THERMAL DESIGN

of

SHELL & TUBE HEAT EXCHANGERS

by

Bill Watson

Further copies available from www.fast-print.net/store.php
at £24.99 + p&p

Shell & Tube Consultancy & Publications Ltd
30 Moatbrook Lane
Codsall
Wolverhampton
West Midlands
WV81DS
UK

E-mail info@satcap.co.uk

ISBN: 978-184426-844-3

Printed in the UK by PrintOnDemand-Worldwide.com
9 Culley Court, Bakewell Road, Orton Southgate, Peterborough, PE2 6XD

SUMMARY OF CONTENTS

The numbers in the left hand column are the format for the way the book is compiled. Those in the right hand column are PDF page numbers for the start of each section to assist in downloading from the company website www.satcap.co.uk.

This page is intentionally left blank.

DETAILED TECHNICAL CONTENTS

LIST OF USEFUL TABLES

THERMAL DATA SHEET & CHECK LISTS

CONTENTS

SOMETHING DIFFERENT

As I grow older I am spending more and more time thinking about what I will be leaving behind. Perhaps that has sparked off a resurgence of my youthful idealism. I have five grandchildren and the world they are growing up in is so different from the one during my formative years.

A global event is taking place as I write this. A volcano erupting in an unpronounceable location in Iceland is causing minor inconveniences by halting air traffic across North West Europe. In 1783 a much larger eruption in the same country at the more pronounceable Laki had more serious consequences. There was widespread destruction of crops and livestock – even the rice crop in Japan failed. The resulting famine in Iceland reduced the population by 25%; all Northern Europe was affected to a lesser degree. Geologists inform us that far greater eruptions have taken place in the past and could occur again at any time.

There is little we can do about natural events but there is a lot we can do about the unnatural way we are treating our planet today. We can learn a lot from history. The people of Easter Island almost exhausted all their raw materials particularly their trees until there was no option but to kill each other until their population was reduced to a sustainable number. The dust bowls in the American Mid West in the 1930's were partly due to the agricultural methods used. The Sahara desert continues to expand as people continue to cut down trees.

The world is now using up resources at an alarming rate whilst increasing its population to exacerbate the problem. Then of course there is the global warming. This is still a highly contentious issue since there are so many variables involved. One thing that does seem to be agreed is that there has been a significant increase in the carbon dioxide content of the air over a relatively short period of time. This is certainly due our activities and is the only issue I have tried to address in this book.

Everyone can do their little bit by recycling, turning off a few lights and driving a sensible car. This certainly helps a bit and gives the participants a 'feel good' factor. Engineers are in a position to do far more, particularly design engineers. For many years now after a particularly good day at the office I have looked up at the sky and said, "I saved you from a few more tonnes of that stuff today". I talk to the sky as well as the trees – often far more satisfying than talking to people who refuse to listen.

Just over two years ago I bought a book which cost me £376.27. That works out at £1.65 for each of the 228 pages. The book is 'Practical Thermal Design of Shell-and-Tube Heat Exchangers by R Mukherjee. The reason that I had to buy it was that the title was exactly the same as the one I had been 'getting round to writing' for the past 15 years.

The book by Mr Mukherjee has clearly been written with full cooperation of HTRI. Many of the diagrams have been taken from its reports and all the case studies involve inputs and outputs from an HTRI program. Some of these case studies are excellent in the way that they show the development of a thermal design but only the one thermal design program is ever mentioned.

The most obvious drawback to this book is the cost – who has ever read it? Companies may buy it but if they do it is unlikely to be more than a single copy. I once worked for a big company that was so tight that several engineers working on different floors had to share one copy of the current TEMA.

The second drawback is that is that it is mainly in US Customary units with a few bits in the old Metric system. Both systems of units used are antiquated and should have been phased out years ago.

I knew it was now or never to make my contribution to heat exchanger design literature. My search for a co-author, who might have spurred me on, had been fruitless. I would have to do it on my own.

I took some time examining my motives for this venture. No one writes highly technical books for the money. It might help their career but mine is almost over. My current royalties amount to about one pint of beer a fortnight, although I often think of it as the most satisfying pint during that period. I had worked hard all my life and had accumulated sufficient pensions to see me through. There was certainly no financial motive.

There is of course the ‘ego trip’ which I would not deny. I still have ambitions well beyond ‘enjoying my retirement’. It has already given me a great deal of personal satisfaction writing this book.

I could see no point in writing a book that few people could afford. I wanted it to be on the desk of everyone who might find it useful; particularly the engineers I worked with in Thailand with incomes far below UK and USA levels. I realised that there was only one way to achieve this aim – **give it away**!

I made the decision to make the book freely available to all the day before I started writing. I knew sufficient about the internet to realise that if I had a web site I could post it on there and let people download it for the cost of the ink and paper. (Even less cost if a memory stick is used - I am a bit old fashioned and still think in terms of a real book.)

As I was nearing the end of the main technical content I realised that I would like some bound copies to give to other people and of course keep one for myself. I found a local printer who would run off copies from a hard original but they could not help with the binding. My search for a binder in the Midlands revealed only one source. On the phone he sounded a charming old guy but he clearly specialised in rebinding old manuscripts. His costs were a minimum of £15 a copy and the maximum he would do was ten – after that he considered repeating the same work to be too boring!

I searched the web for companies that could both copy and bind a book. There were several that looked interesting. The first one I tried looked fine until I got to the point where I read “Before you start writing your book, make sure that the margins at the sides are …. exactly and at the top and bottom are….” A bit late now I thought. I looked at the collection on my bookshelf and realised that they were all different sizes and none of them was A4. I managed to pull myself out of an instant depression and tried another site.

The second company I tried could do almost anything I wanted. All that they required was two PDF files, one for the whole book and another for the cover. The A4 size was not a problem. They even have a program for providing an instant estimate of the cost. Soon I was entering the world of self publishing.

I am expecting to be able to produce 'perfect' bound copies for about £15 per copy. (Perfect in publishing jargon means soft back to most people.) There must be some people out there who would prefer a bound copy rather than downloading and then using a lever arch file. They can send a cheque, payable to 'Shell and Tube Consultancy and Publications Ltd' or cash to cover the production cost and post and packaging. If cash is sent any overpayment will go to one of the many wildlife causes that I support. **Details of the current price will be posted on the web site**. (I have no facilities for dealing with plastic cards. I must have been one of the last people in this country to have one and I have never paid a penny in interest. I have always considered them to be a cunning plot to get economically uneducated people to spend money they have not earned and then pay heavily for it. When the recession came I think I was proved to be right.)

As far as I know there are only two books solely devoted to the thermal design of shell and tube heat exchangers; one cost me £374 for 228 pages; the other is at a nominal cost for 690 somewhat larger pages with almost 240 000 words. The nominal production cost is in reality a subsidised cost. I have already spent a lot of money as well as time on this project.

I started off writing this book with the best of intentions to get permission to publish source material. I ordered a copy of TEMA and after that I found them to be very cooperative. I contacted HEI and they wanted details of publishers etc – not very easy to explain. I tried the manufactures of Kempchen seals to see if they had any better diagrams that those I found on the web – no reply. The latter two examples I thought would be very straightforward.

A lot of the material has been taken from sources that are no longer in existence; companies have changed their names or gone bust. In my last book my publisher and co-author did most of the work in this area and there were no such problems. Anyway how could I get permission to publish a diagram from a design I did 20 years ago?

If seeking permission for publication had involved direct contact with engineers who understood what I was doing there might not have been too many difficulties. But I know such requests would involve administrators or lawyers. They always seem to either provoke me into full combat mode or depression mode when faced with filling in all their forms etc. Either way it would be the sort of distraction I did not have the time for. I therefore made the decision **'publish and be damned'**.

What I have tried to do is acknowledge source material throughout although even this has been difficult at times. My main source of material has been the personal files that I have accumulated over the past 45 years.

This book contains critical reviews of many of the pillars of the world of heat exchanger design – TEMA, HEI, ISO 16812, HTRI and HTFS. Comments are made at appropriate points throughout the book. Any criticism is intended to be constructive and I hope it will be taken as such.

A web search will reveal computer programs from many other sources, some of them remarkably inexpensive. I have never used any of them. Low cost does not necessarily mean low quality and these programs may have severe limitations whilst still producing satisfactory designs for some products. I know how good some low cost programs can be. John Round once wrote a program for tube layout diagrams which was far more flexible and accurate than either the HTRI or HTFS programs available at that time. The first windows program I know of was my own. My old company, Thermodynamic Engineering Ltd was designing a range of high temperature recuperators for glass kilns, (up to 1600°C on the exhaust gas side!). Their design involved a lot of iterative work so I commissioned a final year computing student, (the eldest son of my friend Mac Wright) to produce a program to take some of the drudgery out of the calculations.

Knowing that there are a lot of other computer programs out there I have tried to make general comments which could apply to any program. Inevitably many comments may be more specific to the two main programs used in petro-chemical work. I have used these programs since their earliest versions.

I have tried to make this book more than just a normal textbook. I am among the last generation of thermal design engineers who once did all their work aided only by a slide rule. I have therefore attempted to cover the history of thermal design work from that time up to the present. There is always something to be learned from history.

Tom Eastop, (the co-author of my last book), has looked at a sample chapter which he described as a technical autobiography; an interesting comment. I was certainly making it that without such a succinct description.

I have also tried to 'lighten it up' with a number of stories related to the text. I doubt if such freedom of expression would have been tolerated by an orthodox publisher or even a co-author. To summarise I have tried to make the whole venture – **something different.**

COPYRIGHT

ACKNOWLEDGEMENTS

The first person I would like to thank is Mr R Mukherjee. It was purchasing his book and then reading it that provoked me into getting on to my backside and starting this book, something I had been intending to get round to for many years. I noted that his first acknowledgment was to "Almighty God". I have always found religion to be a fascinating subject ever since the days when I was fortunate enough to gain a place at a grammar school. That is when I started to question the logic of what I was taught in RI. There were no satisfactory answers to many of my questions – unlike mathematics and physics on which engineering is based. I still ask some of the same questions whenever I get the chance. What is certain is that intolerance by many organised religions has caused incredible carnage and suffering throughout history which continues as I write this book.

If I could remember their names I would like to be able to thank many of my teachers from Hutton Grammar School although I expect they will all now be dead. The school was over 400 years old with a large boarding contingency. The Latin motto translated to "work or leave". Discipline was something the modern generation will find hard to believe. Each teacher had his own instrument of punishment; a cane, a belt or a slipper and even the prefects could join in. Just before I left one prefect saw me not wearing my battered school cap on the way to school. He had his moment of sadism but was fortunate that I never saw him again after I left. I had a strange relationship with that school. At 16 I had captained the school water polo team, played at board two in the chess team and always did quite well academically particularly in the science subjects and yet I never quite fitted in and my expulsion was discussed more than once. On reflection my problems originated from my class background. In my year there were only two other pupils whose parents could be described as working class and we all left without even considering entering the sixth form. I did not appreciate it at the time but that school gave me a solid foundation for my future career.

I would like to thank Mr Alan B Turner for provoking me to go to University. We were both cadet engineers for the Blue Funnel Line. Most of the first two years of our apprenticeships were spent at technical colleges. He emerged as the top student from Birkenhead and I did the same at the Riversdale Technical College in Liverpool. That started off a rivalry that also took in sports. He had represented his county at more than one sport, something I had never achieved. Of course he had the advantage of attending the only grammar school in Anglesey; Lancashire was a far bigger county. Later we shared a very small bunkhouse in the half deck of a motor vessel on a trip around the Far East and became firm friends. Before the end of our apprenticeships we had both come to the conclusion that a life on the ocean waves was not for us. I might have settled for a shore job that gave me the opportunity to follow the HNC route to membership of the IMechE; Alan was adamant that he was going to go to university. In July 1965 we both graduated from the University of Newcastle Upon Tyne. I hope Alan is still around and will contact me some day.

I am indebted to Mr James J A Parry and Mr Tom Jupp for introducing me to the petrochemical industry. After graduating I was working as a trainee marine superintendent in Newcastle. Both my parents were living in Liverpool at the time and those of my wife lived relatively close by in Newton-le-Willows so we made regular trips back there. Jim was another good friend from my apprenticeship days but had left the merchant navy after Alan and myself. He had improved his qualifications through the HNC route. On one of my trips back 'home' I met him for a drink on a Saturday lunchtime. Blackburn Rovers, a team I had supported since my school days were playing his team, Everton. The original idea was to have a few drinks and then go to the match. Most of the conversation centred on an advert in the local paper which Jim had brought with him. "New graduates and experienced Engineers required to work in London. Join us and you will be able to afford a house costing over £4 500. Just drop in at Adelphi Hotel for a chat about your career." We were both intrigued and the pints went down as we considered the potential pitfalls. Jim had just bought a small terraced house for about £1000 and I could have bought the place I was living in for £450. (It was in one of the roughest areas in Gateshead and with a small daughter I never seriously considered it.) It was getting close to kick off time when Jim went to the match and I staggered to the Adelphi Hotel for a chat.

The company was Kellogg and the little chat lasted nearly three hours including the sort of IQ tests that I thought I had seen the end of when I passed the entrance exam for the grammar school. In the final face to face Mr Jupp asked me what salary I would expect if I was offered a job. (At that time I was on £1000 a year when a typical salary for a new graduate was more like £750.) I had rehearsed this one in the pub and aimed far higher than I expected to be offered. I gave a little speech about how I was aware of how expensive London was and suggested £1500. He replied that he would make it £1550 and wanted to know how soon I could start. I left that hotel a bit dazed but sober, wondering how my wife would react to the news of a new life in the South. In just over a month I had designed my first heat exchanger.

Kellogg was probably the best place to start thermal design work and I am grateful to them for the training they gave me. I remember Subrata Chatterjee and Harry Rudel being very patient with me as I tried to learn all that I could as quickly as I could.

It would be nice to be able to personally thank the late Gordon Woodhouse who was the head of the process department at Stone and Webster in the sixties. I had an interesting interview with him. After the usual formalities he got rid of the personnel manager and presented me with specifications and drawings for three shell and tube units. He simply asked me what I thought of them. I was glad I was sober for that interview it required quick thinking and I usually like to take my time. The first two were easy - two shells had been used when I could foresee no problems in designing for a single shell and there was nothing on the PDS requiring two shells, my comments on the third one where a bit more involved. When I had finished Gordon replied "I thought so, they were designed on that new in-house computer program which the personnel manager had enthused about. If you want the job you can use whatever design methods you think are appropriate." I never knew how many other applicants there were but Gordon was taking a bit of a gamble is setting up a new section under me when I had less than three years experience.

On my first day at Stone & Webster I was introduced to Mick Smith and Ashoke Banerjee who were to join my group; neither had any experience in thermal design work. I pointed out the imbalance in experience and told the management that we needed another experienced thermal design engineer. I was fortunate enough to recruit Paul Chapman from Foster Wheeler. He was one of the best thermal design engineers

I have ever worked with. He was also very fast and had the unusual ability to carry on crunching numbers on his slide rule whilst discussing the cricket. I always had to pause in my work to respond. We were later joined by Gordon Snashall, who was a new process graduate. I know Paul went to the USA and I have had no contact with him since then. I met up with Gordon again when he was a process manager at Bechtel. I kept in reasonable touch with Mick over the years and he became the engineering manager at Fluor Daniel. Whenever I met them they used to reflect on how much they had learned about heat exchangers and how they had enjoyed their time in our group. I got a lot of satisfaction from their comments which made me think that perhaps I should write a book.

Ashoke became a very special friend. We lost contact for many years. Then one evening I was sitting at an island bar of a hotel in the small village of Liedschendam in the Netherlands when I saw him at the other end of the bar. He was working at Lummus and I was at KTI. It was quite a reunion. It was Ashoke who recommended a walk he had done. I did the 800 km from Saint-Jean-Pied-de-Port to Santiago de Compostela to celebrate my 65th birthday. It was just like he had described, some magnificent mountains and countryside and the chance to meet lots of interesting people. There were lots of conversations about religion and I found it very satisfying to be able to point out to the Christians that the Druids had been doing the same walk long before they started. He also taught me a lot about India and pointed out many things a normal tourist would never see when I was there with him. I taught Ashoke a lot about designing heat exchangers and in return he taught me a lot about life. Sadly he died just as I was starting this book. I know he would have appreciated having one of the first copies.

I would also like to thank David Edge. He took over the heat exchanger section at Stone & Webster and I returned there to work as a consultant many times both during vacation periods whilst I was still in the academic world and after I became a full time consultant. Dave knew Len Lindsall of Foster Wheeler and recommended me to Len. That was the start on another very fruitful relationship and I have been back to Foster Wheeler many times since my first contract. Len was one of the best gaffers I ever worked for. He could be ruthless, like the time when he laid off two contract staff as soon as I became free from the contract I was working on at Stone & Webster. However his staff always judged him to be a very fair man and they knew how well he could defend his department in the higher echelons of office politics. I have seen him once since he retired and I hope he is still enjoying his well earned retirement.

I am not sure now if it was Len or Dave who recommended me to KTI which gave me the opportunity to broaden my experience by designing heat exchangers in another country. All the thermal design work was in a section of the process department just as it had been when I first started at Stone & Webster. The whole company was an amazing conglomeration of races and nationalities and the process department was more varied than most. I worked with Hungarians, Czechs, Germans, Russians, Danes and Indians from four different countries and of course there were some Dutch as well. They never ceased to amaze me with the number of languages they were so fluent in and with their knack of swallowing whole raw fish. Fortunately the one common language was English so that I could learn so much about so many countries just by working in one of them.

I must also mention John Round who I was fortunate enough to work with at Fluor Daniel just before he 'retired'. It was the first time for many years that I had worked with someone who started in thermal design work before I did. It was a great opportunity to kick around a lot of the finer points of thermal design. John also helped with the section on Glycol Injection units.

In 2006 I had a very interesting contract for Foster Wheeler in their office in Thailand. The department I worked in had one mechanical engineer from the UK office that had been there for several years and the rest of the staff were all locals. There were three charming young ladies working on the thermal designs. Phon became my IT crutch; my earnings would have been significantly improved if I had been give a Baht for each time I had to ask her "what's gone wrong with this damned machine now" or "which button do I press next?"

I soon recognised their limitations with some of the more sophisticated aspects of thermal design and held many impromptu tutorial classes. They were the most enthusiastic group I can recall and I have thought of them many times whilst trying to bring out salient points in this book. Outside work I learned a lot about the Buddhist religion and Thai politics. I was very touched by the leaving presents I was given and by the three leaving dinners/lunches. Thank you Khanita, Jeed, Phon and your excellent head of department Korakrit.

From my academic years I would like to thank Roger Briggs and Tom Eastop. Roger is still a good friend and has offered moral support throughout but unfortunately declined my suggestion that he should continue his engineering education by proof reading my work. Tom is the co-author of my other book. Without the experience I gained from writing Mechanical Services for Buildings I could never have started this project.

I would like to thank Ted Saunders, R A Smith, Stanley Yokell, Sadik Kakac, and R T Bott for writing excellent books which should be part of every thermal design engineers library. I have quoted from them many times throughout this book.

Lastly I would like to thank Mac Wright, Peter Hodson and Ian Spittle for their IT assistance. My IT skills now could possibly be described as 'fair to middling', when I started to write this book they could be described very simply – crap!

Mac has been a friend for many years and lives in the same lane. He helped me in compiling some of the tables and did a consistency check on the numbering of the equations, diagrams and tables. I cringed when he pointed out all the mistakes I had made. The photograph is his work although he prefers photographing birds (the ornithological variety). He has given me support and encouragement throughout.

There is no one to thank for any proof reading of the technical content; that will be up to you, the reader.

AUTHOR'S CV

NAME	William Edward Watson
DATE & PLACE OF BIRTH	16/8/1939 Bury Lancashire
EDUCATION	Hutton Grammar School, Preston Riversdale Technical College University of Newcastle Upon Tyne
QUALIFICATIONS	Eur Ing BSc (Hons) FIMechE MCIBSE

ENGINEERING EXPERIENCE IN OTHER INDUSTRIES

MARINE

I completed an apprenticeship as a Cadet Engineer and later spent about five months in the head office of Hunting & Sons as a trainee Marine Superintendent. The total time was about five years.

CIVIL

I spent one of my student vacation summers working with an Irish Tar Gang building the M6 motorway. The machine we used required two men to power the pump which built up sufficient pressure to produce a good spray of tar. I never bothered to go to a Gym while I was on that job. I sometimes reflect on that primitive machine when I am designing exchangers with very viscous fluids.

AIRCRAFT

In between finishing my apprenticeship and going to university I worked as a fitter in the de Havilland factory near Chester. My job was fitting wing sections on to the ill fated Comet planes. It was extremely boring work but I gained some knowledge of aircraft construction.

AGRICULTURAL

While I was a lecture I spent quite a lot of time at what is now Harper Adams University where I had an input to their HND and BSc courses in agricultural engineering. I gained a good insight into the environmental aspects of agricultural engineering including grain drying, crop storage and energy balances in livestock buildings. One of my consultancy projects was the environmental design for the buildings in a snail farm in Scotland.

NUCLEAR

My first involvement with the nuclear industry was whilst I was lecturing and the work was carried out in my own office. The work involved a graphite sample recovery grab and is briefly described in chapter 15 in the section on parallel flow.

After I became a full time consultant I spent about seven months as a thermal analyst working for Thompson Nuclear. This is the only substantial contract I ever did when I could commute from home each day. They were designing a new type of plant for the drying and long term storage of spent fuel elements. A personal contact offered me the job because of my knowledge of both heat transfer and environmental engineering. The latter involved designs with argon/water vapour mixtures rather than the more normal air/water vapour mixtures.

The job was suddenly stopped just as we were about to finalise designs. The whole exercise had been a political/economic scam. The client had agreed new terms for processing spent fuel elements in the traditional way.

All the Engineers I worked with were nuclear engineers and most of them were on short term contracts like me. When we left some of them were very envious that I had already been offered more work in petro-chemicals; their futures were very uncertain; decommissioning work was their only hope because of the myopic vision of UK governments. Fifteen years later our government has woken up to some of the advantages of nuclear power but sadly we no longer have enough engineers to either build or man new power stations.

BUILDING SERVICES

I was first introduced to building services engineering whilst I was lecturing. The National Health Service, (NHS) decided that all its top engineers should have chartered status. At that time very few of them qualified. A crash course was needed rectify this. The crash course route, termed the Academic Test, would only be available for a couple of years and after that a degree would be required. This made the course very unattractive to Universities in the area who should have been in a better position than we were to offer such a course. Our head of department, Tom Eastop, was the only member of staff with any experience in this field. Tom was always up for a new challenge and we soon became the centre for building services in the whole of the Midlands. As the student numbers fell from the crash course we introduced a degree course to replace it.

I was seconded to the NHS design office for six weeks to gain some practical experience. There I designed air conditioning systems for hospitals including the tricky air flow problems involved in operating theatres. During lectures I never stopped learning from my students working in the industry, many of them having very responsible jobs and years of experience. A sad recollection from the first intake to the crash course was a student who was responsible for several hospitals but could not manage the new Academic Test so he took early retirement. On a happier note I have recently received my regular invitation for a reunion with students I taught twenty years ago.

I became involved in consultancy work in this field particularly energy studies where there was the possibility of introducing combined heat and power schemes. After my early retirement I might have carried on in this field. Instead I went back to designing heat exchangers as a full time consultant.

There have been many occasions when I have found my experience in other industries to benefit my work in designing heat exchangers. I am a great believer in cross fertilisation of ideas. This is something I have tried to bring out throughout this book.

EXPERIENCE IN PETROCHEMICALS

CONTRACTORS

I have worked for the following contractors:

Kellogg
Stone & Webster
Foster Wheeler UK
Foster Wheeler Thailand
Procon
Kinetics Technology International (KTI)
Fluor Daniel
Bechtel
John Brown
Snamproghetti

I worked full time for Kellogg and Stone & Webster for a total of about 5 years. The total time of my two contracts at KTI was about eighteen months. My two contracts at Procon covered only the summer vacation period whilst I was still lecturing. I only had the one contract at all the other companies except for Stone & Webster and Foster Wheeler which I have been back to on many occasions.

CLIENTS

Some of the Clients I have designed shell and tube and air cooled heat exchangers for include:

Exxon
Shell
BP
Saudi Aramco
Petropars
Ras Gas
Samref
Texaco
Petrola
Altona

ACADEMIC EXPERIENCE

I lectured at the University of Wolverhampton (formerly Wolverhampton Polytechnic) for almost twenty years. The subjects I taught included thermodynamics, fluid mechanics, heat transfer, environmental engineering in agriculture, building services engineering, energy and power and engineering design. The classes were up to honour degree level. Many of the student projects I supervised were related to heat exchangers and some of them are referred to in the main text.

RESEARCH

Pressure drops in mixtures of lubricating oil and air. This was confidential research for Rolls Royce and was subsequently used in the design of aircraft engine systems.

PUBLICATIONS

Mechanical Services for Buildings –Easton T D & Watson W E, Longman 1992. This book is still available and there have been no revisions.

I started this book on 18th May 2008. My original target for completion was one year; I am now struggling to make it two years. As I am tying up all the loose ends a general election is taking place and I am feeling guilty that I am not doing more to expose our politicians' lack of concrete policies, vision, understanding of the difference between a job which creates wealth and one that consumes it, and of course the usual out and out lies.

LEISURE TIME

When I am allowed to, I like nothing better than to relax in a pub with a pint and a cigar and join in a stimulating conversation. I have strong views on most serious topics particularly the taboo ones of religion, politics and the world's excessive population.

If no one wants to talk I can play most pub games; my darts, snooker and pool are still at a reasonable standard. I would also like to play bridge again with real people instead of just being a 'problem solver' or playing on the net.

I like watching most sports particularly cricket but I would sooner be playing again. If anyone knows ten has-beens who do not mind losing every match I could be their number eleven!

I still enjoy walking, snorkelling and scuba diving. I swim and weight train regularly at a local leisure centre. I often find it very tedious and the only bit I regularly enjoy is the shower and the pint afterwards. The only point of it all is to keep in shape so that I can keep going with all the things I do enjoy.

UNITS

The SI system of units was adopted by the General Conference of Weights and Measures in 1960 and subsequently endorsed by the International Organization for Standardization. By 1970 it was being adopted by the UK, Europe and most other countries in the world apart from the USA. (They are not quite alone; they still have Myanmar and Liberia for company.)

In the UK we had the unfortunate experience of being told by our government that we were *going metric*. Sadly most of the British public still believe this lie which is used to fuel anti-European sentiments. It was never explained that the countries already using the old metric system of units were also undergoing a change. Their calorie and kilocalorie would no longer be official units, they would be checking their tyre pressures in kN/m^2 and not kgf/cm^2, and finally they would be using the Newton for their unit of force/weight. Of course the public in most countries have never adapted to this last change and will quote their weight in terms of kilograms or pounds or stones and pounds. The bureaucrats in my country went madder than usual and started to persecute traders selling goods in our old units of pounds. The more pragmatic Germans respected tradition and their markets still sell produce by the pound! (The German pound is 4.905 Newtons – 0.5 of their old kilogram, and they can still sell goods without any reference to a weight.) Perhaps their government understood that the introduction of a new system of units was to help scientists and engineers not to inconvenience the general public.

The internationally agreed units which we need for thermal design work are as follows:

Quantity	Name	Unit Symbol	Symbol for equations
Length	Metre	m	l (L)
Mass	Kilogram	kg	m
Time	Second	s	t
Temperature	Kelvin	K	T

The definitions of the unit used for each quantity are:

Unit of length – metre

The SI metre is defined as; "the length of the path traveled by light in free space during a time interval of 1/299 792 458 of a second".

This distance can be reproduced at any time and any place without the worry of having a standard which can be damaged or destroyed. In one important respect the US has already accepted this definition as the standard for length since their inch is now exactly 25.4 mm. In the past there was no exact conversion between Imperial and Metric lengths, it was a matter of "how many decimal places do you want it accurate to?"

The old metric and imperial standards both relied on the distance between two points on a metal bar kept in a standards room. The original idea of the Metric meter being one ten-millionth of the distance from the equator to the poles showed inspiration and imagination but was not very practical. In reality the metric meter was the distance between two marks on a bar of platinum in Paris. This distance was then transferred to a small number of secondary standard bars from which it was transferred to a third set and so on until you bought a meter rule in your local store. Each transfer introduced errors, which were cumulative. This did not matter much for buying cloth or carpets but for accurate engineering and scientific measurements it became increasingly unacceptable. With this system of measuring length the basic standard was unique and irreplaceable and all lengths changed slightly when the original standard was once destroyed and had to be replaced by one of the secondary standards.

Unit of mass – kilogram

The SI kilogram is defined as; "the mass of the international prototype of the kilogram".

Unfortunately this still means the mass of a lump of platinum kept in the International Standards Laboratory. Ever since I was a student scientists and engineers have been searching for an alternative definition. As I am writing this I believe we are now getting very close to a universal definition which does not rely on a lump of metal.

Unit of time – second

The SI second is defined as; "the duration of 9 192 631 770 periods of the radiation corresponding to the transition between the two hyperfine levels of the ground state of the cesium 133 atom.

Once again this is a standard which can be reproduced in a laboratory anywhere on earth or on any other planet if ever required. It should be noted that the definition of the second is crucial to the definition of the metre.

Unit of thermodynamic temperature – Kelvin

The SI thermodynamic temperature is defined as; "the fraction 1/273.16 of the thermodynamic temperature of the triple point of water."

I am not entirely happy with this definition and I was tempted to modify it to temperature difference which it clearly what it is. The triple point of water is at 0.01 ^{o}C and 0.0061173 bar pressure. The Celsius or Centigrade temperature scale which we commonly use is derived from the basic definition of the Kelvin. This scale starts close to the triple point of water and is positive as the temperature increases and negative for decreases in temperature.

The second decimal place in the definition of the Kelvin has changed slightly during my studies of thermodynamics. It will be interesting to see if there are any further changes as the physicists get closer and closer to absolute zero.

Once the basic units have been defined all the other units that we use are derived from them. The following table contains some of ones we use in thermal design work:

Quantity	Unit name	Symbol	Basic units
Force	Newton	N	kg m/s^2
Pressure, stress	Pascal	Pa	N/m^2 = kg/ms^2
Energy, work, heat	Joule	J	N m = kg m^2/s^2
Power, heat flow rate	Watt	W	J/s = kg m^2/s^3

Many of the derived units which we use have no unit names these include:

Quantity	Symbol	Basic units
Specific heats, Gas constants	C_p C_v R R_o	J/kg K = m^2/s^2 K
Dynamic viscosity	μ	N s/m^2 = kg/m s
Thermal conductivity	k	W/mK = kg m/Ks^3
Surface tension	σ	N/m = kg/s^2
Heat transfer coefficients	U h	W/m^{2}K = kg /Ks^3

Many of the derived units in the SI system result in either very large or very small values that are difficult to 'take in'. It therefore includes a series of standard prefixes to make the numbers more manageable. The ones we are most likely to use are:

Prefix name	Symbol	Factor
giga-	G	10^9
mega-	M	10^6
kilo-	k	10^3
milli-	m	10^{-3}
micro-	μ	10^{-6}
nano-	n	10^{-9}

An example of the need to use a prefix in thermal design is when specifying the heat load (heat flow rate) which an exchanger needs to be designed for. The kW (kilo-watt) is invariably used and for some units the MW is more appropriate. Another example is when specifying specific heats when we always use kJ/kgK.

The approved prefixes are all in multiples of one thousand. This leaves the centimetre out in the cold. I have no problem with this, 100 mm is just as easy to visualise as 10 cm. The controversial one which concerns us most is the bar not being a recommended multiple of the basic unit for pressure - the Pascal, (N/m^2). The bar is 10^5 Pacsals. The nearest approved multiplier to the bar is the MPa but the kPa is more likely to be used.

The bar is a very special unit being so close to atmospheric pressure, (1 atm = 1.01325 bar). The standard atmospheric pressure is a unit everyone has an appreciation of and operating above or below atmospheric pressure is an important consideration in many aspects of engineering design. For thermal design work we need to work in absolute pressures and a value for the pressure in bar is normally understood to mean absolute pressure but this is sometimes emphasised by adding A to the end to show bar**A**. This distinguishes it from bar**g** which the mechanical engineers use. Many computer programs now use kN/m^2 for pressure. Personally I find this to be a pain in the neck because I immediately move the decimal point two places to 'get a feel' for the magnitude of the pressure I am dealing with and most engineers I have worked with do the same. In thermal design work I think we should all use the bar even though it is a slight deviation from the SI ideal.

At the time of writing this there are two alternative systems of units which a thermal design engineer will have to deal with. These are the old metric/MKS system and the old imperial/US Customary system. Both systems have different units for heat and work which are both forms of energy. In the old metric system the kilogram is a unit of force not mass just as the pound is in the old imperial/US Customary system. This leads to the anomaly that the equation for the velocity of sound which has nothing to do with gravity must have 'g' in it but potential energy which only exists because of gravity does not. The Americans have tried to overcome this problem by introducing g_c as a conversion factor which of course has the same value as g, the gravitational constant. I can remember an alternative approach from my student days when the slug was used as the unit for mass. It never caught on and slugs remained a garden pest or a bit of lead to use in an air gun. A few years ago an American spacecraft that cost over \$100 000 000 crashed on Mars. The failure was reported to be due to a mix up in the system of units used. The problem lay in converting the metric unit for force into the US customary units – the magnitude of the mistake was a factor of 9.81!

It is unlikely that such a major cock up will occur again but anyone who has been involved with thermal design work will be able to recall cases where confusion over units has led to mistakes. The difference in values for some quantities is quite small. For example the thermal conductivity in the MKS system is only about 16% higher than the SI value and the US value is only about 73% higher. These are well within the spread of possible values for the fluids we deal with. For other quantities the differences may be very large and the biggest mistake I can recall finding was a factor of 1000 in a calculation involving surface tension. A table of conversion factors can be found after the main text in this book.

Even when possible mistakes are discounted the cost of maintaining three systems of units (and a few hybrids), must cost far more than the \$100 000 000 for the spacecraft. The worst case I was ever involved in was when I was designing some air coolers at Foster Wheeler. The program they were using was in US Customary units which predicted better results than the programs then available from HTRI and HTFS. The process data was in MKS units which had to be converted before it could be input to the program. The output then had to be converted into the SI system which the client required. This may be an extreme case but just think of all the time involved in using multiple systems of units and of course the extra paper used – don't forget the trees.

We now have a coherent system of units available which has international standing. Surely it is time the alternatives were abolished. Napoleon got rid of the French Customary units and we are all indebted to the French for their pioneering work in the field of measurements. The only downside to their contribution is that we now have an SI system and not a IS system which more people would now find acceptable.

If only the Americans would join the rest of the world I am sure Myanmar and Liberia would follow.

NOMENCLATURE

D_s	Shell internal diameter - m (normally specified in mm)
D_n	Nozzle internal diameter - m (normally specified in mm)
D_b	Bend diameter for U tubes - m (normally specified in mm)
D_{eff}	Effective diameter for cross flow - m
d_i	Tube inside diameter – m (normally specified in mm)
d_o	Tube outside diameter – m (normally specified in mm)
d_e	Equivalent diameter – m (normally specified in mm)
d_d	Diameter of dummy tube or rod m (normally specified in mm)
t	Thickness of tubes, shell etc - m (normally specified in mm)
L	Length – m
L_e	Equivalent length – m
L_{eff}	Effective length for heat transfer- m
A	Surface area – m^2
A_x	Cross-sectional area - m^2
V	Volume – m^3
v	Specific volume m^3/kg
u	Velocity m/s
u_S	Superficial velocity m/s
G	Bulk/mass velocity kg/m^2s
ρ	Density kg/m^3
M	Molecular mass (often wrongly termed molecular weight) or Total mass - kg
$\dot{m}$	Mass flow rate – kg/s
C_p	Specific heat at constant pressure –kJ/kgK
C_v	Specific heat at constant volume – kJ/kgK
γ	Ratio of Cp/Cv - dimensionless
μ	Dynamic viscosity – kg/ms (often specified in cps, 1cps = 0.001 kg/ms)
ν	Kinematic viscosity – m^2/s (normally specified in cks,1cks = 0.001 m^2/s)
σ	Surface tension – N/m (normally specified in dyne/cm,1 dyne/cm=0.001N/m)
k	Thermal conductivity – W/mK, Discontinuity Δp factor – dimensionless
y	Mass fraction - dimensionless
p	Pressure – N/m^2 (or Pa) (1000Pa = 1kPa & 100000 N/m^2 = 1bar) (Pressures are absolute unless a 'g' is added i.e. bar = 1 bar absolute pressure, barg = 1 bar gauge pressure. $p_{absolute} = p_{gauge}$ + atmospheric pressure)
p_C	Critical pressure – normally bar.
Δp	pressure drop – the same as for pressure
f	friction factor - dimensionless
P_t	Tube pitch – mm or m
P_{Lt}	Longitudinal tube pitch – mm or m
P_{tt}	Transverse tube pitch – mm or m
P_b	Baffle pitch – mm or m
P_{row}	Length between tube rows- mm
P_t/d_o	Tube pitch to diameter ratio – dimensionless
Re	Reynolds number = $\rho u d_i$(or d_o or d_e)/μ – dimensionless
Nu	Nusselt number = hd_i(or d_o or d_e)/k – dimensionless
Pr	Prandtl number = $C_p\mu/k$ – dimensionless
P_r	Reduced pressure p/p_C – dimensionless.
Gr	Grashof number = Gr = $\beta g \rho^2 d_o^3/\mu^2$ - dimensionless

Q Rate of heat transfer – W (normally specified in kW or MW)
$\dot{W}$ Work flow rate -- W (normally specified in kW or MW)
H Total Enthalpy - kJ
h Specific enthalpy kJ/kg
h_{fg} Specific enthalpy of phase change kJ/kg
h_i Surface heat transfer coefficient inside tubes – W/m^{2}K
h_o Surface heat transfer coefficient outside tubes– W/m^{2}K
$h_{i@do}$ Inside surface heat transfer coefficient referred to d_o – W/m^{2}K (= $h_i\, d_o/d_i$)
U Overall heat transfer coefficient – W/m^{2}K
$U_{req} = U_{des} = U_{serv}$ the required U value
$U_{calc} = U_{act}$ the actual calculated U value.
U_{clean} the clean U value
f_o Fouling factor outside the tubes - m^{2}K/W
f_i Fouling factor inside the tubes - m^{2}K/W
$f_{i@do}$ Fouling factor inside the tubes referred to d_o - m^{2}K/W (= $f_i\, d_o/d_i$)
T Temperature - °C
T_A Absolute temperature – K
T_C Critical temperature – K
T_r Reduced temperature, T_A/T_C - Dimensionless
ΔT Temperature difference - K
ΔT_m MTD, Mean temperature difference – K
LMTD Log mean temperature difference – K
R_{TT} Total thermal resistance - m^{2}K/W
R_w Thermal resistance of the tube wall - m^{2}K/W
R_{ho} Thermal resistance of the heat transfer coefficient outside the tubes - m^{2}K/W
R_{hi} Thermal resistance of the heat transfer coefficient inside the tubes - m^{2}K/W
$R_{hi@do}$ R_{hi} - referred to d_o - m^{2}K/W (= $R_{hi}\, d_o/d_i$)
F Factor – LMTD Correction, Tube arrangement etc - Dimensionless
BWG Birmingham Wire Gauge
SWG Standard Wire Gauge
PDS Process Data Sheet
TDS Thermal Data Sheet.
OTL Outer tube limit - m (normally specified in mm)
SCD Safe circle diameter - m (normally specified in mm)
NTIW No tubes in the window.
N_T Number of tubes
N_{PP} Number of parallel pass partition lanes
f_N Natural frequency Hz (Cycles/sec)
f_V Vortex shedding frequency Hz (Cycles/sec)
S Strouhal number = $f_V d_o/u$
Ψ Chen number = (Re/S) x Tube bank constant. - Dimensionless

BASIC EQUATIONS & PROPERTIES – CHAPTER 1

INTRODUCTION

I don't know who designed the first heat exchanger but it was probably an Egyptian or a Persian. I designed my first shell and tube heat exchanger in 1965 when all calculations were carried out by hand. Number crunching had already been much improved by the introduction of the slide rule, although I remember one colleague who was still using log tables. I was lucky to start at the Kellogg Company in London. They had an excellent design manual based on up to date equations modified by their own experience. Some other companies at that time were still using McAdams (Ref 5) or Kern (Ref 14) as their manual. McAdams Heat Transmission was first published in 1933 and it is some compliment that the book is still referred to in modern publications (including this one). By 1933 the basic principles of heat transfer were well understood. Since then equations have been modified, refined and extended as more test work has been carried out and this will continue into the foreseeable future.

As computer programs were developed it became practical to carry out more complex calculations. Ideally they should also eliminate arithmetical errors.

One problem with modern computer programs is that they can be too easy to use. In recent years I have come across far too many people who claim to be able to design heat exchangers because they can feed numbers into a program and some sort of design comes out at the other end. There are also too many people in authority who think other people should be able to design exchangers just because they have access to a good computer program. During one of my many contracts with Foster Wheeler I worked with a young novice designer. He carried out designs under supervision but he was not allowed to check other peoples work. His first job in thermal design work had been at a vendor. They had no other thermal design engineers and on his first day he was presented with all the HTRI design manuals. He was given a week or two to read them and after that he was expected to start designing heat exchangers!

The aim of a thermal design engineer is not simply to produce a design but to produce an optimum design. **An optimum design is one which satisfies the process requirements at a minimum cost.** The main factors which effect cost are the type of construction and the amount of metal used. Nowadays I not only think in terms of the money an optimum design can save but the reduction in CO_2 released into the atmosphere by the saving in metal and the improvement in plant performance.

Ideally anyone starting thermal design work should spend at least a few weeks on hand calculations before touching a computer. There are now very few offices were this is likely to happen. One aim of this book is to help the reader to understand what is going on in a computer program. To do this it is essential to get a good grasp of the basic principles involved. This will be emphasised throughout the book.

I have limited this chapter to three fundamental aspects of any design. The first of these is the overall heat transfer coefficient and the method of determining the temperatures through a series of thermal resistances. The second is the method of determining the mean temperature difference for heat exchange. Lastly there is a review of physical properties and their effect on individual heat transfer coefficients.

BASIC EQUATIONS

For heat transfer

$$Q = U\,A\,\Delta T_m \qquad [1.1]$$

Where

Q is the rate of heat transfer in Watts, W
U is the overall heat transfer coefficient in W/m^2K
A is the effective surface area involved in the heat transfer in m^2
ΔT_m is the mean temperature difference between the two streams in K

For most exchanger design work Q will be specified in kW or MW so that a multiplier of 10^3 or 10^6 as appropriate will usually be required on the left hand side of the equation.

ΔT_m is often specified as C or ^oC. Temperatures may be specified in the Celsius or Kelvin scales but it was agreed a long time ago by international conferences on units that temperature differences should be specified as K and I have used this throughout the book.

The effective surface area must only include the portions of the tubes involved in the heat transfer. The lengths inside tubesheets and any length behind a full support plate where there is very little circulation should not be included. There are of course two surfaces involved in the heat transfer, one at the inside of the tubes and another at the outside of the tubes. The outside surface is always used because of the way heat exchanger tubes are specified and manufactured. The outside diameter is fixed (within manufacturing tolerances) but the inside diameter varies depending the gauge of tube used. (The term total surface area is sometimes used. This relates to the full tube length and not the length involved in the heat transfer – this can be confusing when comparing vendor designs.)

It is now necessary to look at U and ΔT_m in more detail.

OVERALL HEAT TRANSFER COEFFICIENTS

A thermal design engineer has to deal with three different values of overall heat transfer coefficient for any heat exchanger. These are as follows:

REQUIRED/SERVICE/DESIGN OVERALL HT COEFFICIENT U_{req}

Rearranging equation 1.1 gives:

$$U_{req} = \frac{Q}{A\,\Delta T_m} \qquad [1.2]$$

U_{req} is often just referred to as U. It may also be called the design coefficient U_{des}, or the service coefficient U_{serv}. It is the heat transfer coefficient which is just required to carry out a specified duty Q, at a specified mean temperature difference ΔT_m in a heat exchanger with an effective surface area of A m^2.

CALCULATED/ACTUAL OVERALL HT COEFFICIENT U_{calc} ,U_{act}

The overall heat transfer coefficient can also be found from an equation which takes into account all the thermal resistances between the two fluids exchanging heat. For a heat exchanger in normal service the basic equation for calculating the overall heat transfer coefficient is:

$$\frac{1}{U_{calc}} = \frac{1}{h_i}\frac{d_o}{d_i} + f_i\frac{d_o}{d_i} + R_w + f_o + \frac{1}{h_o} \qquad [1.3]$$

h_i and h_o are the surface or film heat transfer coefficients in W/m^2K at the inside and outside surfaces respectively. f_i and f_o are the fouling resistances in m^2K/W at the inside and outside surfaces respectively. R_w is the thermal resistance of the tube wall in m^2K/W. d_o/d_i is the ratio tube outside diameter to the tube inside diameter. Since for any cylindrical surface the area $A = \pi d L$ this is also equal to the ratio of the outside surface area to the inside surface area.

Equation 1.3 is unusual in the way that it has some terms expressed as heat transfer coefficients and others as thermal resistances. It is useful to introduce the term total thermal resistance R_{TT} and also treat the surface heat transfer coefficients in the same way, this leads to the following equation:

$$R_{TT} = R_{hi@do} + R_{fi@do} + R_w + R_{fo} + R_{ho} \qquad [1.4]$$

Where $R_{TT} = 1/U_{calc}$, $R_{hi@do} = d_o/h_i d_i$, $R_{fi@do} = fi\ d_o/d_i$, $R_{fo} = f_o$ & $R_{ho} = 1/h_o$

It was explained earlier why the effective surface area in equation 1.1 is always at the outside of the tubes. The same rate of heat flow passes through both surfaces. To find a total resistance to heat flow it is therefore necessary to refer any resistance to heat transfer on the inside of the tubes to the common surface at the outside of the tubes. To maintain the product (Resistance x Area) **a resistance at the inside surface must be increased to have the same effect at the outside surface, hence the multiplier d_o/d_i.**

Tube wall resistance R_w

For a plane surface

$$R_w = \frac{\Delta x}{k} \qquad [1.5]$$

Where Δx is the thickness of the metal in m and k the thermal conductivity of the metal in W/mK.

For a thick cylinder

Consider an elemental cylinder of radius r and thickness δr in a thick cylinder of out side radius r_o and inside radius r_i:

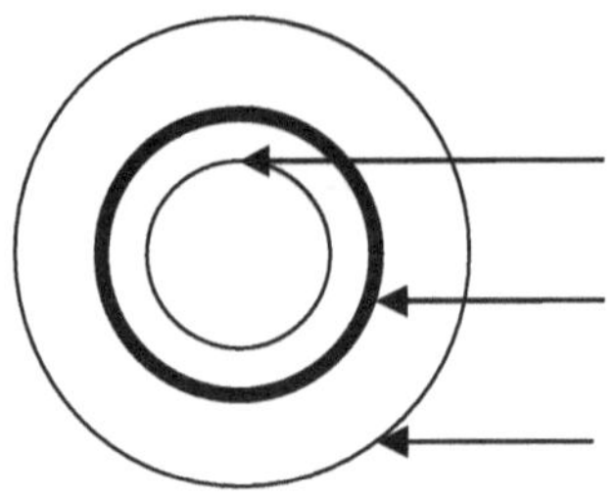

Figure 1.1

For the element, equation 1.5 applies, therefore $\text{Resistance} = \dfrac{\delta r}{k}$

Referring this to the outside surface $\text{Resistance} = \dfrac{\delta r\, r_o}{k\, r}$

r_o and k are constants therefore for a thick cylinder

$$\text{Resistance} = r_o/k \int_{ri}^{ro} \partial r / r$$
$$= r_o/k \ln(r_o/r_i)$$

Using diameter ratio gives

$$\text{Resistance } R_w = \frac{r_o \ln(d_o/d_i)}{k} \qquad [1.6]$$

Example 1.1

Calculate the error involved in using plane boundary theory to calculate the thermal resistance of a 19.05 mm tube of thickness 2.1 mm.

Solution

The correct thermal resistance using equation 1.5 is

$$R_w = \frac{r_o \ln(d_o/d_i)}{k}$$
$$= \frac{19.05}{2}\frac{\ln(19.05/14.85)}{k}$$
$$= \frac{2.37}{k}$$

Using plain boundary theory

$$R_w = \frac{\Delta x}{k}$$
$$= \frac{2.1}{k}$$

The percentage error is (2.37-2.1)/2.37 x100 = **11.4%**

This may seem like a significant error but the thermal resistance of the tube metal is normally only a very small part of the total thermal resistance. Typically the tube wall resistance is less than 4% of the total thermal resistance so that any error in using plane boundary conditions is less than 0.5%. For hand calculations it is normal to use plain boundary theory for the tube thermal resistance. I can only recall one occasion when this would have produced a problem. That case involved thick, none metallic tubes – very unusual. A reasonable computer program will of course use the correct theory.

Note that a value of thermal conductivity for the tube was not necessary to solve this example.

CLEAN OVERALL HT COEFFICIENT U_{clean}

The clean overall heat transfer coefficient is the calculated value without any fouling. Removing the fouling terms from equation 1.3 results in equation 1.7 below.

$$\frac{1}{U_{clean}} = \frac{1}{h_i}\frac{d_o}{d_i} + R_w + \frac{1}{h_o} \qquad [1.7]$$

This is the overall coefficient which would apply when the unit was first commissioned or had just been cleaned. This is the one overall coefficient which has the same name in any literature on heat exchangers.

SUMMARY

The terminology which may be used for any exchanger is:

(a) $U_{req} = U_{des} = U_{serv}$
(b) $U_{calc} = U_{act}$
(c) U_{clean}

The term U can confusingly be used for either (a) or (b). This is because in an ideal exchanger they would have exactly the same value. However an individual could spend their whole working life designing heat exchangers without ever achieving an exact match. Of course rounding to say one decimal place would introduce a few 'matches' and rounding to a whole number would produce regular 'matches'.

Normally the calculated value will be higher than the required value. Tolerances on the discrepancy between the two values are discussed throughout this book.

The clean overall coefficient will always have a higher value than either of the other two. This may sometimes be a factor of two higher depending on the nature of the fouling and the values of the other thermal resistances. Chapter 14 is devoted to fouling.

ELECTRICAL ANALOGY FOR THERMAL SYSTEMS

For the actual flow rate of heat through a series of thermal resistances the U value in equation 1.1 is U_{calc}. Rearranging this equation and introducing $R_{TT} = 1/U_{calc}$ results in the following equation::

$$\frac{Q}{A} = \frac{\Delta T_m}{R_{TT}} \qquad [1.8]$$

At any section through a tube the temperature difference ΔT is the temperature of the hot fluid T_h minus the temperature of the cold fluid T_c. Equation 1.8 can therefore be written as:

$$\frac{Q}{A} = \frac{\Delta T}{R_{TT}} \qquad [1.9]$$

In equations 1.8 and 1.9 the term Q/A is known as the heat flux. This is defined as the heat flow rate per unit area.

For any electrical system (network) the following equation, (Ohm's law) applies:

$$I = \frac{V}{R} \qquad [1.10]$$

In this equation V is the voltage which is also known as the potential difference. R is the electrical resistance and I is the electrical current.

Comparing the three equations, the heat flux is analogous to the electrical current, the temperature difference is analogous to the potential difference and the thermal resistance is analogous to the electrical resistance.

Having established an analogy any rules that apply to electrical networks can be applied to thermal networks. This analogy is a powerful tool used to analyse complex systems in conduction, convection and radiation heat transfer. It can also be applied to hydraulic systems. For most heat exchanger work our system is relatively simple – just a number of resistances in series.

The most important application is in fixing interfacial temperatures in between two thermal resistances in a network as illustrated below.

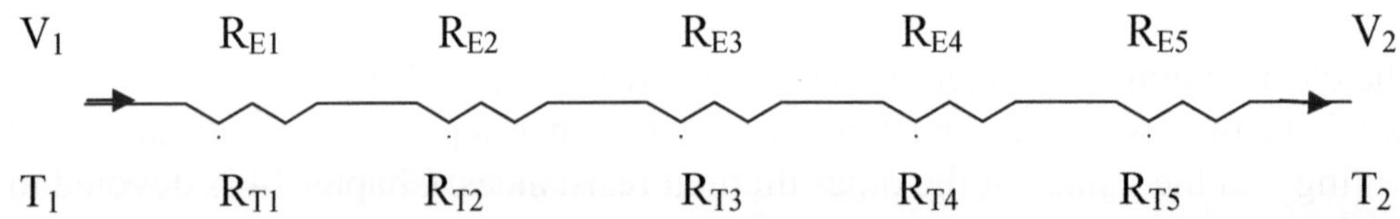

Figure 1.2

The voltage drop across any section of the electrical network is:

$$\Delta V = \frac{\Sigma R_E}{R_{ET}} (V_1 - V_2) \qquad [1.11]$$

Hence the temperature drop across any section of the network is:

$$\Delta T = \frac{\Sigma R_T}{R_{TT}} (T_1 - T_2) \qquad [1.12]$$

For example to find the temperature between resistances 3 and 4 using equation 1.12

$$\Delta T_{1\text{-}3} = \frac{(R_{R1} + R_{R2} + R_{R3})}{R_{TT}} (T_1 - T_2)$$

$$T = T_1 - \Delta T_{1\text{-}3}$$

Example 2

A heat exchanger has condensing steam on the shell side and a hydrocarbon on the tubeside. The heat transfer coefficient for the steam is 10 000 W/m^2K and its fouling resistance is 0.0001 m^2K/W. The heat transfer coefficient for the hydrocarbon is 500 W/m^2K and its fouling resistance is 0.0008 m^2K/W. The tubes are carbon steel with a thermal conductivity of 50 W/mK, an outside diameter of 19.05 mm and a thickness of 2.1 mm. The steam is saturated and condenses at a pressure of 4 bar (absolute pressure). At one section of the heat exchanger the hydrocarbon bulk temperature is 60°C. Establish the temperature profile at this section and calculate the overall heat transfer coefficient and the heat flux.

Solution

From thermodynamic tables such as reference 15 the temperature of the condensing steam is 143.6 °C. The total temperature difference $(T_1 - T_2)$ = 143.6-60 = 83.6 K and the inside diameter is (19.05-2 x 2.1) = 14.85 mm. The ratio of d_o/d_i is therefore 19.05/14.85 = 1.28. The best way forward now is to use tabulation as follows:

SECTION	THERMAL RESISTANCE W/m^2K	ΔT K	T °C
Hot fluid			143.6
Coefficient h_o	$R_{ho} = 1/h_o$ = 1/10 000 = 0.0001	83.6x0.0001/0.00382 = 2.2 K	141.4
Fouling f_o	$R_{fo} = f_o$ = 0.0001	83.6x0.0001/0.00382 = 2.2 K	139.2
Tube wall	$R_w = \Delta x/k$ = 0.0021/50 = 0.00004	83.6x0.00004/0.00382 = 0.9 K	138.3
Fouling f_i	$R_{fi@do} = f_i\, d_o/d_i$ = 0.0008 x1.28 = 0.00102	83.6x0.00102/0.00382 = 22.3 K	116.0
Coefficient h_i	$R_{hi@do} = d_o/h_i d_i$ = 1.28/500 = 0.00256	83.6x0.00256/0.00382 = 56.03 K	60.03
Cold fluid			60
	R_{TT} = 0.00382		

Table 1.1

The first step is to calculate each thermal resistance in the second column from the right. These are then added together to give the total thermal resistance R_{TT}. Equation

1.12 can then be used to calculate the temperature drop across each resistance in the third column. Finally the temperature after each resistance is tabulated in the fourth column. The final value of temperature after the last resistance should be approximately the same as the cold fluid bulk temperature. This of course is a check on the arithmetic from the addition of the resistances to the end. A mistake in calculating a thermal resistance however will not be picked up.

The overall heat transfer coefficient $U_{calc} = 1/\ R_{TT} = 1/0.00382 = 261.8\ W/m^2K$
The heat flux $Q/A\ = U_{calc}\ \Delta T\ =\ 261.8 \times 83.6 = 21\ 885\ W/m^2$
Also, heat flux $Q/A\ =\ \Delta T/R_{TT} = 83.6/0.00382 = 21\ 885\ W/m^2$

Note that the heat flux could have been calculated from any of the thermal resistances

E.g. across the inside fouling layer $Q/A = \Delta T/\ R_{fi@do}\ = 22.3/0.00102 = 21863\ W/m^2$
across the outside coefficient $Q/A = \Delta T/\ R_{ho}\ = 2.2/0.001\ = 22000\ W/m^2$

The slight differences in answer are due to rounding errors. For example the temperature drop across the outside coefficient was rounded up from 2.19 K. Even computers can do this. I have sometimes noticed that an input of say 100 in a HTRI program may be output as 100.001 or 99.999. This is because an SI input is converted to US customary units, the calculations are carried out in US units and then the output is converted back into SI units.

A plane boundary was used in the above calculation for the tube wall which should be justified. From the table above the wall resistance is 100 x 0.00004/0.00256 = 1.56% of the total thermal resistance. The error in not using thick cylinder theory is the same as that calculated in Example 1.1 i.e. 11.4%. The overall error in the U value is therefore 1.56 x 0.114 = 0.18%, this is a very small discrepancy when compared with the tolerances involved in the equations we use to calculate heat transfer coefficients and the uncertainty of fouling factor values. (The error would have been doubled if the tubes had been austenitic stainless steel instead of carbon steel because of the lower thermal conductivity.)

It is useful to have at a sketch of the temperature profile. This is shown in figure 1.3.

This emphasises the low temperature drops across the two external resistances of the steam heat transfer coefficient and the steam fouling resistance. The negligible temperature drop across the tube wall is shown as an almost parallel line.

It is also useful to introduce the term **surface temperature** at this point. There are clearly two possibilities – the surface of the fouling layer which is in contact with the fluid and the surface of the tube underneath the fouling layer. When surface temperature is part of a computer output it is important to know which surface is being referred to – normally it will be the surface of the metal, which will be almost the same as the average metal temperature.

Unless both fluids undergo isothermal processes the temperature profile will vary throughout the length of the exchanger. Normally it will be the profiles at inlet and outlet which will be the important ones. A hand calculation may be necessary to establish a value which is not in a computer output, e.g. surface of the fouling.

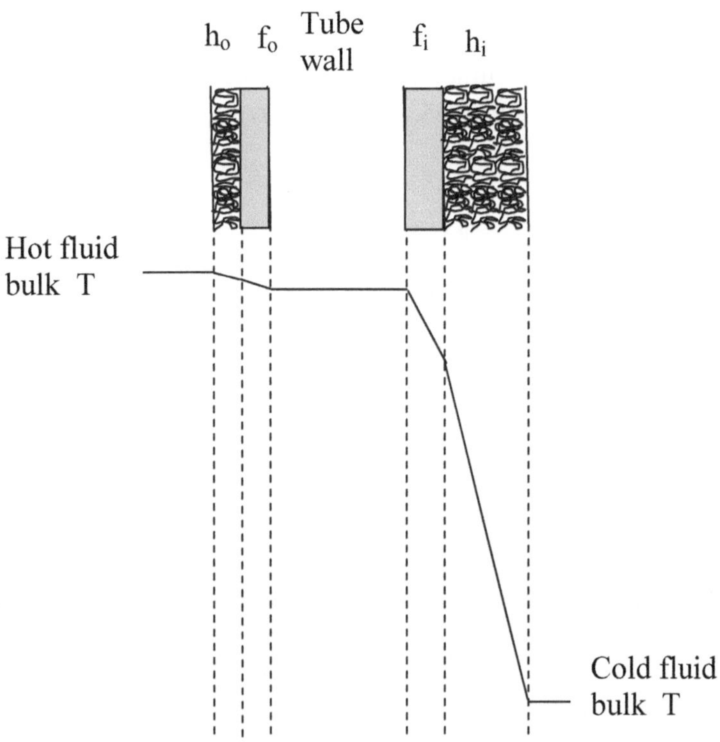

Figure 1.3

For the majority of heat exchanger designs the 'surface' temperature is not important. That still leaves quite a few when they are. For example you may have a fluid which rapidly fouls if it touches a hot surface or one which starts to 'freeze' if it touches a cold one. High or low temperatures can also affect the choice of metal.

ENERGY BALANCE

During normal operation a shell and tube heat exchanger is a steady flow system and therefore the steady flow energy equation applies to each stream. The full steady flow energy equation is as follows:

$$\dot{m}(e_1 + p_1v_1 + u_1^2/2 + z_1g) + Q + \dot{W} = \dot{m}(e_2 + p_2v_2 + u_2^2/2 + z_2g) \qquad [1.13]$$

e is the internal energy, pv is the flow work potential, $u^2/2$ and zg are the kinetic and potential energy terms respectively, Q is the heat flow rate and $\dot{W}$ is the mechanical work flow rate.

There is no mechanical work in a heat exchanger and the changes in kinetic and potential energy are very small (Although they are normally present.)

The internal energy term and flow work potential term are both intensive properties of a fluid and can be combined to form the property enthalpy as follows:

$$h = e + pv \qquad [1.14]$$

The internal energy of a fluid depends only on the temperature of the fluid. For a gas the flow work term remains constant since an increase in pressure produces a corresponding decrease in specific volume. The specific enthalpy therefore depends only on the temperature. Also for a gas $h = C_pT_A$ and a change in specific enthalpy becomes:

$$h_2 - h_1 = C_p (T_2 - T_1) \qquad [1.15]$$

For liquids the flow work potential can be significant. Liquids are relatively incompressible so that increasing the pressure does not produce a corresponding decrease in specific volume.

For example the change in specific enthalpy when water is compressed from 1 bar to 300 bar is approximately $p_2v_2 - p_1v_1 = 0.001(300x10^5 - 1x10^5)10^{-3} = 30$ kJ/kg. The process engineer will normally take care of this but it is important that a thermal design engineer has a good grasp of basic thermodynamics, knows where his equations come from and understands their limitations.

Combining equations 1.12 and 1.13 and eliminating terms which are zero or very small results in the following equation:

$$Q = \dot{m}(h_2 - h_1) = H_2 - H_1 \qquad [1.16]$$

This equation applies to both the hot and cold streams and can be combined with equation 1.1 to give:

$$\mathbf{Q = U_{req}\, A\, \Delta T_m = \dot{m}_H(h_{1H} - h_{2H}) = \dot{m}_C(h_{2C} - h_{1C})} \qquad [1.17]$$

This is one of the most important equations in thermal design work. Any reasonable computer program should flag a warning if there is any imbalance between the heat flow from the hot fluid to the cold fluid.

For two single phase streams equation 1.17 can be written as:

$$Q = U_{req}\, A\, \Delta T_m = \dot{m}c_{pH}\,(T_{1H} - T_{2H}) = \dot{m}c_{pC}\,(T_{2C} - T_{1C}) \qquad [1.18]$$

For two phase streams the specific enthalpies for equation 1.17 must be calculated from the following equation:

$$h = yh_V + (1\text{-}y)h_L \qquad [1.19]$$

Where y is the mass fraction of the gaseous/vapour) component. These terms are defined and discussed later in this chapter under the section on physical properties of fluids.

MEAN TEMPERATURE DIFFERENCE ΔTm

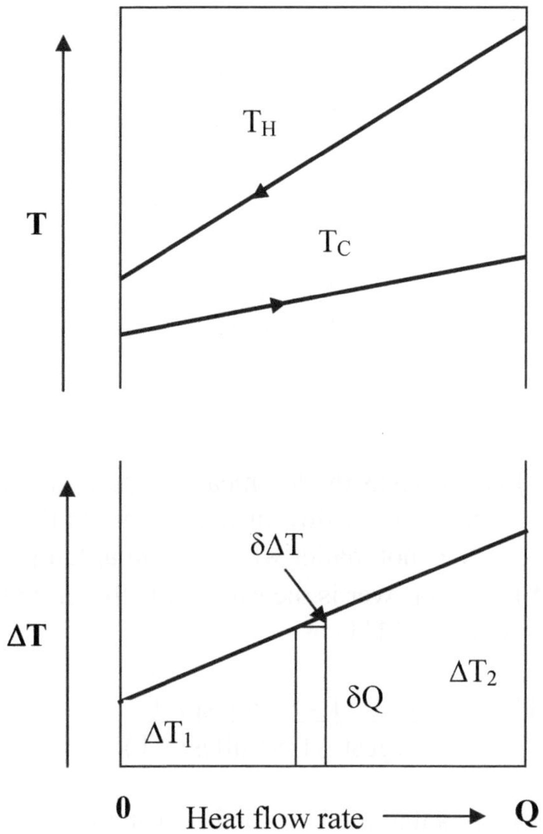

Figure 1.4

There are two graphs on figure 1.4. Both have heat flow rate as the horizontal axis between the limits of 0 and Q, the total rate of heat exchange for the unit. The top graph has a vertical axis of temperature in °C and the lower one a vertical axis of temperature difference ΔT in K. In the lower graph:

$$\Delta T_1 = T_{H2} - T_{C1}$$
$$\Delta T_2 = T_{H1} - T_{C2}$$

Where suffixes 1 and 2 refer to the inlet and outlet temperatures respectively. For an element δQ where the temperature difference is ΔT the change in ΔT is δΔT
The slope of the graph can therefore be found by either considering the element or the overall graph so that:

$$\text{Slope} = \frac{\delta \Delta T}{\delta Q} = \frac{\Delta T_2 - \Delta T_1}{Q} \qquad [1.20]$$

Applying equation 1.1 to the element

$$\delta Q = U\,\delta A\,\Delta T \qquad [1.21]$$

Eliminating δQ from equations 1.20 and 1.21

$$\frac{\delta \Delta T}{U\,\delta A\,\Delta T} = \frac{\Delta T_2 - \Delta T_1}{Q}$$

Rearranging

$$\frac{\delta \Delta T}{\Delta T} = \frac{(\Delta T_2 - \Delta T_1)}{Q} U\,\delta A$$

For the exchanger

$$\int_{\Delta T1}^{\Delta T2} \delta\Delta T / \Delta T = U \frac{(\Delta T_2 - \Delta T_1)}{Q} \int_0^A \delta A$$

Integrating

$$\ln(\Delta T_2/\Delta T_1) = U(\Delta T_2 - \Delta T_1)\,A/Q$$

Rearranging

$$Q = UA \frac{(\Delta T_2 - \Delta T_1)}{\ln(\Delta T_2/\Delta T_1)} \qquad [1.22]$$

LOG MEAN TEMPERATURE DIFFERENCE LMTD

The term $(\Delta T_2 - \Delta T_1)/ \ln(\Delta T_2/\Delta T_1)$ is known as the log mean temperature difference. The above development was for a counter current flow arrangement but the same equation applies to co-current flow. It does not matter which terminal temperature difference is taken to be ΔT_2 and ΔT_1, the answer is the same. It is therefore sensible to avoid negative quantities and remember LMTD as:

$$\text{LMTD} = \frac{\text{Largest } \Delta T - \text{Smallest } \Delta T}{\ln(\text{Largest } \Delta T/\text{Smallest } \Delta T)} \qquad [1.23]$$

This equation is based on the assumptions that the U value is constant and also that both specific heat values are constant. Although specific heats were not directly involved in the derivation, indirectly they were. A graph of temperature against heat flow rate, (a heat release 'curve') is only a straight line for a single phase fluid when the specific heat is constant.

The above derivation of LMTD is the most straight forward I have come across and was published by McAdams in 1933 (Ref 5). Many authors base their derivation on a graph of temperature against area or length. These derivations all directly involve specific heat values and the derivation is more complex. The reader will no doubt have seen one of these and noted that the graphs are not straight lines. This is because different areas of the exchanger transfer different amounts of heat which results in different changes in temperature. (I also like McAdam's approach because it is a lot easier to draw straight lines than curved ones on this bloody computer!)

There are many heat exchangers where the variations in the overall heat transfer coefficient, U and the specific heats are small enough for an overall LMTD to be applied to the whole exchanger. Many other cases are more complex and will be dealt with later.

Example 3

(a) An oil is to be cooled from 50°C to 30°C using cooling water which is heated from 20 °C to 25 °C. Calculate the advantage in using counter current flow compared with co-current flow.

(b) What would be the advantage if the advantage if the oil needed to be cooled to 25 °C and the cooling water outlet temperature was 30 °C?

Solution

(a) Mistakes are less likely to be made in calculating an LMTD if a tabulated procedure is carried out as follows:

	Counter current	Co-current
	50 → 30	50 → 30
	25 ← 20	20 → 25
	25 10	30 5
LMTD	15/ln2.5	25/ln6
	= 16.37	= 13.9

The advantage is therefore (16.37/13.9 - 1)100 = 17.8% and a counter current exchanger would need about 18% less surface area than a co current one. (We must assume that the U values are the same, therefore there would also be a reduction in pressure drops if the length were reduced.)

(b)	Counter current	Co-current
	50 → 25	50 → 25
	30 ← 20	20 → 30
	20 5	30 ?
LMTD	15/ln2.5	
	= 10.8	

The answer to this part of the question is that the counter current arrangement is the only one which could be used. If this had been part of an exam question I had set, I would have given full marks for this statement without any number crunching.

For heat to flow from a hot stream to a cold stream the temperature of the hot stream must be above that of the cold stream and clearly this is not the case for the co-current flow. There is of course no problem designing a counter current exchanger for the new conditions although it would be significantly larger than a unit to satisfy the conditions in part (a).

Conclusion

It is not possible to have a temperature cross in a co-current flow exchanger. (A temperature cross is when the hot fluid leaving the exchanger is at a lower temperature than the temperature of the cold fluid leaving the exchanger).

EXTREME VALUES

We will investigate extreme situations by further examples.

Example 4

The same oil as in the previous example needs to be cooled from 50°C to 30°C. The cooling water is still available at 20 °C but the site is a strange one with lots of cooling water and the temperature rise is only 1 K. Calculate the advantage in using counter current flow compared with co-current flow.

Solution

	Counter current	Co-current
	50 → 30	50 → 30
	21 ← 20	20 → 21
	29 10	30 9
LMTD	19/ln2.9	21/ln(30/9)
	= 17.8	= 17.4

The difference between the two arrangements is now very small – about 2%. In practical design terms there is no longer any significant advantage in using the counter flow arrangement.

If the reader cares to repeat the above calculation with even smaller cooling water temperature rises – say 0.5 K and then 0.1 K the difference will become less and less. For the limiting case of 0 K temperature rise:

	Counter current	Co-current
	50 → 30	50 → 30
	20 ← 20	20 → 20
	30 10	30 10
LMTD	20/ln3	20/ln3
	= 18.2	= 18.2

The cooling water is now undergoing an isothermal process, (not physically possible although it may be approached with a very large flow rate) and the LMTD for co-current flow and counter current flow is exactly the same.

In practical heat transfer work the only process were the cold fluid undergoes an isothermal process is a pure fluid boiling with a negligible pressure drop.

If the hot fluid is a pure substance condensing with a negligible pressure drop the process will also be isothermal and the LMTD will be the same for co-current and counter current flow.

When one fluid undergoes an isothermal process the LMTD is the same for co-current and counter current flow.

What if both fluids undergo isothermal processes? The top line of our equation for LMTD becomes zero and so does the answer. This is clearly nonsense. Changes in temperature difference were basic to the derivation of LMTD so that it can not be applied to this case. **There is no LMTD for a case where both fluids undergo an isothermal process and the mean temperature difference ΔT_m is simply $T_{HOT} - T_{COLD}$.**

Example 5

An oil is to be cooled from 100 °C to 21 °C in a counter current flow exchanger using cooling water with supply and return temperatures of 20 °C and 25 °C respectively. Calculate the change in LMTD when one of the cooling water temperatures is reduced by 1 K.

Solution

	Original Temperatures	T_{out} Reduced	T_{IN} Reduced
	100 → 21	100 → 21	100 → 21
	25 ← 20	24 ← 20	25 ←19
	75 1	76 1	75 2
LMTD	74/ln75	75/ln76	73/ln37.5
	= 17.14	= 17.32	= 20.14

Decreasing the cooling water outlet temperature by 1 K increased the largest terminal temperature difference by 1 K. This made only 1% difference to the LMTD. Decreasing the cooling water inlet temperature by 1 K increased the smallest terminal temperature difference by 1 K. This made 17.5% difference to the LMTD.

Conclusion

Making a change to the smaller of the two terminal temperature differences always has more effect on a design than changing the largest terminal temperature difference by the same amount.

CORRECTIONS TO LMTD

What we have covered so far is fine for co-current and counter current exchangers and there are many such units in addition to all 'pure' double pipe units. However the majority of heat exchangers utilise a combination of co-current and counter current flow. Clearly the overall temperature difference that we must use for these units is some value which lies in between these limiting values.

This is taken care of by introducing a correction factor, F. I am afraid that there are lots of 'F' and 'f' factors in heat transfer; fortunately they do not occur too often in the same section of a calculation. Introducing the F factor leads to the following equation:

$$\textbf{Mean temperature difference } \Delta T_m = F \times LMTD \qquad [1.24]$$

F factors are found from charts such as the one in figure 1.5. This is only one of the 13 charts published in TEMA for correcting the LMTD. Only the first 7 (TEMA figures T-3.2A→ T-3.2G) charts are for arrangements which consist only of combinations of co-current and counter current flow. These will be dealt with first.

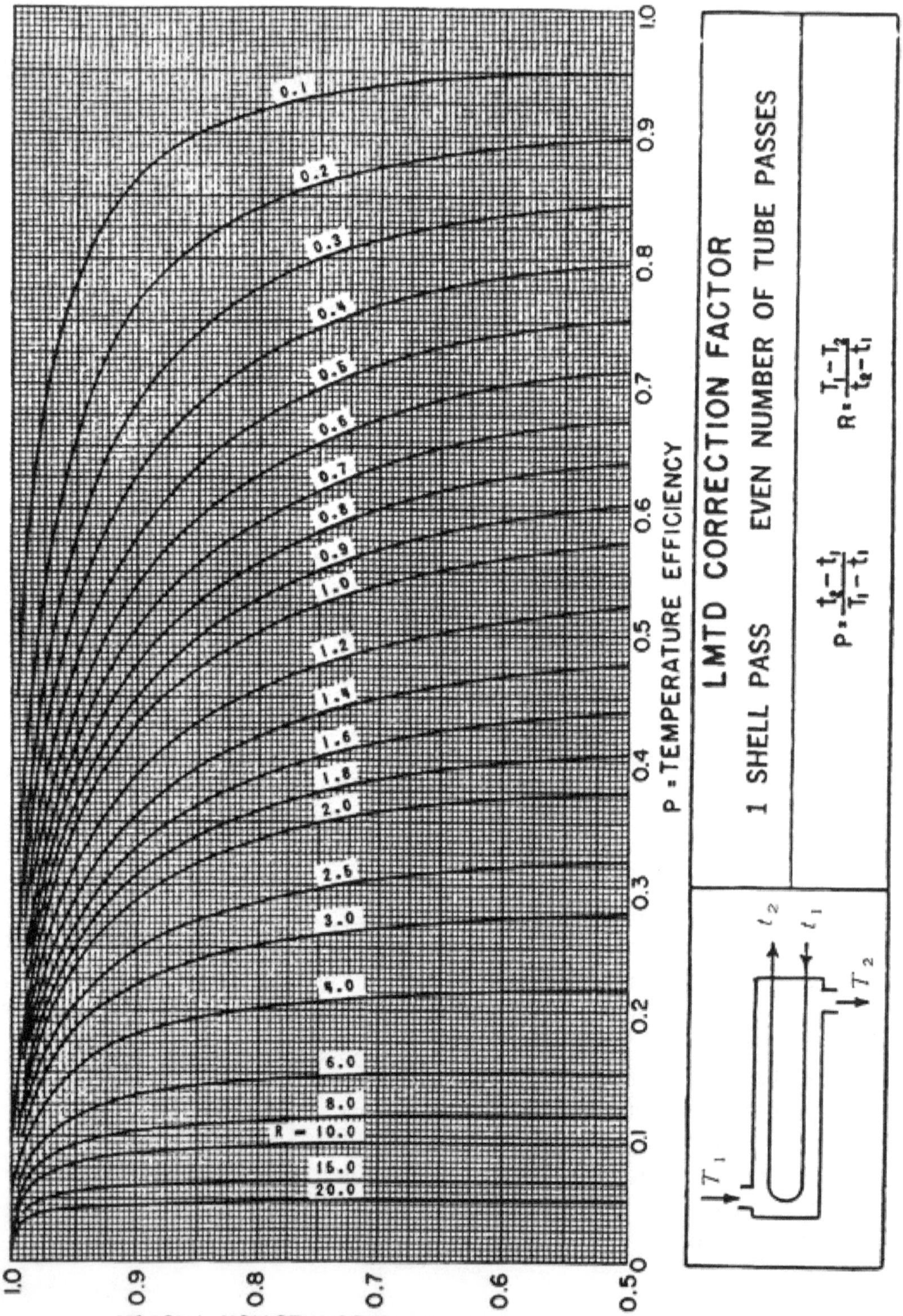

Figure 1.5 (TEMA T3.2A)

From TEMA Ninth Edition – With permission

The **heat capacity** of a fluid stream is defined as the product of the specific heat and the mass flow rate $\dot{m}C_p$. The **ratio** of the heat capacities of the two streams defines the value for **R**. Lines for different values of R are shown on the TEMA chart. Using equation 1.17, R can be expressed in terms of temperature difference as follows:

$$\textbf{Heat capacity ratio R} = \frac{\dot{m}_C C_{pC}}{\dot{m}_H C_{pH}} = \frac{T_{H1} - T_{H2}}{T_{C2} - T_{C1}} = \frac{\Delta T_{HOT}}{\Delta T_{COLD}} \qquad [1.25]$$

Where T_{H1} and T_{H2} are the hot fluid inlet and outlet temperatures respectively and T_{C1} and T_{C2} are the cold fluid inlet and outlet temperatures respectively.

The second term that we need to define is the temperature effectiveness with respect to the cold fluid stream. This is defined as the ratio of the heat transferred from the cold fluid to the heat which would theoretically be transferred if the cold fluid was heated to the temperature of the hot fluid at its inlet. It is designated by the letter **P** and is used as the horizontal axis on the TEMA chart. P is expressed as temperature efficiency on the chart. (I personally do not think that the term efficiency is appropriate to this situation.)

$$\textbf{Temperature effectiveness (efficiency) P} = \frac{T_{C2} - T_{C1}}{T_{H1} - T_{C1}} = \frac{\Delta T_{COLD}}{\Delta T_{MAX}} \qquad [1.26]$$

The use of values of R and P to determine an LMTD correction factor F is best demonstrated through examples.

Example 6

An oil is to be cooled from 100 °C to 90 °C in a shell and tube exchanger using cooling water with supply and return temperatures of 25 °C and 40 °C respectively. The exchanger has one shell side pass and four tube side passes. Calculate the mean temperature difference for heat exchange.

Solution

Once again a tabulation method is useful to solve this problem.

Temperatures °C & Terminal ΔT's	Stream ΔT's	R P & F
100 → 90	10	R = 10/15 = 0.67
40 ← 25	15	P = 15/75 = 0.2
60 65	75	F = 0.99 – from chart

Table 1.2

LMTD = (65-60)/ln(65/60) = 62.48

MTD or ΔT_m = F x LMTD = 0.99 x 62.48 = **61.8 K**

The advantage of this tabulation method is that R is the first number in the stream ΔT's column divided by the number below it and P is the second number in the same column divided by the number below it. Simplifying the presentation of calculations reduces the chance of arithmetical errors.

An observant reader will have noted that the above calculations dealt only with hot and cold stream temperatures without any reference to which fluid was flowing in the tubes and a TEMA chart was then used which identified P and R in terms of shell side and tube side fluids.

Recalculating P and R using tube side and shell side fluids as in TEMA.

$$R = \frac{\Delta T_{\text{shell side}}}{\Delta T_{\text{tube side}}} = \frac{15}{10} = 1.5 \qquad P = \frac{\Delta T_{\text{tube side}}}{\Delta T_{\text{MAX}}} = \frac{10}{75} = 0.13 \qquad \text{from chart } F = 0.99$$

The new values of R and P result in the same value for F. In this example it was quite difficult to read the value for F off the chart and the reader may have come up with a slightly different number.

Example 7

The same oil as in example 6 is to be cooled from 100 °C to 47.5 °C in a shell and tube exchanger using the same cooling water temperatures. The exchanger has one shell side pass and four tube side passes. Calculate the mean temperature difference for heat exchange.

Solution

Using the tabulation method..

Temperatures °C & Terminal ΔT's	Stream ΔT's	R P & F
100 → 47.5	52.4	R = 52.5/15 = 0.67
40 ← 25	15	P = 15/75 = 0.2
60 22.5	75	F = 0.9 – from chart

LMTD = (60-22.5)/ln(60/22.5) = 38.23
MTD or ΔT_m = F x LMTD = 0.9 x 38.23 = **34.4 K**

Recalculating P and R using tube side and shell side fluids as in TEMA.

$$R = \frac{\Delta T_{\text{shell side}}}{\Delta T_{\text{tube side}}} = \frac{15}{22.5} = 0.29 \qquad P = \frac{\Delta T_{\text{tube side}}}{\Delta T_{\text{MAX}}} = \frac{52.5}{75} = 0.7 \qquad \text{from chart } F = 0.9$$

In this example it was a lot easier to read F from the chart. **Also note that that the two methods of selecting F involved quite different regions of the chart.**

Example 8

The same oil as in example 6 is to be cooled from 100 °C to 40 °C in a shell and tube exchanger using the same cooling water temperatures. The exchanger has one shell side pass and four tube side passes. Calculate the mean temperature difference for heat exchange.

Solution

Using the tabulation method.

Temperatures ^{o}C & Terminal ΔT's	Stream ΔT's	R P & F
100 → 40	60	R = 60/15 = 4
40 ← 25	15	P = 15/75 = 0.2
60 15	75	F = 0.8 – from chart

LMTD = (60-15)/ln(60/4) = 16.62

MTD or ΔT_m = F x LMTD = 0.8 x 16.62 = **13.29 K**

Recalculating P and R using tube side and shell side fluids as in TEMA.

$$R = \frac{\Delta T_{\text{shell side}}}{\Delta T_{\text{tube side}}} = \frac{15}{60} = 0.25 \qquad P = \frac{\Delta T_{\text{tube side}}}{\Delta T_{MAX}} = \frac{60}{75} = 0.8 \qquad \text{from chart } F = 0.8$$

Discussion of the results of examples 6 to 8

The LMTD calculated in each case was the LMTD for counter current flow. This is always higher than the value for co-current flow and when both arrangements are present in an exchanger it must be the higher value which is corrected with a factor which is less than one. The only exception to this is when one fluid undergoes an isothermal process when F = 1.

In the first example F had a very high value of 0.99. This is because both fluids had relatively small temperature changes, particularly the hot oil and the terminal temperature differences were both high. The smaller the temperature change in a process the closer it becomes to an isothermal process

In the third example F had a value of 0.8. The value F = 0.8 is always taken as a limiting value when used in hand calculations. It occurs when the temperature of the hot fluid and the cold fluid leaving the exchanger have the same value. This ensures that there can be no point in the exchanger where the cold fluid has a higher temperature than the hot fluid. Some modern computer programs will allow a small overlap in the fluid outlet temperatures which result in an F value slightly below 0.8. The justification for this is that the exchanger is subdivided into many sections for the calculations and in each section the heat is always flowing in the 'right direction'. I always feel a little uneasy about this because I am very conscious that if the overlap is a bit more than the design value then the performance would be significantly reduced. However if a small overlap avoids using two shells instead of one and the resulting F factor was say 0.78 I would just add a bit more oversurface to the design and accept it.

In the second example F had a value of 0.9 which resulted from the hot stream outlet temperature being in between the values in the other two examples.

Using values of R and P calculated from the hot and cold steams resulted in exactly the same value for F as when they were calculated by using tube side and shell side fluids (as in TEMA). It follows that whether the hot fluid is on the shell side or tube side is irrelevant to calculating F. The former approach has the advantage being easier when used in a tabulated calculation.

The two methods of calculating F involved using different areas of the TEMA chart. This did not result in an advantage to either method in these cases. However there may be occasions when one method may be more accurate than the alternative. For example suppose one method results in a value for R which is almost parallel to the value for P. The value for F becomes more of a guess than an estimate. Now is the time to use a second method which may make it easier to estimate F in a different area of the chart.

When reading values off the chart note that the interpolation between lines of constant R are not linear.

Conclusions for F factors in multi pass exchangers with combinations of co-current and counter current flow <u>only</u>.

(1) Base the LMTD on counter current flow.
(2) It does not matter which fluid is in the tubes.
(3) Use the method shown in the tables.
(4) If (3) results in F being in an area of the chart which is difficult to read try the other method.
(5) Before you do any calculations look at the temperatures involved and try to anticipate what you expect the value of F to be.
(6) Your computer program will do most of the above for you but F should be part of the output and studying it may lead you to consider an alternative path to finding an optimum design.

SPECIAL CASE – ONE SHELL PASS & THREE TUBE PASSES

The relevant chart in TEMA (T-3.2G) is the last one (the 7th) for which the flows are all either counter current flow or co-current flow as illustrated in figure 1.6. As such all the equations in the previous section can be applied. What makes this arrangement special is that it is possible to have two passes in counter current flow and one in co-current flow (as shown) or visa versa. There is no chart for two co-current passes and only one counter current pass and I can not think of any good reason for having one. Such a chart could be drawn and for any set of process conditions it would produce a lower value of F than the one from figure 1.6.

The important thing to remember is to **get the shell inlet and outlet nozzles at the right ends in any setting plan or sketch** from which a setting plan may be drawn later. If this is not done correctly the unit will under perform. The same would apply to a one shell pass and five tube pass arrangement. This however is a very unusual arrangement and the difference in F values for the two possibilities would be much lower.

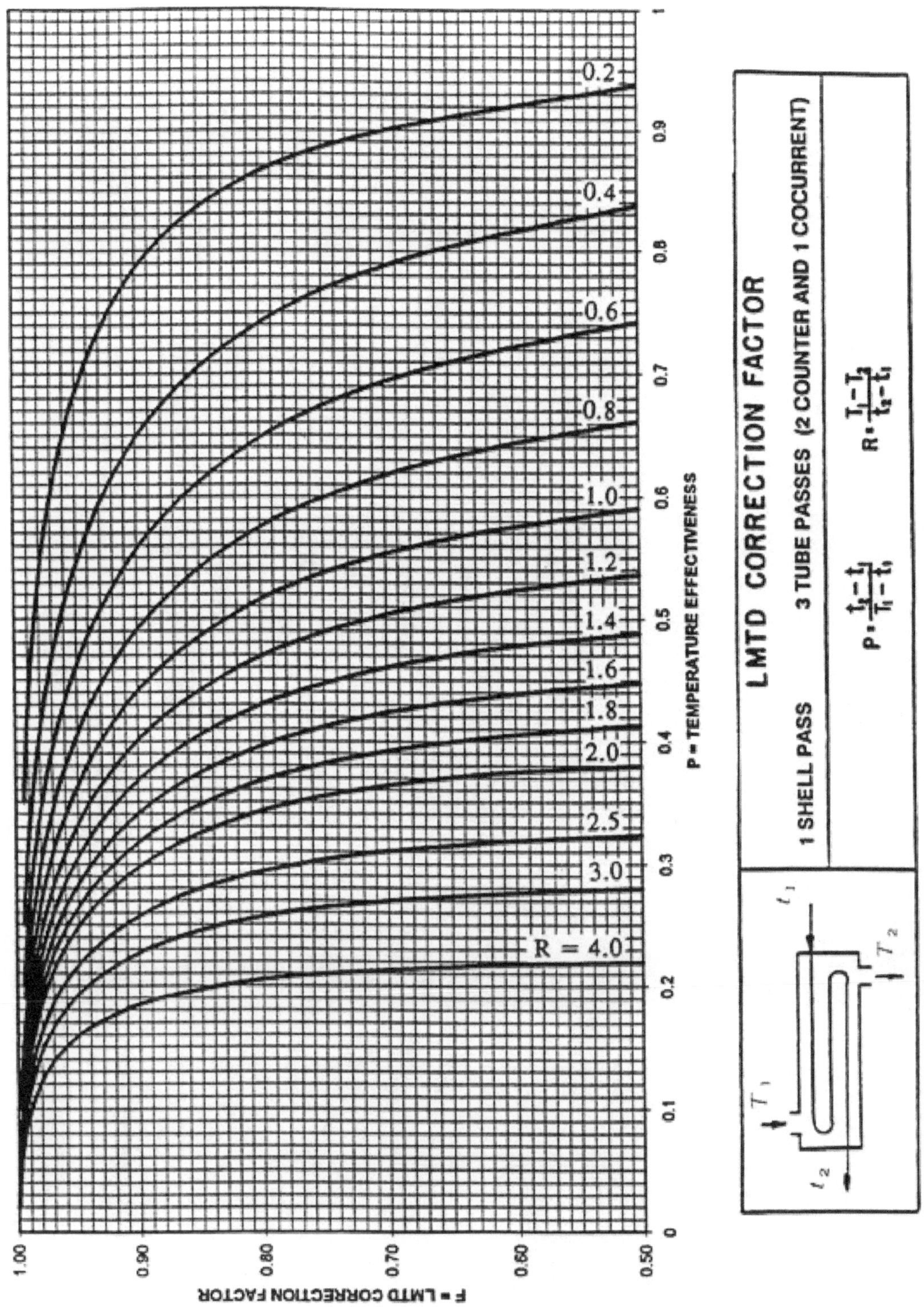

Figure 1.6 (TEMA T3.2G)

From TEMA Ninth Edition – With permission

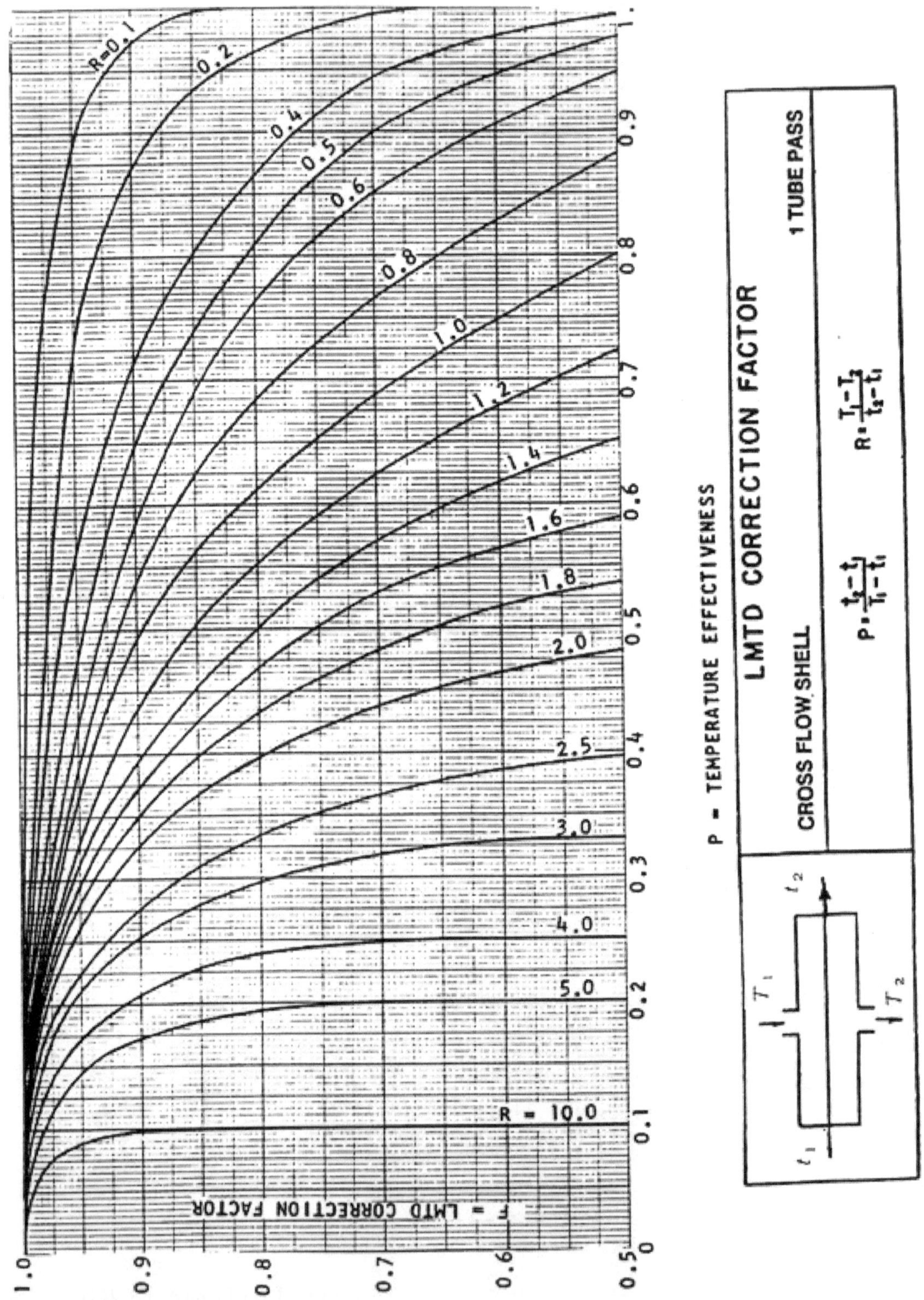

Figure 1.7 (TEMA T3.2H)

CROSS FLOW ARRANGEMENTS

Figure 1.7 (TEMA T3.2H) is the correction factor F for cross flow. This is a pure cross flow case which for a reasonable length of exchanger would normally have more than one shell side nozzle at the top and at the bottom of the exchanger.

Example 9

A gas is to be cooled from 100 °C to 60 °C in a cross flow shell and tube exchanger using cooling water with supply and return temperatures of 25 °C and 45 °C respectively. The exchanger has one tube side pass and the water flows through the tubes. Calculate the LMTD correction factor F using the TEMA definitions of R and P. If the gas had been flowing through the tubes would F be the same?

Solution

$$R = \frac{\Delta T_{\text{shell side}}}{\Delta T_{\text{tube side}}} = \frac{100-60}{45-25} = \frac{40}{20} = 2$$

$$P = \frac{\Delta T_{\text{tube side}}}{\Delta T_{MAX}} = \frac{45-25}{100-25} = \frac{20}{75} = 0.267$$

From the chart **F = 0.955**

Repeating this for the hot fluid in the tubes gives:

$$R = \frac{\Delta T_{\text{shell side}}}{\Delta T_{\text{tube side}}} = \frac{45-25}{100-60} = \frac{20}{40} = 0.5$$

$$P = \frac{\Delta T_{\text{tube side}}}{\Delta T_{MAX}} = \frac{100-45}{100-25} = \frac{40}{75} = 0.533$$

From the chart **F = 0.935**

The difference between the two values for F is hardly dramatic but it is sufficient to demonstrate that the two values are not the same. Defining R and P in terms of tube side and shell side flows as in TEMA is the correct method and defining R and P in terms of stream temperatures is wrong.

Discussion of the results.

This applies to all the TEMA charts from 8 to 13 (TEMA T3.2H→ M). The split flow, divided flow and double split flow all involve a cross flow component and the calculation of the F factor must be with R and P values defined in terms of tube side and shell side flows.

Having a gas on the shell side of a cross flow water cooled exchanger is standard practice. A cross flow design has a very low shell side pressure drop which is essential for units such as the cracked gas coolers on an ethylene plant. I can not think of any situation where the gas would be in the tubes, this is a contrived case to make a point about calculating F factors.

It should now be clear why **TEMA defines R and P in terms of tube side and shell side fluids – for consistency.** The TEMA method is correct for any situation.

However the vast majority of cases involve exchangers where defining R and P in terms of hot and cold streams has an advantage. The reader should understand this and design accordingly. Your computer program will take care of this but you should understand what it going on inside the program.

Conclusions

Defining R and P in terms of shell side and tube side flows is correct every time but for most situations defining R and P in terms of hot and cold streams has advantages.

E STREAM TEMPERATURE PROFILE DISTORTION FACTOR

The HTRI program introduces an additional correction to the LMTD. The HTFS program does not. I am not sure what all the other programs that are now available do. Unfortunately many designers have only ever used one program and may not be aware of this.

The E stream is discussed in detail in chapter 3 and is illustrated in figure 3.17. (The Tinker diagram) It is the stream which flows through the baffle/shell clearance.

My interpretation of the HTRI position is as follows: the E stream flows from inlet to outlet relatively uninterrupted therefore it must distort the LMTD. It is not in contact with the heat transfer surface and there will be no mixing with the other fluid streams. It will arrive at the shell outlet still at the shell inlet temperature. Only then will it mix with the other fluid streams in the outlet nozzle.

My interpretation of the HTFS position is as follows; in reality there can not be an uninterrupted E stream as shown in figure 3.17. The baffle/shell clearances will vary around the circumference of the baffle. This is obvious at the bottom of the shell where baffles must have some contact with the shell in order to support the tubes. With horizontally cut baffles (the norm for single phase fluids) alternate baffles will be in contact with the bottom of the shell. The other baffles may have a larger clearance at the top of the shell, and then again they may not. Shells are not perfect cylinders they vary in diameter depending on the tolerances used. There may be some baffles in contact with the top of the shell. It is not unreasonable to assume that sufficient mixing of the E stream with the other streams takes place throughout the length of the shell to make an E stream correction factor unnecessary.

When using the HTRI program the E stream distortion factor is not significant for the majority of designs. It comes into play:

(1) When the shell side fluid has a high viscosity;
(2) When the baffle pitch is small compared with the diameter, (at/or close to a generally accepted limit of shell diameter to baffle pitch ratio, D_s/P_b, of 5);
(3) When the shell side fluid undergoes a large change in temperature;
(4) Where there is a small difference between the outlet temperatures of the two fluids (The LMTD correction factor, F will be very low – approaching or at a value of 0.8)

Normally most of these conditions all are present when E stream distortion becomes significant and when they all apply the value may fall to as low as 0.8.

Even a value of E stream distortion as low as 0.9 is not normally acceptable and whatever its value is it should be maximised to minimise its effect. Listed below are some of the measures you can take to do this:

(a) One obvious method of reducing E stream distortion is to reduce the shell/baffle clearance. There are practical limits to this manoeuvre which an inexperienced designer may wish to discuss with a mechanical engineer. Table RCB-4.3 in TEMA lists **standard** baffle/shell clearances. Section RCB-4.3 makes clear that a fabricator should take these to be **maximum** values (except where this clearance has no effect on heat transfer – such as condensing steam). The minimum value depends on the fabricator and the length of the exchanger. It is unlikely that you have ever done it but just try and visualise the problems of getting a bundle into a shell. The longer the shell the more difficult it is. Bearing this in mind investigate the effect of reducing the baffle/shell clearance. Just shaving off one millimetre can sometimes have a significant effect. For example if the shell diameter is 1200 mm the standard clearance is 6.4 mm. Most manufacturers could manage with 5 mm. If you use this value in your computer run it is essential to have a note on the thermal data sheet such as: *The baffle/shell clearance should be a **maximum** of 5 mm.*

It can of course be reasoned that since the TEMA values are maximum values the average value must be less than this value and you are justified in using a tighter clearance in your computer run without any note. I know that when bundles are first inserted into a shell there may well be a fitter with a big file at hand. Emphasising the importance of this clearance to the thermal design should encourage his supervisor to make him keep his file in his tool box or at least use it very sparingly.

(b) Any measure which increases the flow rate of any of the thermally active streams will automatically reduce the flow of the inactive E stream. The designer should investigate the effect of increasing the baffle pitch and the baffle cut (individually not both and the same time). Either change will of course reduce the shell side heat transfer coefficient but it is the overall performance that is important. Particularly if fouling is the dominant thermal resistance the reduction in U value may be more than compensated for by the improvement in the E stream distortion factor.

The same principle must of course apply to increasing the tube/baffle clearance. This clearance is very crudely specified in TEMA – the **standard** is either 0.4mm or 0.8 mm depending on the baffle pitch. TEMA states that these values may be reduced when vibration is a problem but says nothing about increasing them. With viscous fluids vibrations are the one thing we do not have to worry about. I can see no reason why the baffle/tube clearance can not be increased. If you do this it is of course essential to have a note on the thermal data sheet.

(c) When the above manoeuvres are not sufficient to give a satisfactory value for the E stream distortion factor (say above 0.85) one simple solution is to use two shells in series.

Changing to two shells in series attacks the problem on three distinct fronts. (Everything in the previous list except of course the first one – the fluid viscosity.)

The shell diameter will be reduced and to keep the pressure drop within the allowable limit the baffle pitch will need to be increased so that D_s/P_b will be reduced. The baffle cut may also need to be increased. This will increase the total flow through the active streams thus reducing the flow through the E stream.

For each shell the fluid will have a reduced change in temperature on the shell side which is so important to the E stream distortion factor.

There will be a substantial increase in the LMTD correction factor F which reduces the required surface area and contributes to a further reduction in the shell diameter.

Overall if in doubt – use two shells.

As with all tricky designs you should discuss them with the process engineer – if he/she is accessible. How a small change in outlet temperatures effects the overall process is crucial to how far you should be prepared to stretch your design into "dodgy regions".

Whatever program you are using the E stream must be inactive for part of its path, (when it crosses the baffle edges), even if it mixes in between baffles. It should therefore always be kept to a minimum. The advantage of the HTRI program is that there is a clear alert to possible problems. However with any program keeping an eye on the E stream as well as just trying to maximise the A and B steams, (the most effective for heat transfer), is very important.

So who is right and who is wrong? I have discussed this many times over the years with people whose opinions I value. I am still not convinced that there is a right and wrong. All modern programs involve complex weighting of the various coefficients with suitable safety factors etc. so that two programs may give the same overall result whilst following different paths to that result. You are the victim of the program at your disposal – unless of course you have access to more than one – which can often be very useful.

If the reader would like more details of the E steam distortion factor Mukherjee – Ref 1 – devotes eight pages to it. I can recommend reading it but at £376, I could not recommend buying it!

TEMPERATURE CROSSES & SHELLS IN SERIES

We have already seen that if an exchanger has a combination of counter flow and co-current flow arrangements the maximum heat that can be exchanged in a single shell is when the temperatures of both fluids leaving the exchanger are the same. (LMTD correction factor is 0.8). There are many process requirements for exchangers where the cold fluid must leave the exchanger at a significantly higher temperature than the hot fluid leaves the exchanger.

Whenever the temperature of the hot fluid leaving the exchanger is lower than the temperature of the cold fluid leaving the exchanger there is said to be a temperature cross. There is no theoretical limit to the number of temperature crosses there can be in a heat exchanger but for each cross it is necessary to add another shell in series. An example is illustrated in figure 1.8.

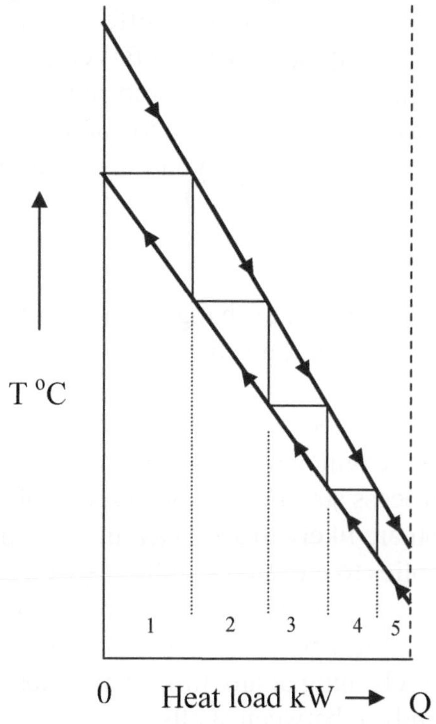

Figure 1.8

The first step is to plot the temperature profiles of the two streams against the heat duty as shown. A horizontal line is then drawn from the cold fluid outlet temperature until it meets the hot fluid profile. From this point a vertical line is drawn downwards until it meats the cold fluid temperature profile. This identifies the limiting temperatures for the first shell. This is the hot shell which would normally be labelled shell 1 or shell A. This process is repeated to identify the limiting temperatures for shells 2, 3 and 4. If this section was the end of the temperature profiles four shells would be required and each shell would have an LMTD correction factor, F of 0.8. However there is still a section of profile left. Taking section 5 (shell 5) in isolation it is clear that the hot stream outlet temperature is well above the cold stream outlet temperature and the F factor for this section would be well above 0.8 In practice with 5 shells required the temperatures would all adjust slightly and each shell could have an F correction factor of slightly above 0.8.

The first six charts of TEMA LMTD correction factors, F are for 1 $\rightarrow$ 6 shells in series with any even number of tubeside passes. You are unlikely to come across a case requiring more than six shells in series. If you did require say eight shells in series your computer program may be able to handle it but if could not you would split the design up into two designs of four shells in series. You would use a diagram like figure 1.8 to establish a break point which would give you the terminal temperatures in between the two banks of four shells. At the end of the design you would of course have to prepare a thermal data sheet for eight shells in series. This will necessitate a few hand calculations to weight the data from your final two computer runs to insert on the single data sheet.

So far we have assumed that all the shells have been identical. This need not be the case. The most common deviation in a series train of shells is when the temperature range requires a change of materials. One or more of the hot shells may require an alloy steel whilst carbon steel is adequate for the colder shells. The difference in thermal conductivity of the tube material will have little effect but any change in tube gauge certainly will. I have designed many units with 16 BWG average wall tubes in the hot shells and 14 BWG minimum wall tubes in the cold shells. Some computer programs may handle this complex input and allow you to carry out the design in a single run others may not. If your program does not you must split up the design into two computer runs as explained above with your break point being the material break point.

There are other complications to designing an exchanger with shells in series which should be taken into account before you start any calculations.

The first is can the shells be stacked one above another? Sorting this one out may involve both process and mechanical engineers. Most cases which require more than two shells in series involve dirty process streams that need removable bundles that can be mechanically cleaned. If you are likely to approach the job spec limit for the weight of a bundle check if this applies to elevated bundles as well as to shells at grade. With smaller shells it is sometimes possible to stack three shells but normally the limit is two. At the other extreme there may be a process requirement that any shell can be taken out of service for cleaning at any time. This rules out any stacking and involves a lot of valves and bends in between shells.

If there are any bends, and worse still valves and bends in between shells who is responsible for the pressure drops in them? If a series train can be accommodated in a single stack there is no problem. The allowable pressure drops are between the inlet and outlet nozzles and there is no loss of pressure which your program will not take account of. If there are discontinuities in between shells is it your responsibility to take the additional losses into account within your allowable pressure drop or have process made allowances for this? If it is your responsibility you will have to sort out what sort of valves will be used and what sort of bends (long or short radius). It is then necessary to take this into account in your allowable pressure drops and carry out the required calculations to confirm your estimate.

Note that because the greatest temperature difference between the two streams is at the hot end, the hot shell transfers the most heat and the contribution of each shell decreases as the temperature differences fall. Looking at this from a different angle it is clear that more surface area is required to transfer a fixed quantity of heat at the cold end of the exchanger. This reinforces a statement made earlier in the section on LMTD that changing the lesser of the two terminal temperature differences has more effect on the design than changing the larger one.

Conclusions

> **(1) The number of shell in series required is always equal to or one more than the number of temperature crosses.**
> **(2) Some trains of series shells can be very challenging and should only be attempted by experienced thermal design engineers.**

MTD, ΔT_m FOR NONE LINEAR HEAT RELEASE CURVES.

The first example is relatively straight forward. Figure 1.9 is typical for a hot stream of mixed hydrocarbons being cooled by cooling water with some of the heavier hydrocarbon fractions condensing. The term cooling curve is also used for convex profiles which increase the area in between straight lines drawn between the inlet and outlet temperatures of the two fluids.

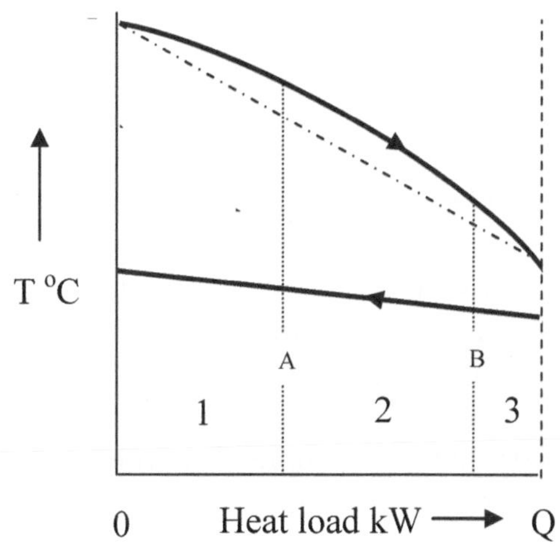

Figure 1.9

If the curvature is not taken into account we would be calculating an LMTD for a hot fluid following the dot – dash straight line. This is clearly not satisfactory because the hot fluid is at a higher temperature than this line except at the inlet and outlet. The first thing to do in this situation is to split the hot curve up into a number of sections which approximate to straight lines. The greater the curvature the higher the number of sections required, for the curve in figure 1.9 three sections are sufficient. The sections do not need to have equal increments of heat duty and the end points may not even be on the heat release curve. To 'straighten' the curve, for each straight line the area above the heat release line should approximately equal the area below the line. I did consider trying to illustrate this on figure 1.9 but my computer graphics are not up to that yet although they are improving 'weakly' as the book progresses.

After straightening the line, temperatures along lines A and B can be read off for the hot and cold fluids. A tabulated calculation is then carried out as follows:

SECTION		1		2		3	
T_{HOT}	T_{IN}	→	$T_{hot\ A}$	→	$T_{hot\ B}$	→	T_{OUT}
T_{COLD}	T_{OUT}	←	$T_{cold\ A}$	←	$T_{cold\ B}$	←	T_{IN}
ΔT	ΔT_{HOT}		ΔT_A		ΔT_B		ΔT_{COLD}
LMTD		LMTD 1		LMTD 2		LMTD 3	

Table 1.3

$$Q_{\text{SECTION 1}}/\text{LMTD 1} =$$
$$Q_{\text{SECTION 2}}/\text{LMTD 1} =$$
$$Q_{\text{SECTION 3}}/\text{LMTD 1} = \underline{\hspace{3cm}}$$
$$\sum Q_{\text{SECTIONS}} / \text{LMTD}_{\text{SECTIONS}}$$

$$\text{EXCHANGER LMTD} = Q_{\text{TOTAL}} / \sum Q_{\text{SECTIONS}} / \text{LMTD}_{\text{SECTIONS}}$$

$$\text{EXCHANGER MTD} = \text{EXCHANGER LMTD} \times F$$

This procedure is known as weighting the LMTD on a duty basis. In theory it can only be used where the overall heat transfer coefficient is constant. In practice an average value has to be used for hand calculations.

The LMTD correction factor, F is calculated in the normal way using the terminal temperatures.

An astute reader will have realised by now that the above procedure is a form of graphical integration. Alternative methods of graphical integration, such as the classical 'Simpson's Rule' are perfectly acceptable alternatives but would be more time consuming. My other book – Ref 4 – contains an example of the application of Simpson's Rule in the section on mass transfer.

We must now progress to investigate more complex heat release curves. Figure 1.10 shows a heat release 'curve' for a pure fluid which starts off as a superheated vapour and leaves the exchanger as a sub cooled liquid.

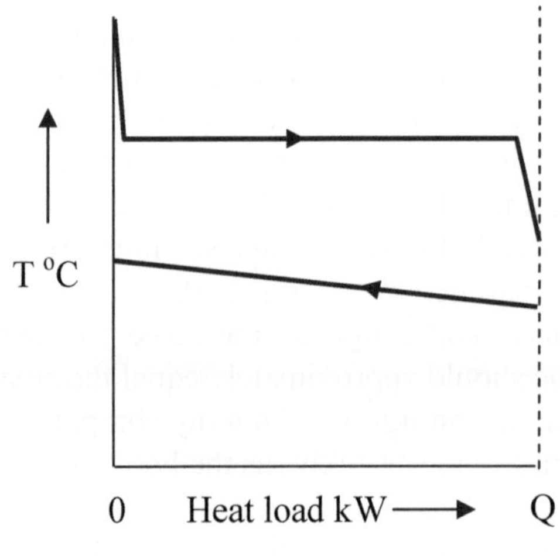

Figure 1.10

To have significant superheating and sub cooling in the same exchanger is not desirable and will be discussed elsewhere. However it could happen. In this situation there will be a high U value in the condensing section, a relatively poor one in the sub cooling section and an even worse one in the desuperheating section. Without drawing any more lines it is clear that desuperheating and sub cooling duties are quite small when compared with the condensing duty (heat load). However because of the drastic differences in shell side heat transfer coefficients they may require a substantial proportion of the surface area. In this situation it is necessary to carry out three design calculations, one for each region. The desuperheating region, next to the shell inlet nozzle will require a close baffle pitch. The isothermal condensing region will require a large baffle pitch. The liquid sub cooling region should ideally have a close baffle pitch which contradicts the requirement for the condensing region and yet the two processes are taking place in the same section of the shell – condensing at the top and sub cooling at the bottom. This will be discussed more fully in the section on condensers.

I think I have written sufficient in this chapter for the reader to appreciate the increase in complexity for this case. **When an area has been calculated for each section and the time comes to fill in the thermal data sheet the MTD should be weighted using the area of each section and not the heat loads.**

Of course nowadays the computer will do all this for you, or will it? Some of the programs I have used are not very good at dealing with large sub cooling loads in condensers; they usually admit it and give you a warning message – take head of it. The most satisfactory way out of course is to have a separate sub cooler but this is not always possible.

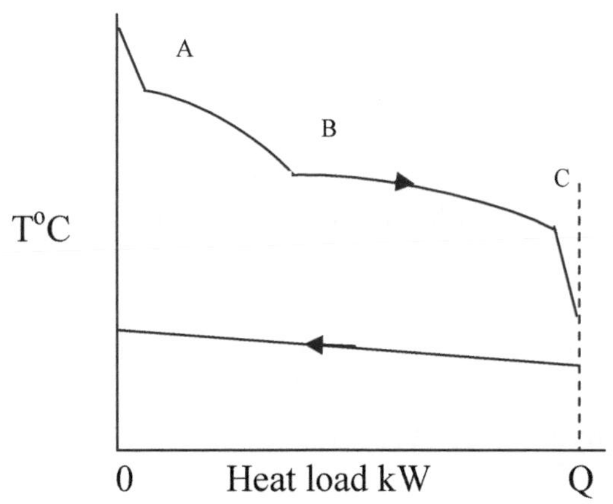

Figure 1.11

Figure 1.11 above shows a curve which has a change of direction in the condensing section as well as at the end of the desuperheating region and the start of the sub cooling region. This is typical of a hydrocarbon stream containing water vapour. At inlet both constituents are in the vapour phase. Then the hydrocarbon starts to condense whilst all the water remains as a vapour. Next the water starts to condense together with some of the lighter fractions of the hydrocarbon. Finally most of the hot stream has been condensed and the main heat load is liquid cooling.

Point A where the hydrocarbon starts to condense is the hydrocarbon dew point. Point B is where the water starts to condense is the water dew point. Quite severe changes in direction of a heat release curve can take place at the dew points. Point C is a bit contrived in the diagram for ease of drawing. There would be a more gradual change of direction and because there may still be some vapour present the last section and the sub cooling load would still be a slight curve.

Points A, B and C are known as **turning points** and with complex streams there may be more than three of them. The definition of **a turning point is a point on a heat release curve is where the profile changes direction from the path it had been following previously.**

In figure 1.11, I assumed that the cooling medium was water again. This of course is not always the case. In a feed/effluent exchanger the hot fluid may start as all vapour and end up mainly liquid whilst the cold fluid starts off as a liquid and finishes as mainly vapour. When a counter current exchanger can not be used these types of units frequently require several shells in series. There is now the extra complication of both streams following irregular paths with a number of turning points in each stream. I have attempted to represent this in figure 1.12. I have had to use a number of straight lines most of which would not be quite straight.

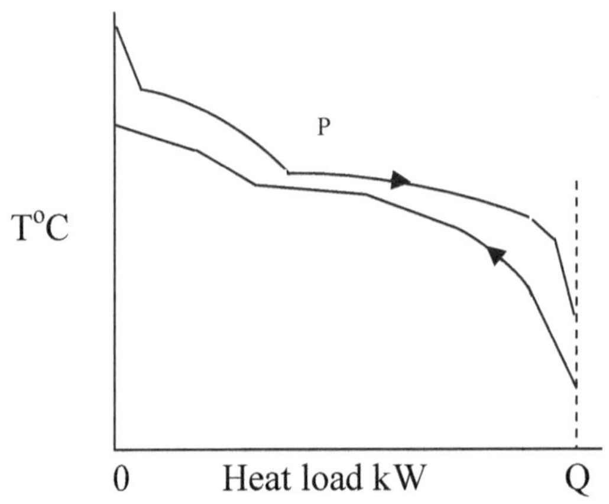

Figure 1.12

I will not bore you with details of how to approach such a problem using hand calculations although I can still vaguely remember one I designed which had to be split up into five different sections. I will never design one of these again using hand calculations and I doubt if anyone else ever will.

There is one vital point that must be understood concerning the input to a computer program. **It is very easy to feed information into the program which will result in a significant error to the MTD.** Remember that curves needed to be approximated to straight lines. The section in between points that you input to a program will be treated as straight lines. The easy way for a process engineer to prepare a heating or cooling curve for the process data sheet is to use equal increments of temperature or heat load (they usually use temperature). This method can miss out all the turning points on a heat release curve. One of the worst examples is where most of the temperature change is in a desuperheating section. Most of the points provided will be for the desuperheating section – which does not need them because it is a straight line – leaving hardly any for the none linear sections.

Conclusion

Heat release curves must include all turning points (dew points and bubble points) and ideally only the end points of any straight line sections.

Under 'check lists' at the end of the book Check List 2 is a hand out which is useful to give to process engineers – ideally before they start issuing process data sheets..

I fear that this problem will get worse as heat transfer design programs are increasingly likely to accommodate a direct input of the process data. There are clearly a lot of advantages to this but a lot of dangers. A conscientious thermal design engineer must be even more vigilant in scrutinising the process data.

If your program will do it; one obvious check is to draw the heat release curves and see where the points on the profile are. It would be nice if the program could plot the heating and cooling curves on the same graph using temperature on the vertical axis as in the illustrations I have used but I have not come across one that is capable of that yet.

If you do feed in an incorrect heat release curve into a computer program you may well get alerts for specific increments such as 'negative specific heat value was calculated'.

The last thing in figure 1.11 which is worth considering is point P. This is the point **where the temperature difference between the hot fluid and the cold fluid is a minimum and is called a pinch point**. A section near a pinch point requires particular attention. Readers who have a chemical engineering background will be familiar with the term, others may not be. I was once involved with a design where process had used average specific heat values for a cold liquid stream which was being heated by a fluid that was condensing. There was quite a small temperature difference at the pinch point. I also noticed that the liquid specific heat varied by about 30% which was sufficient to produce a curve. I redrew a curve for the liquid and the heating and cooling curves crossed. The unit would not have worked and it was critical to the process. When it was all sorted the process engineer became an instant ally and friend – well contractor acquaintance might be a better term. That was when I was working at Procon during one of my vacations from academic work. Procon are no longer with us – perhaps they made too many process errors which were not picked up.

PHYSICAL PROPERTIES

Physical properties were once the responsibility of the thermal design engineer. I can remember spending a lot of time weighting the properties of the constituents of gas streams to determine properties for the mixed stream. This is now the responsibility of the process engineer. Of course he does not have to do any calculations – a process package does it for him/her. However a reasonable understanding of physical properties is still required by both groups of designers. Once again anyone with a chemical engineering qualification will have a head start.

As well as understanding physical properties a thermal design engineer may still have to establish their own. Consider the case where a mixed hydrocarbon process stream starts off as a superheated vapour and ends as a sub cooled liquid. Since there is no liquid at inlet there are no liquid properties on the process data sheet and at outlet there is no vapour and therefore no vapour properties. If this data is input to a computer program it does not matter how cleaver the program is it must assume that the physical properties are all constant throughout the exchanger. A good program will flag up a warning message but will allow you to carry on. Judging by the number of designs I have checked over the years there must be quite a few exchangers in service which have been designed with constant physical properties.

If you receive a process data sheet (PDS) with properties at a single temperature send it straight back. If the process engineer is in the same building pay him a visit. The (PDS) may not have been officially issued yet and it will save a lot of wasted time and paper (think of the trees) if it can be corrected at an early stage. Be nice to him/her. A good relationship between process engineers and thermal design engineers is essential to producing optimum designs.

If you are in the invidious position of working with process data from a remote source such as a process package from a licensee which is sent to the client who passes it on to the main contractor who sends it down to the sub-contractor you are working for and then it spends a couple of days circulation around your office before it reaches your desk you have a problem. Particularly if you are under pressure to get a design out before you set off for the pub on Friday. Now is the time to estimate your own properties just like we had to do many years ago. Provided you know a physical property at one temperature and you know something about the fluid such as its molecular mass it is not too difficult to make a reasonable estimate of the same property at a new temperature.

If you do have to use your own physical properties make sure to cover your backside. Highlight the properties used on the thermal data sheet (TDS) and then write a covering letter. Don't worry about how long it will take for the letter to get back to the data source or if some of the branches fall off the tree on the way back up it – you have done your best. When I have had to do this I have normally received a simple OK and very occasionally a thanks a few weeks later.

PROPERTY DIAGRAMS

Before looking at the physical properties we use in heat transfer it is useful to have a brief overview of property diagrams for a pure fluid. Figures 1.13 and 1.14 are general diagrams which could apply to any fluid. Figure 1.13 is a plot of pressure, p against specific volume, v and figure 1.14 is a plot of pressure against specific enthalpy, h. Lines of constant temperature; isothermals are drawn on both diagrams. Figure 1.14 also has lines of constant dryness fraction, x.

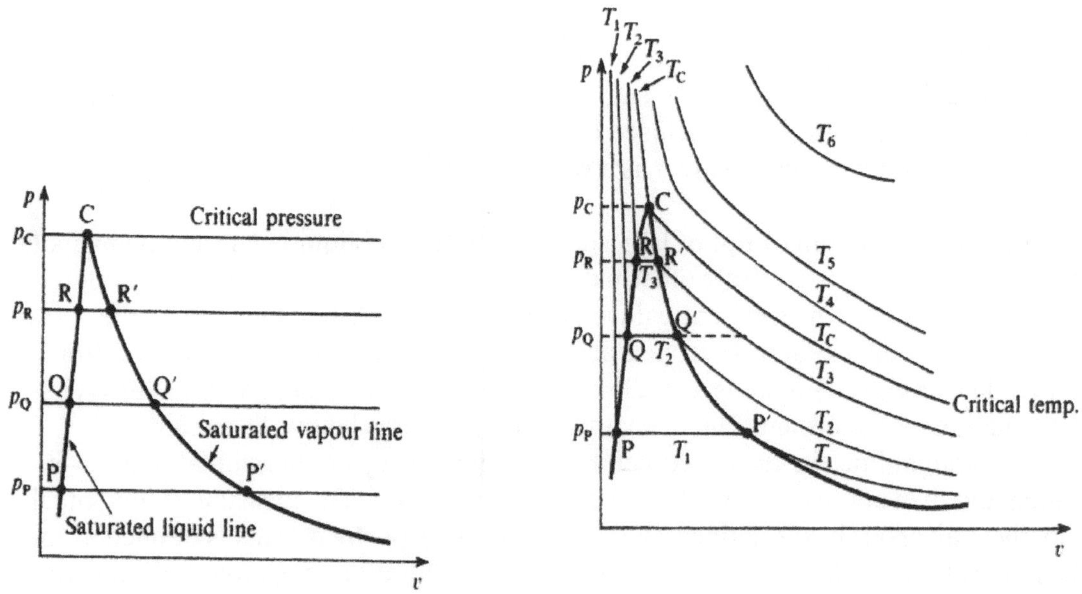

Figure 1.13

When a liquid is heated at constant pressure, an **isobaric process,** its temperature will increase until it reaches a point when bubbles of liquid start to form in the liquid, this is the **bubble or boiling point**. If heating at constant pressure continues there will be no change in temperature until a point is reached when all the liquid has evaporated. During this part of the isobaric process the process will also be **isothermal**. Further heating at constant pressure will increase the temperature of the vapour. If the process is then reversed and the hot vapour is cooled at constant pressure it will pass through the same equilibrium points. The point at which the first drop of liquid is noticed is called the **dew point**. The bottom line on the left hand diagram of figure 1.13 is an isobar at a pressure p which has a boiling point P and a dew point P'. The apparatus for this experiment would be a cylinder with a frictionless piston. The position of the cylinder would be proportional to the specific volume. Readings of temperature and specific volume could therefore be taken along the isobar.

If the process is repeated at higher pressures for isobars at, p_Q and p_R further bubble points can be established at points Q and R and dew points at Q' and R'. It will have been noticed that as the pressure is increased the changes in specific volume between the bubble point and the dew point become smaller and smaller. Eventually a pressure will be reached where they merge. At pressures above this value there is no bubble or dew point.

The point where the bubble and dew points merge is called the **critical point**. This point is identified by a **critical pressure** and a **critical temperature**. There is also a critical specific volume but this term is not in general use. A fluid at a bubble point such as points P, Q and R is known as a **saturated liquid**. A fluid at a dew point such as points P', Q' and R' is known as a **saturated vapour**. (This is sometimes termed a dry saturated vapour to emphasise the point that there no liquid present.) A saturation state is defined as a state in which a change of phase may occur without any change of pressure or temperature. A line drawn through all the bubble points is known as the **saturated liquid line**. A line drawn through all the dew points is known as the **saturated vapour line**. The saturated liquid and saturated vapour lines meet at the critical point.

The fluid to the left of the saturated liquid line is known as compressed or **sub cooled liquid**. The fluid to the right of the saturated vapour line is known as **superheated vapour**. The fluid in the area enclosed by these lines is a mixture of saturated liquid and saturated vapour and is known as a **wet vapour**.

The right hand diagram of figure 1.13 shows the saturated liquid and saturated vapour lines together with a series of isothermal lines. The isotherm which passes through the critical point is known as the critical isotherm and has a temperature T_C. As the temperature of the isotherms increase above T_C there is less and less deviation from a definable curve. At T_6 there is a single curve where the fluid, (gas) will be following the equation pv = constant. The isotherms at temperatures below T_C are almost vertical straight lines in the sub cooled liquid region. They are horizontal in the wet region and then curve downwards in the superheat region. The measure of how much superheat a vapour has is defined by the **degrees of superheat**. This is defined as the difference between the actual temperature and the saturation temperature for the same pressure. For example for pressure P_Q the degrees of superheat are ($T_3 - T_2$) this is illustrated by the dotted line on figure1.13.

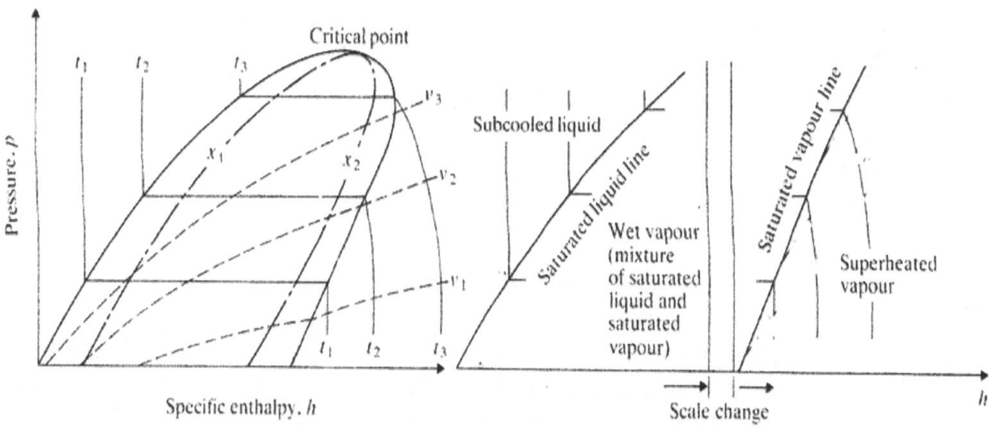

Figure 1.14

There are many property diagrams, (phase diagrams) each with specific applications, for example the reader may have used temperature – entropy diagrams to illustrate thermodynamic cycles such as those used in a vapour power plant. Figure 1.14 is a pressure –enthalpy chart which is widely used in refrigeration work. (I copied this one from the start of the refrigeration chapter in my other book.) The diagram on the left is a full diagram showing a similar range of states to the p – v diagram in figure 1.13. The diagram on the right is a p – h chart which is used to establish property values for practical refrigeration cycles. It only covers the normal working pressures for the refrigerant, it omits most of the wet region and there are different scales for the area around the saturated liquid line and around the saturated vapour line. This produces a significant increase in the accuracy of the chart. This technique is often used on different property diagrams and the reader should understand how these restricted diagrams fit into the 'larger picture'.

The properties of saturated liquids are given the subscript (suffix) f and saturated vapours the subscript g. For example h_f is the specific enthalpy of a saturated liquid and only **either** the temperature **or** the pressure is required to define the precise state. v_g is the specific volume of a saturated vapour and only either the temperature or the pressure is required to define the precise state. Two independent intensive properties (properties independent of the mass) are required to define the state of a fluid. For a sub cooled liquid or a superheated vapour, temperature and pressure are the normal properties used. For a saturated fluid, (liquid, vapour or anywhere in the wet region) temperature and pressure are not independent properties, if you know one of them the other value is fixed.

Specific enthalpy has an additional subscript fg. This is the difference in specific enthalpy between the value for the saturated vapour and the value for the saturated liquid at a specified pressure (**or** temperature). **h_{fg} is the specific enthalpy of phase change.** It is also called the specific enthalpy of vaporisation or the latent enthalpy of phase change. Unfortunately it is also widely referred to as the **latent heat** throughout the petrochemical industry. This is a hang over from antiquated systems of units. **h_{fg}** is a property of a fluid, heat is a form of energy which is transferred from a system to its surroundings whenever there is a temperature difference between them. I assume that young people starting thermal design work will have been taught this. I do not mind the term latent too much it originates from there being no change in temperature during a phase change. Perhaps there is a future for a compromise term – **latent enthalpy.**

$$h_{fg} = h_g - h_f \qquad [1.27]$$

Dryness fraction is the term used to specify the condition or quality of a fluid in the wet region. Dryness fraction is defined as the mass of saturated vapour in one kilogram of a mixture of saturated vapour and saturated liquid. It is therefore dimensionless and has limiting values of 1 when it is all saturated vapour and 0 when it is all saturated liquid.

$$x = \frac{\text{mass sat vapour}}{\text{mass of mixture}} \qquad [1.28]$$

The dryness fraction and either the pressure or temperature specify the state of the fluid in the wet region. Figure 1.13 shows two lines of constant dryness fraction x_1 (about 0.15) and x_2 (about 0.85). Values of specific enthalpy and specific volume are found from the following equations:

$$h = (1 - x)\, h_f + x\, h_g \qquad [1.29]$$

Combining with 1.27

$$h = h_f + x\, h_{fg} \qquad [1.30]$$

$$v = (1 - x)\, v_f + x\, v_g \qquad [1.31]$$

$$v \cong x\, v_g \text{ (neglecting the liquid volume)} \qquad [1.32]$$

Equation 1.29 is the most basic statement for the specific enthalpy of a mixture but I would normally use equation 1.30.

Equation 1.31 is an exact equation. Equation 1.32 is an approximate one which should be used with caution. For low pressure steam the specific volume of the saturated liquid is very small compared with the specific volume of the saturated vapour and if the dryness fraction is also high there will be a very small error in using equation 1.32. However it should never be used for a high pressure hydrocarbon with a high molecular mass when the dryness fraction is low.

GASES & VAPOURS

An academic definition of a gas is a substance which is above its critical temperature and can not be liquefied by increasing the pressure. A vapour is below its critical temperature and can be liquefied by increasing the pressure.

Figure 1.13 showed an isotherm well above the critical temperature where the fluid was behaving like a perfect gas. The constituent gases of dry air are well above their critical temperatures (Nitrogen about – 147°C) and were once thought of as permanent gases. If figure 1.13 had been for nitrogen the specific volume scale would have had to be extended a long way for isotherm T_6 to reach atmospheric pressure.

A useful practical distinction is that gases obey relatively simple equations such as: $pv = RT$ for low and medium pressures. Vapours do not and we need to use tables for properties. Steam of course is a classic vapour and we have all used some form of steam tables to establish properties. However steam is treated as a perfect gas in psychrometry (the study of mixtures of air and water vapour at low pressures – applications air conditioning, cooling towers etc.).

For practical heat exchanger design we do not make any real distinction between vapours and gases. Anything that is not liquid is treated in the same way in a computer input. The gas and vapour component is defined by means of a vapour mass fraction, y as follows:

$$y = \frac{\text{mass flow rate of vapour and gas}}{\text{total mass flow rate}} \qquad [1.33]$$

It may superficially appear that the vapour mass fraction, y and the dryness fraction, x are the same thing. A closer examination will show that this is not the case and there is a good reason for using different symbols. Dryness fraction is used for pure fluids and is concerned with masses of saturated liquids and saturated vapours. Mass fraction of 'vapour' is used with mass flow rates of anything that is not a liquid.

Many data sheets will divide the none liquid components into steam, none condensables and hydrocarbons. These are added together before y is calculated.

None condensables in a condenser are usually air but may contain other none hydrocarbons. A light hydrocarbon with a dew point above the temperatures in the exchanger will behave as a none condensable as far as providing an extra resistance which decreases the condensing coefficient.

The following is a brief overview of the physical properties used in thermal design work.

SPECIFIC HEAT C_p

One simple definition of specific heat is the quantity of heat required to raise unit mass by one degree of temperature during a specified process. In theory there could be a lot of them, in practice only the specific heats for the constant pressure and constant volume processes are of general interest. In this book the specific heat at constant volume is only required for gases chapter 15 on vibrations. Elsewhere it is only the specific heat at constant pressure that is required and is simply referred to as specific heat although the symbol C_p makes clear which one it is.

The basic units are J/kgK but the magnitude of the numbers makes kJ/kgK a more practical unit.

LIQUIDS

Water and weak aqueous solutions have a specific heat value of about 4.2 kJ/kgK. There is little variation over a wide range of temperatures. The specific heat of pure water actually decreases slightly as the temperature increases above 0 ^{o}C to reach a minimum value at 35 ^{o}C and then starts to increase again. This is one of the many unusual features of water properties.

Chemicals and hydrocarbons span a much wider range of values from 1 kJ/kgK to values approaching those for water for very high temperature petroleum fractions. A figure of about 2 kJ/kgK is reasonably typical for most hydrocarbons.

Specific heats of liquid hydrocarbons always increase with increases in temperature. The increase can be over 30% for temperature increases of 100 K. Values are plotted on a linear scale against temperature. At low temperatures the lines are almost straight but curve more at higher temperatures as the rate of increase of specific heat falls off.

There are charts for specific heats of petroleum fractions in TEMA and for specific hydrocarbons in Gallant – Ref 21 as well as many other sources. If you have a value at only one temperature point and need to find a value at a second temperature find a chart for a similar liquid.

This is illustrated on figure 1.15 which is part of a chart with lines of specific heat plotted against temperature for some similar fluids to the one you are dealing with. You have been given a specific heat value C_{p1} at a temperature T_1 and you need a second specific heat value at a higher temperature T_2. Draw vertical lines from the two temperatures. Next draw a horizontal line from the value for C_{p1} until it meets the line from T_1. From this intersection point draw a line parallel to the specific heat lines for the other fluids until it meets the line from T_2. Then read off the second value of specific heat at temperature T_2.

Unfortunately the specific heat values in the two references above are in antiquated units and you may have to do some conversions (see useful table 2) but you will probably have access to properties in modern units.

Liquid specific heats are not affected by pressure.

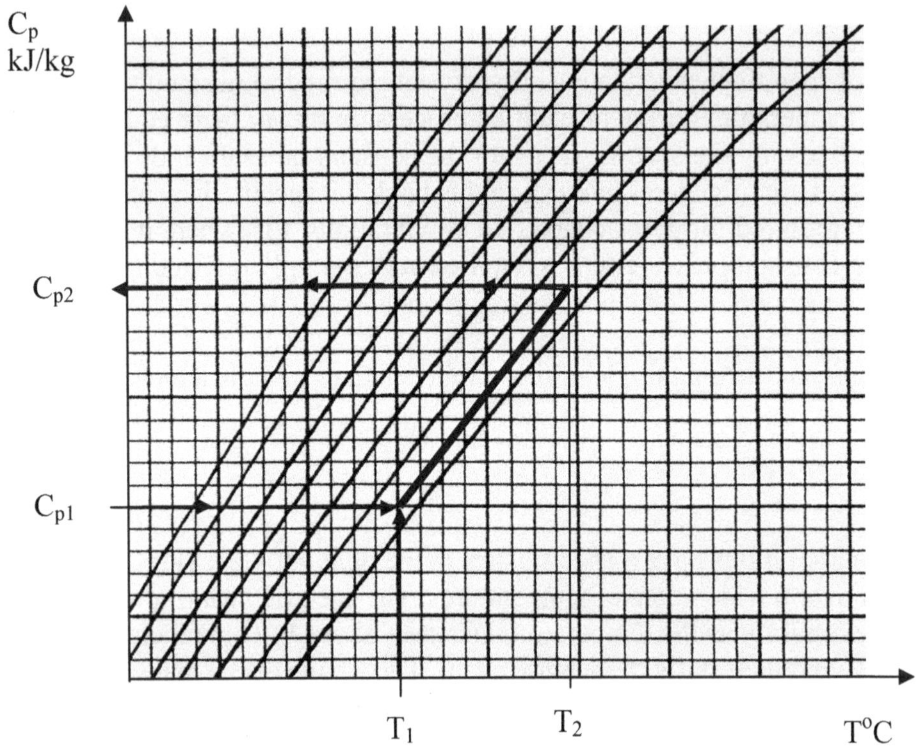

Figure1.15

VAPOURS/GASES

Hydrocarbon vapours have specific heat values which are very similar in magnitude to the values of the liquid states (typical value 2 kJ/kgK). They also vary with temperature in a similar way, so that the above treatment for liquids applies to vapours.

Water vapour/steam specific heats have similar values to those of hydrocarbons and increases with temperature.

For gases hydrogen is the 'odd ball'. It has a value of about 14.5 kJ/kgK which is almost constant (varies less than 4% over a temperature range of 700 K).

ENTHALPY OF PHASE CHANGE (LATENT ENTHALPY) h_{fg}

This has already been defined and a glance at figure 1.14 shows that it rapidly declines as temperatures and pressures approach the critical point. At the critical point it has a value of 0.

Water has a maximum value of just over 2500 kJ/kg at 0°C and falls steadily as the pressure/temperature increase but it still has a value of 2000 kJ/kg at 184°C and 11 bar. It only requires a small mass flow rate of condensing steam to transfer a lot of heat. This is one of the main reasons why it is the working fluid in power plants and a service fluid in many industries.

Ammonia has a value of about 1250 kJ/kg which is one of the main reasons it has always been used in large refrigeration units.

Typical values for hydrocarbons rang between 250 and 600 kJ/kg.

DENSITY & SPECIFIC GRAVITY ρ & SG

Density ρ, is defined as the mass per unit volume. Specific gravity SG, is the ratio of the density to the density of a reference fluid. For all liquids and usually for gases/vapours the reference density is water at 4°C when it has a value of 1000 kg/m3 which is its maximum value. For gases the reference fluid can be air which has a density of about 1.2 kg/m3.

Volumes and the reference state for gases can be very confusing. Two different definitions can be used for the reference state – NTP (Normal temperature and pressure) and STP (Standard temperature and pressure). Both 'normally' use atmospheric pressure of 1.013 bar for the pressure. Normal temperature may be 20°C or 15.5° C (60°F) or 15°C. Standard might be one of these or 0° C. If you ever receive data which uses NTP or STP ask for clarification. If you ever transmit data specify the temperature.

LIQUIDS

Water has a density of 999.8 at 0.01 °C. It then increases until it reaches its maximum value of 1000 kg/m3 at 4 °C. It then decreases as the temperature increases. The difference in density between 0.01 °C and 4 °C may not seem very much but without it life on this planet would be very different. It is this anomalous expansion of water together with the reduction in density when water turns to ice which prevents ponds and lakes from becoming solid ice in winter. Once a layer of ice is formed further cooling causes the remaining water to expand which increases the pressure and lowers the freezing point. At the bottom of deep lakes the temperature is normally 4 °C.

Densities of all other liquids decrease as the temperature increases.

Pressure has practically no effect on liquid densities.

Hydrocarbon densities can vary from about 500 kg/m3 to about 1000 kg/m3. (Specific gravities from 0.5 to 1).

GASES

Gases and vapours have much lower values of density than liquids (atmospheric air about 1.2 kg/m3). For medium and low pressure gases the characteristic equation of state (ideally for a perfect gas) can be rearranged as:

$$\rho = \frac{p}{RT_A} \qquad [1.34]$$

Where R is the specific gas constant for the particular gas, p is the pressure (absolute) and T_A is the absolute temperature.

The density therefore varies directly with pressure and inversely with absolute temperature.

When the terms low and medium have been applied to pressure, as in the paragraphs above, I never even considered attaching numbers to them. A 'high' pressure for one gas may be a 'low' one for another gas. In this situation high and low are relevant to the critical pressure and temperature of the particular gas.

In regions where the equation for a perfect gas can be applied the molecules are so far apart that the volume they occupy is negligible compared with whole volume. The higher the pressure becomes the less valid this assumption becomes and it is necessary to take into account what is called compressibility. Two new terms need to be defined as follows:

$$\text{Reduced pressure} \quad P_r = \frac{p}{p_C} \qquad [1.35]$$

$$\text{Reduced temperature} \quad T_r = \frac{T_A}{T_C} \qquad [1.36]$$

Where p and T_A are the absolute pressure and temperature and p_C and T_C are the critical pressure and temperature respectively. Equation 1.34 is then corrected with a compressibility factor Z as follows:

$$\rho = \frac{p}{ZRT_A} \qquad [1.37\}$$

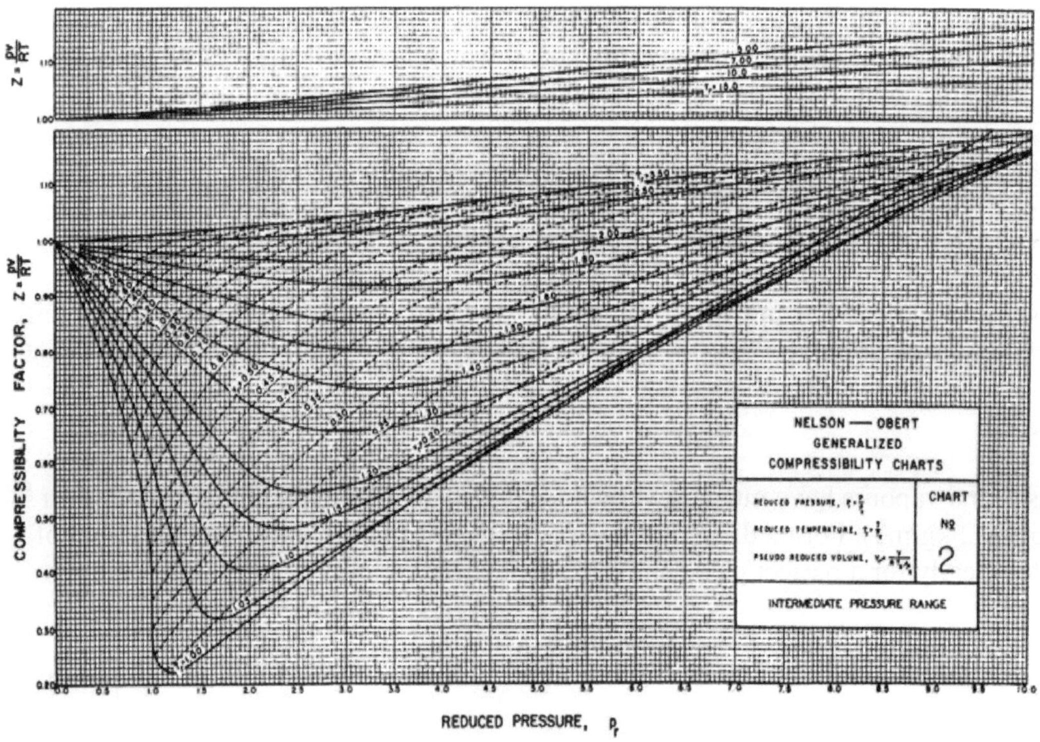

Figure 1.16

A compressibility chart is shown in figure 1.16. The lines of constant reduced temperature all start at the left hand side where the reduced pressure is zero (and hence the pressure is zero) and Z is 1. The line for T_r =1 envelops the lower part of the diagram in a distorted 'v' shape. As T_r increases the lines flatten out and the line enclosing the top is for a T_r value of 3.5. The lines which cross the T_r lines are not necessary for finding Z.

The chart is one of a series originally produced by Nelson & Obert a long time ago and I am sure you can find access to a better copy than figure 1.16 when you need to use one.

VISCOSITY μ & υ

Laminar flow consists of a series of very thin concentric tubes of fluid which slid against each other without mixing and is discussed in more detail in chapter 3.

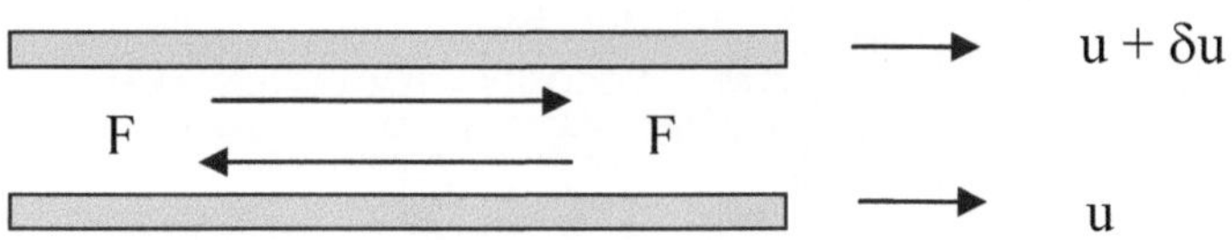

Figure 1.17

Figure 1.17 shows a section through a small area of two adjacent tubes of fluid, (shaded). The tubes have been pulled apart for clarity but in reality are touching. The lower tube is nearest the wall and is moving at a velocity of u. The top tube has a higher velocity u + δu. The upper tube is dragging the lower tube along with a force F and by Newton's third law the lower tube is resisting with an equal and opposite force F. If the distance from the tube wall is y the velocity gradient between the two tubes will be δu/δy which will create a shear stress τ. Since the velocity gradient is causing the shear stress it must be proportional to it therefore:

$$\text{Shear stress } \tau = \frac{\text{Force F}}{\text{Area}}$$

Also $$\text{Shear stress } \tau \propto \delta u/\delta y$$

Therefore $$\tau = \mu \frac{\delta u}{\delta y} \qquad [1.38]$$

Where μ is a constant for a specified fluid at a specified temperature and is known as the absolute or dynamic viscosity. It has the basic units of kg/ms or Ns/m^2 or Pa s but the magnitude of the unit makes it unwieldy for our purposes. We therefore use the unit centipoise. 1centipoise, cps = 0.001 kg/ms. It is easy to relate to the centipoise because our most common fluid, water has a value of 1 at 20°C.

Although it is never used in the input to a design program we also need to define the term kinematic viscosity ν (easily confused with v).

$$\nu = \frac{\mu}{\rho} \qquad [1.39]$$

The practical units for kinematic viscosity are centistokes (cks) although the basic unit is m^2/s. 1 centistoke = 0.001 m^2/s. We only need kinematic viscosities when dealing with liquids which make it easy to transpose equation 1.39 into a more convenient form. Since SG = ρ/1000 equation 1.39 becomes:

$$\nu \text{ in centistokes} = \frac{\mu \text{ in centipoise}}{SG} \qquad [1.40]$$

LIQUIDS

Figure 1.18 shows a range of viscosities for some 'viscous' hydrocarbons. The temperatures are unfortunately in °F. The viscosities are in centistokes. The reason why most charts of viscosity against temperature for liquids use kinematic viscosities is that they are close to being straight lines on a chart with logarithmic scales.

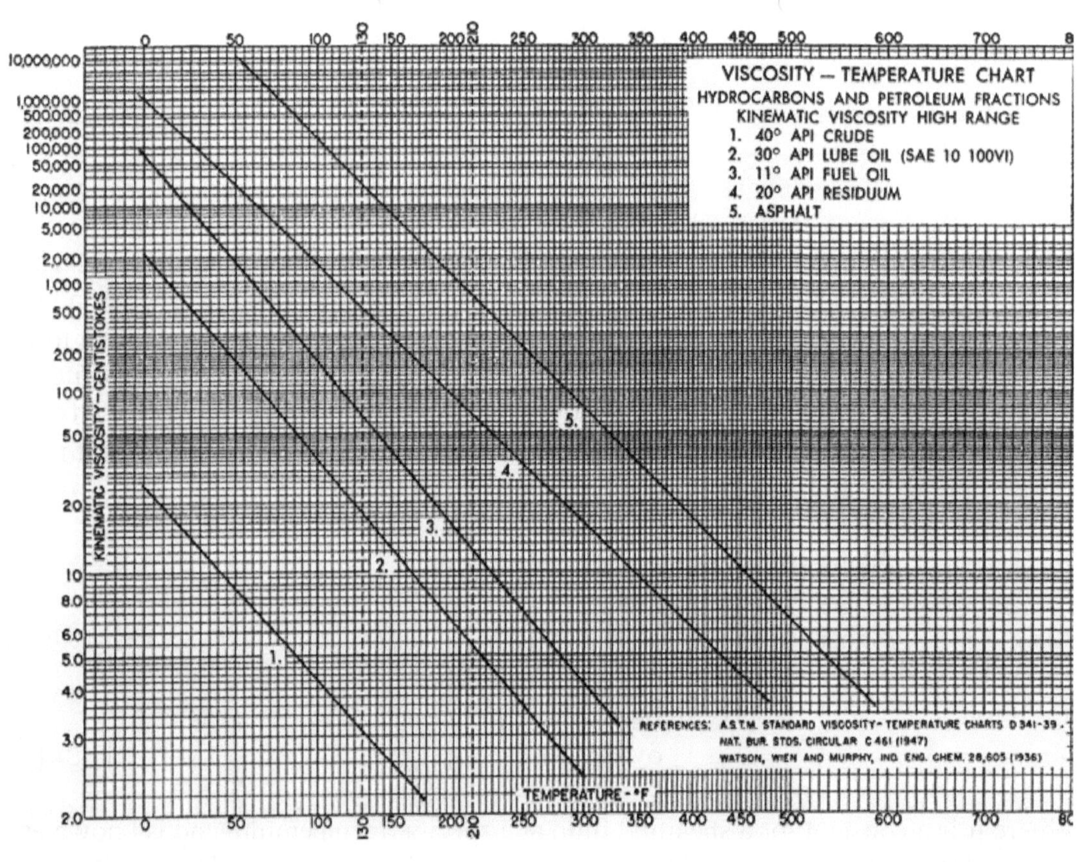

Figure 1.18

These charts are very useful if it necessary to input a third viscosity value into your program. The procedure is as follows:

(1) Calculate the SG values for the two points you have. (divide ρ by 1000)
(2) Change the two viscosity values in cps to cks
(3) Locate these points on the chart
(4) Draw a line between these points and extend it to the temperature required for a third point

(5) Read off the new viscosity in cks
(6) Convert this value back into cps for your computer input.

You can of course find a viscosity value at a second point when only one point is known by drawing a line parallel to adjacent lines as explained in the section on specific heats but this will not be quite as accurate.

Liquid propane has a viscosity of about 0.1 cps at typical room temperatures. The reader has only to glance at figure 1.18 to see the other extremes. Most of the fluids shown will have an SG close to 1 at the lower temperatures so that cks are close to cps in magnitude. Clearly you may have to deal with a fluid with a viscosity greater than 10 000 cps. This results in the highest values being a factor of 100 000 times greater than the lowest values. No other physical property varies so much.

Pressure has no significant effect on liquid viscosities.

GASES/VAPOURS

For low and medium pressures (with respect to the critical pressure) gas viscosities vary between the extremes of about 0.005 cps and 0.1 cps. A typical value to keep in your head is **about 0.015 cps** so that if you see a process data sheet with a value of 0.15 cps it is almost certainly a typing error.

All gases increase in viscosity as the temperature increases. The increase varies between about 20% per 100 K to 60% per 100 K. The values for water vapour and hydrogen are not significantly different from hydrocarbons.

At high pressures (at or above the critical pressure) a correction factor is applied which increases the viscosity value for low pressures. These correction charts require the calculation of the reduced temperature T_r and the reduced pressure P_r which were introduced earlier in the section on compressibility. TEMA and many other sources have an appropriate chart. These corrections are significant and a factor of two would not be unusual and at a reduced pressure, P_r of 3 might be as high as five.

THERMAL CONDUCTIVITY k

The thermal conductivity of a fluid is the property of that fluid which determines how much heat will be transferred through unit area when a temperature gradient exists across a stationary layer of fluid. The units are W/mK.

LIQUIDS

Hydrocarbons have thermal conductivity values which vary between about 0.09 W/mK to 0.17 W/mK. A typical value to remember is 0.13 W/mK. The **thermal conductivity always decreases as the temperature increases**; the variation is typically about 30% per 100 K.

Normally pressure will have little effect on the thermal conductivity of liquids. For very high pressures there is a correction chart in TEMA. However even at pressures well above the critical pressure the corrections are quite small and 'safe' (the factor slightly increases the thermal conductivity).

Water is an 'odd ball' once again. Its average value of about 0.64 W/mK is about five times as high as a typical hydrocarbon value. Not only that but its thermal conductivity starts off with a value of about 0.57 W/mK at 0°C, and then **increases** as the temperature increases to a maximum value of 0.69 at W/mK at 135°C and then it decreases again as the temperature increases above 135 °C.

GASES/VAPOURS

For low and medium pressures (with respect to the critical pressure) gas thermal conductivities for hydrocarbons and steam vary between the extremes of about 0.01 W/mK and 0.04 W/mK. A typical value to keep in your head is about 0.02 W/mK. (Numerically similar to the viscosity value). The **thermal conductivity always increases as the temperature increases**; the variation is typically about 70% per 100 K.

Hydrogen is the 'odd ball' for gases. Its average value of about 0.23 W/mK is about ten times as high as a typical hydrocarbon gas value.

At high pressures (at or above the critical pressure) a correction factor is applied which increases the thermal conductivity value for low pressures. These correction charts are similar to the ones used for correcting gas viscosities. They require the calculation of the reduced temperature T_r and the reduced pressure P_r and there is a chart in TEMA. These corrections are even more significant than the viscosity corrections.

SURFACE TENSION σ

Liquids behave as if their free surfaces were flexible membranes with a constant tension per unit width. This tension per unit of width is known as the surface tension.

It is therefore a force per unit length with SI units of N/m although it is frequently quoted in dynes/cm. 1 dyne/cm = 10^{-3} N/m.

The value of the surface tension depends on:

(a) The liquid.
(b) The temperature and pressure.

A liquid may be in contact with a solid, another liquid or a gas/vapour and may behave differently in different situations. For example in the chapter on condensation it is described how steam will condense in droplets rather than films on some types of surface.

The surface tension values for most hydrocarbons range from about 4 to 25 dynes/cm. Water has values of about 20 dynes/cm at 250°C and about 67 dynes/cm at 20°C. Mercury is an interesting liquid with a value of about 480 dynes/cm at room temperature. This explains why it always forms neat spherical balls when you break a thermometer in the laboratory or at home no matter what surface it falls on. (Younger readers may not have seen this since it has now been banned as a thermometer fluid.) Surface tension always decreases as the temperature increases.

In heat transfer it has been used for many years in correlation used to predict entrainment in kettles. Some modern computer programs now use it for some of their more sophisticated two phase flow correlations. The fact that we managed for many years designing heat exchangers without using it demonstrates that it is the least important of the physical properties we have to deal with.

SUMMARY

Table 1.4 summarises some of the important features of the properties of the fluids we most often need for thermal design work. The values given are typical ones which are worth remembering.

Phase - Property	T increase	HC value	Odd Ball
Liquids – C_{pl} kJ/kgK	increase	2	Water 4.2
μ_l cps	decrease	0.1 → 100 000	0.1 → 1
k_l W/mK	decrease (not H_2O)	0.13	Water 0.64
Vapours – C_{pv} kJ/kgK	increase	2	Hydrogen 14.5
μ_v cps	increase	0.015	0.013
k_v W/mK	increase	0.02	Hydrogen 0.23
Change – h_{fg} kJ/kg	decrease	250 → 600	Water 2000

Table 1.4

The liquid with the unusual properties is water. In the chapter on single phase heat transfer the way in which each property affects the heat transfer coefficient is developed. These are summarised in useful tables 1 at the end of the book. Using these it is interesting to compare the heat transfer coefficients of water and an average hydrocarbon liquid.

$$h \propto C_p^{\,0.33}\ k^{0.67}$$

Therefore $$\frac{h_{water}}{h_{HC}} = (4.2/2)^{0.33}\ 0.64/0.13)^{0.67} = 3.7$$

I have deliberately left out the viscosity since this may have increased or decreased the ratio depending on the hydrocarbon. In easy to remember in general terms to expect water heat transfer coefficients to be about four times higher that the coefficients for light hydrocarbons where the viscosities have similar values. If the comparison had been made with a highly viscous hydrocarbon, the factor would have been much higher.

The water we use in heat transfer calculations is normally cooling water. This may come from a variety of sources: rivers, lakes, the sea, estuaries, cooling towers etc. The fouling factors for different types of water can vary considerably but they would all be classed as fresh water and would require the use of normal water properties.

I once worked on an ethylene plant for a Swedish company which had a very unusual cooling water arrangement. They wanted to keep the cooling water temperatures relatively constant throughout the year. This meant that the velocities through the heat exchangers would be constant. (Control problems with low water velocities in winter are discussed in the chapter on fouling). The site was adjacent to a river flowing into a narrow sea fjord. A barrier across the fjord formed a lake. The relatively constant temperature was achieved by a mixing of river water, fresh sea water and return water from the heat exchangers. I immediately asked for the water properties for the water supplied to the heat exchangers. When I eventually received them I found that the heat transfer coefficients hardly differed from those using fresh water but the pressure drops were 5% lower. This may not sound very much but it was sufficient to 'tweak' some units to achieve that elusive 'optimum design'.

The above was a case for cooling water with a very low salinity value. When sea water is used for cooling the salinity can vary considerably, some areas in the Middle East may have very high values. Always try and get correct values for the sea water you are using.

For gases/vapours hydrogen has the unusual properties.

Therefore $$\frac{h_{hydrogen}}{h_{HC}} = (14.5/2)^{0.33}\ (0.23/0.02)^{0.67} = 9.9$$

This factor of about ten is in practice a bit of an exaggeration. Hydrogen also has the lowest molecular mass of any gas. Pressure drop is inversely proportional to density so that for a fixed allowable pressure drop the bulk velocities would be lower. However the factor is still very high. You may not come across too many units involving pure hydrogen but hydrogen rich streams are far more common and the more hydrogen they contain the higher the heat transfer coefficient will be.

PROPERTIES NEAR THE CRITICAL POINT

It has already been pointed out that gas viscosities and thermal conductivities increase significantly at pressures at and above the critical. However there are significant changes to some properties as they approach the critical point. It has already been demonstrated that the enthalpy of phase change decreases rapidly as the critical point is approached but as h_{fg} decreases the specific heat values of both the liquid and the gas increase. It is quite possible for a liquid hydrocarbon to have a higher C_p value than water and a hydrocarbon gas may have a higher C_p value than hydrogen.

If you ever see unusually high specific heat values for a hydrocarbon on a process data sheet the first thing to check is where you are operating with respect to the critical point.

If you are it is well worth discussing this with process to ascertain the degree of confidence they have in their values on the process data sheet. You may decide that it is prudent to apply a safety factor to the heat transfer coefficient.

ACCURACY

The last time I used the HTRI programs there was a choice of two different sets of properties for *water*! That was something I certainly never expected to see. I would have thought that by now there would be very good agreement on the properties of our most common fluid and in 'steam tables' values are quoted to four significant figures . I never found the spare time to investigate what the differences were in the case of the HTRI options.

If there is still no general agreement on water what about all the other fluids? I had thought about spending some time in getting up to date on this but life is too short. Overall I would say that up to ± 10% is a normal tolerance for most of the fluids we use apart from the service fluids where they should be much tighter. However the tolerances could be much higher for little researched fluids and fluids in 'dodgy regions'.

CONCLUSIONS

The three basic topics covered in this chapter will be part of the design process for any heat exchanger not just shell and tube units.

The term U value has a single straightforward meaning in other industries such as Building Services where the U value of the wall of a building is unambiguous. To design shell and tube heat exchangers it is necessary to understand more than one definition and to know where to apply each one.

In books on general heat transfer the mean temperature difference for heat exchange can also be presented as a relatively easy concept. In many heat exchangers where there are significant changes in heat transfer coefficients it can become quite a complex calculation. Although this calculation will now be carried out by a computer a sound understanding of heating and cooling curves is essential to feed in the right input.

A good thermal design engineer will always be anticipating the heat transfer coefficients expected from any fluid stream. To do this it is essential to appreciate the impact of each of the physical properties on the heat transfer coefficient and pressure drop. Although it is increasingly unlikely that the thermal design engineer will be responsible for providing physical properties there may be mistakes in the data they are given or the fluid may be in a 'dodgy region' and any unusual value should be questioned.

This page has been left blank so that the next chapter starts on a right hand page.

CLASSIFICATION & CONSTRUCTION - CHAPTER 2

Shell and tube heat exchangers are the most flexible of all the possible heat exchanger types. There are no limitations on pressure. I have personally designed units operating from high vacuums to over 400 bar pressure. The only limitations to either very low or very high temperatures are the materials of construction.

The name is derived from the two main components of a cylindrical shell which encloses a number of tubes. One fluid flows though the tubes and the other fluid passes over the tubes in a variety of ways. For readers who are new to thermal design work the obvious starting point is to define and describe some of the terms used.

TERMINOLOGY

UNIT

This term is used for other equipment as well as shell and tube exchangers. A shell and tube unit may consist of a single shell or many shells.

SHELLS

As far as thermal design is concerned the important dimension is the inside diameter. It is now becoming generally accepted that this is what should be specified on the thermal data sheet. This has not always been the case. For small shells it is often more convenient to use a length of standard pipe than it is to roll a shell. When I first started to design heat exchangers it was normal to use pipe sizes up to 24 inches. A 24 inch standard wall pipe has an inside diameter of 23.25 inches (591 mm) but the data sheet would call it a 24 inch shell. Pipe sizes are still used for small shells but as fabricators have refined their techniques to roll smaller and smaller shells the size has been reduced in most job specifications. Currently 16 inches is the normal break point but many fabricators can role even smaller shells. When in the region where there may be a choice, the sensible thing to do is to specify the actual inside diameter of shell you have designed for. So that if a standard 16 inch pipe (15.25" ID) is sufficiently strong for your design specify exactly 387 mm and the vendor knows he has a choice of using a length of pipe or rolling a shell. Do not confuse them by rounding up or rounding down to the nearest 5 mm. (Check in table 3 in useful tables if there is any doubt about the working pressure).

Above the standard pipe sizes it used to be common practice to specify shells in even increments of one inch or two inches (50.8 mm) this was to increase the possibility of using a standard flange. Nowadays shell increments of 10 mm are normally acceptable.

Figure 2.1 contains all the terms normally required to describe a shell and tube heat exchanger, however some of the terminology does not apply to the diagram of the AES shell shown. For example 38, Weir and 39, Liquid level connections are applicable to kettle type exchangers which will be dealt with later. 22→26 apply to a construction which is seldom used and easy to understand if you ever come across it.

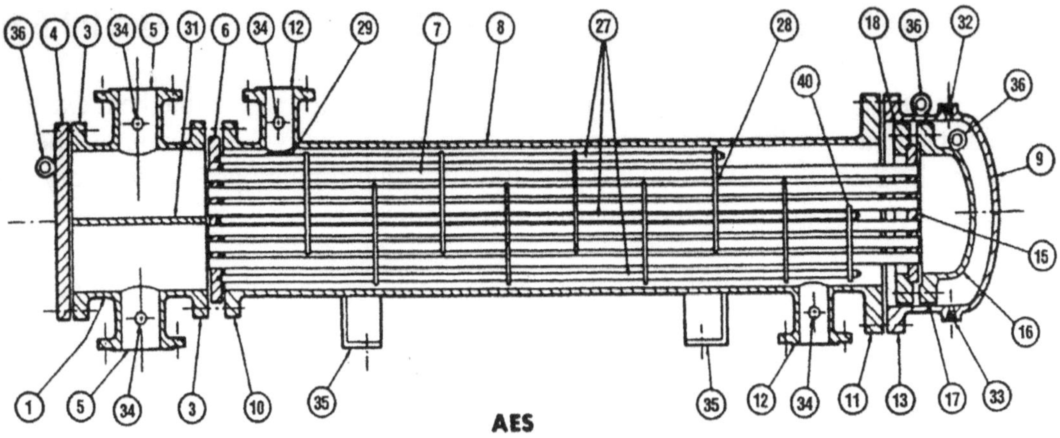

1. Stationary Head-Channel
2. Stationary Head-Bonnet
3. Stationary Head Flange-Channel or Bonnet
4. Channel Cover
5. Stationary Head Nozzle
6. Stationary Tubesheet
7. Tubes
8. Shell
9. Shell Cover
10. Shell Flange-Stationary Head End
11. Shell Flange-Rear Head End
12. Shell Nozzle
13. Shell Cover Flange
14. Expansion Joint
15. Floating Tubesheet
16. Floating Head Cover
17. Floating Head Cover Flange
18. Floating Head Backing Device
19. Split Shear Ring
20. Slip-on Backing Flange
21. Floating Head Cover-External
22. Floating Tubesheet Skirt
23. Packing Box
24. Packing
25. Packing Gland
26. Lantern Ring
27. Tierods and Spacers
28. Transverse Baffles or Support Plates
29. Impingement Plate
30. Longitudinal Baffle
31. Pass Partition
32. Vent Connection
33. Drain Connection
34. Instrument Connection
35. Support Saddle
36. Lifting Lug
37. Support Bracket
38. Weir
39. Liquid Level Connection
40. Floating Head Support

Figure 2.1

STATIONARY HEAD CHANNEL/FRONT END/NEAR END

The stationary head channel, (1) is where the tube side fluid enters the exchanger and also leaves if there is an even number of passes as in the above diagram. Note that there may be a variety of expressions used for the same thing. The term stationary is used because the support saddle (35) on the left hand side is fixed to the concrete plinth (if the unit is at grade) which form the foundations to support the heat exchanger. It is not strictly stationary of course. There must be some contraction or expansion between the fixed saddle and the nozzle through which the fluid enters the channel. This contraction/expansion must be allowed for in the pipework design but the movement is much less that at the other end of the exchanger.

MOVING HEAD/REAR END HEAD/FAR END HEAD

The support saddle (35) at the right hand side of figure 2.1 is free to move and accommodates the main expansion/contraction of the shell. Hence the moving head moves a lot more than the 'stationary' head. This movement is dealt with by the mechanical engineer and then the piping engineers (layout and stress).

The rear head not only experiences the main expansion between the shell and ambient conditions it usually also has to deal with the expansion between the shell and the tubes. This is one reason why there is more variety in the design of rear heads than stationary heads. The other reason for the greater variety is that with an odd number of tubeside passes it must accommodate a nozzle for the tubeside fluid but with an even number of passes it does not.

TUBESHEETS

Both ends of the tubes (only one end for U tubes) are fixed into thick circular plates with pre-drilled holes by either being expanded (usually into grooves) or welded. These plates are called tubesheets, (6) and (15) in the diagram. The tubesheet at the stationary head end (6) is called the stationary tubesheet and the tubesheet at the rear head end is the floating tubesheet when it is free to move independently of the shell. When there are two tubesheets and neither of them can move independently from the shell the unit is termed a fixed tubesheet design (FTS).

TUBE BUNDLE

The tube bundle includes the tubes, tubesheet(s) and means of supporting the tubes. If the tube supports also deflect the shellside fluid across the shell to promote heat transfer they are called baffles (28). If the supports are not involved with promoting heat transfer they are just called supports. There are many types of supports which are detailed later in this chapter.

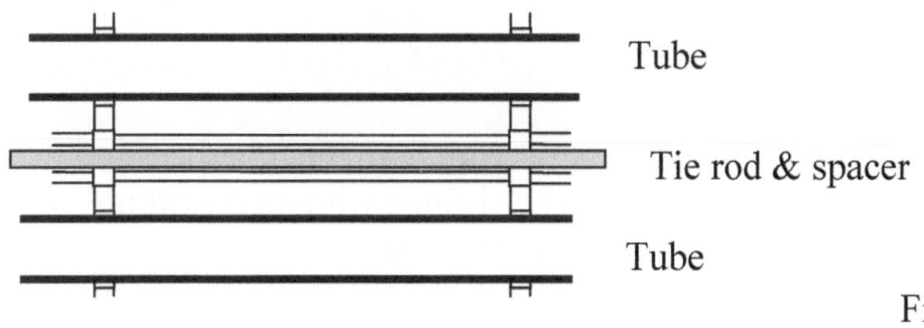

Figure 2.2

The bundle is assembled using a number of tie rods and spacer tubes to keep the baffles and supports in place. Figure 2.2 shows a section through two tubes with a tie rod and spacer arrangement in between them. The spacers are section of tube with a length equal to the baffle pitch minus the thickness of the baffle. The tie rods and the holes through which they pass are smaller than the tube holes. The diameter and minimum number of tie rods required for a particular shell size is specified in TEMA. It is often necessary to use more tie rods than the TEMA minimum. When extra tie rods are required will be demonstrated in later examples.

Tie rods are screwed into the stationary tubesheet at one end and held by a lock nut device at the rear head end. At the rear end the final fixing will be to a support plate which may also be a baffle plate and the last of any baffles containing tubes which do not pass through the final support plate.

When it is possible to take the bundle out of the shell for repair or cleaning it is known as a removable bundle.

TUBULAR EXCHANGER MANUFACTURERS ASSOCIATION (TEMA)

The overview of the main components above is sufficient to move on to the TEMA designations for shell and tube type of heat exchangers. Three letters are used; the first one describes the stationary head, the second the flow path taken by the shellside fluid and the third the rear end head.

The TEMA nomenclature is reproduced in figure 2.3 (with permission). Figure 2.1 was an AES type. We will now look at the full range of options in more detail.

DETAILS OF STATIONARY HEADS

The **A** type head has flanges at both ends. One attaches to a cover which may be removed for cleaning the tubes without disturbing the pipework. The other attaches to the shell flange so that the head can be removed when it is necessary to remove the bundle.

The bonnet, **B** type head has only one flange. This attaches to the shell flange and the pipework must be detached for cleaning the tubes or withdrawing the bundle to clean the shell side.

The **C** type has a channel which is integral with the tubesheet. This head has always puzzled me a bit. Earlier editions of TEMA had no stationary N type head and we used to C for fixed tubesheet exchangers. (A sketch would make it clear that the C was really an N!) I have never come across the C type head as illustrated being used.

Type **N** has a channel which is integral with both the tubesheet and the shell. The removable cover allows the tubes to be cleaned without disturbing the pipework. The elimination of the shell flanges reduces the cost.

Type **D** is a special enclosure for high pressure tubeside fluids. There are many variations of the enclosure illustrated in TEMA which are covered in chapter 11 on special mechanical features.

SHELL SIDE FLOW TYPES

The **E** type is the most common arrangement used in shell and tube heat exchangers. There are two shellside nozzles and each is positioned as close to its adjacent head as is mechanically possible. The shellside fluid may enter from either end and either nozzle may be at the top or bottom, occasionally one nozzle may be at the side and both could be at the top or the bottom depending on the process requirements. The only geometry which identifies an E type is that the shellside fluid enters at one end of the shell and leaves at the opposite end.

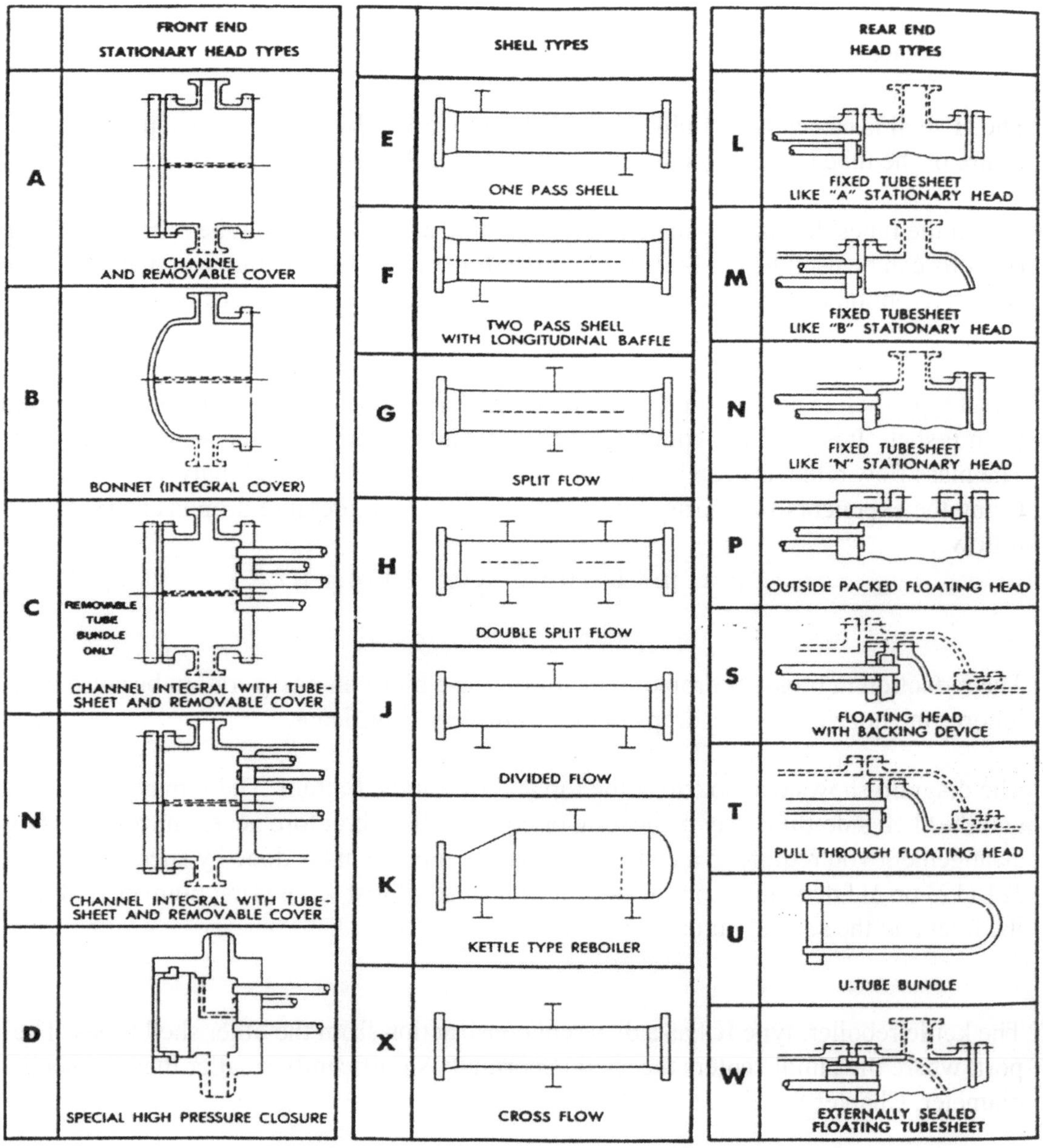

Figure 2.3

The **F** type shell is characterised by both the shell nozzles being at the stationary head end. Normally there are just two passes on the tube side which facilitates an arrangement for true counter flow with all the advantages that brings. However there could be any even number of tubeside passes. If there are more than two tubeside passes the thermal arrangement becomes the equivalent of having two shells in series. The F type shell is in my opinion used far too infrequently and the pros and cons will be discussed later in this chapter.

The **G** type shell has central shell nozzles at the top and the bottom of the exchanger. The shellside flow is split by a horizontal baffle which is close to or on the centre line of the shell. The main application of this shell arrangement is in horizontal thermosyphon reboilers with a short tube length where the inlet is at the bottom and the outlet at the top. The horizontal baffle may be quite short and may be perforated. Its prime function is to ensure that the shellside fluid 'sweeps' the ends and prevents the build up of vapour adjacent to the tubesheets which would create areas which would not be fully effective for heat transfer.

The **H** shell is basically a double G shell. There would always be a full support at the centre of the tube length which is not shown on the TEMA diagram.

When the tubeside fluid is single phase and there are at least two tube passes the G and H shells can be used where there is a temperature cross. This feature can make them an attractive option to using two shells in series for none reboiler applications.

The **J** type shell is normally used when the shellside pressure drop is unacceptably high in an E shell. It is useful to compare an E shell and a J shell with the same baffle pitch. The mass flow rate has been halved and the length of the flow path halved in the J shell. Using the exponents summarised in 'useful tables 1' the pressure drop is changed as follows:

$$\Delta p \propto u^{1.8} L = 0.5^{1.8} \times 0.5 = 0.14 \cong 1/7 \qquad [2.1]$$

The reduction in pressure drop of a factor of 7 is a bit of an exaggeration because it ignores the nozzle pressure drops but it is still very significant.

The diagram shows one nozzle at the top and two at the bottom. These may be reversed and the shell side fluid may be heating or cooling. It is therefore necessary to specify the arrangement you require in a computer program input. This is done by specifying J12 or J21 (12J or 21J depending on the program) instead of just J. The first number is always the inlet and the second number the outlet.

The kettle reboiler, type **K** has a different construction from the other shell types. The port, where the bundle enters the shell is normally significantly smaller than the shell diameter. Chapter 8 is devoted to kettles.

The cross flow, **X** type could be used with the cold fluid on the shellside but its main application is in cooling large volumes of gas (often accompanied by condensation) so that the inlet is normally at the top. Cross flow designs are used when there is a very small allowable pressure drop on the shellside. The diagram shows only one inlet and outlet nozzle. I have to think hard to recall designing a cross flow unit with single nozzles. Normally there will be two or even four nozzles at the top and the bottom to give good distribution.

DETAILS OF REAR END HEADS

The **L** type head is similar, and may be identical to an A type stationary head. It has the same flexibility to be able to clean the tubes without disturbing the pipework. The diagram shows a nozzle in broken lines. For an even number of tube passes there will be no nozzle in this channel.

The **M** type head is similar, and may be identical to a B type stationary head. The pipework must be detached to access the tubes and it would only have a nozzle when there is an odd number of tube passes.

The **N** type rear head is not surprisingly the same as the N type stationary head.

The **P** and **W** types of rear head are both relatively inexpensive ways of accommodating thermal expansion/contraction of the tube bundle relative to the shell. They are however far more likely to leak than alternative ways to overcome this problem and they are not normally used in petro-chemical work.

It is logical to look at the **T** type floating head before the S type. The sketch in figure 2.3 is a bit confusing because if allows for a single pass design with an internal expansion bellows. (Illustrated with broken lines,) I have always thought that a different letter should be used for this special arrangement. The head for a normal T type is very simple and illustrated below in figure 2.4. This one would be for two tube passes if more than two passes are required there would need to be pass partitions in the head.

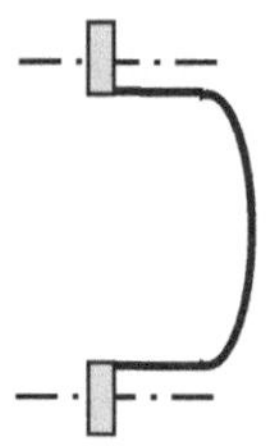

Figure 2.4

The shell has a cover which is aligned with the inside diameter of the shell so that the floating tubesheet is free to expand into the cover if necessary. The **floating head** is bolted to the floating tubesheet. The shell cover and floating head can be removed either to allow the tubes to be cleaned in situ or to allow the bundle to be removed to clean the outside of the tubes.

The shell diameter to outer tube limit clearance is very high for a T type exchanger. For a 1200 mm shell with a design pressure of 27 bar the diametral clearance would be about 140 mm. This high clearance is necessary to accommodate the nuts and bolts which attach the head to the tubesheet.

When a U tube bundle can not be used for a kettle which needs a removable bundle a T type must be used.

A big advantage of the T type is that there is only one relatively simple gasket which minimises potential leakage problems. This is the reason why one client I worked for insisted that all floating head units would be T types.

With the **S** type of rear head the S designation is presumably because of the **split** ring backing devise that is used. The normal S type head is clearer to see in figure 2.1. The S type construction was developed to reduce the bundle/shell clearance for a removable bundle. The ring incorporates holes for bolts and a face for a gasket seal. The split ring is assembled at one side of the tubesheet and the floating head at the other. This enables the tubesheet diameter to be significantly reduced. Section 5 of TEMA shows four possible designs of split rings. For a 1200 mm shell the bundle/shell diametral clearance would be about 70 mm – half the value for a T type.

The advantages of the S type compared with the T type is that more tubes can be fitted into a given shell size and the 'C' stream which by passes most of the heat transfer surface is reduced so that the shellside heat transfer coefficient is improved.

The disadvantages are that the increased complexity and the extra gasket make it more prone to leakage problem and the diameters of the shell flange and the shell cover flange have to be increased.

It is only the tubeside design pressure which is important when deciding if an S type construction can be used. (The shell side pressure is pressing the tubesheet and head together.) The following simple equation can be used to determine the maximum tubeside design pressure for an S type and involves only the shell inside diameter.

$$\text{Maximum design pressure, bar} = 66 - 0.037 \text{ Shell ID in mm} \qquad [\,2.2\,]$$

If the maximum design pressure is below this value there should be no problems with using an S type head. If it is only slightly above you may still be able to use an S type head. This simple equation includes a 20% safety factor and assumes an average allowable stress value for carbon steel bolts. If you are just above the limit - check with a mechanical engineer.

For the 1200 mm shell used to demonstrate a typical clearance required for a T type the maximum tubeside design pressure from equation 2.1 would have been 21.6 bar and a T type was clearly needed.

Finally the **U** type head, although it is arguable if this should be called a head because there is none! In this case the term 'rear end' is more appropriate. The bends in the tubes facilitate the change in direction of the tube side fluid. The majority of U tube exchangers have two tube passes which enables the bend to be in the vertical plane as shown in figure 2.3. However any even number of tubeside passes could be used and four tubeside passes are not uncommon. Except for the two pass arrangement the tube bends are always in the horizontal plane which will be discussed in more detail later.

COMMON COMBINATIONS

Yokell (Ref 9) lists 209 combinations of near head, shell and rear head used to specify different TEMA heat exchanger types. I can not see any in the list that are not practical but there are a lot that I have never seen, or even heard or read about being used. It might be an interesting exercise for the reader to make up their own list! I will concentrate on the ones that are used most frequently in the petrochemical industry.

AES/AET

The AES type is shown in figure 2.1. It is the most versatile of all the shell and tube units. The tubes can be mechanically cleaned in situ or the whole unit can be dismantled to clean the shell side mechanically, renew the bundle or take the bundle to a repair shop. I have had to use a triangular pitch in an AES design. I remember one case which had a clean fluid on the shellside but the client would not accept a fixed tubesheet design even though no expansion bellows was required. Using the triangular pitch avoided using two shells instead of one. However the vast majority of these units will use a square or aligned rotated square tube pitch.

Both the S and T type can only have an even number of tube passes. Except for the special one pass type which has an internal expansion bellows as illustrated below in figure 2.5.

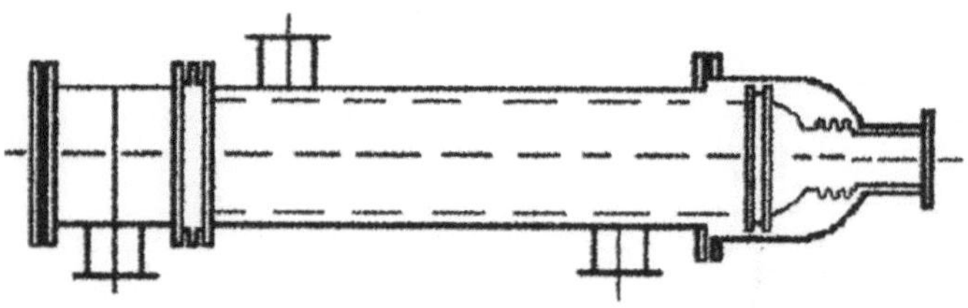

Figure 2.5

Some people would still designate this arrangement as an AES, personally I would specify it as AES Spl. Details of a an expansion bellows can be found in Chapter 11 on special mechanical features.

BEU

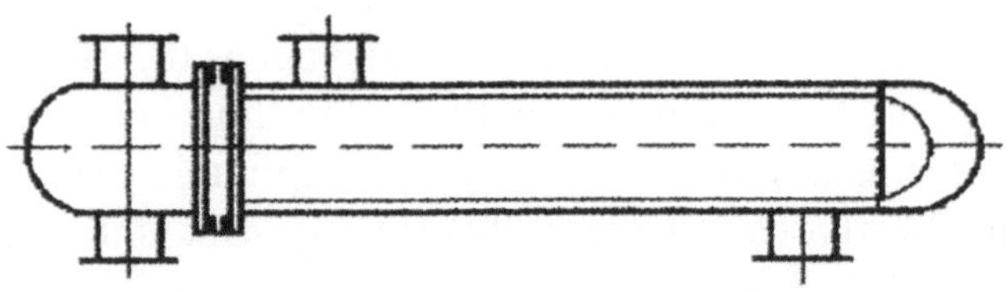

Figure 2.6

The BEU type has a simple construction, a removable bundle and is a widely used exchanger. The tubesheet is trapped between the stationary channel flange and the shell flange so that each fluid is only in contact with one gasket which minimises the chance of leakage. The tubesheet normally has a smaller in diameter than the flanges. An alternative is to have the tubesheet the same diameter as the flanges. This obviates the need for a separate test ring – which can not get lost at the back of a workshop!

It has a typical bundle to shell clearance of 19 mm which minimises the 'C' stream compared with most other types of construction.

The sketch shows an exchanger with the shellside nozzle nearest to the stationary head at the top and at the far end at the bottom. These can be reversed. The arrangement shown is the best one for mechanical design because the two saddle supports are in the central part of the exchanger and the weights are more evenly distributed. If the bottom shellside nozzle is near the stationary head there may be support problems with an exchanger which has a short length compared with its diameter. (If in doubt discuss it with a mechanical engineer.)

If you are running a relatively simple computer design program which calculates overall heat transfer coefficients there is no thermal difference between the two arrangements for the shellside. However if the hot fluid outlet temperature is close to the cold fluid outlet temperature a sophisticated program which divides the exchanger into small sections may detect a slight thermal advantage for one arrangement compared with the other. (Usually no more than 2-3%). If in doubt try both arrangements.

The comments on the arrangement of the shellside nozzles also apply to other TEMA types of exchanger.

There is a further consideration of the shellside nozzle at the far end which applies only to the BEU type.

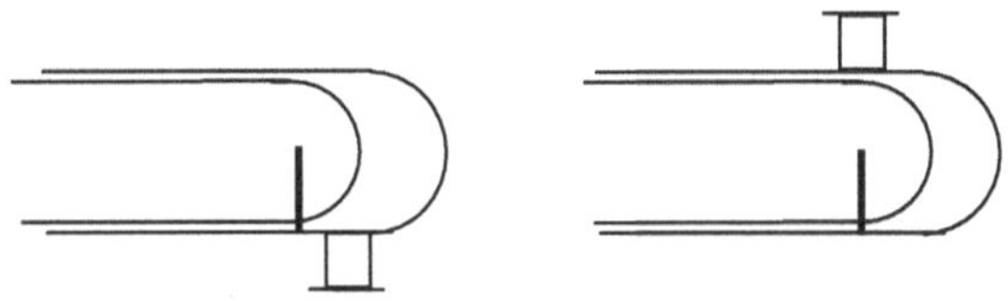

Figure 2.7

Figure 2.6 is the most normal arrangement. The far end nozzle has a full support plate after the nozzle. The shellside fluid does not pass over the U bends and the effective area for heat transfer must not include the bend area. This is the arrangement I would use for most single phase fluid particularly gaseous ones where vibrations could be a problem. Some client specs will only allow this arrangement and the U bend area can only be considered to be effective in kettle reboilers.

The arrangement on the left hand side of figure 2.7 has the nozzle after the support plate. The support plate is acting as the last baffle and should have the same cut as the other baffles assuming they are horizontally cut. This arrangement allows the fluid to pass over the U bends which now take a part in the heat transfer. If you use this option in your program it will no doubt just count the bend area as being fully effective. But is it?

The simple answer is that it may or may not be, and getting it wrong is the reason why some people will not allow it. The first thing to consider is what part of the overall thermal resistance is due to the shellside coefficient. The tube side coefficient and the fouling resistances will be exactly the same in the U bends as in the rest of the exchanger. If the shellside coefficient is only say 20% of the total thermal resistance and this resistance is doubled in the bend region then the bends will still be 80% effective. The next thing to look at of course is the velocity over the tubes compared with the average velocities in the rest of the exchanger. The top half of the tubes are mainly in the window above the support baffle and the velocities in this region will be quite reasonable. The velocities in the lower half of the bends will be much lower because of the by passing around the end of the bundle. This by passing could in theory be reduced by welding a shaped horizontal deflecting plate at the centre of the integral shell cover. This would of course involve some detailed expansion calculations to make sure it did not clash with the bundle. This together with the extra cost of extending the shell would lose most of the advantage over a simple design with a full support baffle. If you ever use this option you are probably safe to assume that the U bend surface is 50% effective and go a bit higher to say 70% when the shellside film resistance is a small part of the total thermal resistance but you should never assume it to be fully effective.

The option on the right hand side of figure 2.7 is more common and the shell does not need to be lengthened as much as for the option on the left. This can be useful when it is an outlet nozzle and the shellside fluid is boiling or when it is an inlet nozzle and the shellside fluid is condensing. In either case the support should be cut 5 mm above the top tube in the lower part of the bundle and there should be a large notch at the bottom of the support to allow the fluid to circulate. If the heat transfer is mainly condensing or boiling, velocities are not as important and may even be irrelevant, the most important thing is to make sure the fluid comes into contact with the tubes. For a condensing service and probably for a boiling service the normal baffles would have a vertical cut.

For a condensing service I would always consider the U bend area to be 100% effective. For a boiling service, particularly with a big bundle where there may be some 'dry' patches at the centre I might cut this back a bit but no lower than say 80%.

This arrangement will increase the maximum unsupported length compared with using a full support plate and extra U bend supports may be needed.

BEM

The BEM type is a fixed tubesheet design with bonnet covers at both the stationary end and the far end as shown in Figure 2.8. The stationary channel has to be removed to access the tubes which involves disturbing the pipework.

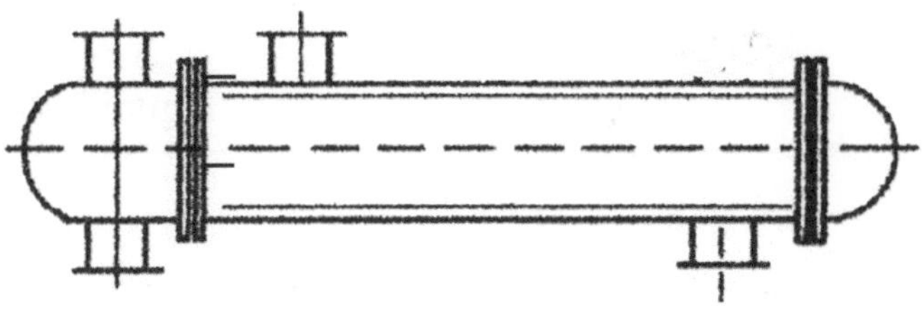

Figure 2.8

The potential expansion problems with fixed tubesheet exchangers are dealt with in chapter 11 on special mechanical features.

The shell side can only be chemically cleaned. Its applications are therefore limited to exchangers with clean shellside fluids or situations where chemical cleaning is available. It is normal to use a triangular or rotated triangular tube pitch to maximise the number of tubes in a fixed diameter shell.

A BEM may be used for a counter flow design with one pass on the tubeside in which case the channels at both ends would have a nozzle. To minimise the pressure drop on the tubeside of such designs the nozzles are often axial as shown in figure 2.9.

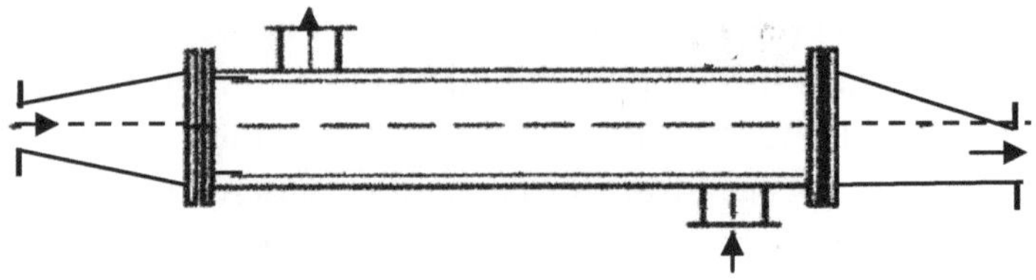

Figure 2.9

The nozzle at the inlet channel would be positioned on the shell axis but it is normally desirable to have the outlet nozzle offset, as shown in the diagram, to make the tubes free draining. If condensation is taking place inside the tubes offsetting the outlet nozzle is essential.

I would designate the exchanger in figure 2.9 as a BEM Spl.

AEM

The only difference between the AEM and the BEM is the flat cover plate at the stationary head end to enable the tubes to be cleaned without disturbing the pipework.

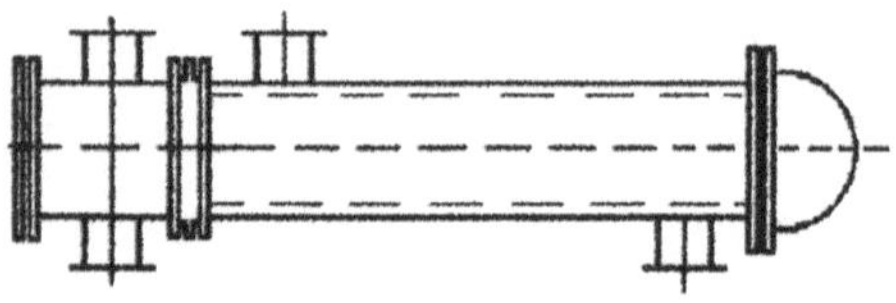

Figure 2.10

AEL

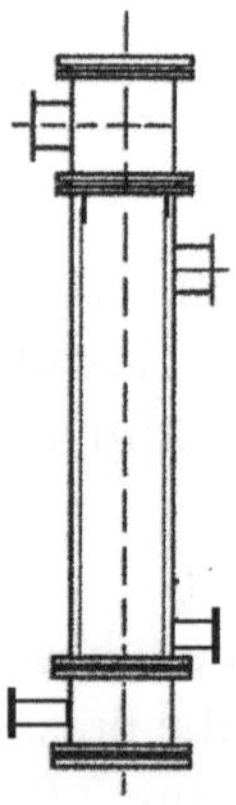

Figure 2.11

The AEL is a fixed tubesheet design with removable covers at both ends. One of the main applications is in vertical reboilers which require frequent cleaning of the tubes.

NEN

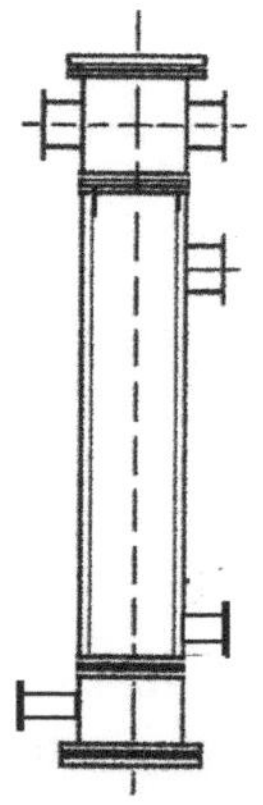

Figure 2.12

The NEN is a cheap version of the AEL and in the past I have used this construction for many vertical thermosyphon reboilers. The cost is reduced because four flanges have been eliminated. It is just like the AEL as far as cleaning the tubes is concerned but with the channels being integral with the tubesheets and the shell the tubes are much more difficult to access if any maintenance is required. It is a long time since I used one of these; nowadays clients seem prepared to spend the extra cost for the more flexible BEM or AEL types.

BXS

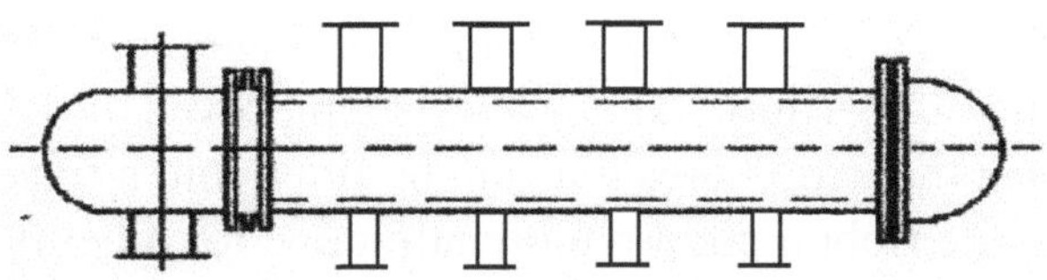

Figure 2.13

Figure 2.13 shows a BXS design with four nozzles at the top of the shell and four nozzles at the bottom of the shell. The nozzles at the top are larger than the nozzles at the bottom which is typical of a large volume gas stream being cooled with some condensation taking place. It would be more usual to have an A type stationary head but the B type trims the cost a bit.

One of the most important things in a cross flow design is to try and get an evenly distributed flow across the tube banks. I have seen distributor plates recommended to achieve this. I have always thought that this is a crazy waste of precious shellside pressure drop. (For a distributor to be effective it must use up a significant portion of the allowable pressure drop.)

A far better solution is leaving a good space above and below the tube banks. For most exchangers with an impingement plate at inlet we would need a height of $D_n/4$ at the top of the exchanger, and where there is no impingement plate at the outlet, a height of $D_n/6$. I always increase the height under the inlet nozzle to at least $D_n/3$ and above the outlet nozzle to $D_n/4$ for cross flow units. This is illustrated in figure 2.14.

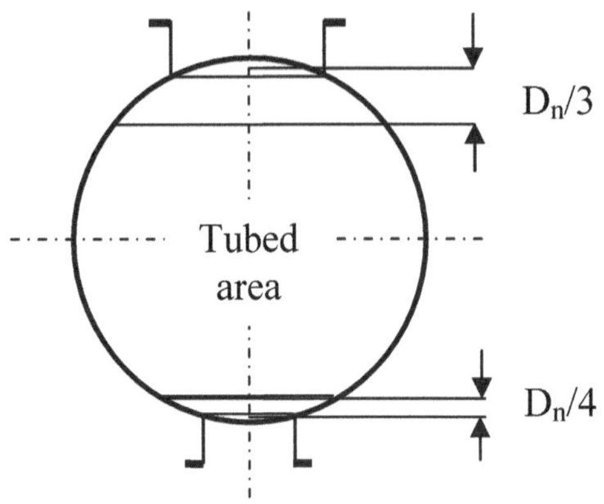

Figure 2.14

The nozzle diameters are the inside diameters and further details of how the top and bottom rows of tubes are positioned in heat exchangers can be found in the section on impingement protection later in this chapter.

The other thing I would never have in a cross flow unit is a solid impingement plate. With large nozzles and a plate at least 10% larger than the nozzle inside diameter they are likely to develop stagnant areas below them as well as increasing the pressure drop. I would always use impingement rods. If you have a client who insists on a solid impingement plate add an extra 50 mm to the height above the top tube row and keep the impingement plate 50 mm above the top tubes.

It is often more important in cross flow units than in most other units to keep the shellside velocity as high as possible. The shellside coefficient is invariably controlling the overall heat transfer coefficient and any by passing on the shell side should be minimised. For example if more than two tube passes are used they should be arranged in a ribbon pattern to avoid any F stream in a pass partition lane. This also improves the MTD. (This mainly applies to cross flow shells; the extra tubes that can be fitted into a fixed shell make alternative arrangements to the ribbon pattern more attractive for other types). The only other variable that you have any control of is the C stream which passes around the shell in the bundle/shell clearance. If you do not input any sealing strips into your program it may well come up with a number. The number it comes up with is seldom an optimum for this type of unit. You should keep on increasing the number of sealing strips as long as there is a worthwhile improvement in the shellside coefficient. I have gone up to a pair of sealing strips ever two tube rows. Sealing strips are cheap so don't be afraid to use them.

When dealing with large volume flows of gas think of the rest of the plant. Somewhere in the stream there must be a very big expensive compressor which is absorbing a lot of power. Making the best use of any pressure drops is very important for this compressor.

The other area in which the thermal design engineer should get involved is the pipework. You may have noticed that the inlet nozzle in figure 2.14 was quite large compared with the shell. This was not simply to illustrate the points I was trying to make. In trying to maximise the pressure drop across the tubes the nozzles on this type of unit can become very large and so can the associated pipework. With four nozzles the pipework should ideally be symmetrical as shown in figure 2.15. The symmetry should apply to both the inlet and the outlet pipework.

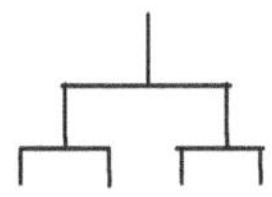

Figure 2.15

There may be times when this is not possible in which case there are two options as shown in figure 2.16.

Figure 2.16

The left hand connections should never be used, you should insist on the arrangement on the right hand side. With the arrangement on the right at least each 'section of the fluid' has the same length of flow path. This alone will not ensure correct distribution. The header pipes (the horizontal lines) should be as large as possible which will almost certainly mean swaging up from the line size. You should work closely with the process and piping engineer on this one. If they are not on your wavelength just say that you will have to put a note on the thermal data sheet commenting that performance can not be guaranteed because of possible maldistribution from the connecting pipework.

There will always be some uneven distribution in any heat exchanger, unless it is a in a laboratory test rig with a rectangular duct, and some variation in velocity is inevitable and acceptable but it should always be minimised.

A cross flow design is often regarded as a vibration free exchanger, unless it is single phase and there is the possibility of an acoustic problem. This is true because it is possible to add as many support plates as necessary without disrupting the thermal performance. This is one aspect of thermal design I am a bit conservative on. I add supports until the last stars have disappeared from view. I then add one more. This may seem like a needless expense when trying to design the most economic exchanger. However I am always conscious that today's grass roots design is tomorrows revamp when the flow rate will be increased by 20% and an extra support plate or two does not add much to the cost.

DEU

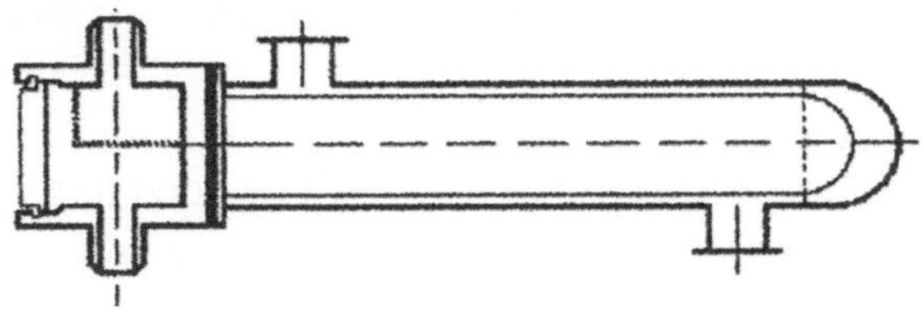

Figure 2.17

The D type stationary head is used for high pressure enclosures. The high pressure is only on the tubeside, the shellside is relatively low pressure. These special units are considered in more detail in chapter 11 on special mechanical features.

BKU

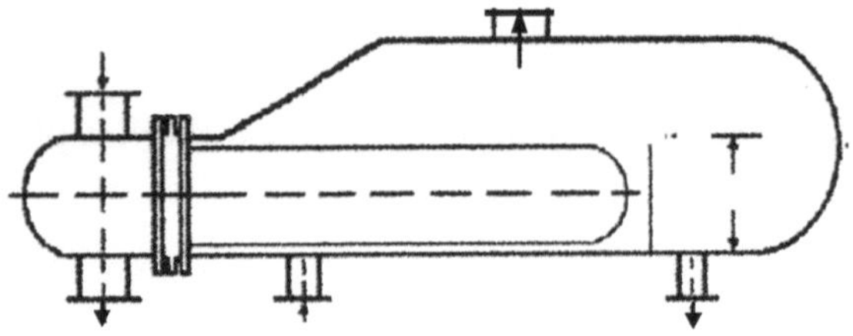

Figure 2.18

This is a kettle reboiler and kettles are dealt with in detail in chapter 8. An AKU would have a flat cover at the stationary end to access the tubes for cleaning without disturbing the pipework. However if the tubes needed regular mechanical cleaning an AKT construction would probably be used.

BFM

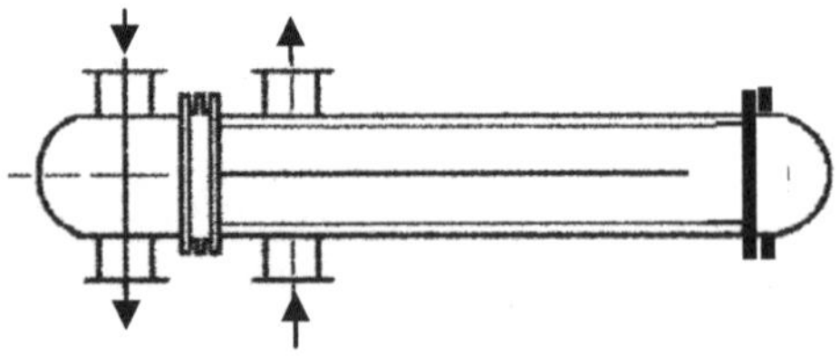

Figure 2.19

The BFM type has two tubeside passes and two shellside passes. The fluids are therefore in true counter current flow and are used when there is a temperature cross. (The cold fluid outlet temperature is higher than the hot fluid outlet temperature). Reasonable temperature crosses may be accommodated in shells involving cross flow such as the X shell or the G shell. For substantial temperature crosses the F shell can be a very inexpensive alternative to using multiple shells in series.

I did once design what I called a BFM Spl. It was on a revamp job with limited space which fixed the tube length. Using E shells would have meant three shells in series so I immediately opted for an F type shell. With two passes on the tubeside and two passes on the shellside I soon found that I was 'chasing my tail'. As I increased the shell size the coefficients fell off so much that the shell size had to be increased again. It should have only been a small unit but it was fast becoming a much bigger one. Without any length restrictions there would have been an easy solution – use a smaller diameter shell and a longer tube length.

The shellside fluid was clean so that I could use a fixed tubesheet and I decided to try a three pass shell with three tube passes. There were no problems with the computer program because there were no computers in those days. I quickly found an optimum design which satisfied everyone's requirements. I then sat back to think about the practicality of what I had come up with. The horizontal baffles could be welded so there would be no leakage problems and eventually I saw a way that it could be assembled. Unfortunately I forgot to write a procedure down and this was one of the last units I designed before leaving to join the academic world.

The story has a happy ending because the vendor did cry 'help' and Harry Rudel who took over from me as head of heat exchanger section at Stone and Webster got a nice trip out of it. The unit worked fine when the plant was restarted. A few years later I was looking through a Patterson and Kelly catalogue and there was a nice picture of an exchanger with three shell side passes! Yorkell (Ref 9) refers to some highly specialised exchangers being built with 16 shellside passes!)

It would be an interesting challenge to work out the necessary fudges to design another three shellside pass unit using a modern computer program.

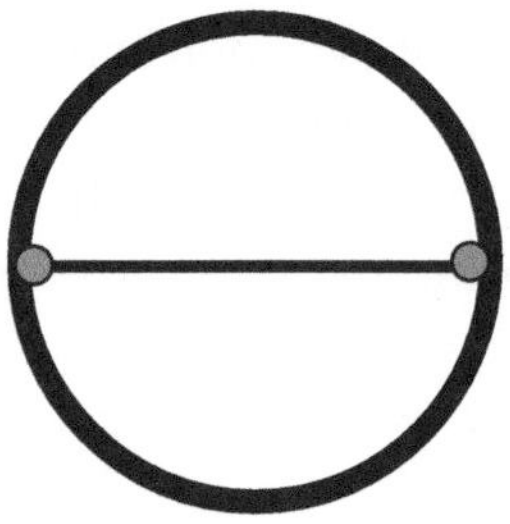

Figure 2.20

The BFM (or AFM if the tubes need to be cleaned without disturbing the pipework) have no sealing problems. The horizontal baffle can be welded to the shell as shown in figure 2.20. (The type of weld shown can only be used when the shell is large enough to give access. For smaller diameters the shell must be cut in two and then welded again integral with the horizontal baffle.)

BFU

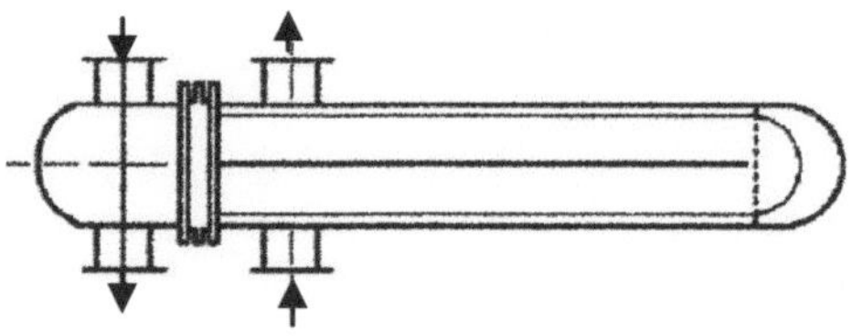

Figure 2.21

The BFU has the advantage of having a removable bundle so that it could be used with a dirty shellside fluid requiring mechanical cleaning of the outside of the tubes although I would only consider using it with a dirty fluid if the tube length was quite short minimising bundle pulling problems. This advantage is also its weakness because there are always potential sealing problems with the horizontal baffle. Some of the ways of providing a seal are as follows.

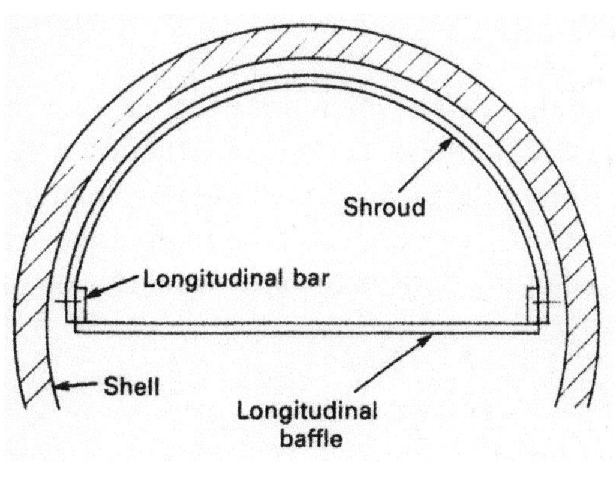

Figure 2.22

Figure 2.22 shows a shroud which completely encloses the tubes above the horizontal baffle. The only leakage point is where the fluid enters, (or leaves) the shroud adjacent to the stationary head. To minimise this leakage the nozzle is fitted with a removable sleeve.

The nozzle sleeve protrudes a few mm into the shrouded area so that the leakage area is approximately the sleeve/shroud clearance multiplied by the sleeve outside diameter. To remove the bundle the pipework must be broken and the sleeve removed from the nozzle.

Apart from the complicated construction the main disadvantage with this arrangement is that the bundle/shell clearance must be increased by about 25 mm – to about 45 mm. This not only reduces the number of tubes that can be fitted into a particular shell it also has thermal implications because the top half is not symmetrical with the bottom half. I doubt if computer programs will ever be able to handle this one in a single run. You will need two design runs one with clearances appropriate to the top half and another to suit the bottom half and then a bit of 'fudging' to bring the two runs together for the overall performance.

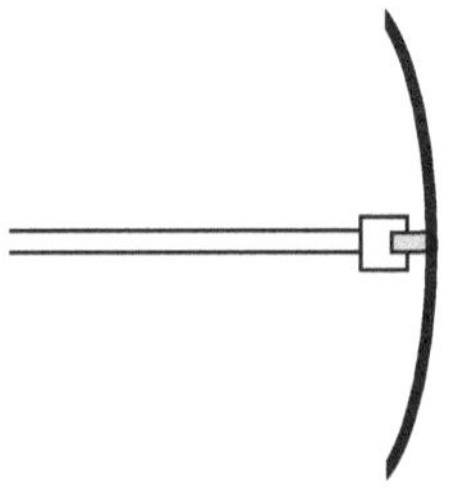

Figure 2.23

Figure 2.23 shows a seal with a braided packing strip. I can see lots of potential problems with this device and I have never personally seen it used.

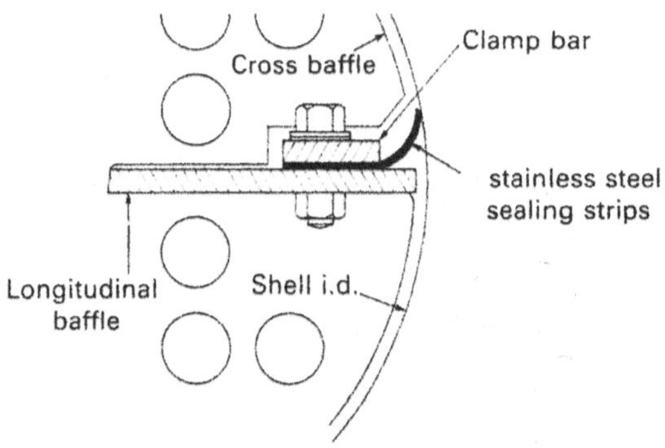

Figure 2.24

This is traditionally the most popular sealing device which has been used for many years. The seal is provided by one or more flexible stainless steel strips held down by a clamp bar which is bolted to the horizontal baffle. In the arrangement shown the high pressure side would be at the top so that the pressure difference would be forcing the seal against the shell. If the shellside fluid had entered at the bottom the seal would be on the underside of the longitudinal baffle.

The most recent development in seals is shown in figure 2.25. The left hand side is a cross section and the right hand side a pictorial view. This type of seal is supplied by Kempchen I do not know if anyone else supplies a similar type of seal.

The construction consists of lamellae, (thin strips of 316 SS), which are spot welded into a holder section. I have seen a sample section and they are very robust. The holder just clips over the edge of the horizontal baffle which may be up to 25 mm thick. There are two ways the seal can be used. One involves cutting grooves in the seal holder to fit around each baffle. I do not like the idea of compromising the integrity of the seal. Only the second method is shown in figure 2.25. This has slots cut in the baffles which the holder slips into. (Except for the last baffle which acts as a stop and may need to be thicker than the rest of the baffles.) Unless the exchanger is exceptionally long just one continuous seal is required.

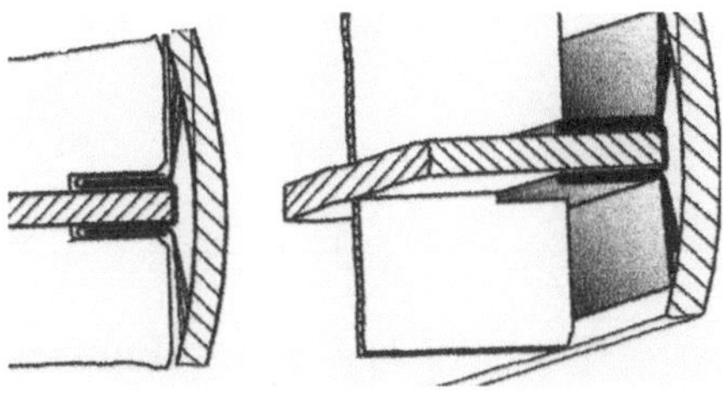

Figure 2.25

This arrangement provides a seal above and below the horizontal baffle so that where there are irregularities in the shell diameter there are two narrow gaps in series for the by pass fluid to cross. This must halve the flow compared with a seal on one side only. Also the lamellae vary in length, being shorter towards the centre. This results in multiple contact points or leakage points. Any leakage must therefore pass over a series of gaps and this effect will be similar to the effect of the labyrinth glands used to seal the shafts of turbines and compressors.

Some client's specs require the seals to be replaced every time the bundle is removed. This is a very quick and simple procedure with this type of seal.

If you visit the Kempchen website do not take too much notice of the diagram which shows the semicircular baffles which would completely block the flow path. I am very impressed with their seals but I do not think they should branch out into designing heat exchangers.

BFM & BFU

There are some features which apply equally to both types of construction. There is no theoretical restriction on the type of baffles which may be used in an F type shell other then the ones which would apply an E shell. (e.g. not using horizontal baffles for condensing fluids.) In practice you may find a more restricted selection in your program and you will have to live with it. When vertically cut baffles are used it is important to arrange for an offset adjacent to the turning space where the shellside fluid crosses over from the top to the bottom or visa versa. This is shown in figure 2.26.

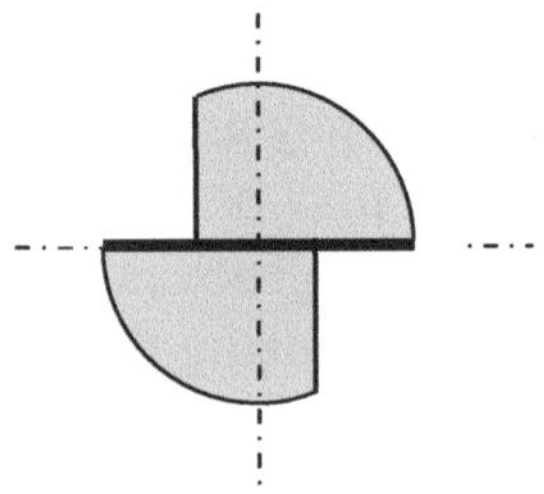

Figure 2.26

The sketch shows only the two end baffles and the horizontal baffle. If the two baffles had been at the same side most of the turning space would have become a dead pocket and the tube surface area in this region ineffective.

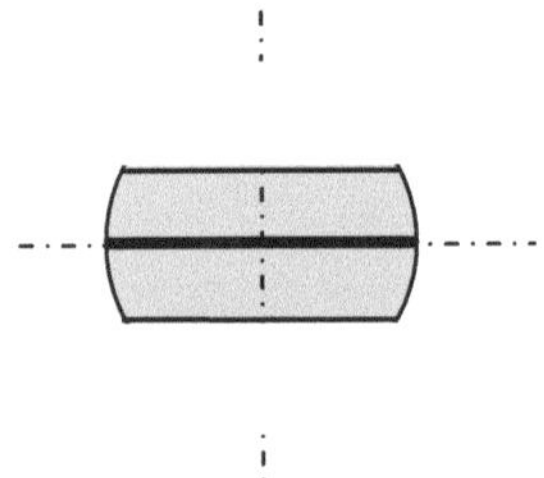

Figure 2.27

The same thing when horizontal baffles are used as shown in figure 2.27 above.

Insulated horizontal baffles

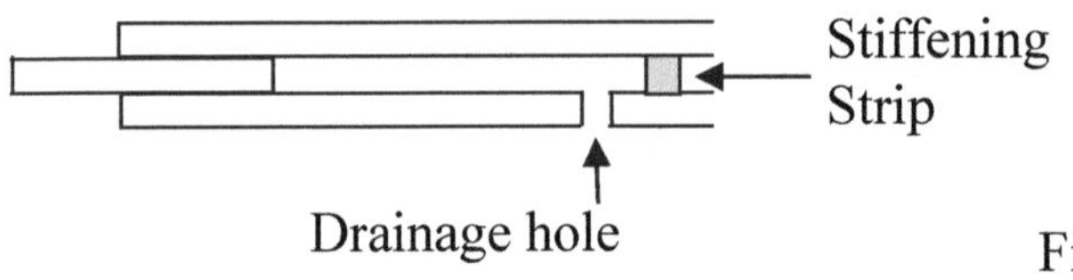

Figure 2.28

The 'insulation' in an insulated baffle is provided by the shellside working fluid. There are in effect two horizontal baffles separated by a gap. For a fixed tubesheet design this could simply be two horizontal baffles welded separately to the shell. For removable bundles the construction is shown in figure 2.28. The two baffles are welded to a strip of metal which protrudes sufficiently for a sealing device to be attached to it. If necessary stiffening strips can be welded to the two baffles either longitudinally as shown or across the diameter. These will greatly increase the second moment of area of the combined section (like an 'I' section) and provide stiffness against bending. These strips should be as thin as practical to minimise 'fining' effects. The thermal design engineer should simply state that these are acceptable and leave the details to the mechanical engineer or vendor. It is important to state that each compartment of the horizontal baffle should have a drain hole at each end – say 10 mm diameter.

When vertically cut baffles are used with U tubes the insulated baffle has a useful side effect. Suppose the sections of metal shown in the diagram are all 10 mm. The total thickness of the combined horizontal baffle is 30 mm and if the U bend gap is 38.1 mm (19.05 mm tubes on a 3 x d_o bend) the by pass gap between the inner tubes and the horizontal baffle is only about 4 mm which is quite acceptable. With an uninsulated baffle you may need to use sealing strips to block off this by passing, in which case the sealing strips would also help stiffen the horizontal baffle but only by a small amount compared the effect of stiffeners in an insulated baffle.

Well after all that, just how effective is an 'insulated' baffle? Consider a stationary hydrocarbon liquid in a 10 mm gap. With a typical thermal conductivity of 0.13 W/mK the thermal resistance is:

$$\frac{\Delta x}{k} = \frac{0.01}{0.13} = 0.08 \text{ m}^2\text{K/W}$$

Or if you prefer to think in terms of coefficients 13 W/m^{2}K, and if you add on the thermal resistances of two slabs of metal and the heat transfer coefficients at either side of the baffle (which will be lower than the values across the tubes) it is easy to conclude that there will be very little heat flow across the baffle. With a gas the insulating effect is even better.

Materials

It is an easy one to miss but we should be certain that 316 SS is compatible with the shell side fluid. I can not remember a case when it was not. I know Kempchen can provide seals in other materials but these are regarded as specials and will cost more. If you are dealing with a fluid that contains chlorides 316 SS will not be acceptable. You should of course have an alert from the materials specified for the shellside. If in doubt discuss it with your metallurgist, (sorry technologist!).

Pressure drop and temperature difference limitations

The pressure drops allowable on the shell side are often limited by the client's specs. These tend to vary between 0.7 bar, (10 psi) and 0.35 bar, (5 psi). It should be remembered that these limits are per shell and if you have two shells in series it will be the hot shell which will control. It is also often unclear if this applies to the shellside, (nozzle to nozzle), or just across the horizontal baffle, (excluding nozzle losses), which logically it should do.

Any limitations should take into account the diameter and the length of the shell. A small diameter shell with long tubes has a relatively large ratio of potential leakage area to normal flow path areas in each baffle space. As the shell size increases this ratio will decrease. Large diameter shells with a short tube length have the least potential problems. Also a gas is more likely to leak than a liquid particularly if the gas contains hydrogen. Of course blanket limitations do not take these issues into account.

Some specs may have temperature difference limitations – typically 194 K (350°F). Whether there is a spec limitation or not you always need to be careful of large changes in temperature of either fluid but especially the shellside fluid with large diameter shells. The reason is that one half of the tubesheet will be exposed to the highest temperature fluid and the other half to the lowest temperature fluid. This of course will be the same at the tubeside; the hottest fluid in the channel will be adjacent to the hottest fluid on the shellside. This produces a temperature gradient across the tubesheet which will tend to bend it. If there is a large temperature difference consult a mechanical engineer before you carry out too much work on an F type shell. (Note that in an E type shell the tubesheets are exposed to the shellside fluid at or close to the inlet, or outlet temperatures across the full height of the tubesheet.)

Summary

Some modern programs will allow an option of an insulated or uninsulated baffle so it is easy to run both options and see if one is worthwhile. There may also be options of percentage leakage. With fixed tubesheet designs this is of course zero; with removable bundles a judgement must be made. Once again it is possible to try a few different values and observe the effect. If the shell diameter is a reasonable size compared with the length and if I intended to specify Kempchen type seals I would consider 10% leakage to be quite sufficient.

An F type shell involves a lot of complication and potential pitfalls compared with an E type shell. Experience of bad designs is probably why some clients ban them altogether. More enlightened clients will ask for them to be submitted for approval. I hope that what I have covered will help you in making out a case for using one where appropriate. An F type shell instead of a number of E type shells in series can save a lot of money, and of course a lot of tonnes of CO_2 being released into the atmosphere.

COST COMPARISONS OF TEMA TYPES

I had intended to try and write a substantial section on costs but quickly gave up. Ted Saunders – ref 6, has 13 pages on detailed costing methods. The following table is an extract from his work.

	S type head	T type head	FTS	U tube
Cost ratio	1.0	1.04	0.78	0.86

Table 2.1

The table is based on a surface area of 310 m^2, and a tube length of 6.1 m.

I would recommend ignoring the small difference between the S and T types of floating head. Cost is never a factor in deciding which one to use.

For shorter tube lengths the cost ratio of the U tube and FTS items would fall. I would therefore recommend thinking in rounded terms of a 25% reduction in cost for FTS constructions and a 20% reduction for U tube constructions. (The cost ratio for the FTS is presumably for one which does not require an expansion bellows.)

For any type of construction there are other considerations which affect the cost. Clearly an A or L type head will cost more than a B or M type head. This choice has an impact on maintenance and will therefore normally be in any clients spec.

I will not venture into the field of absolute costs. Heat exchangers are like any other commodity their cost depends on what a vendor is prepared to sell for and prices can vary widely depending on the vendor's situation. If they already have lots of work in hand the cost will go up, if their shop floor is empty and they are thinking of laying off staff the price will be rock bottom. I have seen lists of quotes where the highest price was 50%

higher than the lowest one. Most reasonably sized companies will have a specialist purchasing department and this work should be left to them.

It is worth emphasising that the one variable which the thermal designer will normally have some control over is the length and diameter of the unit so I will close this section with a little ditty which applies to any type of unit.

Long and thin is sure to win.
Short and squat can cost a lot.

ADDITIONAL STANDARDS

There are two standards in addition to TEMA which are relevant to thermal designs work.

ISO 16812 / API 660 (Ref 40)

I can remember when API standards were only used for air cooler designs where they are the basic standard (API 661) since there is no equivalent of TEMA for air coolers. Nowadays they are normally called up as well as TEMA for shell and tube job specs. API has now been combined with the European Standards as EN ISO 16812.

This International Standard specifies requirements and gives recommendations for the mechanical design, material selection, fabrication, inspection, testing and shipment of shell and tube heat exchangers for the petroleum, petrochemical and natural gas industries.

It is quite a small document compared with TEMA having only 46 pages and TEMA is referred to throughout demonstrating that it is not intended to be an alternative standard. Most of this standard is mainly of interest to the mechanical engineer and vendor rather than the thermal design engineer.

The one big exception is the heat exchanger data sheet. The previous edition of API 660 only had a reproduction of the TEMA data sheet which has been around since I designed my first heat exchanger. The current data sheet is 5 pages and I have reproduced it at the end of this book under Check Lists and Data Sheet.

The first page is very similar to the TEMA data sheet and my only comment is that the momentum flux on line 29 is written as rV^2 when clearly ρV^2 was intended. I found it very encouraging to find a basic typing error in an international document when I have been worrying about the errors I must be making.

The second page starts with a nozzle schedule which the thermal design engineer would certainly fill in. The rest of the page is listed as optional and could be left to a mechanical engineer except that the thermal designer should fill in a preliminary estimate of weights.

The third page is a blank space for additional comments and sketches. In my experience it will often be necessary to add additional sheets at this point, say 3A, 3B etc.

The fourth sheet contains room for additional process data. The data list is very comprehensive but I have some strong comments on the layout. The spaces for the heat release data have only 7 spaces for one fluid and 8 spaces for the other fluid when process normally provide 10 or 11 sets of values. There are also two very small areas where heat release curves can be plotted. It would be much more useful if the two curves were on the same graph and the temperature was on the vertical axis. The enthalpy scale on the graph is also confusing. The units are stated to be enthalpy, (actually specific enthalpy) but the numbers are for a specific enthalpy ratio which would be dimensionless. There is an unnecessary space at the top of the page which could be used to address these issues. If this page were redesigned and a computer wiz kid arranged for the graphs to be plotted as the data was fed in it could be very useful.

The last page is for fixed tubesheet exchangers only. I had already written a section on FTS exchangers in chapter 11 before writing these comments on the ISO standard. I will leave it to the reader to compare the two.

One general comment on the overall document is that I can not see the point in repeating lines 1, 2 and 3 on every page. After page 1 the job number and item number should be sufficient and pages like page 4 need the extra space.

There are some points in this standard which affect the thermal design and are not covered in TEMA. For example “transverse baffles and support plates shall have notches that are 10 mm high to facilitate drainage”. This has always been standard good practice, (except of course in kettles where it is pointless), calling up ISO is a way of ensuring that it is done.

There is the following comment on impingement plates “When used, an impingement plate ‘baffle’ shall extend at least 1 inch (25 mm) beyond the projection of the nozzle bore.” There is no guidance in TEMA on impingement plate size. However ISO might be a bit restrictive with a 43 mm inlet nozzle (2 inch nominal bore).

There is a section on sealing strips which concurs with standard good practice and avoids the need to add some notes on the thermal data sheet. (e.g. “Peripheral bypass seal strips shall not restrict the inlet and outlet flows”) There are also some notes on bundle skid bars which I can not find mentioned in TEMA.

There is a useful section on the details which should be submitted on vendor drawings (which should be checked by a thermal design engineer as well as a mechanical one), but I can find no other sections which have a direct impact on the thermal design. Even if they are not called up for a particular job the reader should have access to these standards and be aware of what they contain.

HEAT EXCHANGE INSTITUTE STANDARDS – HEI

These standards, (unlike ISO), provide **an alternative** means of designating a shell and tube heat exchanger. They are the standards which are used throughout the power industry to classify heat exchangers. I had been designing heat exchangers for a couple of years using the TEMA standards before I became aware of their existence. That was my loss because they do have a lot to offer thermal designers outside the power industry particularly in the design of high pressure units. At the opposite end of the pressure scale they are also involved in vacuum equipment.

It is interesting to compare their system of designating units with the TEMA system. TEMA uses a three letter system; HEI uses five symbols which are a combination of letters and numbers. The first two symbols specify the front or stationary head. The first one is a letter and the second a number. The letter specifies the type of enclosure and the number the stationary tubesheet arrangement.

The third symbol specifies the shellside flow path and is always S with an appropriate suffix. S_1 is a one pass shell (TEMA E type), S_2 is a two pass shell, S_d is a divided flow, S_s is a split flow and S_k is a kettle.

The last two symbols specify the rear head. The first of these is a number (which may be the same as the number used to specify the stationary head but 6 for example is for a split ring floating head. The final symbol is a letter which specifies the far end enclosure for example U is for U tube – the same as in the TEMA system.

An AES TEMA type would be specified as $C1S_16F$ in the HEI system a TEMA BGM would be specified as $B3S_s3B$ in the HEI system. Although five symbols instead of three may seem more complicated than TEMA some things are easier to remember such as the shellside flow path.

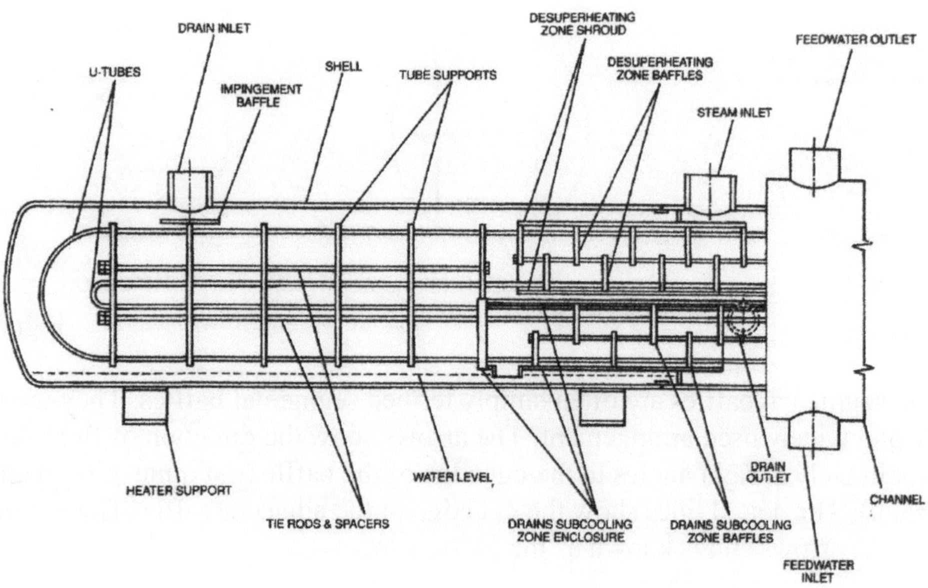

By permission of the HEI Figure 2.29

Figure 2.29 is an interesting example of a high pressure feed heater. The first point of interest is that neither the TEMA nor HEI designation systems would provide a comprehensive description. As well as having standards for general heat exchangers the HEI has additional standards for closed feed water heaters (in an open feed heater the steam mixes directly with the feed water), without offering guidance for specifying them (Yokell – Ref 9).The best TEMA description would be a DFU Spl and the type of high pressure enclosure and the baffle arrangement would need fully explaining.

High pressure enclosures are dealt with in chapter 11. Desuperheating and sub cooling are dealt with in detail in chapter 5 on condensing and figure 2.29 will be referred back to in that chapter..

Yokell – ref 9 has an excellent section on the HEI standards. It is well worth reading if the reader does not have access to the original standards.

DETAILS OF CONSTRUCTION

Having looked at the basic terminology and classification it is appropriate to now look at some of the details of construction.

BAFFLES

Baffles must satisfy two requirements. Their heat transfer function is to divert the shellside fluid across the tubes. Their mechanical function is to support the tubes.

SINGLE SEGMENTAL BAFFLES

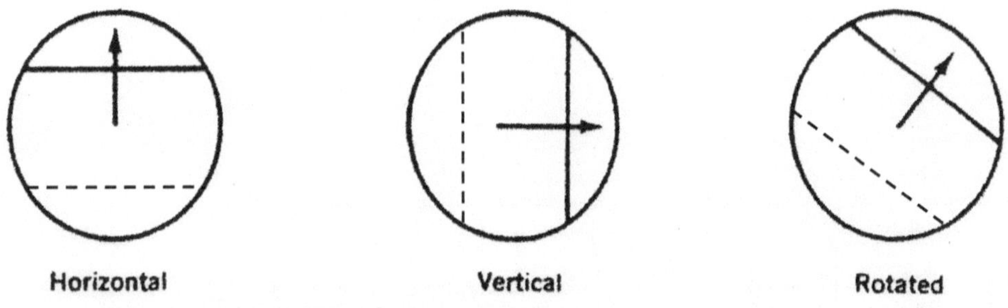

Figure 2.30

Single segmental baffles are often simply termed segmental baffles. They are the most basic and widely used arrangement. The arrows show the direction of fluid flow which is approximately at right angles to the cut edge of the baffle (not quite at right angles near the shell). The dotted lines show the cut edge of the adjacent baffle. The segmental area where the arrows end is known as the baffle window.

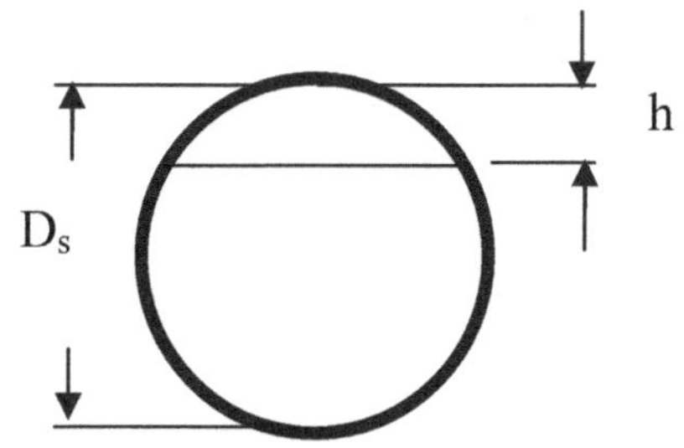

Figure 2.31

BAFFLE CUT

An important feature of a baffle is the baffle cut. In figure 2.31 h is the maximum distance from the cut edge of the baffle to the inside of the shell of internal diameter D_s. The **baffle cut** is defined as the ratio **h/Ds** although this ratio is normally multiplied by 100 to express it as a percentage. The normal range of values for the baffle cut is between 15% and 45%. For single phase flow the optimum utilisation of pressure drop usually corresponds to a cut of about 25%. However it may be necessary to significantly deviate from this for a particular design. 'Opening up' the cut – increasing it, will reduce pressure drop and decreasing it, will increase the pressure drop. Adjustments to the baffle cut normally have the same effect on the heat transfer coefficient, increasing it reduces the shellside coefficient and reducing it increases the coefficient. However reducing the cut does not always increase the overall performance of the exchanger. Below about 18% cut, the increased flow in bypass streams, particularly the E stream, may reduce the overall performance.

The baffle cut as defined above is the only one which is used as an input to (or an output from) a computer program. The baffle cut may also be defined as the ratio of the long flow area in the window to the long flow area for the whole shell. The window long flow area is the area of the segment minus the area unavailable to shellside flow in the window. The main areas unavailable to shellside flow in the window are the sections occupied by tubes or tie rod spacers. For the whole shell areas may also be lost to bundle runners, sealing strips and dummy tubes as well. A baffle cut defined in this way is known as a baffle cut on an area basis. The difference between the values of baffle cuts on a dimension and an area basis can be quite significant particularly for low values of cut in units with a high bundle/shell clearance. The area basis cut is the most fundamental for heat transfer because it represents the proportion of the flow in the window and was normally used in hand calculations (although it was ideally expressed as a dimensional ratio on the thermal data sheet for the mechanical engineers to use). A computer program will do its own internal calculations to work on an area basis.

There is an important practical limitation to the value you should specify for a baffle cut. This is illustrated in figure 2.32 The figure shows three sections through a baffle each with a tube hole adjacent to the baffle cut edge. The section on the left shows a hole which is completely surrounded by metal. The middle section shows a semicircular hole – the cut is on the centre line of a row of tubes. The last section on the right shows a cut that is off centre. The edges at the top of the hole with the offset cut form nasty sharp points. These are a hazard for people assembling the bundle and must be avoided.

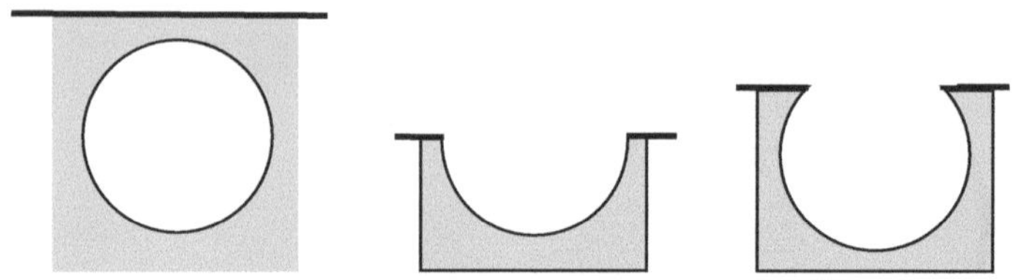

Figure 2.32

Tube holes in baffles must always be either surrounded by a minimum of about 2.5 mm of metal or have baffle cuts on the centre line of tube rows. It is normally only possible to surround the tube hole with metal with a square pitch layout. To fix the edge of the baffle correctly will result in a precise number for the cut – say 25. 23%. This should not be rounded down to 25%. Look at the tube layout and the position of the cut for your last computer run.

ASYMETRICAL BUNDLES

Bundles are normally symmetrical about the vertical centre line. The vertical centre line either passes through a vertical row of tubes or midway between two vertical rows of tubes. Bundles are seldom completely symmetrical about the horizontal axis because of the difference in the areas unavailable for tubes at the top and bottom of the exchanger adjacent to the nozzles. Tube rows however are normally spaced symmetrically about the horizontal centre line so that if horizontally cut baffles are used the cut will be the same at the top and bottom.

There are sometimes situations where the tube rows are not symmetrical about a centre line. For example if U tubes are used in a shell with large nozzles it may be possible to fit in more tubes by having the bundle slightly offset from the shell horizontal centre line. If horizontal baffles are used the baffle cut will pass through a different section of a tube row at the top and bottom of the shell. It is still essential to have the baffle cut passing through the centre of a tube row, (or in between a tube row for a 90° layout). The cut will therefore need to be slightly different at the top and the bottom of the shell. The arithmetic mean of these two cuts should be used for the computer run.

BAFFLE PITCH

The other variable for all baffle types is the baffle pitch. The baffle pitch, P_b is the distance between the centres of the baffles as illustrated in figure 2.33.

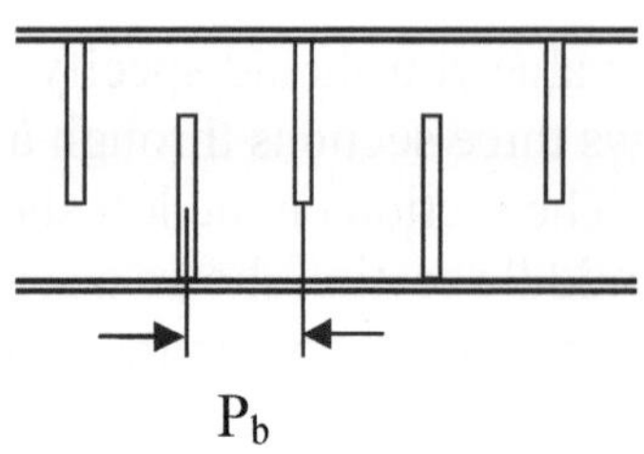

Figure 2.33

The distance between the baffle faces used to calculate the cross flow velocity is the baffle pitch minus the thickness of one baffle. The difference between this and the baffle pitch is relatively quite small for large units with a wide baffle pitch but is more significant for small shells with a close baffle pitch.

The minimum 'normal' baffle pitch in TEMA is $D_s/5$ or 51 mm, (2 inches) whichever is greater. The normal maximum baffle pitch is the shell diameter, D_s. In TEMA the maximum is only governed by considerations for maximum unsupported length. These are listed for different tube diameters and different materials in table RCB-4.52. An example from this table is that for a 19.05 mm tube a maximum span for a carbon steel or alloy steel is 1524 mm but for most none ferrous materials 1321 mm. The maximum unsupported span is in the window and therefore has a length of twice the baffle pitch minus the thickness of one baffle. In practice, in large shells, maximum baffle pitches are usually limited by vibration considerations.

The baffle pitch is the main factor controlling the cross flow velocity. Decreasing the baffle pitch increases the pressure drop and normally increases the heat transfer. However as with a small baffle cut using a very small pitch may cause sufficient flow distortions, particularly increasing the E stream, to reduce the heat transfer. This is most likely to occur with viscous fluids.

In general the baffle cut and pitch should be modified until the cross flow and long flow velocities are reasonably balanced to achieve an optimum design.

WHEN TO USE EACH TYPE OF BAFFLE

HORIZONTAL & 45° (ROTATED CUT)

For single phase flow the baffle cut is invariably horizontal. With horizontal baffles it is essential to get the number of shellside crosses correct this is illustrated in figures 2.34, 2.35 and 2.36.

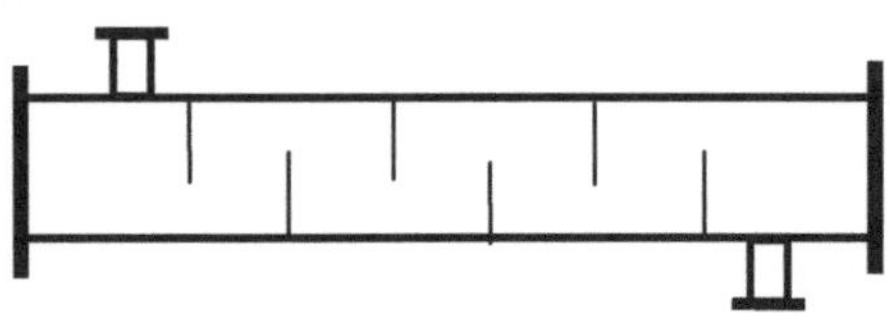

Figure 2.34

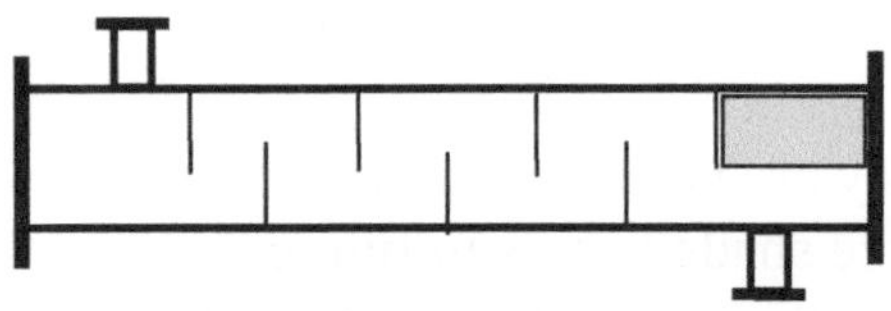

Figure 2.35

In figure 2.34 the shellside fluid crosses the shell 7 times, once across the inlet space, once across the outlet space and five times between the end baffles. There are six baffles and the number of spaces is always one more than the number of baffles.

In figure 2.35 there are 7 baffles and therefore eight spaces. If you follow the path of the fluid, very little of it will pass over the shaded area which becomes a 'dead space'. This is a loss of fully effective heat transfer area.

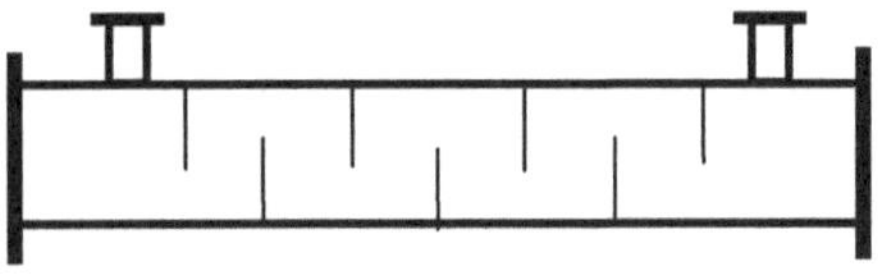

Figure 2.36

When the shellside nozzles are on the same side, (either top or bottom or at one side in a vertical unit), an odd number of baffles resulting in an even number of passes is necessary to avoid any dead spaces.

For horizontal baffles use an odd number of crosses when the shellside nozzles are on the opposite side and an even number of crosses when the shellside nozzles are on the same side.

This also applies to 45° cut baffles, which are termed rotated cut in TEMA.

The main use for 45° cut baffles is with viscous fluids where a 45° tube pitch may have a thermal advantage over a 90° tube pitch. The baffles are rotated instead of the tube pitch so that visually and mechanically the tube pitch is at 90° but thermally it is at 45°. This arrangement allows cleaning lanes to be maintained without increasing the width of the pass partition dislocation lanes. This is explained more fully in the section on pass partition lanes later in this chapter.

VERTICAL CUT

Vertical cut baffles are usually used for two phase flow. Figure 2.37 is a plan view of a shell with vertical cut baffles. The first thing to note about vertically cut baffles is that it does not matter if there is an odd or even number of shellside crosses. This is an extra degree of freedom compared with horizontal baffles where the number of baffles must be changed in multiples of two.

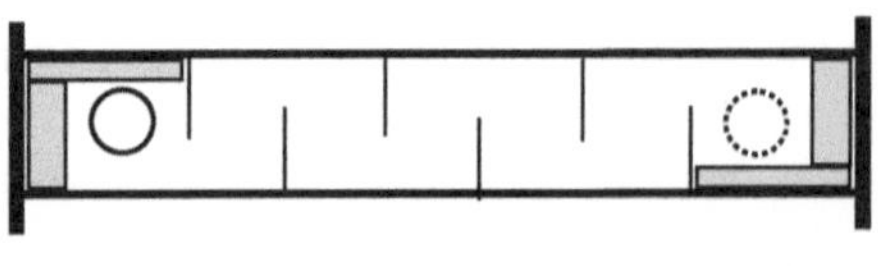

Figure 2.37

The second thing to understand is why they are hardly ever used for single phase fluids. If they were used for single phase fluids the shaded areas in figure 2.37 would be relatively ineffective for heat transfer. Also the untubed areas under the top nozzle and above the bottom nozzle create potential by pass areas which will increase the C stream. (In a horizontally baffled exchanger only the baffle/shell clearance is available to the C stream.)

Some specs will insist on vertically cut baffles for all two phase flows on the shellside but this too restrictive. They are invariably used for condensers but there is occasionally a good case for using horizontally cut baffles such as when there is only a very small percentage of the shellside fluid condensing – this is discussed in chapter 5 on condensers. There are more cases where horizontal baffles are the best option for boiling services – this is discussed in chapter 7 on boiling. However I can not think of a good case for using vertically cut single segmental baffles for a single phase fluid.

For two phase fluids the shaded spaces in figure are not ineffective although they may be less effective than the spaces between the end baffles. Boiling can take place in a stagnant fluid and condensation does not rely on fluid velocity although it may be assisted by it.

It does not matter if the shellside fluid is boiling or condensing the largest volume flow rate and hence the largest nozzle will always be at the top of the shell. Sometimes the nozzle at the top of the shell can be very large – up to 50% of the shell diameter. This creates a large untubed area at the top of the shell. It is therefore often necessary to add 'ears' to the top of vertically cut baffles to prevent axial bypassing as illustrated in figure 2.38. Sealing strips above the tubed area to reduce the C stream by passing will also be required

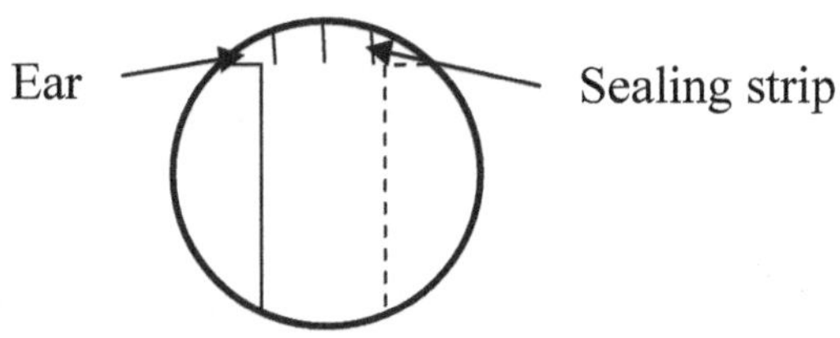

Figure 2.38

MULTI SEGMENTAL BAFFLES

There are two types of multi segmental baffles; double segmental and triple segmental. They are both used to either reduce the shellside pressure drop or reduce the baffle pitch and hence the unsupported length to avoid a vibration problem. I have tried to use triple segmental baffles a couple of times without finding a satisfactory design. It is very difficult to keep all the cut edges of the baffles on a centre line of a tube row and to keep the different longitudinal flow paths reasonably equal. I will therefore just concentrate on the much more widely used double segmental type.

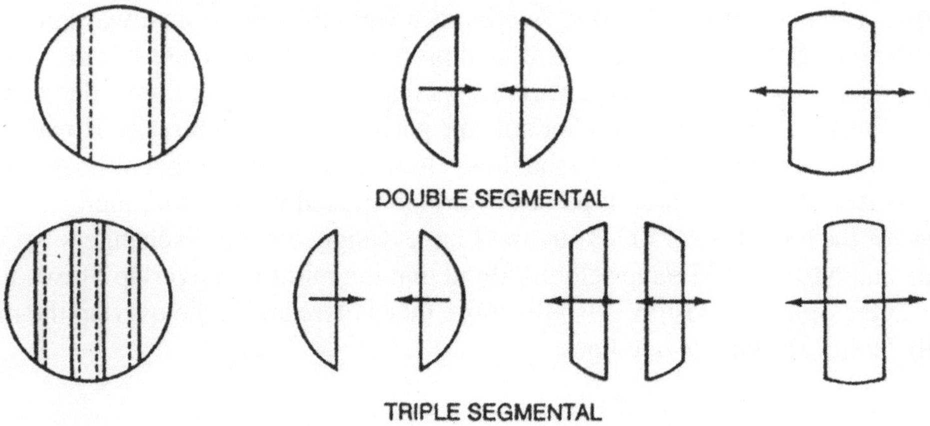

Figure 2.39

These are shown along the top of figure 2.39 in a sectional view and in figure 2.40 in a plan view of a shell.

There are now two different types of baffle, a central baffle adjacent to two outer baffles. These must overlap to keep the bundle together. The overlap is shown on the left hand sketch in figure 2.39 and should cover a minimum of two vertical tube rows. The baffle cut is defined by the segment removed from the central baffle and to ensure that there is sufficient overlap this is normally limited to a maximum value of about 25%.

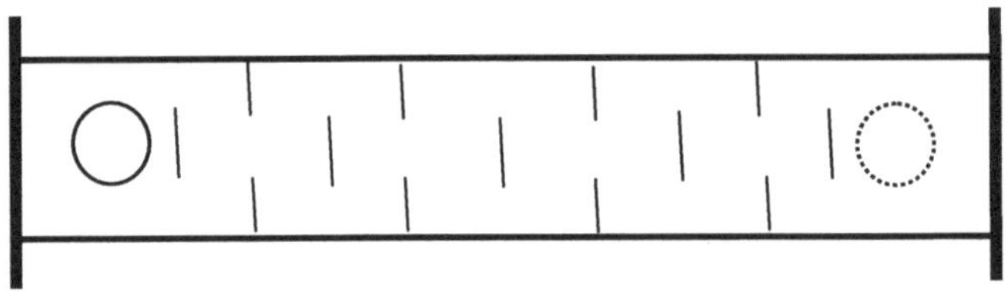

Figure 2.40

The plan view shows that there are central baffles adjacent to each nozzle. If there were side baffles adjacent to the nozzles there would be potential dead spaces next to the shell in both of the end spaces. Unlike single segmental baffles double segmental baffles may be used for single phase flows as well as for two phase flows. To maintain this arrangement there must always be an odd number of baffles and there is always one more central baffle that there are pairs of outer baffles.

It is useful to compare a double segmental baffle arrangement with a single segmental arrangement with the same baffle pitch. In the double segmental arrangement the cross flow velocity has been halved and the cross flow path has also been halved. This is a similar situation to changing from an E shell to a J shell which was discussed earlier in this chapter where the change was shown to reduce the pressure drop by a factor of about 7 – equation 2.1. The reduction in long flow pressure drop is not so dramatic, the length of flow path is unchanged and the velocity in the window is usually not quite halved. The overall effect is still a large reduction in pressure drop, at the expense of a reduction in the shellside heat transfer coefficient of course. In practice some of this reduction in pressure drop will be utilised to reduce the baffle pitch.

When double segmental baffles are input into a computer program it will normally just work out the baffle cut required to give the minimum of two vertical tube rows overlap. The stream analysis will look very good because the B stream will be high. This can be very deceptive. The number of tube rows overlap is a variable which can always be input. As this is increased above two rows the B stream proportion will decrease but the shellside heat transfer coefficient will invariably increase. The B stream must decrease because more fluid is forced through other streams particularly the A stream. However the number of rows of tubes involved in the cross flow is increased so that although the B stream is decreased the surface area it passes over is increased which more than compensates for the reduction of the B stream. I have designed many exchangers with double segmental baffles and I have seldom found that the minimum overlap of two tube rows results in the optimum design. Also the more tubes there are in the overlap the more mechanically 'solid' the bundle becomes.

To optimise the shellside of a double segmental baffle arrangement vary the number of baffles/spaces in increments of two and the number of tube rows in the overlap, (leave out the baffle cut this is dependant on the tube rows in the overlap).

There is sometimes one exception to the baffle arrangement shown in figure 2.40 which is when a J12 is used particularly if the spec. requires a solid impingement plate. If central baffles are used next to the central inlet nozzle the resulting velocities and momentum fluxes off the impingement plate may be unacceptably high. It may then be necessary to have a pair of side baffles adjacent to the inlet nozzle as shown on figure 2.41.

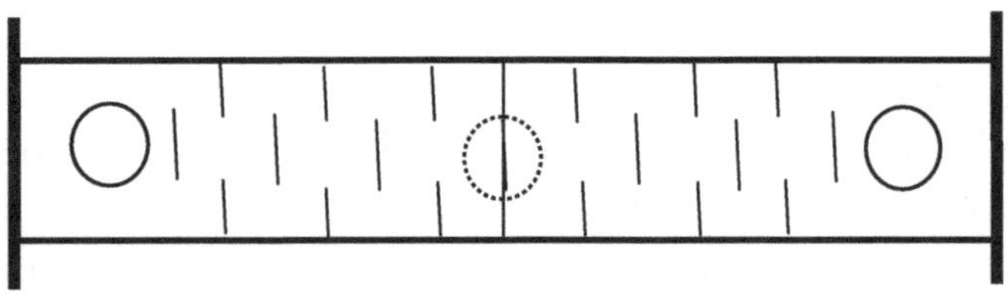

Figure 2.41

NO TUBES IN WINDOW – NTIW

This is a special single segmental baffle arrangement when the baffle cut is always perpendicular to the nozzle axis. Normally the nozzles are at the top and bottom of the exchanger so that the baffle cuts are horizontal as shown in figure 2.42. The name is obvious from the fact that only the central section contained within the cuts of all the baffles is tubed. The nozzle at the top will always be either the same size or larger than the bottom nozzle because it carries the highest volume flow rate.

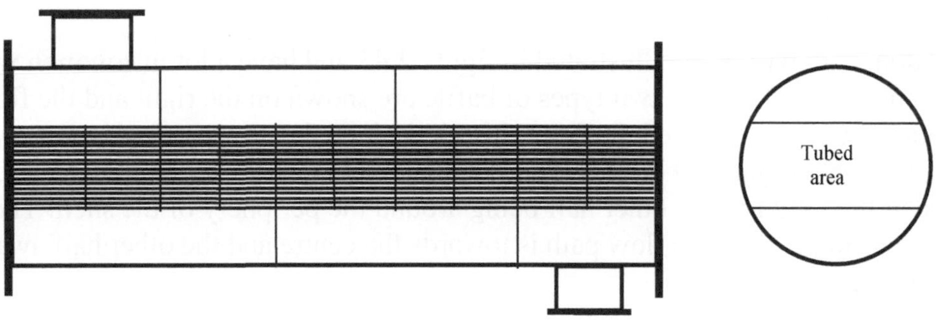

Figure 2.42

All the normal flow streams are present but the A stream is significantly decreased because the baffle segments in the windows do not have any holes. The B stream is therefore very high and the heat transfer coefficients are high compared with any other segmental baffle arrangement apart for a cross flow unit.

This should be a vibration free exchanger (although there may be acoustic problems but these will not lead to tube failures). As far as fluid elastic instability is concerned this can be eliminated by simply adding support plates in between the baffles in the tubed area. The diagram shows only one support plate in each cross flow space but any number can be used without significantly affecting the heat transfer or pressure drop. As with a cross flow unit you can just run your program adding support plates to your design until all the vibration stars fade away and then add one more for when the flow rate is increased at the future revamp.

Normally a NTIW design will have the spaces at inlet and outlet the same as in between the end baffles. There are no dead spaces, the full length of all the tubes are normally fully effective. One exception to this is when U tubes are used and the bend area is not utilised. Another is with a split ring floating head design when some allowance may have to be made for the loss of effective area behind the support plate. Particularly with a floating head design there should be lots of sealing strips to minimise the C stream by pass. (Probably more than your computer program will suggest.)

A vibration free design with excellent heat transfer characteristics must have some drawbacks. The first is the increase in shell size for a specified number of tubes. This is sometimes not as bad as it may seem. If the shellside fluid is mainly gaseous there will be a large volume flow rate and consequently large nozzles at inlet and outlet. If this is the case there are already large untubed spaces adjacent to the nozzles and the shell size may only need to be increased a couple of sizes to accommodate a NTIW design. The minimum baffle cut for a NTIW design is 15% but most designs have a higher cut than this. The other main problem is that when the allowable shellside pressure drop is relatively low the number of cross passes may be quite low. There is a very large change in pressure drop between a three and a five cross design. From useful tables 1 the change is a factor of $(5/3)^{2.8} = 4.2$. Even changing from 5 to 7 crosses involves a factor of $(7/5)^{2.8} = 2.6$. If your design is over on pressure drop and increasing the number of cross passes is not practical it becomes necessary to increase the baffle cut and increase the shell diameter.

DISC AND DOUGHNUT BAFFLES

Disc and doughnut baffles are illustrated in figure 2.43 and have a lot in common with double segmental baffles. The two types of baffle are shown on the right and the flow path through the shell is on the left. From this view alone the arrangement is exactly the same as for double segmental baffles. The longitudinal flow is axial; half its path being at the centre of the shell and the other half being around the periphery of the shell. The cross flow is always radial; half the flow path is towards the centre and the other half away from the centre.

Whilst disc and doughnut baffles could be used for horizontal shells they are more common in vertical shells.

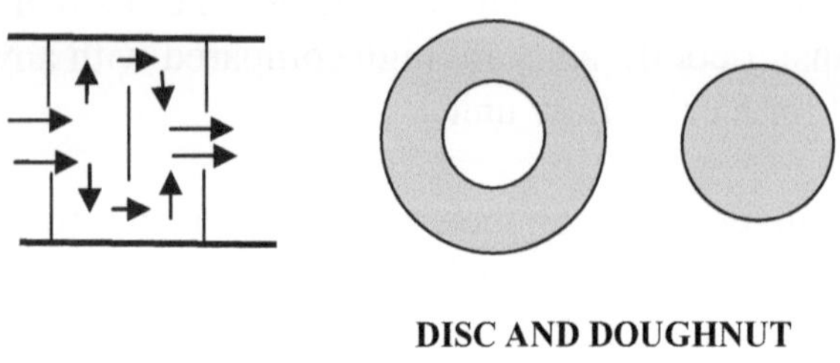

Figure 2.43

DISC AND DOUGHNUT

It is interesting that HTRI had this arrangement as an option for many years and then dropped it from the menu. It was of course not on the menu when I had my only first hand experience of disc and doughnut baffles.

I had just started work at Foster Wheeler where I was replacing two contract thermal design engineers who had just left. One of them had spent a couple of weeks looking at about half a dozen disc and doughnut designs. The designs were part of a process package and the client had asked Foster Wheeler to check them. The designer who had checked them came to the conclusion that they were all under surfaced – the worst one by over 30%. The company who had carried out the designs had a lot of experience in designing with disc and doughnut baffles and presumably had their own program. The head of the heat transfer section was therefore reluctant to report back to the client that the designs were unacceptable so I was asked to check the checks that had been done.

There were detail drawing of all the units and the computer runs on CST using double segmental baffles were clearly the closest computer option. My first reaction was that I should just check one using hand calculations. Then I did what I usually do when faced with an unusual challenge – nothing! I just thought about it for a few hours – going outside to have more cigars than I normally would to stimulate the old brain and then it came to me that there were several fundamental errors in the computer runs I had in front of me.

The first and most obvious one was that the computer program had been allowed to default to the minimum of two tube rows overlap when clearly there were a lot more tube rows between the edges of the two types of baffle. As I was counting the tube rows the second mistake became clear.

All the computer runs had been carried out for a 30° triangular tube pitch (the exchangers were all vertical FTS designs). This was correct for one row of tubes only – the one on the horizontal centre line. The vertical row at 90° to this row on the vertical axis was at exactly 60° to the fluid flow. With a radial flow path most of the tubes were somewhere in between these limiting values. I corrected the tube rows crossed for the original run and then set up a second input with the tubes at 60° with the appropriate number of tube rows crossed for this configuration. An average of these two runs predicted a design that was no longer under surfaced. Then as I examined the stream analysis for these runs one last detail occurred to me.

There is no C stream in a disc and doughnut baffle. I could not eliminate this because it would interfere with the E stream which was still present. I just minimised it by inserting lots of sealing strips.

Finally I decided to carry out some hand calculations to compare the velocities with those on the CST output. If drawings are available it is easy to work out an effective diameter for cross flow and there are two values; one corresponding to the outer edge of the cross flow area and another corresponding to the inner edge. My calculations predicted 4 m/s in and 4.5 m/s out giving an average value of 4.25 m/s. The CST runs showed 4.49 and 4.52 m/s for the 30° and 60° runs respectively for the mid point values and 4.2 and 3.75 m/s for the X flow values. This was a better agreement than I had expected and it did not take long to check the rest of the units.

I could find nothing wrong with any of them and even the one which had been estimated to be over 30% under surfaced was now only 4% under surface. This would be acceptable for any type of check on other people's designs, and in this case I was sure sub contractor would have used their own specialised computer program. So there was no embarrassing report – just a simple "the designs have been checked and found to be acceptable".

ROD BAFFLES

Rod baffles consist of a combination of vertical and horizontal rods which have the same diameter as the clearance between the tube rows. The tube pitch is therefore normally 90°; a triangular pitch could be used but the tube pitch to diameter ratio has to be increased which loses the main advantage of the triangular pitch. The rods are always in groups of four so that there is one rod above a tube, one rod below and one at either side as illustrated in figure 2.44 for a single tube.

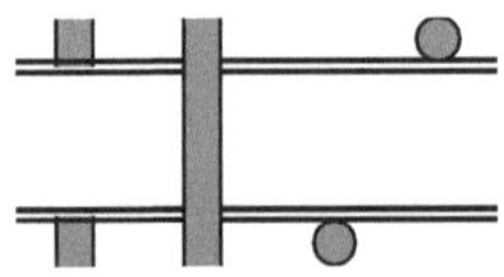

Figure 2.44

The rods are held in position relative to the tubes by peripheral support rings or plates. The support rings are then welded to longitudinal bars to prevent any axial movement. The longitudinal bars are then welded to the stationary tubesheet.

Rod baffled exchangers may be used for fixed tubesheet, U tube or floating head designs. They must have distributors at both the inlet and outlet. Except for areas around the inlet and outlet distributors the flow is purely longitudinal.

The rod baffled exchanger was developed in the early 1970's by Phillips Petroleum Co. They had had failures of tube bundles on a butadiene plant due to flow induced vibrations. Fortunately the bundles had a square pitch so that they could carry out emergency repairs by inserting rods between the tubes to reduce the unsupported lengths. They must have put quite a lot of these in to be sure that the vibration problem was eliminated. The rods solved the vibration problem and it was noticed that the extra turbulence they created improved the shellside heat transfer coefficient. That started a research and development program using only rods instead of segmental baffles. The result was the rod baffled exchanger which they patented. It could only be used under licence from Phillips and they naturally wanted to recoup their development costs so that they were relatively expensive.

I have dabbled with patents on a number of occasions – all have failed to date! My understanding is that patents only last for 20 years so that the Phillips monopoly should be well over. There is now a rod baffle option in some commercial computer programs.

ORIFICE BAFFLES

Orifice baffles have no baffle cuts; they occupy the full shell diameter (minus the bundle/shell clearance.) The flow path is created by enlarging alternate tube to baffle clearances as shown in figure 2.45. The fluid flows through the enlarged clearances of one baffle pass diagonally across the tubes to the enlarged clearances of the next baffle.

The inlet and outlets to the shell require distributors the same as the rod baffled exchangers.

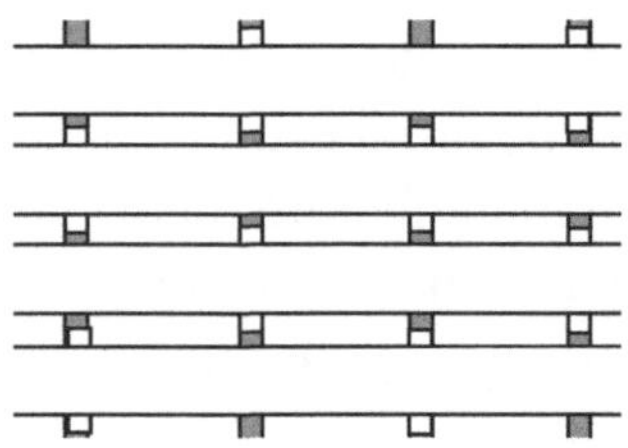

Figure 2.45

The effective flow stream is therefore very similar to the A stream in a conventionally baffled exchanger. The only bypass streams are the E stream and an F stream if there are any pass partition lanes.

My only experience of an orifice baffled exchanger was in the academic world. I once supervised a student project to compare orifice baffles with conventional segmental baffles. My motive was to find a vibration free exchanger which could be used as an alternative to the expensive Phillips rod baffle design.

Our test facilities were very limited and the only fluids we could use were water, low pressure steam or air. The results showed that the heat transfer was very similar to a segmental baffled exchanger using the same pressure drop.

It would certainly eliminate vibration problems in the same way as the rod baffled exchanger. The one disadvantage would be that it would be limited to relatively clean fluids because the orifice holes might block up.

The idea of the orifice baffled exchanger has been around a long time: it is mentioned in Kern (ref 14). I have never heard of one being used commercially but a net search uncovered some recent academic work on them!

HELICAL BAFFLES

I am not even going to consider an illustration for this arrangement. There is a very pretty multicoloured picture on the ABB Lummus website. Instead of having baffles perpendicular to the exchanger axis this arrangement has quadrant shaped baffles sloped away from the perpendicular to give a spiral effect. The main flow path of the shellside fluid spirals through the exchanger rather than passing from side to side. This clearly has a pressure drop advantage compared with a conventionally baffled exchanger where the direction of the fluid is abruptly changed at each baffle. The B stream is increased which results in a reduction of the other flow paths particularly the A steam (which is of course effective for heat transfer.) Overall the claim is that there is a substantial increase in overall heat transfer coefficient which seems to be a reasonable claim.

Unlike the rod baffle or orifice baffle arrangement the "Helixchanger" does not need distributors at inlet and outlet so that tubes will normally need to be omitted adjacent to the nozzles to limit the momentum fluxes.

This is a reasonably recent development (late 1980's) and when I first heard of them my first question was "how do you drill holes in an angled baffle so that you can insert a round tube through them?" The answer of course is that they are not drilled but are made with a patented stamp forming technique. Some of the patents for the helical baffled exchangers may still be effective and they are only be produced under licence. There are therefore only a small number of vendors who can manufacture them. This together with the complicated fabrication makes them relatively expensive. Although possible vibration problems are claimed to be reduced they are not effectively eliminated as they are with a rod baffled exchanger.

EMbaffle

This is a very recent development, (this centaury) by Shell Global. The baffles are made of a plate material that has been slit and expanded. They look a bit like a garden trellis designed for climbing plants – visit their website to see the pretty pictures. They are similar to the rod baffles in covering the whole cross section of the shell (minus the baffle/shell clearance) and having peripheral plates to stiffen them and attach them to bars which prevent axial movement. Since the holes are not round the fluid can pass through the parts of the hole which is not in contact with the tubes. The flow through the exchanger is therefore longitudinal and the fluid passes from the holes in one baffle to the holes in the next baffle in a diagonal path which is the same flow path as for orifice baffles.

These units must have distributors at inlet and outlet and there will be some cross flow adjacent to the distributors. As with all long flow units there should be no vibration problems but momentum fluxes must be limited and/or some form of tube protection introduced adjacent to the inlet distributor.

The baffle plates are made from quite thin material – usually a stainless steel. They do not therefore comply with TEMA requirements in this respect. Perhaps the next edition of TEMA will have an exception clause to cover this.

There is a very interesting variant to having an exchanger fully baffled with EMbaffles and that is to have EMbaffles combined with conventional segmental baffles. For large diameter shells the limiting maximum value for baffle pitch is normally the unsupported length in the window and/or a potential vibration problem. EMbaffles can be fitted at the mid point between each baffle. This halves the unsupported length and changes the natural frequency of the tube span by a factor of four. This of course increases the pressure drop but the increase in turbulence particularly in the window will increase the heat transfer coefficient a bit. There is no theoretical reason to limit the number of EMbaffles to one per baffle space. The big advantage of this arrangement is that it is no longer necessary to have expensive distributors at inlet and outlet. I can see a promising future for the EMbaffle when used in this way.

There is now an option in the HTRI programs for a design with EMbaffles. I have not personally tried it at the time of writing this book.

SUMMARY OF BAFFLE TYPES

The difficulty in designing an optimum shell and tube heat exchanger is invariably the design of the shell side. Ten different arrangements of baffles and baffle types have been covered above.

The long flow arrangements and the helical baffle type all have their originators making all sorts of claims that their products are the best innovation to hit the market. The main one of course is the elimination of flow induced vibrations. This must be correct for the longitudinal flow types. The helical baffle advocates do not claim to eliminate the possibility of vibration only to substantially reduce it. I am not sure how 'substantial' the reduction is.

Another claim that they all make is that fouling is reduced. I have devoted a separate section to fouling in this book which I hope makes clear what a complex topic that is. The elimination of dead pockets must reduce the fouling of some heavy fouling streams however there are other fluids where I doubt if they will make any difference. I would not want to reduce a fouling factor without some very good data to justify this for a particular stream.

The other claim is that they are more economical. This certainly needs to be examined critically. Whenever there is a proprietary product there are a limited number of suppliers, this always increases the cost. Distributors are expensive and increase the shellside pressure drop. There may well be a number of situations were they are the most economic solution but there will be a lot more were they are not. It is up to the thermal design engineer to recognise where using one of these designs could be the right solution. If a computer program is available to the designer it is reasonable to have a few runs to see how these are turning out. However I doubt if it is really worthwhile to try and optimise a design (unless it is a feed job and one is known to have been used on a similar plant). The licensor will have teams of engineers who are specialists and at the end of the day you will have to go to them for a price and importantly a thermal guarantee. It is simpler to just go to them with the process data in the first place. Normally of course there will be an alternative solution using a convention design, perhaps a NTIW if vibrations are a potential problem. An alternative design should be optimised and immediately sent to vendors for a comparative quote. Unless there are other considerations such as pipework limitations the optimum design is then the cheapest one.

BUNDLE SUPPORTS

The baffles provide support for the tubes as well as directing the flow of fluid across the tubes in most shell and tube exchangers. However there are cases were an extra support is required and when mechanical support only is required.

An obvious case where support only is required is in kettle reboilers. The fluid flow is basically in a vertical direction through the bundle and from a purely thermal viewpoint the bundle would be better off without them. Each support includes all the tubes which are therefore full circle except that it may be advantageous to cut the segments off the top of the supports 5 mm above the top of the tubes in the top tube row. These segments are superfluous material and just add to the weight of the bundle. The number of supports is fixed by TEMA or spec limits for the maximum tube span. Increasingly specs are only allowing maximum spans of 80% of the TEMA maximum. (Very, very occasionally extra supports may be used in a kettle because of a 'possible' vibration concern.)

A historically interesting, but now irrelevant case where only supports were required, is when saturated steam (or steam with a small degree of superheat) is used on the shell side. When I first started designing heat exchangers we never calculated a coefficient for condensing steam on the shellside of an exchanger – we just used a value of 2000 Btu/h ft^2 ^oF (11356 W/m^2K). Thermodynamically there is no need for baffles and the pressure drop should ideally be zero during isothermal condensation so we simply put a note on the thermal data sheet 'supports only required'. The mechanical engineers would then design supports/baffles with a maximum allowable pitch and a baffle cut that ensured a minimum of two tube rows overlap. It is interesting how little inaccuracy this resulted in. Just try it out with one of your designs from a modern computer program and see if it makes any difference to the size!

Back to more conventional support plates in current use. Figure 2.46 shows some supports that are used with floating heads (S and T types). The left hand sketch is known as an anchor support plate for obvious reasons the other two are alternatives.

Figure 2.46

The anchor design is the one we normally used at Kellogg in the early 1960's and is still used today. The aim in all these designs is to allow fluid to circulate around the section of tubes behind the support. The circulation in the right hand sketch is improved by an oversized drain notch. Note that the top has been cut off the two supports on the right hand side. Any metal above the top tube row is not providing any support and is inhibiting circulation. Supports should therefore have a segment cut off about 5 mm above the top of the top tube row unless some of it is required for tie rods when 'ears' should be left in place. The area between the support plate and the floating tubesheet is not normally counted as effective area for heat transfer even though with a condensing stream and the inlet adjacent to the floating head end it may well be fully effective. Even though the area may not be counted for heat transfer it is good practice to allow circulation around the support. Any circulation even with a single phase fluid provides a little bit of safety factor.

Figure 2.47 shows the type of supports that can be used for a U tube bundle with vertical U bends.

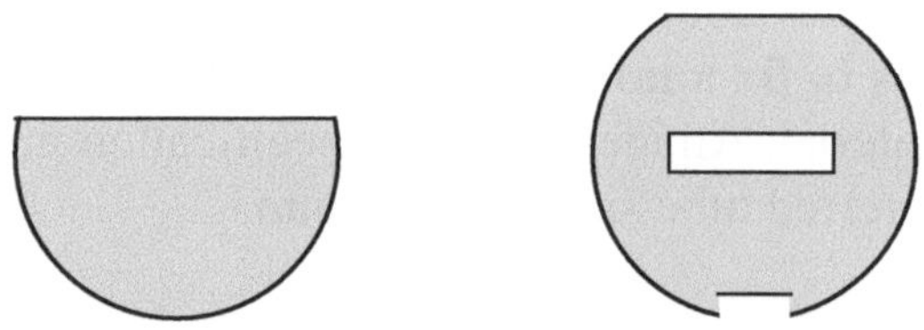

Figure 2.47

The figure on the left hand side is very much a support/baffle and is used when the when the bend area is to be included in the effective area for heat transfer as explained earlier in this chapter in the section on BEU units.

The figure on the right hand side is a full support plate when the bend area is not counted as effective area for heat transfer but it still allows some fluid circulation. The hole in the centre should be about 5 mm from the two inner rows of tubes.

Figure 2.48

Figure 2.48 shows possible supports for horizontal U tubes. The figure on the left will give a very good circulation. However it increases the unsupported length which may be acceptable for small shells and may require extra U bend supports on larger shells.

The figure on the right hand side is the equivalent of the right hand side of figure 2.47. In this case the vertical central hole is contributing very little to circulation and may as well be omitted.

There are many variations of the tube supports shown above and ideally a decision as to what is most appropriate should be made after studying the tube layout diagram. The other use of supports is where they are used to prevent flow induced vibrations and these will be discussed in chapter 15 on vibrations.

TUBES

Tubes are always specified by the outside diameter d_o and information on the thickness from which an inside diameter d_i can be established to use in the calculations for the heat transfer and pressure drop inside the tubes. This may sound like a very straightforward and simple procedure but it continues to cause problems with interpretations of how the thickness is defined. The confusion arises in the way the terms; minimum thickness, minimum gauge and minimum wall are used. At the time of writing the clients specifications of some very big companies are still confusing in this respect.

The first thing to understand is that there should be two distinct stages in specifying the thickness of a tube. **The first stage is to fix a minimum gauge** to be used for a specified outside diameter and a particular material. An ideal client's specification might be as shown in table 2.2 below for the preferred tube size of 19.05 mm.

TUBE MATERIAL	BWG	THICKNESS	
		mm	inches
Carbon and low alloy steels	14	2.11	0.083
Aluminium, Al brass, Al bronze	14	2.11	0.083
Austenitic stainless steels	16	1.65	0.065
Ni-Fe-Cr alloys	16	1.65	0.065
Admiralty brass	16	1.65	0.065
Cupro-Nickles	16	1.65	0.065
Copper	16	1.65	0.065
Monel/Zirconium/Hastelloy	18	1.25	0.049
Titanium	20	0.89	0.035

Table 2.2

The tube gauges are all in even numbers which is standard practice. Tube gauges with odd numbers are very seldom used. I am not sure of the historical origin of why the system developed where the higher the number of the gauge the thinner the tube wall becomes.

For 25.4 mm tubes it used to be standard practice to decrease the gauge number by two for each material. This is still the conservative approach for many companies but others are starting to use the same minimum gauges for 25.4 mm tubes as for 19.05 mm tubes. This trend seems to have been started with air coolers. Normally the minimum tube diameter for an air cooler is 25.4 mm and for many years there was a minimum tube gauge of 12 BWG in API 661 (air cooler specification) for carbon steel. This was then changed this to allow 14 BWG for 25.4 mm carbon steel tubes. This trend now seems to be spreading to shell and tube units so that the same minimum gauges may be used for the two most popular sizes of tube. This trend presumably reflects a general improvement in the quality of tubes. Whatever is used it must be made clear in the specifications before a new job is started.

One point worth noting in this table is that for titanium the minimum gauge is 20 BWG for a 19.05 mm tube. This is not a problem and I have often used 20 BWG tubes for titanium. However there is a very brief and rather confusing table on tube wall thicknesses in ISO 16812. This table requires a minimum gauge for titanium to be 19 (in earlier editions it was 18). If a clients spec allows a 20 BWG tube and also calls up API this conflict must be resolved at an early stage.

The term '**minimum** ' for this stage is used to allow for the fact that heavier gauge tubes may be required for mechanical strength if the operating pressures require a thicker tube. This should be checked for each exchanger if there is any doubt about a particular pressure. This can be done quickly using table 5 in useful tables at the end of the book. The figures for the internal pressures have been adopted from TEMA where a much more extensive list can be found. The TEMA table D-9 has a range of tube diameters from ¼ inch to 3 inch and have a wider range of allowable stresses. I have only included values that you are likely to use regularly. The other data I have included in table 5 are values for external pressures. TEMA no longer has this information. I have found this data for external pressures very useful over the years **but it should be treated with caution**. If you are 'well inside the limits' there should be no problems but if you are in the slightest doubt then you should calculate a thickness or consult a mechanical engineer.

Thicknesses for external pressures depends on the code being used e.g. ASME. These contain external pressure charts for different material groups.

The second stage is to look up the tube specification for the material being used. Table 6 in useful data and tables at the back of this book contains extracts from the ASTM tube specifications for most of the materials you are likely to use. The important extracts required for a thermal design are the tube wall tolerances which are all included.

Tube tolerances are specified as either **minimum wall** or **average wall**. The tube thicknesses in table 2.2 are **nominal values**.

For example a common specification for carbon steel tubes is A179. This is a seamless tube manufactured to a wall tolerance of - 0 to + 20%. (This is known as a minimum wall tolerance.) For a 19.05 mm, 14 BWG tube the minimum thickness must be 2.11 mm and the maximum thickness must be 2.11 x 1.2 = 2.53 mm. The mean thickness is therefore 2.32 mm.

The mean thickness is just the arithmetic average of the limiting values. The average thickness of the tubes is not necessarily the same thing. About three years ago, before the price of steel started to rocket, I had just finished checking some designs which had been carried out by another company and concluded that the tube side pressure drops where all too high. Before I wrote my report I decided to carry out a very brief survey of what tube manufacturers actually supplied for A179 carbon steel tubes. I was conscious of the fact that I had been using the same tubes with the same tolerances ever since I had been designing heat exchangers. Elsewhere tolerances had tightened up significantly – it seemed a long time since I was very careful in 'running in' a new car engine, so I made a few phone calls. One salesman told me that they usually managed to keep to +10% and - 0%. Another salesman told me that they usually managed to keep to +15% and -0%. I also rang up someone I knew at a small vendor who I had worked for in the past. He obligingly went down to the workshop and took sample measurements some tubes in stock. They averaged + 12.5%.

.

A 179 CS tubes are produced by drawing over a die which has to be regularly changed. The tubes will be at one end of the tolerance range at the start of a run and at the other end of the range at the end of run. A typical run might be 5000 m which is sufficient for 833, 6 m tubes. This corresponds to the requirements of a shell of about 900 mm diameter depending on the shell internal construction. Medium and large diameter shells will therefore have a mix throughout the tolerance range. Small shells could of course be fitted with tubes which are all close to one of the limits. SS tubes are produced by cold rolling thick cylinders over a mandrel which needs regular adjustments and there is not the same pattern to the tolerances

So which tube thickness should we input to the program we are using? Clearly to use the nominal value is quit wrong and unsafe. The pressure drop through a tube varies inversely with the inside diameter to the power of 4.8 (see useful tables 1). If all the tubes were close to the maximum tolerance the error in pressure drop is as follows:

$$\text{Error ratio in } \Delta p = \{\ (19.05 - 2.11)/(19.05 - 2.53)\ \}^{4.8} = 1.13$$

If they were at the arithmetic mean value the error ratio becomes:

$$\text{Error ratio in } \Delta p = \{\ (19.05 - 2.11)/(19.05 - 2.32)\ \}^{4.8} = 1.06$$

It is clearly quite wrong to use an inside diameter based on the nominal wall thickness for a minimum wall tube. This is one of the mistakes made by the company that provoked my little bit of research above. I have come across many companies who regularly do this – particularly vendors. It is one of the, tricks, that is done to gain a 'competitive edge'. They will usually get away with it because of the tradition in the industry of using cumulative safety factors. (A safety factor on the pump/compressor, a safety factor on the pipe pressure drop and a safety factor on the heat exchanger pressure drop.)

Recommendations.

(1) Use the tube specification arithmetic mean tolerance value for minimum wall tubes. For medium and large exchangers this will be correct. A HTRI engineer I once rang up agreed with this approach.

(2) If you know where the tubes are to be supplied from at the start of a job carry out a survey of the actual tolerances suppliers will be using for minimum wall tubes and use the arithmetic mean of the one with the widest tolerance. This is a sensible and justifiable way to gain a 'competitive edge'. On some jobs political considerations may restrict sourcing to a specific country so there may not be too many potential suppliers.

(3) For average wall tubes just use the nominal wall thickness. (Any 'tightening up' of the positive tolerance to save metal will result in a much smaller change in internal diameter which will be 'safe'.)

U TUBES

The points made above apply equally to U tubes but there are further considerations for these.

When a U tube is bent there must always be some thinning of the tube. The thinning will be negligible for the outer rows of tubes but is much more significant for the inner rows, this may be over 10% and TEMA will allow a maximum of 17%. To avoid excessive thinning there are restrictions on the minimum bend which can be used. This minimum bend is expressed in terms of a bend diameter to tube outside diameter ratio as illustrated below in figure 2.49

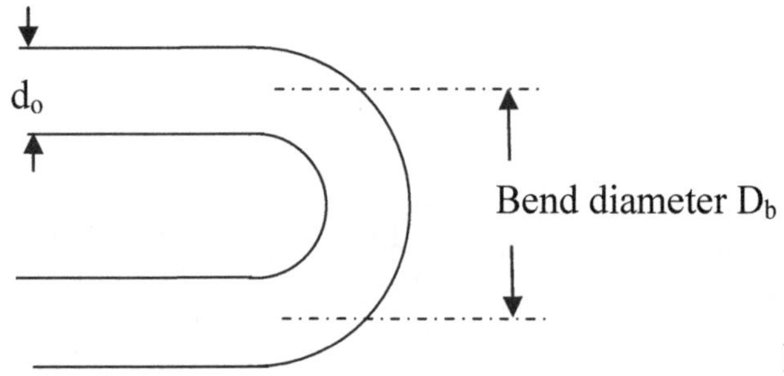

Figure 2.49

$$\text{Bend to diameter ratio} = \frac{D_b}{d_o}$$

The bend to diameter ratio depends on the tube material some typical values are shown in table 2.3 below.

Tube material	Bend/diameter ratio D_b/d_o
Carbon, low alloy and S Steel	3
All other materials	3.5
Titanium	4

Table 2.3

This information will normally be found in the clients specs. If it is not then something must be agreed and inserted in the 'Job Notes' (see check lists at the end of the book) before starting on any designs.

Having agreed on D_b/d_o ratios the inner rows of tubes will still be weaker than the outer rows and unlike straight tubes they can not be replaced. (In theory the tubes in the outer row could be replaced but I have never come across this being done. The normal course of action if a U tube fails is to plug the tube – losing the effective area from two tube lengths.)

The vulnerability of the inner tubes may be acceptable in a low pressure exchanger but normally it is sensible to use thicker walled tubes. **It is good practice to specify the two inner rows of tubes two tube gauges lower than the rest of the bundle for 90° tube layouts and the inner three rows of tubes two tube gauges lower for 45° and either triangular pitch layout.**

The use of U tubes clearly results in a significant untubed area at or near the centre line of the bundle. (In most U tube bundles it will be on the centre line but when there is a large difference between the sizes of the two shell nozzles it may be possible to fit in more tubes by having a small offset between the centre line of the shell and the centre line of the bundle.) The untubed space has a negligible effect on heat transfer coefficients in two tube pass units when the U bends are vertical and the baffles are horizontal. However if both the U bends and the baffles are vertical, as in figure 2.50 this can introduce a substantial bypass stream (F stream).

This bypass stream can be reduced by using dummy tubes or blocked completely using a series of sealing plates. A more satisfactory approach is to use inclined or crossed U tubes in the pass partition lane area.

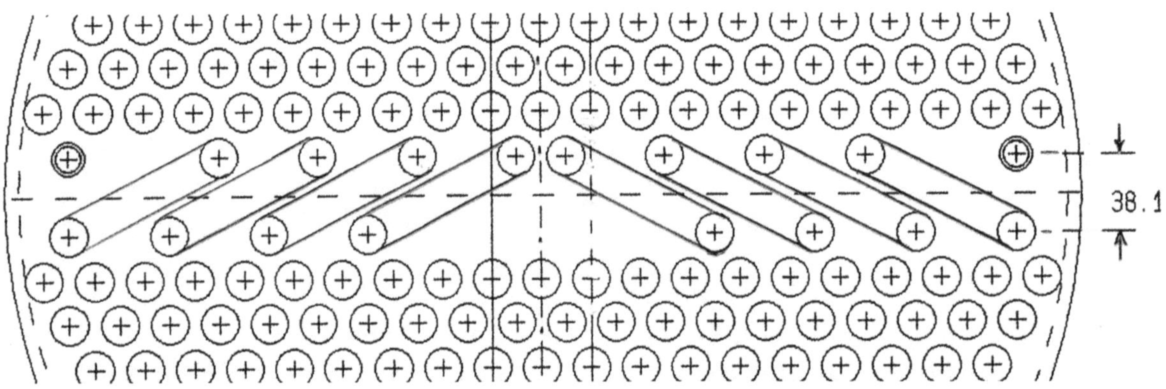

Figure 2.50

Figure 2.50 shows the central section of a design with inclined U tubes in the area which is not available for vertical U tubes. The unit is a small condenser with the minimum of two tube rows overlap resulting in a vertical baffle cut of about 45%. The tubes are 19.05 mm on a 25.4 mm triangular pitch, (30° 'as viewed' but 60° in terms of fluid flow for heat transfer). The curved broken lines adjacent to the lines of the shell inside diameter are the outer tube limit. Two tie rods are shown. The distance between the centre lines of the inclined tubes is 38.1 mm so that the gap through the centre is reduced to 19.05 mm.

As well as reducing bypassing the extra surface area of about 5% was very important in this case because it was a high pressure unit – DEU – and minimising the shell diameter resulted in a greater cost saving than would have been the case with a low pressure unit.

An alternative to using inclined tubes is to use crossed tubes. Two cross tube arrangements are shown in figure 2.51. These arrangements are more effective in blocking the F stream flow path. (With inclined tubes the fluid will flow at an angle to the centre line rather than flow straight across through the gap required for the partition plate in the stationary head.)

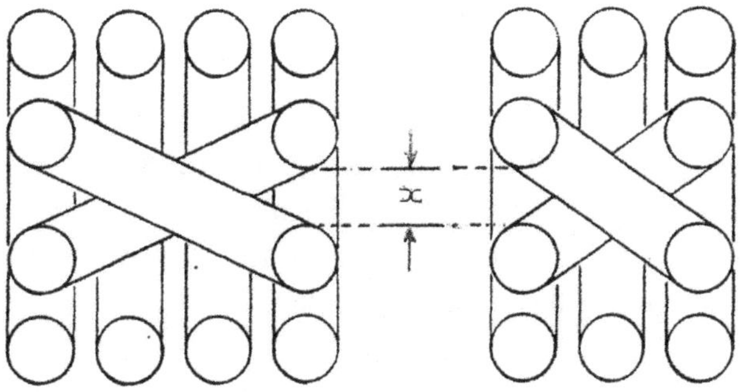

Figure 2.51

The distance x must be equal to or greater than the distance required for the pass partition lane. (It was 19.05 mm in figure 2.50.)

Inclined or crossed tubes should have the same gauge as the inner two or three rows of tubes.

If reducing the F stream is very important to the design dummy tubes may also be introduced to a bundle with either inclined or crossed tube arrangements. Dummy tubes may be in any part of the pass lane including the central area. It is normally only worthwhile fitting dummy tubes across the baffle overlap and perhaps two or three vertical rows at either side.

It is not mechanically possible to fix the support/last baffle plate right up to the bend of U tubes. This is illustrated in figure 2.52

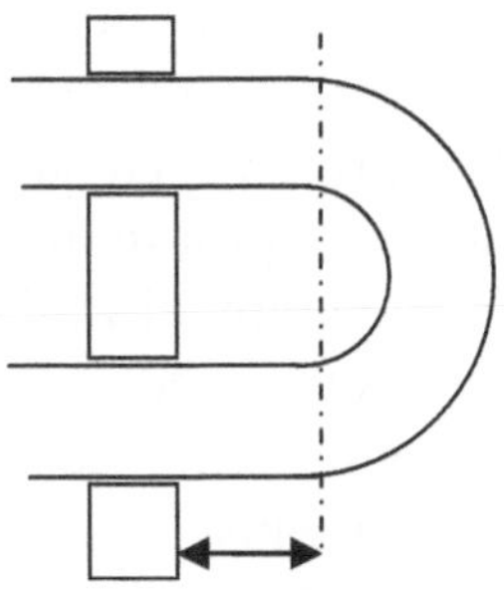

$2d_o$ mm

Figure 2.52

Typically about twice the tube outside diameter is allowed between the tangent line of the U bend and the support. The dimension should be rounded to the nearest 5 mm so it becomes 40 mm for 19.05 mm tubes. This must be allowed for in the thermal calculations as well as the thickness of the tubesheet. Particularly if crossed tubes are used this distance needs to be increased for the tubes in the main tube bank. If the U bend area can be considered to be fully effective for heat transfer this is not a problem. If the U bend area is not to be counted for heat transfer then this extra length of ineffective straight tube is an extra cost particularly for the cross tube arrangements.

PASS PARTITION DISLOCATION LOSSES

In the previous section on U tubes it was emphasised that there must always be a minimum gap across pass partition lanes. It is important to understand how this gap is established for all types of tube. Figure 2.53 shows two tubes adjacent to a pass partition. The top tube is attached to the tubesheet by expanding the tube into grooves. The bottom tube is attached to the tubesheet by welding the tube to the face of the tubesheet.

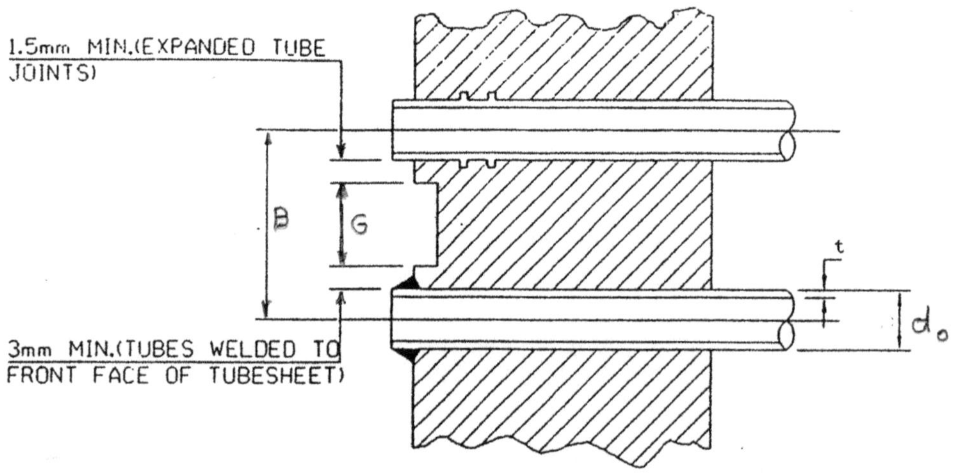

Figure 2.53

For expanded joints a minimum of 1.5 mm must be allowed between the groove and the adjacent tube. To accommodate the thickness of the weld this distance has to be increased to 3 mm for welded tubes. The thickness of a weld depends on the thickness of the tubes and 3 mm is satisfactory for tube gauges of 14 BWG which have a nominal wall thickness of 2.108 mm and a maximum value of 2.53 for minimum wall tubes with a 20% tolerance. 12 BWG tubes are more marginal. The nominal wall thickness is 2.7 mm but with minimum wall tubes with a 20% tolerance some tubes could be 3.3 mm thick. This would be a point worth discussing with a mechanical engineer if in doubt be safe and use 3.3 mm. For tubes thicker than 12 BWG the 3 mm must be increased depending on the nominal thickness of the tube and the specification tolerance.

The dimension G in figure 2.53 is the width of the groove in the tubesheet which accommodates the gasket used to seal pass the partitions in the channel head. This dimension depends on the thickness of the partition plate but is not necessarily the same dimension. This is because the partition plate is normally tapered to fit into the partition groove.

SHELL ID mm	GROOVE, G mm	GAP –Expanded Tubes mm	GAP –Welded Tubes mm -min
→ 610	10	13	16
610→ 1200	13	16	19
>1200	16	19	22

Table 2.4

The thickness of partition plates increase as the shell diameter increases. Section RCB-9.13 of TEMA gives details of how to calculate the thickness of partition plates. This is not necessary for a thermal design and table 2.4 is sufficient to establish the gap between the adjacent tube rows.

Your computer program should be using these values – probably defaulted to the gap for expanded tubes. It is therefore only necessary to remember to increase this gap when welded tubes are to be used. (You should check the program you are using to verify this.)

So far in this section we have been looking at how to establish minimum gaps for partition lanes. It is now necessary to see how these gaps affect the tube count and to understand why there are situations where the minimum gap should not be used.

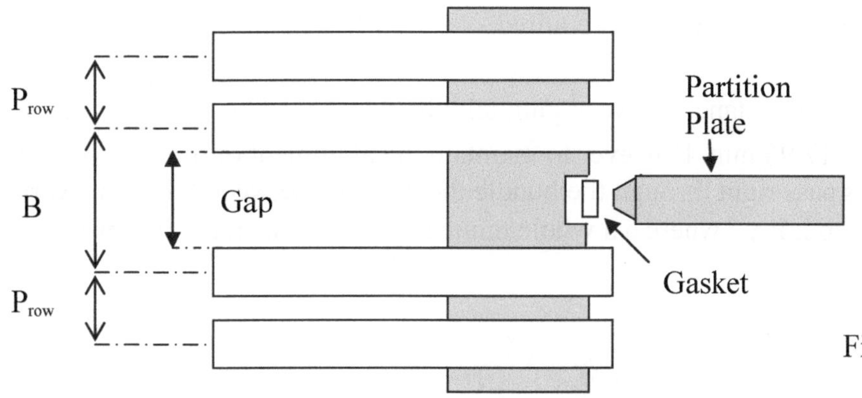

Figure 2.54

Figure 2.54 shows a section through a tubesheet adjacent to a pass partition in one of the heads of an exchanger. Dimension B is the length between the centre lines of the tubes at either side of the pass partition lane. Dimension P_{row} is the length between the tube rows parallel to the gap. Therefore the length which is not available for tubes is ($B - P_{row}$). When designing exchangers using hand calculations it is useful to have tables which cover all the tube (d_o) and pitch (P_t) combinations commonly used. Now that a computer program is used for nearly all designs it is not worthwhile including such a table. The reader should of course understand why they vary because this can affect the tube count for alternative orientations e.g. 30° via 60° tube pitch. Some examples follow:

Example 2.1

An exchanger with a 1000 mm shell has 19.05 mm tubes on a 25.4 mm, 90° pitch. The tubes are expanded into the tubesheet. Calculate the width of the area which is unavailable for tubes because of a pass partition lane.

Solution

From table 2.3 a gap of 16 mm is required. For a □ pitch the distance between parallel tube rows is clearly the tube pitch of 25.4 mm.

Therefore:

$$B = Gap + d_o = 16 + 19.05 = 35.05 \text{ mm}$$

Width of unavailable area = $B - P_{row} = 35.05 - 25.4 = 9.65$ mm

This is a relatively small loss because of the large distance between the parallel tube rows.

Example 2.2

Use the same data as in Example 2.1 to calculate the % change in the width of the area which is unavailable for tubes for a 45° pitch - 'As viewed'. (This is still a 45° pitch with either vertical or horizontal baffles.)

Solution

From table 2.3 a gap of 16 mm is required.

For a ◇ pitch the distance between parallel tube rows is P_t Cos 45° (or $P_t\, 0.5\sqrt{2}$). = 25.4 x 0.707 = 17.96 mm. However to maintain the alignment of the tube bank so that the cleaning lanes pass right through the bundle the distance between tubes in a vertical row must be equal to $n\, P_t\sqrt{2}$ where n a whole number. This is illustrated in figure 2.55.

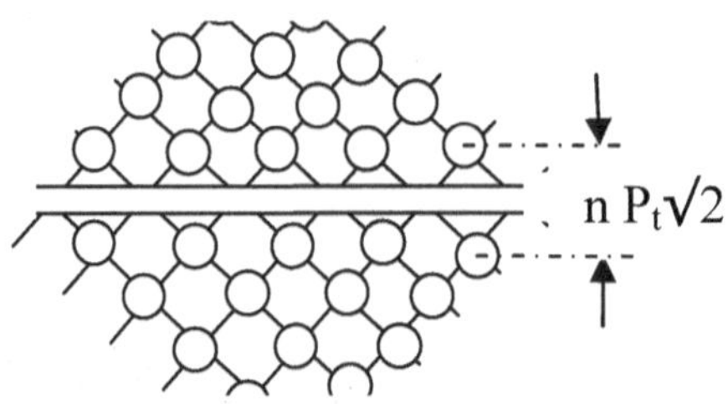

Figure 2.55

If n = 1

$$B = P_t\sqrt{2} = 25.4 \times 1.414 = 35.92 \text{ mm}$$
$$\text{Gap} = B - d_o = 35.92 - 19.05 = 16.87 \text{ mm}$$

The min minimum gap is 16 mm so this is satisfactory. It is worth noting that if the tubes had been welded particularly with thick walled tubes it would have been necessary to use n = 2.

The ratio of the gap compared with the gap for the 90° arrangement is 16.87/9.65 = 1.75 or a 75% increase in the area unavailable for tubes for each partition lane.

Nowadays the 45° tube arrangement is normally only used for viscous fluids with low Reynolds numbers when it creates a mechanical turbulence which results in higher heat transfer coefficients than a 90° arrangement. Because of the difference in areas unavailable for tubes in the pass partition lanes of the two alternative arrangements the most satisfactory solution is to use a 90° pitch 'as viewed' and 45° cut baffles. This maximises the tube count and the heat transfer.

Example 2.3

An exchanger with a 1000 mm shell has 19.05 mm tubes on a 25.4 mm, 30° pitch - 'as viewed'. (With horizontal baffles it is 30° and with vertical baffles it is 60° to the direction of flow for heat transfer.) The tubes are expanded into the tubesheet. Calculate the width of the area which is unavailable for tubes because of a pass partition lane.

Solution

From table 2.3 a gap of 16 mm is required.

For a $\triangle$ pitch the distance between parallel tube rows is the height of the triangle i.e. P_t Cos 30° (or P_t Sine 60°) = 25.4 x 0.866 = 22 mm. Therefore:

$$B = Gap + d_o = 16 + 19.05 = 35.05 \text{ mm}$$

$$\text{Width of unavailable area} = B - P_{row} = 35.05 - 22.0 = 13.05 \text{ mm}$$

Example 2.4

Use the same data as in Example 2.3 to calculate the % change in the width of the area which is unavailable for tubes with a 60° pitch - 'As viewed'. (With horizontal baffles it is 60° and with vertical baffles it is 30° to the direction of flow for heat transfer.)

Solution

From table 2.3 a gap of 16 mm is required.

For a rotated triangular pitch ('as viewed') the distance between parallel tube rows is half the tube pitch as illustrated in figure 2.56.

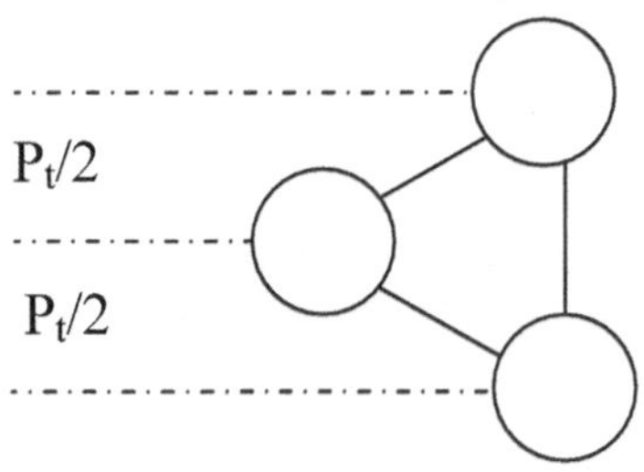

Figure 2.56

$$B = Gap + d_o = 16 + 19.05 = 35.05 \text{ mm}$$

$$\text{Width of unavailable area} = B - P_{row} = 35.05 - 25.4/2 = 22.35 \text{ mm}$$

The ratio of the change is therefore 22.35/13.05 = 1.71 or 71%.

As far as the area lost to any parallel pass partitions are concerned this gives the triangular pitch an advantage. However the rotated triangular arrangement has the advantage of a shorter distance between parallel rows so that it is often possible to fit in more tube rows. If there are no other considerations try both tube orientations to see which results in the higher tube count.

Example 2.5

A 1400 mm diameter BEU exchanger has 19.05 mm carbon steel tubes on a 45° pitch. Calculate the width of the partition lane which is unavailable for tubes.

Solution

The minimum bend/diameter ratio for carbon steel tubes is normally 3. This fixes a minimum value for B as 3 x 19.05 = 57.15 mm. However to maintain cleaning lanes B must be a multiple of 1.414 P_t. The next multiple is two, which increase the value of B to 2 x 1.414 x 25.4 = 71.83 mm. This is the value which should be used for B so that the bend/diameter ratio must be increased to 71.83/19.05 = 3.77 for the tubes with bends in the vertical plane.

The width of the partition lane unavailable for tubes is therefore:

$$B - P_t = 71.83 - 25.4 = 46.43 \text{ mm}$$

This is a significant loss and the pass partition gap is $(B - d_o)$ = 71.83 – 19.05 = 52.78 mm. It is therefore worth considering using inclined or crossed tubes to increase the tube count taking care to maintain the cleaning lanes. (Use a piece of graph paper to see where inclined or crossed tubes could be fitted.)

PASS PARTITION LANE ARRANGEMENTS

Up to this point we have been looking at how to establish the width of partition lane gaps and the subsequent loss of area available for tubes. This is all that is necessary for two tube pass units and the two pass exchanger is the most commonly used. It is of course possible to have a lot more than two tube passes.

ODD NUMBER OF PASSES

Figure 2.57 shows the pass partition lanes for a three pass exchanger. The diagram on the left shows the pass partition in the stationary head and on the right the pass partition in the rear head.

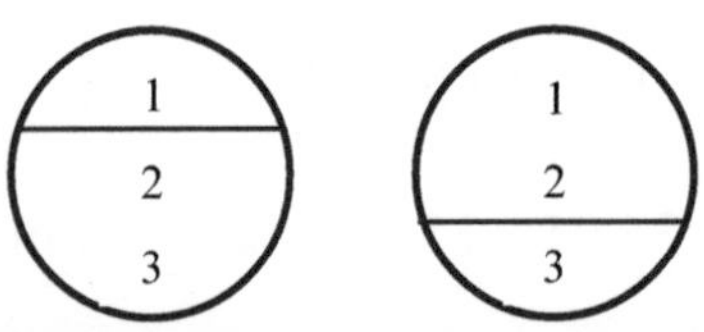

Figure 2.57

There are two pass partition dislocation lanes. A five pass exchanger would have 4 pass partition dislocation lanes.

EVEN NUMBER OF PASSES

Figure 2.58 illustrates arrangements of pass partitions for exchangers with an even number of passes. Type 1 and type 3 are alternative arrangements for straight tubes.

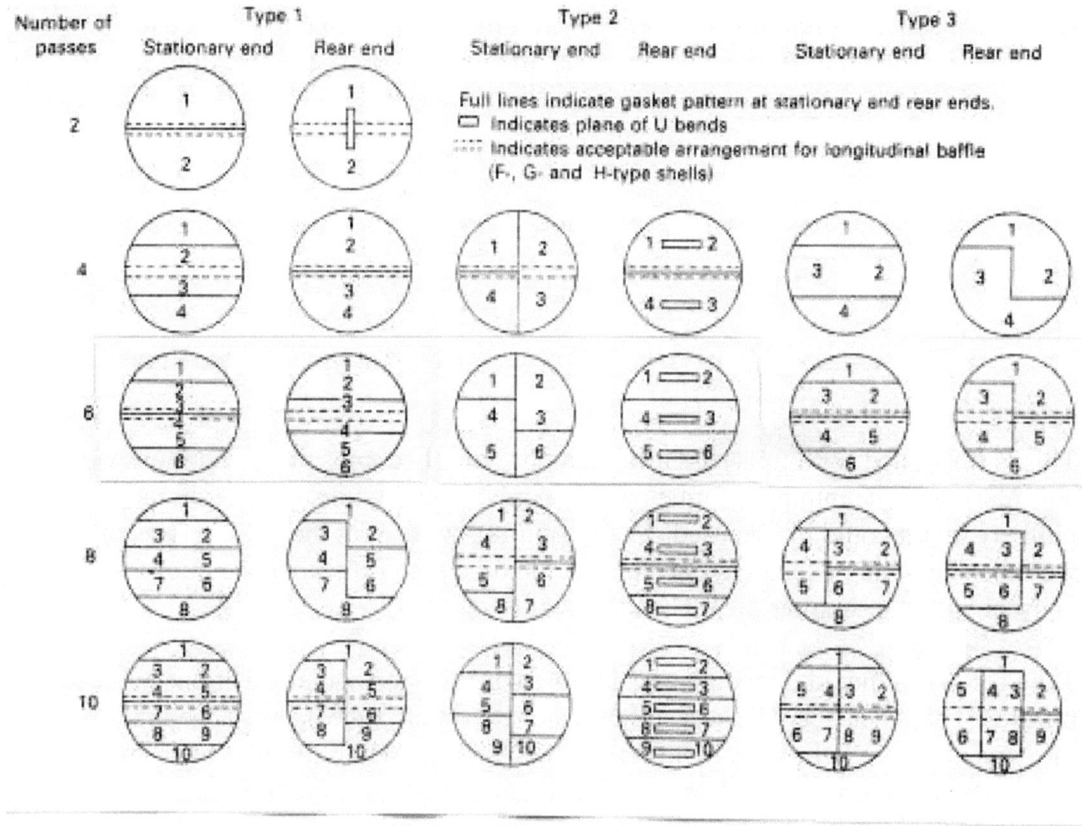

U TUBE PASS PARTITION LANES

The type 2 partition lane arrangements only apply to U tube exchangers. For two passes the partition lane is the same as for any other type of unit and the U bends are in the vertical plane. For all other arrangements the U bends are horizontal. This means that there will be a large pass partition lane on the vertical centre line and a number of smaller horizontal ones. The width of these will depend on the tube arrangement. One disadvantage of U tube bundles with more that two passes is that many people offset the channel nozzles as shown in figure 2.59.

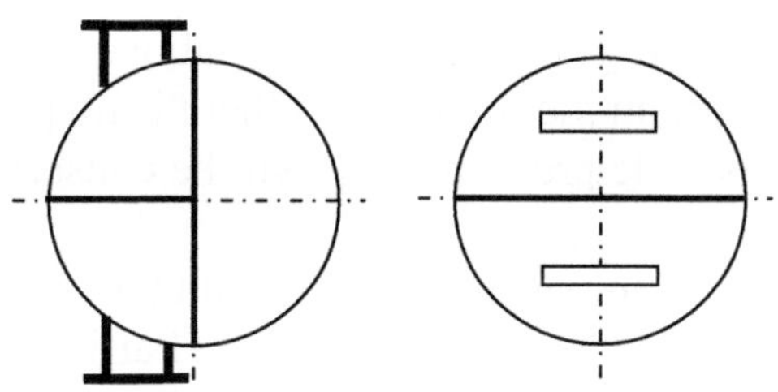

Figure 2.59

This complicates the construction of the channel and therefore increases the cost of the exchanger. It also has an impact on the connecting pipework which increases costs outside the exchanger. It is often assumed that with more than two passes in a U tube exchanger the nozzles must be offset. This is not always true. Whenever the shell side nozzles are relatively large compared with the shell inside diameter there will be significant spaces at the top and bottom and it may be possible to arrange the pass partitions as shown in figure 2.60 and keep the nozzles on the centre line.

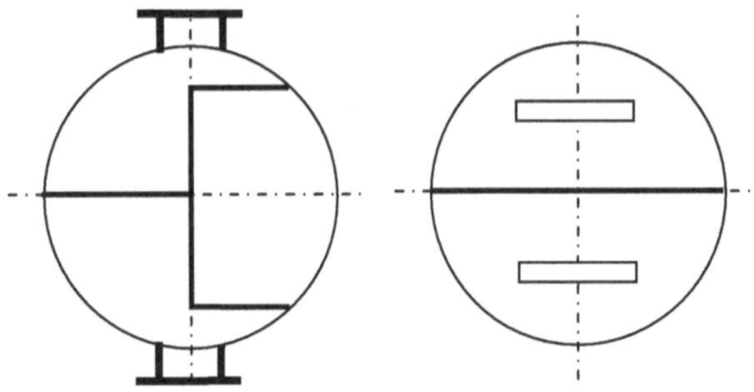

Figure 2.60

This is illustrated for a four pass arrangement but it becomes more and more worthwhile considering as the number of tube passes increases and with say an eight pass bundle it is unlikely that it could not be used. The reason for this is that the more passes there are on the tubeside of any exchanger the smaller the tube side flow area becomes and consequently the smaller the tube side nozzles.

Take a copy of the tube layout and add the two short horizontal partitions adjacent to the nozzles. The inner line will be say 2 mm (for expanded tubes) above the top tube row and the same distance below the bottom row, then add the width of the groove from table 2.3. Add the vertical grooves and draw in the inside diameter of the tube side nozzles using data from useful table 4. This will result a diagram looking like figure 2.61.

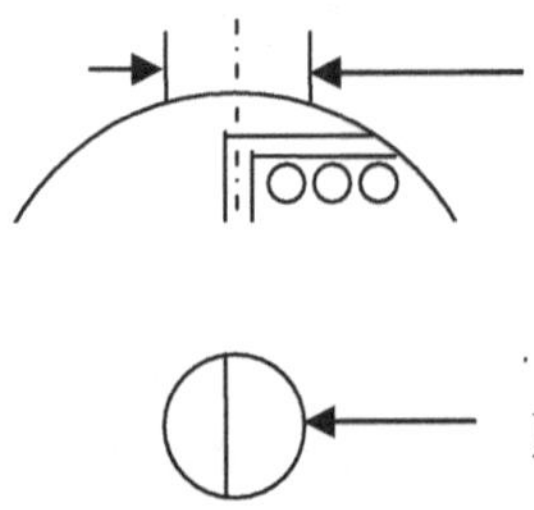

Figure 2.61

It is then a matter of comparing the cross sectional area of the nozzle with the minimum flow area into the first compartment in the channel. There is no obstruction to entry through the left hand side of the plan view. The flow area at the right hand side is the length of the arc multiplied by the average height of the arc above the partition plate. If the flow area is the same, or greater than the nozzle area there is no problem and adding one nozzle velocity head to the tube side pressure drop will be conservative.

If the flow area into the first channel compartment is less than the channel nozzle cross sectional area it may still be possible to use an in line nozzle. Particularly with small nozzles there is a big jump between sizes. For example the change in cross sectional area, and hence velocity, between a 4 inch nozzle and a 6 inch nozzle is about 2.8 so that the change in momentum flux is $2.8^2 = 7.8$. If a 4 inch nozzle had been just too small there would still be a lot in hand in a six inch nozzle to justify the flow area into the compartment being less than the nozzle area. The only real limitation is the maximum momentum flux for the fluid which is based on erosion criteria. (See table 2.5) Because the partition plate immediately below the nozzle has fluid impinging on it, it is sensible to specify this short section bit thicker than the rest of the partition plates, say an extra 3 mm.(Which you must allow for in your drawing and calculations).

One technique I have found very useful over the years for changing or adding details on an existing drawing is to use a scale 'ruler'. In the above exercise we started with a print of the tube layout which would show the shell nozzle inside diameter but not the tube nozzle inside diameter. If you are very lucky – and you can find it – you may have a piece of suitable plastic which you could use. Mine are always missing or not quite right for the job so I create a purpose made scale rule. Just place the edge of a piece of paper across a section of the tube bank and mark off a scale in divisions equal the tube pitch. If necessary subdivide the first section into four or even ten subdivisions. (You could use your 'O' level geometry to do this accurately but in most cases I just judge it by eye.) You now have a purpose built scale rule. If the ID of your nozzle had been 146 mm and the tube pitch was 25.4 mm this becomes 146/25.4 = 5.75 divisions on your scale rule. This is easy to insert on your diagram. To take dimensions off the diagram just read them off your scale rule and multiply them by the tube pitch.

It is of course not satisfactory to just show in line nozzles on your data sheet or preliminary setting plan, any one who know a little bit about exchangers will smugly point out that you have made a mistake. A note on the TDS may be sufficient for a rough feed job but ideally you should include a sketch similar to figure 2.60 with a note that the modified partition lane should be as close to the top tube row as possible to maximise inlet flow area. Unless the bundle is symmetrical about the horizontal axis another sketch for the bottom nozzle is required.

This method of avoiding using offset nozzles may take a bit of time but there is no impact on the construction costs since there is no change in the length of the channel or position of the channel nozzle. If this method is not practical there is another way.

EXTENDED CHANNEL TO AVOID OFFSET NOZZLES

This method of avoiding offset nozzles with U tube bundles can be used for any number of passes and is independent of the size of the shellside nozzles. It is illustrated in figures 2.62 and 2.63. The former is an elevation through the channel and the latter shows end views of cross sections at three points.

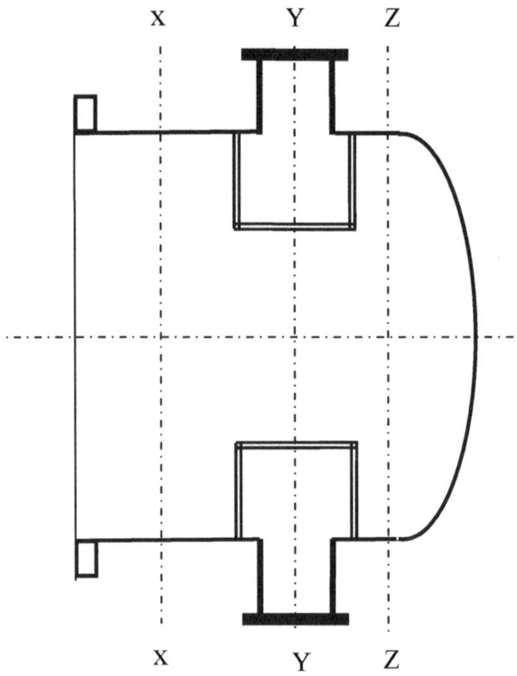

Figure 2.62

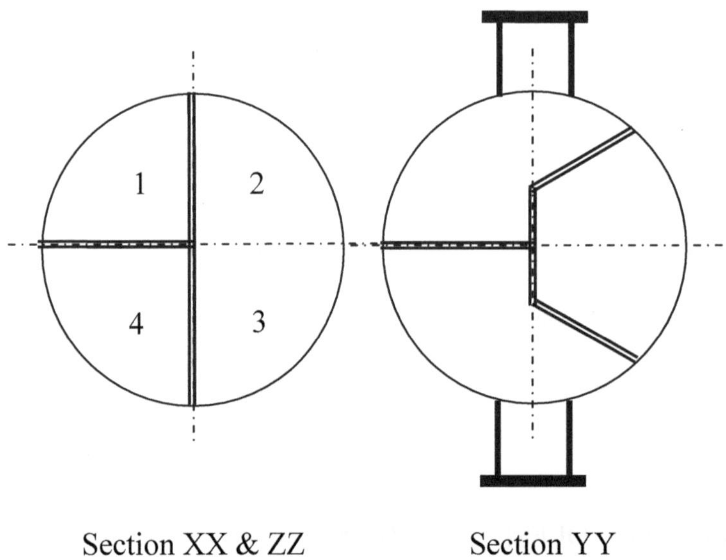

Figure 2.63

A four pass arrangement is shown. The partitions at section XX and ZZ are exactly the same as they would be if offset nozzles were to be used. In the central section YY the fluid from the top nozzle passes into a conical section which diverts it into the normal quadrant section ready to flow into pass one. The fluid returning to the channel through pass two crosses over as normal to enter pass three before finally returning through pass four where it enters the conical section directing the fluid into the bottom nozzle.

The total cross over area at any point should be at least 1.3 times the flow area through the tube passes as normal and the length of the channel has to be increased to accommodate this. This arrangement is particularly advantageous for six and ten passes which would otherwise require a nozzle to be offset at each side of the channel adding further complication to the pipework. (Note – the conical deflectors are on opposite sides of the channel for 6 and 10 passes.)

The only negatives with this approach are the extra cost a short length of channel, a more complex partition plate arrangement and remembering to allow for the extra length between the shell and tube nozzles on the setting plan. I consider the advantage of eliminating offset nozzles more than worth it and I can not understand why more companies do not do it.

STRAIGHT TUBE PARTITION LANES

Type 1 and type 3 pass partition arrangements are used for exchangers with straight tubes. There are other possible variations of the arrangement of the partition lanes with the higher numbers of tube passes. Ten passes is not a limiting factor I have designed units with up to 16 passes. Your limitation is what has been included into your computer program. Your control over which type you want is by filling in the number of vertical and horizontal lanes. For example if you wanted four passes you would fill in 3 horizontal (two partitions in the stationary head and one in the rear end) and zero vertical for type 1. Fill in 2 horizontal and 1 vertical for type 3. The program will probably default to one of the options if you omit this input.

When selecting a type 1 or a type 3 arrangement there are three considerations:

(1) Does either arrangement have a thermal advantage?
(2) What effect does the choice have on the F bypass stream?
(3) Which arrangement maximises the tube count?

Types 1 are known as ribbon pass arrangements. The 4 and 6 are true ribbon passes, The 8 and 10 are modified ribbon passes. A true ribbon arrangement for an 8 pass would have four horizontal partitions in the stationary head and three in the rear end. If the shell side fluid is not undergoing an isothermal process and is in cross flow there can be a significant advantage in the higher MTD of the ribbon pass arrangement. This is because the hottest fluid on the shell side is only exposed to the tube pass with the hottest fluid and the coldest fluid on the shell side is only exposed to the tube pass with the coldest fluid. This applies to X, G and H type shells. These shells can accommodate a temperature cross so that a single shell can be used instead of two shells in series. Where there is a thermal advantage use only ribbon passes.

The effect on the F stream depends on the baffle orientation. With horizontal baffles a true ribbon pass arrangement has no F stream bypass. This can be very important in X shells handling large volumes of gas on the shellside when it is often a struggle to maximise the velocity to optimise the heat transfer coefficient. For other units it is always desirable to minimise the number of bypass lanes perpendicular to the baffle cut if there are no other considerations.

Finally the arrangement which gives the highest tube count should be used if nothing else overrules this. This will normally be a type 3 arrangement. The areas unavailable for tubes depend not only on the width of the pass partition lanes but on their total length. For example the total length of pass partition lane losses for a six pass type 1 design is about 5 x 0.9 x SCD (Safe circle diameter) = 4.5 SCD. For a type 2 it would be about 4 x 0.9 x SCD = 3.6 SCD. The F stream bypass in partition lanes can of course be reduced by using dummy tubes or other obstructions which will be covered next.

SEALING PASS PARTITION LANES

Tubes, solid rods or rectangular strips of metal can be used to seal pass partition lanes. When tubes are used in this way they are called dummy tubes. Dummy tubes must be sealed at one end and open at the other. If both ends were open there would be a potential bypass problem and sealing both ends could be dangerous and would add a small extra cost.

The first thing to consider of course is, is it really necessary. Often it is not. Run your program with and without sealing and observe the effect this has not only on the shell side heat transfer coefficient but on the overall heat transfer coefficient.

Assuming sealing is necessary you then have to decide what length of the pass partition lane requires sealing. It is unlikely that your computer program will do this for you but it must be specified in some way in your TDS ideally on the tube layout diagram. There are no fixed rules, it is a matter of judgement – just imagine the fluid flowing through the bundle as discussed in the stream analysis section of chapter 3.

Figure 2.64 shows a section through a bundle with a single horizontal pass partition lane and vertical baffles. The only variable involved is the baffle cut which determines the baffle overlap. For bundles with a low baffle cut, say below 26% it should only be necessary to seal the section of the lane covered by the baffle overlap. For higher baffle cuts I will sometimes extend this into the window area, the higher the baffle cut the more I might extend it. I would only extend beyond the baffle cut when using dummy tubes or rods but not when using strips. Sealing the ends of the pass partition lanes would be just another cost and there would be hardly any effect on the heat transfer.

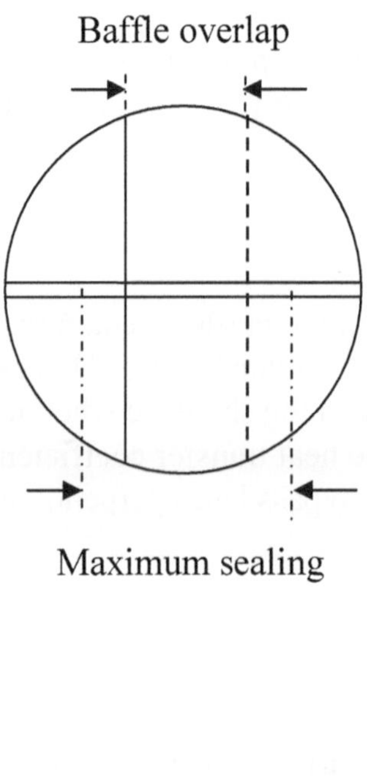

Figure 2.64

If any sealing device is to be affective the clearance between it and adjacent tubes should not be more than the clearance between the tubes ($P_t - d_o$). It should not however be so small that there is no fluid flow to one side of the nearest tubes. The ideal distance would be half the clearance between the tubes and this should be specified when strips are used. For dummy tubes and rods select the nearest standard size which is the most suitable. Some examples follow.

Figure 2.65 shows an in line square pitch arrangement. If the partition lane gap is 19.05 mm and the tubes are also 19.05 mm on a 25.4 mm pitch the nearest standard diameter is 9.525 mm (0.375 inches) which you would use for blocking the pass partition. The clearance would therefore be half of this or 4.762 mm which is less than the clearance between the tubes of 6.35 mm. If the diameter was 10 mm the clearance would be 4.52 mm which would be even better.

For small diameters, say below 12.7 mm, rods are more satisfactory than tubes in this case I would specify 10 mm rods. (If a sensible vendor had no 10 mm rods he would just use 9.525 mm rods.)

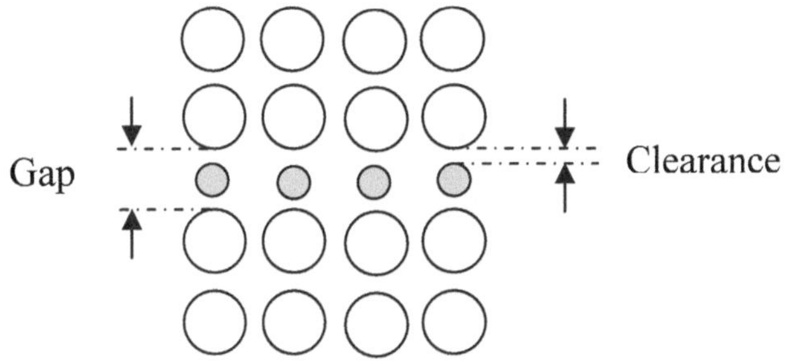

Figure 2.65

One disadvantage of sealing the lanes on an in line square pitch arrangement is that part of the tubes exposed to the lane can not be mechanically cleaned.

Figure 2.65 shows an in line square tube pitch arrangement with U tubes which have a bend to diameter ratio of three resulting in a gap equal to two tube diameters. It is now necessary to use two dummy tubes in a vertical row and if the tubes are 19.05 mm on a 25.4 mm pitch the nearest standard size is 12.7 mm. There are three clearances in the lane and the average clearance is only (2 x 19.05 – 2 x 12.7)/3 = 4.23 mm which may seem satisfactory but the total of the three clearances is 12.7 mm which still allows a significant by pass stream. (The flow through the central clearance is totally ineffective for heat transfer.)

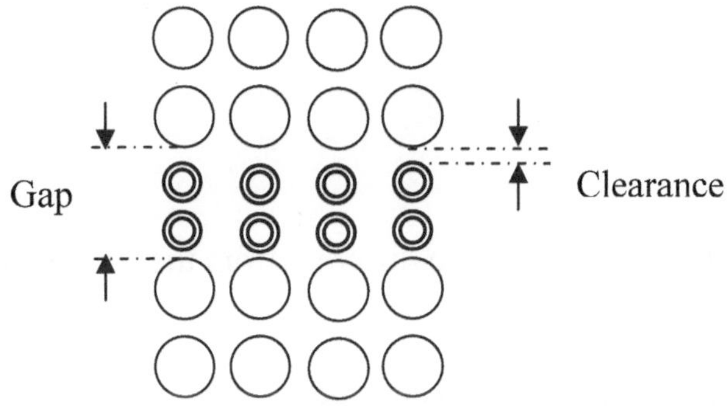

Figure 2.66

In this situation I would use sealing strips rather than dummy tubes as shown in figure 2.67. The clearance between the strip and the tube can be specified as half the tube pitch which is ideal. With this arrangement you are fully justified as considering the pass partition lane as being fully sealed in the input to your program.

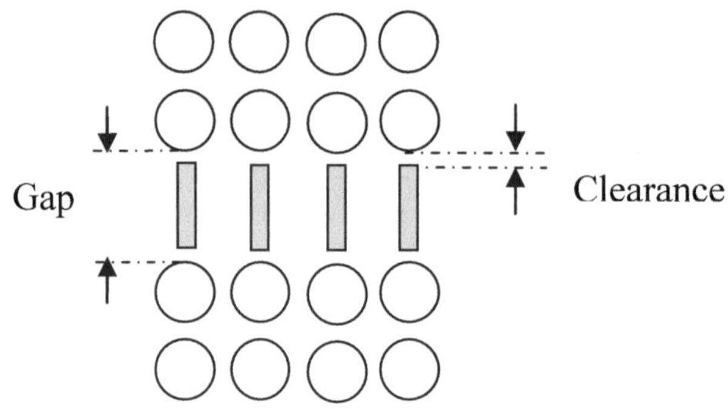

Figure 2.67

Figure 2.68 shows a pass partition lane with the tubes on a rotated square pitch. The gap has been fixed to maintain the cleaning lanes through the tube bank as explained previously. This is the simplest of all cases to deal with – the dummy tubes are the same size as normal tubes. As far as the shell side fluid is concerned there is no pass partition lane.

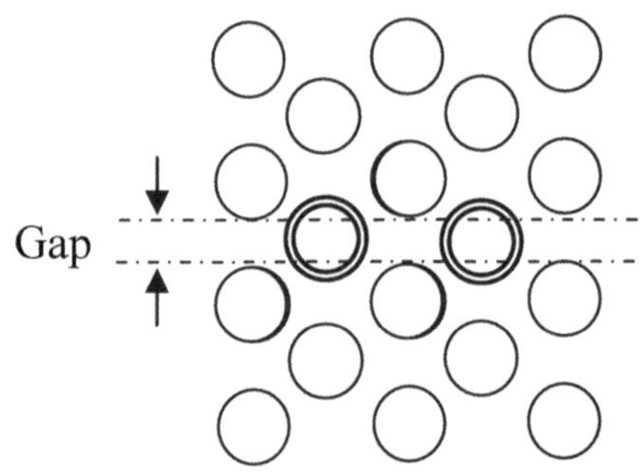

Figure 2.68

Figure 2.69 is a U tube bundle with a rotated square pitch. It is now necessary to have three rows of dummy tubes to achieve the same effect.

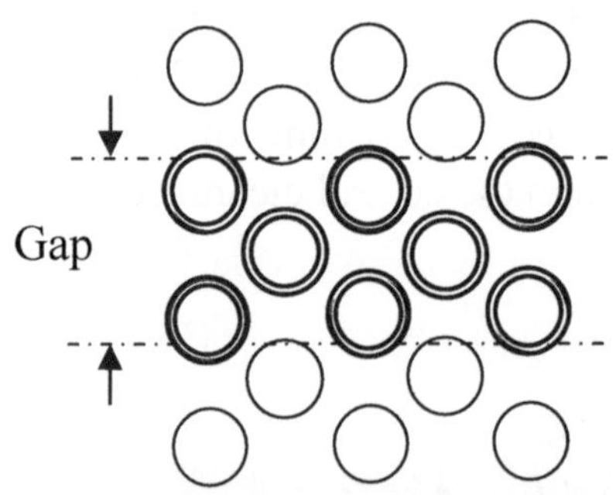

Figure 2.69

The triangular tube pitch arrangements are different from the square tube pitch in two respects:

(1) There are no cleaning lanes to consider
(2) The pattern at either side of the partition lane may not be symmetrical.

Figure 2.70 shows a triangular (30° 'as viewed') tube layout. The baffle cut may be any of the possible orientations. In this case the tube layout is symmetrical about the pass partition lane.

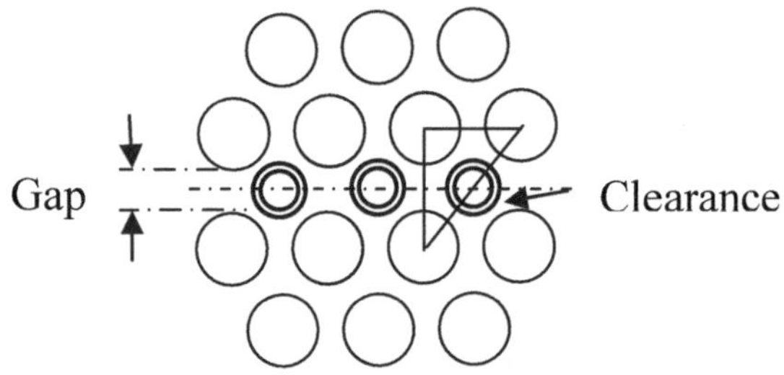

Figure 2.70

To determine the clearances for this arrangement a little basic geometry is necessary. The triangle in figure 2.70 is shown again in figure 2.71. The two sides adjacent to the right angle can then be used to calculate the diagonal line, (the hypotenuse).

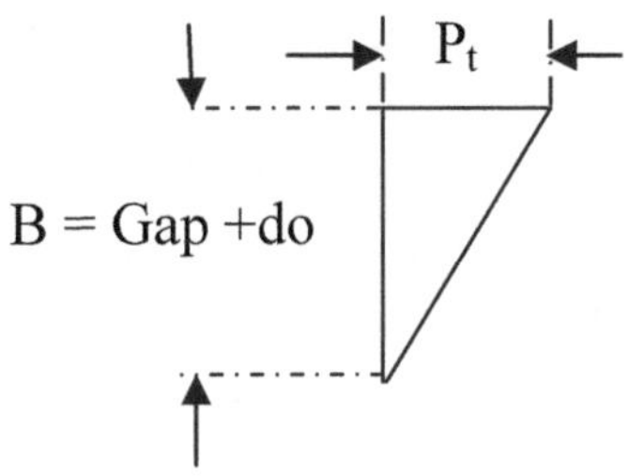

Figure 2.71

$$\text{The diagonal} = (P_t^2 + B^2)^{0.5}$$

$$\text{Clearance} = (\text{Diagonal} - d_o - d_d)/2$$

Where d_d is the diameter of a suitable dummy tube or rod.

Example 2.6

A bundle with straight tubes inside a 1000 mm shell has 19.05 mm tubes with a 25.4 mm triangular pitch, 'as viewed' adjacent to a horizontal pass partition. The tube pitch is symmetrical about the partition lane and the tubes are welded into the tubesheet. What is a suitable diameter for dummy tubes or rods to reduce bypassing through the partition lane?

Solution

From table 2.4 on page 48 the gap required for welded tubes is 19 mm. The distance between the centre lines of the tube row, B, is therefore (19 + 19.05) = 38.05 mm.

$$\begin{aligned}\text{The diagonal} &= (P_t^2 + B^2)^{0.5}\\ &= (25.4^2 + 38.05^2)^{0.5}\\ &= 45.75\text{ mm}\end{aligned}$$

Try 22 mm (7/8 inch) dummy tubes.

$$\begin{aligned}\text{Clearance} &= (\text{Diagonal} - d_o - d_d)/2 \\ &= (45.75 - 19.05 - 22)/2 \\ &= 2.35 \text{ mm}\end{aligned}$$

This is less than half the clearance between the tubes [(25.4 – 19.05)/2 = 3.2] and is therefore a bit too tight.

Try 19.05 mm (3/4 inch) dummy tubes.

$$\begin{aligned}\text{Clearance} &= (\text{Diagonal} - d_o - d_d)/2 \\ &= (45.75 - 19.05 - 19)/2 \\ &= 3.9 \text{ mm}\end{aligned}$$

16 mm dummy tubes would have resulted in a clearance of 5.5 mm. These would have also just satisfied the criteria of being below the tube clearance of 6.4 mm but above half that value. There would be significantly more bypassing with these.

Use 19 mm dummy tubes – the same size as the active tubes in this case.

In Table 7 in useful tables at the back of this book Table 7A contains dummy tube sizes for the most commonly used combinations of tubes and pitches covering both expanded and welded tubes with a triangular orientation.

Figure 2.72 shows a triangular (30° 'as viewed') tube layout. In this case the tube layout is not symmetrical about the pass partition lane. It is displaced by half a tube pitch. This results in a significant complication of the geometry. There are now two possible values for the clearance. Clearance C1 is the clearance between the dummy tube and the tube below the partition pass which is in a vertical plane to the dummy tube. Clearance C2 is the clearance between the dummy tube and the adjacent two tubes in the first tube row above the partition lane. The dummy tube is on an inclined plane to the tubes above it.

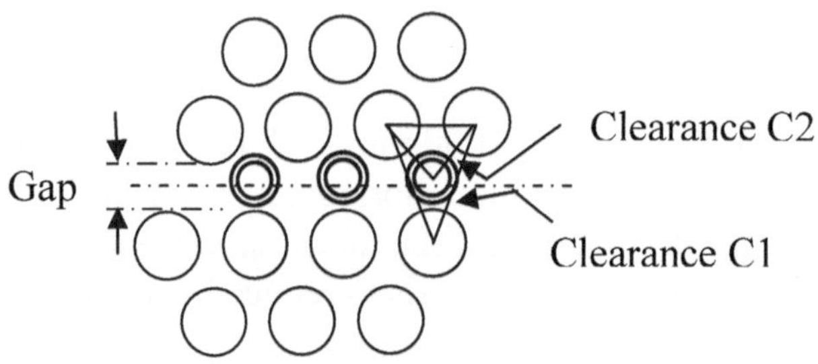

Figure 2.72

To analyse the geometry it is now necessary to consider the two triangles shown in figure 2.72. These two triangles are reproduced and enlarged in figure 2.73

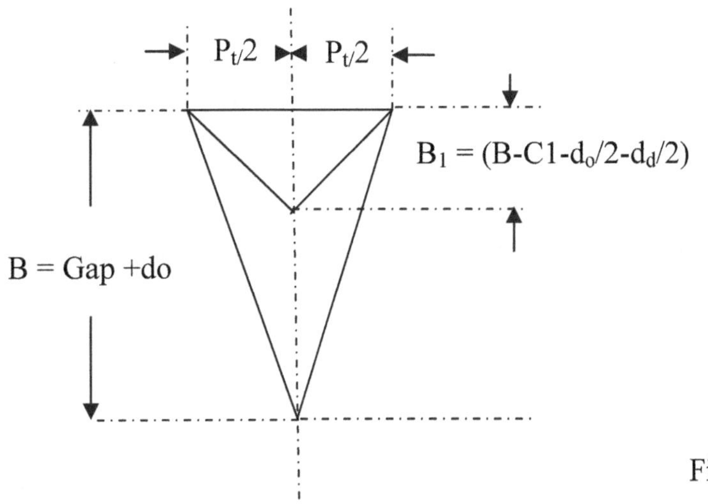

Figure 2.73

The larger of the two triangles is used to establish the height of the two smaller triangles. For one of the small triangles:

$$\text{Diagonal} = [B_1{}^2 + (P_t/2)^2]^{0.5}$$
$$\text{Clearance C2} = (\text{Diagonal} - d_o/2 - d_d/2)$$

To proceed with this calculation it is necessary to assume a value for C1 before C2 can be calculated. The mean value is then the arithmetic average of these two numbers. If this value is then specified it is not strictly mathematically correct. This is because a change in C1 of say 1 mm does not result in a corresponding change of 1 mm in C2. The mathematically correct procedure is to then use the mean value for C1, recalculate C2, and carry on this iteration until there is agreement to whatever degree of accuracy is required.

Fortunately the error in the first calculation is usually in the second decimal place. In the days when tube layouts were produced on a drawing board it was normal practice for the thermal design engineer to simply state 'space equally in the tube nest' and the mechanical designer would do this 'by eye'. Now that tube layouts are electronic and mechanical engineers may not be involved the simple accurate way is to just specify the clearance C1 knowing that that C2 will be OK.

To decide how this should be done it is necessary to look at the fluid flow past the dummy tube. The flow must divide into two streams, one passing through the clearance C1 and the other passing through two clearances C2. If the two clearances are the same there will be a much higher flow rate past the single resistance. Ideally they should be the same which requires C1 to be smaller than C2. The two clearances could be optimised which could be an interesting but purely academic exercise. This is not necessary but there is now a simple guide as to how we should specify C1.

Look at the mean clearance value and then round this down to a sensible dimension and use this to specify the clearance between the dummy tube and the adjacent tube on the same vertical centre line – C1.

Example 2.7

A bundle with straight tubes inside a 1000 mm shell has 19.05 mm tubes with a 25.4 mm triangular pitch, 'as viewed' adjacent to a horizontal pass partition. The tube pitch on the row of tubes below the partition lane is displaced by half a tube row with respect to the tube row above the partition lane. The tubes are welded into the tubesheet. What is a suitable diameter for dummy tubes or rods to reduce bypassing through the partition lane and a suitable value for the clearance between the dummy tube and the top of the tubes in the bottom row?

Solution

From table 2.4 on page 48 the gap required for welded tubes is 19 mm.

For the large triangle in figure 2.73 the distance between the centre lines of the tube row, B, is therefore (19 + 19.05) = 38.05 mm.

Try a dummy tube diameter of 19 mm (same as the tube) and estimate C1 as half the tube pitch which is (25.4 - 19.05/2 = 3.2 mm

For the small triangle in figure 2.72

$$B_l = (B - C1 - d_o/2 - d_d/2)$$
$$= (38.05 - 3.2 - 19.05/2 - 19.05/2)$$
$$= 15.8 \text{ mm}$$

$$\text{Diagonal} = [B_l^2 + (P_t/2)^2]^{0.5}$$
$$= [15.8^2 + (25.4/2)^2]^{0.5}$$
$$= 20.3 \text{ mm}$$

$$\text{Clearance } C2 = (\text{Diagonal} - d_o/2 - d_d/2)$$
$$= (20.3 - 19.05/2 - 19.05/2)$$
$$= 1.3 \text{ mm}$$

This is far less than half the clearance between the tubes (C1 = 3.2 mm) and is therefore unacceptable.

Try a dummy tube diameter of 16 mm and estimate C1 as half the tube pitch as before.

For the small triangle in figure 2.72

$$B_l = (B - C1 - d_o/2 - d_d/2)$$
$$= (38.05 - 3.2 - 19.05/2 - 16/2)$$
$$= 17.35 \text{ mm}$$

$$\text{Diagonal} = [B_l^2 + (P_t/2)^2]^{0.5}$$
$$= [17.35^2 + (25.4/2)^2]^{0.5}$$
$$= 21.5 \text{ mm}$$

$$\text{Clearance C2} = (\text{Diagonal} - d_o/2 - d_d/2)/2$$
$$= (21.5 - 19.05/2 - 16/2)/2$$
$$= 4 \text{ mm}$$

$$\text{Mean clearance} = (\text{C1} + \text{C2})/2$$
$$= (3.2 + 4.0)/2$$
$$= 3.6 \text{ mm}$$

This is acceptable and **16 mm dummy tubes** are appropriate. The clearance C1 could be specified as the minimum we assumed of 3.2 mm. However this 'minimum' is not a hard and fast rule from anywhere it was just a guide I suggested. In this case I would **use 3 mm**, a practical round number which would help to even out the flow streams around the dummy tube.

In Table 7 in useful tables at the back of this book Table 7B contains dummy tube sizes for the most commonly used combinations of tubes and pitches covering both expanded and welded tubes with a displaced triangular orientation about the partition lane.

The reader will have noticed by now that this table like all the others in this section are an Excel spreadsheet. If you find you are designing exchangers with tubes or pitches outside the ranges in the tables it will take you a lot less time than it took me to make up your own spreadsheet. You can then insert any values you like and see the effect on the clearances around the dummy tubes.

Figure 2.74 shows a rotated triangular (60° 'as viewed') tube layout. The baffle cut may be any of the possible orientations. In this case the tube layout is symmetrical about the pass partition lane.

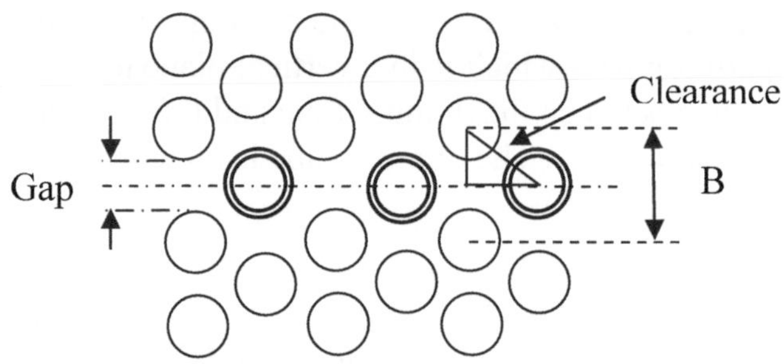

Figure 2.74

To determine the clearances for this arrangement the basic geometry is relatively simple and both clearances have the same value. The triangle in figure 2.74 is shown again in figure 2.75. The two sides adjacent to the right angle can then be used to calculate the diagonal line.

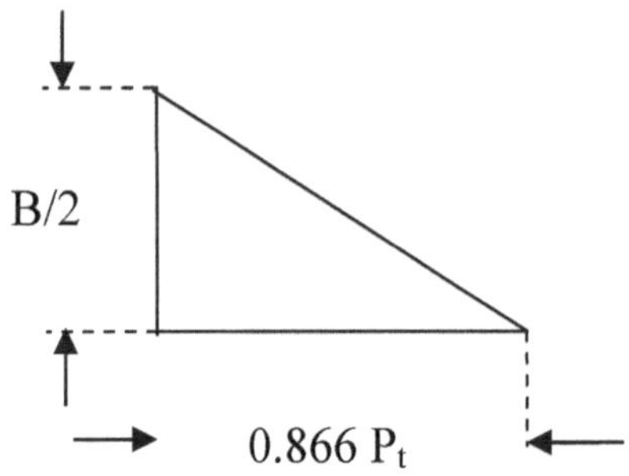

Figure 2.75

$$\text{The diagonal} = ([0.866P_t]^2 + [B/2]^2)^{0.5}$$

$$\text{Clearance} = (\text{Diagonal} - d_o/2 - d_d/2)$$

Table 7 in useful tables at the back of this book Table 7C contains dummy tube sizes for the most commonly used combinations of tubes and pitches for this arrangement.

As for the symmetrical triangular pitch the dummy tube is always on the gap centre line so that only the diameter of the dummy tubes is required. There is no need to specify any clearance so long as you are sure that it is OK.

If you look at table 7C the clearance for the 25.4 mm tubes on a 31.75 mm pitch with a gap of 16 mm is a bit high at 5.8 mm. 29 mm (1.125 inches) dummy tubes would have been ideal but there is no standard tube of this diameter in TEMA.

Figure 2.76 shows a rotated triangular (60° 'as viewed') tube layout. In this case the tube layout is not symmetrical about the pass partition lane. It is displaced by the height of the tube pitch triangle. This again complicates the geometry. There are once again two possible values for the clearance. Clearance C1 is the clearance between the dummy tube and the tube below the partition pass which is in a vertical plane to the dummy tube. Clearance C2 is the clearance between the dummy tube and the adjacent two tubes in the first tube row above the partition lane. The dummy tube is on an inclined plane to the tubes above it.

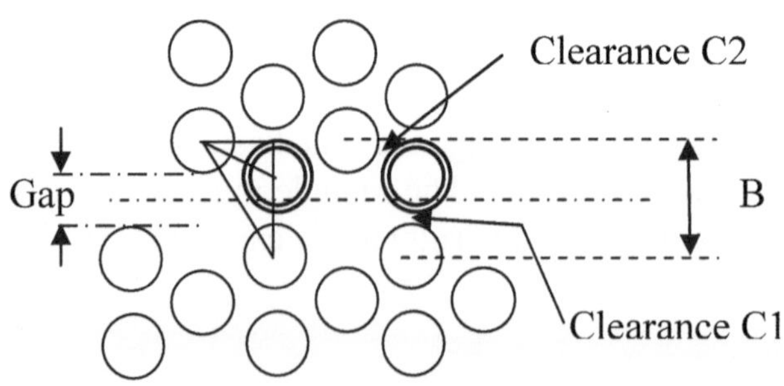

Figure 2.76

To analyse the geometry it is now necessary to consider the two triangles shown in figure 2.76. These two triangles are reproduced and enlarged in figure 2.77

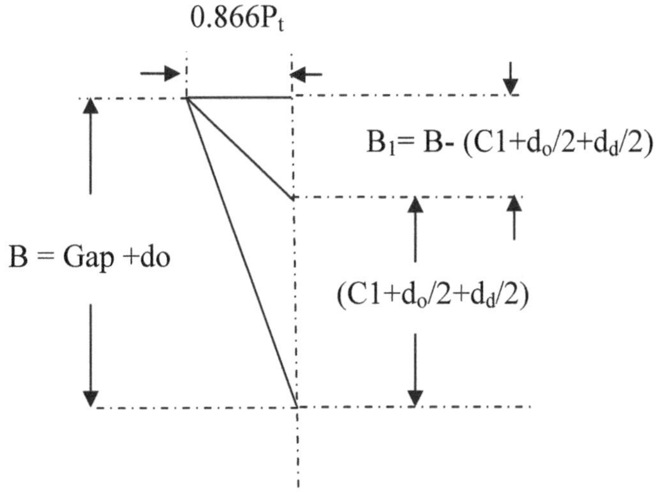

Figure 2.77

The larger of the two triangles is used to establish the height of the smaller triangle.

For the small triangle:

$$\text{The diagonal} = ([0.866P_t]^2 + [B/2]^2)^{0.5}$$

$$\text{Clearance} = (\text{Diagonal} - d_o/2 - d_d/2)$$

To proceed with this calculation it is again necessary to assume a value for C1 before C2 and the mean value can be calculated. Table 7D in useful tables at the back of the book after the main text has the results of the calculation for arrangements you are most likely to use. Because the dummy tubes are not on the centre line it is necessary to specify the diameter of the dummy tubes and the clearance C1.

All the calculations in useful table 7 have been carried out on an excel spreadsheet. I found it necessary to use a lot more columns than the ones printed, for example columns for $[B/2]^2$ and $[0.866P_t]^2$ in the equation above. I am sure it is possible to just write in much larger equations so that all that is necessary are the columns shown. Attempting this was driving me to drink earlier than usual in the evenings. I am writing this book in splendid isolation without being able to call in an IT whiz kid every time I get stuck. The reader will no doubt find basic computing as easy as I still find basic geometry.

You should also have noticed some inconsistencies in the decimal points. For example I have used 19.05 mm for the tube diameter but 19 mm for the dummy tube diameter when they have the same value. This is because so much data was once only in the outdated US customary system of units, (which was then the British Imperial system). I thought about going through them all again making minor changes, (which would not affect the result,) for constancy. Then I thought it was more satisfactory to leave them as they are as a demonstration of the irritations we will all have to endure for many years to come. (OK you win – it was also the lazy way out!)

BUNDLE/SHELL CLEARANCE, OTL & SCD

BUNDLE/SHELL CLEARANCE

This could be defined as the distance between the inside diameter of the shell, D_s and the outside of the nearest tube to it. However it is more convenient to define it on a diametral basis which is twice this value.

For the input to any calculation a **minimum** value is specified. This is dependant on the construction of the bundle and the diameter of the shell.

The **actual** bundle/shell diametral clearance is twice the distance between the inside diameter of the shell and the outermost tube as calculated from the tube layout diagram.

The difference between the two values may be zero or very close to zero for a large bundle but several millimetres for a small bundle. A sophisticated computer program will observe the minimum value but then use the value it calculates from the tube layout it has established.

OUTER TUBE LIMIT, OTL

This is defined as the diameter of a circle equal to the inside diameter of the shell minus the bundle/shell diametral clearance.

$$OTL = D_s - \text{bundle/shell diametral clearance.} \qquad [2.3]$$

SAFE CIRCLE DIAMETER SCD

This is defined as the diameter of a circle which encompasses the centres of all the tubes.

It is therefore the diameter of a circle equal to the inside diameter of the shell minus the bundle/shell clearance minus the tube outside diameter.

$$SCD = D_s - \text{bundle/shell diametral clearance} - d_o \qquad [2.4]$$

The OTL and SCD will therefore have **maximum** and **actual** values.

The OTL is always shown on electronically generated tube layout diagrams. This is the maximum value. It is easy to confirm this by looking at a tube layout for a small shell when it is unlikely that any of the tubes will touch the OTL.

The SCD is an essential to a hand calculation for a tube count. For electronically generated tube layouts it is no longer necessary.

As far as heat transfer is concerned the bundle/shell clearance should always be a small as possible to minimise C stream bypassing.

U TUBE BUNDLE CLEARANCES

U tube bundles only need sufficient clearance to cater for the ligament thicknesses of the baffles between the holes for the outer tubes and the baffle edge, and the baffle/shell clearance which allows for irregularities in the shell diameter.

For shells up to about 750 mm diameter 13 mm clearance is sufficient. For larger shells 19 mm is required.

FIXED TUBESHEET BUNDLE CLEARANCES.

The clearance required for a fixed tubesheet design depends on the method used to attach the shell to the tubesheet.

Figure 2.78 shows the simplest and least expensive method. The sketch shows the same joints at each side of the tubesheet so that the channel is also fixed to the tubesheet. It is therefore a detail from a TEMA type NEN exchanger.

For this form of construction the same bundle/shell clearance can be used as for a U tube bundle i.e. up to about 750 mm diameter shell a 13 mm clearance is sufficient and for larger shells 19 mm is required.

This type of construction is only suitable for pressures up to about 25 bar. It is also unsuitable for low temperature exchangers. For this purpose low temperature can be defined as a temperature for which standard carbon steel would not be suitable. The other drawback to this simple construction is that the welds can not be radiographed so that if a spec or code specifies full radiography it can not be used.

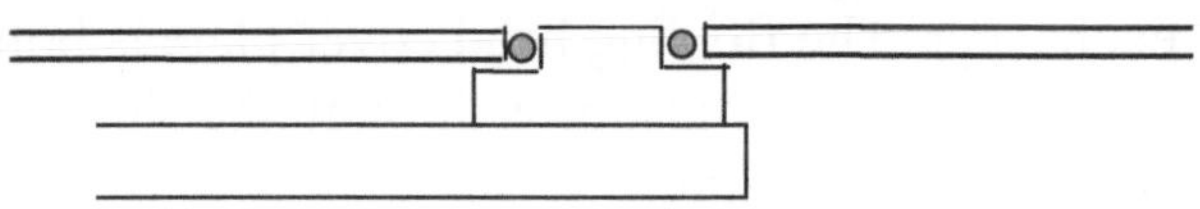

Figure 2.78

For low temperature work and pressures above 25 bar it is necessary to eliminate the sharp corner between the shell and tubesheet and replace it with a radius to avoid a stress concentration point. This is shown on figure 2.79 for one half of the tubesheet only. The other half might be the same for an NEN construction or there could be a flange if the stationary head was an A or B type and the rear head an L or M type.

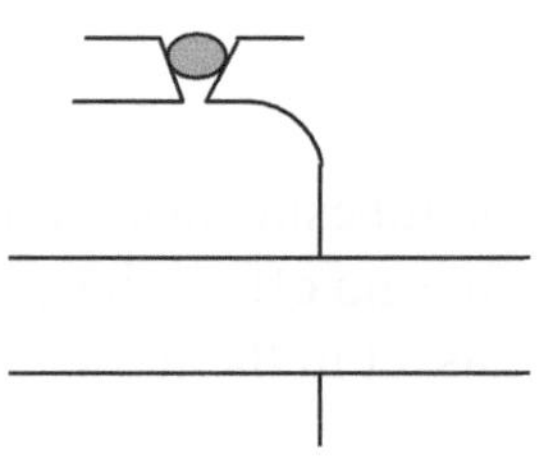

Figure 2.79

The addition of the radius increases the bundle/shell clearance. The minimum diametral clearance is now twice the radius used. If the tubes are to be welded to the tubesheet, (which I would always recommend for fixed tubesheet exchangers), it is also necessary to add twice the thickness of the weld say a further 3 mm x 2 = 6 mm to ensure that the welds do not overlap the radius for NEN shells only. (With other types of head there is no radius on the tubesheet in the channel or rear head.)

The radius required depends on the thickness of the shell. By the ASME code it must be a minimum of 10 mm for shell thicknesses up to and including 38 mm. Above 38 mm it 0.25 times the thickness of the shell up to a maximum value of 19 mm.

FLOATING HEAD TUBESHEET BUNDLE CLEARANCES

Figures 2.1 and 2.3 illustrate the construction of a typical floating head with a split ring backing device. The tubesheet diameter has to be increased to accommodate the inner face of the split ring and its gasket which results in a high bundle/shell clearance. This increases as the diameter of the shell increases.

Values for the clearance and OTL for different shell diameters are shown in table 8A in useful tables at the back of this book.

For welded tubes twice the thickness of the weld, say 2 x 3 = 6 mm (0.25 ins) should be added to the bundle/shell clearances in this table. This decreases the OTL by the same amount. (For 16 BWG tubes a 2 mm weld would be sufficient and for very thick tubes more than 3 mm may be required but 3 mm will cover over 90% of the exchangers you are likely to design.)

PULL THROUGH FLOATING HEADS – TEMA T TYPE

This construction requires the greatest bundle/shell clearance of any of the TEMA types of shell and tube exchanger. The floating head cover plate is bolted directly to the tubesheet and the whole assembly must pass through the shell, (or port for a kettle type of unit.) This is illustrated in figure 2.80.

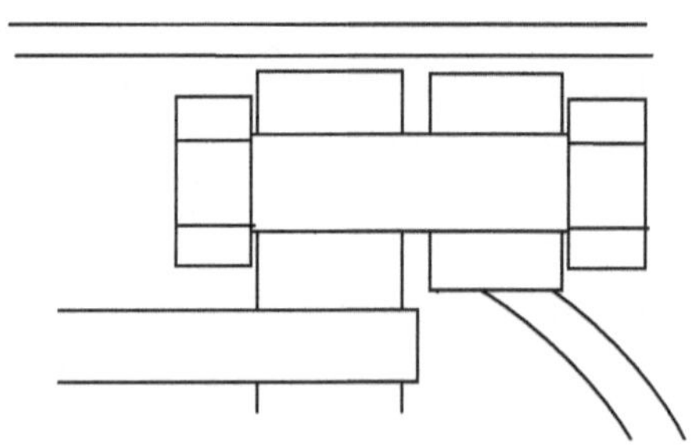

Figure 2.80

A particular design is limited by the tubeside design pressure which is trying to push the joint apart. The shellside pressure has no effect since this is pushing the cover onto the tubesheet and helping to seal the gasket in the joint.

To accommodate higher pressures in a fixed shell size it is necessary to increase the diameter of the bolts. This of course increases the shell/diameter clearance and reduces the OTL.

Table 8B in useful tables has values for maximum design pressures for different bolt sizes. The table is not exhaustive, there is no limit to the pressures which may be used for any size of shell, but the bolts must get bigger and bigger and consequently the clearances larger and larger. The table should cover most of your needs. As with the S type head it is necessary to increase the clearance when welded tubes are used.
When the shell side fluid undergoes convective heat transfer the T type construction has potentially severe bypassing in the C stream and it is necessary to use lots of sealing strips to minimise this. For kettle reboilers a T type may even have thermal benefits by having a good circulation path under the bundle. It does of course increase the cost by increasing the size of port and channel.

NOZZLES

The function of the main nozzles in a heat exchanger is to deliver the working fluids to and from the exchanger, from and to the adjoining pipework.

Someone new to thermal design work may consider them to be a detail which can be left to a mechanical engineer or vendor. A more experienced engineer will appreciate that a basic understanding of nozzles is often essential to producing that elusive optimum thermal design.

Two types of nozzle are shown in figure 2.81. On the left is a nozzle with a weld neck flange. It consists of a forged flange welded to a length of pipe. Pipe sized flanges are widely used elsewhere on any plant and are therefore mass produced and relatively cheap. The length of pipe depends on the nozzle projection from the shell and is dealt with later. It is the standard type of nozzle used for most exchanger work in the construction industry. A cheaper variation from this design is to use a slip on flange. This does not have the tapered neck. The flange fits over the pipe before being welded to it. The flange shown has a raised face which fits inside a recessed space in the connecting pipe to confine the gasket. Another variation is to have a flat face. These are only ever used for low pressure work with innocuous fluids where minor leaks could be tolerated. As far as the thermal design is concerned they are all treated in exactly the same way.

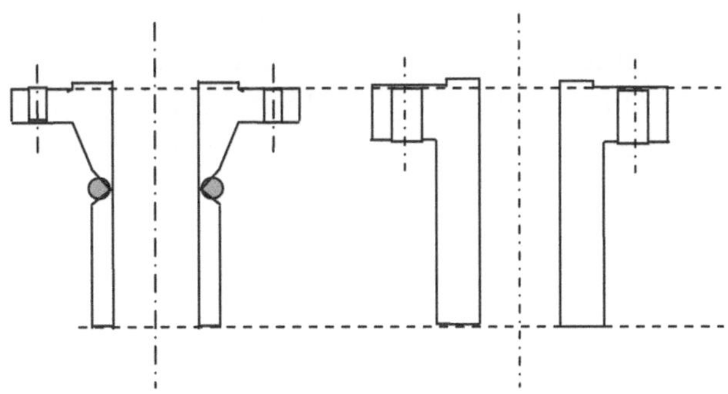

Figure 2.81

On the right hand side is a long forged flange nozzle. The whole nozzle is produced from a single forging. It is therefore much more expensive. How much more expensive it is depends on the nozzle projection. Typically it will cost about three times as much.

If there is no reason not to use it we would always use the nozzle on the left and simply refer to it as a raised face flange nozzle or simply raised face, RF. The alternative on the right is usually termed a long forged welding neck nozzle.

To understand when we resort to using a long forged welding neck it is necessary to consider one of the basics of mechanical design. Whenever it is necessary to make a hole in an exchanger, (or any other pressure vessel), the structure is weakened. To restore the strength of the vessel it is normally necessary to replace this metal in some way. The only occasions when it is not necessary is when a standard thickness has been used for the vessel which is greater that that required by the appropriate code. The reduction in strength may then be acceptable. This will normally only occur with small nozzles in low pressure units. It can be left to the mechanical engineer or vendor to decide if this is allowable. The sensible and safe assumption for a thermal design engineer is to assume that the metal lost in the hole will be replaced.

Figure 2.82 shows how the metal is replaced for our normal nozzles.

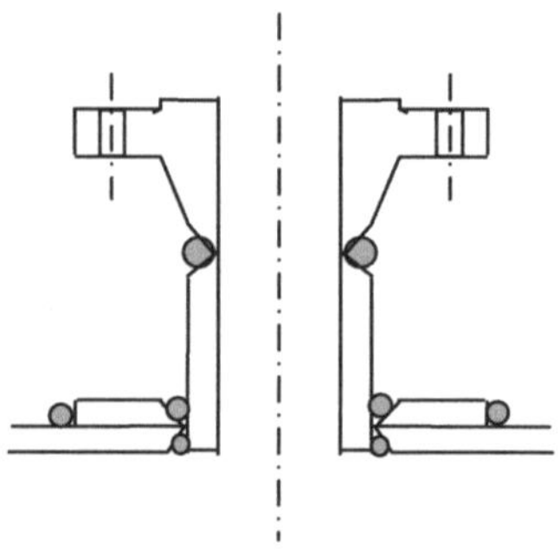

Figure 2.82

A reinforcing pad is used which has a maximum diameter equal to twice the outside diameter of the nozzle. A small hole is drilled in the pad to prevent any trapped air from blowing the welds. All the thermal design engineer needs to know is that nozzles are likely to have such a pad and take account of this when positioning nozzles.

The long forged welding neck nozzles do not need reinforcing pads. The extra thickness in the nozzle barrel is sufficient to replace the metal lost in the hole. They are therefore used when it is thermally necessary to minimise the distance between a shell nozzle and the adjacent tubesheet. An example of such a situation is in a vertical thermosyphon reboiler when having the bottom nozzle too far from the tubesheet will leave a stagnant pool of fluid at the bottom of the shell which will reduce the effective length of the tubes for heat transfer.

The last type of nozzle is one you would only use in very high pressure exchangers.

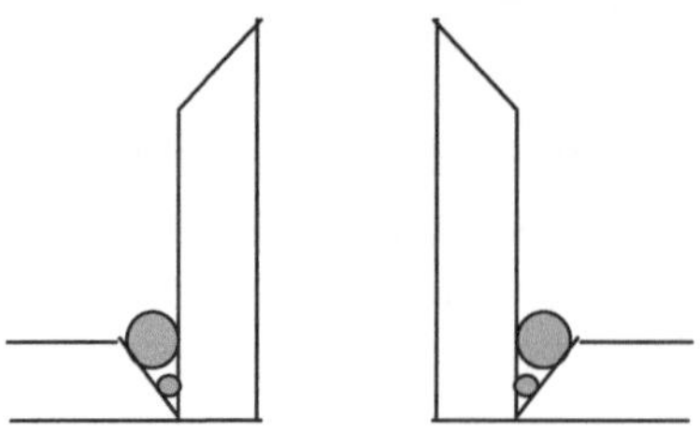

Figure 2.83

This is illustrated in figure 2.83 above. The nozzle is a forging with no flange. The nozzles are welded directly to the adjoining pipework. With no gasket the possibility of leakage is eliminated..

NOZZLE INSIDE DIAMETER, D_n

For thermal design work the most important dimension is the nozzle internal diameter.

For long forged welding neck flanges this is relatively simple. The inside diameter is the internal diameter of the nominal pipe size. These are listed in TEMA and are also shown in table 4 of useful tables. There are two tables of nozzle internal diameters one in SI units and one in US customary units. The dimension you need is the one headed 'Actual ID' in the section of the table under zero corrosion allowance.

For our 'normal' nozzles it is a bit more complicated. The inside diameter depends on the corrosion allowance and the flange rating. Standard flange ratings which cover most exchangers are 150 lb, 300 lb, 600 lb and 900 lb. They depend on the temperature and pressure of the system. They affect not only the heat exchanger but the adjoining pipework and whatever is at the other end of a connecting pipe. They are therefore normally specified by process and provided on the PDS.

150 lb and 300 lb ratings cover most of the heat exchangers you are likely to design. Table 4 at the back of the book lists nozzle inside diameters for different values of corrosion allowance for these ratings. For higher ratings the inside diameter should be reduced. Note that for very small nozzles it is normal to use long welding neck flanges which increase their overall mechanical strength.

NOZZLE MOMENTUM FLUX, ρu^2

The nozzle inside diameter controls the momentum flux of the fluid entering and leaving the exchanger which is why it is important to get it right rather than just using a nominal value.

A good starting point for nozzle momentum fluxes is table 2.5. I have used the values in this table for many years whilst working for several different companies in more than one country.

		ρu^2 kg/ms^2	$\rho u^2/2$ bar	ρu^2 lb/fts^2	psi
Channel nozzles	Liquid	9660	0.048	6492	0.7
	Gas	4830	0.024	3246	0.35
	2-Phase	2760	0.014	1855	0.2
Shell nozzles	Liquid	4830	0.024	3246	0.35
	Gas	4830	0.024	3246	0.35
	2-Phase	2346	0.012	1577	0.17

Table 2.5

I have included velocity head values as well as momentum fluxes since these are more directly related to pressure drop. In the SI system of units one velocity head, is $\rho u^2/2$ which in table 2.5 has been adjusted from N/m2 to bar so that the numbers in column four are the numbers in column 3 divided by 2 times 10^{-5}. Pressure drops due to nozzles are expressed in terms of velocity heads. For a sudden enlargement, (inlet nozzle), the pressure drop is one velocity head (about two velocity heads for a shell inlet with an impingement plate). For a sudden contraction, (outlet nozzle) the pressure drop is half a velocity head.

The reason for limiting momentum fluxes is to prevent erosion. For liquids the momentum flux of any individual particle is the same as the momentum flux of the bulk of the fluid. This is not the case whenever a gas/vapour is involved. Even with a gas there may be times when there are particles of liquid, (on start up or after hydraulic testing) and even the odd solid particle left in the pipes. Gases, particularly low pressure gases may have very high velocities resulting in very high momentum fluxes for an individual particle. In two phase flow the momentum flux of the liquid particles is always far higher than the momentum flux of the bulk of the fluid. This is reflected in the relative values in table 2.5.

For tubeside nozzles TEMA gives no guidance on momentum fluxes. (There are momentum flux guidelines for flow through tubes which is dealt with in Chapter 3) I would therefore recommend that you use the values shown in table 2.5.

For the shellside TEMA has two standards. The first concerns the need for impingement protection under shell inlet nozzles. It states that impingement protection is required whenever the nozzle momentum flux of a single phase non abrasive fluids exceeds 2232 kg/ms^2 (1500 lb/fts^2) or 744 kg/ms^2 (500 lb/fts^2) for all other fluids. These values are well below the values in table 2.4, which assumed that impingement protection would be provided at the inlet nozzles. The value of 744 kg/ms^2 is quite low and I can see no reason for not using it provided account has been taken of any overcapacity. I am less happy with the 2232 kg/ms^2 for non abrasive fluids. This is because I am often not sure which fluids, particularly liquids, are none abrasive. It is much easier to give examples of abrasive fluids, for example any cooling water except possibly for closed circuit systems. I look at each case individually taking into account the fluid, the tube metal, and how close the value of the momentum flux is to the limit and you should do the same. If in doubt use impingement protection. (It is a decision you may not have to make too often because many specs insist on some form of protection.)

The second statement in TEMA is that in no case shall the shell or bundle entrance or exit area produce a momentum flux in excess of 5953 kg/ms^2 (4000 lb/fts^2). The first thing to take note of here is that this is not the same as the nozzle momentum flux; it refers to the shell entrance and exit areas. This is followed by a series of statements which I find to be rather confusing. This is discussed in detail in the section on impingement protection later in this chapter.

A momentum flux of 5953 kg/ms^2 is far higher than any of the values in table 2.4 and there is no point in having a much lower momentum flux in the nozzle and then accelerating the fluid as it enters the critical area where it is about to make contact with the first tubes. This figure also takes no account of the phase of the fluid and there is no doubt that two phase fluids are far more likely to cause erosion problems than single phase fluids. The values in table 2.4 may be a bit conservative but the areas around nozzles are the most vulnerable areas in any heat exchanger. I therefore still use these values as a guide, (but not fixed maximums), and I would certainly never use a value of 5953 kg/ms^2 for a two phase fluid. You are of course still fully complying with TEMA by being more conservative.

NOZZLE PROJECTION

The final topic on nozzles is the nozzle projection. This is something the thermal design engineer will only get involved in if his/her brief includes a preliminary setting plan. The projection is the distance between the face of the flange and the outside diameter of the shell as shown in figure 2.84.

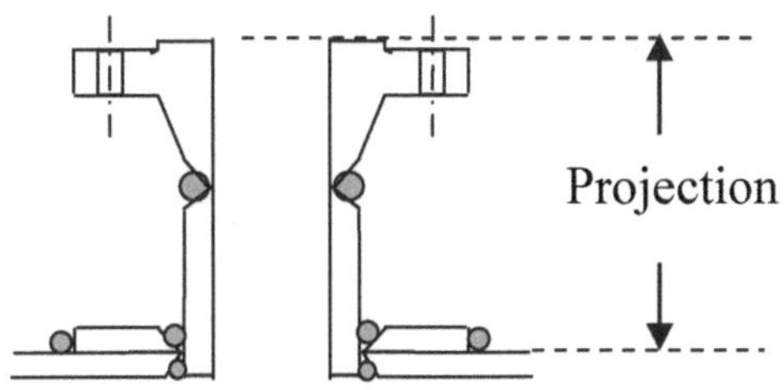

Figure 2.84

On a setting plan the projection is normally added to the radius of the outside of the shell to give a dimension from the exchanger centre line. The projection depends on the flange rating, the size of the nozzle and the thickness of any insulation. Nozzle projections apply to other pressure vessels as well as heat exchangers and most companies have their own standards. I have noticed that standards do seem to vary for no apparent reason indicating that judgement rather than precise calculations have been used to compile them. If in any doubt about which value to use the largest one is the most conservative.

If there are no standards for the job you are working on table 9 in 'useful tables' after the main text has some typical values.

BUNDLE ENTRANCE AND EXIT SPACES

The effective length of a shell and tube exchanger consists of a central baffled section enclosed by end spaces. For an F type of shell the end spaces are both adjacent to the stationary head, for E type shells there is one space at the stationary head and the other at the rear head. The inlet/outlet can be at either end and the length of the spaces may be quite different.

The J type of shell is a special case. The mechanical construction may suggest significantly different lengths at each end. However the divided flows should have paths with approximately the same hydraulic resistance. Therefore it may be necessary to increase the smaller of the two lengths.

The length of the end spaces is normally significantly larger than the baffle pitch in the central section. For an exchanger with a large diameter relative to the tube length end spaces may occupy a substantial proportion of the available space.

For any exchanger it is important to fix the length of these spaces before starting to manipulate the baffle details to optimise the shell side design. Some of the best design programs now make a reasonable job at fixing these lengths but you may be working with one which does not. It is well worth checking some of the values your program is using.

If the values for the end lengths are too small the first time you may discover this is when the vendor points it out. You will then have to rerun the unit with modified baffling. The worst case scenario is that you may have to increase the shell size and the vendor can then start to mark up his 'extras'. If your buyer has 'screwed him right down' from his original price the 'extras' may be substantial.

If your end spaces are too large and convective heat transfer is involved the reduced velocities and hence reduced heat transfer coefficients, in the end sections, means that you may have already designed a unit with a larger shell than was necessary.

The length of an end space is dependant on the type of head used in the construction. The main heads used are as follows.

STATIONARY HEAD L1

This length applies to all exchangers. It has therefore been designated as L1. A fixed tubesheet exchanger will have an L1 value at each end. These values will not be the same unless the outlet nozzle has the same diameter as the inlet nozzle. All other types will have an L1 length at one end only.

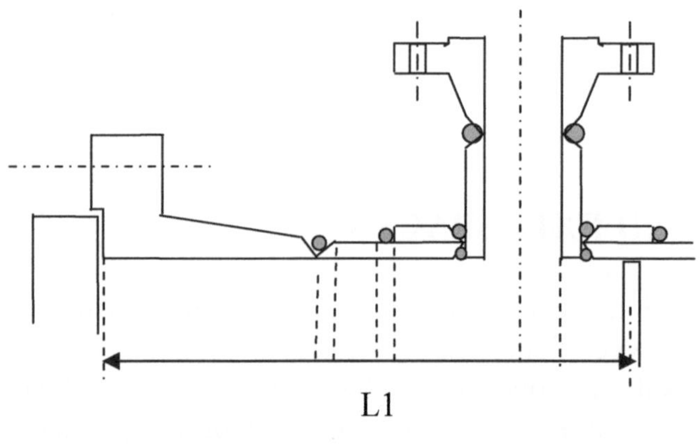

Figure 2.85

Figure 2.85 illustrates L1 which starts at the inner face of the stationary tubesheet and ends at the centre line of the first baffle. The vertical broken lines indicate the individual sections which are added together to establish this length.

Starting from the tubesheet at the left hand side these are:

(a) The length of the shell flange.
(b) The length of the weld preparation on the shell which is the same as the thickness of the shell, t.
(c) 50 mm or t if this is greater.
(d) The length of the weld fixing the reinforcing pad to the shell which is the same as the thickness of the shell, t. (This might be a bit less if the pad is thinner than the shell.)
(e) The outside diameter of the nozzle. (This might be a bit less if the diameter of the pad is less than its maximum value of twice the outside diameter of the nozzle.)
(f) The radius of the inside of the nozzle.
(g) A distance between the inside edge of the nozzle and the face of the first baffle. This distance should be just sufficient to allow unrestricted flow between the nozzle and the space. If it is an inlet space account should be taken of any impingement device.
(h) Half the thickness of the first baffle.

Some of the qualifications in brackets in the above list will not be known until the detailed mechanical design stage. It is therefore necessary to assume maximum values.

Most of the above data is determined by the shell side design pressure and temperature and is independent of what is taking place on the tubeside. The one exception is the length of the shell flange which must match up with the rating of the flange on the channel. If the tubeside design pressure is higher than the shell side design pressure this length may need to be increased.

Table 10 in useful tables at the back of the book lists values for L1 for a range of shell side pressures from 150 psig (10.34 barg) to 750 psig (51.7 barg). They allow for a reasonably sized round impingement plate. The tables are not fully extensive in so much as they do not allow for every possible nozzle size in every possible shell size, (now sometimes in increments as small as 10 mm). Tables to cover every possibility would have required more paper than you will need to print off this book. They must therefore be used with sensible, (safe) values. Always use values for the pressure and nozzle size which fully cover your data. This is best illustrated by means of examples.

Example 2.8

An AES exchanger has a 1200 mm shell with 14 inch nominal bore nozzles at inlet and outlet. The design pressure for both sides is 10 barg. Determine the end space length required at the stationary head end.

Solution

The table for 10.34 barg covers the pressure range for both sides of the exchanger and the shell diameter of 1219 mm is safe for the diameter. However there is no 14 inch nozzle in the table for this diameter. There is a value for a 14 inch nozzle in the next shell size of 1270 mm. This is the value which should be used i.e. **622 mm**.

Example 2.9

The exchanger in example 2.8 has a design pressure on the shell side of 10 barg but the tubeside design pressure is now 45 barg. Determine the end space length required at the stationary head end.

Solution

The length of the flange will now need to be increased to match the flange on the tubeside. The pressure which covers the new tubeside pressure is 51.7 bar and for a 1219 mm shell the flange length is 203 mm. For a pressure of 10.34 barg the flange length is 105 mm. Therefore:

$$L1 = 622 + (203 - 105) = \mathbf{720\ mm}$$

SPLIT RING FLOATING HEAD L2 S

Figure 2.86 illustrates L2 S which starts at the inner face of the stationary tubesheet and ends at the centre line of the first baffle. The vertical broken lines indicate the individual sections which are added together to establish this length.

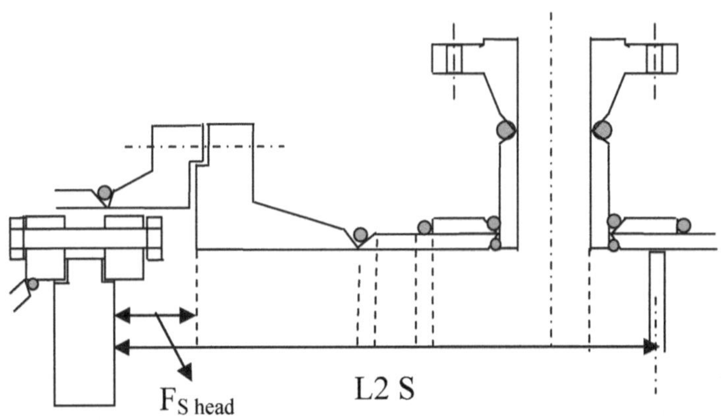

Figure 2.86

All the lengths listed for the stationary head inlet space are also present in the space required for a split ring floating head. There is an additional space required to accommodate the split ring and bolts and to allow for thermal expansion.

The flange is unaffected by tubeside pressure because both the shell flange and the shell cover flange are only exposed to the shell side pressure. The split ring and bolts are affected by the tubeside pressure and whenever the tubeside pressure is greater than the shell side pressure the length allowed for these parts, $F_{S\ head}$, must be increased.

Table 10 in useful tables has values for L2 S. An example of their use follows.

Example 2.10

An AES exchanger has a 1090 mm shell with 16 inch nominal bore nozzles at inlet and outlet. The design pressure is 10 barg for the shell side and 43 barg for the tube side. Determine the end space length required at the floating head end.

Solution

The table for 10.34 barg covers the pressure range for the shell side of the exchanger, a shell diameter of 1118 mm is safe for the diameter and there is a 16 inch nozzle in the table for this diameter. From the table a value of 864 mm is satisfactory for the shell side pressure but this needs to be corrected for the tube side pressure because of the increase in length of the split ring and bolts.

The length of $F_{S\ head}$ in the 10.34 barg table is 108 mm. The pressure which covers the tubeside is 51.7 barg which has an $F_{S\ head}$ value of 219 mm. The corrected value to use for L2 S is therefore:

$$L2\ S = 864 + (219 - 108) = \mathbf{975\ mm}$$

PULL THROUGH FLOATING HEAD L2 T

Figure 2.87 illustrates L2 T which starts at the inner face of the stationary tubesheet and ends at the centre line of the first baffle. The vertical broken lines indicate the individual sections which are added together to establish this length.

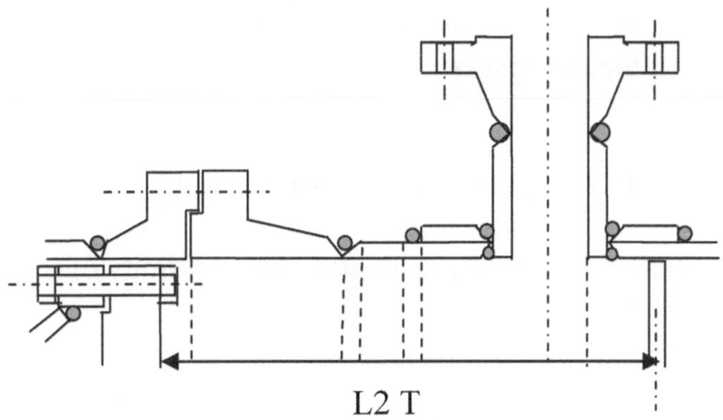

Figure 2.87

All the lengths listed for the stationary head inlet space are also present in the space required for a pull through floating head. There is an additional space required to accommodate the bolts and access to the bolts so that the floating head cover can be removed for cleaning the tubes without pulling the bundle.

The extra space required is affected by the tube side pressure which is often well above the shell side pressure for this type of head. The extra length required for the bolts and access to the bolts needs to be taken into account. The $F_{T\ head}$ column in table 10 can be used to take this into account in the same way that $F_{S\ head}$ was used to correct for tube side pressure with S type heads.

Example 2.11

An AET exchanger has a 1090 mm shell with 16 inch nominal bore nozzles at inlet and outlet. The design pressure is 10 barg for the shell side and 43 barg for the tube side. Determine the end space length required at the floating head end.

Solution

The table for 10.34 barg covers the pressure range for the shell side of the exchanger, a shell diameter of 1118 mm is safe for the diameter and there is a 16 inch nozzle in the table for this diameter. From the table a value of 716 mm is satisfactory for the shell side pressure but this needs to be corrected for the tube side pressure because of the increase in length of the bolts.

The length of $F_{T\ head}$ in the 10.34 barg table is 63.5 mm. The pressure which covers the tubeside is 51.7 barg which has an $F_{T\ head}$ value of 101.6 mm. The corrected value to use for L2 T is therefore:

$$L2\ T = 716 + (101.6 - 63.5) = \mathbf{754.1\ mm}$$

This value would be sensibly rounded up to 755 mm.

By the time this book is published the most sophisticated programs should be predicting reasonable values for the end spaces. However my limited searches on the internet revealed an astonishing number of programs which presumably are being used by someone somewhere which are very unlikely to do this. (This includes the ones I use when I do work for vendors.) In any case it is important to understand how end spaces are calculated for different types of construction. Some types of head do not need any tables or detailed calculations to fix values for end spaces.

U HEADS AND PULL THROUGH FLOATING HEADS L2 T Spl

These units have adjacent nozzles which are not dependant on the proximity of a shell flange to establish the length of the end space.

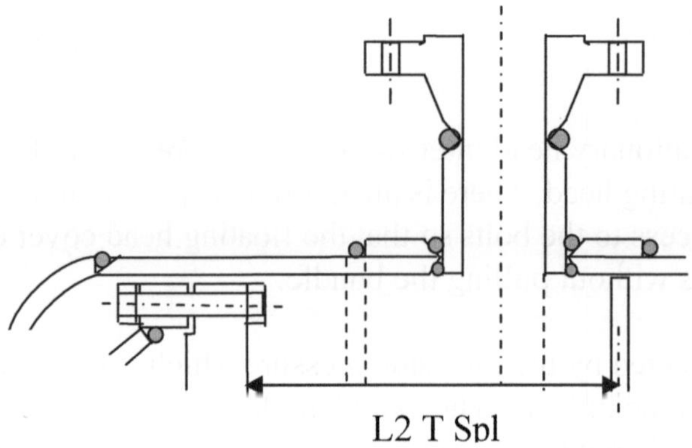

Figure 2.88

The L2 T Spl is a cheaper version of the normal T type. The reduced cost is due to the elimination of a shell flange and a shell cover. The disadvantage is that the tubes can not be cleaned without pulling the bundle.

For both these types of design the nozzle can be positioned at the centre of the adjacent space – the optimum position for fluid flow

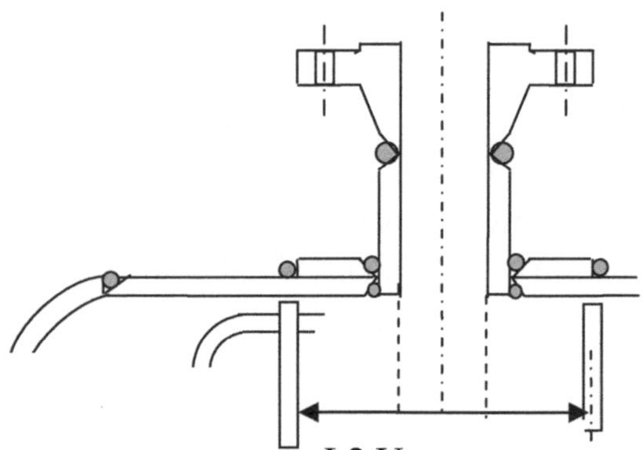

L2 U

Figure 2.89

The L2 T Spl is an unusual one but the comments on the U type illustrated in figure 2.89 apply to both types.

It is only necessary to start with the inside diameter of the nozzle and make any allowance necessary for any impingement device. For an outlet nozzle such as illustrated in figure 2.89 the flow path at any section before the nozzle is clearly going to be more than the flow path through the nozzle. For this arrangement the length of the end space can normally be taken to be the same as the pitch in the main baffled section.

If an impingement device is fitted under an inlet nozzle this must be taken into account. For a solid flat plate just make sure that there is sufficient space around the plate to avoid any flow restriction. For most sizes of nozzle this can be done just by looking at the tube layout, occasionally with a large nozzle you may have to think a bit longer but I can not recall ever finding it necessary to do any detailed calculations.

Even with an impingement plate it is sometimes possible to use the same length of space as the pitch in the baffled section. If the minimum end space is actually less than the pitch in the baffled section the space itself would be a restriction and the length must be increased to the same as the baffle pitch. (Watch out for computer programs doing this!)

Once the length of the end space has been fixed the nozzle should be at the centre but you will want to round to a sensible number when you fix the distance between nozzles on any setting plan.

IMPINGEMENT PROTECTION

Impingement protection is a topic that all thermal design engineers should have a thorough understanding of. Badly designed or inappropriate devices are one of the main causes of vibration problems.

OUTER TUBE ROW HEIGHTS

Figure 2.90 illustrates the geometry used to determine the maximum distance for the outer row of tubes from the centre line of the shell. At the top of the diagram is a nozzle with an impingement plate with the centre line of the outer row of tubes y_1 mm from the exchanger centre line. At the bottom of the diagram is a nozzle with no impingement plate and the centre line of the outer row of tubes is y_2 mm from the exchanger centre line. Either nozzle could be at the top depending on whether the shell side fluid is being heated or cooled.

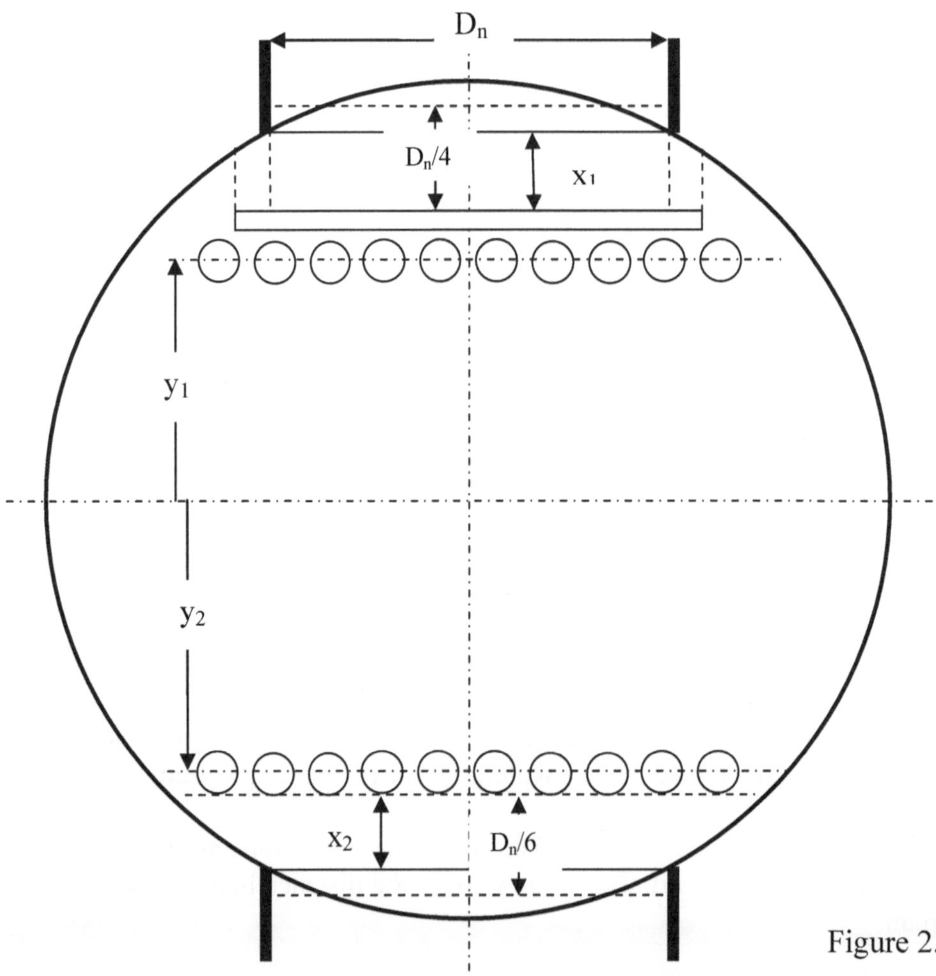

Figure 2.90

For the nozzle at the top of the diagram there is a segment of a circle enclosed by the diameter of the shell and a line drawn across the points where the inside diameter of the shell meet the inside diameter of the nozzle. The height of this segment may be established from two right angle triangles as follows:

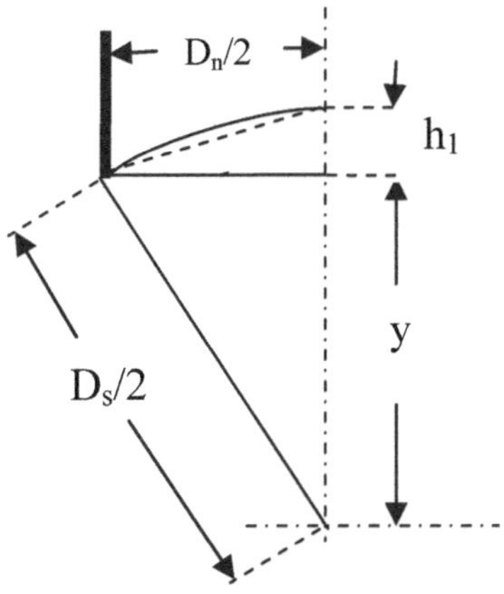

Figure 2.91

$$y = [(D_s/2)^2 - (D_n/2)^2]^{0.5}$$

$$h_l = D_s/2 - y$$

$$h_l = D_s/2 - [(D_s/2)^2 - (D_n/2)^2]^{0.5} \quad [2.5]$$

For the fluid to enter the exchanger with the same velocity, (and hence the same momentum flux), as in the nozzle the area where the fluid enters the exchanger must be the same as the nozzle cross sectional area. The area through which the fluid passes from the nozzle to the exchanger is cylindrical. The diameter of this cylinder is the same as the nozzle internal diameter and the average length is the distance x in figure 2.90 plus $h_l/2$. Therefore:

Cylindrical area = Nozzle area

$$\pi D_n (x_l + h_l/2) = \pi D_n^2/4$$

$$(x_l + h_l/2) = D_n/4 \quad [2.6]$$

$D_n/4$ is normally taken to be a minimum dimension. The minimum distance to the first tube row can now be calculated from the top of the shell by adding together the following dimensions:

$h_l/2$
$D_n/4$
The thickness of the impingement plate.
A clearance between the plate and the first tube
Half the diameter of the tubes d_o.

There are no generally accepted standards for the thickness of the impingement plate. Some companies use a standard plate of 9.5 mm irrespective of the nozzle size. This is a reasonable thickness for most nozzles. However 6.3 mm should be sufficient for nozzles up to 6 inches nominal bore and with small shells this may help the tube layout. For nozzles of 18 inch nominal bore and above I would either increase this to 12.7 mm or add an extra support under the plate.

It is good practice with most companies to allow 3 mm clearance between the bottom of the impingement plate and the adjacent tubes.

Typically therefore:

$$D_s - y_1 = h_1/2 + D_n/4 + 9.5 + 3 + d_o/2 \qquad [2.7]$$

The dimension ($D_s - y_1$), the height under the nozzle is an optional input to most modern computer programs. Since ($D_s - y_1$) is a minimum dimension the distance y_1 is a maximum dimension.

For a nozzle with no impingement plate the height of the segment can be calculated in the same way as for the nozzle with the impingement plate. However to maintain the same velocity as in the nozzle there is now more area for the fluid to flow through. There is still a cylinder with the same diameter as the nozzle inside diameter and a length of ($x_2 + h_2/2$) but there is now an extra area due to the clearances between the tubes in the row adjacent to the nozzle. Part of this extra area available can be 'seen' looking through the nozzle at the tubes as illustrated in figure 2.92.

Figure 2.92

This extra area depends on the ratio of the clearance between the tubes and the tube pitch. Values for this ratio for the tube outside diameter d_o and tube pitch, P_t are shown for some common tube and pitch combinations in table 2.6.

Tube d_o mm	Pitch P_t mm	Clearance mm	Ratio
19.05	23.81	4.76	0.2
19.05	25.4	6.35	0.25
25.4	31.75	6.35	0.2

Table 2.6

The smallest extra available area is with a ratio of 0.2. For this ratio the extra area available becomes:

$$\text{Extra area} = 0.2\ \pi/4D_n^2$$

Taking account of this extra area 'seen' through the nozzle leads to the following relationship:

Cylindrical area + extra area = Nozzle area

$$\pi D_n (x_1 + h_1/2) + 0.2\ \pi/4D_n^2 = \pi D_n^2/4$$

$$(x_1 + h_1/2) = D_n/5 \qquad [2.8]$$

This is another simple equation comparable to equation 2.6 for the nozzle with the impingement plate and would result in a very conservative positioning of the adjacent tube row. However there is some extra area which this simple analysis omits. At the end of each lane created by the clearance between the tubes there is an additional area allowing the fluid to flow in the same plane as the fluid flowing across the imaginary cylinder. A section of part of this additional area is illustrated in figure 2.93.

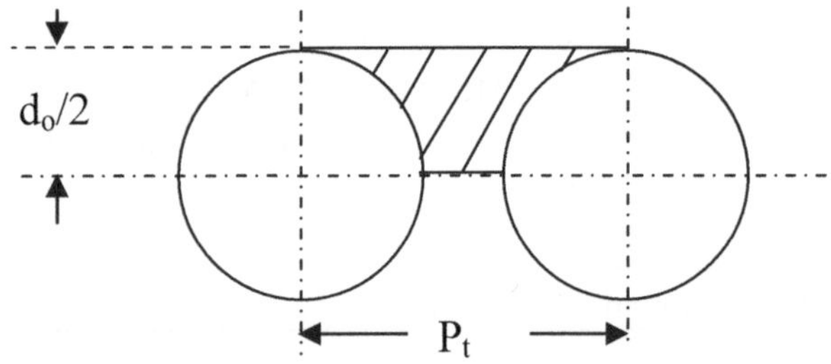

Figure 2.93

This additional area between two tubes is shown shaded in the sketch. There is one of these areas at each end of a flow path 'lane' formed by the clearance between the tubes. For this section:

$$\text{Area} = P_t\, d_o/2 - \pi d_o^2/8$$

There is a lane every tube pitch so that across the section of a nozzle there are D_n/P_t 'lanes', therefore:

$$\text{Total area} = 2(P_t\, d_o/2 - \pi\, d_o^2/8)\, D_n/P_t$$

The full equation now becomes:

Total flow area into the exchanger = Nozzle area

$$\pi\, D_n\, (x_2 + h_2/2) + (P_t - d_o)\, \pi\, D_n^2/4\, P_t + 2(P_t\, d_o/2 - \pi\, d_o^2/8)\, D_n/P_t = \pi D_n^2/4$$

Rearranging and 'simplifying' gives:

$$(x_2 + h_2/2) = D_n/4 - (P_t - d_o)\, D_n\, /4P_t - 2(P_t\, d_o/2 - \pi\, d_o^2/8)\, /P_t\, \pi \qquad [2.9]$$

When I first started in thermal design work I did the easy check on the $D_n/4$ for the nozzle with an impingement plate but questioned the $D_n/6$ for the nozzle with no impingement plate. I was told just to use it but decided to do a couple of spot checks to convince myself. 1/6 = 0.167 and the checks I did resulted in values very close to this, but not exactly the same. I have used $D_n/6$ ever since without any problems. Then when I came to write this section of the book I thought it was worthy of further investigation.

Equation 2.9 clearly shows that the distance we are interested in, $(x_2 + h_2/2)$ depends on the diameter of the nozzle, the tube pitch and the tube outside diameter. However h_2 also depends of the shell diameter Ds. This results in the dimension we require depending on four variables.

I set up a spread sheet using only the worst case of tube clearance/tube pitch ratio of 0.2 (25.4mm tubes on a 31.75 mm pitch) and three shell diameters of 381 mm, 1000 mm and 1800 mm. I used the full range of nozzles compatible with the shells (only four in the case of the 381 shell but up to 24 inches for the other two).

For the 381 mm shell the ratio of 0.167 was never exceeded and a value of 0.137 for a 6 inch nozzle was the highest ratio. (A ratio of $D_n/7 = 0.143$ would have been sufficient for this shell size.)

For the 1000 mm shell the ratio of 0.167 was never exceeded and a value of 0.161 for a 6 inch nozzle was the highest ratio.

The data for the 1800 mm shell was more interesting and the results are reproduced in table 2.7. The ratio of 0.167 was exceeded for five of the nozzles but not by very much – maximum of only 2% over.

			RHS equation 2.8						ratio
D_n	d_o	P_t	term1	term 2	term 3	$(x+h_2/2)$	x_2	$D_n/6$	x_2/Dn
50.8	25.4	31.75	12.7	2.54	3.005	7.2	7.0	8	0.137
97.2	25.4	31.75	24.3	4.86	3.005	16.4	15.8	16	0.162
146.3	25.4	31.75	36.58	7.315	3.005	26.3	24.8	24	0.169
193.4	25.4	31.75	48.35	9.67	3.005	35.7	33.1	32	0.171
242.9	25.4	31.75	60.73	12.145	3.005	45.6	41.5	40	0.171
303.3	25.4	31.75	75.83	15.165	3.005	57.7	51.2	51	0.169
333.3	25.4	31.75	83.33	16.665	3.005	63.7	55.9	56	0.168
381	25.4	31.75	95.25	19.05	3.005	73.2	63.0	64	0.165
428.7	25.4	31.75	107.2	21.435	3.005	82.7	69.8	71	0.163
477.8	25.4	31.75	119.5	23.89	3.005	92.6	76.4	80	0.160
574.6	25.4	31.75	143.7	28.73	3.005	111.9	88.4	96	0.154

Table 2.7

To complete the exercise I repeated the calculations for the 1800 mm shell with 19.05 mm tubes on a 25.4 mm pitch (a clearance/pitch ratio of 1.25). As expected all the values were now under the 0.167 ratio for a height of $D_n/6$.

			RHS equation 2.8						ratio
D_n	d_o	P_t	term1	term 2	term3	$(x+h_2/2)$	x_2	$D_n/6$	x_2/Dn
50.8	19.1	25.4	12.7	3.18	2.49	7.0	6.9	8	0.135
97.2	19.1	25.4	24.3	6.08	2.49	15.7	15.1	16	0.155
146.3	19.1	25.4	36.6	9.14	2.49	24.9	23.5	24	0.160
193.4	19.1	25.4	48.4	12.09	2.49	33.8	31.2	32	0.161
242.9	19.1	25.4	60.7	15.18	2.49	43.1	38.9	40	0.160
303.3	19.1	25.4	75.8	18.96	2.49	54.4	47.9	51	0.158
333.3	19.1	25.4	83.3	20.83	2.49	60.0	52.2	56	0.157
381	19.1	25.4	95.3	23.81	2.49	68.9	58.7	64	0.154
428.7	19.1	25.4	107.2	26.79	2.49	77.9	64.9	71	0.151
477.8	19.1	25.4	119.5	29.86	2.49	87.1	71.0	80	0.148
574.6	19.1	25.4	143.7	35.91	2.49	105.2	81.7	96	0.142

Table 2.8

An interesting feature of all the tables is that the smallest and largest nozzles all require less space under the nozzle than provided by $D_n/6$. This is mainly of interest for the larger nozzles where more tubes could be fitted into a shell by reducing the height under the nozzle. For example in table 2.7 the nozzle velocity into or out of the shell could be maintained if the adjacent tube row was positioned (96 – 81.7) = 14.3 mm closer to the nozzle. The odds are strongly in favour of being able to fit in an extra row to tubes.

Many times in the past I have 'infringed' the $D_s/6$ 'rule' by the odd mm or two to fit in an extra row of tubes but only when I knew that the nozzle momentum flux was well below any limiting value. I would never have considered moving it by as much as 14.3 mm. From now on I will.

To enable the reader to check out the exchanger he/she is working I originally included a spread sheet in the tables at the back of the book. When this was in Excel format it could be copied and the values input into another spread sheet where the value for the shell diameter could be changed. I was very pleased with my new IT skills. Unfortunately now that I am close to publication I find that everything will have to be in PDF format and I may as well remove it.

This approach to positioning tubes adjacent to nozzles with no impingement plate could easily be incorporated into any modern design program. If any of the organisations writing programs cares to do this I would take it as a compliment. The extra row of tubes in even a small percentage of the thousands of heat exchangers which are designed each year would really make a contribution to cutting down on the CO_2 being released into the atmosphere.

Finally on this topic, the D/6 had been around for a long time before I started work in this field. I have no idea who first introduced it but it proved to be a useful contribution to the thermal design of shell and tube exchangers. Before spreadsheets and computer programs no one would have been able to spend the time going through the basic geometric calculations necessary for a more accurate positioning of the first row of tubes.

'NORMAL' IMPINGEMENT PLATES

The previous section described how to position an impingement plate relative to the inlet nozzle and the adjacent row of tubes. The thickness of the plate was also discussed. It is now necessary to consider some of the other characteristics of impingement plates.

The first of these is the shape. When I am asked what the question "which shape would I recommend" my usual sarcastic reply is that I always use round impingement plates with round nozzles but rectangular plates for square nozzles. I then go on to explain that there may be situations where a rectangular or square plate would be acceptable. An important factor to consider is how the plate is going to be supported. Impingement plates should be held in place by being welded to the spacers of two tie rods. Two possible positions are shown in figure 2.93

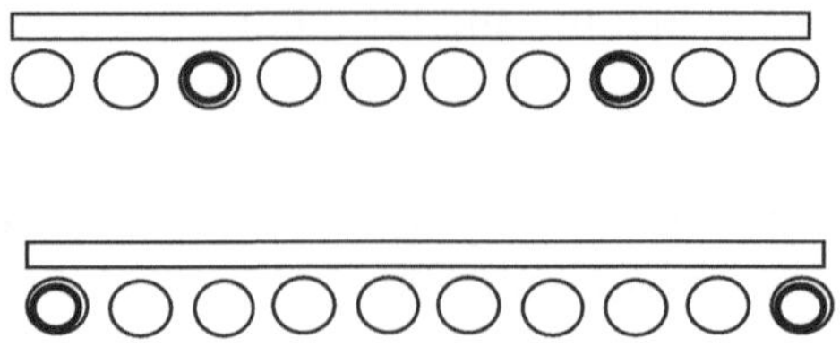

Figure 2.94

The top sketch in figure 2.94 shows the supporting tie rods and spacers ideally positioned for a round plate at about ¼ and ¾ of the diameter of the plate. The bottom sketch shows tie rods in the worst possible positions for a round plate. This is because there would only be a very short weld retaining the plate in position which could easily fail. This is illustrated in the left hand sketch of figure 2.95.

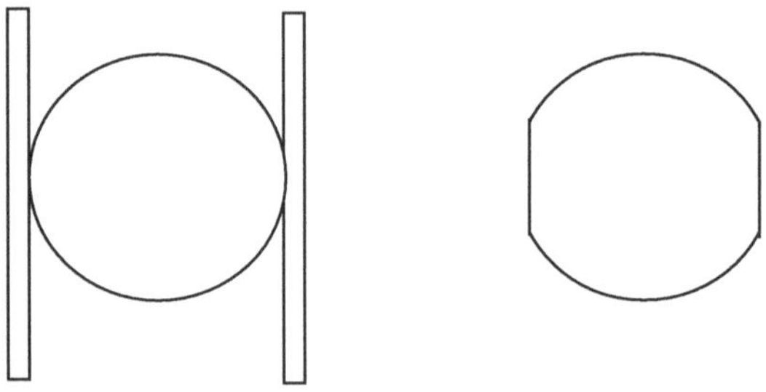

Figure 2.95

This problem can be overcome by slightly increasing the diameter of the plate and trimming off the edges to give an acceptable length of weld as shown in the right hand sketch. However when the supports are at the edges of the plate it is much simpler and normally quite acceptable to just use a square or rectangular plate. The disadvantage of the rectangular plate is that it covers more area; and the more area the impingement plate covers the higher will be the velocities and momentum fluxes into the bundle. This is discussed in more detail later.

The second decision about impingement plates is whether to perforate or not to perforate. Some specs require a series of small holes to be drilled in the plate immediately above the gaps between the tubes particularly for two phase flow units. The theory behind this is that it prevents the formation of a 'dead space' immediately below the impingement plate. I am not convinced that this additional complication is worthwhile and unless a spec insists on it I would just use a solid plate. (Since writing this section I note that ISO now ban the use of perforated plates.)

The final decision about impingement plates is the diameter. I have come across several different guides to fixing this dimension. A simple one is just the inside diameter of the nozzle plus say 50 mm. (This is now the ISO recommendation.) For a 2 inch, 43 mm ID, nozzle in a shell with a small shell to bundle clearance this might result in the impingement plate touching the inside of the shell and for very large nozzles it may produce a plate which is too small to avoid direct impingement on some of the tubes.

A more reasonable approach is to make the impingement plate diameter a fixed ratio of the nozzle inside diameter. The usual ratio used is 1.1. This again falls down with the smallest nozzle we normally use, (2 inch), because the plate would only be 47 mm in diameter.

I found an attempt at a more scientific approach to establishing the diameter of an impingement plate in an old German standard (or Dutch? – my linguistic abilities go no further than being able to order a beer in many different languages). This is illustrated in figure 2.96. The only changes I have made are those necessary to adapt the nomenclature to that used in the rest of this book.

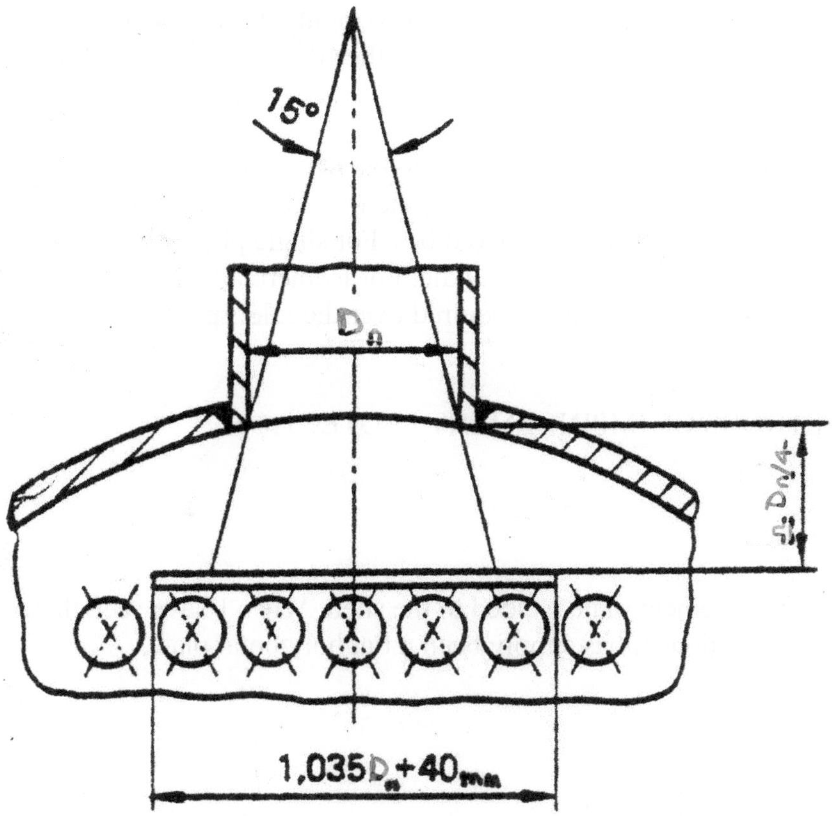

Figure 2 .96

I like the 15° angle. It 'looks right' and with many aspects of thermal design if it looks right it probably is right. However I have never come across any scientific basis for it. The $D_n/4$ is qualified by an approximation sign probably to indicate that the actual dimension to the intersection of the shell and nozzle inside diameter is ($D_n/4 – h_l/2$) as shown earlier in figure 2.90.

I did a lot of geometric calculations to try and establish the constant of 1.035 without success, (if a reader manages it please let me know). I used both the approximate and actual distance from the top of the plate. (When $h_l/2$ is taken into account it is necessary to assume a ratio for the shell diameter to the nozzle diameter – see equation 2.5.) The constant of 40 mm still results in a very large plate for a 2 inch nozzle.

If I was asked to write a standard for a computer program I would use the same form of equation as shown in figure 2.94 modified as follows:

$$\text{Impingement plate diameter} = 1.05\ D_n + x \text{ mm} \qquad [2.10]$$

Where x would be found from the following table:

Nozzle NB ins	2 → 6	8 → 14	16 → 20	24
x mm	20	30	40	50

Table 2.9

I have never used an equation such as 2.10 in the past and I would not recommend it to the reader. I have always just looked at the tube layout, taken into account the supports for the plate and tried to picture the path the fluid will be taking. It is particularly important to try and visualise the flow past the tubes at the edge of the plate since these are the ones most susceptible to vibrations.

Whenever possible the tie rods and spacers supporting the plate should also be picking up the corners or edges of the baffles. Often this is not practical and it becomes necessary to add two extra tie rods and consequently lose two tubes. For single phase fluids the extra tie rods will extend over the full length of the bundle. For isothermal condensing – particularly condensing steam, they are only required over the inlet space.

ALTERNATIVE METHODS OF IMPINGEMENT PROTECTION

Plates welded to the shell

Figure 2.97 shows an impingement plate welded to the shell. When I have seen this arrangement in other publications it has simply been a flat horizontal plate.

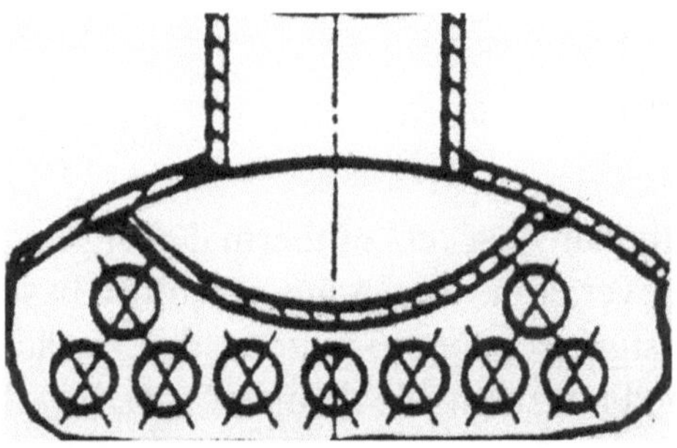

Figure 2.97

These could clearly only be used with fixed tube sheet items or perhaps BEU units with the entrance at the rear head end. Many clients' specs specifically ban their use and I would never use them.

'Wavy' Plate

This is a 'normal' impingement plate which is shaped to the profile of the tubes and is illustrated below in figure 2.98.

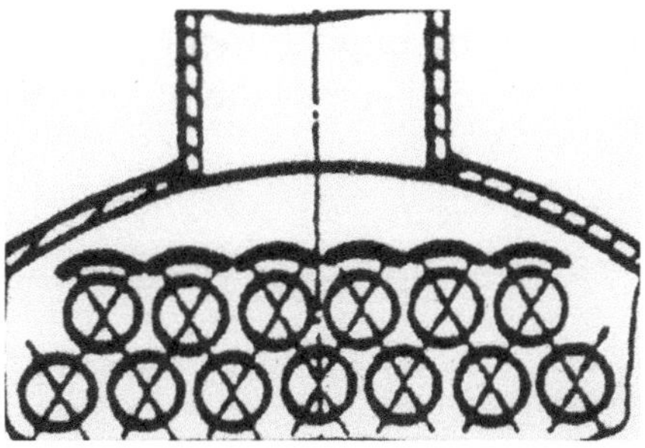

Figure 2.98

I can see advantages for this design, particularly the way in which the fluid is directed away from the outer tubes. However I am not convinced that the added complication is worthwhile and I have never seen it used in petrochemical work. It could be justified for small batch produced units such as oil coolers for diesel engines.

Impingement strips

Figure 2.99 shows an arrangement with a series of metal strips. A variation on this type of design is to have each tube covered by an inverted V shaped strip the same as the strip covering the tube on the centre line in the sketch.

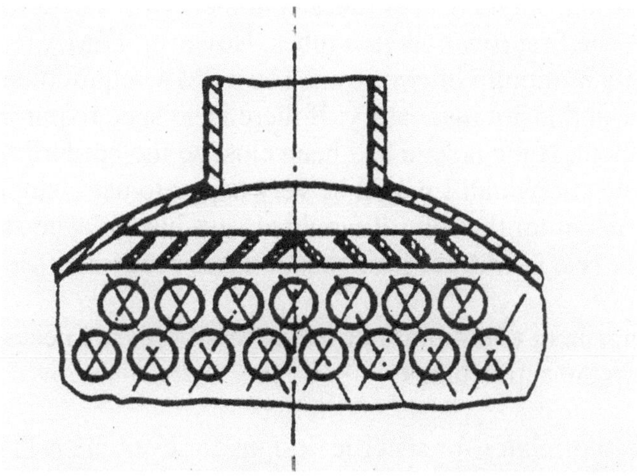

Figure 2.99

I can not see that this arrangement provides as good a protection as a conventional impingement plate. I have never used it and would not recommend it.

Impingement rods

This method utilizes bars or thick tubes, sealed at one end like dummy tubes, instead of a plate. The rods normally have the same diameter and pitch as the active tubes. This is illustrated in figure 2.100.

The top sketch in the diagram shows a single row of impingement rods for a bundle with an inline square pitch. The bottom sketch shows two rows of impingement rods for a 30° (as viewed), triangular pitch. For either arrangement the first row of rods adjacent to the nozzle are positioned in exactly the same way as the first row of tubes would be for an outlet nozzle or an inlet nozzle which did not require any impingement protection as discussed earlier. Impingement rods normally only extend across the inlet space.

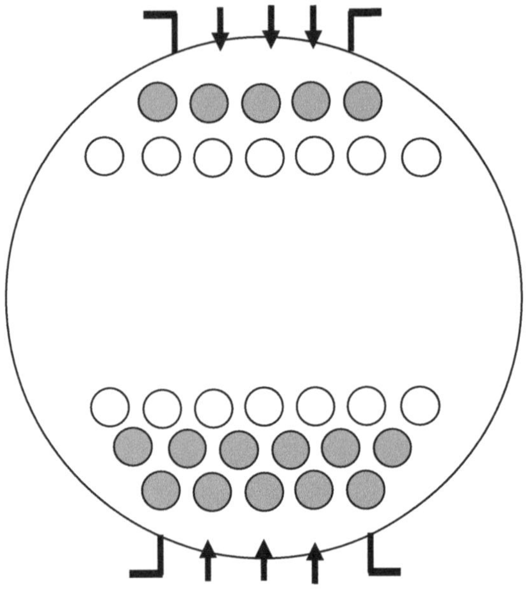

Figure 2.100

For the square pitch arrangement a single row of tubes is in theory sufficient to prevent direct impingement on any of the first row of active tubes. However TEMA RGP-RCB-4.625 and some specs require a minimum of two rows. The TEMA requirement is 'recommended good practice' and is not mandatory. If there is no spec requirement it is once again a matter of judgement. If the nozzle had been close to the borderline of requiring no impingement protection at all I would be very happy to use a single row. If the nozzle was close to its momentum flux limit I would always have played safe and used two rows. Now that there is a TEMA recommendation it is prudent to use two rows.

For any triangular pitch arrangement a minimum of two rows of rods are necessary to prevent direct impingement on the active tubes.

I have always used rods instead of plates for specific designs for example NTIW, (No tubes in the window), units with large shell nozzles where a plate would occupy a very high proportion of the inlet space. If these have a square pitch I will often use one row of rods across the full width and a second row across the central section extending to the same width as an impingement plate would have. This helps to even out the cross flow distribution and of course there is no possibility of any 'dead patches'. For the same effect with a triangular pitch I would use two rows across the full width and a third row across the central section. In either case it is not normally necessary to lose any active tubes.

Over the years I have found myself increasing the proportion of the units that I designed which utilised rods rather than plates. In recent years some clients have stated a preference for rods rather than plates. This is no doubt because of experience with tubes adjacent to impingement plates failing because of flow induced vibrations. These may be due to badly positioned or badly sized plates but I expect the trend to use rods rather than plates to continue.

BUNDLE ENTRANCE & EXIT FLOW AREAS

These are normally the most important flow path areas in any shell and tube heat exchanger. They are where the fluid is in contact with the active tubes for the first and last time. They are the areas where the tubes are most likely to suffer from flow induced vibration problems.

In section 10 of recommended good practice in TEMA there is now a page of text and a page of diagrams on this topic. These correctly refer to approximate and minimum entrance and exit areas. They are probably incorporated into most modern computer programs. At the time of writing this I do not have access to these to check this out – perhaps an inquisitive reader with time to spare between 'bookable work' might like to try. For now I would simply like to point out how the accuracy of the equations in TEMA can be improved.

To demonstrate this I have modified and reproduced an earlier diagram as figure 2.101. The top nozzle is now an inlet nozzle with an impingement plate and the bottom nozzle is the outlet nozzle.

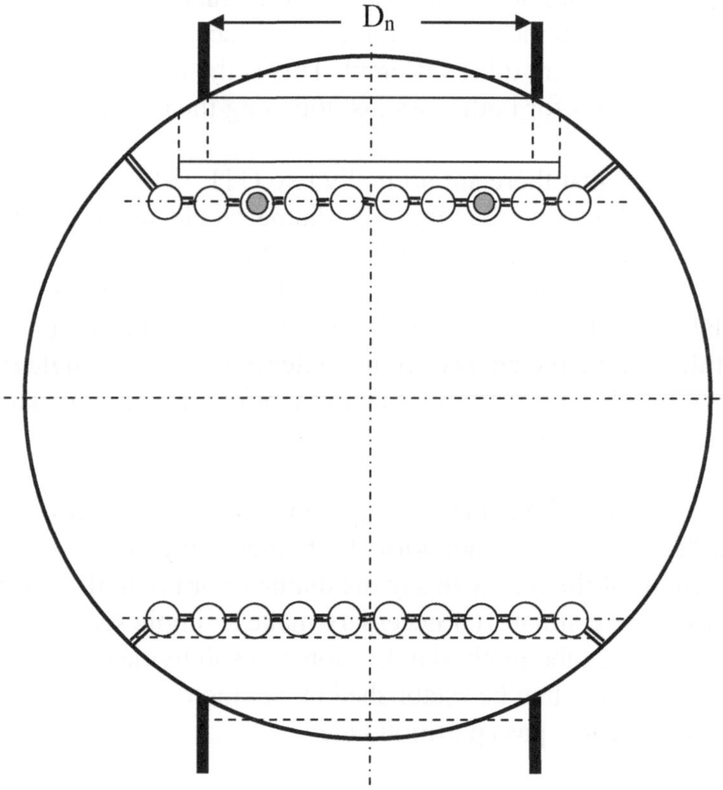

Figure 2.101

For the inlet nozzle the bundle entrance area would be found from the following equation:

$$\text{Entrance Area} = W_B \times L_{BE} - \text{Plate Area} + A_L \qquad [2.11]$$

Where

W_B = The width of the flow path across the bundle illustrated by the double line.
L_{BE} = The length of the space at the bundle entrance.
A_L = Any unrestricted longitudinal flow area.

For the outlet nozzle the bundle exit area would be found from the following equation:

$$\text{Exit Area} = W_B \times L_{BE} + A_L \qquad [2.12]$$

Where

W_B = The width of the flow path across the bundle illustrated by the double line.
L_{BE} = The length of the space at the bundle exit.
A_L = Any unrestricted longitudinal flow area.

I think it is easier to understand the calculations by using two equations instead of one combined equation as TEMA does. It is also easier to remember two simple equations rather than one complicated one and if you have a good mental picture of what is going on the equations are obvious.

The calculation for the plate area clearly depends on the shape of the plate.

The first term in the TEMA equation includes (D_s – OTL); which is the sum of the two clearances between the outer tubes of the row adjacent to the nozzles and the shell. This is the same for either an inlet or an outlet nozzle. The diagrams in TEMA have been conveniently drawn so that the outer tubes do just touch the OTL which make this correct. However a glance at figure 2.100 indicates that the clearances between the shell and the outer tubes, for the two nozzles, are quite different. This is the normal case. It would be very unusual to have the tubes for both rows just touching the OTL.

In an earlier section it was explained that there are normally two OTL values. The maximum OTL value is fixed by subtracting the minimum diametral clearance for the type of bundle being used from the shell inside diameter. The actual OTL is then calculated from the actual tube layout diagram. The difference between the two is normally insignificant for a large bundle but is not always insignificant for a small bundle. Whatever the size of the bundle the chances of the outer tubes of any single row just touching the OTL are very small. These clearances will therefore normally be larger than those calculated from the TEMA formula.

If you have a tube layout in front of you the geometry required to accurately calculate these clearances is simple but tedious. I would not normally bother doing any calculations. Comparing the length of the line with say the diameter or pitch of the tubes is good enough to make a reasonable estimate. (If necessary make a simple scale rule on the edge of a piece of paper using one tube pitch as a division and subdivide one division by eye.) The width of the flow path can then be established by adding on the clearances between the tubes and across any vertical pass partition lanes.

The unrestricted longitudinal flow area depends on the type of bundle. The TEMA diagrams are for a fixed tubesheet exchanger so that there is clearly no longitudinal flow in the direction towards the rear head. For U tube and floating head units there must be some flow past the support plate. How much flow there will be is difficult to calculate. If I ever found it necessary to allow for flow across this area I would work out the area and then make an estimate of what proportion of this area it was reasonable to use. In the other longitudinal direction – past the first baffle towards the fixed head end of the exchanger there is a small area which could be fully counted which is illustrated in figure

2.102. This area is approximately equal to the length of the arc above the bundle multiplied by half the baffle/shell diametral clearance.

Figure 2.102

In practice I would normally ignore the longitudinal flow areas altogether. This provides some safety factor.

Once a value for the bundle inlet or exit area has been established the velocity and momentum flux can be calculated by comparing them with the values for the appropriate nozzle as follows:

$$\text{Bundle velocity} = \text{nozzle velocity} \times \frac{\text{bundle area}}{\text{nozzle area}} \qquad [2.13]$$

$$\text{Bundle momentum flux} = \text{nozzle } \rho u^2 \times \frac{(\text{bundle area})^2}{(\text{nozzle area})^2} \qquad [2.14]$$

The bundle entrance or exit velocities and momentum fluxes calculated in this way will unfortunately not be homogeneous values. More fluid will flow into paths with low hydraulic resistance than paths with relatively higher hydraulic resistance. For example in figure 2.100 more fluid will flow through the spaces between the outer tubes and the shell than into or out of the main bank of tubes. Once again judgement has to be used. If you are close to a limit assuming homogeneous values then there will be local areas where the limit will be exceeded and the design may need to be modified accordingly.

TUBE COUNT PROCEDURE

A hand calculation method of obtaining a tube count is more than just something of historical interest because it also increases ones understanding of a tube layout. The procedure is to establish the area available for tubes and then divide this are by a factor which depends on the tube pitch and layout.

Figure 2.103 shows a simple bundle with two tube side passes. The areas inside the safe circle diameter, SCD, which are unavailable for tube centres are show shaded. The safe circle diameter was defined earlier as the circle within which the centres of all the tubes must lie. The area available for tubes is the safe circle area, SCA, minus these areas.

Tubed area = SCA – Inlet & outlet areas lost – pass partition lane losses. [2.15]

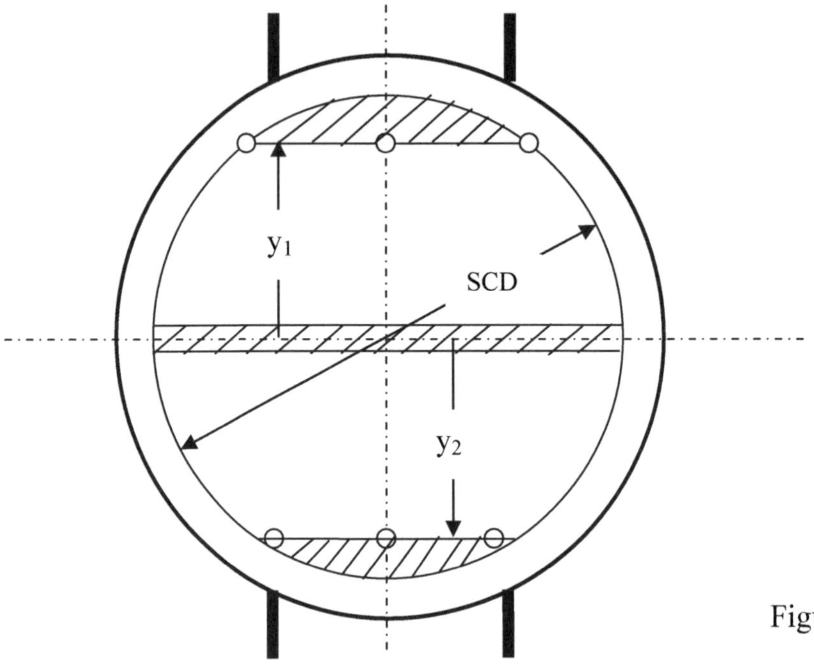

Figure 2.103

The inlet and outlet areas unavailable for tubes are the segments bounded by the SCD and a line drawn through the centres of the outer tube rows. The geometric calculations necessary to establish the positions of the outer tube rows were explained earlier in this chapter. The height of the segments in figure 2.103 can therefore be calculated as follows:

$$h_{inlet} = (D_s/2 - y_1) - (D_s - SCD)/2 \qquad [2.16]$$

$$h_{outlet} = (D_s/2 - y_2) - (D_s - SCD)/2 \qquad [2.17]$$

The areas of the segments can then be calculated exactly but it is much easier to use a table of chord lengths & areas of circular segments such as the one in TEMA – table D-14 in section 9.

The width of pass partition lanes unavailable for tubes was covered earlier in this chapter with examples 2.1 to 2.5 demonstrating how they can be calculated. In figure 2.103 the length of the partition lane is the SCD. The area lost is almost exactly the width multiplied by the SCD.

In more complex pass arrangements there may be both vertical and horizontal pass lanes which may be of different widths. (e.g. a 4 pass U tube bundle.). If the length of a lane is not along a diameter it is sufficiently accurate to estimate what proportion of the SCD the length of a lane is.

Having calculated the area available for tubes the number of tubes can be found from the following equation:

$$\text{Number of tubes} = \frac{\text{Area available for tubes}}{FP_t^2} - \text{Tie rods} \qquad [2.18]$$

Where F is 1 for any square pitch and 0.866 (Cos 30°) for any triangular pitch, P_t is the tube pitch. The reader should satisfy himself/herself that the term FP_t^2 is the area occupied by one tube. The minimum number of tie rods can be found from table R-4.71 in TEMA. These are minimum values and it is prudent to add a few extra.

This method is very accurate for all but the very smallest pipe size shells. For many years the HTRI programs gave very poor tube counts and I often used to do a hand calculation and input this into the computer program. Some companies developed their own in house programs. The first commercial tube count program which gave good results was the OPTU program by HTFS.

Modern programs not only give good results but will produce the tube layout for you to look at. Do not forget to look at any program option of the vertical axis passing through or between vertical tube rows, it can make a difference to the tube count. These programs do not go through the calculations shown above. They plot out each tube row as a draughtsman used to do. They still have minor problems of course and it can still be challenging to get exactly the layout you require for a particular design.

MISCALLANEOUS ITEMS

SEALING STRIPS

Sealing strips are used to minimise the amount of fluid bypassing the tubes in the bundle. Figure 2.104 show a sealing strip adjacent to the edge of a vertical baffle. The clearance between the sealing strip and the tube should ideally be about half the clearance between the tubes and never more than the clearance between the tubes.

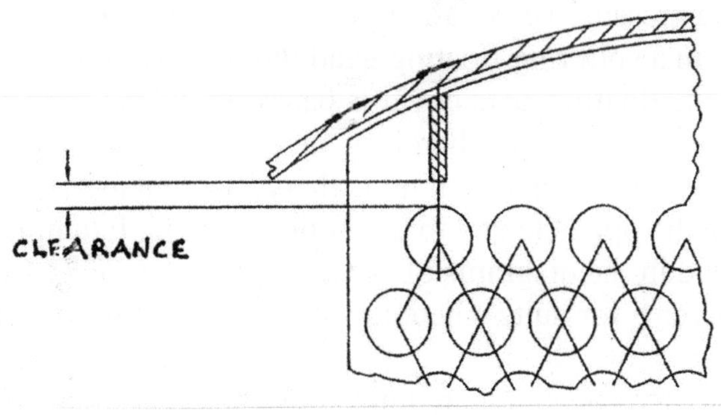

Figure 2.104

Sealing strips should never block cleaning lanes in square pitch arrangements.

If the number of pairs of sealing strips is not input into modern computer programs there is often a default setting and it is easy to be lazy and rely on this. The default may be two pair – one near the edge of each baffle (as illustrated in figure 2.103) or one pair every four or five rows. I would recommend that the designer should take into account the bundle/shell clearance for horizontal baffles and the heights between the shell and the top row of tubes for vertically cut baffles. Sealing strips are used to reduce the C stream bypassing. A glance at the first computer run will show how significant this is. Experimenting with the number of sealing strips will then show if the design can be improved by adding more. Particularly with cross flow and NTIW units adding more sealing strips can significantly improve the design and I have often used a pair of sealing strips for every second tube row.

For horizontally cut baffles it is important to extend the sealing strips for the full tube length between the tubesheets or tubesheet and support plate. For vertical baffles they should only extend as far as the end baffles so that they do not interfere with the flow into or out of the bundle.

For units such as condensers which will normally have a large inlet nozzle at the top and a relatively small one at the bottom there may be no need of a bottom strip to 'make up the pair'. Any bundle runner will give a seal at the bottom and there will be a liquid level which will reduce vapour by passing. Adding very short sealing strips at the bottom is pointless, (although the program may assume that they are present).

VENTS & DRAINS

These miscellaneous items apply to any pressure vessel on a plant and there will normally be a standard for the site. A vent is necessary whenever there is a space in a vessel which is above the highest nozzle to allow air to be expelled particularly at start up. A drain is necessary in any vessel whenever there is an area below the lowest nozzle. Vents can vary from ¾ inch for small vessels up to 1½ inch for the largest. Drains are liable to blockage and are therefore normally larger the vents and range from a minimum of 1 inch to up to 3 inch for very large vessels. The only guidance in TEMA is that they should be 'a minimum of ¾ inch nominal pipe bore'. Some vents and drains may need to be connected to pipework so that they do not discharge to the ambient air. This is a process consideration.

A shell and tube exchanger is a pressure vessel and must have vents and drains as explained above. For example in a split ring floating head unit the area enclosed between the rear head shell flange and the floating head cover is partly above the shell and partly below the shell. The floating head cover must therefore have a vent and drain as illustrated in figure 2.1. For most other horizontal units there is a vent at the opposite end to the top nozzle and a drain at the opposite end to the bottom nozzle. Examples of where they are positioned can be found in the diagrams of the first few pages of TEMA. It is often prudent to provide extra vents for units where the shellside fluid is condensing.

Adequate venting and draining of vertical units requires more thought. Some methods of venting the shellside are shown in figure 2.105.

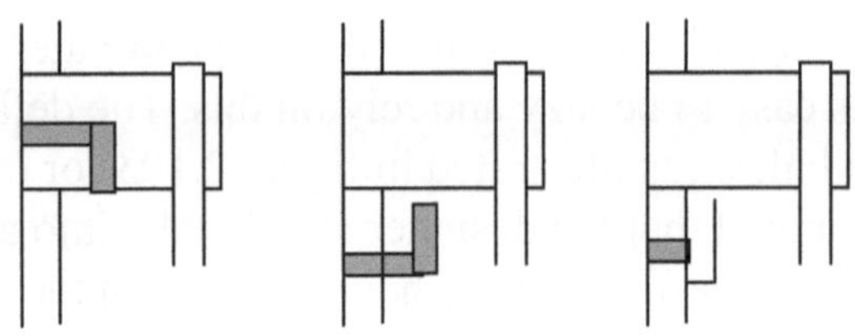

(a) (b) (c)

Figure 2.105

Sketch (a) shows the most satisfactory way of venting but can only be used with relatively thick tubesheets. The shaded section shows the venting path provided by drilling two holes in the tubesheet.

Sketch (b) shows the venting path provided by a pipe elbow welded to the inside of the shell.

Sketch (c) shows an alternative to (b). A 'conventional' vent hole is surrounded by metal except at the very top adjacent to the tubesheet.

When the shellside fluid is condensing in a vertical unit there will always be some liquid level at the bottom. This may need to be taken into account in the thermal design (fudge the input with a very thick tubesheet). It also raises the interesting question of venting. Venting is normally only associated with clearing air or inerts from the top of a space. However if inerts are cooled they may fall to the bottom of a unit to lie above the liquid level of a condenser. Just in case this takes place I would always recommend fitting an extra vent above the liquid level in large vertical units. (Say 150 mm above the top of the outlet nozzle).

The only 100% way of draining a vertical unit is through the tubesheet which would have a flow path as shown in sketch (a) of figure 2.106. Types (b) and (c) in figure 2.106 will provide extra drainage compared with a simple drain fitted to the shell by siphoning out some of the liquid but will never completely drain the fluid.

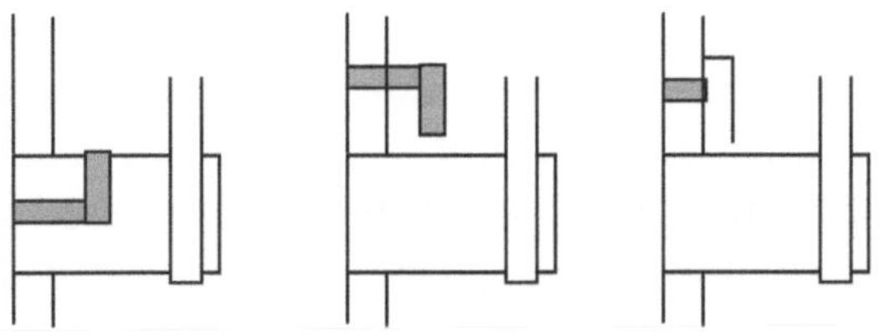

(a) (b) (c) Figure 2.106

If you have any doubt about what type of vent or drain is practical for a vertical unit discuss it with a mechanical engineer.

It should not be necessary to specify vents and drains on the TDS for a standard design. It is of course essential to specify any extra items and any none standard items such as those shown in figures 2.105 and 2.106. They must also be on the check list for the thermal designer's check of any vendor's drawings.

BUNDLE RUNNERS / SLIDING STRIPS / SKID BARS

I can not find any mention of these in TEMA and yet they are fitted to most shell and tube exchangers with removable bundles and may even be used in large fixed tubesheet items to assist in assembly.

ISO 26812 is more forthcoming and section 7.6.6 states that all removable bundles with a mass of more than 5450 kg (12000 lb) should have continuous sliding surfaces to facilitate bundle removal. If skid bars are used:

(a) A minimum of two skid bars shall be provided.
(b) They shall not be situated in the projected area of the nozzles.
(c) They shall be welded to the transverse baffles and support plates.
(d) They shall protrude 0.8 mm (1/32 inch) beyond the baffle/support OD.
(e) Their leading and trailing edges shall be provided with a radius or a bevel to prevent damage to the shell when inserting or removing the bundle.

Section A.2.4 provides further guidance to section 7.6.6 by suggesting that alternatives to skid bars should be considered if the bundles are too large for conventional pulling devices. Examples are where the bundle mass exceeds 18150 kg (40000 lb), the diameter exceeds 1220 mm (48 ins) or the length exceeds 7.3 m (24 ft). Alternatives for consideration are listed as:

(a) Bundle rollers
(b) Skid bars on a rail
(c) Removable shell.

It is interesting to note that ISO correctly refers to the kg as a unit of mass not weight and that precedence is given to the SI units over the US customary units. It is also interesting that runners are only considered essential for bundles with a mass of over 5450 kg. They are normally used on all but the smallest pipe sized shells. The protrusion of 0.8 mm should I believe be taken to be a minimum value since I have worked on jobs with up to 2 mm in the specs. The further guidance section strangely lists a diameter limit of 1220 mm. This is regularly exceeded without any problems. The length and mass limits are the important ones particularly the latter. Removable shells are covered in the section on high pressure units.

The function of bundle runners is to make it easier to insert and withdraw removable bundles to and from the shell. For horizontal units there are normally two. They are usually strips of metal although round bars may also be used. Both are illustrated in figure 2.107.

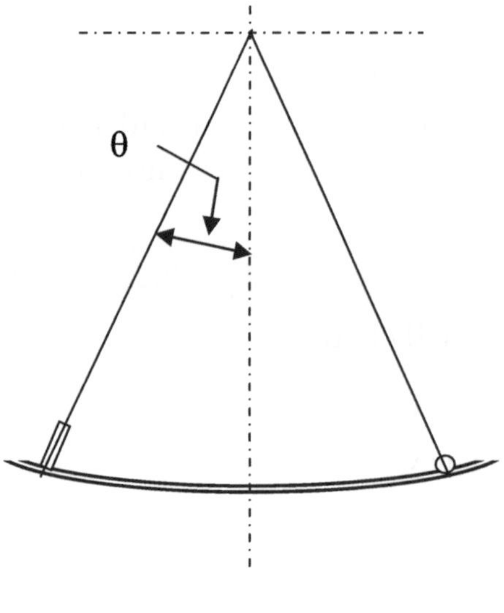

Figure 2.107

Bundle runners are positioned at an angle θ to the vertical centre line of the exchanger. The ideal angle for θ is between 25° and 30° and I would suggest a minimum of 20° and a maximum of 35° although some specifications allow up to 40° maximum.

Bundle runners are only necessary between the first and last baffle/support plate. They should not restrict the flow into or out of the bundle nor obstruct the cleaning lanes on square pitch arrangements.

Some specs may give details of the dimensions of rectangular runners. A typical example is shown in table 2.10 below.

Shell D_s mm	Height mm	Thickness mm
150 → 390	30	6.5
400 → 690	40	10
700 → 890	50	13
900 → 1290	60	13
Above 1300	70	16

Table 2.10

Such detail is unusual in a spec and it is normally up to the thermal designer to decide on the size of the bundle runners. If the bottom shell nozzle is relatively small it may be necessary to take out several tubes to comply with table 2.10. For small shells – up to say 400 mm some companies do not regard runners as being necessary. For very deep runners it is worthwhile considering providing some drain notches at the mid points between the baffles.

I have worked several times with a good mechanical engineer who had spent many years working for a vendor. We once discussed the issue of bundle runners and he said that the company he had worked for had always used cylindrical bars. 19.05 mm diameter bars were satisfactory for shells of up to 1270 mm and above that 25.4 mm. He had been responsible for inserting many bundles without any problems. I pointed out how flexible such small bars would be compared with the much stiffer runners in table 2.10 for the larger shells. He countered by suggesting I work out the contribution of hefty runners to the second moment of area of the whole bundle. I did not even bother putting finger to calculator to realise that he was right and the effect would be negligible.

The only purpose of bundle runners is to prevent the edges of the baffle catching on the edge of the entrance to the shell when the bundle is being moved. Without runners the baffles could be severely damaged. Provided the runners are firmly welded to the baffles and remain in place they will fulfil that function. It still 'feels right' to use larger runners for larger shells. However unless a spec includes a table similar to table 2.10 there is no good reason to over engineer bundle runners. I rely on the old adage that 'if it looks right it probably is right'. I look at the tube layout and put in runners which look right. One situation where hefty runners 'look right' is in cross flow units and NTIW units with very large baffle pitches. Special consideration should be given to such units. Once again this is essentially a mechanical not a thermal problem and it should be discussed with a mechanical engineer.

The orientation of rectangular runners is normally along a radial line as shown in figure 2.107. This is ideal because the edge of the runner is parallel to the edge of the baffle. There will be cases where this is not convenient particularly if trying to fit in large runners to comply with a spec where the angle will inevitably infringe on cleaning lanes with square pitch bundles. I can not see any reason for not using vertical runners provided the edge which protrudes from the baffles is rounded to avoid a sharp edge contact with the shell as required in ISO 16812.

The above text mainly refers to horizontal shells which are the most common. Vertical shells with removable bundles also need runners. With vertical shells there is no clear area of contact as there was with the bottom of the shell for horizontal units. It is therefore necessary to fit four runners ideally spaced at about 90° around the circumference. I would always use cylindrical bars in this situation since they would have a negligible effect on the thermal design. (Make out a special case if you are using an onerous spec.)

Any bundle runner will also act as a sealing strip which should be taken into account in the thermal design. They may also be counted as tie rods since they clearly support the edges of the baffles.

TUBE LAYOUT DIAGRAMS

Tube layout diagrams were once produced on a drawing board by a mechanical engineer and checked by the thermal design engineer to make sure that they satisfied the thermal requirements. There are now many computer programs which can do the drawing and tube layouts should now be an integral part of any TDS (thermal data sheet) and are the sole responsibility of the thermal design engineer.

The details used in the construction of tube layout diagrams have already been covered. The following checklist is a summary of what to look out for.

(1) The OTL should be shown either on the diagram or in an adjacent table.
(2) The position of tube rows adjacent to nozzles.
(3) The position, size and method of attachment or any impingement device.
(4) The width of any pass partition lanes. Are cleaning lanes unobstructed for square pitch arrangements? Are U bends satisfactory? Have welded tubes been catered for? Does the lane require sealing to minimise by passing? If it does is the sealing arrangement adequate to restrict the F stream to the value predicted by the thermal design run?
(5) For multi pass exchangers are the tubes per pass in reasonable agreement? For most designs they should ideally be the same which is seldom possible. The deviation from the mean value should not normally exceed 10%. There are exceptions such as example 2.18
(6) Have sufficient tie rods been used? Are they correctly positioned to pick up the edges of the baffles? If in doubt add more and remember that tie rods have a smaller diameter than the tubes so that they may be placed where a tube could not.
(7) Are sealing strips required? If they are required are they in the right place? They should be placed in the widest gaps around the bundle not necessarily in a row

next to a partition lane. Looking at the diagram do you think it will produce a C stream compatible with the one on the thermal design run?

(8) Are bundle runners required? If they are, are they in a sensible position?

(9) Have drain notches been provided at the bottom of the support plate and baffles? The normal drain notch is 10 mm and is required by ISO16812, larger notches may be required for some designs.

(10) Do baffle cuts pass through the centre line of a row of tubes or in between rows so that there are no sharp edges?

(11) Does the final tube count agree with your final computer run? This question is not as daft as it may appear. You may have replaced some tubes with tie rods or eliminated some tubes around an inlet or outlet area although it is unlikely to affect the overall design it is embarrassing to have it pointed out to you that the number of tubes you have specified on the TDS is not the same as on the tube layout.

There are now a series of examples of actual tube layout diagrams. These have been taken from a variety of sources produced by different methods and programs. Some examples will contain all the information needed; others include only sufficient information to illustrate points in a particular design.

Example 2.12

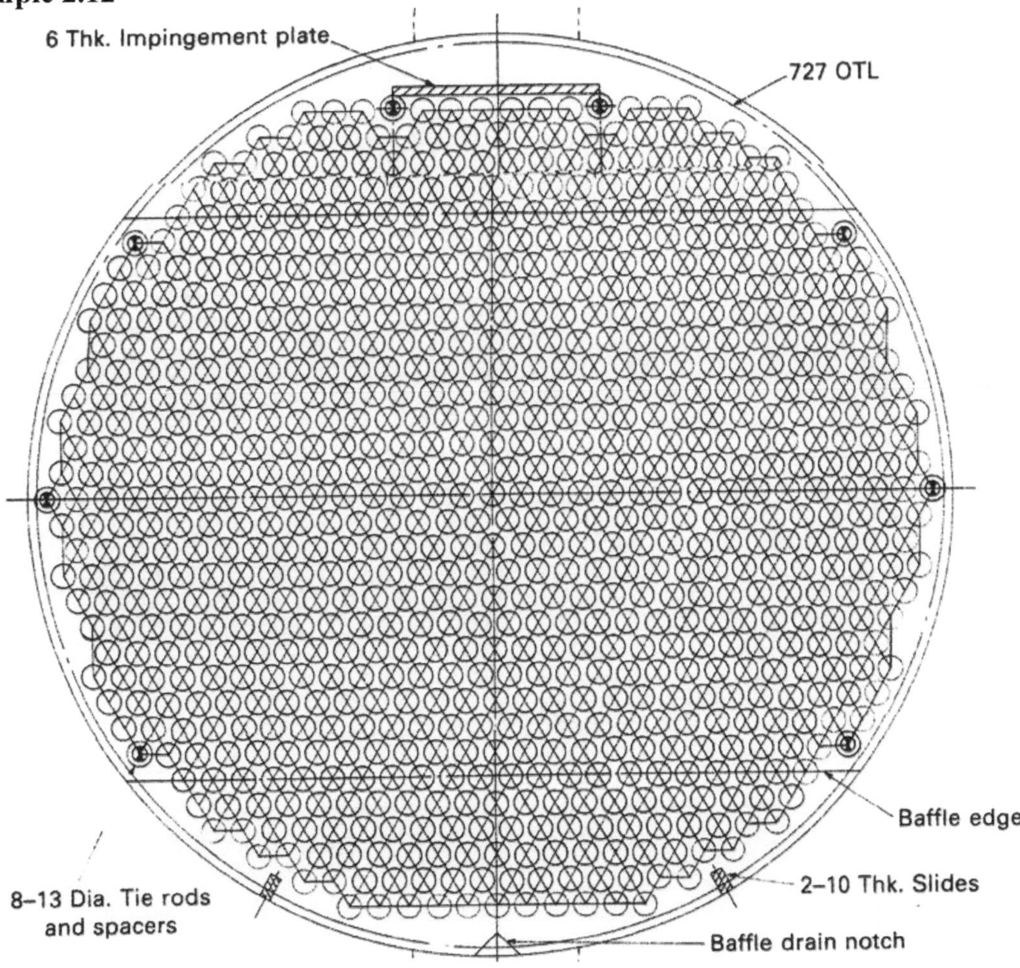

Figure 2.108

This is a very simple layout for a 740 mm shell, fixed tubesheet exchanger with horizontal baffles. There are 780, 19.05 mm tubes on a 23.81 mm triangular pitch. The diametral clearance between the tubes and shell is only 13 mm. It was not therefore considered necessary to add any sealing strips to avoid C stream bypassing.
The size of the drain notch is not indicated but it would be assumed to be the normal 10 mm.

There is clearly adequate clearance under the nozzles and the impingement plate which would be square judging by the supports. The size of the impingement plate should either be on the diagram or mentioned elsewhere.

Despite being a fixed tubesheet item small bundle runners have been used for ease of assembly. Note that they are simply termed slides which are yet another description to add to bundle runner, sliding strip and skid bar.

Example 2.13

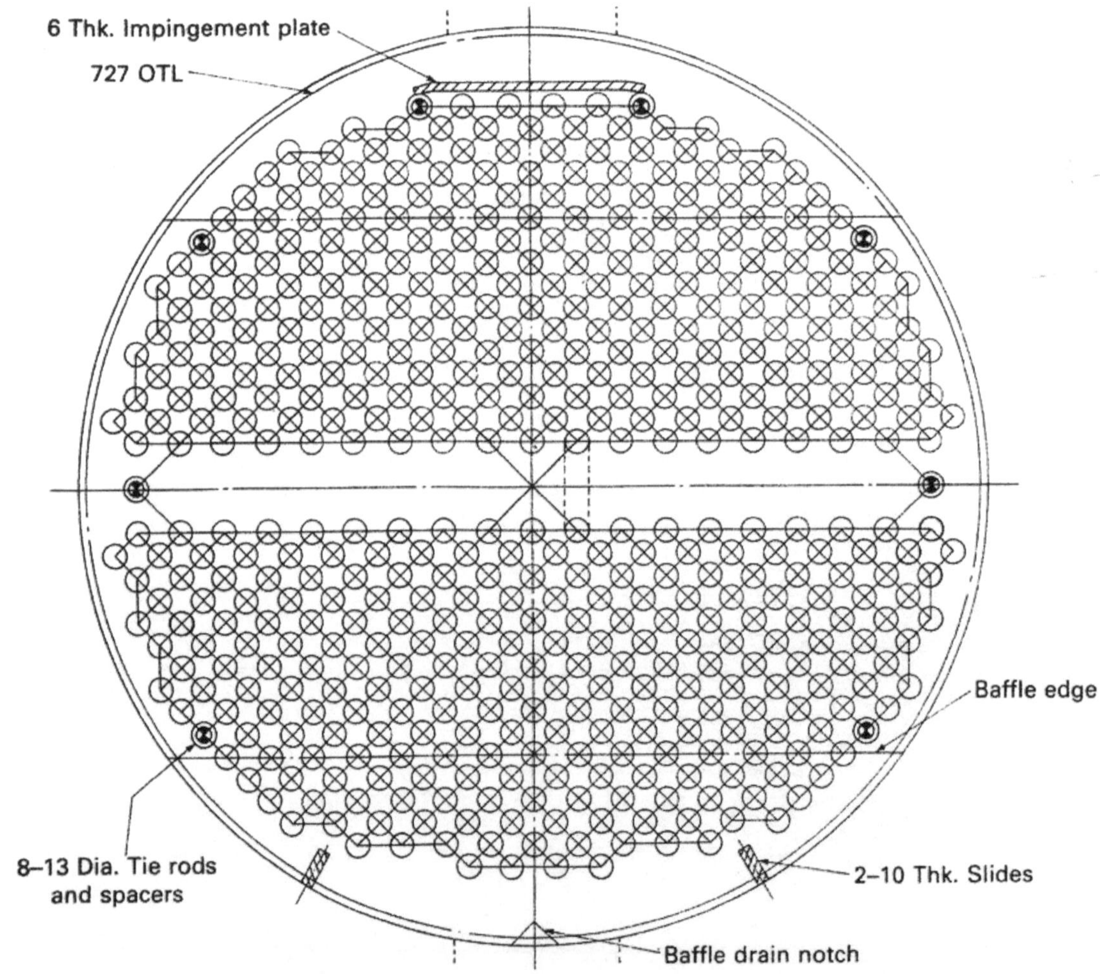

Figure 2.109

The shell size, OTL and baffle arrangement are the same as in the previous example. However this time there are 19.05 mm U tubes on a 25.4 mm rotated square pitch. The tube to shell diametral clearance is the same so there are no sealing strips.

The shell side flow passes across the pass partition lane so there is no need for any sealing. The bundle runners are at about 30° to the vertical which is ideal.

It is important to maintain cleaning lanes in between the tubes and the cross at the centre of the bundle confirms that this has been done. Note the tie rods along the shell horizontal centre line. They have also been positioned so that they do not restrict the cleaning lanes.

Example 2.14

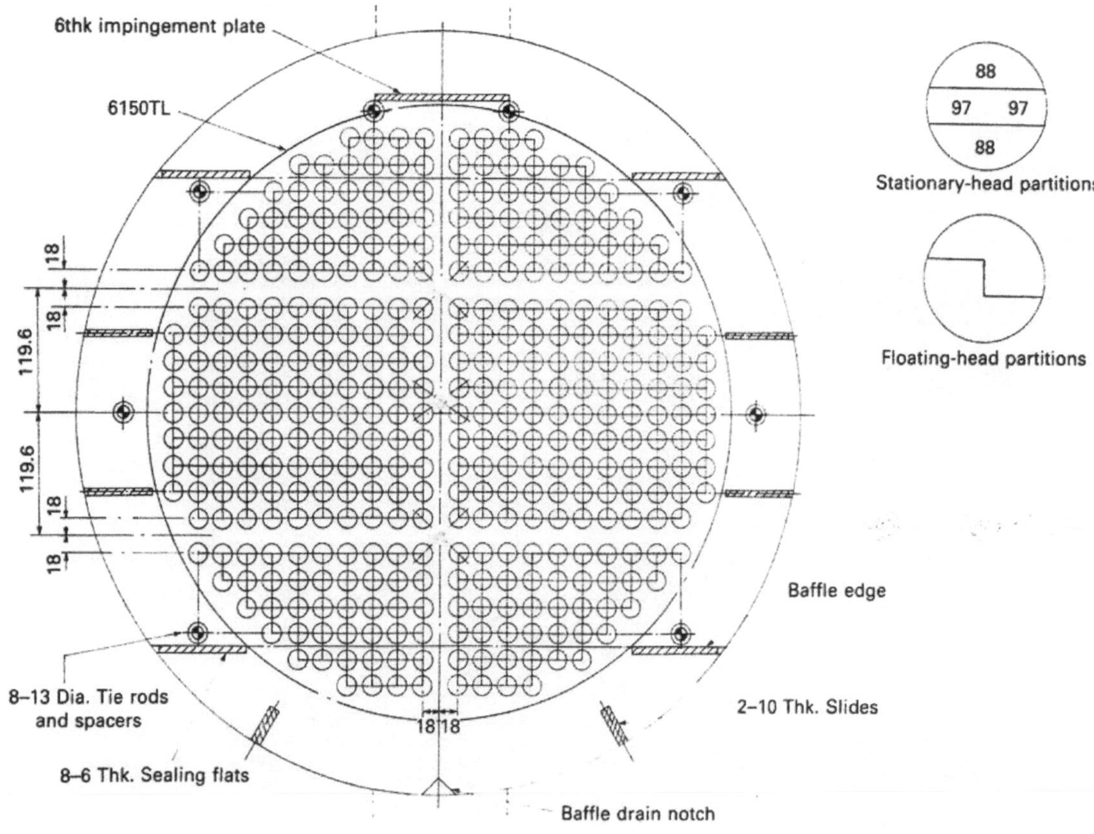

Figure 2.110

The shell size is the same as for the previous examples and the baffles are still horizontal but this is now a pull through floating head unit with 19.05 mm tubes on a 25.4 mm in line square pitch.

The diametral clearance between the shell and outer tubes is now a minimum of (740 – 615) = 135 mm. The actual clearance is a bit more because none to the tubes quite touch the OTL. This creates an enormous C stream by pass potential and the sealing strips are very wide to reduce this. Looking at this diagram I would probably have moved the outer sealing strips to be opposite the third tube rows from the horizontal pass partition lanes and add an extra pair of strips along the shell horizontal centre line. Note that all the tie rods are outside the OTL which is normal practice for this type of construction.

Because of the relatively small gap (compared with the flow path across the tubes and around the C stream) in the vertical pass partition lane it was not considered necessary to provide any sealing.
The aside sketch on the right hand side shows the tubes per pass. The average tubes per pass is 92.5 which results in a maximum deviation of under 5% which is quite acceptable.

Example 2.15

Shell I.D.	1350.00	(mm)
OTL	1331.00	(mm)
Tube O.D.	19.05	(mm)
Tube Pitch	24.00	(mm)
Pass Part. Vert.	19.00	(mm)
Pass Part. Horiz.	19.00	(mm)
Tube Layout	30.0	degrees
Alignment	Semi	
Baffle Cut	34.1	%
Inlet Nozzle I.D.	294.00	(mm)
Outlet Nozzle I.D.	294.00	(mm)
Number of Passes	3	
Total Tubes	2508	
Tie Rods/Spacers	12	

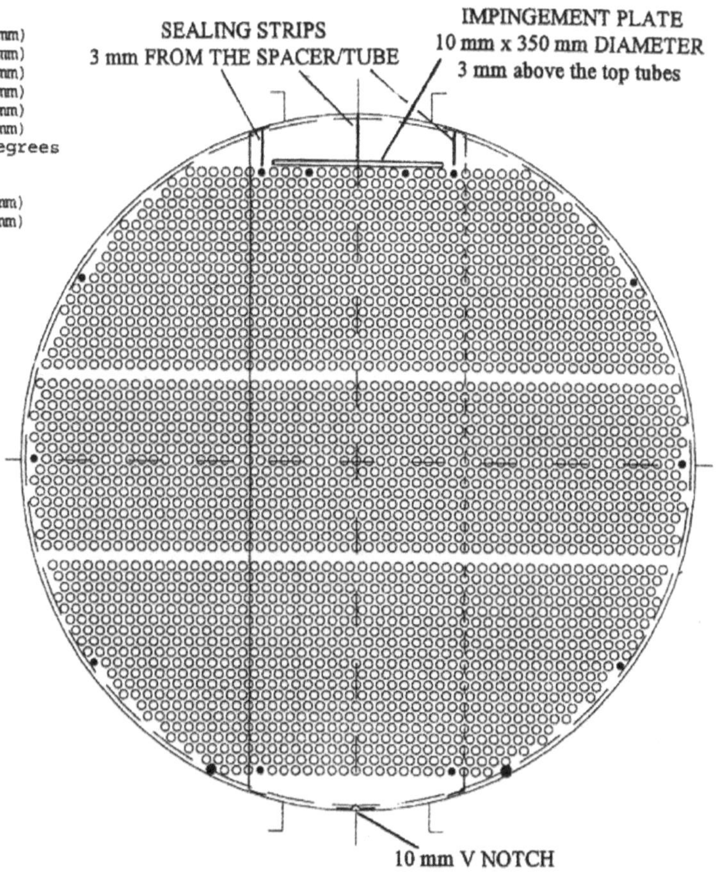

Figure 2.111

This example is a propylene trim condenser which I designed many years ago. It is a three tube side pass, fixed tubesheet exchanger with vertically cut baffles. The baffle cut has been slightly modified by hand to ensure that the cut is either on a tube centre line or in between tubes.

Three sealing strips have been used at the top of the bundle to avoid by passing but there are none at the bottom of the exchanger where the liquid level would provide a seal.

The impingement plate is round and fully specified. The tie rods supporting the plate are close to the ideal position. The shell inside diameter, OTL, etc are specified in the table on the top left hand side. I will leave it to the reader to check that the distribution of tubes per pass is satisfactory.

Example 2.16

Run Number:		
Shell I.D.	1067.00	(mm)
OTL	1047.00	(mm)
Tube O.D.	19.05	(mm)
Tube Pitch	25.40	(mm)
Pass Part. Vert.	15.88	(mm)
Pass Part. Horiz.	40.00	(mm)
Tube Layout	90.0	(deg.)
Alignment	U-Tube	
Baffle Cut	0.00	(%)
Inlet Nozzle I.D.	363.00	(mm)
Outlet Nozzle I.D.	478.00	(mm)
Number of Passes	2	
Total Tubes	1064	
Tubes Removed	6	
Tubes Remaining	1058	
Tie Rods/Spacers	10	

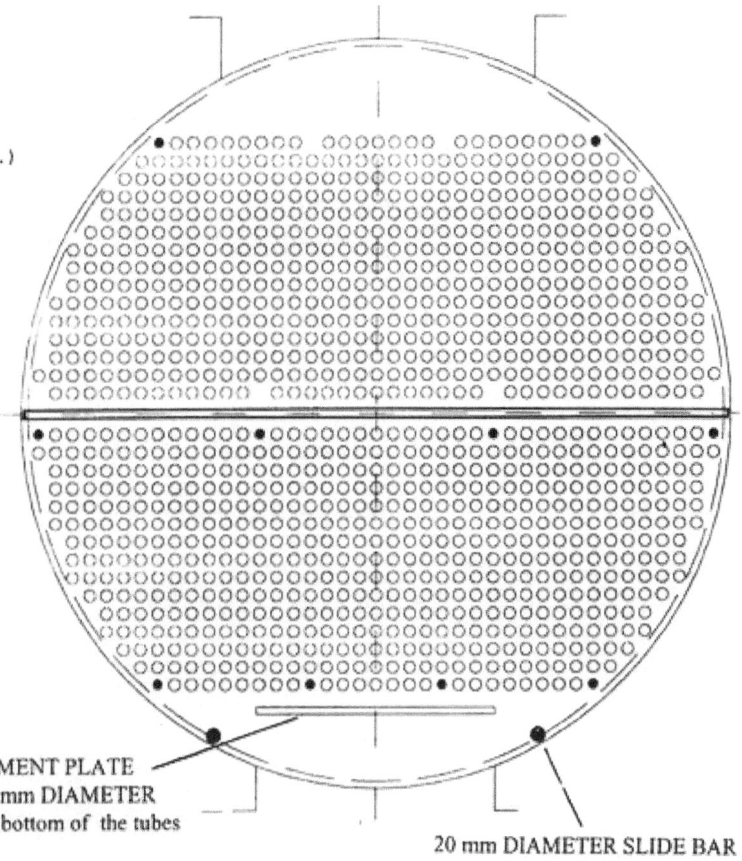

BUNDLE DETAILS

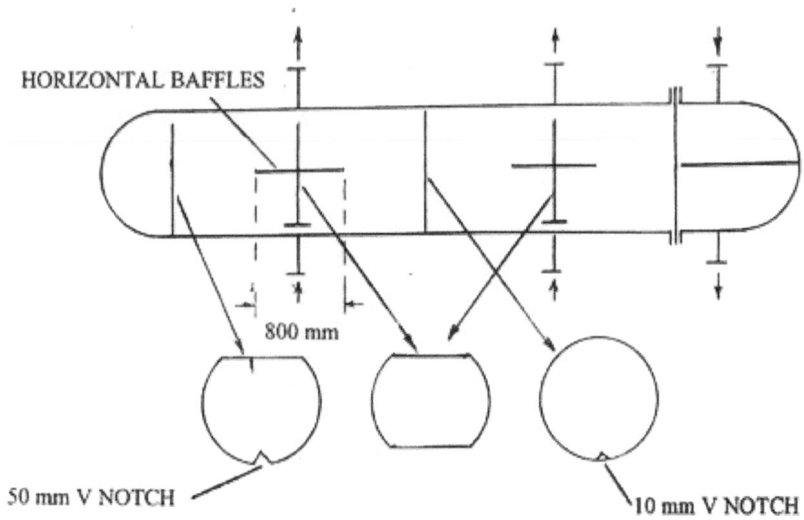

Figure 2.112

This unit is a BEU horizontal thermosyphon reboiler. A sketch of the baffles has been included to help explain some of the features of the tube layout. The horizontal baffle shown in the layout is to help circulate the boiling fluid around the full length of compartments. This example is typical of many reboilers where it is debatable whether horizontal baffles are really necessary. I have come to the conclusion that they do not cost very much, there is no loss of tubes because they fit comfortably into the U bend pass partition lane and they may help so I always use them. The impingement plates are fully specified but are clearly not to scale with respect to the distance below the first row of tubes. This is acceptable.

No notches have been indicated on the tube layout because they are all different as shown in the sketch below. The central baffle is the only one with a normal 10 mm notch. (Perhaps this baffle should also have had the top cut off?) The support plate has a much larger notch to provide an easy path for liquid to reach the U bends. With this design of support plate the bend area should be fully effective. Whether you are allowed to count this area in the design depends on the spec you are working to. If you are not allowed to count this area you can confidently push the oversurface to the absolute minimum.

Example 2.17

Row	Tubes
1	10
2	14
3	15
4	16
5	17
6	18
7	19
8	20
9	21
10	8
11	8
12	21
13	20
14	19
15	18
16	17
17	16
18	15
19	14
20	10

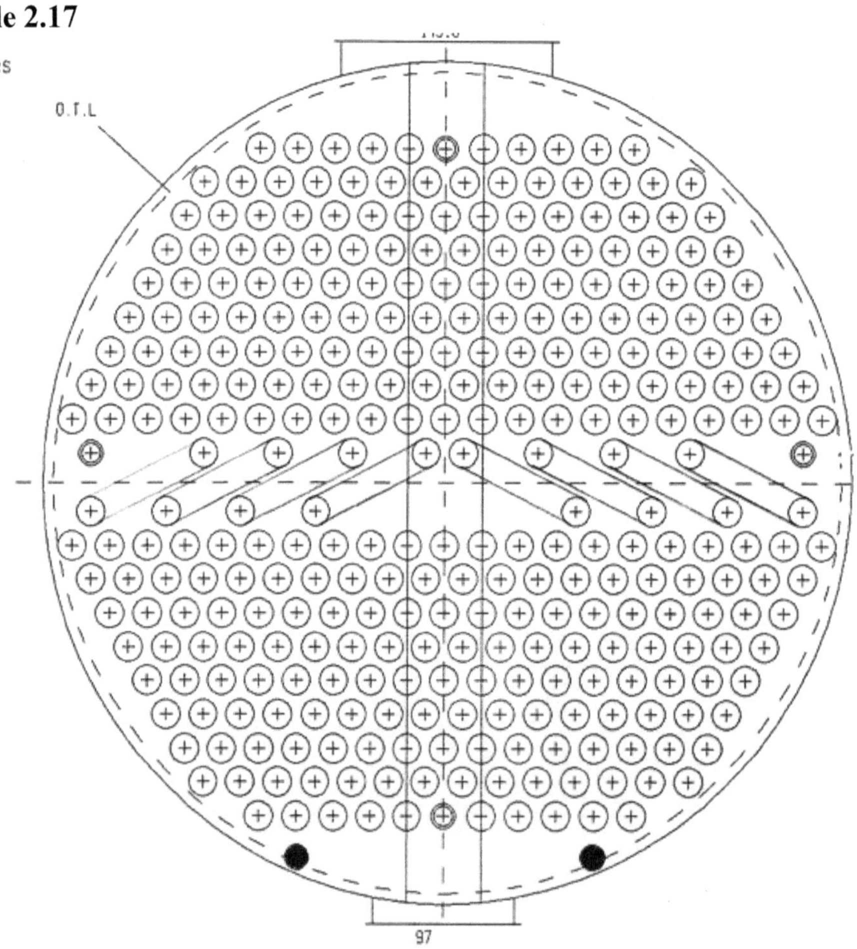

TEMA type	DEU
Shell Diameter	547.7
Outer tube limit	533.6411
Top clearance to tube	47.3
Bottom clear' to tube	47.3
Tube Diameter	19.05
Tube pitch	25.4
Flow arrangement	60
Number of tubes	316
Number of tube passes	2
Horizontal pass lane	38.1
Baffle cut % diameter	45.36

TUBE PASS DETAILS

Pass	No.Rows	Tubes
1	10	158
2	10	158

Figure 2.113

The tube layout program used to draw this diagram, (figure 2.112), shows the number of tubes in each row as well as the details shown in previous programs. The unit is a DEU which has high pressure on the tubeside and a relatively low pressure on the shell side. For an economic design it was therefore necessary to keep the shell diameter to a minimum to minimise the diameter of the expensive channel. The tube count has been maximised by fitting inclined tubes into the space created by the bends of the other tubes in the bundle.

No impingement protection is shown in this diagram because the inlet nozzle was beyond the U bend. The shellside fluid is condensing steam so the baffles are at their maximum cut and there is no need for sealing strips or sealing in the pass partition lane. The U bend area was counted as being fully effective.

Example 2.17

Figure 2.114 is an overhead cooler with a large volume flow rate of vapour with some condensation. The inlet and outlet nozzle being the same diameter is an indication that the amount of condensation is relatively small. The design is a NTIW which was necessary to give a low pressure drop without any potential vibration problems other than acoustic ones which could not be avoided. (There were several intermediate supports in each cross space to keep well clear of any fluid elastic instability).

This is an example of an exception to the 'rule' that any unit with condensation on the shell side should have vertically cut baffles. I did of course investigate the possibility of using vertically cut baffles but the designs were horrendous and I knew before I started that I would finish up using horizontal baffles.

The sizing of the liquid notches was critical and varied for each baffle. Too large a notch would allow excessive bypassing and too small a notch would allow the liquid level to build up excessively. The pressure drop across each baffle was taken from the computer run and applied to the appropriate notch. The pressure drop through the notch was then calculated, by hand of course, and compared with the pressure head of any liquid back up. The calculations took up some time but the results were worth it, I was finally confident that the unit would work OK. If any reader is faced with such a design they should check with process at an early stage that there are no off design conditions which need to be considered.

An acoustic baffle was considered necessary and is positioned at about 40% of the shell inside diameter. The positioning of acoustic baffles is fully discussed in the chapter on vibrations. Note that the acoustic baffle extends beyond the tube bundle. It should extend as far as possible without interfering with the flow path of the shell side fluid. The bundle runners are quite hefty which was probably required in the spec. I decided that they should be vertical rather than radial because the end of a radial runner on the left hand side of the diagram would be closer to the acoustic baffle thus reducing the flow path to the tubes at the left hand side of the acoustic baffle.

The position of the sealing strips is worth noting. For the two sections of the tube bundle adjacent to the horizontal centre line they are opposite the outer tube row which is 'normal'. For the outer sections of the tube bank they are positioned along the line of the first and third tube rows from the centre line. The reason for not putting one of the pair of sealing strips in line with the first or second row of tubes adjacent to the baffle edge is because the effective diameter for cross flow is already a minimum for the bundle at this point. (The spacers for the tie rods are just over the OTL on the top and bottom tube rows.)

Shell I.D.	1676.40	(mm)
OTL	1625.60	(mm)
Tube O.D.	19.05	(mm)
Tube Pitch	25.40	(mm)
Pass Part. Vert.	19.00	(mm)
Pass Part. Horiz.	19.00	(mm)
Tube Layout	90.0	degrees
Alignment	Semi	
Baffle Cut	22.00	%
Inlet Nozzle I.D.	590.00	(mm)
Outlet Nozzle I.D.	590.00	(mm)
Number of Passes	4	
Total Tubes	2032	
Tubes Removed	34	
Tubes Remaining	1998	
Tie Rods/Spacers	4	

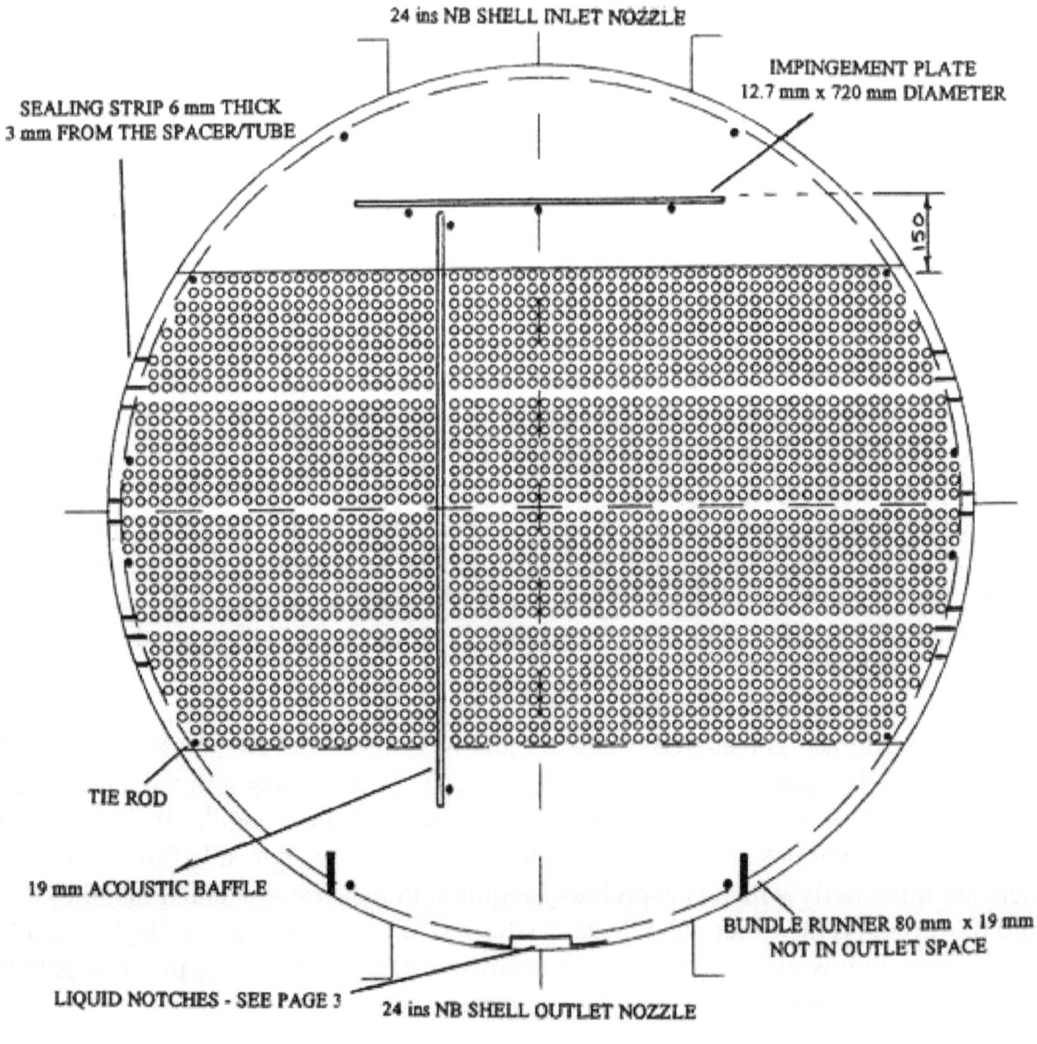

Figure 2.114

Example 2.18

Row	Tubes
1	15
2	21
3	25
4	27
5	31
6	33
7	35
8	37
9	37
10	39
11	43
12	43
13	45
14	45
15	45
16	47
17	47
18	49
19	49
20	49
21	49
22	49
23	49
24	49
25	49
26	49
27	49
28	49
29	49
30	49
31	47
32	47
33	45
34	45
35	43
36	43
37	43
38	41
39	41
40	39
41	37
42	33
43	31
44	31
45	27
46	23
47	19
48	13

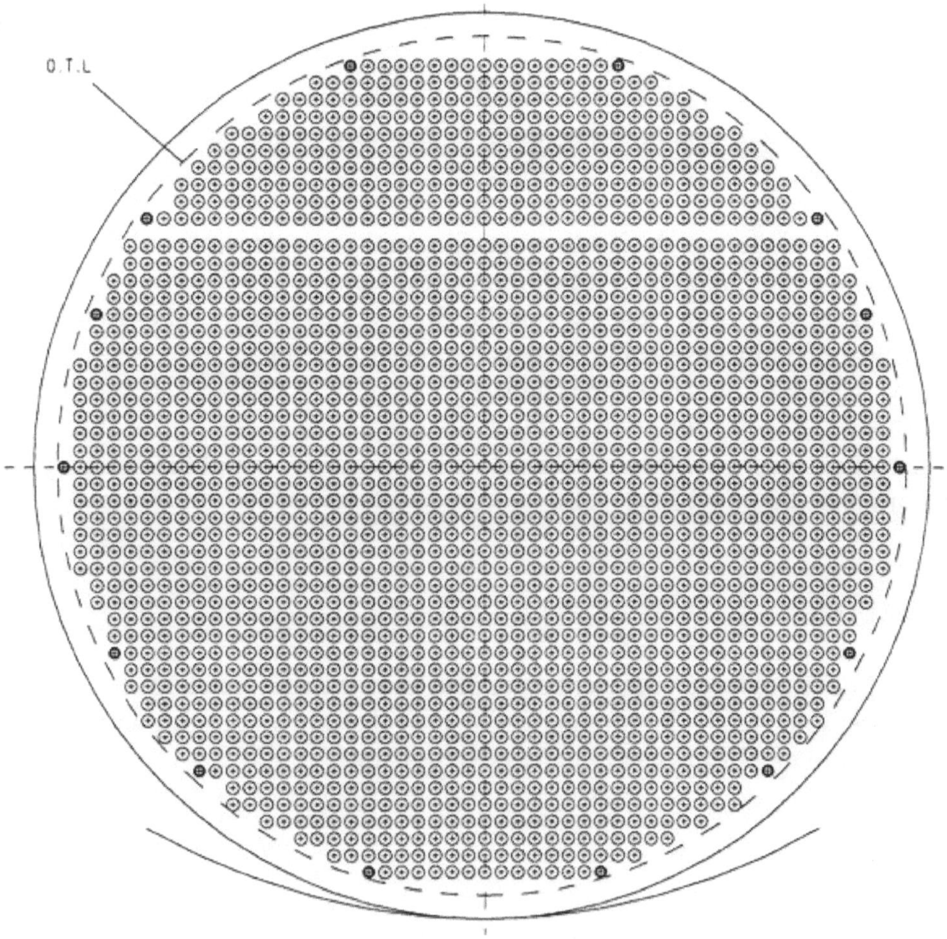

TEMA type	BKM
Bundle Diameter	1360
Outer tube limit	1289.17
Top clearance to tube	70.5
Bottom clear' to tube	60.8
Tube Diameter	19.05
Tube pitch	25.4
Flow arrangement	90
Number of tubes	1900
Number of tube passes	2
Horizontal pass lane	41.28

TUBE PASS DETAILS

Pass	No.Rows	Tubes
1	10	300
2	38	1600

Figure 2.115

This is an interesting tube layout for a kettle reboiler. It is a two pass exchanger but the distribution of tubes per pass is a long way from being reasonably equal. The type of tubes in each pass is also different. The first pass at the top has normal plane tubes; the second pass at the bottom has low finned tubes.

The unit is part of a refrigeration system and the MTD is quite low. The tube side fluid is a superheated vapour which then fully condenses. The shell side fluid is a pure hydrocarbon so that the boiling is close to isothermal.

A run with equal tubes per pass showed that the desuperheating coefficient was strongly controlling and required a large percentage of the surface area. The simplest way out of such a situation is to have a separate desuperheating unit. This is not what process required and it would not have been the most economic solution in this case.

Two thermal design runs were required, one for each pass. For the first pass the vapour cooling coefficient is controlling and it is therefore pointless to use expensive low finned tubes.

For the second thermal design the tubeside fluid is condensing and the condensing coefficient is far higher than the boiling coefficient due to the low temperature difference across the boiling film. Hence the use of fine pitched low fin tubes to increase the number of nucleation points as well as increasing the surface area to reduce the shell side thermal resistance.

The two designs were then combined with appropriate weightings to give overall performance figures for the TDS.

SUMMARY

I hope I have provided sufficient examples to give a reasonable overview of tube layout diagrams. If the reader has an interesting case that brings out further points then please forward it to me for inclusion in the next revision.

CONCLUSIONS

The title of this chapter suggests that it is concerned only with mechanical features and that was my intention when I started to write it. However I soon found that it made no sense to describe a mechanical feature without discussing its impact on the thermal design. For example it is essential to understand momentum fluxes to asses the different types of impingement protection that are available.

I found this a very challenging chapter to write because I had the time to investigate some of the 'standard good practices' which I had taken for granted in the past, for example the $D_n/6$ for an outlet nozzle. I also found I had some explaining to do for so many mechanical features which I had always checked with a mechanical engineer or used table values which had been used for many years. I did find some mistakes in some of these which I have corrected before reproducing them in a new format in the SI system of units whenever possible.

There have been many occasions when I have had to discuss some of the discrepancies I came up with by checking with colleagues I have worked with in the past and are still in the industry. I am very grateful for their help.

CHAPTER 3 – SINGLE PHASE HT & Δp

INSIDE TUBES

LAMINAR AND TURBULENT FLOW

Fluid flow inside tubes is classically divided into laminar flow and turbulent flow.

In laminar flow the fluid moves as if it consisted of a series of very thin concentric tubes of matter which slide against each other, maintaining their integrity, without any mixing with adjacent layers. The velocity of the fluid is zero at the wall of the tube and increases in a curved profile to reach a maximum at the centre as shown in figure 3.1. The temperature profile follows a similar curved path with a minimum value at the wall of the tube and a maximum at the centre for a fluid being cooled and a maximum at the wall and a minimum at the centre for a fluid being heated.

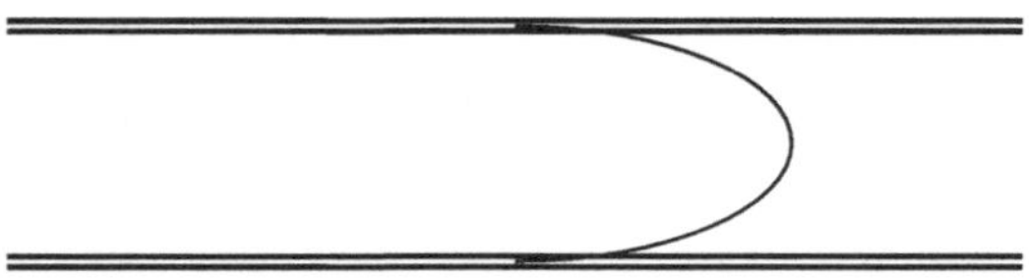

Figure 3.1

Turbulent flow is a much more complex situation. Figure 3.2 shows a section through a tube containing a fluid in turbulent flow. To the right of the section is a typical profile for the fluid velocity and temperature.

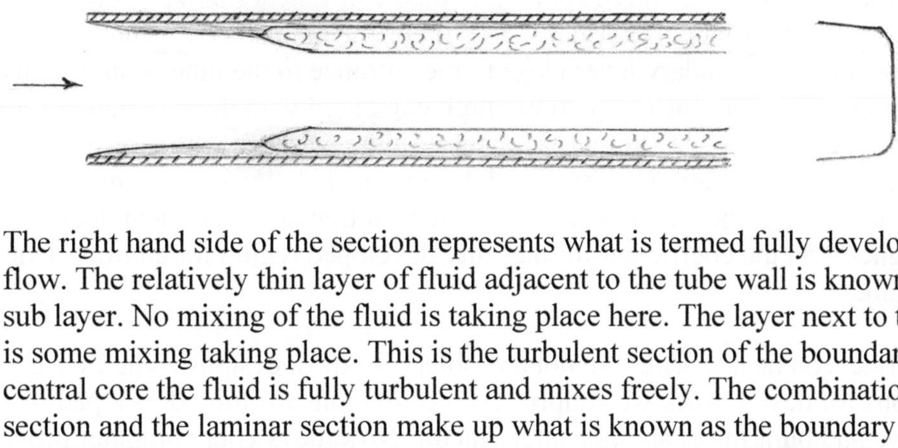

Figure 3.2

The right hand side of the section represents what is termed fully developed turbulent flow. The relatively thin layer of fluid adjacent to the tube wall is known as the laminar sub layer. No mixing of the fluid is taking place here. The layer next to this is where there is some mixing taking place. This is the turbulent section of the boundary layer. In the central core the fluid is fully turbulent and mixes freely. The combination of the turbulent section and the laminar section make up what is known as the boundary layer for turbulent flow. (Before academics jump on me from a great height I am fully aware that this analysis is a simplification. For example it is now generally accepted that there is a thin buffer layer between the laminar sub layer and the turbulent layer. However a practical thermal designer requires a simple mental picture of what is going on to evaluate a specific situation.)

The velocity and temperature profiles in a turbulent boundary layer are quite different to the profiles for laminar flow. Most of the changes take place across the laminar sub layer. There are then further, smaller changes across the turbulent layer. Finally the core of the fluid is at almost constant velocity and temperature. The velocity and temperature profiles are not identical but it is only the general shape that should be firmly fixed in the mind of a thermal design engineer.

Conduction heat transfer is a topic that most students will have met at a very early stage of their education. It is a very useful analogy to think about convection as being conduction through a layer of fluid. As we have seen above there are two distinct thermal resistances in series. However these can be added together to give a single resistance and any change in a variable that effects one layer will effect the other. It is therefore strongly recommended that **convection should be thought of as conduction through a thin layer of fluid.**

It is then clear that any quantity that changes either the composition or thickness of the boundary layer will change the heat transfer coefficient. If the thickness of a boundary layer is reduced the thermal resistance will be reduced and hence the heat transfer will be increased. An increase in velocity will reduce the thickness of the boundary layer and therefore increase the heat transfer coefficient. An increase in the physical property viscosity will increase the thickness of the boundary layer and therefore reduce the heat transfer coefficient. How each variable effects heat transfer is summarised in table 1 in useful tables at the end of this book.

ENTRANCE EFFECTS

The text above only described the fully developed flow of a fluid through a tube. The boundary layer at the entrance to the tube with turbulent flow was clearly quite different. For the first section of the tube there was only a laminar layer which became the laminar sub layer further down the tube as the boundary layer developed. In practice the distance between the entrance to the tube and the point at which a turbulent boundary layer is fully developed can be quite substantial (normally > 10 times the diameter). It is not practical to draw this to scale. The boundary layer close to the entrance to the tube is quite thin and the heat transfer coefficient is correspondingly high compared with the coefficient for the fully developed region. Since the laminar layer starts at the entrance with a zero thickness there is no thermal resistance at this point and the heat transfer coefficient is infinitely high. As the distance from the entrance increases the heat transfer coefficient decreases but remains higher than the coefficient for the fully developed region for an important section of the tube.

The entrance effects on heat transfer are not a problem for most designs. When we used hand calculations for design work we simply used correlations for fully developed flow knowing that we had built in yet another safety factor. Nowadays your computer program will probably take them into account. To take account of entrance effects an additional dimensionless group – diameter to length ratio is introduced into equations of the form shown under dimensional analysis later in this chapter.

However entrance effects should be understood because there are occasions where a good knowledge is essential to produce a satisfactory design. One such case is where a hot gas is used to generate steam in a reboiler. The overall metal temperatures will be near to that of the boiling water and a carbon or low alloy steel is perfectly acceptable. A closer examination will show that with a normal construction the metal temperatures at the entrance to the tubes are well above an acceptable limit for a low grade material. One way to overcome this and still use the low grade material is to fit an insulated ferrule at the entrance to the tube. The disadvantage of using insulated ferrules is the significant increase in pressure drop.

There is an entrance effect in laminar flow which has been exhaustively covered in many academic books. It increases the heat transfer coefficients in exactly the same way but I have never come across a case where this could cause a problem. In general laminar flow inside tubes should be avoided wherever possible because of the very low heat transfer coefficients associated with it.

REYNOLDS NUMBER Re

A Reynolds number is a dimensionless group. All dimensionless groups are the ratio of two quantities. There is no need to get involved in dimensional analysis to develop an equation for this group. The Reynolds number is defined as the ratio of inertia forces to viscous forces and it can be shown that:

$$Re = \frac{\text{Inertia Forces}}{\text{Viscous Forces}} = \frac{\rho u d_i}{\mu} = \frac{G d_i}{\mu} \quad [3.1]$$

Where ρ is the density, u is the velocity, d_i is the inside diameter of the tube and μ is the dynamic (absolute) viscosity.

It is assumed that the basic form of the Reynolds number will be well known to the reader. In the right hand side expression ρu has been replaced by G. G is known as the bulk (or mass) velocity and is an alternative way of expressing the Reynolds number.

In liquids density will show small variation with temperature. However in gases there may be quite significant variations. Viscosity exhibits the opposite behaviour. In gases there will be small variations with temperature but with some liquids there may be very significant changes.

For a constant mass flow rate through a tube, if the density falls (when it is heated) the velocity will increase and if the density increases, (when it is cooled) the velocity will fall. This is why we always used the bulk velocity form of the Reynolds number when we did all these calculations by hand. It also clearly shows that density variations have no effect on the Reynolds number. As you will see later in this chapter that means that density variations have no effect on the heat transfer coefficient. You will also see that density does effect the pressure drop (pressure drop is inversely proportional to density.)

The Reynolds number is so important to a heat transfer engineer that he/she should spend some time thinking about it. For a given process fluid it is clear that the only control we have is to optimise the velocity for a given allowable pressure drop to maximise the heat transfer coefficient. (Or when this is not realistic consider switching the fluids.)

TRANSITIONAL FLOW

We have looked at the 'clear cut ' flow regimes of laminar and turbulent flow. The problem now faced is how to predict the flow regime in a particular heat exchanger tube. This is where the Reynolds number is so vital. For a practical thermal design engineer it is our only way to decide if likely to be laminar or turbulent.

When I was first introduced to Reynolds number as a student I seem to recall a magic number of 2100 (or was it 2300). Below this value the flow was laminar and above it the flow was turbulent. This was analogous to Mach number where velocity in a gas increases until a point is reached, (the sound barrier) where a sudden change to supersonic velocity takes place. Fluid flow regimes in a tube are far less predictable.

I have skimmed through the many books and papers I have in my collection to search for any new 'magic numbers' – there are none. Laminar flow can start to break up at a Reynolds number as low as 1800 and fully developed turbulent flow may not be established until the Reynolds number is over 10000. In between we could be in an intermediate regime termed transitional flow. Clearly this possible intermediate regime covers a large span of Reynolds numbers and will occur in many heat exchanger designs.

There is a wide variety of fluids to be dealt with in heat transfer with the most basic division being between liquids and gases. However within a single phase there can be sub-divisions. For example water behaves differently than a hydrocarbon. Even when I did my first designs in 1965 there were separate equations for water and hydrocarbons. The flow regime is also very dependant on the surface of the tube. Even for a specified fluid flowing through a specified tube there can be some odd effects. Laboratory tests have been undertaken in which the mass flow rate was slowly increased to the point at which laminar flow started to break down and the Reynolds number noted. The test was then repeated with a high mass flow rate which was then slowly reduced until laminar flow was re-established and the Reynolds number noted again. The Reynolds number was significantly higher when the flow was slowly increased. It has also been observed that in laminar flow a sharp tap on the tube may be sufficient to change the flow regime into the transitional zone.

The spread of Reynolds numbers in transitional flow can be much less for particular fluids in commercial tubes than the extremes quoted above. Fortunately the thermal designer does not have to worry too much about this. Modern programs will incorporate different equations and the program will select the appropriate one for each situation. However it is essential that the thermal designer understands what is going on and looks at the Reynolds numbers throughout the exchanger (if the program provides this information). Reynolds number can also be the key to looking for ways to improve the first design a computer program will come up with.

CHANGING FLOW REGIMES TO IMPROVE DESIGN

As a simple example suppose a viscous oil is to be heated inside tubes by steam. The tube length is fixed and with two tube passes. there is plenty of pressure drop to spare but there is insufficient pressure drop available to increase the number of passes to four. An inexperienced designer (or one pushed for time or just a lazy one?) may accept the first solution the computer comes up with. A more experienced and conscientious designer may observe that the first tube pass is in laminar and transition flow and the second pass in turbulent flow.

Your program may not allow an input of variable tubes per pass (but one day it will). Now is the time for a composite design. Note or estimate the oil outlet temperature from the first pass and use this to design a single pass unit. You will use up a lot more of the allowable pressure drop for this design. It may be necessary to adjust your first estimate of outlet temperature to keep this unit close to the required oversurface. You will find that the surface area required for this part of the duty has been significantly reduced due to the improvement in flow regimes and hence surface heat transfer coefficients. Then feed the terminal conditions for this unit into another design which will require a lot more tubes. When a point is reached that the required oversurface is OK for both units within the allowable pressure drop then the sum of the surface areas should be sufficient for the overall design. This type of composite design is much quicker to optimise if you design one shell on your computer and the other one on that of a colleague who is just reading specs or away at a meeting.

This example is relatively simple because the tube side coefficient would totally dominate the total thermal resistance and the flow path on the shell side would not be important. (See the section on condensing steam as a service fluid in chapter 5). You would still need to draw a tube layout for the actual two pass design with variable tubes per pass and carry out a few hand calculations for the information required on thermal data sheet.

Some programs already allow a simpler method to solve this problem. Just design two shells in series with different numbers of tubes in each shell. If you do this be sure to specify ridiculously large intermediate nozzles. There is no point in having 'none existant' nozzles taking up pressure drop. It may also help to have much larger shells than required for the tube count to keep the shell side pressure drop to what it would be in the single shell. (A significant pressure drop with steam will reduce the MTD – see the section in chapter 5 on condensing steam.)

The larger the exchanger the more worthwhile it is two carry out a complex calculation for an apparently simple design. It is always very satisfying to do this and the saving in metal means a reduction in the CO_2 contribution to global warming.

EQUATIONS DEVELOPED FROM DIMENSIONAL ANALYSIS

I have no intention of going through this in any detail. Anyone reading this book will surely have been introduced to the topic as a student. If they have not there is a plethora of books which have already covered it. Dimensional analysis is resorted to when a direct mathematical approach fails to come up with practical design equations.

It involves writing down the variables which are know to affect the quantity of interest, in this case surface heat transfer coefficient. For convective heat transfer the variables are velocity u, the inside diameter of the tube di, the specific heat at constant pressure Cp, the dynamic (absolute) viscosity μ, thermal conductivity k and the density ρ. A dimensional analysis results in the dimensionless group Nusselt number Nu, being a function of the

Reynolds number Re and the Prandtl number Pr. That is the end of the dimensional analysis and the start of the hard work in the laboratory.

A lot of research is required to establish constants which change an equation of proportionalities into a practical design equation. One relatively simple design equation for fully developed turbulent flow is:

$$Nu = 0.023\ Re^{0.8}\ Pr^{0.333} \qquad [3.2]$$

or

$$\frac{hd_i}{k} = 0.023\ \frac{(\rho u d_i)^{0.8}}{\mu^{0.8}}\ \frac{(Cp\mu)^{0.333}}{k^{0.333}}$$

or

$$h = 0.023\ \frac{k}{d_i}\ \frac{(\rho u d_i)^{0.8}}{\mu^{0.8}}\ \frac{(Cp\mu)^{0.333}}{k^{0.333}} \qquad [3.3]$$

There are literally dozens of equations of similar form for convective heat transfer, each with different values for the constants and there may be additional dimensionless groups. For example many use a power constant of 0.4 for the Prandtl number. At one stage I thought I might try and count them all but soon gave up. The reason that there so many equations is that some give better results than others for a limited range of situations. Different equations have been developed for ranges of Reynolds number, ranges of Prandtl number, heating and cooling etc. In general the more limited the application is the more accurate the equation will be.

Once again selection of appropriate equations is not a normal problem for a thermal design engineer. Your program will contain many such equations and will make a selection for you without you being conscious of it. There may of course be the very rare occasion when you are dealing with an unusual fluid and it may be worth checking up on what the computer program is using and doing a bit of research to try and find if there is a more accurate equation. In such situation it is not too difficult to carry out hand calculations using both equations and then introduce a correction factor for h in the input to your program.

It is not necessary to resort to dimensional analysis to develop equations for laminar flow although it is necessary to make assumptions for the shapes of the temperature and velocity profiles. However it is useful to transpose these equations into a dimensionless format so that they can be easily compared with a typical equation for turbulent flow. This results in the equation below.

$$\frac{hd_i}{k}\ \alpha\ \left(\frac{\rho u d_i}{\mu}\ \frac{Cp\mu}{k}\ \frac{d_i}{L}\right)^{0.333} \qquad [3.4]$$

The diameter length ratio, d_i/L takes account of the distance from the tube inlet.

USEFUL RELATIONSHIPS

It is important that a thermal design engineer understands how changing a single variable will affect the heat transfer coefficient. From the simple equations above it is clear that the heat transfer coefficient h, is proportional to the velocity to the power of 0.8 for turbulent flow. If the velocity is doubled (by say changing from 2 tube passes to 4 tube passes without changing the number of tubes) the heat transfer coefficient will increase by a factor of $2^{0.8} = 1.74$.

It is worthwhile for the reader to work out the other relationships for himself/herself. They are all summarised in table 2 of useful tables at the end of this book.

PHYSICAL PROPERTIES

As discussed in the earlier sections of this chapter convective heat transfer can be thought of as conduction through a layer of fluid. It is therefore logical that the physical properties in the relevant equations should be at the average conditions in the boundary layer.

This presents a problem. Suppose a fluid inside the tubes is being heated by steam. If the pressure drop on the steam side is negligible, which it should be, if the steam is on the shell side, the process will be almost isothermal (constant temperature). This together with the very high heat transfer coefficients and low fouling factors typical for steam may well result in the wall temperature, and even the average boundary layer temperature, being higher throughout the exchanger than the outlet bulk temperature of the fluid being heated. Since the process data sheet will normally just specify physical properties at inlet and outlet they will all be outside the range of the temperatures in boundary layer. How to establish a property at a second temperature when only one value is given has been explained in Chapter 1.

A sophisticated modern design program will overcome this problem by extrapolation but this is never as satisfactory as interpolation. Ideally physical properties should be specified at two temperatures which cover the range of temperatures in the boundary layer as well as the bulk temperatures. In practice we normally manage with bulk temperatures – except for occasions when we need a viscosity reading at an extra point.

When we used hand calculations to design heat exchangers it was not practical to use properties in the boundary layer. For liquids we only took account of viscosity - the property most susceptible to significant changes. This was done by introducing another dimensionless group – viscosity ratio where μ_b is the viscosity at the bulk temperature and μ_w the viscosity at the wall temperature. This group had an exponent of 0.14 so that $(\mu_b/\mu_w)^{0.14}$ was added to our practical design equations. (The wall temperature was usually just a conservative estimate.)

FRICTIONAL PRESSURE DROPS

Frictional pressure drop Δp through a tube can be found from the following equation which is normally credited to Darcy:

$$\Delta p = \frac{4fL\rho u^2}{2di} \qquad [3.5]$$

Once again this is an equation which the reader will already be familiar with and if he/she is not there are many textbooks on heat transfer and fluid mechanics to refer to. The basic SI unit for pressure drop is N/m^2 which is often referred to as a Pascal, Pa. The other terms in the equation have been used earlier in this chapter except for f which is a dimensionless friction factor, often termed the Fanning friction factor. A word of caution if you ever do any hand calculations using this equation – make sure of your data source for f. Some researchers have tried to 'simplify' the Darcy equation by eliminating the 4 and using friction factors four times as high as the Fanning values. This has of course just produced an unnecessary complication and source of a potential serious error.

For laminar flow f = 16/Re irrespective of the nature of the surface of the tube. For transitional flow and turbulent flow fixing a value for f is much more difficult. Early researchers have tried to develop a single complicated equation to predict f for turbulent flow but gave up long ago. As with heat transfer your computer program will contain many equations for different situations and will select the most appropriate one. You don't even have to worry about the magnitude of the friction factors it is using.

Looking at the Darcy equation it is easy to make the mistake of assuming that frictional pressure drop is proportional to velocity to the power of 2. However many of the equations for the friction factor contain the term $Re^{-0.2}$ which is proportional to the velocity $u^{-0.2}$. Frictional pressure drops in a tube are therefore proportional to $u^{1.8}$.

F varies significantly within different ranges of Reynolds number it is also dependant on the surface finish of the tube. A chart which demonstrates this is shown in figure 3.3 and is usually referred to as a Moody diagram.

Moody Diagram

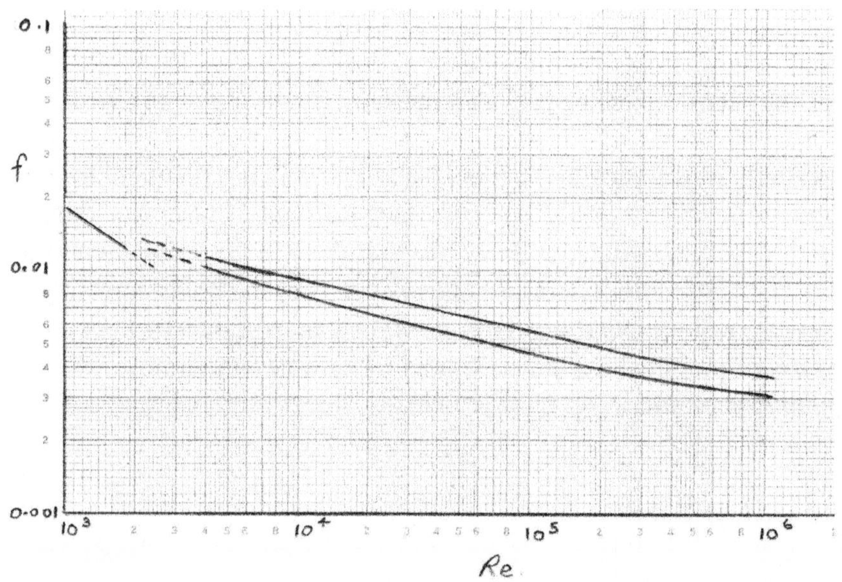

Figure 3.3

The first thing to note about this chart is that both the vertical and horizontal axes have a logarithmic scale. The second thing is that I have simplified it compared with the ones you will find in most textbooks. They often contain a much more complicated family of curves which allow for surface roughness. (Defined as the height of surface irregularities divided by the diameter of the pipe/duct.) However I have never designed a heat exchanger with concrete tubes or even riveted ones so I decided to simplify the diagram to include only the range of interest to heat exchanger designers.

The short 'curve' on the left hand side is the one used for laminar flow. Curve is perhaps not the correct word to use because it is a straight line. In the turbulent region the lowest curve is for a smooth surface. The curve above it is for commercial pipes (not tubes). In between are curves appropriate for normal commercial heat exchanger tubes. Your computer program will default to a commercial tube option but there may be situations when you would want to change this. (If the program allows you to). For example if there is a corrosive fluid inside the tubes and it has been decided to use thicker tubes rather than a superior metal it would be sensible to use higher friction factors. (Or just a higher safety factor).

The HTRI program has a default of commercial tubes but it also has the option of smooth tubes. I recently checked a series of design by a Korean vendor. I was in agreement with their heat transfer calculations and consequent surface areas but could not understand why their tube side pressure drops were so low compared with mine. I assumed that they had incorrectly been using an average diameter for the tube d_i when the tubes were specified as being minimum wall. (See tube specifications in Chapter 2). This gave a closer agreement but knowing that we were using the same computer program there should not have been any difference. It finally transpired that they had opted for the smooth surface option for the friction factors.

Before I wrote my report on the designs (which was rather damming as far as tube side pressure drops were concerned.) I contacted HTRI to confirm my estimation of the situation regarding the selection of the tube surface. A smooth tube surface is what you might get in laboratory experiments or when a heat exchanger is new and there has been no time to build up any fouling layer. This option should never be used to design a heat exchanger.

DISCONTINUITY PRESSURE DROPS

A discontinuity pressure drop is basically a pressure drop caused by anything other than friction or acceleration of the fluid. In a thermosyphon reboiler design there are discontinuity losses due to bends, valves etc in the connecting pipework. These are normally dealt with in terms of equivalent lengths. In all other types of heat exchanger discontinuities are dealt with in terms of velocity heads.

A velocity head is the rise in pressure a moving fluid would experience if it were brought to rest. It is the kinetic energy term in Bernouilli's equation expressed in the units of pressure. (In fluid mechanics Bernouilli's equation may be expressed in terms of energy pressure or pressure head depending on the application.)

$$\text{Velocity head} = \frac{\rho u^2}{2}\ \text{N/m}^2 \qquad [3.6]$$

For a specific discontinuity:

$$\text{Pressure drop} = k\,\frac{\rho u^2}{2} \qquad [3.7]$$

k is a constant for the discontinuity Any basic text book on fluid mechanics will show that for a sudden contraction k = 0.5 and for a sudden enlargement k = 1. The inlet and outlet to a tube are close to these standard cases. The inlet and outlet velocity heads will of course be different, particularly for a gas when there may be significant changes in density.

TOTAL PRESSURE DROP

The total pressure drop through the tube side of a heat exchanger is therefore the sum of the frictional pressure drops in each tube pass together with the sum of all the discontinuity losses. (Acceleration losses are negligible in single phase flow).

As well as the discontinuity losses in and out of the tube passes there are losses in the nozzles at inlet and outlet with the same k factors. A small allowance may also be made for 'turnaround' loses particularly if the unit has a floating head. In a floating head the passages between tube passes should always have a cross sectional area at least 1.5 times the cross sectional area of the tube passes. With other types of head construction the flow area between passes will be well above this figure.

The overall pressure drop on the tube side of a heat exchanger therefore proportional to $u^{1.8}$ for the friction component and u^2 for the discontinuity component. Allowing about 20% for the discontinuity component, results in the total pressure drop being proportional to velocity to the power of 1.85 which I recommend in useful tables 1 at the end of this book.

VELOCITY & MOMENTUM FLUX LIMITS IN TUBES

TEMA states that consideration shall be give for special devices to prevent erosion of the tube ends when:

(1) The inlet nozzle is axial.
(2) The fluid is a liquid with a momentum flux of more than 8928 kg/ms^2 (6000 lb/ft s^2).

AXIAL NOZLES

It must be emphasised that TEMA only recommends *considering* special devices for axial nozzles. No protection is needed for the majority of axial nozzles. The simplest type of protection is an impingement plate as shown in figure 3.4. This has the disadvantage of having a relatively high pressure drop – somewhere between one and two velocity heads.

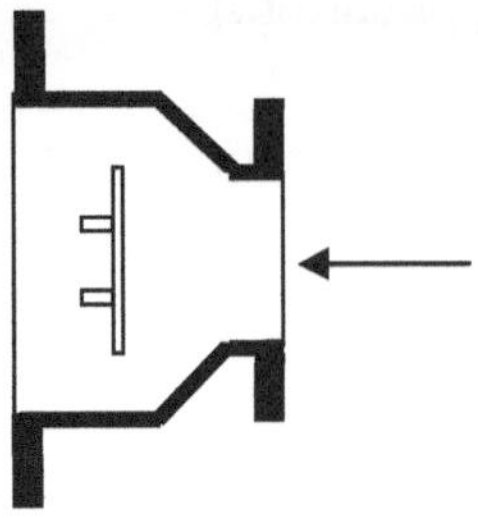

Figure 3.4

Direct impact with the tube ends can also be avoided by using deflecting plates. One possible arrangement of deflecting plates is shown in the figure 3.5. This arrangement has a far lower pressure drop and with a long tapered channel acting as a diffuser there may not be any need for protection.

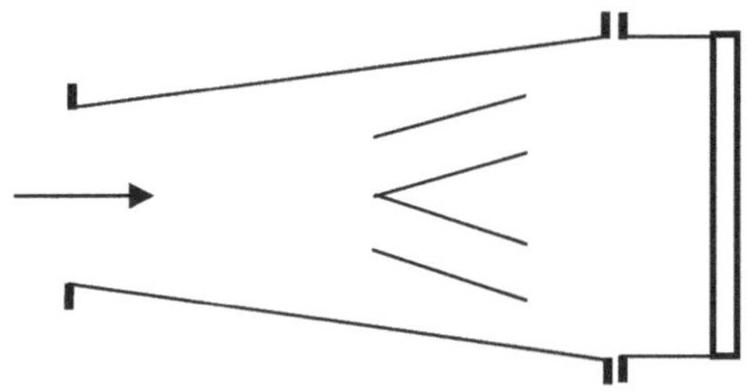

Figure 3.5

Deflecting plates are most likely to be used where there is two phase flow with a small volume of liquid compared with the volume of vapour. In this situation the momentum flux of the liquid particles may be much higher than the average momentum flux of the fluid. You will often find that deflector plates are 'traditional' for a particular item on a specific plant.

Deflector plates will always increase the tube side pressure drop. This increase may be small – certainly less than one nozzle velocity head, but sometimes the pressure drop is critical for single pass units. Faced with a new situation I would prefer to use some form of tube end protection provided there was a reasonable pressure drop through the tubes. The higher the pressure drop through the tubes the better the distribution will be to each tube.

TUBE END PROTECTION

Protection plates

A simple protection plate is shown in figure 3.6. It is a flat plate with holes drilled to match up to the tube holes. It is normally attached to the tubesheet around the periphery but for large diameter shells consideration should be given to a few more central fixing points even if this means losing an odd tube or two. (Purely a mechanical problem so discuss it with a mechanical engineer.)

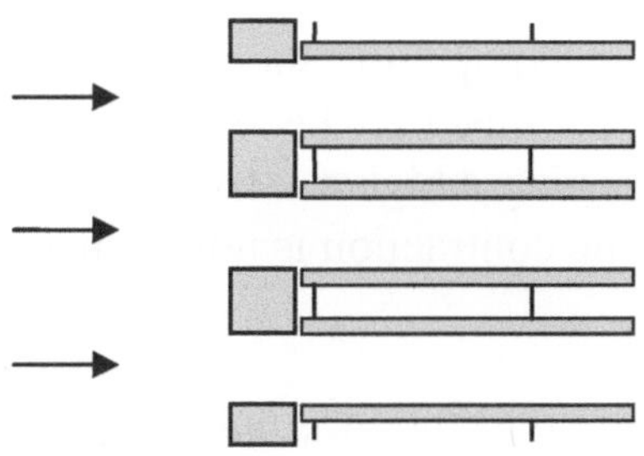

Figure 3.6

If the plate is made from the same material as the tubes and tubesheet there will be no expansion problems. In this situation it is sensible to specify a spare plate. However it is normally desirable to make the plate out of a more erosion resistant material than the tubes. If this is done the expansions and stresses should be checked out.

Whenever I specify erosion protection plates I never use a plain plate. There are significant advantages in using a plate with a more gradual opening such as the one shown in figure 3.7.

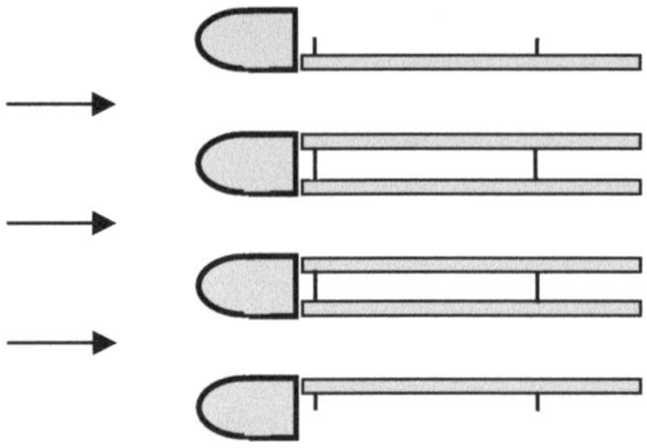

Figure 3.7

The pressure drop due to a sudden contraction from a large space is half a velocity head which we normally think of as being a discontinuity k factor of 0.5. There is a very neat treatment of contractions in Fluid Mechanics for Engineers by Barna – Ref 22. Figure 3.8 is one of his diagrams.

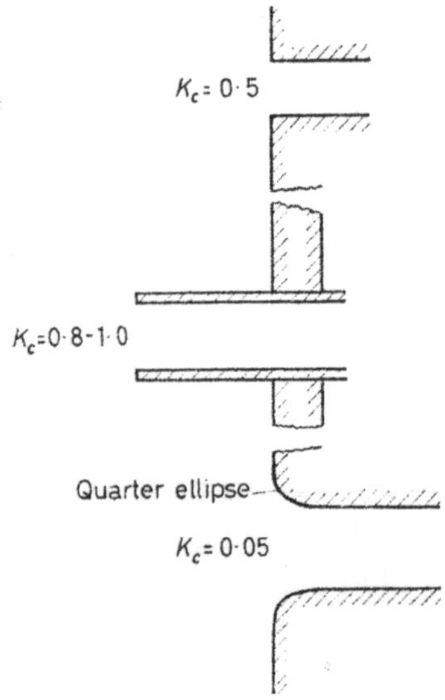

Figure 3.8

Whilst the contraction factor for the sharp edged opening should be well known to the reader the contraction factor for the protruding tube may be less well known. In a normal heat exchanger the tube does protrude suggesting a higher k factor. However the protrusion is normally only a few mm and the contraction is not from a very large space so that a factor of 0.5 is not unreasonable.

What is very interesting is the very low k factor of 0.05 for the quarter ellipse. This is used in nozzle designs (not the heat exchanger type). I am not sure how a vendor would respond to a request for quarter elliptical holes but it may be worth discussing if the pressure drop is very critical and impingement protection is required. I have always specified a radius profile as shown in figure 3.13 and assumed that a k factor of about 0.25 would be about right.

Ferrule Protection

An alternative to using a plate is to use ferrule protection. These are relatively thin inserts made from a material with superior erosion properties than the tube material. The sketch in figure 3.9 illustrates this. The ferrules are lightly rolled into each tube.

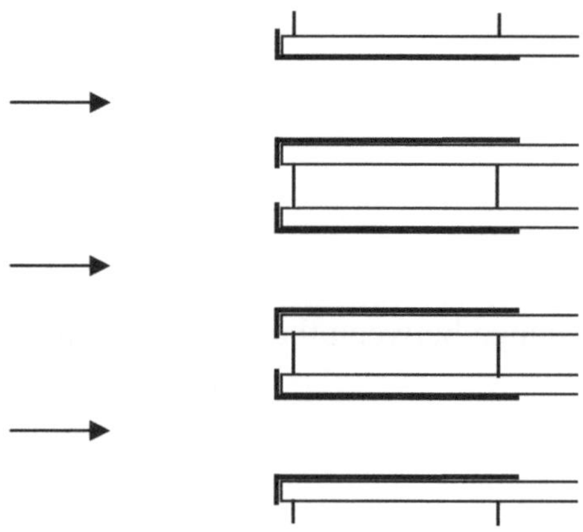

Figure 3.9

I have shown ferrules with a sharp edge in figure 3.15 but it is clearly more satisfactory if the inlets are rounded as shown in figure 3.10.

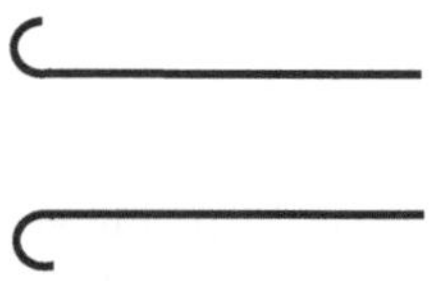

Figure 3.10

If the ferrules are sharp edged there must clearly be some increase in pressure drop. How much extra pressure drop is required depends on the thickness of the ferrule and the diameter of the tubes. If the ferrule is quite thin and the face rounded there will be a very little if any increase in pressure drop.

As an example of what can be achieved with ferrules I remember a unit I once worked on for a well known client who was known for having a conservative approach to many heat exchanger problems. The unit was a steam heated pump through reboiler with no vaporisation. (The heated liquid was flashed off to separate the liquid and vapour components in the tower.) The liquid being heated in the tubes was a fairly heavy hydrocarbon with a density of about 900 kg/m^3 and a viscosity of 1.5 cps at inlet. **The momentum flux in the tubes was about 20000 kg/ms^2.** The tubes were made of carbon steel and the ferrules were made of Incoloy 825. It was a four pass unit and the ferrules were fitted to the tubes at the inlet to each pass.

Ferrules may also be used to protect the ends of the tubes and the tubesheet from high temperatures. A simplified insulated ferrule is shown in figure 3.11 below.

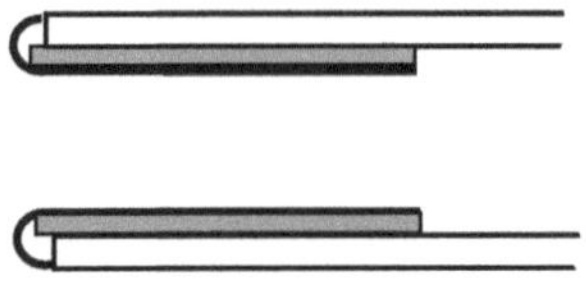

Figure 3.11

These ferrules are of a much thicker construction with a double skin containing an insulating material which is the shaded section in the sketch. The pressure drop from this type of ferrule is often a large proportion of the tubeside pressure drop. The most common application is in waste heat boilers.

OUTDSIDE TUBES

BASIC DEFINITIONS

The **tube pitch P_t** is defined as the distance between the centres of adjacent tubes of outside diameter do.

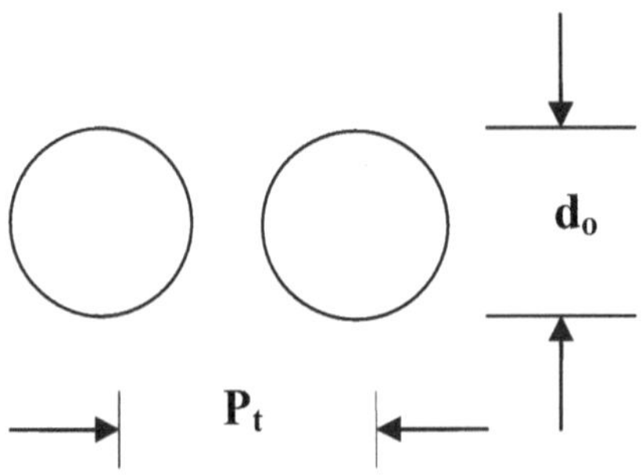

Figure 3.12

Another term which the reader should become familiar with is the **pitch to diameter ratio P_t/d_o** and the **clearance between tubes is $(P_t - d_o)$.** The minimum pitch which may be used for a specified tube is controlled by the ligament thickness. The ligament thickness is the thickness of the metal in the tubesheet between two tubes and therefore always less than (Pt – do) when tolerances have been applied. When I first started designing heat exchangers it was quite common to use a 23.81 mm (15/16 inch) triangular pitch with 19.05 mm (3/4 of an inch) tubes. The clearance was therefore 4.76 mm. There then followed many years when many clients and contractors insisted on a minimum clearance of 6.35 mm (1/4 inch). In recent years the 23.81 mm pitch seems to be making a comeback. I remember one of the first times I used it again was on a revamp job. The new duty was of course higher than on the original design. I found that by reducing the tube pitch I could cram in just sufficient extra surface in a new bundle to avoid the cost of a new shell and disruption to the pipework – a significant saving.

The two factors which should be taken into account when deciding on the clearance between tubes are the method of attaching the tube to the tubesheet and the thickness of the tubes. If a tube is welded to a tubesheet it should have a weld preparation on the tubesheet which is at least as thick as the tube. It is not good practice for the welds to overlap. For expanded tubes the thicker the tube is the greater the force becomes during expansion and the more susceptible the ligament is to distortion. Table RCB-7.22M in TEMA gives recommended tube gauges for different tubesheet thicknesses it does not distinguish between expanded and welded tubes.

Personally I would not hesitate to use a clearance of 4.76 mm with 16 BWG tubes if they are expanded and they would be just OK for welded tubes. However I would not even consider using a clearance of less than 6.35 mm with a 12 BWG tube that required welding to the tubesheet. If in doubt about a particular case consult a good mechanical engineer. If your company does not have one then phone up a few vendors. (Hopefully they will all give you the same answer!)

TUBE PITCH ARRANGEMENTS

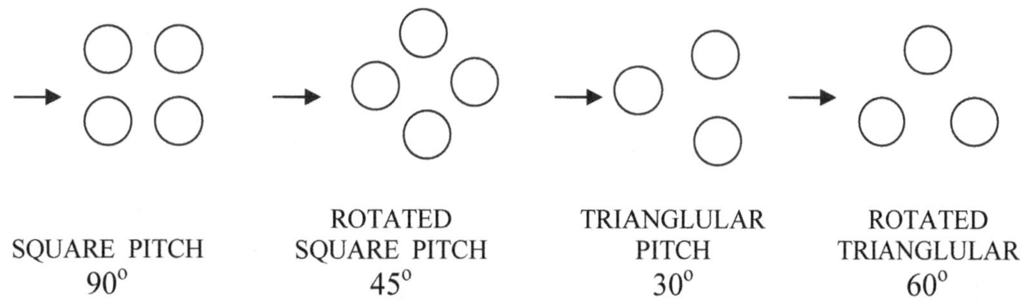

Figure 3.13

One simplification for shell and tube heat exchangers is that there is normally only one tube pitch (unlike air coolers for example where there may be more than one.) The arrows indicate the direction of the fluid flow which will be perpendicular to the edge of the baffle cut.

Unfortunately tube pitches can be misinterpreted. For thermal design work the direction of fluid flow defines the arrangement. However a mechanical engineer may not see it this way. Take away the arrow and the rotated triangular arrangement above looks like a triangular pitch. There are of course no arrows on a tube layout diagram so you can understand his viewpoint. A thermal designer should be aware of this when discussing tube pitch arrangements with people outside the thermal design group.

The four standard arrangements illustrated above in figure 3.13 will now be considered in more detail.

SQUARE PITCH

This is the simplest and on many plants the dominant tube pitch arrangement. It caters for any method of cleaning so that the minimum clearance is always 6.35 mm (1/4 inch). This results in $Pt/d_o = 1.333$ for 19.05 mm tubes and $Pt/d_o = 1.25$ for 25.4 mm tubes. $Pt/d_o = 1.25$ is the minimum allowed by TEMA for class R exchangers. Some relaxation is allowed for class B and C exchanger using smaller tubes or shell diameters below 305 mm.

For single phase fluids the baffles are normally horizontally cut with an odd number of crosses so that there are no ineffective areas in the end spaces. (Unless the shell nozzles are required to be on the same side in which case there will be an even number of passes.)

ROTATED SQUARE PITCH

This was once a very popular arrangement for single phase units when we used hand calculations. It resulted in higher heat transfer coefficients than a square pitch. There is one potential disadvantage in this arrangement and that is that it can result in more potential tube space being lost to pass partition lanes and the greater the number of tube side passes the greater the loss.

The minimum height of a pass partition lane is determined by the groove in the channel which seats the pass partition. The subsequent loss in available space for tubes depends on the tube arrangement. Details of these were shown in the section on pass partitions in chapter 2. The pass partition area losses are much greater for a 45° arrangement than for a 90° arrangement because of the need to maintain the alignment of the tube clearances which become the cleaning lanes. The distance between the centres of tubes in a vertical row P_{row}, of a 45° arrangement is $P_{row} = 2^{0.5} P_t = 1.414\ P_t$. For a 90° arrangement it is only $P_{row} = P_t$.

We had a neat way of overcoming the potential problem of the extra pass partition losses. The tube layout was 90° as far as a mechanical engineer was concerned but the baffles were cut at 45° so that the fluid flow arrangement was rotated square pitch.

In general the rotated square pitch does not indicate any significant advantage in heat transfer compared with a square pitch for most ranges of Reynolds number in modern computer programs. The one exception is when the Reynolds number is very low. Then there may be significant increases in heat transfer coefficient with a rotated square pitch. This is the only situation now when rotated square pitches are now regularly used.

It is debateable if there is any true laminar flow on the shell side of an exchanger with a rotated square (or any orientation of triangular pitch). This is because there is a natural mechanical turbulence created by the fluid continually changing directions. **Whenever you are dealing with a viscous fluid on the shell side try both arrangements.**

If I am designing a unit where there is a clear advantage of using a rotated square pitch and there are only two tube passes I would just accept the extra pass lane dislocation loss and design with horizontal baffle cuts. However if there were twelve tube passes I would certainly go back to an arrangement with vertical cleaning lanes but with the baffles cut at 45°. (Provided the program allows for this). It may initially raise a few eyebrows but it will be easy for you to demonstrate that this is the best design.

TRIANGULAR PITCH

Triangular pitch tubes are only used when the shell side does not need to be cleaned or if a chemical cleaning method is satisfactory. Any triangular arrangement with only one value for the pitch forms an equilateral triangle with all the angles being 60°. For a triangular pitch the angle between the direction of flow and the sloping sides of the triangle is therefore half of this i.e. 30°.

The advantage of a triangular pitch is that it is possible to fit more tubes into a fixed shell size than when using the square pitch arrangements. The triangular pitch is far more widely used than the rotated triangle.

ROTATED TRIANGULAR PITCH

For this arrangement the angle between the direction of the fluid flow and the sloping side of the triangle is 60°. This arrangement is now far less common than it once was. When only gravitational effects were taken into account during condensation calculations it had the advantage of more condensing streams. (The condensing streams are equal to the number of vertical tube rows.) Now that modern programs also take account of shear forces the calculations are more complicated and there is often little difference between the heat transfer for the two arrangements.

Some specifications now even ban this arrangement. The reason for this is that the rotated triangular pitch is more susceptible to vibration problems than the triangular pitch. (See chapter 15 on vibrations.) This is not a very good reason for banning it – just a very sensible reason one not to use it when an exchanger is susceptible to vibrations.

The main situation nowadays when rotated triangular pitch should be used is with vertically cut baffles. It is then often possible to fit more tubes into a particular diameter shell than with a triangular pattern. This is particularly true when there are a large numbers of tube passes or when dealing with small diameter shells. The reason for this is that the tube rows adjacent to any horizontal pass partition lanes contain more tubes. (Remember 30° is 60° to a mechanical engineer so if you look back at the diagrams in Figure 3.5 in a horizontal row there is a tube every pitch in the far right sketch but only one every 2 x 0.866 = 1.732 pitches in the sketch next to it).

CROSS SECTIONAL AREAS REQUIRED FOR TUBES

Each tube in a bundle occupies a cross sectional area A_X, calculated by:

$$A_X = FP_t^{\,2} \qquad [3.8]$$

Where F is the tube arrangement factor. This was first introduced in chapter 2 as equation 2.16 but it will now be looked at in more detail.

F = 1 for a square (or rotated square) pitch
F = 0.866 (Cos30^{o}or Sine60^{o}) for any triangular pitch

These factors can be deduced by studying figure 3.5. If a square is drawn by linking the centres of four tubes the enclosed area is P_t^2 and there are four 90° segments of tube which have the same sectional area as one tube. If a triangle is drawn for either of the right hand sketches linking the centres of three tubes encloses a triangle with an area of 0.5 x Pt x $0.866P_t$. The triangle encloses three 60° segments of a tube which is equal to the sectional area of half a tube. The area a full tube will occupy is therefore:

$$A_X = 2 \times 0.5 \times P_t \times 0.866P_t = 0.866\ P_t^2.$$

For a given cross sectional area the ratio of the number of tubes in a triangular pattern to those in any square pattern with the same pitch is 1/0.866 = 1.15. **It is therefore possible to fit in 15% more tubes with a triangular pattern.**

For a practical heat exchanger shell the full cross sectional area is not available for tubes. There is an annulus of area lost due to the clearance between the shell and the outermost tubes. There may be areas lost due to pass partition lanes there are normally areas lost adjacent to the inlet and outlet nozzles. (See the section on tube layouts in Chapter 2 for details).

For a given area available for tubes, if a triangular pitch is acceptable **changing from a 25.4 mm pitch to a 23.81 mm pitch** produces an increase in tube count in the ratio of the tube pitch squared. This is $(25.4/23.81)^2 = 1.138$ so that **the percentage increase is almost 14%.**

With modern computer programs the designer has a lot of flexibility and should not think in terms of a pitch of either 25.4 mm or 23.81 mm. Having studied a particular problem (looking at ligament thicknesses etc) and possibly having discussed it with a mechanical engineer **there are a lot of tube pitches in between these limiting values**. Just changing from a 25.4 mm pitch to a 25 mm pitch will fit in an extra 3% tubes. If you are on a maximum shell size limit and this means using one shell instead of two there is a big saving to be made for a very small change in pitch.

FLOW ACROSS TUBE BUNDLES

Heat transfer and pressure drop on the shell side of a heat exchanger are much more complicated than for a fluid flowing through tubes. We will start by looking at two ideal situations which are only approached during laboratory tests.

IDEAL CROSS FLOW

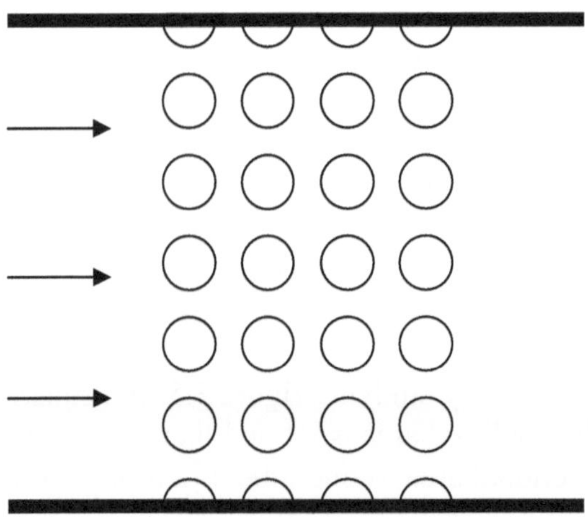

Figure 3.14

The above sketch shows a laboratory type of set up for a square pitch arrangement. The tubes are in a rectangular duct and not a cylindrical shell. The fluid approaches the first row of tubes at a free stream velocity which is calculated from the cross sectional area of the duct. As soon as it reaches the first row it must accelerate as it passes through the flow area created by the clearance passages in between the tubes. It then slows down when it reaches the flow path in between the tube rows. This is repeated throughout the tube bank.

The equations governing heat transfer and pressure drop have a similar form to those for flow inside tubes. The characteristic dimension in this situation is clearly tube outside diameter d_i. The velocity is the maximum value as the fluid passes through the clearances in between the tubes. For the other possible tube arrangements the section of minimum area where the fluid reaches a maximum velocity is shown in figure 3.15. The full length of the section used to calculate the maximum velocity is called the effective diameter for cross flow.

For the two **in line** arrangements (**90° and 30°**) the effective diameter is the number of tubes in a vertical row minus one multiplied by the clearance in between two tubes. For the **rotated pitch** arrangements (**45° and 60°**) it is not quite so simple. There is a flow through two clearances for each length between any two tubes in a vertical row. For the rotated square arrangement the distance between two tubes in a vertical row is $2^{0.5} P_t = 1.414 P_t$ so that there is one clearance for every 0.707 tube pitches. For the rotated triangular arrangement there is a clearance every 0.866 of a tube pitch.

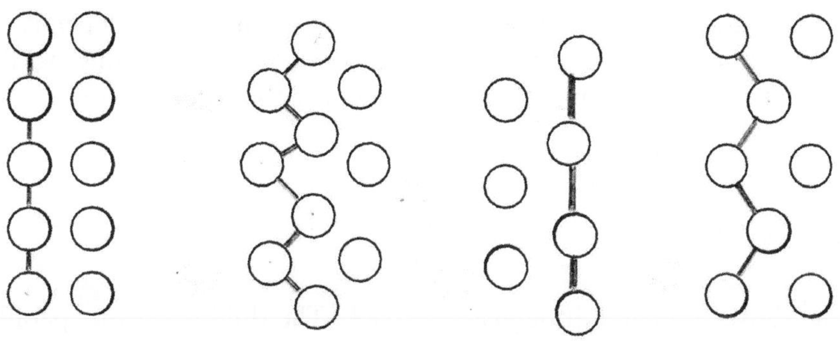

SQUARE PITCH 90° | ROTATED SQUARE PITCH 45° | TRIANGLULAR PITCH 30° | ROTATED TRIANGLULAR 60°

Figure 3.15

DIMENSIONLESS EQUATIONS

The following basic equation applies to flow across tube banks and as for the equation for flow through tubes I have transposed it to clarify how each variable affects the heat transfer coefficient.

$$Nu = \text{a constant} \quad Re^{0.6} \quad Pr^{0.333} \qquad [3.9]$$

or

$$\frac{hd_o}{k} = \text{a constant} \quad \frac{(\rho u\, d_o)^{0.6}}{\mu^{0.6}} \quad \frac{(C_p\mu)^{0.333}}{k^{0.333}}$$

or

$$h = \text{a constant} \; \frac{k}{d_o} \; \frac{(\rho u d_o)^{06}}{\mu^{0.6}} \quad \frac{(C_p\mu)^{0.333}}{k^{0.333}} \qquad [3.10]$$

Comparing this basic equation with the one for flow through tubes leads to some very important deductions. Firstly the exponent of the Prandtl number is the same for both equations. Secondly the exponent of the Reynolds number has been reduced from 0.8 to 0.6. This means that changing the velocity in cross flow has less effect on the heat transfer coefficient than changing the velocity when a fluid is flowing through tubes. Finally I decided to just leave the term 'a constant.' (I could have inserted a figure of about 0.3). This is to emphasise the point that another variable has been introduced compared with flow through tubes. The constant in this equation depends on the tube arrangement particularly the difference between an in line pitch and a rotated pitch.
As well as the modifications in the basic equation due to tube pattern all the other factors which affected flow through tubes still apply. Therefore the basic equation may be modified for heating and cooling, gases and liquids and different ranges of Reynolds number. How each variable affects heat transfer is summarised in table 1 of useful tables at the end of this book. The reader should work these out for himself/herself as an aid to remembering them. A sophisticated modern computer program will contain many equations which fortunately a designer does not have to worry about. He/she should of course appreciate the basics of the equations.

FRICTIONAL PRESSURE DROPS

The basic equation has a similar form to the one used for frictional pressure drops inside tubes. The additional complication is that a value for the friction factor depends on the tube arrangement. If a Moody type of diagram is drawn there will be additional lines to allow for different tube arrangements. Once again the designer need not worry about this but he/she should be aware of it.

IDEAL LONG FLOW

The figure below shows a section through a tube bank. The fluid is flowing parallel to the tubes i.e. at 90° to the plane of the page.

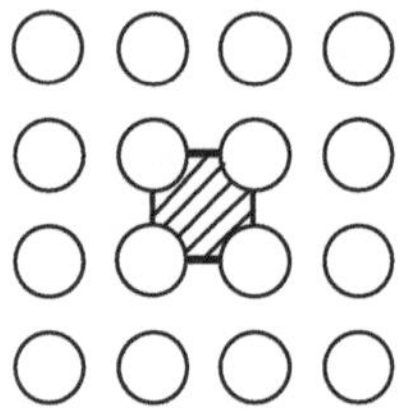

Figure 3.16

For heat transfer and frictional pressure drop calculations the relevant equations require a characteristic dimension. The tube outside diameter is not suitable because the fluid does not move in a path around this dimension. This is a problem for any none circular conduit for example elliptical tubes or the annulus in a double pipe heat exchanger. This difficulty is solved by introducing the concept of an equivalent diameter. The basic definition of equivalent diameter d_e is:

$$d_e = \frac{4 \text{ x Cross sectional area}}{\text{Wetted perimeter}} \qquad [3.11]$$

The wetted perimeter is the length of the surface involved in the process being considered. Applying this equation to the circular cross section of the inside of a tube:

$$d_e = \frac{4 \times \pi/4\, d_i^2}{\pi\, d_i} = d_i$$

Consider the section formed by linking four tubes in a tube bank shown in figure 3.16.

$$d_e = \frac{4 \times (Pt^2 - \pi/4\, d_o^2)}{\pi\, d_o} \qquad [3.12]$$

If this section had a tube pitch of 25.4 mm and the tubes were 19.05 mm the equivalent diameter, de would be 24.07 mm. This would of course be exactly the same for a rotated square pitch. It is a useful exercise for the reader to work out for himself/herself that the equivalent diameter for the same tubes and same tube pitch in any triangular arrangement is 18.29 mm.

For ideal long flow the equations for both heat transfer and frictional pressure drop are essentially the same as for flow through the inside of a tube provided the equivalent diameter is used in those equations.

ANNULUS EQUIVALENT DIAMETERS

I had originally thought in terms of writing a separate chapter on double pipe and multi tube items in which case this section would have been there. I decided against it because most double pipe items are vendor designs and the book was already taking a lot more time than I had estimated. The only other reference to double pipes is in chapter 11 where some alternative types of construction are covered. However the heat transfer and pressure drop through an annulus raises some interesting issues which are worth including.

A true double pipe is simply one pipe inside another pipe as illustrated in figure 3.17. Equation 3.11 applies to the annulus formed between the two pipes.

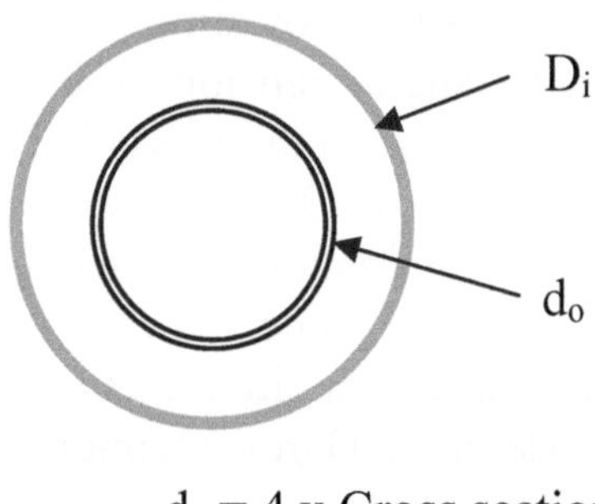

Figure 3.17

$$d_e = \frac{4 \times \text{Cross sectional area}}{\text{Wetted perimeter}}$$

For pressure drop.

The wetted perimeter is the length of the surface involved in the process being considered. For pressure drop both the inside surface of the outer pipe and the outer surface of the inner pipe will result in frictional losses therefore:

$$d_e = \frac{4 \times \pi/4\ (D_i^2 - d_o^2)}{\pi\ (D_i + d_o)}$$

$$d_e = \frac{(D_i - d_o)(D_i + d_o)}{(D_i + d_o)}$$

$$d_e = (D_i - d_o) \qquad [3.13]$$

This is the diametral clearance between the two surfaces and is an easy dimension to remember.

For heat transfer.

For heat transfer it is only the outside surface of the inner pipe, d_o which is active therefore:

$$d_e = \frac{4 \times \pi/4\ (D_i^2 - d_o^2)}{\pi\ d_o}$$

$$d_e = \frac{(D_i^2 - d_o^2)}{d_o} \qquad [3.14]$$

For a double pipe unit with a 4 inch schedule 40 pipe inside a 6 inch schedule 40 pipe the two values for equivalent diameter become:

For pressure drop; $d_e = (D_i - d_o) = (6.055 – 4.5) = 1.555$ inches

For heat transfer; $d_e = \frac{(D_i^2 - d_o^2)}{d_o} = \frac{(6.055^2 - 4.5^2)}{4.5} = 3.65$ inches

The difference between the two values is a factor of 2.35! Having two different equivalent diameters leads to two different values of Reynolds number, one used to calculate a heat transfer coefficient and the other one pressure drop. When I started to do thermal design work we used the same equations as we did for flow through tubes using different values of Reynolds number. However Reynolds number also serves the other important function of identifying the flow regime. This raises an interesting conundrum that one value of Reynolds number might predict turbulent flow and the other one laminar flow?

I have noticed that some authors dodge this issue by simply saying that the diametral clearance should be used as the characteristic dimension for both heat transfer and pressure drop which eliminates the dilemma with the Reynolds number. The two main providers of design programs differ on this one. One recommends the 'traditional', 'correct' approach and the other one the simple method of using the diametral clearance for both. Of course they may both result in the same answer by using different constants in the empirical equations used. (This is a similar situation to the problem of having two different values of friction factor for use in the Darcy equation which was referred to earlier.)

COMBINED IDEAL CROSS FLOW AND LONG FLOW

Taking the idealized situations one stage further try to imagine a tubular exchanger in which there is only long flow and cross flow as illustrated in the diagram below.

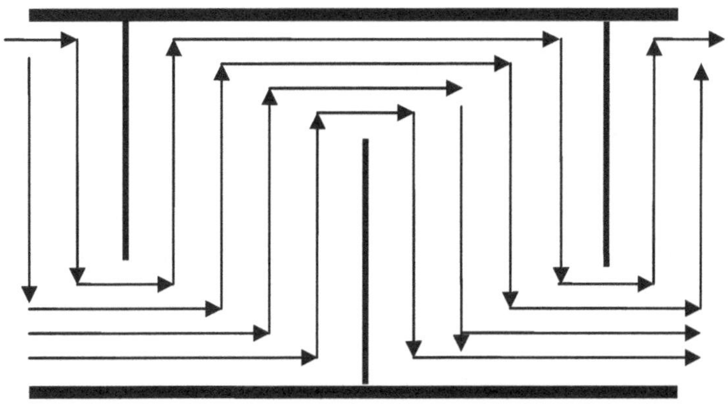

Figure 3.18

The exchanger would have to be a fixed tube sheet design since the baffles would have to be welded to the shell to avoid any baffle to shell clearance. The tubes would also have to be seal welded to the baffles to avoid any flow through the tube to baffle clearances. (I can imagine a construction which would have very little leakage but would be expensive and impractical – can you?) If you are on good terms with a vendor ask him to quote for an experimental exchanger which he must guarantee will be 'leak proof' – the best day of the year to do this is April 1st.

Assuming your vendor carries on with the fun and sends you a sensible looking quote would you expect to have a shell and tube exchanger with only ideal long flow and cross flow as discussed previously? The answer I am afraid is no. When ideal flows were first introduced I guardedly used the expression 'approached during laboratory tests'. Even with the cross flow in a duct illustrated in figure 3.14 there must be end effects. Some of the fluid in the top and bottom flow paths is in contact with the wall of the duct. This effect can be minimised by fixing dummy half tubes to the sides of the duct but it is still there. With a cylindrical shell and baffles the end effects are magnified. Part of the long flow path is in contact with the inside of the shell and another part crosses the edge of the baffle. Part of the cross flow stream is in contact with the shell and another part in contact with the face of the baffle. Another complication is that the cross flow velocity changes as the chord length of the section of the shell changes.

SHELL SIDE PRESSURE DROPS

Frictional pressure drops in both long flow and cross flow have been introduced earlier. There is another important contributor to shell side pressure drop which will now be introduced. Looking a figure 3.18 it is clear that the fluid is changing direction every time it comes up against a baffle or the side of the shell. Changing the direction of a fluid produces a momentum pressure loss which is proportional to the velocity head of the fluid. These loses are often termed 'window turn around' losses or simply window losses. The total pressure drop on the shell side of an exchanger is therefore:

Total Δps = Frictional losses + window losses + Nozzle losses

The nozzle losses are a bit more complicated than the nozzle losses on the tube side because the loss at the inlet nozzle depends on whether an impingement protection device is fitted or not and if it is what type of device. (See the section on impingement protection in Chapter 2.)

PRACTICAL TUBE BUNDLES

More realistic flow paths through the shell side of a heat exchanger are shown below in figure 3.19. This diagram was first proposed by Tinker and has been around since before I started designing exchangers using hand calculations.

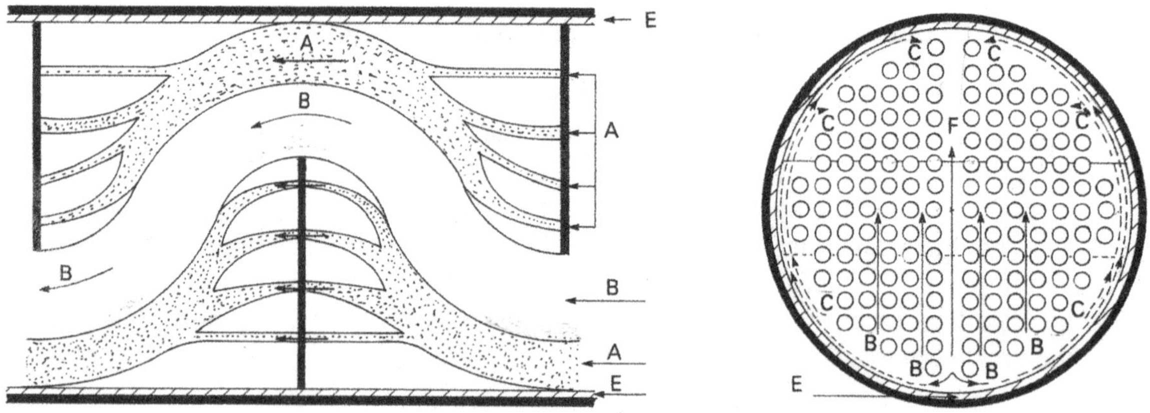

Figure 3.19

The Tinker diagram identifies all the normal leakage streams as well as the long flow and cross flow streams. We should first consider each of these in detail.

B STREAM

This is the closest we can get to ideal cross flow in a heat exchanger with segmental baffles. In the section of the window next to the baffle this stream starts in long flow but soon start to cross the tubes. The nearest to pure cross flow is in the region of the baffle overlap where the stream is closest to being perpendicular to the tubes. It is the most effective stream for heat transfer.

A STREAM

The A stream is the closest any of the streams involved in the heat transfer gets to being pure longitudinal flow. Fluid flows through the annuli formed by the tube/baffle clearances. The velocity is a maximum as it passes through the annuli; it then slows down as it crosses over to the next baffle and the thickness of stream increases.

E STREAM

Fluid flows through the annulus formed by the clearance between the baffles and the shell. Although it is illustrated as a stream with a fixed thickness it will also have a maximum velocity as it passes the edge of the baffles and will slow down in between baffles so that the stream will be thicker there. However it is still unlikely to have significant contact with the tubes and therefore is not involved in the heat transfer process.

C STREAM

It can be confusing to try and show a path for this stream in a side view. In the cross sectional sketch it is shown passing around the bundle in the space between the outermost tubes and the section occupied by the E stream. It must also be passing along the shell so that its flow path is at an angle to the plane of the paper. This stream will always contribute something to the heat transfer however the greater the clearance between the outer tubes and the shell the less effective it is.

F STREAM

The sectional sketch shows a tube layout with one pass partition lane perpendicular to the baffle cut. The fluid is passing along this lane and both sides of the stream are in contact with tubes. If the width of the pass partition lane is minimal this stream can be quite effective for heat transfer. As the width of the pass partition lanes increase the contribution to heat transfer falls.

SUMMARY

Anyone new to thermal design work may find it useful to jot down an aid to remembering these steams since a computer output will just identify a stream by a letter. I offer the following suggestions:

A	Annulus hole flow
B	Best (cross flow)
C	Circumferential flow
E	Extrusion flow (extruded like a tube)
F	Flat flow (the only one taking a linear path)

As with all mnemonics if you make something up for yourself they are more effective.

An observant reader will have noted that there are areas of Tinkers diagram which have not been allocated to any particular stream. These are located in the corners formed between the baffles and the shell. These are relatively stagnant areas. There is some fluid movement due to eddies formed by adjacent streams but they are not as effective for heat transfer as the areas closer to the centre of the bundle. The Tinker diagram is still a simplification of what takes place on the shell side of a heat exchanger. There is always interaction between streams in contact with each other.

It is essential that a thermal design engineer has a vivid picture of what is taking place in the shell side of an exchanger. The complexity will be to easier visualise by someone who has been down white water rapids in an inflatable, swum and snorkelled in rivers or scuba dived in adventurous waters such as those in Scapa Flow in the Orkneys. One moment you may be almost stationary and yet only one stroke away from a current you would be foolish to try and fight against.

I recommend the reader to try and imagine that he/she in a small boat or better still in scuba gear. Start from the inlet nozzle of the exchanger and work your way through to the outlet nozzle by as many different paths as you can find. Going via the C stream is an interesting one and just try going via the E stream at the top of the exchanger and then again at the bottom. Are the two E streams the same? Don't forget the F stream and try and include the end compartments adjacent to the nozzles which are normally larger than the spaces between the baffles.

CALCULATING LONG FLOW AND CROSS FLOW AREAS

The first step for a shell side calculation is to establish long flow and cross flow areas from which the appropriate velocities can be calculated. An equivalent diameter also needs to be calculated for the long flow.

For long flow:

Total flow area with no baffles, $A_{X\,tot} = \pi/4\ D_S^2 - N_T\ \pi/4\ do^2$

Flow through the window only = baffle cut ratio x $A_{X\,tot}$

A stream by pass = (1 – baffle cut fraction) πd_o(tube/baffle radial clearance)

E stream by pass = Length of arc around the baffle x Baffle/Shell clearance

The baffle cut ratio must be on an area basis. This is the most fundamental definition of baffle cut and must always be used for calculations. This is then transposed to a height to diameter ratio or percentage for the data sheet and the mechanical design. Transposing from an area to a dimension basis is straightforward geometry but it is far quicker to use a chart.

The flow area for the equivalent diameter calculation excludes the untubed areas adjacent to the nozzles.

$$\text{Equivalent diameter} = \frac{4 \text{ x Flow area}}{\text{Wetted Perimeter}} = \frac{4\,(A_{X\,tot} - \text{ nozzle entrance \& escape areas})}{N_T\,\pi d_o} \qquad [3.15]$$

For a practical bundle the equivalent diameter will always be larger than the values calculated earlier for an ideal bundle.

For cross flow:

It is necessary to calculate an effective diameter for flow across the bundle. This is relatively simple for an ideal tube bank as illustrated earlier in figure 3.16. For a practical tube bank it may be quite complicated. The gaps between the tubes will depend on the tube arrangement. All by pass streams must be taken account of. The value will be quite different for vertical and horizontal cut baffles in the same shell. Lastly there is the effect of sealing strips. This is very important for pull through bundles. It is also very important for any shell with large nozzles and vertically cut baffles.

Cross flow with vertically cut baffles.

Figure 3.20 shows a section through an exchanger with vertically cut baffles.

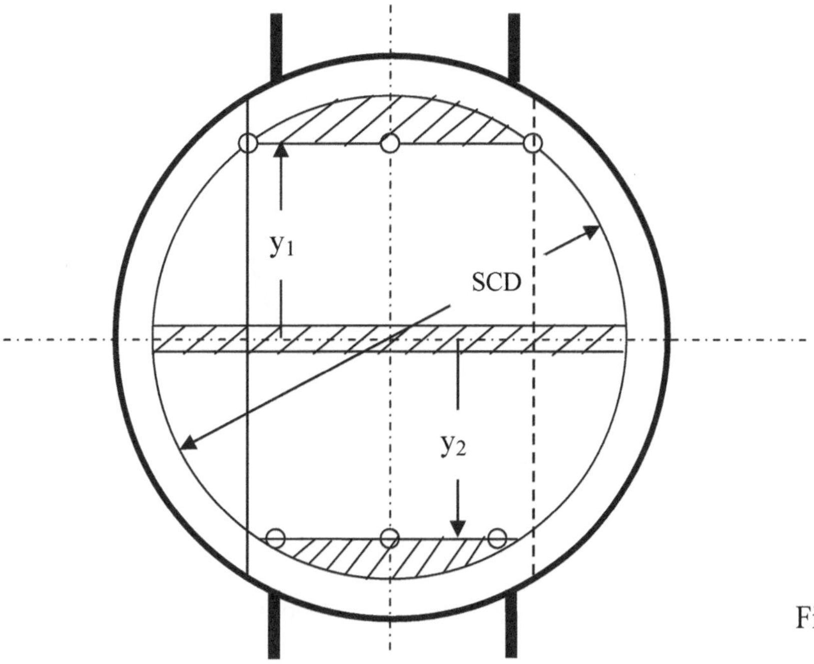

Figure 3.20

The equations used to calculate values of effective diameter, D_{eff} for vertically cut baffles are as follows:

For heat transfer and pressure drop with no sealing strips

$$D_{eff} = \frac{(P_t - d_o)(y_1 + y_2 - BN_{pp})}{F\,Pt} + (Ds - [y_1 + y_2 + d_o]) + N_{pp}(B - d_o) \qquad [3.16]$$

For heat transfer with sealing strips

$$D_{eff} = \frac{(P_t - d_o)(y_1 + y_2 - BN_{pp})}{F\,Pt} + 0.4(Ds - [y_1 + y_2 + d_o]) + N_{pp}(B - d_o) \qquad [3.17]$$

For pressure drop with sealing strips

$$D_{eff} = \frac{(P_t - d_o)(y_1 + y_2 - BN_{pp})}{F\,Pt} + N_{pp}(B - d_o) + 10\text{ mm} \qquad [3.18]$$

When sealing strips are used they are assumed to be partially effective for heat transfer but fully effective for pressure drop. This results in different values of velocity and hence Reynolds number for heat transfer and pressure drop calculations.

Cross flow with horizontally cut baffles.

Figure 3.20 shows a section through an exchanger with horizontally cut baffles.

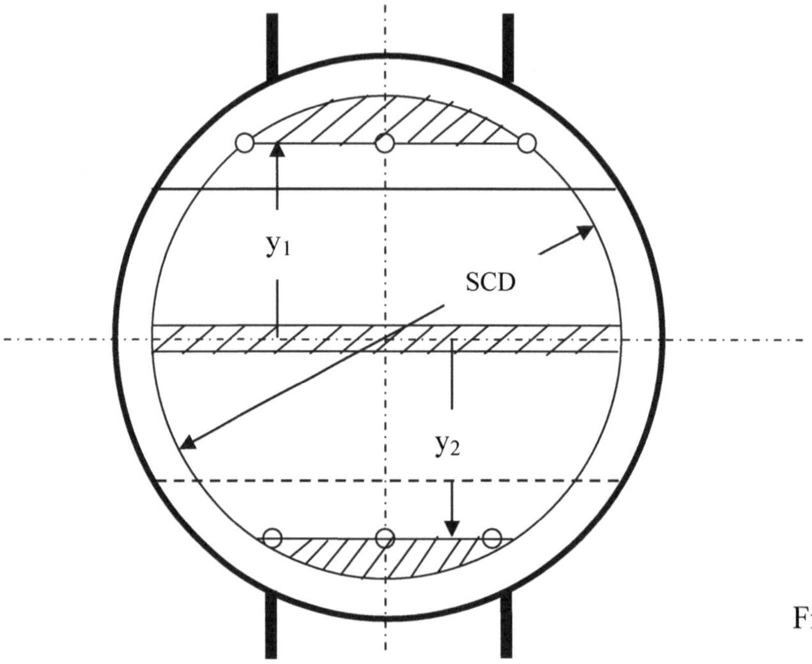

Figure 3.20

The equations used to calculate values of effective diameter, D_{eff} for horizontally cut baffles are as follows:

For heat transfer and pressure drop with no sealing strips

$$D_{eff} = \frac{(P_t - d_o)(SCD - BN_{pp})}{F\,Pt} + (Ds - SCD - d_o]) + N_{pp}(B - d_o) \quad [3.19]$$

For heat transfer with sealing strips

$$D_{eff} = \frac{(P_t - d_o)(SCD - BN_{pp})}{F\,Pt} + 0.4(Ds - SCD - d_o) + N_{pp}(B - d_o) \quad [3.20]$$

For pressure drop with sealing strips

$$D_{eff} = \frac{(P_t - d_o)(SCD - BN_{pp})}{F\,Pt} + N_{pp}(B - d_o) + 10\text{ mm} \quad [3.21]$$

The terms not illustrated in the diagrams are:

F - Tube arrangement factor

1.0 For in line square pitch
0.707 For a rotated square pitch
1.0 For triangular pitch
0.866 For a rotated triangular pitch

B - The distance between the tube rows adjacent to a partition lane.

N_{pp} The number of pass partition lanes parallel to the direction of flow.

N_{pp} is one in figure 3.20 for the vertical cut baffles and zero for the horizontally cut baffles in figure 3.21. For a 4 pass U tube bundle the value would be 1 for both arrangements but the value of B would be quite different.

These equations were used for hand calculations before the advent of computer programs. They were also used in some of the earliest computer programs. Over the years they have been refined and the latest programs will enable you to investigate the effect of varying the number of sealing strips for example. They are also considered to be sufficiently accurate that it is no longer necessary to use different equations for heat transfer and pressure drop.

CONCLUSIONS

My original intention was to give some worked examples of hand calculations and then compare them with designs from a modern computer program but as I am writing this I do not have access to a modern program. Perhaps this is something to be included in a later edition? I have at least given the reader an insight into how we tackled these problems in the sixties and as always the principles remain the same only the sophistication and refinement of the methods used change.

There will also be times when it is necessary to do some hand calculations as an addition to a computer run particularly the calculation of an effective diameter to determine a cross flow velocity. A good example of such a situation is the design of subcooling sections in condensers which is covered in chapter 5. For such a case it is very unlikely that any computer program will be able to handle the geometry for an optimum design. It then becomes necessary to fudge a computer run to simulate the actual geometry as close as possible. A hand calculation is then necessary to calculate the velocities in the actual design and compare them with the values from the computer run.

This page has been left blank so that the next chapter starts on a right hand page.

CHAPTER 4 – DESIGN of SINGLE PHASE UNITS

PRELIMINARY SIZING

This section applies to any type of heat exchanger not just single phase shell and tube items.

PART 1

The surface area of a heat exchanger is found from equation 1.1:

$$A = \frac{Q}{U\,\Delta T_m}$$

The heat load Q will be known but ΔT_m and U need to be estimated.

ESTIMATION of ΔT_m

If+ both fluids are single phase the log mean temperature difference LMTD can be calculated quite accurately from equation 1.23:

$$LMTD = \frac{\text{Largest } \Delta T - \text{Smallest } \Delta T}{\ln(\text{Largest } \Delta T / \text{Smallest } \Delta T)}$$

If one of the fluids undergoes a phase change using the terminal temperature values in the LMTD equation will not be accurate. If you have a graph of the curve it is a simple matter to draw a straight line through it so that the area between the new fictitious straight line and the other fluid profile is approximately the same as between the two actual heat release lines. If you are given a table of values but no plot then look at the temperature drops at either side of the 50% heat release point this will give you an indication of how much the actual heat release line deviates from a straight line. Figure 4.1 shows the temperature profiles for an exchanger where the cooling fluid is single phase but the hot fluid process involves some condensation.

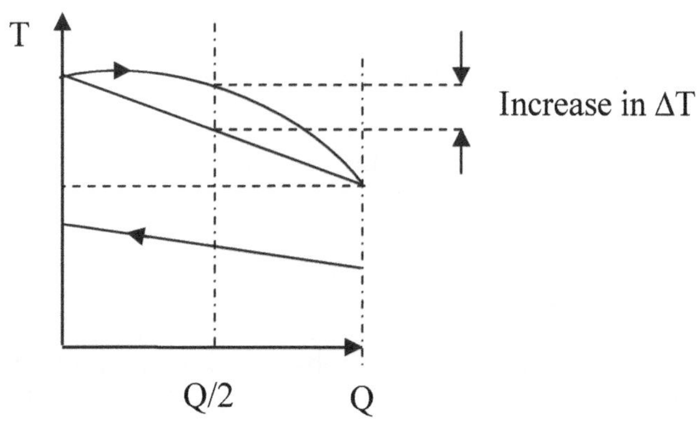

Figure 4.1

The straight line between the hot fluid terminal temperatures is the line assumed in any LMTD calculation. The temperature at the mid point of this line is half the difference between the terminal temperatures added to the inlet temperature. This is compared with the value from the heat release data. In this case the increase in ΔT will increase the LMTD value. The deviation from the calculated LMTD will normally be a bit less than half the value of ΔT at the mid point. Jot down this deviation.

Figure 4.2 shows the temperature profiles for an exchanger where the heating fluid is steam and there is some liquid heating of the cold fluid before it starts to boil.

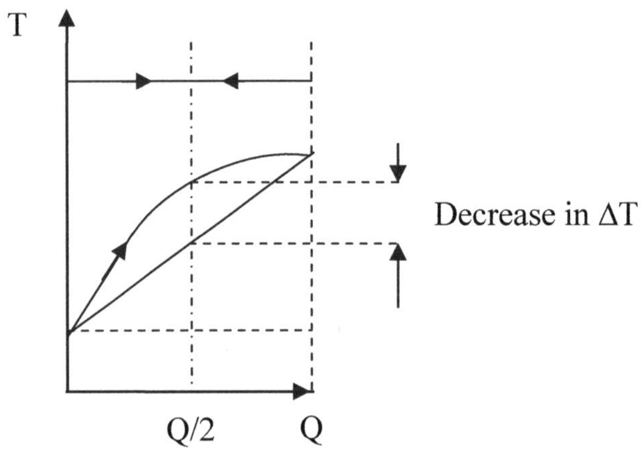

Figure 4.2

The same procedure as above is used to show that there is an increase in the mid point temperature of the cold fluid compared with the mean value for a straight line process. In this case the increase in ΔT will result in reducing the LMTD. As before the deviation in LMTD will be a bit less than half the value of ΔT. Jot down this deviation noting that it is a negative.

Unless one of the fluids undergoes an isothermal process or you are sure that the flows will be counter current the LMTD will require a correction factor, F. You know that the value of F will lie between 0.8 and 1. If the outlet temperatures of the two fluids are close to the same value the correction factor will be close to 0.8. If there is a substantial difference between the two outlet temperatures the correction factor will be approaching 1. Just look at the terminal temperatures and make an estimate of what you think is F will be. If you are starting to take too much time thinking about it just use 0.9.

Then calculate the mean temperature difference, ΔT_m from:

$$\Delta T_m = (\text{LMTD} +/- \text{Any deviation})F \qquad [4.1]$$

ESTIMATION of U_{calc}

If you have access to similar items from past jobs then it is obvious to use these. They may need to be modified for example if the cooling water is from a different source the fouling factors may be different. If there are none or you can not find them it is worthwhile to estimate the individual thermal resistances and then calculate the overall heat transfer coefficient, U_{calc}.

U_{calc} can be calculated from equation 1.2:

$$\frac{1}{U_{calc}} = \frac{1}{h_i}\frac{d_o}{d_i} + f_i\frac{d_o}{d_i} + R_w + f_o + \frac{1}{h_o}$$

Fouling factors will normally be included in the process data, if they are not; there are many sources which the reader will probably have access to. TEMA has a good selection but unfortunately they are only in the outdated US customary units and need to be converted. (The latest additions no longer include the Chicago sanitary canal with its famous fouling resistance of 0.00175 m^2K/W [0.01 in US units] presumably because something has now been done to clean it up a bit!.) Many of the largest client companies have their own lists of fouling resistances. The Shell DEP on fouling resistances has over 30 pages covering many different types of plant and they are in SI units.

The wall resistance for a carbon steel tube is about 0.00005 m^2K/W. If the tubes are stainless steel the thermal conductivity will be less but this will be compensated for by the tubes being thinner. For copper or brass tubes it will be less but a fixed value of 0.00005 m^2K/W is quite accurate enough for estimate work.

The ratio of the tube diameters d_o/d_i is about 1.3 for carbon steel tubes and 1.2 for none ferrous tubes.

Equation 1.2 can now be modified to:

$$\frac{1}{U_{calc}} = \frac{1.3 \text{ or } 1.2}{h_i} + f_i\, 1.3 \text{ or } 1.2 + 0.00005 + f_o + \frac{1}{h_o} \qquad [4.2]$$

The values of h_i and h_o depend on the fluid and the process it undergoes and need to be estimated. Tables of heat transfer coefficients are not as readily available as tables of fouling resistances so I have provided some in table 11 of useful tables. These have been compiled from a variety of sources. In many cases you will not find an exact match with what you require so just select a value for the nearest fluid.

When I started to compile table 11 I realised that once you have selected a value for a surface heat transfer coefficient the first thing you do with it is to take the reciprocal to determine the thermal resistance. I have therefore provided thermal resistances along with the coefficients. The relationship between coefficients and resistances is an interesting one; each provides one of the terms in equations 1.2 and 4.2 and there is no logical reason for them being in different units. To overcome this I have noticed some clients in recent years quoting fouling coefficients in place of fouling resistances. This is more logical but it also means more work since it is now necessary to take the reciprocal of two more terms. If we were starting again from scratch the sensible thing might be to think and work throughout in terms of thermal resistances. (No need for h or U values!). I know I will continue to think in terms of heat transfer coefficients and fouling resistances even though it may be illogical.

The values for the heat transfer coefficients in table 11 have been compiled assuming typical allowable pressure drops; say 0.7 bar for most liquids other than very viscous ones inside tubes which invariably require more. Having selected coefficients, and if necessary fouling resistances, the U_{calc} value can then be established. With the U_{calc} value and the ΔT_m the surface area can now be estimated. This is best illustrated by means of an example.

Example 4.1

15 kg/s of diesel are to be cooled from 80°C to 50°C whilst heating up 35 kg/s of an oil mixture with an average viscosity of 4 cps from 35 °C to 47 °C. The duty is 1350 kW and the diesel is inside the tubes of an AES shell and tube heat exchanger. The fouling resistances for the diesel and oil are 0.001 m^2K/W and 0.0015 m^2K/W respectively. What is the approximate surface area of the unit?

Solution

Calculate the ΔT_m as recommended in chapter 1

$$\begin{array}{ll} & 80 \rightarrow 50 \\ & \underline{47 \leftarrow 35} \\ & 33 \quad\ 15 \\ \text{LMTD} & 18/\ln(33/15) \\ & = 22.8\ \text{K} \end{array}$$

There is only a 3K temperature difference between the two exit temperatures so that the correction factor F will be closer to 0.8 than 1.0 – say 0.85 hence:

$$\mathbf{\Delta T_m} = 22.8 \times 0.85 = \mathbf{19.4\ K}$$

Select appropriate heat transfer coefficients from table 12A. There is a value for cooling diesel of 570 W/m^2K (Thermal resistance 0.00175 m^2K/W). Interpolating for the viscous oil being heated gives a coefficient of about 580W/m^2K (Thermal resistance 0.00172 m^2K/W). Assuming carbon steel tubes the thermal resistances can now be added together and U_{calc} calculated from equation 4.2 as follows:

$$\begin{array}{lll} 1.3R_i = 1.3 \times 0.00175 & = 0.00228 \\ 1.3\ f_i = 1.3 \times 0.001 & = 0.0013 \\ R_W = & 0.00005 \\ f_o = & 0.0015 \\ h_o = & \underline{0.0015} \\ R_{TT} = & 0.00663\ m^2K/W \end{array}$$

$$U_{calc} = 1/R_{TT} = 1/0.00663 = \mathbf{150.8}\ W/m^2K$$

The surface are is therefore:

$$\mathbf{A} = \frac{Q}{U\,\Delta T_m} = \frac{1350000}{150.8 \times 19.4} = \mathbf{461.4}\ m^2$$

This may be all that is necessary for a preliminary design. The magnitude of the answer shows that it is a reasonably sized unit but not too big to require more than one shell. (A typical 15 tonne bundle made out of carbon steel will have about 750 m^2

of surface area.) Note also that for this type of work it is not necessary to distinguish between U_{calc} and U_{req} .

An estimated surface area can be used to:

(a) Estimate the probable number of shells.
(b) See if any units are sufficiently small to be worthwhile sending out for alternative quotes for proprietary double pipe designs. The breakpoint for this does not seem to have changed much over the years and 20 m^2 is good guide. If a lot of items are involved this is probably the most economic path to take. If there is just a single item it is probably better to just design a small shell and tube unit or your own special double pipe design – see chapter 11.
(c) Estimate an approximate cost.

PART 2

The second stage of preliminary sizing is to calculate the number of tubes and fix a shell size. Once again this is easier to explain using an example.

Example 4.2

Estimate a suitable length and diameter for the unit in example 4.1

Solution

The type of unit and material of construction have been established and we now know that it is a reasonably sized unit in terms of surface area. The most economic unit will be the one with the maximum tube length – say 6 m.

Calculate the number of tubes as follows:

$$\text{Number of tubes} = \frac{A}{\pi\, d_o\, L} = \frac{461.4}{\pi\, 0.0195 \times 5.7} = \mathbf{1353}$$

Round this up to say **1360** to allow for some tubes being lost for tie rods.

Note that if the tube length is 6 m the effective length will be reduced by the thickness of the two tubesheets and some loss of effective length behind the floating head support plate. I have estimated this to be about 0.3 m.

The cross sectional area required for a single tube can be found from equation 3.8:

$$A_x = FP_t^{\ 2}$$

Where F is the tube arrangement factor which is 1 for any square pitch and 0.866 for any triangular pitch and P_t is the tube pitch. For an AES unit with a high fouling resistance on the shell side the pitch will be either 90° or 45°. If 19.05 mm tubes are used the pitch will be 25.4 mm.

X Sect Area for tubes = Number of tubes x 1 x $P_t^{\ 2}$
= 1360 x 25.4^2
= 877418 mm2

Figure 4.3 shows a section through a shell with the shaded areas being unavailable for tubes. The sketch is of a two pass unit but at this stage we have no idea how many passes will be required.

It is now necessary to make an allowance for the area unavailable for tubes. In this example there is a liquid on the shell side so the nozzles will be relatively small compared with the shell and about 10% should be sufficient. If there had been a low pressure gas on the shell side you would have allowed more and if you had reason to believe it would be a NTIW design quite a lot more – say 30%.

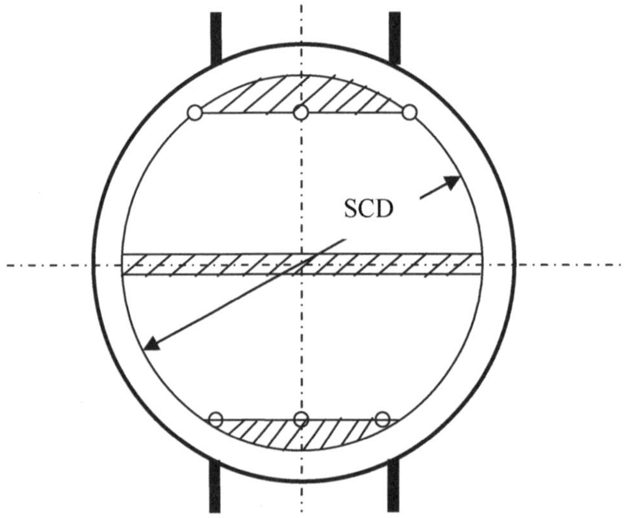

Figure 4.3

The safe circle diameter, SCD can now be estimated as follows:

SCD = (Tubed area x Factor x $4/\pi)^{0.5}$
= (877418 x 1.1 x $4/\pi)^{0.5}$
= 1108 mm

The shell ID, D_s is then calculated from:

D_s = SCD + Tube d_o + Bundle/Shell clearance.
= 1108 + 19 + 44
= 1171 say **1170 mm**

The 44 mm can be found in table 8A of useful tables but an experienced designer would probably just have estimated something between 40 and 50 mm without looking it up.

An estimate of the length and shell diameter of a unit:

(a) Improves the estimate for the cost.
(b) Is useful for plant layout purposes.

In the days when all calculations were done by hand with the help of a slide rule to do the number crunching this preliminary design process would continue. The next step would have been to calculate a tube side velocity. A velocity would be worked out assuming one or two passes and then the number of passes would be increased until the velocity looked about right for the allowable pressure drop (for example 1-2 m/s for water or low viscosity liquids.)

If the preliminary sizing is to be continued into detailed hand calculations, which should still occur during training or academic courses, the nozzles would have been sized before estimating a shell size because these are unlikely to change. The ΔT_m would also have been worked out accurately by calculating the LMTD correction factor. We also had pages of tube count tables which covered most of the units we designed and could be used to select a shell diameter. A lot of books still contain such tables and I am sure that my collection is more comprehensive than most. I could not see any point in reproducing them in this book. For training it is better to work things out from first principles and for detailed work to use a computer program.

Preliminary sizing of a shell and tube unit will only result in a very approximate size. If a lot of units are sized in this way some will be underestimated and others overestimated. The overall accuracy for a project can be surprisingly high.

An interesting story was told to me by an old colleague who was a first class thermal design engineer. He had once been in the situation where a job, (not an estimate or a feed), had come in for the design of a large chemical plant and he was already snowed under with work on other jobs. To make matters worse he had just lost one of his staff. The deadline he was given made it impossible to carry out detailed designs on the heat exchangers. His solution was to use approximate sizing for all the units except for a couple of very critical ones. He used coefficients from a similar plant and made sure that the vendors would be responsible for the thermal guarantees so that someone would be doing some detailed design work. The job went off smoothly and by the time he told me this story the plant had been built and had been working OK for a couple of years.

Returning to the present time, my advice to any novice designer is to develop a competence in approximate sizing. With experience it can take less than ten minutes to calculate a surface area and it is still the first thing I do with any unit. To fix a diameter takes longer and if it is necessary to check velocities then turn to the computer. If all the necessary data such as fluid properties is not available just estimate them to produce a preliminary design,

FLUID ALLOCATION – Tube side or Shell side?

This section also applies to any phase of fluid not just single phase fluids.

The decision as to which fluid should be on the shell side and which fluid should be on the tube side of a shell and tube exchanger will normally have already been made by a process engineer by the time the process data sheet, PDS reaches a thermal design engineer. However this is not always the case particularly for single items which are not part of a large plant. There will also be cases when the process engineer did not find the decision to be an easy one and he/she might have got it wrong. For cases where the decision is not clear cut the process engineer should add a note on the PDS that a switch of fluid sides is acceptable. Frequently they do not and the thermal design engineer should recognise these cases and bring it to their attention. (This is a lot easier if they are in the same department or at least working in the same building.) A thermal design engineer should therefore fully understand all the issues involved in fluid allocation. These are as follows:

MATERIALS OF CONSTRUCTION

A fluid on the tubeside of an exchanger is in contact with the tubes, the channel, channel cover, channel nozzles, partition plates, and the stationary head tubesheet in all units. This is all it will be in contact with for U tube items. It will also be in contact with the rear head and rear head tubesheet in fixed tubesheet items and the rear head tubesheet and cover in floating head items.

A fluid on the shell side of an exchanger is in contact with: the tubes, the shell, the shell nozzles, the bundle internals such as baffles supports runners etc, and the stationary tubesheet in all units. For U tube items it will be in contact with a shell cover. It will also be in contact with the rear head tubesheet in fixed tubesheet items and the rear head tubesheet and cover in floating head items.

Nozzles have different sizing criteria for the shell side and tubeside based the momentum fluxes, (Table 2.5 in chapter 2), but this small change is not worth taking into account. Contact with tubesheets and floating head covers is the same with the fluid on either side. This leaves the main contact difference being between the shell, shell cover and bundle internals compared with the channels and channel covers.

If the material required for one of the streams is significantly more costly than the material required for the other stream then **the most economic solution is to have the fluid requiring the more expensive material on the tube side.** The difference in cost ratio will be a maximum for U tube items and a minimum for fixed tubesheet items.

PRESSURE

If one fluid is at a very high pressure, say over 100 bar and the other fluid is at a relatively low pressure, say below 10 bar this will probably be **the overriding factor and the high pressure fluid should be inside the tubes.**

If a shell is under high pressure not only does this increase the thickness of the shell which is much longer than the channel but thicker tubes will be required. (Tubes can withstand about twice as much internal pressure as external pressure – see useful table 5.) A very high pressure on the shell side would result in a substantial increase in cost.

If the difference in pressure between the two streams is less than 10 bar it is unlikely to have any impact on the allocation of the fluids. Above this it becomes one of the many factors which have to be considered.

FOULING

Fouling of heat exchanger surfaces is discussed in detail in chapter 14. As far as fluid allocation is concerned it is only important for fluids with high fouling factors (say 0.00035 m^2K/W and above) which require regular cleaning of the surfaces they are in contact with.

To clean the inside of tubes, the channels or channel covers must be removed and in the case of floating head items the shell cover and floating head cover. Cleaning is then a relatively straightforward forward exercise. (Cleaning methods are covered in detail in chapter 14 on fouling.) Not only is the process relatively easy but it is a simple matter to inspect that it has been done (except for the bends of U tubes!). U tubes are interesting in so much as it used to be strictly forbidden and still is in most specs, to put a fouling fluid inside U tubes. Improved cleaning methods have encouraged some clients to now allocate a fouling fluid such as cooling water to the inside of U tubes.

To clean the outside of tubes it is necessary to pull the bundle. The cleaning process is then more difficult that cleaning inside tubes and it is not as easy to be sure when it is clean. **If regular cleaning of the heat transfer surfaces is required it is better to have a fouling fluid inside the tubes.**

HIGH VISCOSITIES

If one fluid has a high viscosity, say over five centipoise, it may well be at least partially in laminar flow when it is inside the tubes. This will result in a very low heat transfer coefficient. If a viscous fluid is on the shellside there is a mechanical turbulence which results in much higher heat transfer coefficients. Having the fluid with the highest viscosity on the shellside invariably results in a heat exchanger with a lower surface area than when the viscous fluid is inside the tubes.

Unfortunately high viscosity streams are usually high fouling streams so that there may be a conflict with the cleaning requirements. **Considering viscosity alone the most viscous fluid should be on the shellside**.

LARGE TEMPERATURES DIFFERANCES

If one fluid has a very large temperature difference between inlet and outlet it may cause mechanical problems when it is on the tubeside of a heat exchanger. It is most likely to be a problem when there are only two tube side passes. Half of the tubesheet is exposed to the highest temperature and the other half is exposed to the lowest temperature. This effect is tempered by the shell side of the tubesheet being exposed to a fluid where there is only a very small change in temperature. However the difference in temperature between the top and bottom of the tubesheet will result in bending and the associated stresses which may not be acceptable. The temperature difference above which tubesheet bending may be a problem can vary between as little as 100 K to up to 150 K in different client's specs.

If there is a possible problem it is necessary to provide the mechanical engineer with heat transfer coefficients for both faces of the tubesheet and inside the tubes – where the coefficient will be very high even if the fully established coefficient is quite low. This is not easy and involves a lot of engineering judgement, some pro ration and some hand calculations etc. To err on the safe side the surfaces exposed to the hot fluid should be allocated a heat transfer coefficient as high as can be realistically expected and the coefficient on the surface exposed to the cold fluid as low as can be realistically expected. A further complication is that the heat transfer coefficient on each side of the pass partition plate may be significantly different if the tube side fluid undergoes a change of phase.

One solution is to use insulated ferrules which were covered in chapter 3. These will protect most of the tubesheet surfaces from the hot fluid and result in a more even temperature distribution. A protection plate could be added if there are still any doubts.

Where there are four or more tubeside passes, tubesheet bending is less likely to be a problem. However there may still be problems with differential expansions of the tubes between passes. This is most likely when there is a large change in the tubeside heat transfer coefficient between passes such as when the cold fluid starts off in laminar flow and finishes in turbulent flow. It also requires a relatively poor heat transfer coefficient on the shell side. For example if the shell side fluid was condensing steam and the tubeside fluid heat transfer coefficient was relatively low then the tube metal in both passes would be close to the steam temperature and there would be no problem. A metal temperature difference above about 35 K between adjacent tube passes is worth checking out with a mechanical engineer. (The calculations are similar to those necessary for fixed tubesheet items – see chapter 11.)

If there are more than two passes expansion problems are less likely. Also if there are concerns with expansion problems in a four pass unit it is possible to split the floating head arrangement so that the two hottest and the two coldest passes can expand independently. Problems with differential expansion of the tubes in adjacent passes will not be identified until you have had your first computer run.

Problems with tubesheet bending or differential expansion between tube passes can both be solved by switching the fluid with the large temperature range to the shell side. (Another alternative might be to use two shells in series even if LMTD considerations did not require this.)

LMTD DISTORTION FACTORS

The LMTD distortion factor is used in some computer programs to allow for the **possibility** of the E stream passing along the length of the shell side of an exchanger without mixing between baffle spaces and was discussed in detail in chapter 1. This is not a factor in all computer programs. If you are running a program such as the HTRI program and E stream distortion is a problem it may sometimes be worth considering switching the problem fluid to the tube side.

VOLUME FLOW RATES.

As far as pressure drops in heat exchangers are concerned the designer should think in terms of volume flow rates rather then mass flow rates. Unless the pressures are very high, (approaching the critical point), the specific volumes of gases and vapours are far higher than those of liquids. It is often worthwhile to check out the volume flows at the start of a design if there are any doubts about the relative values. If nozzles have been sized these will give a good indication of the relative volume flow rates of the two streams. (You may even have connecting line sizes on the PDS.)

Low volume flow rates are easy to handle on the tubeside. The number to tube passes can be increased until the allowable pressure drop is used up. High volume flow rates are not and there must be a minimum of two passes for many types of unit. If the pressure drop is too high with two passes and the most economic tube diameter then either the tube diameter must be increased or the tube length shortened.

The shell side has the opposite characteristics. Low volume flows can be a problem. If the allowable pressure drop has not been used up using the minimum baffle cut and pitch there is not much that can be done about it. On the other hand there is a lot of flexibility in dealing with large volume flow rates. The options range from using double segmental baffles to cross flow exchangers with multiple nozzles for very high volume flow rates. The basics of all the possible options have been covered in chapter 2 and will also be discussed in more detail later in this chapter.

As far as volume flow rates are concerned **the stream with the highest volume flow rate should be on the shell side**.

SUMMARY

To allocate fluids to the tube side or shell side consider:

(a) Material cost - expensive - tubeside.
(b) Pressure - high - tubeside
(c) Fouling - high - tubeside
(d) Volume flow rate - high - shellside.
(e) Viscosity - high - shellside

Check for:

(a) A high temperature difference in a tubeside fluid.
(b) A high E stream distortion factor in a shellside fluid.

Having listed the requirements for deciding which fluid should be allocated to which side of an exchanger there will not be many cases where all the factors point in the same direction. Normally however some factors will be more important than others for a particular situation and it will be clear which fluid should be on the tubeside.

Cooling water is an interesting example of fluid allocation. The cooling water will invariably be on the tubeside of a shell and tube exchanger. Many sources of cooling water such as the sea or an estuary will require expensive materials such as brass, copper nickel or titanium. Water from any 'open' source will have a high fouling factors and will require regular cleaning. The temperature ranges are very low and so are the volume flow rates. The pressure may be considerably higher on the shellside but all the other factors point in the same direction and are usually overriding.

If the cooling water is from a closed or 'semi closed', (system with a cooling tower), source there is no need for expensive materials. There will also be reduced fouling factors and less cleaning will be required, possibly none at all with a fully closed system. Although it is no longer so clear cut the cooling water will still probably be on the tube side. Then we come to the situation where the fluid being cooled requires an expensive material and the balance will almost certainly swing in favour of having the cooling water on the shell side.

There will be cases when the most experienced process and thermal design engineers will not be sure on which sides the fluids should be allocated to produce the most economic design. It is important to recognise these cases as early as possible in a thermal design. After a couple of computer runs it is normally clear if it worthwhile looking at a design with the fluids switched.

TUBE SIDE SINGLE PHASE DESIGN

For the most economic exchanger the aim is to fully utilise the allowable pressure drop for each fluid. This is normally more difficult to do on the tubeside than on the shellside. The variables available to us will now be looked at in detail.

CHANGE THE NUMBER OF PASSES

If two tubeside passes are not using up very much of the allowable pressure drop the next step for most units is to try four passes. For the allowable pressure drop not allocated to the nozzles and a fixed number of tubes the effects of this are:

(a) A doubling of the tubeside velocity.
(b) The length of tube the fluid passes along is doubled.
(c) The number of discontinuity pressure losses at the inlet and outlet to each tube pass is also doubled and the turn around losses in the headers is increased.

Using the relationship between velocity and pressure drop from useful tables 1 the pressure drop is changed by the following factor:

$$\Delta p \text{ change factor} = 2^{1.85} \text{ x } 2 = 7.2$$

To change from 4 tubeside passes to 6 tubeside passes the change is reduced to:

$$\Delta p \text{ change factor} = (6/4)^{1.85} \text{ x } 6/4 = 3.2$$

To change from 6 tubeside passes to 8 tubeside passes the change is reduced to:

$$\Delta p \text{ change factor} = (8/6)^{1.85} \times 8/6 = 2.3$$

To change from 8 tubeside passes to 10 tubeside passes the change is reduced to:

$$\Delta p \text{ change factor} = (10/8)^{1.85} \times 10/8 = 1.89$$

The change from two passes to four passes is the most common one and a thermal designer should remember that this changes the pressure drop by a factor of about seven. It is also worth remembering that even a change from eight passes to ten passes will still almost double the pressure drop. If the above exercise had been carried out for a fixed shell size the total number of tubes would have fallen as the number of passes was increased and the factors would have been even more dramatic.

It is quite common to be in situation where one tube pass arrangement uses up a small proportion of the allowable pressure drop but the arrangement with the next highest number of passes is well over the top on pressure drop. (An advantage of fixed tubesheet items is that the number of tube passes can be varied in odd numbers of tube passes as well as even numbers of tube passes therefore you are much more likely to establish a design for a particular length of tube within the allowable pressure drop.) If a sensible proportion of the allowable pressure drop can not be used up by varying the number of passes then the next variable to look at is the tube length.

CHANGE THE TUBE LENGTH

You would start most designs using the maximum allowable tube length since this will result in the most economic unit if a design is possible. (*Long and thin is sure to win. Short and squat can cost a lot!*) For many years the maximum tube length has been limited to 24 ft for fixed tubesheet items and 20 ft for items with a removable bundle by handling and cleaning limitations. These were then reduced in increments of 4 ft resulting in 16, 12 and 8 ft becoming standard lengths. The idea behind using standard lengths is to minimise stocks of spare tubes. 10 ft tubes were often also acceptable because a 20 ft tube could be cut in half. (If cutting is acceptable it is only necessary to keep stocks of 24, 20 and 16 ft with modern stock control methods). Now that most people have moved on to the SI system these standard lengths have been maintained but converted to the new system to become; 7315, 6096, 4877, 3658 and 2438 mm. The tube lengths to be used on a job are normally in the job specs and should always be in the job notes. It is important to establish if 6096 mm is to be used for the most common length or whether this is to be rounded to 6100. When I write job notes I will write this dimension in bold to emphasise this point. Sometimes now there are jobs with a maximum tube length of 6m and increments of 1m which gives more flexibility to the thermal designer.

A U tube exchanger is a special case. It is becoming increasingly popular because of its low capital cost and improved methods of cleaning tubes around the bends. Only the outer row of tubes could possibly be replaced and I have never come across a case where spare outer tubes were kept in stock. **The tube length can therefore be specified in sensibly rounded increments of say 5 or 10 mm.** I have never heard a reasoned argument against this although I have worked on many jobs were I have been told that I must still stick to one of the standard length systems described in the previous paragraph. (Clearly a past committee decision and one from a committee with either no thermal designer on, or only an inexperienced or very quiet one present.) When you are allowed to do it small changes in tube length will often allow you to fully optimise the tube side pressure drop.

It is now necessary to investigate the effect of changing a standard tube length. If the length of a heat exchanger is reduced it is necessary to increase the shell diameter to accommodate more tubes which will maintain the surface area. The effect on pressure drop of changing the tube length from 6096 mm to 4877 mm is as follows:

Decrease factor due to length change = 4877/6096 = 0.8

Decrease factor due to velocity change = $(4877/6096)^{1.85}$ = 0.662

Net decrease factor = 0.8 x 0.662 = 0.53

The effect on pressure drop of changing the tube length from 4877 mm to 3658 mm is as follows:

Decrease factor due to length change = 3658/4877 = 0.75

Decrease factor due to velocity change = $(3658/4877)^{1.85}$ = 0.587

Net decrease factor = 0.75 x 0.587 = 0.44

As a rough guide to keep in ones head both these changes can be thought of as halving the pressure drop, (excluding the nozzles).

CHANGE THE TUBE DIAMETER

Our last variable is the tube diameter. This is only considered as a last resort since smaller diameter tubes normally result in the most economic unit. The most common change is from 19.05 mm tubes to 25.4 mm tubes so this change will be looked at in detail. If we start by assuming the same surface area, the number of tubes will decrease in proportion to the ratio of the two outside diameters i.e. 1.333. If both tubes are 14 BWG minimum wall the thickness will be 2.3 mm resulting in inside diameters of 14.45 and 20.8 mm respectively. Using the approximate relationships in useful table 1 the change in pressure drop can now be calculated as follows:

Decrease factor due to d_i change = $(20.8/14.45)^{-4.8}$ = 0.174

This assumes that all other parameters including the velocity remain the same but if the surface area is the same the velocity will increase in the ratio of the change in the number of tubes.

Change in Δp due to velocity change = $1.33^{1.85}$ = 1.67

Net change in Δp factor = 0.174 x 1.67 = 0.29

This change in pressure drop by a factor of about 0.3 may be sufficient to go back and increase the number of tube passes. (Note that if the spec had required the 25.4 mm tubes to be 12 BWG the change in diameter and hence pressure drop would have been a bit smaller.)

Unlike the two previous variables the tube diameter also affects the heat transfer coefficient as well as the pressure drop. In turbulent flow the relationship is $d_i^{-0.2}$ which for the above example would reduce the heat transfer coefficient by about 7%. If the flow is laminar the relationship is $d_i^{-0.67}$ which for the above example would reduce the heat transfer coefficient by about 27%.

The lower heat transfer coefficients together with the increase in shell diameter required for a fixed surface area per unit length of exchanger make the use of larger tubes the least attractive option.

ASK FOR MORE ALLOWABLE PRESSURE DROP

The above options are based on the assumption that the allowable pressure drop must be strictly adhered to. This is not always the case. If you have an otherwise good design where the pressure drop limit is exceeded by a small amount say 10-15% it may be worth looking at the full circuit the heat exchanger is in. Increasing the allowable pressure drop will ultimately be a decision for the process engineer but often a good thermal design engineer will want do his own preliminary investigation before he makes out a case to justify it.

Process flow diagrams, such as the process and instrument diagrams (P&ID's) and plant layouts should be available throughout any job. Find the appropriate diagrams and look at:

(a) The full circuit and all other items of equipment in the circuit – particularly any other heat exchangers.
(b) The means of providing the fluid flow.

There may be some very simple solutions to your problem such as:

(a) There is another heat exchanger in the circuit. If it happens to be one you have already designed and have some pressure drop left over the problem may be solved. If you have not designed the other unit yet suspend the calculations you are doing until you have done sufficient work on the other unit to see if some pressure drop can be reallocated.
(b) Check if the circuit is driven by a pump or compressor or by a hydraulic head. If it is a simple circuit driven by a hydraulic head margins will already be tight and it is unlikely that the allowable pressure drop can be increased.

Other solutions may not be so obvious. An interesting and important case is a cooling water system. Pumps adjacent to the source will deliver cooling water to heat exchangers all over the plant. The heat exchangers served will be at different distances from and at different elevations to the pumps. (It is very similar to a hot water radiator system used for central heating in winter in northern latitudes.) Often the same allowable pressure drop will be allocated to all the units. Central to the design of any cooling water system is to identify the index init. This is the one with the greatest hydraulic length from the pumps and this unit must have the required flow rate and allowable pressure drop available to it. After the index circuit has been sized the flows to the other items are controlled by the pipe sizes and valves.

If the unit you are investigating is on the index circuit or a circuit which is close to it (there may be several circuits with similar hydraulic lengths) then you can forget about asking for more pressure drop. If however the unit you are investigating is hydraulically much closer to the pumps than some of the other units you will have a good case for asking for more pressure drop.

For non cooling water circuits with lots of other items of equipment particularly valves it may be worthwhile going straight to the process engineer and discussing the problem. This assumes that he/she is available. If you do not have access to the process engineer you will need a very strong case. One situation where it is still worth pursuing is when the only alternative to a bit more pressure drop is to use two shells.

MULIPLE SHELLS

The above has been written assuming that the unit will be a single shell. This is not always the case. It will be evident at an early stage if more than one shell is required. If a large number of shells are required it is important to look at all possible combinations. For example twelve shells may be arranged as; 2S6P, 3S4P, 4S3P or 6S2P. Design where the arrangement of shells becomes another variable can be very time consuming but it is time well spent because these are very expensive items.

SUMMARY

To optimise the tube side pressure drop:

(a) Vary the number of tubeside passes.
(b) Vary the tube length.
(c) Consider using larger diameter tubes.
(d) Vary the arrangement for multiple shell items.

Consider asking for the allowable pressure drop to be increased.

SHELL SIDE SINGLE PHASE DESIGN

You are more likely to be able to fully utilise the allowable pressure drop on the shell side than on the tube side. This is because there are more variables which can often be changed in smaller steps. These are as follows:

(a) The main variable – the baffle pitch can sometimes be changed by smaller increments than either the number of tube passes or the tube length, (except for U tubes).
(b) The type of baffle can also be changed.
(c) The shell side flow path can be changed by changing the TEMA type.
(d) The tube pitch can be changed.
(e) The C stream, (**C**ircumferential flow) and any F stream, (**F**lat flow) can be modified.

Also

Heat transfer coefficients are less sensitive to changes in velocity, ($u^{0.6}$ compared with $u^{0.8}$), in turbulent flow and there are no sudden changes in flow regime with viscous fluids because of the mechanical turbulence.

As with tubeside design the first thing is to establish an allowable pressure drop excluding the nozzles. Nozzles should be sized during an early computer run. The program you are running may size nozzles for you by allocating say 10% of the allowable pressure drop to the nozzles. These sizes may not be the optimum ones and may need adjusting as explained in chapter 2. Each variable will now be looked at in more detail.

BAFFLE PITCH AND CUT

Baffles were discussed in detail in chapter 2 and some of the following text is a resume of some of the points made there.

Baffle pitches can be varied between a diameter to pitch ratio, D_s/P_b of 1 and 5. This will normally be called up by a spec or standard but even if it is not there is very unlikely to be an optimum design outside these limits.

Baffle cuts can vary between about 15% and 45%. The maximum figure is governed by the requirement to have a baffle overlap of two tube rows and it may be higher for large shells. Very large baffle cuts are most frequently used when the shell side fluid is condensing. An optimum baffle cut is unlikely to be more than about 35% for a single phase unit. Increments of baffle cut are limited by the requirement for the edge of a baffle to be either through the centre of a tube row or in between two tube rows as explained in the section on baffles in chapter 2. You would not bother about this requirement during early computer runs but it is essential that it is correct on your final run.

Baffle pitch and baffle cut are in theory independent variables and most computer programs will allow you to change either one while the other remains constant. However they are not totally independent. A good shell side design should have a reasonable balance between the cross flow and long flow velocities. To achieve a good balance, short baffle pitches, (high D_s/P_b ratios) will require a low baffle cut and as the length of the pitch increases so will the value of the baffle cut.

Figure 4.4 shows a typical single segmental baffle arrangement for a single phase fluid on the shell side of a heat exchanger. The baffles are horizontal which eliminates any dead spaces in the end compartments. One disadvantage of the horizontal baffle is that there must be an even number of baffles to give an odd number of flow paths across the unit.

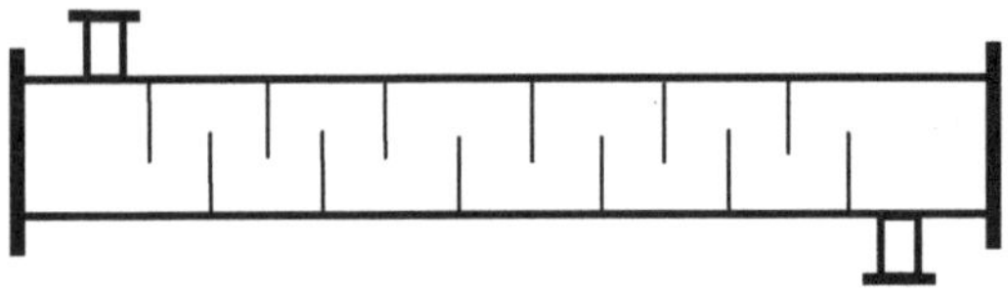

Figure 4.4

At an early stage of the design you should fix the length of the end spaces as explained in chapter 2 although these may change a bit as you change the shell diameter or re-size a nozzle. You will have input or allowed the computer to calculate the thickness of the tubesheets so you will now have a tube length between end baffles to work with. Some very old hand calculation methods assumed a baffle cut of 25% for single phase fluids. I still find this a good starting point for computer runs until it needs adjusting to balance up the long and cross flow velocities.

For a fixed effective length between end baffles the relationship between baffle pitch and pressure drop is $P_b^{2.85}$ from useful table 1. The power of this expression consists of 1.85 for the velocity change plus 1 to allow for the increase in the number of cross flow paths.

The first target for any shell side design is to achieve a solution with single segmental baffles in an E type shell. The baffle pitch for the section between end baffles is therefore reduced to try and fully utilise the pressure drop. The computer input will probably be the total number of cross flow paths which is the number of crosses in the baffled section plus two for the end spaces. The resulting baffle pitch will not be a sensibly rounded number, this will later be adjusted.

During this process you may decide to deviate from a cut of 25% if it becomes clear that this is the way to go to balance the long and cross flow velocities but at this stage you will be using increments of say 5%. If there is a solution which uses up most of the pressure drop the final stage is to optimise the baffle cut. You are probably looking at a design where the baffle cut is neither through the centreline of a tube row or in between tube rows and this needs to be corrected on your final run. You may also find that a % cut slightly off the nearest practical one will improve the design. For example if your last run was with a 30% cut and the nearest practical cut is 30.2% try practical cuts at either side of this value as illustrated in the following table:

Baffle Cut %	Oversurface %
34.2	0
32.2	0.5
30.2	0.9
28.2	1.7
26.2	1.2

Table 4.1

Provided the pressure drop is still OK the optimum cut is clearly 28.2%. The variation shown in table 4.1 is typical. In this example it may be argued that a 30.2% cut would have done the job so why bother? My counter to this is that one should always try and optimise any design, and what if the 30.2% cut had resulted in 0.6% undersurface? A designer who was unaware of the changes a small change is baffle cut can produce might have rejected the design.(This is of course most important when the shell side coefficient is the controlling one.)

Optimising the baffle cut in this way may appear to be a lengthy tedious exercise but for an experienced designer it only takes a few minutes and is just a back of a cigar packet job. When I am supervising a novice designer I insist that they produce a table for the checker to look at.

If you are just over on pressure drop and can do nothing about it with horizontal single segmental baffles I have come across cases where switching to vertical cut baffles where the baffle pitch can be varied in increments of one will just produce a design. This is only likely to occur when the end spaces are not too much larger than the baffle pitch and the thermal resistance of the shell side stream is a small proportion of the total thermal resistance. (Of course the spec may not allow you to try this.)

If the pressure drop is too high with single segmental baffles the next thing to try is changing the type of baffle.

CHANGE THE TYPE OF BAFFLE

The standard change of baffle type is to use double segmental baffles instead of single segmental baffles. This means switching from horizontal to vertical cut baffles as illustrated in figure 4.5, which is a plan view.

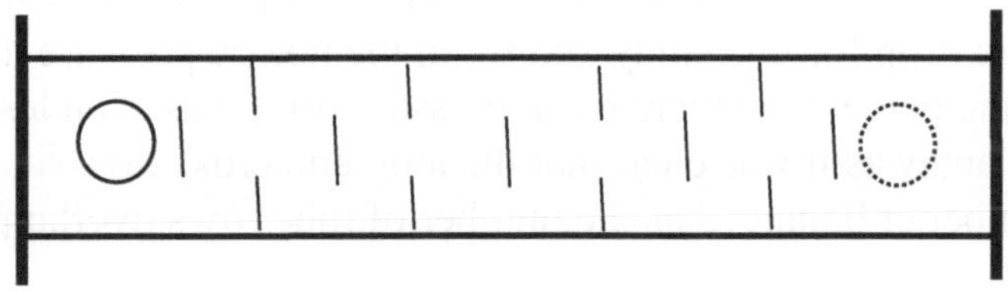

Figure 4.5

Figure 4.6 shows the flow paths across a double segmental baffle and a single segmental baffle with the same pitch.

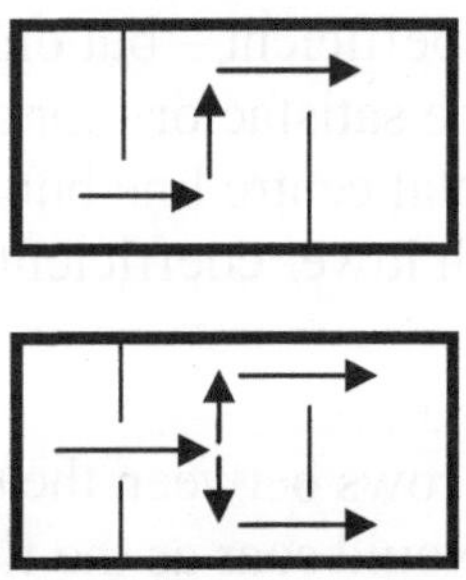

Figure 4.6

The cross flow velocity has clearly been halved. The cross flow pressure drop has therefore been reduced by a factor of $2^{-1.8} = 0.29$ assuming the same number of tube rows has been crossed. However the number to tube rows crossed will also have been reduced so the factor will be even smaller. The long flow has not been affected so dramatically but the cross flow is the main contributor to shell side pressure drop. The reduction of pressure drop of course comes at the expense of a reduced heat transfer coefficient.

When double segmental baffles are input into a computer program the usual output will be for design with the minimum of two tube rows overlap and a baffle cut of about 25%. This will give the minimum pressure drop for a particular baffle pitch. This is very seldom the optimum solution as table as table 4.2 below demonstrates.

INPUT	OUTPUT					
Rows Overlap	Baffle cut %	Overlap mm	S Area %	Δp_S kPa	B Stream	A Stream
None	27.26	52	- 12.4	30.9	0.67	0.066
5	24.46	130	- 1.91	32.7	0.54	0.102
10	19.61	260	+ 12.8	34.3	0.46	0.155

Table 4.2

Table 4.2 above is part of a study I did in the early nineties to alert other members the group I was working with of the importance of using the number of tube rows overlap as one of the input variables. I had just checked a second item where there had been no input and the program had defaulted to the minimum of two rows overlap. Neither run had produced an optimum design.

The shell side heat transfer coefficient was strongly controlling the design I selected – between 50% and 60% of the total thermal resistance. The table shows that changing the number of tubes overlap from the minimum to 10 resulted in a change in the surface area required of over 25% for an increase in pressure drop of only 11%. I had isolated the bundle pressure drop for this study and when the nozzle pressure drops where taken into account the percentage increase in pressure drop was even less. Looking at the full stream analysis it was clear that the long and cross flow components became much better balanced as the number of tube rows overlapped was increased.

A double segmental baffle design with only two tube rows overlap will have a very high B stream. To an inexperienced designer this may appear to be a very good feature but like many aspects of life appearances can be very deceptive. A high B stream value will result in a high heat transfer coefficient – but only across two tube rows at either side of the exchanger. This may be satisfactory for a very small bundle with a small number of tubes across the horizontal centre line but as the bundle size increases more and more tubes will have a much lower coefficient and the design will diverge from the optimum solution.

There is of course a limit to the number to tube rows between the edges of adjacent baffles. Too many rows and the pressure drop would soar as the fluid was squeezed through the diminishing window areas. A balance is required between the long flow and cross flow just as with single segmental baffles. However it is only occasionally with very small shells that I have found a two row overlap produces the optimum design.

Returning to figure 4.5 it is clear that the number of baffles must be varied in steps of two, (the same as horizontally cut single segmental baffles) to minimise dead areas in the end spaces. The number of cross passes and the number of tube rows overlapped can then be varied to optimise the design.

If the shell side pressure drop is still too high the next logical step might seem to be to consider triple segmental baffles which were briefly described in chapter 2. The disadvantages of these are that to maintain symmetrical baffling it is necessary to work in increments of three baffle pitches and it becomes more difficult to balance the long and cross flows. We never used them with hand calculations. For many years these were not an option in the programs I was using which has limited my experience of them. If you are only just over on pressure drop they may be worth investigating. Otherwise it is time to change the type of shell.

USE A TEMA J TYPE SHELL

Changing the type of shell is the last resort in reducing the shell side pressure drop. This is because not only is the cost of the shell going to increase if more than two nozzles are used but the complications and costs of the pipework outside the exchanger are going to increase. A typical J type of shell is shown in figure 4.7.

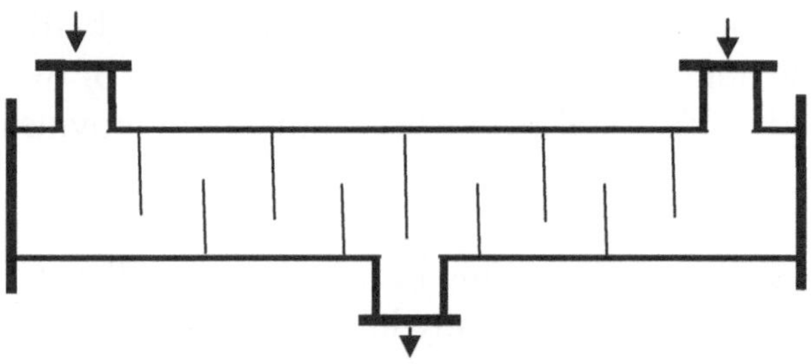

Figure 4.7

For many years a J type shell was just a J type shell, a sketch on the thermal data sheet would indicate which way the shell side fluid was flowing. With the advent of computer programs it became necessary to be more specific without any need for a sketch. Figure 4.7 would now be termed a 21J, (or J21 – check which your program is using). The first number indicates the number of inlet nozzles and the second number the number of outlet nozzles. Reverse the direction of the arrows in diagram 4.7 and the unit would be designated a 12J.

The first consideration is how a J shell compares with an E shell. Assuming the same shell size with the same number of tubes and the same baffle pitch it is clear that for a divided flow:

(a) The velocity has been halved.
(b) The length of the flow path has been halved.

The change in pressure drop, excluding the nozzles, will be governed by the following factor:

$$\Delta p \text{ change factor} = 0.5^{1.85} \times 0.5 = 0.139 \text{ (about } 1/7^{th}\text{)}$$

The pressure drop change factor is the inverse of the pressure drop factor for changing from two passes to four passes on the tube side.

The second consideration is should we use a 21J or a 12J for a particular design? The answer is that the fluid with the highest volume flow rate should pass through the two nozzles since this will minimise the area unavailable for tubes adjacent to the nozzles. For single phase fluids this means that the two nozzles will always be at the top irrespective of whether the shell side fluid is being heated or cooled. The only exception I can think of is on a re-vamp where there may be piping constraints. (An exception in two phase flow is a horizontal reboiler where having a single outlet nozzle has considerable advantages – see chapter 9)

Note that for a shell with a divided flow the central "baffle" is a full diameter support which is only cut about 3 mm away from the tube row adjacent to the nozzle.

Although you will have already tried using double segmental baffles in an E type shell you should switch back to single segmental baffles for your first runs on a J type shell. Figure 4.6 shows horizontal baffles which are the most suitable for single phase flow. Unfortunately they do have a disadvantage. Cross passes must be changed in steps of two on either side of the central support. This means changing the number of cross flows for the full length of the exchanger in steps of four. I can not think of any way of predicting how well this will work out until you have your first computer run. You may sometimes have the option of trying vertical single segmental baffles where it is only necessary to change the number of cross passes in increments of two. However the increase in ineffective areas makes this very unlikely to provide a satisfactory solution.

Another limitation of the J type shell is with floating head units where the end space required for the floating head end will be significantly longer than that required for the fixed head end. Whilst there is no theoretical reason why the flows in each direction must be the same it is unlikely that your computer program will be able to handle this and it will be necessary to increase the shorter length to maintain symmetry.

When all the sensible combinations of tube pitch, (and %cut) have been exhausted and the pressure drop is still too high it is time to investigate using double segmental baffles in a divided flow shell. The changes in pressure drop will be the same as they were for an E type shell and once again it will be a matter of varying the number of cross passes and the number of tube rows in the baffle overlap. For the optimum flow path it is necessary to work in steps of four for the number of crosses as it was with the horizontally cut single segmental baffles.

One case where the ideal flow path created by having a central baffle adjacent to each nozzle may cause problems is when the bottom nozzle is the inlet nozzle and the baffle pitch is relatively small. Having central baffles adjacent to the nozzle may restrict the flow path and cause unacceptably high local velocities and hence momentum fluxes. This is most likely to be a problem when a plate must be used for impingement protection. One solution to this is to increase the length of the spaces adjacent to the central support another is to relax the requirement for a central baffle adjacent to the nozzle as illustrated in figure 4.8.

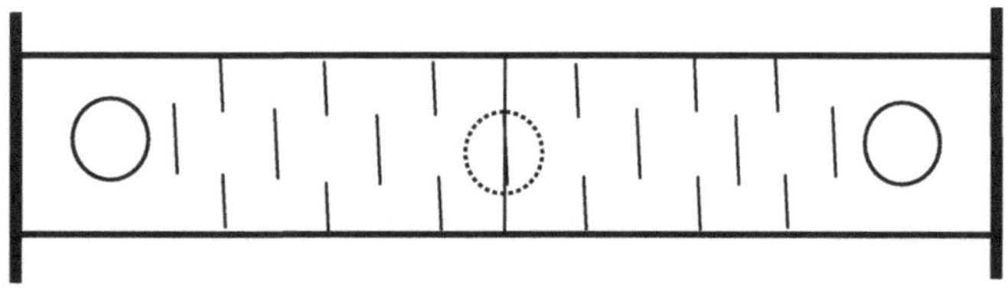

Figure 4.8

OTHER TEMA TYPES

With a liquid on the shell side you should have found a solution which utilised most if not all of the shell side pressure drop long before reaching the divided flow with double segmental baffle stage. The later part of the above procedure will only have been necessary with a gas.

With low pressure gases you will probably have already been experiencing vibration warning messages unless the shell is a very small one. There are ways of dealing with many of these which are discussed in the appropriate sections in chapter15. With very severe problems it may be necessary to use vibration free designs, (except for acoustic vibrations) such as NTIW and cross flow types with multiple nozzles. It will normally be clear at an early stage in the design if this is the way forward.

TUBE PITCH

It is easy to forget that the tube pitch is now a possible variable particularly for fine tuning of the shell side. This was not always the case. From the days when calculations were aided only by a slide rule through the development of the first computer programs we would always stick to standard tube pitches. This was because of the way vendors produced heat exchangers. Their templates and drilling machines would all be set up to deal with the standard pitches. A none standard tube pitch presented problems which could only be overcome by measures which would increase the cost. With the advent of numerically controlled machines it is now a relatively simple matter to produce heat exchangers with any tube pitch required without increasing the cost.

The first thing the designer should be aware of is the effect a small change in tube pitch has on the pressure drop and heat transfer coefficient. The following tables have been compiled by just taking pressure drop to be proportional to the velocity to the power of 1.85 and the heat transfer coefficient to the power of 0.6:

Increase in P_t	Δp factor	h_s factor
1 mm	0.77	0.92
0.5 mm	0.87	0.96
0.1 mm	0.97	0.99

Table 4.3

I started with a standard gap between tubes of 6.35 mm in each case for both tables. They show that even quite small changes in tube pitch can have a significant effect on pressure drop while having a much smaller effect on the heat transfer coefficient

Decrease in P_t	Δp factor	h_s factor
1 mm	1.36	1.11
0.5 mm	1.16	1.05
0.1 mm	1.03	1.01

Table 4.4

Decreasing a tube pitch may of course infringe a specification but on a number of occasions I have trimmed down a 25.4 mm pitch to 25 mm when using 19.05 mm tubes with a square pitch arrangement. I would not go any further with a square pitch arrangement and would not even try that if the fluid was a heavy fouling one or the client an immovable one. There is often a much better case for a small reduction in tube pitch when a triangular arrangement is being used with a 25.4 mm pitch and 19.05 mm tubes. Tube pitches, ligament thicknesses and welded via expanded tubes are discussed in more detail Chapter 2. If I were dealing with a situation when I felt that a 23.81 mm pitch would not have been unreasonable I might try trimming the tube pitch down to say 24.5 mm.

Any reduction in tube pitch should be discussed with all concerned before you go too far and you will have to present a good case. The best cases are where you can stay with an E type shell instead of a J type shell or when you can keep to one shell within other parts of the spec instead of having to use two shells.

I can not think of any technical objections to increasing a tube pitch except that the number of tubes which can be fitted into a fixed shell size will be reduced.

The greatest change I ever made was from a 25.4 mm pitch to a 28 mm pitch. The shell side fluid was cooling water so the tubes were on a square pitch arrangement. It was a revamp job and there was only room for an E type shell. The volume flow rate was high which resulted in not only excessive pressure drops but vibration problems. (Vibration problems are unusual with liquids but not unknown particularly with cooling water.) Another designer had worked on the unit and reached the conclusion that the only solution was using expensive rod baffles. This meant sending off for a proprietary design and a subsequent delay. When it was passed over to me I went through the calculations he had done and could find no mistakes. I went outside for an inspirational cigar and returned to investigate the effect of changing the tube pitch. Within a couple of hours I had a design which was just on the allowable pressure drop and all the vibration warning stars had disappeared. The pressure drop factor for the change in tube pitch I used was 0.54 and the heat transfer factor was 0.81. The shell diameter had to be increased which reduced these values even more. The shell side thermal resistance was so low compared with the total thermal resistance that the reduced heat transfer coefficient had little effect on the overall design. The dramatic reduction in pressure drop meant that I could fit in more baffles which together with the reduced velocity eliminated the vibration worries.

The above example can hardly be described as 'fine tuning' but it dose demonstrate how important a variable the tube pitch can be. Most of the changes I have made to a tube pitch have only been by a fraction of a millimetre.

MODIFY BY PASS STREAMS

This is not a separate step or topic to consider when optimising a heat exchanger but something that would have been carried out throughout all the above procedures. I have left it until the end to avoid distractions from the main points I have been trying to make in each section.

A C stream, (**C**ircumferential flow) would be present in all the types of unit we have looked at. This stream can be modified by changing the number of sealing strips used. It is unlikely that sealing strips would not be required for any type of floating head design particularly for a pull through floating head. At the other extreme a U tube bundle, particularly with a large shell and tubes on an inline square pitch, may not require any at all. If you are using a computer program which defaults to inserting sealing strips for you let it have its way during your early runs and then look to modify these as you approach a final design.

There may not be any F stream, (**F**lat flow) particularly with horizontally cut baffles. At the other extreme with vertically cut baffles and a lot of tube passes or with a U tube design with horizontally cut baffles and more than two tube passes the F stream might be quite important. Pass partitions and the methods of sealing them were discussed in detail in chapter 2. If the F stream is low – say below 10% it is probably not worth sealing it because both sides of the stream are in contact with the tubes and therefore making some contribution to the heat transfer. (Unlike the C stream where only one side of the stream is in contact with the tubes.) With a large F stream you may have pressure drop to spare and need to improve the heat transfer coefficient. Then fully blocking it may be the answer. I have noticed in between the limits of no seals or fully sealed F streams some computer programs are a bit dodgy and the designer should be aware of this.

The E stream, (**E**xtruded flow) is the only one which contributes absolutely nothing to the heat transfer and ideally it should always be minimised. However every aspect of design on the shell side is a balancing act and any change in one flow stream will affect all the others. I am still working on a relatively cheap but reliable seal to drastically reduce the E stream for units where it can become a major problem. I am sure I am not alone and if the reader can come up with anything he should keep it to him/herself until he has seen a patent agent.

SUMMARY

For an optimum design you should have used up most if not all of the allowable pressure drop on both sides of the exchanger. I have been through the steps you would take to optimise the tubeside and shellside separately when of course you would be carrying out both processes simultaneously. Dealing with so many variables which interact with each other is what makes thermal design work so challenging.

I have only dealt with single phase streams but many points will apply equally to two phase streams. Other points would not. For example it is often not ideal to try and use up the entire allowable pressure drop for a pure fluid that is condensing, particularly steam.

Lastly I have not mentioned data checks which apply equally to both fluids. We have all been there at some time but there is nothing worse than finalizing a design and then having it pointed out by the checker that part of the input is either wrong or that some aspect of the design does not comply with the job specifications. Try and get it all right before you start.

SAFETY FACTORS

This is always an important and often controversial issue with any item of equipment but particularly for heat exchangers. This section applies to any heat exchanger not just single phase ones.

PROCESS SAFETY FACTORS

The first safety factor which is applied to many heat exchangers is the sole responsibility of the process engineer. . Process may use overcapacity factors for critical circuits or just critical items. The PDS, (process data sheet) may have:

(a) An overcapacity factor which applies to the flow rates of both streams and the duty.
(b) A requirement for a specified value of oversurface.
(c) An overcapacity factor for the duty only.

The first item in the above list is very easy to deal with. The computer program you are using may even have an input for overcapacity which saves you the trouble of reaching for the calculator. Although it should not be necessary it is advisable to check that the allowable pressure drops apply to the overcapacity case.

The second item is also relatively easy to deal with. The thermal design engineer simply comes up with a design with the required oversurface which overrides whatever value may otherwise have been used. It may of course in effect temper the value which may otherwise have been used. For example if 20% oversurface had been required and the 'optimum' design was 18% oversurface, (2% undersurface with no process factor) this should be acceptable.

Both the process engineer and the thermal design engineer should be aware that for example a 10% oversurface will not necessarily produce a 10% increase in duty. If the mass flow rates and inlet temperatures are maintained for single phase fluids the percentage increase in duty will always be less than the percentage oversurface. For such a case the outlet temperature of the hot stream must decrease and the outlet temperature of the cold stream increase. This reduces the Δt_m by decreasing both the LMTD and the correction factor, F. In some cases the percentage increase in duty may be significantly less than the increase in oversurface. On the other hand if the two mass flow rates are increased with the temperatures constant the duty will increase by more than the percentage oversurface because both heat transfer coefficients will increase.

The last item which may occur on a PDS is a requirement for a percentage increase in the duty only. If you try to input this into a reasonable computer program you will at least be warned about the imbalance between the duty and the heat flows calculated for the two streams. If you receive a PDS with a requirement for an overcapacity on duty only take it or send it back and point out the at the only practical options are (a) or (b).

HEAT EXCHAGER SAFETY FACTORS.

Safety factors for the actual heat exchanger design can be much more complicated than process safety factors as far as the thermal design engineer is concerned. Also there will often be many items on a plant with no process safety factors. A novice designer may not he aware of the considerations which have been taken into account to arrive at a safety factors for the exchanger. He/she may just follow the job notes and select the appropriate values.

Ideally a safety factor should be allocated to both the heat transfer coefficient and the pressure drop on each side of the heat exchanger. It is normally only necessary to specify safety factors for the pressure drops. This is because an excessive pressure drop in a heat exchanger could cause serious problems as soon as a plant is commissioned. If a heat exchanger underperforms on heat transfer it may not be so serious because other items in the circuit may be able to compensate fore this. Often safety factors are part of clients' specifications and may change for different types of plant. Sometimes there will be no values given therefore an experienced designer should understand all the variables which need to be taken into account to allocate sensible values.

Before looking at the variables it is important to mention the tube inside diameter. Tube wall thicknesses were discussed in detail in the section on tube specifications in chapter 2. I recommended using the mean value of the possible extreme values allowed by the specified tolerances. (Resulting for example in using the nominal wall value for stainless steel tubes which have an average wall specification tubes and +10% for the thickness of carbon steel tubes which have a minimum wall specification.) This is just a sensible approach to fixing a tube inside diameter and if +10% has been used for a carbon steel tube it should not be thought of as a safety factor.

The variables we need to consider when allocating safety factors are as follows:

(a) Physical properties
(b) Shell side or tube side.
(c) Fouling
(d) Heat transfer & pressure drop correlations.

PHYSICAL PROPERTIES

Physical properties were discussed in detail in chapter 1. Properties for the commonest fluids such as air or water are well established although I am still cynical about the four and five figure accuracy of the data in sources such as the Rogers and Mayhew tables (Ref 15).

The accuracy of some of the physical property data for a lot of the fluids we deal with is probably no better than plus or minus 10%. The exception is the specific heat which should be much more accurate because:

(a) It is the easiest property to measure in a laboratory.
(b) Weighting a property for a mixture from its constituents is a precise procedure.
(c) It is so important to get it right because of its effect on the heat balance.

Occasionally it is necessary to deal with a fluid which is new to you are you are cautious about, perhaps because of unusual property values. If in any doubt you should discuss the properties of the fluid with the process engineer. If the fluid is confirmed as being a bit dodgy I would use a safety factor of 0.9 or even less for the heat transfer coefficient. This would be noted in the calculation file for the checker. Any action on the pressure drop safety factor would depend on the value of the viscosity and the side of the exchanger the fluid was on.

FOULING FACTORS – EFFECT ON HEAT TRANSER

The effect of fouling on the overall heat transfer coefficient is to reduce it and the differences between clean and dirty or fouled heat transfer coefficients are well documented and referred to throughout this book.

The relationship between fouling factors and individual heat transfer coefficients is quite different and less well known. Any fouling will increase the individual surface heat transfer coefficients. This is best illustrated by numerical examples

Example 4.3

A shell and tube exchanger has 20 mm tubes which are 2 mm thick on a 26 mm in line square pitch. What is the effect of a 1 mm layer of fouling inside the tubes on the heat transfer coefficient assuming?

(a) Turbulent flow and a clean coefficient of 1000 W/m^{2}K.
(b) Laminar flow and a clean coefficient of 100 W/m^{2}K

Solution

(a) The inside diameter for the clean design is 16 mm and this is reduced to 14 mm when the tubes are fouled. Assuming the mass flow rate is constant the velocity change will be inversely proportional to the ratio of the diameter change squared. The change in the characteristic dimension, (the inside diameter) will also affect the coefficient. Using the relationships from useful table 1:

$$h_{fouled} = h_{clean} (d_{ic}/d_{if})^{0.2} (d_{ic}/d_{if})^{2 \times 0.8}$$
$$= 1000 (16/14)^{0.2} (16/14)^{2 \times 0.8}$$
$$= \mathbf{1272\ W/m^2K}$$

When this fouled coefficient is referred to the outside diameter of the tube it will be reduced by a factor of (d_{if}/d_o) instead of (d_{ic}/d_o) for the clean coefficient resulting in a net reduction of (d_{if}/d_{ic}). The fouled coefficient above therefore needs to be corrected for a true comparison with the clean coefficient.

$$\text{Corrected } h_{fouled} = 1271\ (d_{if}/d_{ic}).$$
$$= 1271\ (15/16)$$
$$= \mathbf{1191\ W/m^2K}$$

This is effectively a 19% improvement in the coefficient although it may be more mathematically sound to think of it as a 19% reduction in the thermal resistance due to the inside heat transfer coefficient when it is referred to the tube outside diameter.

(b) Again using the relationships from useful table 1 at the back of this book.

$$h_{fouled} = h_{clean}\ (d_{ic}/d_{if})^{0.67}\ (d_{ic}/d_{if})^{2 \times 0.33}$$
$$= 100\ (16/14)^{0.67}\ (16/14)^{2 \times 0.33}$$
$$= \mathbf{119\ W/m^2K}$$

$$\text{Corrected } h_{fouled} = 119\ (d_{if}/d_{ic}).$$
$$= 119\ (15/16)$$
$$= \mathbf{112\ W/m^2K}$$

This is effectively a 12% improvement in the coefficient.

Example 4.4

Use the data in example 4.3 to calculate new effective heat transfer coefficient when the fouling is on the shell side and the flow is turbulent.

Solution

The outside diameter for the clean design is 20 mm and this is increased to 21 mm when the tubes are fouled. Assuming the mass flow rate is constant the velocity in cross flow will increase in the ratio of the gap between the tubes. The change in the characteristic dimension, (the outside diameter) will also affect the coefficient. Using the relationships from useful table 1:

$$h_{fouled} = h_{clean}\ (gap_c/gap_f)^{0.6}\ (d_{oc}/d_{of})^{0.4}$$
$$= 1000\ (6/4)^{0.6}\ (20/21)^{0.4}$$
$$= 1251\ W/m^2K$$

When this fouled coefficient is referred to the outside diameter of the tube it will be **increased** by a factor of (d_{of}/d_o).

$$\text{Corrected } h_{fouled} = 1251\ (d_{of}/d_o).$$
$$= 1251\ (21/20)$$
$$= \mathbf{1313\ W/m^2K}$$

This is effectively a 31.3% improvement on the coefficient.

The calculations for the tubeside only relied on the approximate relationships from useful table 1 and are therefore reasonably accurate. The overall improvement in the heat transfer coefficients are quite significant and on the safe side.

On the shell side the improvement in heat transfer coefficient is even more significant. However the calculations I have used are a simplification. They may not me too far off for a cross flow unit but for other types of the design they take no account of the changes in the stream analysis which fouling may induce. These effects may of course increase the coefficient even more if some of the bypass streams become blocked.

There is no way that a computer program can take advantage of the increases in heat transfer coefficients unless a value for the thickness of the fouling layer in input which is very seldom the case. **For the vast majority of designs fouling remains a hidden safety factor for the heat transfer coefficient in single phase flow**.

FOULING FACTORS – EFFECT ON PRESSURE DROP

Fouling always increases the pressure drops and can be taken into account in a number of different ways. The first thing to look at is the effect of allocating a thickness to the fouling layer in the same way as in the previous examples for heat transfer.

Example 4.5

A shell and tube exchanger has 20 mm tubes which are 2 mm thick on a 26 mm in line square pitch. What is the effect on pressure drop of a 1 mm layer of fouling?

(a) Inside the tubes.
(b) Outside the tubes.

Solution

The increase in pressure drop inside the tubes is found from one simple relationship:

$$\begin{aligned}\Delta p_{\text{fouled}} &= \Delta p_{\text{clean}} (d_{ic}/d_{if})^{4.8} \\ &= \Delta p_{\text{clean}} (16/14)^{4.8} \\ &= \boldsymbol{\Delta p_{\text{clean}} \times 1.9} \text{ – A 90\% increase in } \Delta p\end{aligned}$$

On the shell side the pressure drop will change due to the change in velocity in the gaps in between the tubes.

$$\begin{aligned}\Delta p_{\text{fouled}} &= \Delta p_{\text{clean}} (gap_c/gap_f)^{1.8} \\ &= \Delta p_{\text{clean}} (6/4)^{1.8} \\ &= \boldsymbol{\Delta p_{\text{clean}} \times 2.07} \text{ - A 107\% increase in } \Delta p\end{aligned}$$

As for the heat transfer case, the B stream velocity would change if the flow through the by pass streams change so that this is a very approximate estimate.

For fouling on either surface the results for increases in pressure drop are both interesting and disturbing. The 1 mm thickness assumed is not unrealistic.

There has been an option for inputting a fouling layer thickness on the inside of tubes for many years in some of the programs I have used. Some programs will now allow a fouling layer thickness outside the tubes to be input. I have on occasions used the option on the tubeside but never on the shell side. The problem of course is getting the data. A thermal design engineer can only to use what he/she is given.

Using thickness values, (together with fouling thermal conductivity values), is clearly the most scientific approach to dealing with fouling. I assume that some enlightened clients must be now using them but I have not been lucky enough to work on one of their jobs. This topic is discussed in more detail in chapter 14 on fouling.

One of the more traditional approaches has just been to use different safety factors for different values of fouling factor as typified in tables 4.5 and 4.6 which have been taken from different specs I have worked with in the past..

Fouling factor m^2K/W	Safety factor
0.00009 - 0.00033	1.1
0.00034 – 0.00085	1.2
0.00086 & above	1.5

Table 4.5

Fouling factor m^2K/W	Safety factor
< 0.00017	1.05
0.00017 – 0.0004	1.1
> 0.0004	1.2
Heavy fouling	1.3

Table 4.6

For high fouling fluids on the shell side there is a more scientific way to investigate the increase in pressure drop due to fouling than just selecting a high safety factor. There has been an option in some programs for many years now to block the clearances between the baffles and shell and the baffles and tubes, (The E and A streams.) This gives a good indication of how fouling will effect the shell side pressure drop. This is sometimes found in a spec but I have often used it in a high fouling situation when there has been no mention of it in a spec. I can not visualise a situation where all the clearances are blocked. However some will be and there will also be layers of dirt outside the tubes which will compensate for any unblocked clearances.

HEAT TRANSFER & PRESSURE DROP CORRELATIONS

The final reason for applying safety factors to heat transfer coefficients and pressure drops is to take account of the inaccuracies in the correlations used to calculate them.

Table 4.7 is an extract from a spec I once had to work to in the early nineties. The table takes account of the tolerances for the correlations being used in one particular computer program. It is only for heat transfer coefficients hence all the safety factors are less than one.

	Accuracy %	Safety factor
Tubeside – Water turbulent	± 10	0.9
– Other fluids turbulent	± 15	0.85
– Laminar/Transition	± 30	0.7
Shellside – All single phase fluids	± 30	0.7
Two phase – Either side	± 35	0.65

Table 4.7

Extracts from another spec of the same era are shown in table 4.8

	Safety factor
Tubeside – Water turbulent	1.0
– Other fluids turbulent	0.826
Shellside – All single phase fluids	0.79
Two phase – Either side (No inerts)	1.0

Table 4.8

Since the nineties the accuracy of the correlations used in computer programs has no doubt improved but any relationship which relies on test data will always have some tolerance.

Comparing the two tables I remember being surprised at any safety factor for water as in table 4.7. It would have little effect because the very high heat transfer coefficients of water normally contribute a very small resistance to the total thermal resistance.

The most interesting comparison of course is the safety factors for the changes of phase. It is generally accepted that the correlations used with changes of phase have a much wider spread of data than those for single phase. This is particularly true for boiling fluids where the accuracy for some situations is still no better than about ± 50%.

In recent years most of the specs I have worked to did not require any safety factor to be applied to the heat transfer coefficients. This has not stopped me occasionally using one when I considered it to be necessary.

It is very bad practice not to apply any safety factor to pressure drops. Despite this I have know some companies who do it – particularly vendors who are under pressure to be very competitive. (They are more likely to do this if they know that process have already applied a safety factor – particularly a significant factor on the flow rates and duty.)

Typical standards for pressure drop safety factors which I have worked with in recent years are shown in table 4.9

	Safety factor
Tubeside – Single phase	1.1
– Two phase	1.15
Shellside – Single phase B stream >0.6	1.1
– Single phase B stream <0.6	1.15
– Two phase	1.15

Table 4.9

These are sensible minimum safety factors which could be increased if the designer considered it to be necessary for a particular unit. One interesting feature is the shell side factors for single phase flow which depend on the magnitude of the B stream. This makes sense since far more tests have been carried out for cross flow over tube banks than for any of the other shell side streams.

COMBINED SAFETY FACTORS

Some specs have safety factors for pressure drops which combine more than one variable such as the ones shown in table 4.10.

Tube di mm	Wall t mm	Carbon steel	Non-Ferrous
25.4	1.65	1.24	1.1
25.4	2.11	1.28	1.12
25.4	2.77	1.35	1.15
19.1	1.25	1.28	1.14
19.1	1.65	1.34	1.17
19.1	2.11	1.41	1.2
19.1	2.77	1.53	1.26

Table 4.10

This table was designed to take account of the difference in specifications between minimum wall and average wall tubes and the fouling inside tubes made of different materials.

It is worthwhile to isolate these two components. The tolerance for carbon steel tubes is -0 to + 20%. The arithmetic mean of the tolerance is therefore + 10%. The difference in pressure drop between using the nominal wall value and a mean minimum wall value can be calculated for the 2.11 mm thick 19.1 mm tube as follows:

$$\text{Difference in } \Delta p = \{(19.05 - 2.11) / (19.05 - 2.32)\}^{4.8} = 1.06$$

The factor of 1.41 in table 4.9 therefore allows 0.06 for the variation in tube wall and 0.35 for fouling. This compares with 0.2 for the none ferrous tube which would have an average wall specification and therefore no adjustment to the tube wall thickness. The 15% difference between these two figures reflects the tendency of carbon steel tubes to foul more readily than non ferrous tubes. (This introduces yet another variable which will be discussed more fully in the chapter on fouling.)

Ironically for the shell side in the same spec there was a 1.15 safety factor for liquids and boiling services and no safety factor for other fluid states – no mention of fouling!

SUMMARY

(1) If you are tight on pressure drop remember that it is not necessary to apply any safety factor to the nozzle losses. Even heavy fouling should not affect these. (If a reasonable sized nozzle has a 1 mm layer of

fouling the new diameter will probably still be within the tolerance of the inside diameter.)

(2) There are several variables which can affect safety factors:
 (a) Physical property tolerances.
 (b) Fouling (& the surface being fouled.)
 (c) Heat transfer and Δp correlation tolerances.

(3) It would not be unreasonable to include (a) in (c) but it was more useful to analyse them separately.

(4) The effect of the variables on the safety factor may be different on each side of the exchanger.

(5) Many client specs concentrate on only one variable.

(6) Tube wall tolerances should ideally not be included in any safety factor.

(7) **The effect of fouling on pressure drop can be grossly underestimated**. I have seen situations where heavily fouled units have had a lot more than 1 mm of deposits on the surface of the tubes. This applies to both sides of an exchanger.

(8) I have never worked with a spec which covered all the variables and I probably never will. A good thermal design engineer must work to the spec he/she is given but should recognise situations where it is necessary to deviate from the spec. (Conservative deviations from a spec are not normally too much of a problem if you have a reasonable case.)

(9) Safety factors are probably the most dodgy aspect of thermal design work

CONCLUSIONS

There has always been a wide range of variables available to the thermal design engineer. In the days of hand calculations there was seldom time to investigate all the possibilities and producing a good design relied heavily on the designers experience in identifying the critical variables.

With the advent of computer designs it is now possible for a novice to look at the effect of far more variables provided he/she can identify them. The number of variables has also increased. For example I would once have simply assumed what I considered to be a sensible number of sealing strips for a design, now I can investigate the effect of varying the number.

I have focused on the design of single phase units in this chapter but most topics are equally relevant to the design of other units.

CONDENSERS – CHAPTER 5

INTRODUCTION

Condensation occurs whenever a saturated vapour comes into contact with a surface at a temperature below the saturation temperature of the vapour.

It is sometimes presented as a topic under the general umbrella of convection. I have always thought of it as a separate topic. As far as the practical design of heat exchangers is concerned there are some similarities but there are also very significant differences between convective heat transfer and condensing heat transfer. For thermal design work the most important difference is the allowable pressure drop allocated and the way that this should be treated.

Ideal condensation of a pure fluid is not only isothermal but also isobaric. Process will allocate a relatively low pressure drop to a condensing stream. Some pressure drop is necessary, particularly for condensing inside tubes. For condensing on the shell side an ideal design would only require some pressure drop to cover the nozzle losses. Any pressure drop in the tube bundle will reduce the saturation temperature and hence the MTD as explained in chapter 1. This is quite different from the requirement to always try and use up the entire allowable pressure drop in convective heat transfer.

NUSSELT'S THEORY

The most important contributor to our knowledge of condensation was Nusselt who published his paper in 1916. It is encouraging to remember that someone was carrying out such work at a time when most scientists and engineers were fully engaged in developing more effective methods of killing people.

Nusselt made the following assumptions in his analysis of filmwise condensation:

(1) The temperature of the surface is constant.
(2) All physical properties are at the mean film temperature.
(3) The condensate film is so thin that the temperature gradient across it is linear.
(4) The curvature of the film is neglected.
(5) The downward flow is laminar and heat is transferred across the film by conduction only.
(6) There is no sub cooling of the condensate. The heat delivered depends only on the enthalpy of phase change of the fluid, h_{fg}, at the condensing temperature.
(7) The thickness of the condensing film at any section if a function of the mean velocity and the mass of condensate passing through the section.
(8) The mean velocity at any section is a function only of the viscous forces resisting flow and the gravitational forces causing flow.

He started with the most straight forward geometric shape – a flat plate. This is illustrated in figure 5.1 below.

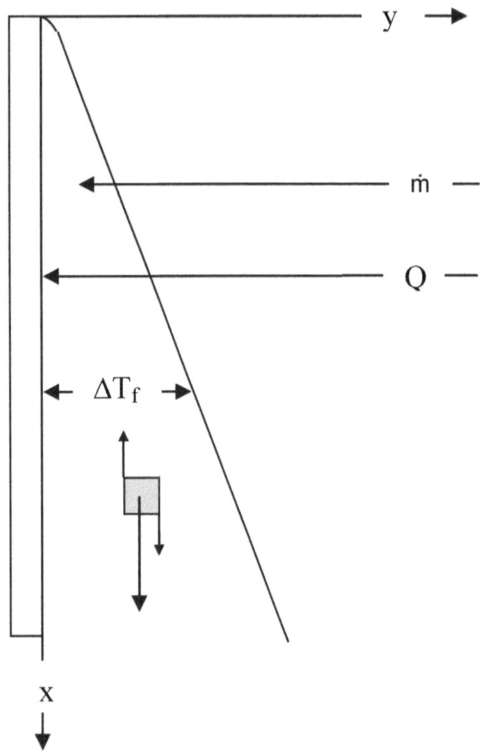

Figure 5.1

The theory analyses the forces on an element of fluid in the film with dimensions δx and δy. The gravitational force causes the element to flow downwards and is resisted by a viscous force. This is used to establish a velocity profile across the element. The velocity profile has a maximum value at the interface between the liquid and vapour and a value of zero at the surface of the plate. The other fundamental equations used are a mass balance and a heat balance.

The details of the analysis take up several pages and are covered in many standard textbooks such as reference 14 by Kern. The result is the following equation for the average heat transfer coefficient between the top of the plate and a distance L from the top of the plate:

$$h = 0.943 \frac{(k^3 \rho^2 g\, h_{fg})^{1/4}}{(\mu L\, \Delta T_f)^{1/4}} \qquad [5.1]$$

ΔT_f is the temperature difference across the liquid film and the physical properties are for the liquid at the mean film temperature except for h_{fg} which is the enthalpy of phase change. g is the acceleration due to gravity.

The constant in the equation is from one of the integrations and is not an empirical value. (My apologies to the reader for the presentation of this equation, I never found a way of extending brackets over two lines of text so on editing this section I added the power to the bottom of the equation as well as the top to avoid confusion in the early part of this chapter. Later on I have omitted the power from the bottom line when the spacing should make clear what is intended.)

For the outside of a tube the Nusselt analysis is more complex because of the geometry which is illustrated in figure 5.2

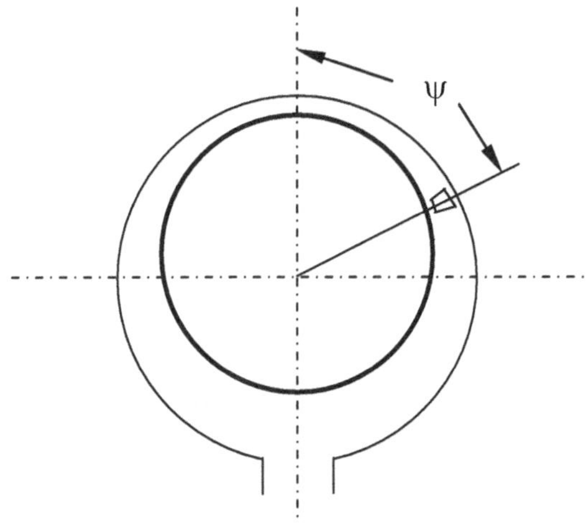

Figure 5.2

In this case the element of fluid is at an angle ψ from the vertical. The dimensions of the element are now $\delta\psi$ and δr. Although the mathematics is more involved the procedures followed are similar to those used for the analysis of condensation on a flat plate. The final equation for the average heat transfer coefficient during condensation outside a single horizontal tube is:

$$h = 0.725 \frac{(k^3 \rho^2 g\, h_{fg})^{1/4}}{(\mu d_o \Delta T_f)^{1/4}} \qquad [5.2]$$

The only differences between equations 5.2 and 5.1 are the numerical constant and the tube outside diameter; d_o replacing the length from the top of the plate, L.

The values for the average heat transfer coefficient during condensation predicted by the Nusselt equations are in good agreement with experimental data. Equation 5.1 was developed for a flat plate but it can also be applied to the outside of a vertical tube. The following points should be noted from Nusselt's work:

(a) For a flat plate or a vertical tube the heat transfer coefficient approaches infinity at the top where the condensate film approaches zero thickness. As the length from the top increases the heat transfer coefficient decreases. This is the same phenomena as in convective heat transfer at the entrance to a tube which was discussed in chapter 1. For the horizontal tube there is always a high but finite value at the top and a very low minimum value at the bottom.

(b) Condensing heat transfer can be thought of as conduction through the condensate film and any variable which affects the thickness of the condensate film will affect the heat transfer coefficient. This again is similar to thinking of convective heat transfer as being conduction through the boundary layer.

(c) The equations predict the effect of specific physical properties. For example the heat transfer coefficient is proportional to $k^{3/4}$ and $\mu^{-1/4}$. (These exponents are modified for flow across banks of tubes.)

(d) The equations can be used to compare the condensing coefficients on a vertical tube with those on a horizontal tube. Consider a 19 mm diameter tube 1000 mm long in both orientations condensing the same fluid with the same temperature difference across the film. Dividing equation 5.2 by 5.1 results in:

$$\frac{h_{horizontal}}{h_{vertical}} = \frac{0.725}{0.943}\left(\frac{L}{d_o}\right)^{0.25}$$

$$= 0.769\,(L/d_o)^{0.25} \qquad [5.3]$$

$$= 0.769\,(1000/19)^{0.25}$$

$$= \mathbf{2.07}$$

The heat transfer coefficient for the horizontal tube is more than twice that for the vertical tube. If the tube length had been 5 m the ratio of the two coefficients would have been more than three. **Condensing heat transfer coefficients are always higher for horizontal tubes than vertical tubes.**

Example 5.1

Compare the heat transfer coefficients for condensation outside a single horizontal tube 19 mm in diameter when the fluid is:

(a) Steam at a pressure of 10 bar.
(b) Butane at a temperature of 40°C.

Assume the temperature drop across the film is 2K for the condensing steam and 6K for the condensing butane.

Solution.

The physical properties for steam/water have been taken from reference 15 and for the butane from reference 21. These are as follows:

	Steam/Water	Butane
k – W/m^{2}K	0.676	0.1
ρ – kg/m3	885	560
h_{fg} - kJ/kg	2015	330
μ - cps	0.15	0.15

The units above are not all basic units but they are the most convenient ones to remember values for. When they are used in a basic equation all the units must be the fundamental ones. The two condensing coefficients can now be calculated from equation 5.2.

For steam

$$h = 0.725\,\frac{(k^3\rho^2 g\, h_{fg})^{1/4}}{(\mu d_o \Delta T_f)^{1/4}}$$

$$= 0.725\left(\frac{0.676^3 \text{x}\ 885^2 \text{x}\ 9.81 \text{x}\ 2015000}{0.00015 \text{x} 0.019 \text{x} 2}\right)^{0.25}$$

$$= \mathbf{21943}\ \text{W/m}^2\text{K}$$

For butane
$$h = 0.725 \left(\frac{k^3\rho^2 g\, h_{fg}}{\mu d_o \Delta T_f}\right)^{1/4}$$
$$= 0.725 \left(\frac{0.1^3 x 560^2 x 9.81 x 330000}{0.00015 x 0.019 x 6}\right)^{0.25}$$
$$= \mathbf{2012}\ W/m^2K$$

The difference in the order of magnitude of the two coefficients for steam and butane is typical for practical heat exchangers as well as for a single tube.

One last thought before moving on to practical design equations. The Nusselt equations provide a reasonable starting point for a bar room discussion – after a suitable number of drinks of course - how would you tackle designing a condenser for use in a space ship?

DEVELOPMENT OF EQUATIONS USED FOR DESIGN

There are clearly limitations to equations 5.1 and 5.2 when an average condensing heat transfer coefficient for a bank of tubes is required. These developments will now be considered.

INTRODUCING VAPOUR DENSITY

If a net buoyancy force is used instead of the gravitational force on the element of liquid in the development of equations 5.1 and 5.2 the density of the vapour becomes a variable. Using the suffixes L and V refer to the liquid and vapour properties respectively this modifies equation 5.1 for a vertical tube to:

$$h = 0.943 \left(\frac{k_L^3 \rho_L \{\rho_L - \rho_V\} g\, h_{fg}}{\mu_L d_o \Delta T_f}\right)^{1/4} \qquad [5.4]$$

And equation 5.2 for a horizontal tube to:

$$h = 0.725 \left(\frac{k_L^3 \rho_L \{\rho_L - \rho_v\} g\, h_{fg}}{\mu_L d_o \Delta T_f}\right)^{1/4} \qquad [5.5]$$

.For condensing steam the density of the vapour is so small compared with the density of the liquid that the difference between using equation 5.2 and 5.5 would hardly be noticed. However for a high pressure hydrocarbon condensing at a temperature close to its critical value there are much smaller differences between the liquid and vapour densities and the enthalpy of phase change is reduced. For such a case the increased accuracy of equation 5.5 is more significant.

CONDENSING STREAMS & CONDESATE LOADING

Figure 5.3 shows the two top tubes of a vertical row of tubes in a tube bank. Equation 5.5 could be applied to the top tube but not the lower tube. The layer of liquid is thicker because condensate from the top tube is added to any newly condensed vapour. This tube will therefore have a lower heat transfer coefficient. The third tube down would have an even thicker layer of condensate and the heat transfer coefficient would be further reduced.

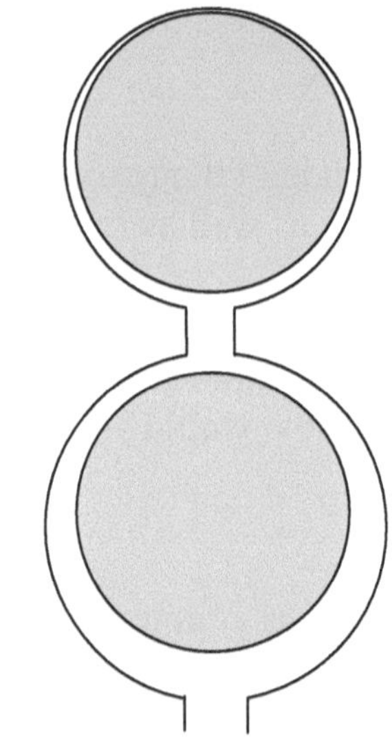

Figure 5.3

The progressive thickening of the condensate layer and subsequent lowering of the heat transfer coefficient will continue down the row of tubes as long as the flow around the tubes remains laminar. If there are sufficient tubes in the row a point will be reached when the layer becomes too thick to adhere to the tubes and it will start to break up. Just before this happens the heat transfer coefficient will have reached a minimum value. Once the laminar layer starts to break up and become turbulent the heat transfer coefficient will start to increase again and will carry on increasing until the bottom tube in the row is reached.

The same phenomena occur outside a vertical tube. The heat transfer coefficient at any point will decrease as the condensate falls down the tube. It will reach a point where the surface of the film starts to become wavy and then breaks up into turbulent flow and the heat transfer coefficient will start to increase again.

It is necessary to find some way of determining the point at which laminar flow will end and turbulent flow will begin. As with convective heat transfer a Reynolds number is used. This is defined as:

$$Re = \frac{\rho u d_e}{\mu}$$

Where u is the velocity of the condensate film, all the physical properties are for the liquid and d_e is the characteristic dimension. The characteristic dimension is the equivalent diameter (hydraulic diameter) which was first introduced in chapter 3 and is defined as:

$$d_e = \frac{4 \text{ x Cross sectional area}}{\text{Wetted perimeter}}$$

The velocity can be expressed as:

$$u = \frac{\text{Volume flow rate}}{A_X} = \frac{\dot{m}}{\rho A_X}$$

For a vertical tube the wetted perimeter is the circumference of the tube therefore the Reynolds number can be transposed as follows:

$$\begin{aligned} Re &= \frac{\rho u d_e}{\mu} \\ &= \frac{\rho\ \dot{m}\ 4\ A_x}{\rho\ A_x\ \mu\ \pi\ d_o} \\ &= \frac{4\ \dot{m}}{\mu\ \pi\ d_o} \end{aligned} \qquad [5.6]$$

For a horizontal tube the wetted perimeter is the length of each side of the tube therefore the Reynolds number can be transposed as follows:

$$\begin{aligned} Re &= \frac{\rho u d_e}{\mu} \\ &= \frac{\rho\ \dot{m}\ 4\ A_x}{\rho\ A_x\ 2\ L} \\ &= \frac{4\ \dot{m}}{\mu\ 2\ L} \end{aligned} \qquad [5.7]$$

It has been established from experimental work that the change from laminar flow to turbulent flow takes place at a Reynolds number of about 1800 on a vertical tube. For a horizontal tube the break point is about 2700.

Once it is necessary to rely on experimental data is becomes convenient to further transpose original Nusselt equations and use empirical equations. One such transposition is:

$$h = \text{A Constant}\ \ Re^{-1/3}\ \left(\frac{k^3 \rho^2 g\ h_{fg}}{\mu^2}\right)^{1/3} \qquad [5.8]$$

Note that ΔT_f is no longer in the equation, the effect of buoyancy is assumed to be negligible and the exponents of the physical properties have been modified as follows: For the liquid viscosity $\mu^{1/3}$ x $\mu^{-2/3}$ = $\mu^{-1/3}$ and the liquid thermal conductivity is now to the power of 1.

These variables now have more effect on a condensing stream than they did for condensing on a single tube.

The next step is to consider a bank of tubes. The terms condensing stream and condensate loading now need to be introduced. A condensing stream is just a vertical row of tubes where the condensate from one tube falls on the tube below it until it all drains off from the bottom tube in the row. The condensate loading is the mass flow rate which runs off each condensing stream. Equations 5.6 and 5.7 can now be modified as follows:

$$\text{For vertical tubes } Re = \frac{4\ \dot{m}}{\mu\ \pi\ d_o\ N_S} \qquad [5.9]$$

Where N_S is the number of condensing streams and is equal to the number of tubes.

For horizontal tubes $Re = \frac{4\dot{m}}{\mu\, 2LN_S}$ [5.10]

The $\dot{m}$ in both equations is now the total mass flow rate of fluid condensed in the shell.

The number of condensing streams is more complicated for horizontal tubes since it depends on the tube pitch arrangement. Figure 5.4 shows the tube pitch arrangements used in the design of shell and tube exchangers.

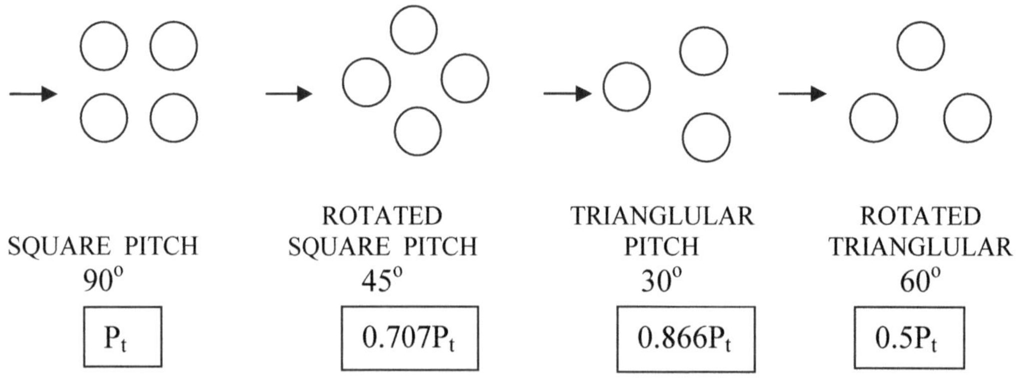

Figure 5.4

Baffles are always vertically cut in condensers so the arrows show the direction of fluid flow. The figures in the oblong boxes are the distances between vertical tube rows. For a fixed width of bundle it is clear that there will be about 40% more condensing streams in a rotated square pitch bundle than in an inline square pitched bundle. The rotated square pitched layout will therefore have a correspondingly reduced condensate loading. There is an even bigger difference between the triangular pitch and the rotated triangular pitch. The figure is now over 70%.

Provided there are no spec or vibration problems you should always **try the rotated triangular pitch arrangement when designing a condenser**. It frequently gives significantly higher heat transfer coefficients. For a square pitch there is not as much of an advantage in using the rotated arrangement since other considerations, (such as the decrease in space available for tubes due to pass partition lanes) makes it less likely to lead to an optimum design.

The development of equation 5.8 can be taken a step further for practical design work. Equation 5.11 is one I used for years for designing condensers using hand calculations.

$$h = 6540 \times j \times k \times \frac{(SG)^{2/3}}{(\mu)} \qquad [5.11]$$

SG is the specific gravity which took the place of density. The viscosity is in centipoise. There is no g, this is incorporated into the constant to minimise the arithmetic. (Remember we used slide rules in the days of hand calculations.) The units used were the old imperial units. The j factor was taken from the charts shown in figures 5.5 and 5.6.

For vertical tubes

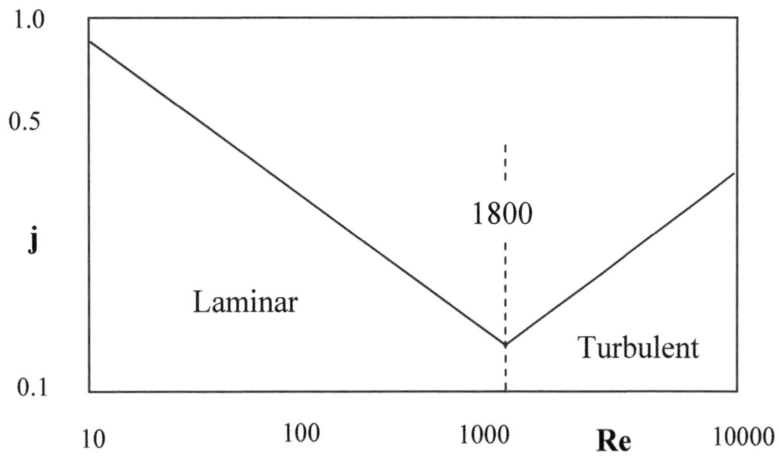

Figure 5.5

The point at which the heat transfer coefficient is a minimum for vertical tubes corresponds to a Reynolds number of 1800 and a j factor or about 0.16.

For horizontal tubes

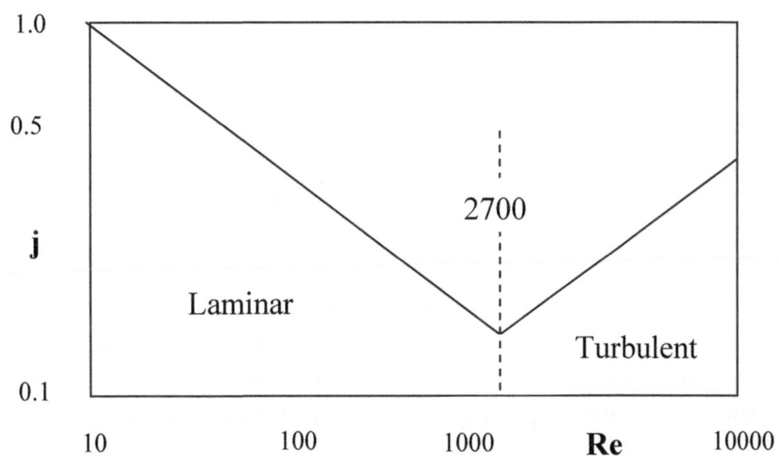

Figure 5.6

The point at which the heat transfer coefficient is a minimum for horizontal tubes corresponds to a Reynolds number of 2700 and a j factor of about 0.18.

The charts have not been accurately reproduced. Their purpose is to demonstrate the order of magnitude of the changes in condensing heat transfer coefficient for a particular fluid with changes in the condensing Reynolds number.

THE EFFECTS OF VAPOUR SHEAR

All the previous analysis in this chapter has been based on the important assumption that condensation is taking place in still surroundings. This is always a very reasonable assumption for the outer tube rows in a bundle which are adjacent to the shell. For an outer tube row the only velocities will be the very small ones of the vapour migrating from the bulk of the fluid to the surface of the tube.

There are other parts of a bundle where the assumption of a still fluid is totally unrealistic. Figure 5.7 shows a vertical tube row subjected to a strong horizontal vapour velocity.

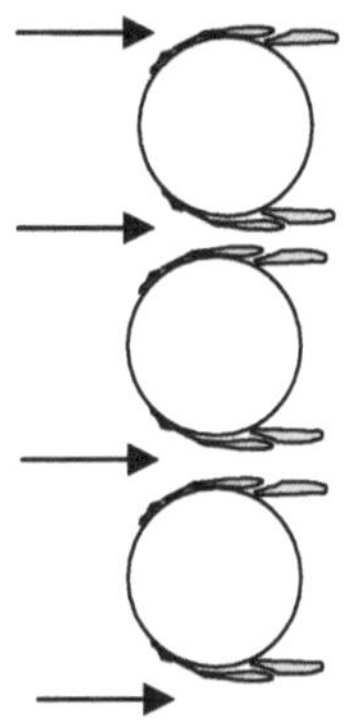

Figure 5.7

There is no longer any dripping of the condensate from one tube down to the tube below it. As soon as any liquid appears it will be wiped off the tubes and splashed on to the tubes in the next row. There will always be some areas in a shell and tube exchanger where the velocities are high enough to wipe the tubes of condensate particularly around the inlet nozzle where the velocity is downwards as illustrated in figure 5.8.

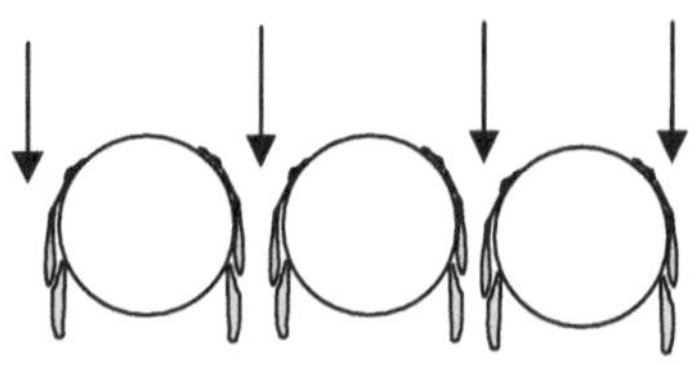

Figure 5.8

If condensate is being wiped from the tubes the average thickness of the condensate layer around the tubes will be quite small. Condensation can still be thought of as conduction through a layer of liquid. The condensing coefficients in shear controlled condensation are therefore correspondingly high compared with filmwise condensation.

This phenomenon has been known for many years but because it leads to higher heat transfer coefficients it is safe to ignore it for practical design work. It was never taken into account in the days of hand calculations. The incredible growth in computing capacity has meant that modern programs can take advantage of the higher heat transfer coefficients in the shear controlled areas of a bundle.

I have no intention of trying to sort through any of the correlations used for shear controlled condensation. I have never calculated a coefficient in a shear controlled situation and I sure I never will. The important thing for practical design work is to be aware of where it is likely to be taking place and to recognise this on a computer output.

DROPWISE CONDENSATION

The mechanism of dropwise condensation is quite different to filmwise condensation. The surface never becomes coated in liquid. Droplets of liquid form as illustrated in figure 5.9. They increase in size and then fall off taking any small droplets with them leaving a dry area of tube behind them.

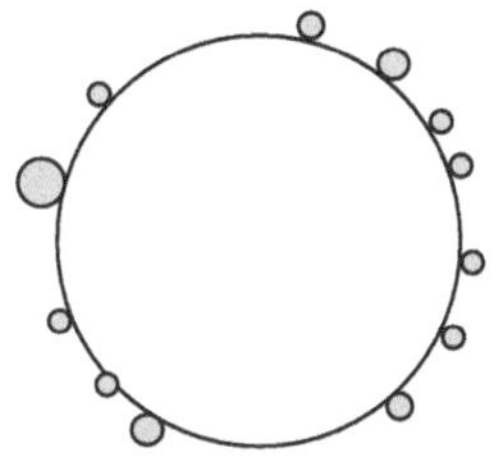

Figure 5.9

Because there is no liquid film on the tubes the thermal resistance is very low and the condensing coefficient very high. Research papers have reported dropwise condensing coefficients of over five times higher than the corresponding filmwise coefficient.

A similar phenomenon is easy to observe in a none stick frying pan and Teflon coated surfaces are one of the ways which have been used to promote dropwise condensation in the laboratory. The surface finish is clearly important and highly polished or very clean surfaces will promote some dropwise condensation. There has been some success with surfaces coated in one of the noble metals. I remember a long time age reading an article suggesting that we would all soon be using gold plated tubes in condensers. It would have made a nice change to do something useful with the metal instead of hiding it back in the ground after taking all that trouble to dig it out and process it.

Another approach which has been used to promote dropwise condensation is to introduce chemicals which coat the surface. Once again these work for a while in the laboratory but then break down and filmwise condensation takes over.

There is another very practical problem to utilising dropwise condensation in our industry – it has only been observed for one fluid – steam. I am not sure why this is but the very high surface tension value for water compared with other fluids seems to be the most likely reason. The steam they use in laboratory tests will also normally have a higher level of purity than the steam we use industrially.

Coefficients for filmwise condensation of steam are normally at least 10000 W/mK and the corresponding thermal resistance is normally a very small part of the total thermal resistance. Increasing the coefficient would make very little difference to the size of most exchangers. If refrigerants or hydrocarbons could be persuaded to condense in a dropwise manner there would be more to gain when the other fluid in an exchanger had a high coefficient and both fouling factors were very low.

In conclusion filmwise condensation possibly occurs for a while in some areas of a commercial heat exchanger while the unit is clean. It is an interesting topic but I can not see it ever contributing to practical thermal designs in the petrochemical industry.

Research is mainly directed at the surface condensers used in the power industry where the steam is very pure and zero fouling factors may be used on the water side by using a continuous cleaning system with flexible brushes or balls passing through the tubes. (See the chapter 14 on fouling.)

PRESSURE DROP

The pressure drop for condensing on the shell side is complicated by the fact that any section along the shell will have condensing vapour occupying most of the space at the top and there will be some liquid which has already condensed occupying a smaller space at the bottom.

Figure 5.10 shows a section through a condenser with the liquid running along the bottom of the shell.

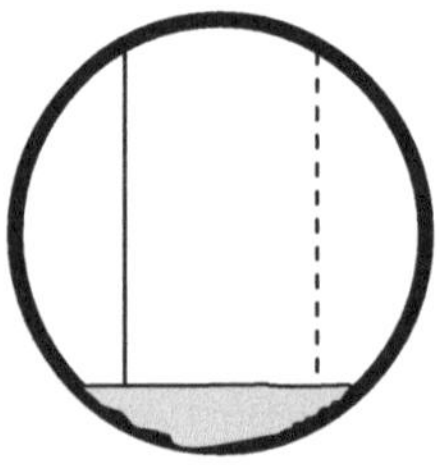

Figure 5.10

Figure 5.11 shows a side view. The solid line, the dashed line and the dotted line all represent liquid levels. The lower dashed line represents a liquid level in a condenser with no pressure drop through the bundle. There are only very small changes in the liquid level through the bundle. The top solid line is another idealised line illustrating what 'should' take place when condensing a single pure fluid i.e. all vapour and no liquid at the face of the inlet channel tubesheet and all liquid at the face of the outlet channel tubesheet.

The middle dotted line represents a more practical liquid level in a condenser with a pressure drop through the bundle. In this situation the liquid level will progressively increase from the inlet to the outlet but never reach the top of the exchanger. It may be argued that it will increase in steps across each baffle rather than a smooth line as shown. The height of the liquid in the outlet space is affected by the size of the outlet nozzle which will be discussed later.

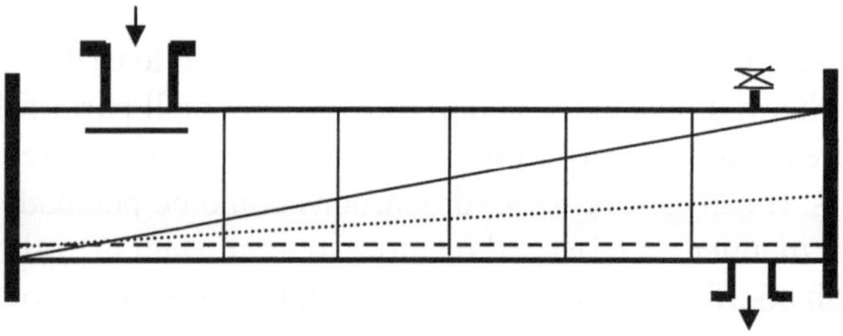

Figure 5.11

For any design it is necessary to establish the average volume fractions occupied by the vapour and the liquid before any pressure drop calculation can be carried out. The equation we used for hand calculations was:

$$\frac{1}{F_L} = \left(\frac{\mu_V}{\mu_L}\right)^{1/9} \left(\frac{\rho_L}{\rho_V}\right)^{5/9} \left(\frac{\dot{m}_V}{\dot{m}_L}\right) + 1 \qquad [5.12]$$

The mass flow rates are the average mass flow rates of the vapour and liquid. The ratio of the mass flow rates is 1 for a pure fluid which starts out as all vapour and finishes as all liquid. However the same equation was used for condensation of multi-component fluids and partial condensers which could include inerts. For such cases the ratio of the mass flow rates would not be 1. After determining a value for the fraction of liquid, F_L the fraction of vapour, F_V was simply $(1 - F_L)$.

The next step was to work out the cross flow and long flow areas including all the by pass streams in the same way as for a single phase fluid as explained towards the end of chapter 3. For condensers equations 3.16, 3.17 and 3.18 would be modified because there is no longer any by passing at the bottom of the bundle because of the liquid seal. The long flow and cross flow areas were then reduced by multiplying by the vapour fraction. The pressure drops for the vapour in long flow and cross flow were then calculated. This should ideally be the same as the pressure drop for the liquid flow but this was never calculated as a check. This was not unreasonable since the liquid fraction was normally quite low compared with the vapour fraction.

SHELL SIDE CONDENSING STEAM.

I have never calculated either a heat transfer coefficient or a pressure drop for condensing steam on the shell side of a heat exchanger.

For hand calculations I always assumed a coefficient of 11356 W/m^2K (2000 BTU/h ft^2 ^oF). I knew that some people were using a more conservative figure of 1500 BTU/h ft^2 ^oF at the time. They are both clearly rounded figures and when I later started to use the SI system I used 10000 W/m^2K. It is a useful exercise for the reader to check the condensing coefficients on some designs from the latest computer programs. Then work out the effect of substituting 10000 W/m^2K for the computer value and see the effect on the overall design. I can not remember a case where it would have been worthwhile changing the design.

Pressure drop calculations can be more complex and controversial. I was trained to accurately calculate the nozzle losses and then assume a nominal value of say 0.015 bar for the bundle. The baffling was then left to the mechanical engineer who would use the maximum pitch and cut allowable by the codes. It must be remembered that in those days the maximum size of shells was smaller than we now use and hence there were less likely to be any vibration problems. Of course we never even checked for vibrations we just relied on good design practices and there were very few problems which could not be attributed to bad design practices. The most common example of bad practice was filling up the space under an inlet nozzle with tubes which created very high local velocities. (This normally occurred at companies where a thermal design engineer never checked the vendor drawings.)

Shell sizes have grown and even with 'normal' good design practices vibrations are often predicted for exchangers with condensing steam on the shell side. I have come across designs were people have used small baffle pitches and even double segmental baffles to try and fade away the 'stars'. The pressure drops necessary for such designs must result in a fall in the saturation temperature which will lower the MTD. This is all completely unnecessary. I maintain that **I can still design a shell and tube unit with a low pressure drop using maximum baffle pitches and cut which will be vibration free** – without using a computer..

Potential problems always occur around the inlet space where the tubes have a maximum unsupported length and the volume flow rate of steam is also a maximum. The solution is quite simple - just reduce the maximum velocities into the bundle and reduce the maximum unsupported length at the same time. The velocities can be reduced by clipping the tops off the first few baffles and if necessary moving the inlet nozzle further away from the tubesheet.

I first came across the idea of moving the inlet nozzle in the late sixties. I was working on a phenol plant for a well known chemical company. There were a lot of steam heated vertical thermosyphon units and the clients spec clearly stated that the steam inlet nozzles should be positioned approximately a third of the way down the shell. I was new to the position of being in charge of a thermal design department and wary about questioning specs of such a big company. None the less this was something that went against current practice so I decided I must make a good case for pointing out that this was wrong. I thought about it at work, I thought about it over the weekend, and I had still not come up with anything to suggest a potential problem. There would have been a vent close to the top tubesheet irrespective of where the inlet nozzle was positioned so there would be no problem with venting inerts. The nozzles were positioned as required by the spec. As far as I know they worked OK, (feedback from site is usually negative) and I reverted to more normal practice for the next job.

Years later when I first came across potential vibration problems in a unit with steam on the shell side I thought again about the designs I had done on that phenol plant. Having the nozzle a third of the way down the shell had meant that a third of the steam would have flowed upwards and two thirds downwards. The net effect is that the velocity of the steam entering the bundle is significantly reduced. From then onwards I started to apply these principles to all condensing steam items even those which did not have vibration warnings.

Figure 5.11 is a full 'belt and braces' job for a unit with warnings about severe vibration problems with a 'conventional' design.

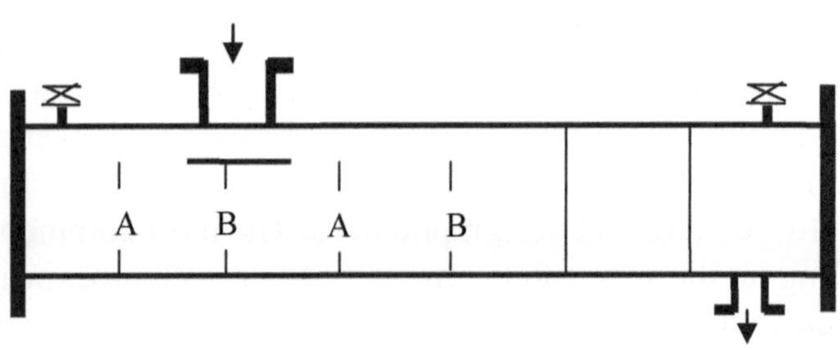

Figure 5.12

The first four baffles are as illustrated in figure 5.13. The remaining baffles, only two in this case, are conventional vertical cut baffles with ears if necessary to prevent the venting of two much vapour whilst air is being expelled.

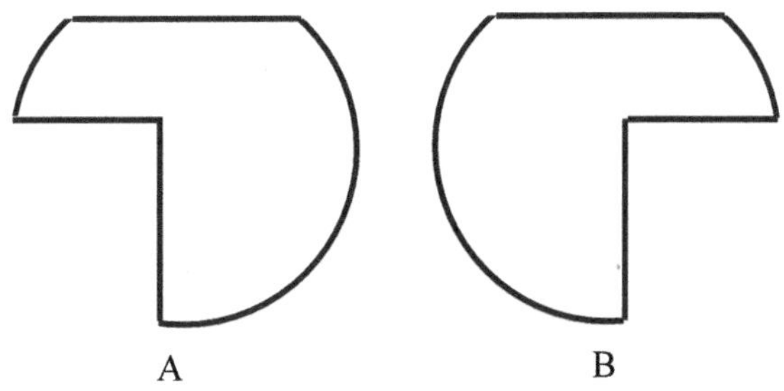

Figure 5.13

The advantages of this arrangement are:

(a) The velocity across the first row of tubes has been drastically reduced.
(b) The top rows of tubes are supported every baffle pitch in the vulnerable area. It is a matter of engineering judgement as to how many rows of tubes at the top of the bundle should have the support of the 'overhang'. I just look at the tube layout. Take into account any pass partition lanes and the position of the tie rods, (adding extra ones if absolutely necessary) and specify X rows of tubes to be fully supported by type A and B baffles. For any bundle a minimum of 6 rows is necessary to make this measure worthwhile and it would normally be a lot more in a large bundle.
(c) The overall bundle pressure drop will be very small and consequently there will be a negligible loss of MTD.

It is not very often that I have moved the inlet nozzle far enough down the shell to justify the second vent shown in figure 5.12. If a second vent is required it is sensible to 'cover' it as illustrated in figure 5.14.

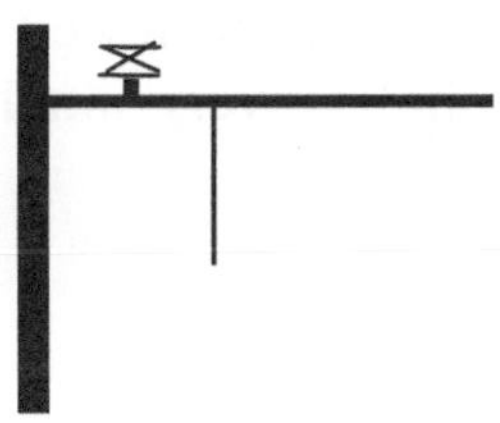

Figure 5.14

The vent is covered by a segmental section which covers the same tubes that are fully supported by baffles type A and B. It will probably need a short extra support at the top. This is just to prevent excessive loss of steam during venting.

In previous chapters for single phase fluids the importance of keeping the inlet nozzle as close to the adjacent tubesheet as practical has been emphasised. For condensing steam the opposite may be required.

For most designs with very minor vibration warnings I normally just take a very conservative view of the inlet length adding a couple of hundred millimetres to the minimum value. I then position the first baffle immediately below the inlet nozzle and specify the first three baffles as type A/B.

For a steam heated item with 'conventional' baffles and no vibration warning I would still specify that the tops should be clipped off the top of the first couple of baffles just to minimise the pressure drop.

The downside to being a bit creative in a design is that your computer program will not like it. There is unlikely to ever be any way that you could input some of the modifications suggested above. That is why it is important to understand the basics used for hand calculations and have a 'feel' for how a change in a variable will affect the overall design. There should be no need for any detailed calculations. Just remember that this measure will say more than halve the velocity and it will change the natural frequency of the tube by a factor of about four.

Reducing the steam velocity in the first part of the exchanger will of course remove any shear controlled condensation but the condensate loading will also be reduced and with such high coefficients the overall effect will be negligible. If you use a computer value for the pressure drop you know it will be very safe even if you have only just clipped the top off a couple of baffles.

If you are a vibration worrier and take everything you read too literally you could support the top rows of tubes for the full length of the 'belt and braces' design in figure 5.12. Lastly to convince yourself and others you could try running this design as two shells in series with the first one being a cross flow unit and the second one being a conventional unit. (Only if you have lots of time to kill and no deadline to meet!)

SURFACE CONDENSERS

Steam surface condensers are a feature of every vapour power plant. They are designed by power plant engineers or vendors. My only reason for introducing them here is because there are some features of their design which we can learn from and apply to more general condensers. Figure 5.15 shows a very simplified section through a surface condenser.

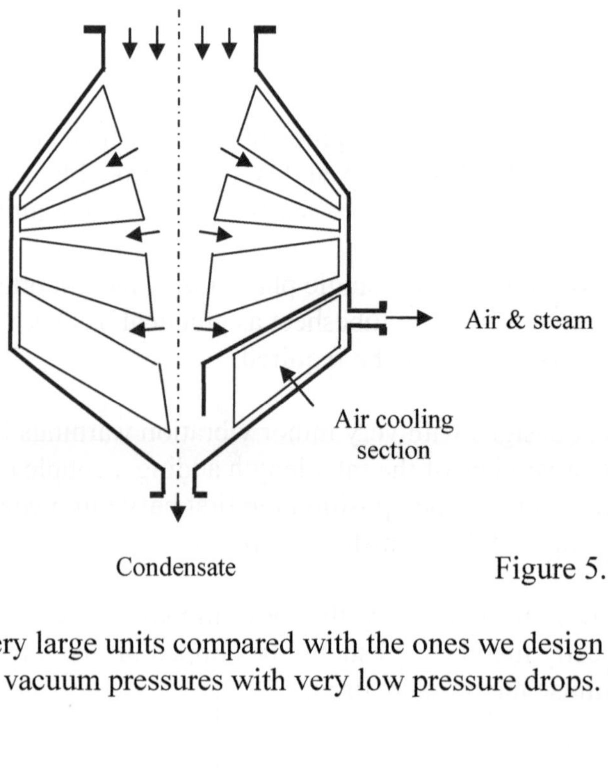

Figure 5.15

Surface condensers are often very large units compared with the ones we design and they always operate under high vacuum pressures with very low pressure drops. The higher the vacuum the lower the saturation temperature becomes which results in increasing the overall thermal efficiency of the plant. (Remember the Carnot and Rankine cycles from the early years of your course in thermodynamics.)

The sketch shows the steam entering the unit through a very large nozzle which is likely to be more complex than the simple round nozzle illustrated. The steam then enters the main condensing tube banks through a series of tube free passages. This arrangement allows the steam to reach all the tubes. (If one solid bank of tubes were to be tried the tubes at the centre of the bank would be ineffective because they would be starved of steam.) Condensate leaves at the bottom and should not be over sub cooled which would reduce the plant efficiency. Air, with some steam, leaves at the side after passing over an air cooling section. This section is in series with the main condensing section.

The passages through the main tube banks not only allow access to the steam but causes a 'regenerative effect'. As the steam moves along a passage the condensate from the bank above passes through the steam which reduces any sub cooling. There are also baffles in between the upper tube banks to remove condensate and reduce the condensate loading on the banks below.

The high vacuum conditions are present in the LP turbine as well as the condenser and although measures such as labyrinth glands are used along the drive shaft to minimise this there will still be a continuous leakage of air into the system. This requires continuous venting. When steam is used as a service fluid the pressures are always positive with respect to ambient pressure and venting is an intermittent process ideally only required for start up.

It is not possible to expel the air from a surface condenser without losing some valuable steam with it. The steam loss must be minimised. To achieve this, the air must leave the condenser at as low a temperature as possible. The air cooler tubes are therefore supplied with cooling water at the lowest possible temperature..

The effect of this air cooling section is best demonstrated by means of an example.

Example 5.2

20000 kg/h of steam enters a surface condenser as a saturated vapour at 38°C. The steam contains 0.3 kg of air for every 1000 kg of steam. The pressure throughout the unit may be considered to be constant and the condensate leaving the unit is sub cooled by 2 K. The air leaving the air cooler section is at 27°C.

What are the savings made by having the air cooler section?

Solution

Properties for steam and the gas constant for air are taken from tables such as reference 15.

At entry p_s is 0.06624 bar and v_g is 21.63 m^3/kg. The total volume flow rate of steam is therefore 20000 x 21.63 m^3/h and 6 kg/h of air are present in the same volume.

$$\text{Air partial pressure } p_a = \frac{m_a R_a T}{V} = \frac{6 \times 287 \times 311}{20000 \times 21.63} = 1.2 \text{ N/m2} = 0.000012 \text{ bar}$$

This is negligible compared with the partial pressure of the steam and may be ignored.

Without any air cooling section the air would have to be extracted at the condensate temperature of 36°C. The steam associated with the air would have a saturation pressure of 0.0594 bar and a specific volume of 23.97 m^3/kg therefore:

$$p_a = p - p_a = 0.06624 - 0.0594 = 0.00684 \text{ bar}$$

The volume flow rate of air removed is therefore:

$$V = \frac{m_a R_a T}{p_a} = \frac{6 \times 287 \times 309}{0.00684 \times 10^5} = 778 \text{ m}^3/\text{h}$$

The mass of steam associated with volume V of air is:

$$m_s = \frac{V}{v_g} = \frac{778}{23.97} = \mathbf{32.45\ kg/h}$$

With an air cooling section the air is extracted at a temperature of 27°C. The steam associated with the air has a saturation pressure of 0.03564 bar and a specific volume of 38.81 m^3/kg therefore:

$$p_a = p - p_a = 0.06624 - 0.03564 = 0.0306 \text{ bar}$$

The volume flow rate of air removed is therefore:

$$V = \frac{m_a R_a T}{p_a} = \frac{6 \times 287 \times 300}{0.0306 \times 10^5} = 168.9 \text{ m}^3/\text{h}$$

The mass of steam associated with this air is:

$$m_s = \frac{V}{v_g} = \frac{168.9}{38.81} = \mathbf{4.35\ kg/h}$$

The saving in condensate is (32.45 – 4.35) = **28.1 kg/hr**. Without the air cooling section this loss would have had to have been compensated for with treated water at ambient temperature which would have required an input of heat to bring it up to the condensate temperature.

Just as important is the drastic reduction in the capacity of the air ejector units which maintain the vacuum. They now only need to handle 168.9 m^3/h compared with 778 m^3/h.

Points which are useful to apply to other types of condenser.

(a) Small changes of temperature can have a big effect on the amount of condensable fluid carried out of an exchanger by none condensable gases. This will be discussed further in a later section on condensing with inerts.
(b) For very large tube bundles it may be necessary to introduce vapour passages to reach the more remote tubes in a bundle. One of our main applications of this is in boiling rather than condensing where lanes may be necessary to prevent vapour binding in the centre of the bundle.
(c) There will always be some sub cooling in any full condenser.
(d) Because of very low density values of the steam, (high specific volumes), there are very high velocities in a surface condenser. However the momentum fluxes are very low. In the above example if steam at the inlet conditions had a velocity of 100 m/s the moment flux would only be 462 kg/ms^2. This will be referred to in the chapter on vibrations.

SUB COOLING IN CONDENSERS

Sub cooling is also referred to as under cooling. There are likely to be a few degrees of sub cooling in any full condenser. In surface condensers measures are taken to minimise it but the condensate will still leave a few degrees below the saturation temperature.

Two basic terms need to be defined:

Degrees of sub cooling

Degrees of sub cooling is defined as the number of degrees of temperature difference between the saturation temperature and the actual temperature of the fluid. The units are K.

A sub cooled liquid must therefore be at a higher pressure than the saturation pressure corresponding to the stream temperature.

Flashing

Is the production of vapour from a sub cooled liquid stream which occurs when the pressure of a sub cooled liquid stream is reduced to the saturation pressure corresponding to the stream temperature

There are times when process requires some sub cooling of the condensate. I tend to classify these situations into reasonable sub cooling and excessive sub cooling. There is no break point between the two in terms of degrees of sub cooling. The more important design feature is how much of the total effective surface area is required for sub cooing. Anything above about 20% is becoming excessive.

Process normally requires the sub cooling to prevent flashing in the lines or elsewhere downstream of the condenser. When the sub cooling becomes excessive there is normally an additional reason such as lowering the temperature down to a safe storage temperature.

Even with excessive sub cooling, the liquid cooling duty is still normally a small percentage of the total heat duty.

REASONABLE SUBCOOLING

The first essential is to make sure that you input the data correctly into your computer program. If you are dealing with a common fluid the most satisfactory way is to use the automatic properties stored in the program. It is then only necessary to input the inlet and outlet temperatures and the program will do the rest for you. If it is not a common fluid or process require you to use their properties because the data bank they are using is different from the one in your program then ask for cooling curves at the inlet and outlet pressures. If it is a pure fluid make sure to tick the box to input this, there is no point in unnecessarily penalising the condensing coefficient. The default in most programs is to include a convective vapour resistance, (this is covered in detail later in this chapter.) This includes steam of course although it is unlikely to make much difference to steam coefficients except possibly in vertical units where all the other thermal resistances are quite low.

An ear on the last baffle.

Having input data you are happy with, a sophisticated program will indicate how much area is required for sub cooling. You will also probably get a warning that the calculation for the liquid cooling is a bit dodgy. Even with no warning assume it is a bit dodgy. This is because the heat transfer will be a mixture of natural convection and very low velocity forced convection. When sub cooling is required you should have sealing strips at the bottom as well as the top to encourage the latter.

The simplest way to achieve a small amount of sub cooling is to specify an ear on the last baffle before the outlet space. This is illustrated in figure 5.16.

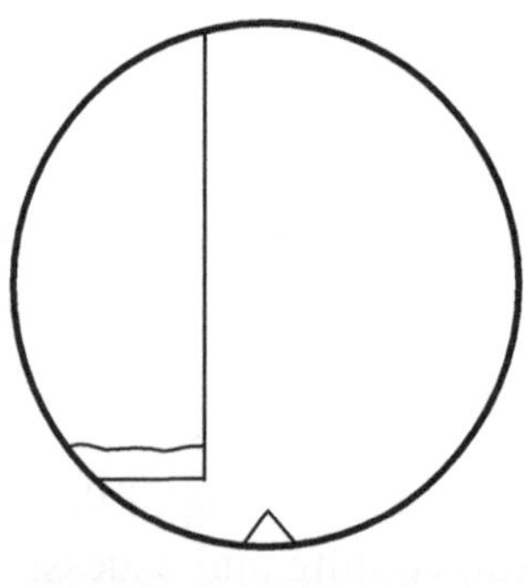

Figure 5.16

The ear will act as a weir which will establish a liquid level along the bottom of the bundle.

Once you know how much sub cooling area is required calculate out how many tubes this represents and look at the tube layout and work out how many rows need to be covered. Like all baffle edges the ear must be cut either on the centre line of a row of tubes or in between two rows. Suppose your calculation predicts that 5.3 rows of tubes are required for sub cooling. It might be reasonable to make your ear cut between the fourth and fifth row from the bottom. This assumes that the distance between the liquid level and the cut would cover the fifth row from the bottom and the liquid level would lie along the bottom of the sixth row.

A detailed calculation would be very involved and require assumptions regarding the A and E by pass streams and there will always be some flow through the drain notch which could be important for small shells. You would then need to calculate the flows resulting from the pressure drop across the last baffle. This could take more time than the rest of the design and is not justified. Just make a sensible judgement as to where you think the cut on the ear should be.

Side outlet nozzle

This is a possibility that I have never used. To be honest I would never have thought of it until I studied figure 2.29 in chapter 2. This is an HEI diagram of a feed heater and it clearly has a side outlet nozzle. I can not see any thermo fluids argument against using it in other types of condenser. It would certainly establish a liquid level at the bottom of the shell in a more positive way than using an ear on the last baffle.

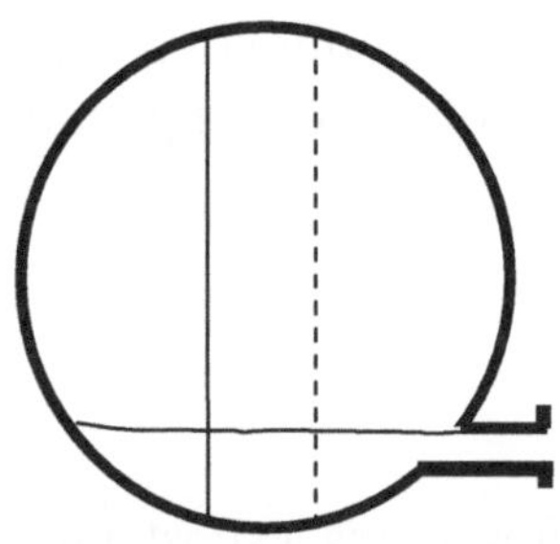

Figure 5.17

Offset nozzles are never popular with either manufacturers or piping. However liquid outlets are normally quite small and I can not see this being too much of a problem.

Having never used it I am reluctant to recommend it, there may be a problem I have not thought of. I will therefore leave it as something for the more experienced reader to consider.

A gooseneck

This arrangement is illustrated in figure 5.18. The condensate outlet is at the bottom of the shell as normal and there is a second small nozzle on the shell above the liquid level.

The gooseneck is part of the pipework. The liquid level in the shell is maintained by the top of the 180° bend in the pipe. (The bends of course would be standard pipe bends and would not have sharp corners as illustrated.)

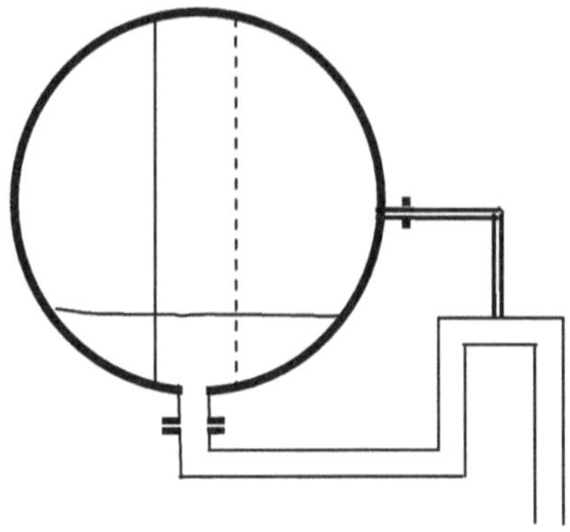

Figure 5.18

The pipe from the shell to the top of the bend in the pipework is necessary to break the liquid seal. Without this small pipe the gooseneck could become ineffective by allowing the liquid to siphon out of the exchanger thus lowering the liquid level. To prevent the liquid level becoming too high in the shell the pressure drop in the pipework should be minimised by using low velocities and long radius bends. During design operation the liquid level in the shell will always be above the top of the gooseneck to give sufficient head to overcome the pressure drop in the pipe. (The pressure at the surface of the liquid and the top of the gooseneck are the same.) This should be taken into account when fixing the height of the gooseneck.

I have been using goosenecks ever since I started designing heat exchangers. They are clearly more expensive than just adding an ear to the bottom of a baffle but are safer if the sub cooling is critical. The greater the amount of sub cooling required the more worthwhile it becomes to use this arrangement.

EXCESSIVE SUBCOOLING.

Sub cooling becomes excessive not only when it requires a substantial part of the total surface area but when it is starting to dominate the overall design. As discussed earlier the heat transfer coefficients are very low and the MTD is also lower than in the condensing section above the liquid level. A point is often reached where adding surface area just makes the heat transfer coefficient worse and requires more surface area to be added. When this happens more drastic measures are required than those covered under ‘reasonable sub cooling’. It is necessary to find some way of increasing the velocity and hence the heat transfer coefficient of the liquid without affecting the condensing section.

Separate sub cooler.

The most effective way to deal with excessive sub cooling is to use a separate liquid cooler. This means another item of equipment and extra pipework because the coolant must now be supplied and controlled to two units. It may not mean any extra cost of heat exchangers because the condensing shell will be smaller and the sub cooler will require quite a small surface area. As far as the thermal design is concerned it is now straightforward to produce two optimum designs without any worry of ‘adding a bit extra to be safe’ on a design which is ‘dodgy’ in places.

The downside of course it that using a separate sub cooler may require a bit of extra pressure drop. Once you realise that the sub cooling is going to be excessive this option should be discussed with the process engineer.

Vertical orientation

This is an option which is easily missed. It may not be practical for a large unit but for a small unit it may be acceptable. For a vertical unit such as illustrated in figure 5.19 there is no problem in dealing with liquid cooling. The baffles at the top can have a wide pitch with a high cut where the condensing is taking place and a smaller pitch and cut at the bottom for the liquid cooling.

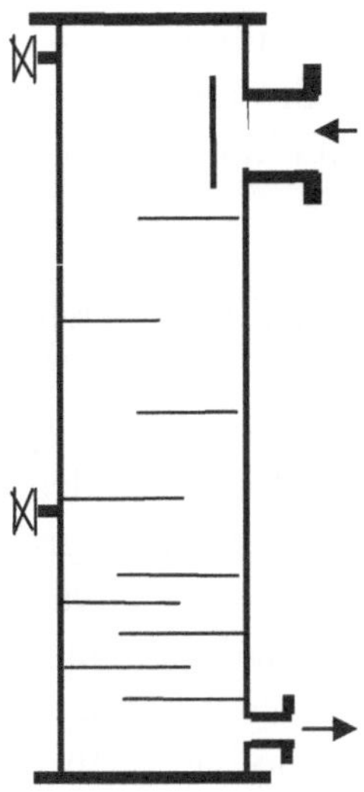

Figure 5.19

The baffles are now horizontally cut with respect to the inlet nozzle so there may no longer be any need for sealing strips. The orientation of the outlet nozzle is flexible. It is shown on the same side as the inlet nozzle in the sketch. This means that the only tubes lost are those under the inlet nozzle. (Check with piping to see if they have any preference for a particular orientation.)

For a vertical design a second vent under the last baffle in the condensing section is a sensible precaution to prevent cooled inerts lying above the liquid level.

The more sophisticated modern programs should be able to handle this design in a single run. If your program can not do this it may be necessary to design the condensing section and the sub cooling section in separate runs.

Tubeside condensing.

If the condensing fluid is at a reasonable pressure and the cooling medium relatively clean it may be worthwhile investigating switching the streams.

Sub cooling is not a problem when the condensing is inside the tubes. The exchanger is physically split into two sections, as shown in figure 5.20. There is no longer any need to carry out calculations or make assumption as to how many tubes will be required for the sub cooling.

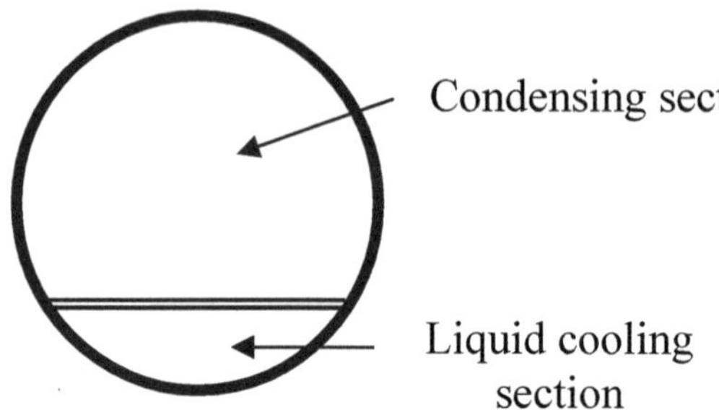

Figure 5.20

The first pass on the tubeside is through the condensing tubes and the second pass it through the sub cooling section. A computer program is unlikely to be able to handle this in a single run but this is one of the most straight forward of combination run designs. The tube side coefficient and pressure drop will be spot on and the shellside coefficient from the condensing run can be input into the run for a small single pass liquid cooler.

The condensing stream pressure drop is the critical feature which is why it is only likely to be possible with medium or high pressure streams. You may of course wish to investigate using larger diameter tubes to keep the pressure drop to a minimum. If you do this remember that the larger tubes are unlikely be ideal for the sub cooling section. It will probably be a long time before computer programs will be able to produce a tube layout with different sized tubes. A clear advantage of the old days when a good draughtsman, (women were also called draughtsmen in those days), would not have had any problem with this. An easy way out is to keep the same tube pitch in both sections your program may be able to handle this. More challenging is to use the ideal pitch for each section. If the large tubes are on say a 1.25 inch pitch and the smaller ones on a 1.0 inch pitch you could fit in approximately $(1.25/1)^2 = 1.56$ more tubes in the lower section. Use the computer tube layout for the larger pitch and then manually modify the bottom part.

Use a horizontal baffle

An F type shell and a G type shell can both be used to maintain a physical separation between a condensing section and a sub cooling section on the shell side of an exchanger. These are illustrated in figures 5.21 and 5.22 respectively.

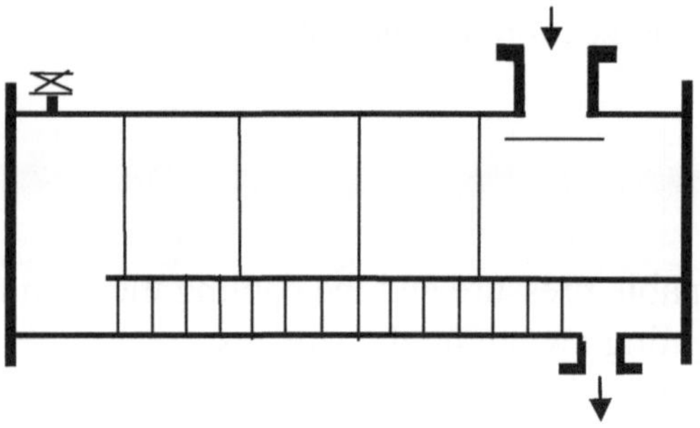

Figure 5.21

Figure 5.21 for the F shell is very similar to the HEI diagram figure 2.29 in chapter 2 without the desuperheating section. Two computer runs would be necessary to design this type of unit. Each design would be for an E type shell.

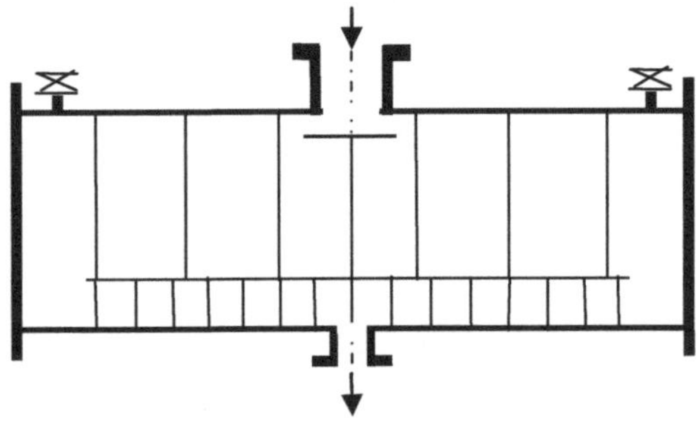

Figure 5.22

The G type in figure 5.22 has the advantage of a much lower pressure drop. The two computer runs would be for J type shells. The condensing section would be a 12J and the sub cooling section would be a 21J. In chapter 4 it was shown that the difference in pressure drop between an E type shell and a J type shell was a factor of about 1/7.

For either of these designs the liquid cooling will be in contact with the lowest temperature coolant. A break point in the coolant temperature range for the two duties is the first thing to work out. If you are very lucky the break point may coincide with the first tubeside pass. The break point temperature is then the coolant outlet temperature for the liquid cooling section and the inlet temperature for the condensing section. It is more likely that the first tubeside pass will include the liquid cooling section and the lower rows of tubes in the condensing section. This is best illustrated by means of an example.

Example 5.3

1 kg/s of a pure hydrocarbon is condensing at a temperature of 100°C and process requires it to be cooled down to 40°C in the same shell. The hydrocarbon has an enthalpy of phase change of 500 kJ/kg and an average liquid specific heat of 2 J/kgK. Cooling water is available with supply and return temperatures of 25°C and 35°C respectively and has an average specific heat value of 4.18 kJ/kgK.

You have tried a single run design with the intention of using a gooseneck liquid seal which shows that over 60% of the surface area would be required for liquid sub cooling. You have now had preliminary runs for a combination design which suggest that less than 25% of the total surface area is required for sub cooling provided you can achieve a reasonable forced convective coefficient for the liquid cooling. You have also established that for an optimum cooling water velocity the design will have two tubeside passes so that the first pass will supply water to both sections, Determine the cooling water inlet temperature which should be input to the design for the condensing section.

Solution.

$$\text{Condensing duty} = \dot{m}\, h_{fg} = 1 \times 500 = 500 \text{ kW}$$

$$\text{Liquid cooling duty} = \dot{m}\, c_p\, \Delta_T = 1 \times 2 \times 60 = \underline{120 \text{ kW}}$$

$$\text{Total duty} = 620 \text{ kW}$$

Total cooling water flow rate = $Q/C_p\ \Delta_T$ = 620/4.18 x 10 = 14.83 kg/s

It is anticipated that about 25% of the surface area will be required for the sub cooling therefore half of the total water flow rate will pass through the liquid cooling tubes on its first pass.

$$T_{out} \text{ from cooler} = T_{in} + Q/\dot{m}\,C_p = 25 + 120/7.42 \times 4.18 = 28.87\ ^{o}C$$

Imagine this water is then mixed with inlet temperature cooling water.

$$T_{mix} \times 1 = 28.87 \times 0.5 + 25 \times 0.5$$

$$\mathbf{T_{mix} = 26.9\ ^{o}C}$$

This is the temperature which you should input into your computer run for the condenser design. It is not quite thermodynamically correct but the error will be quite small and as with most combination designs you will probably add and extra 5 or 6% extra oversurface to allow for minor distortions in the design procedure. (You might have allowed less extra oversurface for four tube side passes when the outlet temperature from the sub cooling section would have been the inlet temperature to the condensing section.)

You may of course need to modify this mixed temperature a bit as you approach your final runs for the combined design and find perhaps that only 23.5% percent of the tubes are needed for the sub cooling.

Even allowing for the extra oversurface, the total surface area required has been reduced by about 30% compared with a single run cooler with a gooseneck seal. This is a big cost saving. If you spot one of these designs coming up make out a case for at least double the hours required for a single run design.

Lastly you are designing an F type or G type shell which involves a horizontal baffle. A lot of specs and individuals do not like these. This was discussed in detail in the section on F types of shell in chapter 2. For a typical condenser with a sub cooler section the overall allowable pressure drop is unlikely to be more than about 0.15 bar so there is no problem with the pressure difference across the baffle. If it is a fixed tubesheet exchanger the baffle can be welded to the shell which eliminates another main objection. That leaves a removable bundle which requires a seal between the baffle and shell.

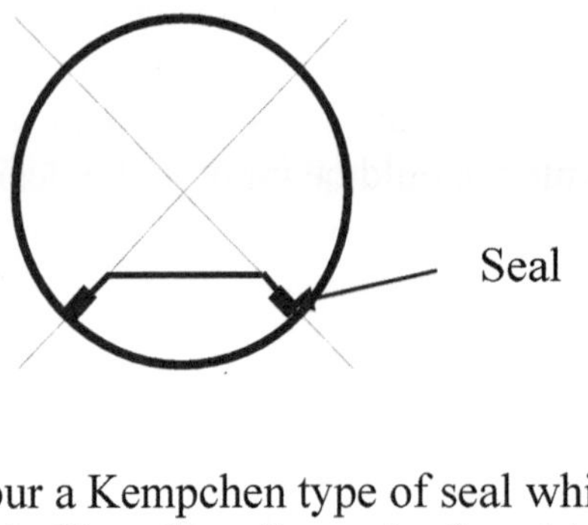

Figure 5.23

I now always favour a Kempchen type of seal which I am sure is very effective for normal horizontal baffles. (Details can be found in chapter 2). However a 'normal' horizontal baffle would be on the centre line of the shell which is not the case for the above example. This can be overcome by modifying the horizontal baffle as shown in figure 5.23 above.

This modification allows the seals to be fitted along radial lines as they would be for a horizontal baffle. It also stiffens the horizontal baffle. A final bonus is that the channels formed on either side of the baffle would be running full of condensate and a liquid is less likely than a vapour to leak across a seal.

Computer runs

How you fudge your computer runs will depend on the program you are using. For the condensing section you can probably use the actual shell diameter and input a height above the outlet nozzle to keep the bottom row of tubes just above your horizontal baffle. Your computer will assume a layout as shown in figure 5.24.

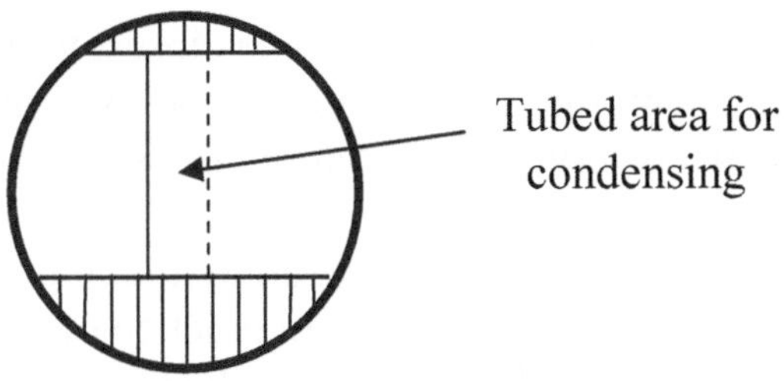

Figure 5.24

It may be necessary to use a fictitious large outlet nozzle to achieve this. In the computer run use lots of sealing strips to avoid bypassing as illustrated in the right hand diagram. The important feature for the condensing section is that the condensate loading is the same for your computer run as for the final design.

For the sub cooling section life is not quite so easy. Computer programs will always assume round shells and you have no option but to run with a very artificial input to try and simulate the segment at the bottom of the shell. You need to select a much smaller shell size which has a cross flow velocity as close as possible to your final design. This run will require a fictitious large inlet and outlet nozzle as shown in figure 5.25.

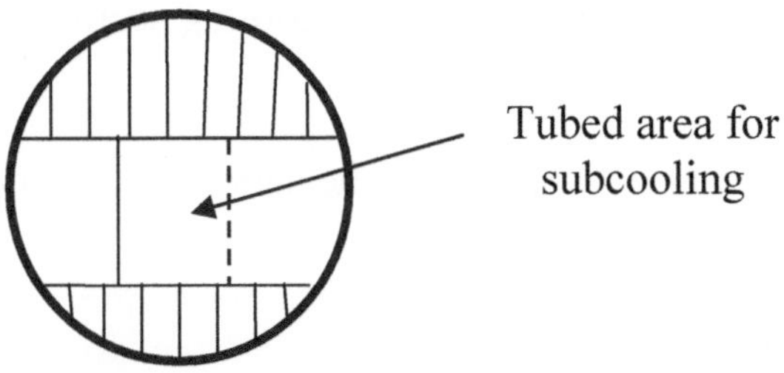

Figure 5.25

As with the condensing run add lots of sealing strips to get the cross section and hence the cross flow velocity as close as possible to the segmental section in the final design. You are unlikely to ever achieve a match and it will be necessary to do some hand calculations to pro rate between your computer run and the actual design for both heat transfer and pressure drop. You can afford to err on the 'safe side' for both calculations. (It is probably simpler to just do a hand calculation for this section!)

Horizontal baffles in the subcooling section are normally a better option. The narrow segmental section you are trying to simulate is the central section of a horizontally baffled shell with very large inlet and outlet nozzles as illustrated in figure 5.26. The shell diameter will be approximately equal to the average width of the actual segmental section.

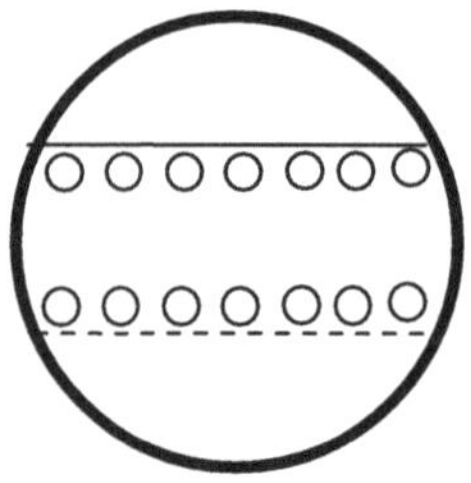

Figure 5.26

For horizontal baffles a NTIW design as illustrated may come closest to the section you are trying to imitate although this will mean losing a few tube rows in the bottom segment it is worth it. There will now be very little pro ration necessary.

The large outlet nozzle for the condensing section and the large inlet nozzle for the subcooling section will result in very low pressure drops. In practice of course there is no pressure drop between these two sections. However is will be necessary to add on the pressure drop of the actual liquid outlet nozzle used.

There is no problem with the heat transfer and pressure drop on the tubeside. You need a run with a full tube bundle which will give the overall pressure drop and a heat transfer coefficient which can be used as input to the individual section designs.

Two final points:

(a) Be conservative with the pressure drop – a safety factor of say 30% for the sub cooling section would not be unreasonable.
(b) I quoted an overall extra oversurface of about 5-6% earlier in this topic. You may have complete confidence in the condensing section run and decide to add an extra 20% to the dodgy sub cooling section which has the same overall effect.

These may not be necessary with a NTIW design for the subcooling section.

SIZING OUTLET NOZZLES

The last thing on liquids leaving a condenser is the sizing of the outlet nozzle. For any condenser with or without any requirement for sub cooling it is important to keep the velocities and hence the momentum fluxes low. This minimises both the chance of flashing in the line and backing up of condensate where sub cooling is not required.

When I first started to design heat exchangers I was told to use a maximum velocity of 3 ft/s and I have never seen any good reason to change this approach. This translates into about 0.9 m/s. The only time I might increase this a bit is to match up to a line size which I know is a final process line size.

SUPERHEATED VAPOUR ENTERING A CONDENSER

The degree of superheat in a vapour is defined as the temperature difference between the actual temperature and the saturation temperature at the same pressure. The units are K.

A highly superheated vapour is best dealt with in a separate desuperheater just as a separate sub cooler was thermodynamically the best solution to excessive sub cooling. However many condensers will have a few degrees of superheat and a few will a more significant amount of superheat.

For most pure condensers the way to deal with superheat is very simple – do nothing – except to remember to tick the box for wet wall condensing which is now in most modern programs. Without this your program will work out a very low convective heat transfer coefficient for vapour cooling until the bulk of the vapour is at its saturation temperature (**dew point**). This may require a substantial part of the total surface area of the unit whilst carrying out a very small part of the total heat duty.

Condensation will take place whenever a vapour is in contact with a surface at a temperature below the vapour saturation temperature. An example is the ideal way to investigate the effect of a desuperheating coefficient on the heat transfer surface it is in contact with and to illustrate if condensation will or will not take place.

Example 5.4

A condenser has 19 mm carbon steel tubes with a wall thickness of 2.3 mm. The thermal conductivity of carbon steel is 50 W/mK. At a point in the exchanger the cooling water temperature is 40°C and the condensing hydrocarbon is at 120 °C. The saturation temperature for the hydrocarbon is 80 °C. The fouling factors for the water and hydrocarbon are 0.00035 and 0.00015 m^2K/W respectively.

Show that condensation will take place if the vapour cooling coefficient is:

(a) 50 W/m^2K
(b) 250 W/m^2K

What vapour cooling coefficient would be necessary to keep the surface temperature one degree above the saturation temperature?

Solution

The ratio of d_o/d_i = 19 / (19 – 2.3x2) = 1.32. The total thermal resistance excluding the resistance of the vapour coefficient can be calculated at the tube outside surface as follows:

For the cooling water $1/h_o \times d_o/d_i = 1/5000 \times 1.32 = 0.000264$
Water fouling $f_i\, d_o/d_i = 0.00035 \times 1.28 = 0.000462$
Wall resistance $\Delta x/k = 0.0023/50 = 0.00045$
Hydrocarbon fouling $= \underline{0.00015}$
Total (excluding R_{ho}) $= 0.000922\ m^2K/W$

The temperature drop across a series of resistances can be found using equation 1.12.

$$\Delta T = \frac{\Sigma R_{RT}}{R_{TT}}(T_1 - T_2)$$

Where ΣR_{RT} is the sum of the thermal resistances up to that point and R_{TT} is the total thermal resistance.

For a vapour cooling coefficient of 50 W/m^2K

$$R_{ho} = 1/h_o = 0.02$$
$$R_{TT} = 0.000922 + 0.02 = 0.020922\ m^2K/W$$
$$\text{Surface Temperature} = 40 + \frac{\Sigma R_{RT}}{R_{TT}}(T_1 - T_2)$$
$$\text{Surface Temperature} = 40 + \frac{0.000922}{0.020922}(120 - 40)$$
$$= \mathbf{43.5\,^oC}$$

For a vapour cooling coefficient of 250 W/m^2K

$$R_{ho} = 1/h_o = 0.004$$
$$R_{TT} = 0.000922 + 0.004 = 0.004922\ m^2K/W$$
$$\text{Surface Temperature} = 40 + \frac{\Sigma R_{RT}}{R_{TT}}(T_1 - T_2)$$
$$\text{Surface Temperature} = 40 + \frac{0.004922}{0.020922}(120 - 40)$$
$$= \mathbf{55\,^oC}$$

For a surface temperature of 81 ^oC

$$81 = 40 + \frac{0.004922\,(120 - 40)}{1/h_o + 0.000922}$$
$$\text{Hence } h_o = \mathbf{1140\ W/m^2K}$$

50 W/m^2K is a very low vapour cooling coefficient. 250 W/m^2K is a more typical value. 1140 W/m^2K is a ridiculously high coefficient which would require a very high pressure drop and probably result in vibration problems. Both the lower coefficients result in a surface temperature well below the saturation temperature showing that condensing would be taking place. This example is typical of many units you will come across where it can be shown that there is no need to be concerned about quite a high degree of superheat and there is no need for a separate desuperheating section in the exchanger.

One very important point arises from the above calculation is that **the lower the vapour convective coefficient becomes the lower will be the surface temperature and the more likely it is that condensation will take place.**

This demonstrates the folly of using a desuperheating section when one is not necessary. An unnecessary desuperheating section will introduce the following penalties to a design:

(a) More surface area will be required because of the lower U value in the desuperheating section.
(b) The pressure drop in the desuperheating section will lower the condensing pressure and hence the saturation temperature for the condensing section. This will reduce the MTD and therefore increase the area required for the condensing section.
(c) The baffle arrangement is more complex.

To check if you may need a desuperheating section just try a run with desuperheating and check out the surface temperatures. If your program does not give you a surface temperature near the inlet you may need to carry out a quick hand calculation similar to the one above. If it is a close call remember that lowering the convective coefficient will reduce the surface temperature. This can be done by clipping the top off the first couple of baffles as discussed earlier in this chapter under the section on shell side condensing steam.

So is there ever a case for a desuperheating section with a pure fluid? This is a question I am still asking myself. I refer the reader back to the HEI feed heater design illustrated in figure 2.29 in chapter 2 which clearly has a desuperheating section. A power plant will have a train of many feed heaters. Each one increases the feed waster temperature using steam at increasing pressures and temperatures until it finally enters the economiser. The economiser is basically the last feed heater which uses exhaust gas instead of steam to heat the feed water up to just below the boiling section saturation temperature.

The steam used for the feed heaters is bled off from in between the train of turbines so that the steam for the highest pressure feed heater is extracted after the HP turbine and may be highly superheated. The degree of superheat is far higher than when steam is used as a service fluid. I am reluctant to contradict colleagues who are specialists in a different industry and I may have missed something. However I still can not see why reducing the velocity of the steam entering the bundle to a point where the vapour coefficient is low enough to ensure condensation would not be a better option. If you ever meet a power plant specialist be sure to bring this point up.

If it was necessary to design a pure condenser with a desuperheating section you could use a horizontal baffle at the top as they do in the power industry or try something like figure 5.27 which would be much more straightforward to design.

The desuperheating section would be closely baffled to give a reasonable vapour coefficient and the rest of the unit would have more widely spaced baffles for the condensing section. A sophisticated modern program will allow you to use a single run for this type of design. You may need a combination design, run as two shells in series, with a more primitive program.

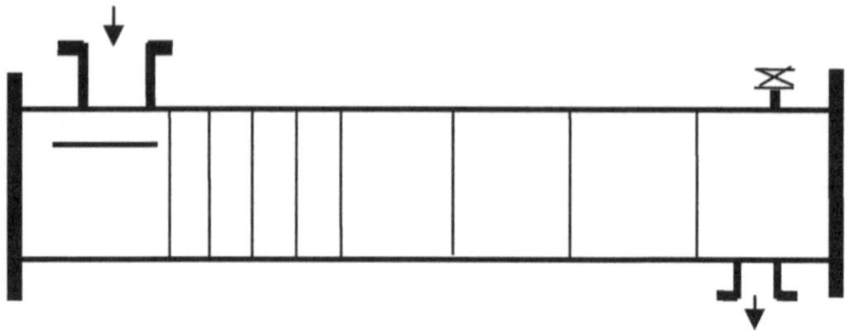

Figure 5.27

The inlet space in figure 5.27 is worth looking at closely. It would have a very low vapour cooling coefficient which would seem to justify using an inlet distributor to reduce the length of this space. On the other hand the 'dead pockets' at the corners would have such low coefficients that you might get some condensation!

I am scratching my head to think of a situation where such a design might be necessary with a pure fluid condensing. Maybe if the cooling fluid were a gas or a high viscosity liquid the tubeside coefficient would be very low which would increase the temperature at the outside surface of the tubes to above the stream dew point. I can not remember having come across such a situation.

A dual baffled exchanger is far more likely to be used for partial condensers with mixed vapours except that the close pitch baffles would be at the outlet end.

PARTIAL CONDENSERS / MIXED VAPOURS

Partial condensers with a mixture of more than one vapour or gas are as common as condensers with a single pure fluid. They can vary from units which are basically gas coolers with some condensation to units which have only two vapours both of which are condensable. (The rare occurrence of just two vapours both being fully condensed could be classed as a full condenser but it could not be dealt with in the same way as for a pure single fluid condensing.)

Partial condensers are all characterised by having an extra thermal resistance to heat flow in at least some part of the exchanger. Figure 5.28 shows a section through a tube which is covered in condensate. Out side the condensate film is a relatively stationary layer of vapour which is the boundary layer for the vapour being cooled.

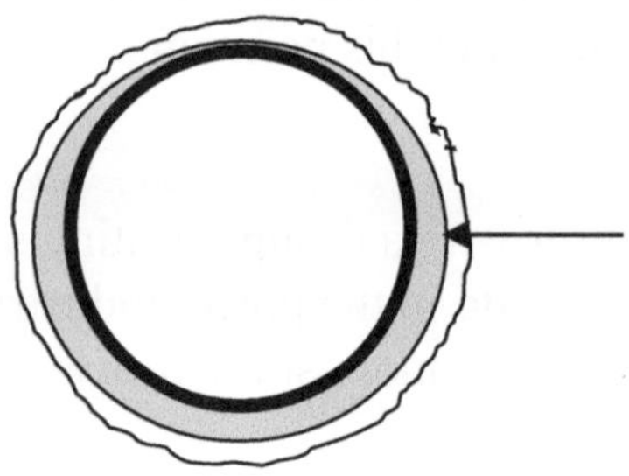

Figure 5.28

This boundary layer will be mainly composed of gas or vapour molecules which are above their dew point temperature. For a molecule of condensable vapour to reach the surface of the tube or condensate film it must pass through the none condensable layer first. This introduces a mass transfer resistance which normally has only a small effect on the condensing coefficient. The important resistance is the vapour cooling resistance which is the inverse of the vapour cooling coefficient. This is always a very low coefficient compared with the condensing coefficient.

It is again worthwhile to briefly look back at how we used to tackle this situation in the days of hand calculations. A condensing coefficient would be calculated in exactly the same way as for a pure fluid condensing. Equation 5.12 would then be used to calculate the fraction of the shell occupied by the average vapour flow, Fv.

Long flow and cross flow areas would then be calculated as explained towards the end of chapter 3 in exactly the same way as they would be for a single phase fluid. The areas calculated would then be reduced by multiplying by the vapour fraction before the cross flow and long flow heat transfer coefficients were calculated. The amount of cross flow and long flow clearly depends on the baffle cut. The coefficients were therefore weighted as follows to determine the vapour cooling coefficient, h_o.

$$h_o = 2 \times \%cut/100 \times h_{long\ flow} + (1 - 2 \times \%cut/100) \times h_{cross\ flow} \qquad [5.13]$$

The vapour cooling coefficient calculated in this way had been found to be very conservative as long ago as 1937 when Colburn and Drew published a paper suggesting a correction factor. This correction factor was used which could increase the heat transfer coefficient by up to 80%. The correction factor involved the use of a chart which I do not see any point in trying to reproduce here. The dominant variable used in establishing the correction factor was the condensate loading. If very little condensation was taking place there would be a negligible correction. For a full condenser with a high condensate loading there would be a significant correction factor. It is not difficult to see the logic of this correction. Figure 5.28 is a simplification which would apply to the top row of tubes. Lower down the bundle streams of condensate would also be contributing to the vapour cooling.

Having established a corrected vapour cooling coefficient it is clear from figure 5.28 that for any vapour cooling there is also the thermal resistance of the condensate film surrounding the tube which must be taken into account.

There are two resistances in series as illustrated in figure 5.29 which must be added together to find a combined thermal resistance.

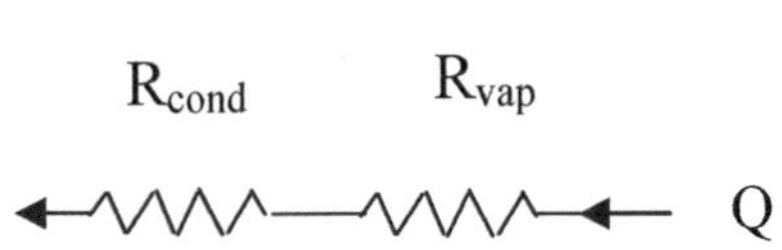

Figure 5.29

In terms of heat transfer coefficients where h_{vap} is the corrected vapour cooling coefficient:

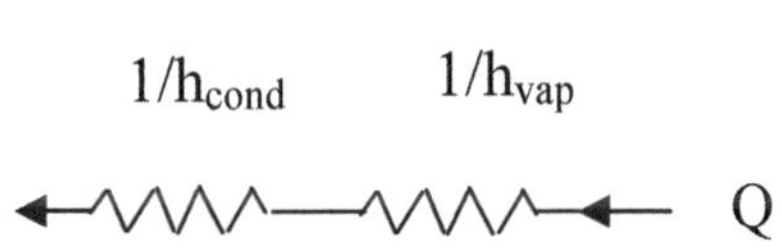

Figure 5.30

A combined heat transfer coefficient, h_{comb} was then calculated from the following equation:

$$h_{comb} = \frac{h_{vap} \times h_{cond}}{h_{vap} + h_{cond}} \qquad [5.14]$$

The shell side heat transfer coefficient, h_s is then found by weighting the condensing and vapour cooling duties with their respective coefficients. Hence:

$$h_{(cond + vap)} = \frac{Q_{(cond + vap)}}{Q_{cond}/h_{cond} + Q_{vap}/h_{comb}} \qquad [5.15]$$

A similar parallel calculation was carried out for the liquid cooling at the bottom of the shell using the liquid fraction, F_L to establish a liquid cooling coefficient h_{liq}.

An MTD would have been calculated at an early stage for the liquid cooling and the condensing plus vapour cooling regions. The latter would have involved a cooling curve which was dealt with as explained in the section on none linear heat release curves in chapter 1. If the coolant was water there were charts to minimise the calculations for establishing a tubeside heat transfer coefficient which could be applied to both regions.

There was now sufficient information to calculate U_{calc} values and hence the surface area required for each region. These were then compared with the surface area available. The balance between the surface area in each region also had to be checked. If there was insufficient surface area for the liquid cooling, the liquid level would need to be raised – normally by using a gooseneck. The calculations would then have to be repeated with an increased value of F_L and a correspondingly reduced value of F_V. For most designs there was no need for a recalculation for the liquid surface area. If there was too much surface area there was nothing that could be done about it and some extra liquid cooling was inevitable.

Most of the above calculation would have to be repeated several times to achieve a satisfactory design. We had standard calculation sheets for many sections not only to minimise errors but to make it easier for a checker to follow. We used very soft pencils and a lot of rubbers, (erasers to any American readers).

The hand calculation procedure outlined above is the same as was used in some of the earliest computer programs which is why a basic understanding of it is important to a present day thermal designer. Since then programs have been developed and refined to increase their accuracy. For example some now take account of the mass transfer effect which we used to ignore.

It is interesting that it would have taken a couple of days to design a partial condenser using hand calculations and the time we now take using a computer has not changed much and if anything may have increased. Of course the thermal data sheet has grown from a single page to four or more pages and now includes a tube layout and some form of preliminary setting plan. However we should also be looking at options which were too time consuming to consider with hand calculations.

MULTI PITCHED BAFFLES.

A partial condenser which condenses over 90% of the total shell side flow rate would ideally be baffled as shown in figure 5.31.

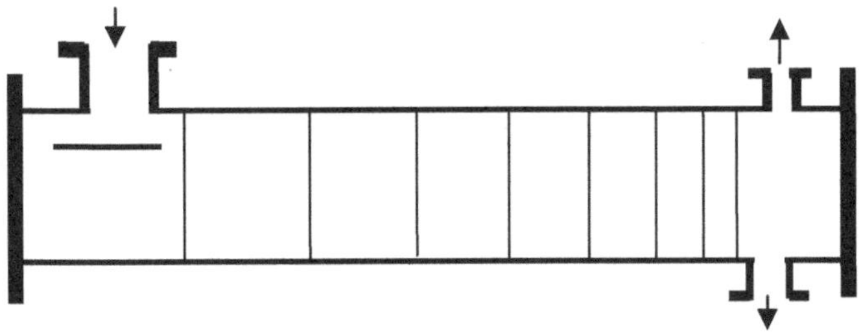

Figure 5.31

If the unit had been fully baffled with the wide pitch used at the beginning it would have worked OK. However towards the outlet the vapour cooling coefficient would have been so low that it would have dominated the thermal resistance and a lot of surface area would have been used up to cater for a small part of the duty. I have noticed that a lot of people will design such a unit in the same way as they would a partial condenser where only say 50% of the fluid condenses i.e. they would use a single pitch which would use up the entire allowable pressure drop. Given a free hand you computer program will do the same thing.

We have seen earlier in this chapter that for a pure condenser a pressure drop is not required and will have a negative effect on the MTD. For the first part of a partial condenser the heat load is dominated by condensation and needs very little pressure drop. Towards the outlet pressure drop is needed to maintain a reasonable vapour cooling coefficient and that is where most of the allowable pressure drop should be utilised.

I can not ever remember having the time, (or the program?) to vary each pitch in the ideal way illustrated in figure 5.30. I can recall many cases where there has been up to a 20% saving in surface area by using more than one baffle pitch. Today's sophisticated programs will allow you to input this data quite quickly. If you are running a more basic program you may need a combination design run as two shells in series. An experienced designer may spot the likelihood of a multi pitch design just by looking at the cooling curve data. But for an experienced designer or novice the first run would be with evenly spaced baffles. The changes in the shellside coefficient are then studied to see how they change along the shell and what effect the shell side coefficient is having on the local overall coefficient, (U value). It then becomes clear if multi pitched baffles will improve the design. Often using just two pitches is sufficient to make a worthwhile saving in surface area. The final design would then look something like figure 5.27 in the section on desuperheating except that the closely baffled section would be near the outlet nozzle.

MISCIBLE & IMMISCABLE CONDENSABLES

Most hydrocarbons in either the vapour or liquid phases will readily mix together and for many calculations can be treated as a single fluid with properties weighted from the constituent components. The term miscible literally means mixable.

For condensing heat transfer it is the way that liquid water behaves with most liquid hydrocarbons that is important because a liquid film on the tubes determines the condensing heat transfer coefficient. In general water is immiscible with most hydrocarbons. (The one pleasant exception is a mixture of ethanol and water!)

The worst case scenario is if the hydrocarbon forms the film on the tubes and any water vapour condensation takes place on the surface of the hydrocarbon film. It would clearly then be wrong to use mixed liquid properties.

Faced with this situation, which frequently occurs, I would recommend playing it safe and using the more conservative hydrocarbon liquid properties. The most important property of course is the thermal conductivity which should always be the value for the hydrocarbon. I would also use the hydrocarbon viscosity and density. The specific heat is different. Using a value for one fluid only may upset the heat balance particularly if there is also significant liquid cooling section at the bottom of the shell. You will have already built in some conservatism with the assumptions for the other properties so use the mixed value for the specific heat.

If you only have mixed values on the process data sheet you should discuss this with the process engineer. If no process engineer is available make sensible estimates and use your own fudged properties. (They may not need to be completely fudged if you know the mass flow rates of liquid hydrocarbon and water at outlet you could look up the properties for water in tables and work back to predict each hydrocarbon property from the mixture property.)

VAPOUR COOLING SECTIONS

In a partial condenser anything which does not condense behaves like, and is treated as an inert even though the same fluid may be condensed in another exchanger further downstream. In this section I will consider 'almost full' condensers which contain some 'true inerts'. The inert is invariably air which has leaked into the hydrocarbon stream either by being dissolved in a feed stream or leaking in at a point in the system which was at a negative pressure while it passed through rotating machinery.

All condensers need to be vented at start up and may need intermittent venting to expel very small quantities of inerts. However in some cases the flow rate of inerts is sufficient to require continuous venting. Some large units will justify the use of a separate vent condenser other units will not.

It is the units which have a small but continuous flow rate of inerts and no vent condenser which will be considered here. The first thing to appreciate is what happens to the inerts once they leave the condenser. Their normal destiny is a flare stack. This can of course be checked on a process flow diagram. In the flare stack any hydrocarbons are burnt producing carbon dioxide and contributing to global warming.

The second thing to appreciate is that we have a situation which is the same as in a surface condenser. In the surface condenser it was necessary to continually expel air but also to minimise the amount of steam which was expelled with this air. In a process condenser we need to expel inerts while minimising the loss of hydrocarbons.

In the surface condenser a separate air cooling section was used to achieve this and this section of the exchanger was supplied with the lowest temperature cooling water. These principles can be applied to a process condenser.

The temperature of the coolant – invariably cooling water, is the first thing to look at. Having the cooling water inlet at the top of the exchanger would certainly help to condense hydrocarbons under the vent connection but this is overruled by the requirement for the coldest water to be in the liquid cooling part of the exchanger. However there is still a subtle choice to be made which may not have too much effect on the liquid cooling provided substantial sub cooling is not required. Consider the two possible nozzle arrangements shown in figure 5.32.

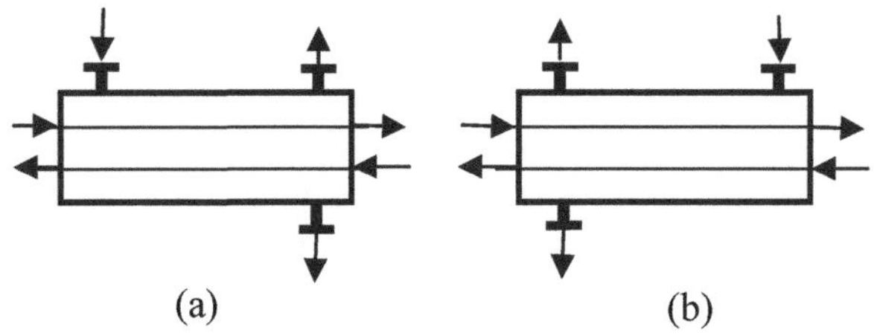

Figure 5.32

For the nozzle arrangement in case (a) the vapour about to leave the exchanger through the vent is in contact with tubes where the cooling water is at its highest temperature. For case (b) the vapour about to leave the exchanger is crossing tubes where the cooling water is at a lower temperature. One of the wonders of using modern computer programs is that it takes very little time to try both arrangements. If there is no clear thermal advantages in using (a) then use (b).

With either arrangement it is not difficult to add at least one baffle near the vent nozzle to increase the vapour heat transfer coefficient and prevent by passing. A more elaborate but more effective baffle arrangement is illustrated in figure 5.33. This is a vapour cooling section – the equivalent of an air cooling section in a surface condenser.

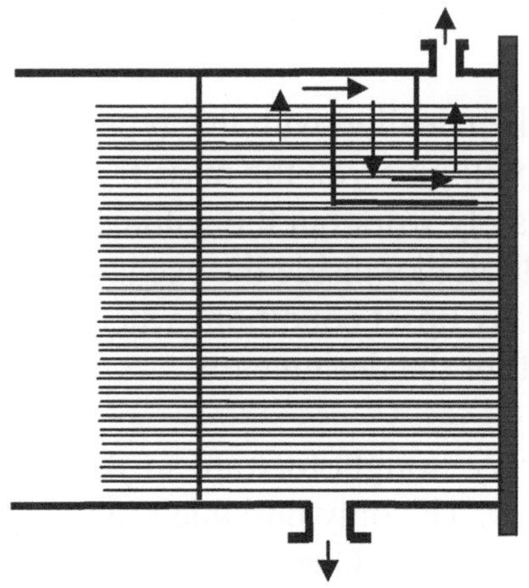

Figure 5.33

If there is a convenient pass partition lane or if the tubes are on an inline square pitch a short horizontal baffle can be fitted together with two horizontally cut baffles as shown. The horizontal baffle does not need to be attached to either the shell or the tubesheet. The clearances around the edges will allow condensate to drain from the vapour cooling section down the sides of the shell and the face of the tubesheet.

If the tubes are on a triangular pitch and there is no convenient pass partition lane a corrugated horizontal baffle could be used. This is an added complication but it would not cost very much. The lazy way of course is to just use a single vertical baffle adjacent to the vent nozzle as normal.

There is no need for any detailed calculations for a vapour cooing section. Just roughly check that the long and cross flow paths are well above the area of the vent nozzle and the pressure drop will be negligible.

These measures will not affect the size of the exchanger but they will reduce the mass of hydrocarbons being burn off in a flare stack.

There are many ways in which the simple design in figure 5.33 might be improved. For example use more baffles than the minimum illustrated. More sophisticated vapour cooling sections might require another computer run. This will only happen when a vapour cooling section is requested on the process data sheet. Process engineers are under increasing pressure to be environmentally friendly and the more enlightened ones may soon be specifying vapour cooling sections for condensers. This would be extra and very challenging work for the thermal design engineer.

HORIZONTAL BAFFLES IN CONDENSERS

No, no, no is the usual cry. My normal response is never say never. I have used horizontal baffles in exchangers where some of the shell side fluid condenses on many occasions. Such units are normally termed gas coolers and the amount of condensation is very small. The fluid condensing is normally water so that the exchanger is acting as a dryer as well as a cooler. Most of the units I have been involved with where I have used horizontal baffles have also involved large volume flow rates of gas so there were also potential vibration problems. The one certain way out of vibration problems, (other than acoustic ones), is to use a NTIW design. NTIW designs are not very suitable for designs with vertically cut baffles.

One of the early questions one should ask when dealing with this type of exchanger is how small is very small? The cooling curve will give you the mass fractions so you know what the mass condensed is. I immediately work out the volume flow rates at outlet to give me a picture of what it would look like inside the exchanger. It is also necessary to check with process if there are any upset conditions where the condensation would be higher than specified for the design case.

The computer run for this type of design is very straight forward so you should have something close to a final design quite quickly. The only possible problem is the condensate. Will it build up and flood the condenser? One of the first things I do is to work out the velocity of the liquid flow at outlet through a standard 10 mm drain notch which will be at the bottom of each baffle. This gives a good feel for the potential problem you are dealing with. If the velocity is very low I might just specify a larger than normal notch for the last baffle just to show that I had thought about the problem and leave it at that.

The cases which are not immediately obvious are the interesting ones. Too small a notch could lead to build up of condensate. Too large a notch would increase the E stream and allow the gas to by pass the tube bundle. This is another opportunity for some hand calculations.

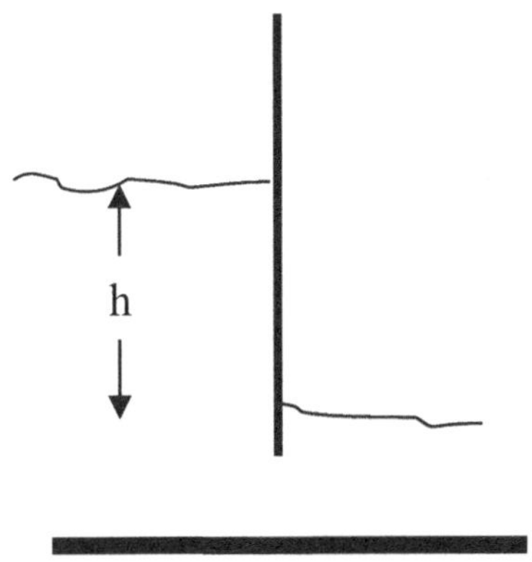

Figure 5.34

The equation $u = \sqrt{(2gh)}$ will be familiar to the reader. It is the equation used to calculate the discharge through an orifice in the vertical wall of a tank with a liquid level h above the orifice. The equation was first developed by Torricelli in 1643. (He was a pupil of Galileo.) The reader is more likely to remember deriving the expression from Bernoulli's equation. Massey, (ref 17) shows that this equation also applies to a system with two liquid levels when h becomes the difference between the two liquid levels. It seems reasonable to apply the same equation to the flow of condensate through a notch in the bottom of a baffle as shown in figure 5.34.

The velocity predicted by Torricelli refers to the vena contracta which is a short distance downstream of the orifice. To refer this to the orifice area it is necessary to introduce a coefficient of discharge, C_d , which has a value of about 0.6. The equation can now be expressed as:

$$u = 0.6 \sqrt{(2gh)} \qquad [5.16]$$

In a heat exchanger the condensate flow rate across a baffle will not only depend on the pressure head of the build up of liquid but on the frictional pressure drop, Δp_F across the bundle at either side of the baffle. This can be read off your computer output. This pressure drop can be expressed in terms of head of fluid, h_F as follows:

$$h_F = \frac{\Delta p_F}{\rho g} \qquad [5.17]$$

Equation 5.16 can now be modified for flow across a baffle.

$$u = 0.6 \sqrt{2g\{h + h_F\}} \qquad [5.18]$$

It is convenient to use a notch with a right angle at its apex and a base length of L as shown in figure 5.35.

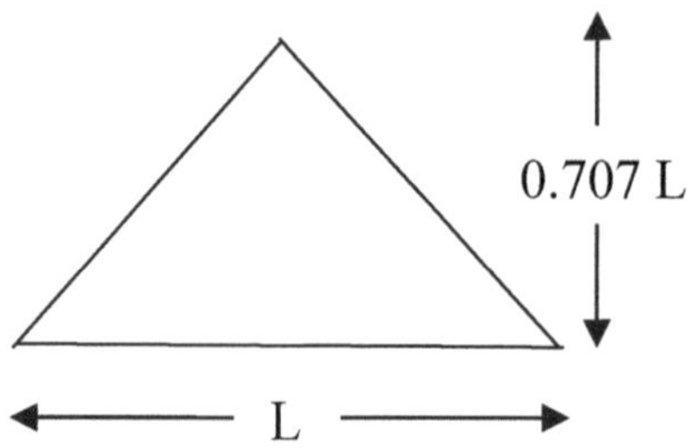

Figure 5.35

The area of this triangle is 0.3535 L^2. For a volume flow rate, V the velocity through the triangular section is therefore:

$$u = \frac{V}{0.3535\, L^2} \qquad [5.19]$$

The bottom line of the triangle should of course be a slight curve because it represents the bottom of the shell but the small extra area involved is not worth taking into account. The velocity through the notch must satisfy both equations 5.18 and 5.19 therefore:

$$L = \left(\frac{V}{0.6 \times 0.3535 \sqrt{2g\{h + h_F\}}}\right)^{0.5} \qquad [5.20]$$

For any computer run the volume flow rate is the amount of fluid condensed up to the baffle in question and h_F is known. That leaves only a value of h being necessary to size the notch.

So what value should be selected for h? The answer of course is zero. The ideal liquid level is just at the top of the notch. Having sized the notch the effect a build up of liquid would have on the flow through the nozzle can be investigated by inserting values for h in equation 5.20.

Alternative theoretical approach.

The theory above applies to a sharp edged orifice. The opening through the notch in a baffle is not a sharp edged orifice. Over a third of the contact area is the smooth bottom of the shell and the contact area with the baffle is not sharpened.

An alternative approach is to treat the notch as a very short tube. The frictional pressure drop through the thickness of a baffle is negligible. The pressure drop is therefore half a velocity head at inlet and one velocity head at outlet making a total of one and a half velocity heads. For the pressure drop across the baffle:

$$\Delta p_F = 1.5 \frac{\rho u^2}{2}$$

Rearranging

$$u = \sqrt{(2\, \Delta p_F / 1.5\rho)} \qquad [5.21]$$

The velocity in equations 5.19 and 5.21 must be the same therefore;

$$\text{Length of notch } L = \left(\frac{V}{0.3535 \sqrt{\{2\, \Delta p_F / 1.5\rho\}}}\right)^{0.5} \qquad [5.22]$$

The obvious way to investigate the use of equations 5.20 and 5.22 is through a worked example.

Example 5.5

A gas cooler condenses 1200 kg/h of vapour before it reaches the last baffle prior to the outlet nozzle. The liquid has a density of 950 kg/m^3. The computer output shows that the pressure drop across the last baffle is 0.05 bar. What is the minimum size of a 90° notch to prevent the build up of condensate? What effect would a 100 mm head of liquid above the nozzle have?

Solution.

$$\text{Volume flow rate } V = \frac{1200}{3600} \times \frac{1}{1000} = 0.000333 \text{ m}^3/\text{s}$$

Using equation 5.20

$$\text{Fluid friction loss head } h_F = \frac{\Delta p_F}{\rho g} = \frac{5000}{950 \times 9.81} = 0.527 \text{ m}$$

$$\text{Length of notch } L = \left(\frac{V}{0.6 \times 0.3535 \sqrt{2g\{h + h_F\}}}\right)^{0.5}$$

h is zero therefore

$$L = \left(\frac{0.000333}{0.6 \times 0.3535 \sqrt{2g \times 0.527}}\right)^{0.5}$$

$$= 0.\ 022$$

$$\mathbf{= 22\ mm}$$

Using equation 5.22

$$\text{Length of notch } L = \left(\frac{V}{0.3535 \sqrt{\{2\, \Delta p_F / 1.5\rho\}}}\right)^{0.5}$$

$$= \left(\frac{0.000333}{0.3535 \sqrt{\{2 \times 5000/1.5 \times 950\}}}\right)^{0.5}$$

$$= 0.\ 019$$

$$\mathbf{= 19\ mm}$$

Although different assumptions were made in developing the two equations the results are in reasonable agreement.

To investigate the effect an extra head of fluid equations 5.19 and 5.18 can be used. Applying these to the more conservative 22 mm notch gives:

With no liquid static head

$$u = \frac{V}{0.3535\, L^2} = \frac{0.000333}{0.3535 \times 0.022^2} = 1.95 \text{ m/s}$$

With a static head of 0.1 m

$$u = 0.6 \sqrt{2g\{h + h_F\}}$$

$$u = 0.6 \sqrt{2g\{0.1 + 0.527\}} = 2.1 \text{ m/s}$$

The difference is only about 8%. This is because the pressure drop across the baffle is much higher than the static head pressure. If the pressure drop across the baffle had been much lower there would have been a much higher increase in velocity. This would mean that the fluid would be far less likely to build up a significant head on the face of the baffle. If this had been the case and the velocity had been reasonable I would have rounded the notch size up 25 mm. In this example a small build up of liquid does not provide much of a safety factor.

My other concern about this design is the magnitude of the velocity. It does not 'feel right' that the velocity through the notch should be so much higher than the velocity I would use to size the liquid outlet nozzle which would be a maximum of 1 m/s.

If the velocity had been less than 1 m/s and a small static head of liquid had predicted a significant increase in velocity I would have just rounded up the size of the notch to 25 mm.

In this example the size of notch required to give a velocity of 1 m/s needs to be investigated.

Rearranging equation 5.19

$$L = \sqrt{(V/0.3535u)} = \sqrt{(0.000333/0.3535)} = 0.0307 = 30.7 \text{ mm}$$

The theory we have used assumed a triangular notch. The bit of area that was missed off the bottom will be sufficient to reduce the velocities slightly. I would therefore use a 30 mm notch.

The danger of over sizing a notch is that it will increase the vapour by pass stream. However if the actual level is only half of the level at the top of the notch as illustrated in figure 5.36 only one quarter of the notch area would be contributing to the vapour by pass streams. This ratio is independent of the angle chosen for the notch.

Figure 5.36

This is the advantage of using a triangular notch rather than a rectangular section which would expose half the notch area to vapour by passing if the liquid level was half way down the notch.

This calculation can be repeated for a sample of the other baffles. The pressure drop across each baffle will not of course be the same. If there are only two or three baffles at the bottom of the shell you should do it for each one. If there are a lot of baffles, (which is unlikely), take samples and group baffles together. Make sure on the data sheet that the largest notches are at the outlet end! There is clearly a case for a spreadsheet to take care of the arithmetic and make this a relatively quick procedure. I must get round to doing this next time I am involved with one of these designs.

Any academic reading this who is looking for a good student project might like to consider flow through the type of notch used in commercial heat exchangers. It would be very interesting to sort out which of the theories I used above is the more accurate. It would not involve any health and safety issues to be industrially useful because the most usual fluid involved is a laboratory favourite - water. I suspect the most useful outcome would be a modified value of C_d to use with the first theory. (Why did I not think of it when part of my job was coming up with a new undergraduate project each year?)

TUBESIDE CONDENSATION

The first thing to consider is a pure fluid. There is an important thermodynamic difference between condensing inside tubes and condensing outside tubes. For condensing a pure fluid outside tubes it was possible to approach the ideal of an isothermal, (hence also isobaric), process. When a fluid flows inside tubes there is always some pressure drop and the process is no longer isothermal. The pressure of the fluid will fall as it passes along the tube resulting in a corresponding fall in the saturation temperature. This fall in saturation temperature will result in a lower MTD for a fluid condensing inside tubes than would have been the case for the same fluid, using the same coolant, condensing outside the tubes. The smaller the MTD is the greater this effect will be.

This applies to steam the same as any other fluid. Condensing steam on the shellside requires hardly any pressure drop above the unavoidable nozzle losses. Condensing steam inside tubes will require a lot more. I remember items when I could not avoid using about 0.5 bar, (0.1 bar would be more normal.)

BASIC HEAT TRANSFER THEORY

Condensation inside tubes can still be thought of as conduction through a film of liquid of varying thickness and a Nusselt analysis applied to it. This is illustrated in figure 5.37.

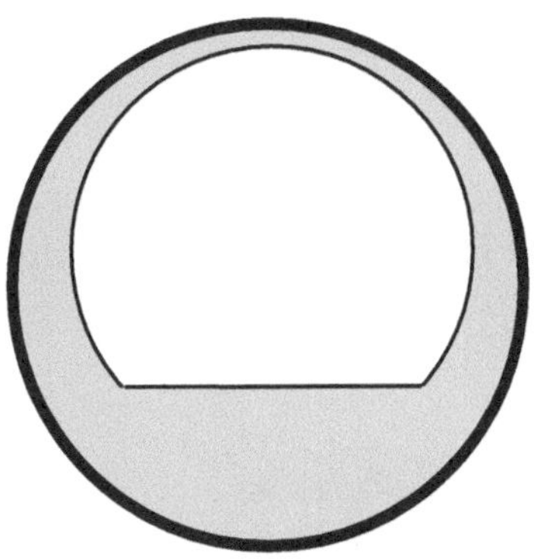

Figure 5.37

The Nusselt theory applies to the liquid film in the top section of the tube but not to the lower section where there is a liquid level which has built up from condensation upstream in the tube. For condensing inside tubes there is no need to worry about condensing streams these are just the number of tubes in each tube pass. I do not intend to get involved with any of the theory in detail. I consider it to be sufficient and more useful to just outline the way calculations were done before the age of the computer.

One of the first tasks was to establish an average value for the volume occupied by the vapour and the liquid. The equation we used is not surprisingly very similar to equation 5.12 which was used to establish a liquid level during shell side condensing. This equation is:

$$F = \left(\frac{\mu_V}{\mu_L}\right)^{1/9} \left(\frac{\rho_L}{\rho_V}\right)^{5/9} \left(\frac{\dot{m}_V}{\dot{m}_L}\right) \qquad [5.23]$$

The similarity with the equation for condensing outside tubes is that the ratios of the same variables are raised to the same power. However in this case F did not refer to either the liquid or vapour fractions. A chart with logarithmic scales with F on the vertical axis and F_L on the horizontal axis was then used to establish a value for F_L, the liquid fraction. After determining a value for the fraction of liquid, F_L the fraction of vapour, F_V was simply $(1 - F_L)$.

The tube was then treated in two sections as shown in figure 5.38. Below the liquid level is a liquid cooling section and above the liquid level a condensing and vapour cooling section. Even in a pure condenser there is always likely to be some vapour cooling because of the fall in saturation temperature due to pressure drop. For a pure condenser this effect would be very small. However the same calculations were used for partial condensers where the effect of vapour cooling is far more important.

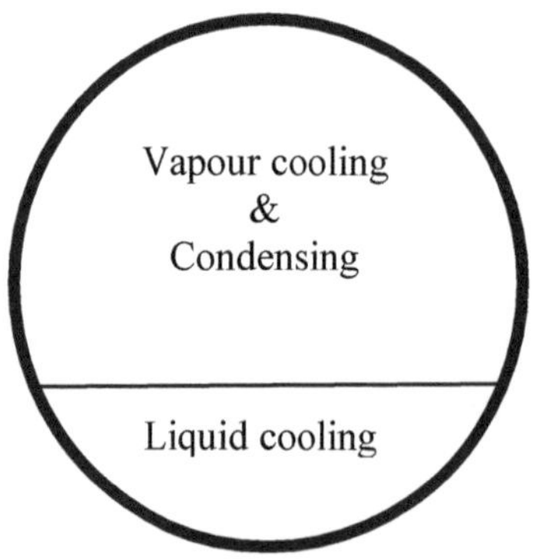

Figure 5.38

The two sections were treated as two independent parallel streams which were assumed to have the same pressure drop. (The basis of the derivation of equation 5.23).
For the condensing heat transfer:

$$\text{Load Factor} = \frac{\dot{m}_{condensed}}{L_{eff} \times N_T \times F_V} \qquad [5.24]$$

A Reynolds number was then calculated from the load factor which was used to look up a value for j to use in equation 5.11 which is repeated below:

$$h = 6540 \times j \times k \times \left(\frac{SG}{\mu}\right)^{2/3}$$

The main difference in the treatment of the shell side and tube side condensing coefficient was therefore in the calculation of the condensate loading factor.

For the liquid and vapour cooling we used to fill in a table which took account of the geometry of the shapes involved. This is illustrated below in table 5.1.

	Fluid Fraction	Area fraction for HT	d_e/d_i	d_e
Liquid	F_L			
Vapour	F_V			

Table 5.1

The area fractions for heat transfer are the length of the arc in contact with each fluid divided by the total circumference of the inside of the tube. For any value of F_L or F_V it is a basic but complicated geometric calculation to establish the length of the arc. We therefore used a chart with arc fraction plotted against fluid fraction. It was only necessary to take the value for one fluid from the chart but it was a good check to read off both and make sure that they totalled 1.0.

Values for the hydraulic or equivalent diameter ratio could also be looked up from a chart or taken from tables such as the one in 'Handbook of Hydraulics,' by King (Ref 23). These values were then multiplied by the tube inside diameter to establish the equivalent diameters. The reader should be able to quickly check that for a fluid fraction of 0.5 the equivalent diameter is the tube inside diameter the same as for a fluid fraction of 1.0. It is a more challenging exercise to show that the equivalent diameter has a maximum value of 1.217 at low fluid fractions between 0.0134 and 0.0245.

The fluid fractions were then used to calculate the velocity for each fluid. These were then used together with the equivalent diameters to calculate Reynolds numbers to use in the same equation as would have been used for a single fluid occupying the whole tube.

Exactly the same correction factor was applied to the vapour cooling coefficient as was used for condensing on the shellside. Equation 5.14 was then used to calculate a combined vapour cooling coefficient and finally equation 5.15 was used to weight this with the condensing coefficient to establish a heat transfer coefficient for the top section of the tubes.

The surface areas required for the two sections were then compared with the available surface area. The calculation was then repeated until a design with an acceptable oversurface was established.

Comparing condensing inside tubes with condensing outside tubes two further points have emerged from the calculation method outlined above. Firstly for a full condenser there is likely to be more sub cooling because the liquid is in continuous contact with the tubes. Secondly for a partial condenser it is highly unlikely that the flow regime will be stratified throughout the length of the tube. The justification for using a stratified flow model is that this is a conservative assumption as far as heat transfer is concerned. Deviations from stratified flow will increase the heat transfer coefficient in the same way that vapour shear increased the heat transfer coefficient outside tubes.

PRESSURE DROP

The subject of two phase flow is sufficiently complex to justify the contents of an entire book such as reference 24 by Butterworth and Hewitt which has over 500 pages. Many people have spent a substantial part of their working lives on research work. All I will attempt to do in this chapter is explain the method we once used to design heat exchangers using hand calculations. This should give the reader an appreciation of some of the problems of two phase flow pressure drops. I will start by defining some of the basic terms required:

Superficial Bulk/Mass Velocity

This is the bulk velocity which each component of the two phase flow would have if it alone occupied the full cross sectional area of the tube. Therefore:

$$G_{LS} = \frac{\dot{m}}{A_X} \quad [5.25]$$

And

$$G_{VS} = \frac{\dot{m}}{A_X} \quad [5.26]$$

Where G_{LS} and G_{VS} are the superficial bulk velocities of the liquid and vapour respectively. In the SI system of units bulk velocity are kg/m^{2}s but we were still using the old imperial system, (now the US customary system) and the units were lb/ft^{2}s.

The bulk velocities used in the pressure drop calculation were therefore quite different from the bulk velocities used in the heat transfer calculation which were derived using the vapour and liquid fractions. The superficial bulk velocity of the vapour used in the pressure drop calculation was normally only slightly smaller than the bulk velocity used in the heat transfer calculation but the difference between the two values for the liquid was much more significant. I have just looked up a hand calculation for a demethanizer 1st stage precooler from January 1970. The bulk velocities used for heat transfer and pressure drop for the vapour were 31.9 and 28.4 lb/ft^{2}s respectively. For the liquid component the values were 46.9 lb/ft^{2}s and 6.35 lb/ft^{2}s respectively.

Superficial Pressure Drop

This is the pressure drop for each component of the two phase flow calculated using the superficial bulk velocity and the individual properties of each component. For the liquid it is therefore the pressure drop for the liquid alone occupying the full cross sectional area of the tube. The Reynolds number for each component was calculated using the inside diameter of the tube and the appropriate viscosity value.

Martinelli Multiplier ϕ_{Ltt}^2

This is an empirically determined ratio of the two phase pressure drop to the liquid superficial pressure drop. Therefore:

$$\text{Two phase pressure drop, } \Delta p_{TP} = \phi_{Ltt}^2 \times \Delta p_L \qquad [5.27]$$

The suffix L in the Martinelli parameter refers to the liquid and the suffix tt indicates that it is based on the assumption of turbulent flow in both individual phases. The empirical equation for ϕ_{Ltt} is:

$$\phi_{Ltt} = \left(\frac{1-y}{y}\right)^{0.9} \left(\frac{\rho_V}{\rho_L}\right)^{0.5} \left(\frac{\mu_L}{\mu_V}\right)^{0.1} \qquad [5.28]$$

The history of the Martinelli multiplier is interesting and should help to understand its complexity and limitations.

This form of multiplier was first introduced by Lockhart and Martinelli shortly after the Second World War. According to Butterworth, (Ref 24), the square on the multiplier when used in equation 5.27 serves no useful purpose. It was convenient to use because of the graph paper available to them at the time. However the power of 2 continues to be used.

The Lockhart/Martinelli experiments were carried out mainly using air and water as the fluids at ambient temperatures with pressures up to 3.5 bar. The only other combination of fluids used was oil and air. Most of the data was close to ambient pressure. The use of empirical data to produce a chart automatically takes account of slip due to the vapour component having a higher velocity than the liquid component.

Martinelli then started to work with Nelson using a single pure fluid – water, to provide the two phases. The water/steam data extended the pressure range up to 220 bar – the critical pressure of water. Their paper was published in the ASME transactions in 1948.

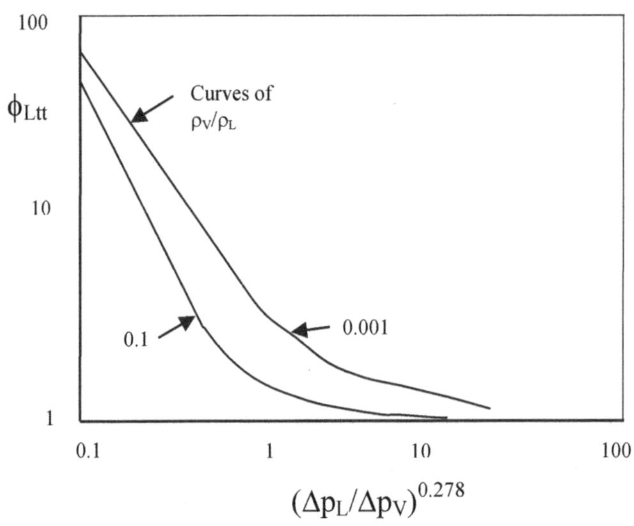

Figure 5.39

We used modified Martinelli/Nelson curves in the days of hand calculations. I have very roughly reproduced a simplified version of the chart we used in figure 5.38.

There were eight curves of constant, vapour density/liquid density; only the limiting values are shown. It should be noted that using ϕ_{Ltt} instead of $(\phi_{Ltt})^2$ on the vertical axis increased the accuracy of reading off values.

The superficial pressure drops were then calculated in the same way as for any single phase fluid. The ratio of these pressure drops, $\Delta p_L/\Delta p_V$ was then raised to the power of 0.278. The ratio of the average vapour and liquid densities was then calculated. These two quantities allowed a value of ϕ_{Ltt} to be read from the chart. There was then sufficient information to calculate the two phase frictional pressure drop from equation 5.27.

The discontinuity losses in the nozzles and at inlet and outlet to the tubes were then calculated assuming homogeneous flow to calculate two phase densities.

Limitations

At the time we used this chart we were aware that it was mainly based on data for water/steam and we were mainly using it for designs with hydrocarbons despite the fact that we used a different equation for water and hydrocarbons for convection! The only alternative at the time would have been to use homogeneous properties. The density would have been reasonable although it would have ignored slip but what about the viscosity? The difference between the liquid and vapour values for hydrocarbons could be a factor of between 10 and 100. How should viscosity be weighted? Methods using homogeneous properties had been be tried and found to be less accurate than using the Martinelli/Nelson method.

A very reasonable question for the reader to ask is why was there no test data on hydrocarbons? The answer of course is health and safety. The rules in those days would not have been as onerous as the present day ones in the UK but they were there and rightly so. I remember going on a course at the National Engineering Laboratory in East Kilbride in the early seventies. One of my many questions was "what fluids do you use for experimental work"? The answer was steam/water, air and that 'very safe and stable' compound refrigerant 12. (Also known as Arcton 12 or Freon 12 depending on which company flag you fly.) CFC refrigerants are given a two or three digit number with the first digit representing one less than the number of carbon atoms, the second digit representing one more than the number of hydrogen atoms, and the third digit representing the actual number of fluorine atoms: if one digit is zero it is omitted, hence R12 is CCl_2F_2.) Less than ten years later the stability of the fluorinated hydrocarbons proved to be their Achilles heel when it was found that the one molecule they would react with was ozone.

There are frequently situations were existing correlations to predict two phase pressure drops are not satisfactory. I was involved in one such situation while I was working as a senior lecturer at what is now Wolverhampton University. I was asked to do some confidential research for Rolls Royce. They had found that the pressure drops in the scavenge lines from the bearings in aircraft engines could not be accurately predicted by any of the existing correlations. I ran tests on different mixtures of air and lubricating oil for each of the pipe sizes being used throughout the range of temperatures encountered in practice. The data I recorded was used to produce a family of design curves for each pipe size which could then be used to accurately predict the pressure drop.

I repeated many runs particularly those which showed a distinct step in the curves which was no doubt due to a change in flow regime. I wanted to analyse the results to see if I could come up with a correlation to fit the curves. I was told that this was not necessary, they were very happy with some practical design charts. I made a brief start in my own time but never got very far. It would be interesting to know how the latest correlations in the computer programs we now run would predict the data I produced. Only someone at Rolls Royce might know. There is a poignant ending to this story. A few months ago I took some of my grandchildren to the aircraft museum at Cosford. There was an RB211 engine, a great beast of engineering ingenuity. It was the first time I had ever seen one. My first reaction was to try and explain my humble contribution to the youngsters but it remained as a thought.

The Martinelli/Nelson correlations, as we used them for hand calculations, were used in the early computer programs. Developments since then have been to refine them as more test data, particularly on hydrocarbons has become available. It is also now possible to estimate a flow regime at each stage in the process and some modifications will only apply to a particular flow pattern. These developments will continue. To find out where the program you are using is up to you will have to read the appropriate manual. It is hoped that my brief outline will help you to understand your manual.

CONTROL OF CONDENSING STEAM UNITS

The control of heat exchangers is the responsibility of the process engineer. However when steam is used as a service fluid it is important that the thermal design engineer is involved since it may have an impact on the design of both the exchanger and equipment used to control it.

When a unit is clean or operating at part load there will be a lot more surface area available than is required to satisfy the duty. There are two fundamentally different methods used to control steam heated items:

THROTTLING THE STEAM

This method requires a control valve in the steam inlet line and is the ideal method of control. As the pressure of the steam is reduced the saturation temperature will fall which reduces the MTD. This is continued until a thermal balance is reached. That is provided a thermal balance is possible. To check if a thermal balance is possible it is necessary to work out a minimum condensing pressure.

Minimum condensing pressure

This is defined as the pressure to which steam must be throttled to achieve a stable operating state utilising all the effective surface area when a unit is clean and operating at minimum load.

Some examples will help to understand this:

Example 5.6

Saturated steam at a pressure of 10 bar is used to heat a pure fluid which is boiling isothermally at 70°C. An exchanger has been designed which has an overall heat transfer coefficient of 820 W/m^{2}K. The total fouling for the unit is 0.0006 m^{2}K/W. Calculate the minimum condensing pressure assuming the boiling heat transfer coefficient remains unchanged and there is no overcapacity or minimum load requirement.

Solution

For the design case:

From property tables the saturation temperature of steam at 10 bar is 179.9°C, therefore the design MTD is (179.9 – 70) = 109.9 K.

When the unit is clean:

$$R_{TT} = 1/\,U_{calc} = 1/820 = 0.00122$$
$$\Sigma ff = \underline{0.0006}$$
$$R_{TT}\text{ clean} = 0.00062\ m^2K/W\ \&\ U_{clean} = 1613\ W/m^2K$$

Q/A = U ΔT_m = constant. Therefore:

$$\text{New MTD} = 109.9 \times 820/1613 = 55.9\ K$$
$$\&\ \text{New } T_{sat} = 70 + 55.9 = 125.9°C$$

From property tables this corresponds to a saturation pressure of about 2.4 bar this is the minimum condensing pressure.

As the steam was throttled it would have become superheated. The effect of this on the calculation would have been negligible.

The assumption that the boiling heat transfer coefficient remained constant when the temperature drop across the coefficient changed is not correct but it is safe since any change would reduce the overall heat transfer coefficient and increase the minimum condensing pressure.

A computer program will take into account any changes in heat transfer coefficients and.all that is necessary is to set the fouling factors to zero and then progressively reduce the steam pressure until the oversurface is approximately zero.

Example 5.7

Repeat example 5.6 assuming the unit had been designed with 10% overcapacity.

Solution

The MTD now needs to be reduced further in the ratio of the two heat loads. The new MTD is therefore 129.5 x 1/1.1 = 117.7 K. From property tables the new minimum condensing pressure is about 1.85 bar and this will still be satisfactory.

The only change in the input to a computer program is the reduced duty.

Example 5.8

Repeat example 5.6 using saturated steam at a pressure of 2 bar.

Solution

From property tables the saturation temperature of steam at 2 bar is 120.2°C, therefore the design MTD is (120.2 – 70) = 50.2 K. The clean coefficient will remain the same.

$Q/A = U\,\Delta T_m$ = constant. Therefore:

$$\text{New MTD} = 50.2 \times 820/1613 = 25.4\ \text{K}$$
$$\&\ \text{New}\ T_{sat} = 70 + 25.4 = 95.4^{\circ}\text{C}$$

From property tables this corresponds to a saturation pressure of about 0.85 bar this is the minimum condensing pressure.

In this example the minimum condensing pressure is not realistic and is unacceptable. The condensate must leave the exchanger under positive pressure. How much positive pressure is required depends on the pressure drop in the steam trap and the lines taking the condensate to the flash chamber before it is returned to the hot well. This is something for process or systems, if they operate as separate departments, to calculate.

It only takes a few minutes to calculate a minimum condensing pressure using a computer program. If control through throttling the steam is not possible or if it has been decided to cut costs by eliminating the control valve in favour of a cheaper shut off valve then the exchanger must be controlled by back flooding.

BACK FLOODING

The aim of back flooding is to reduce the effective surface area available for heat transfer. The steam pressure and saturation temperature remain constant therefore the MTD is constant, hence:

$$Q/\Delta T_m = U_{calc}A_{eff} = U_{clean}A_{clean} \qquad [5.29]$$

The surface area required in the clean condition can be found using equation 5.29 above. The difference between the design effective area and the clean effective area is the area which needs to be flooded. Process may require connections to a level gauge to check this.

Equation 5.29 is for the design condition. If a unit is required to operate at a part load condition the basic equation still holds but the value of Q is reduced. The reduced value of Q divided by the design value of Q is termed the **turndown** and is normally expressed as a percentage.

There are occasions when process may require a very low turndown – say 20%. This is not a problem with steam condensing on the shellside of horizontal units where the theoretical limit is zero because all the tubes could be covered in condensate. For steam condensing on the shellside of vertical units the only practical limit is when the condensate starts to cover the steam inlet nozzle.

Condensing inside horizontal units with the steam inside U tubes is a more interesting situation. All steam heated units have two tubeside passes and the U bends are normally vertical.

There should be no problem with such units if the required back flooding covers less than half the surface area. Once half the surface area needs to be covered all the tubes will be blocked with condensate. As explained earlier in this chapter the pressure drop with condensing steam on the shellside is negligible but for condensing inside tubes it is not. As the control valve in the condensate line closes there may no longer be sufficient pressure drop available to condense the required amount of steam and the tubes will flood more than required. The valve will then open again to compensate leading to an unstable operation.

To overcome this problem the U tubes can be turned so that the bends are in the horizontal plane. Some specifications will require this arrangement whenever back flooding control is required even when it is only necessary to cover less than 50% of the tube surface area. This is a sensible precaution and I would always use horizontal U tubes in this situation.

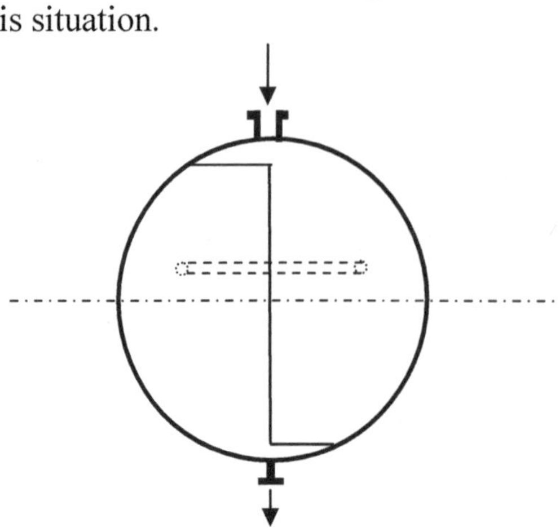

Figure 5.40

For two pass horizontal U tubes the channel inlet nozzle must be at the top as it would be for a vertical U tube unit. It is tempting to use a side inlet nozzle with a simple vertical pass partition lane along the vertical centre line and I have heard about a case where this was used. Of course the back flooding was limited by the liquid level reaching the inlet nozzle. This was not sufficient and some very costly modifications were needed.

Figure 5.40 shows a channel partition arrangement for horizontal U tubes. It assumes that there is sufficient untubed area at the top and bottom of the bundle to allow unrestricted flow into and out of the channel. If there is not sufficient space an extended channel or offset nozzles can be used as discussed in chapter 2 pages 53/7.

When back flooding is used with the steam on the shell side there is large volume of liquid which passes around a series of baffles which have a calming effect resulting in a stable liquid level. Connections to a level controller can therefore be taken from the shell.

For condensing on the tube side a separate condensate pot is fitted adjacent to the channel. This is a slim vertical vessel which extends from the top elevation of the channel to below the bottom elevation of the channel. The condensate passes from the channel to the side of the pot and the top of the pot has a vent line back to the top of the channel. The condensate leaves the bottom of the pot via a valve which is operated by a level controller with connections to the pot.

COMBINED THOTTLING AND BACK FLOODING

For low turn down units requiring fine control a combination of throttling and back flooding may be used. This is of course a very expensive option and process will only tend to use it for units with a history of problems. One example where it is often used is for fractionating tower reboilers

CONCLUSIONS

I have only given an overview of the development of the basic equations used in condensation; there are many textbooks which give a detailed analysis. My journey through the hand calculation methods we once used was the easiest way for me to highlight the important variables. This is not covered in many other text books and as with so many topics a good understanding is best arrived at by digesting views from a variety of sources.

Most of the practical points I have developed have been the standard good practices I have been brought up to use. Some others are ones I have developed myself over the years and many of these have been have been successfully tried by other people. These have usually been instigated as a response to a particular problem. As always writing provides time for reflection and I have suggested ways out of potential problems that I have never faced. I hope I have made it clear which category each section belongs to.

This page has been left blank so that the next chapter starts on a right hand page.

FLOW MAPS – CHAPTER 6

INTRODUCTION

Matter can exist in three states or phases: solid, liquid and vapour/gas. Most people will have experienced flow systems where all three have been involved at the same time. Leaving a beach as a sudden storm breaks can leave you covered in liquid and irritating particles of solid. In the UK it is not unusual to have rain and hail at the same time. In this case water is present in all three phases.

There are important industrial processes involving solids such as fluidised beds and the transport of granulated particles. However for the thermal design of most shell and tube heat exchangers we only need to be concerned with combinations of liquid and gas/vapour. Some of the basic terms used in two phase flow are as follows:

Mass fraction/quality y

This is the ratio of the mass flow rate of vapour to the total mass flow rate. It has been used in earlier chapters and for steam is normally termed dryness fraction.

Void fraction F_V

This is the fraction of the total volume that is occupied by the vapour/gas. It can also be defined as the fraction of the cross sectional area of a tube occupied by the vapour which is the way it was used in the previous chapter on condensers

Homogeneous or mixed density

This is the density weighted on a mass basis therefore:

$$\rho_{mix} = \frac{\dot{m}_L + \dot{m}_V}{\dot{m}_L/\rho_L + \dot{m}_V/\rho_V}$$

Or in terms of mass fraction

$$\rho_{mix} = \frac{1}{(1-y)/\rho_L + y/\rho_V}$$

Homogeneous Velocity

This is the velocity calculated for the mixed flow using a mixed density weighted using the mass flow rates or the mass fractions of the two phases.

Slip ratio/slip factor

This is the ratio of the actual vapour velocity to the actual liquid velocity.

$$\text{Slip ratio} = u_V/u_L \qquad [6.1]$$

In two phase flow the vapour will always be moving at a higher velocity than the liquid. The progress of any particle of fluid will be restricted by its viscosity and density. Any particle of liquid will have shear forces resisting motion due to its viscosity which will be much higher than the shear force acting on a particle of vapour. It also requires more energy to move a heavy particle than a light one. The movement of a liquid particle can be compared with someone walking through marshy ground carrying a heavy rut sack while the vapour particle is moving along a parallel path of firm ground with no rucksack.

Superficial Velocity

This is the velocity which each component of the two phase flow would have if it alone occupied the full cross sectional area of the tube. Therefore:

$$u_{LS} = \frac{\dot{m}_L}{\rho_L A_x} \qquad [6.2]$$

And

$$u_{VS} = \frac{\dot{m}_V}{\rho_V A_x} \qquad [6.3]$$

Where u_{LS} and u_{VS} are the superficial velocities of the liquid and vapour respectively.

We will now look at the range of range of flow patterns/flow regimes which can be found in tubes and pipes.

HORIZONTAL TUBES

Some of the flow regimes which have been identified in horizontal tubes are as follows:

BUBBLE FLOW

Figure 6.1

The volume flow rate of the vapour is much smaller than the volume flow rate of the liquid. The individual bubbles of vapour vary in size and will tend to flow mainly towards the top of the tube due to gravitational effects.

PLUG FLOW

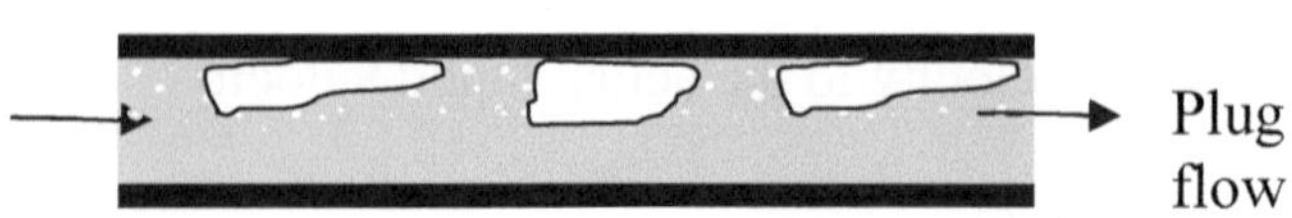

Figure 6.2

If the volume flow rate of vapour is increased while remaining less than the volume flow rate of the liquid some of the bubbles will coalesce to form plugs of vapour. These plugs of vapour will have a minimum cross sectional area at the front in the direction of the flow and a minimum cross sectional area towards the back. They are therefore often described as **bullet shaped**.

STRATIFIED FLOW

Stratified flow Figure 6.3

If the volume flow rate of vapour is further increased whilst the velocities remain relatively low gravity will cause the two phases to completely separate and the liquid will flow along the bottom of the tube and the vapour at the top.

This is the classical flow regime for which a Nusselt analysis applies resulting in the equations used for filmwise condensation.

WAVY FLOW

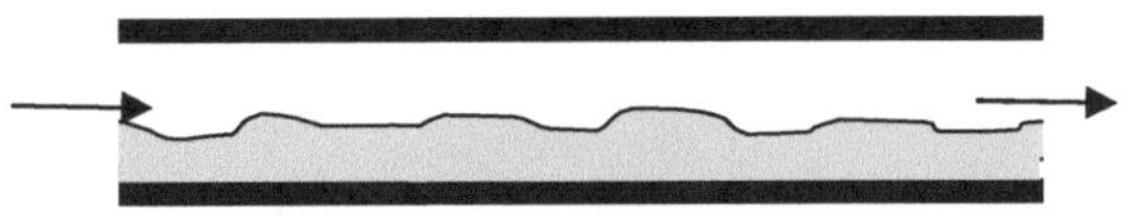

Wavy flow Figure 6.4

Stratified flow can be compared to a calm sea when a yacht would be making little progress without cheating and switching on the engine. As the wind picks up waves start to develop and sailing can once again be fun. In wavy flow in a tube the size of the waves increases as the vapour velocity increases.

SLUG FLOW

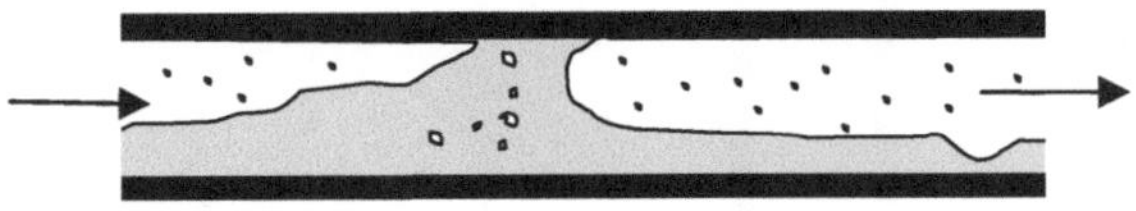

Slug flow Figure 6.5

The velocity of the vapour is now so high some of the waves reach the top of the tube. There are bubbles of vapour in the waves and the vapour space if full of liquid spray. This is known as slug flow. If you are still yachting it is time to reef in the main sail and be prepared to get very wet and cold.

In horizontal pipes slug flow can cause vibrations when the slugs of liquid repeatedly hit the next bend upstream. It is therefore a flow regime which should be avoided whenever possible. If it is unavoidable for example for a part load case then the pipe supports will need special attention.

ANNULAR FLOW

Figure 6.6

The velocities are now so high that you would have no more worries if you had been in that yacht. The vapour space is full of spray and its velocity is much higher than the liquid velocity. Most of the liquid clings to the surface of the tube. Gravity is still having an effect and the liquid layer will be thicker at the bottom than at the top.

MIST FLOW

Figure 6.7

This will occur at high velocities when the volume flow rate of liquid is very small compared with the volume flow rate of vapour. (Back to the starting point of bubble flow except that the volume fractions of the two phases has been reversed.)

VERTICAL TUBES

Some of the flow regimes which have been identified in horizontal tubes are as follows:

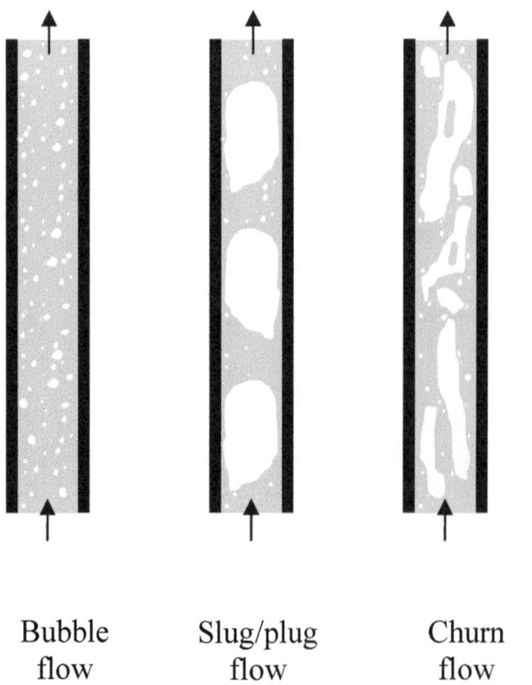

Figure 6.8

BUBBLE FLOW

The volume flow rate of vapour is small compared with the volume flow rate of liquid. This is very similar to bubble flow in a horizontal tube except that the bubbles are evenly spaces across the tube because the gravitational force is acting in the same plane as the direction of flow.

SLUG/PLUG FLOW

When the volume fractions of the liquid and vapour have similar values the bubbles coalesce to form plugs of vapour. These are even more recognisable as bullet shaped than the ones in horizontal flow. Confusingly the terms slug flow and plug flow are interchangeable in vertical tubes. I tend to think of plugs of vapour sandwiched in between slugs of liquid to remember this.

In vertical pipes the unevenness of the flow can cause the pipes to shake or vibrate particularly when the flow hits a bend. It is therefore a flow regime which should be avoided.

It is not normally a problem in heat exchanger tubes where the flow regime will change as the fluid is heated or cooled and the slugs/plugs will be at different heights in different tubes.

CHURN FLOW.

If the gas flow is increased the plugs of vapour in the slug/plug flow regime will break up and the new flow regime is termed churn flow.

This flow regime can also cause problems is vertical pipes because it is associated with oscillations of the liquid which can move up and then down the pipe. It is therefore another unstable regime which should be avoided.

ANNULAR FLOW

At higher vapour velocities the liquid will cling to the wall of the tube as it did in annular flow in the horizontal tube. Now gravity is in the same plane as the direction of flow and although the liquid film may vary in thickness as some of the liquid is torn away the variation in thickness will be random. The vapour core will be full of liquid droplets.

WISPY ANNULAR FLOW

If the liquid flow rate is increased from an annular flow pattern the liquid droplets in the core of vapour will coalesce to form thin streaks of liquid. This is known as wispy annular flow.

High velocities are required to establish either form of annular flow.

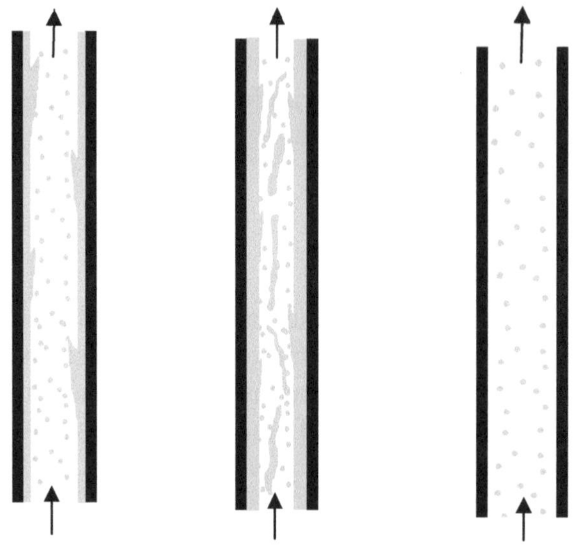

Figure 6.9

MIST FLOW

This is the same as for the horizontal tube. The volume flow rate of the vapour is much smaller than the volume flow rate of the liquid. The individual small particles of liquid are dispersed throughout the stream of vapour.

HEAT EXCHANGER TUBES

During two phase flow through a heat exchanger tube the proportions of liquid and vapour will be continually changing. If the fluid is being heated the mass flow rate of vapour will normally steadily increase as the mass flow rate of the liquid decreases. If the fluid is being cooled the mass flow rate of the liquid will steadily increase and the mass flow rate of the vapour will correspondingly decrease.

The flow regime of the fluid is therefore likely to change at some point along the tube and in some cases there may be more than one change. This could not be taken into account when we used hand calculations. All we could do was to make conservative assumption to come up with a practical design which would satisfy the duty required. In some cases the assumptions were very conservative. Revamp work can be very interesting because it gives an opportunity to compare old designs with the predictions made by modern computer programs.

Some modern programs have the ability to check a flow regime and than use the most applicable correlations for heat transfer and pressure drop. This is something over which the designer has no control and it is all too easy to treat the computer as a black box without worrying about it. Design work is much more interesting with a good understanding of the fundamentals and some knowledge of just how sophisticated the program being used is. To find out where your program is up to you just have to read the manual, although I know some of these can be very boring.

In the above description of the various flow regimes there was no mention of heating or cooling although they will be referred to in other sections of the book. However one of the most important situations when it is necessary to identify flow regimes does not involve any heating or cooling. That is in the design of the pipework for thermosyphon reboilers. This is an integral part of the thermal/hydraulic design and it is essential to minimise the possibility of unstable flow in the outlet lines. Flow maps are a useful aid to identifying the flow regime in a pipe.

FLOW MAPS

A flow map gives a picture of the possible flow regimes for a particular flow configuration. They are based on empirical equations to predict the change of one flow regime to another. All empirical equations are limited by the test data used to formulate the correlations. A good chart must be relatively simple and easy to use. However simplification can reduce accuracy and any flow map is a compromise solution. Different charts take account of different variables and it is not surprising that they may predict different flow regimes for the same set of conditions. Some of the maps the reader may find useful are described below.

HEWITT & ROBERTS MAP FOR UPWARDS FLOW IN A VERTICAL PIPE

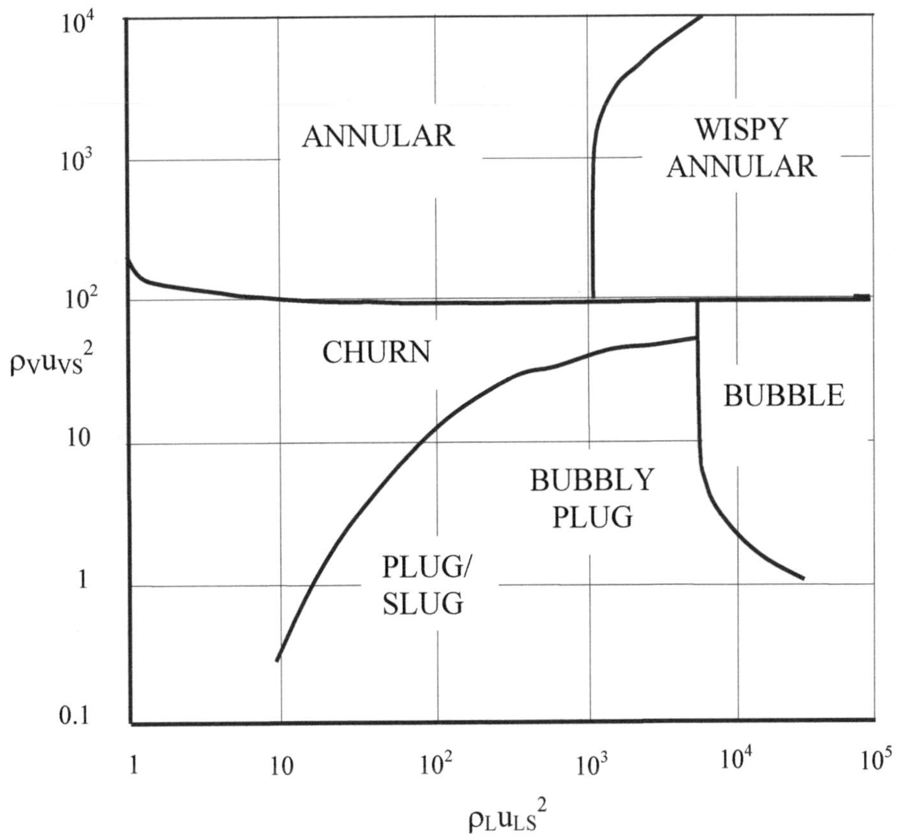

Figure 6.10

This is a relatively simple chart to use and covers a very wide range of possible flow regimes. To use the chart it is necessary to define yet another superficial quantity – the superficial momentum flux, (ρu^2).

This is the momentum flux which each component of the two phase flow would have if it alone occupied the full cross sectional area of the tube. Superficial velocities have been introduced earlier in this chapter as equations 6.2 and 6.3. The superficial momentum fluxes are therefore calculated using the superficial velocities and actual densities of each phase.

$$\rho_L u_{LS}^2 = \frac{\dot{m}_L^2}{\rho_L A_x^2} \quad \text{kg/ms}^2 \qquad [6.4]$$

And

$$\rho_V u_{VS}^2 = \frac{\dot{m}_V^2}{\rho_V A_x^2} \quad \text{kg/ms}^2 \qquad [6.5]$$

The thermal designer is most likely to use this map to check the flow regime in a vertical pipe in a reboiler system.

I can not recall a situation where the liquid momentum flux was above about 5000 kg/ms^2. Nor one where the vapour momentum flux was above 1000 kg/ms^2 or below 5 kg/ms^2. The majority of cases you are likely to investigate will hover around the almost straight line dividing the annular flow from the churn flow at a vapour momentum flux of about 100 kg/ms^2. Taking values from a map with a logarithmic scale can be very tricky. I would advise the reader to add a lot more intermediate points to the chart in the 'working area'. This is a simple geometric procedure which the reader will have covered at some time in a course on engineering drawing.

The aim is to select a pipe which results in a stable annular flow. A word of warning, your computer program may just output the flow regime for a particular pipe and if this is OK it is easy to just leave it at that. However the boundaries between different flow regimes are quite blurred. Changing the diameter of the pipe diameter by a few millimetres will give some indication of how probable the predicted flow regime will be. However it is much more satisfactory to check where you are on a flow map where you can see how close to a boundary you are.

Limitations

The data for this chart was taken from tests on air-water at atmospheric pressure and steam-water over a range of pressures. We are normally dealing with mixtures of hydrocarbons. The chart has been used for hydrocarbons for many years and therefore must predict reasonable results for most situations. However there will always be cases where it is found to be unsatisfactory. (Remember the problems with the two phase pressure drops in tubes which were discussed towards the end of chapter 5?)

VELOCITIES FLOW MAP FOR UPWARDS FLOW IN A VERTICAL PIPE

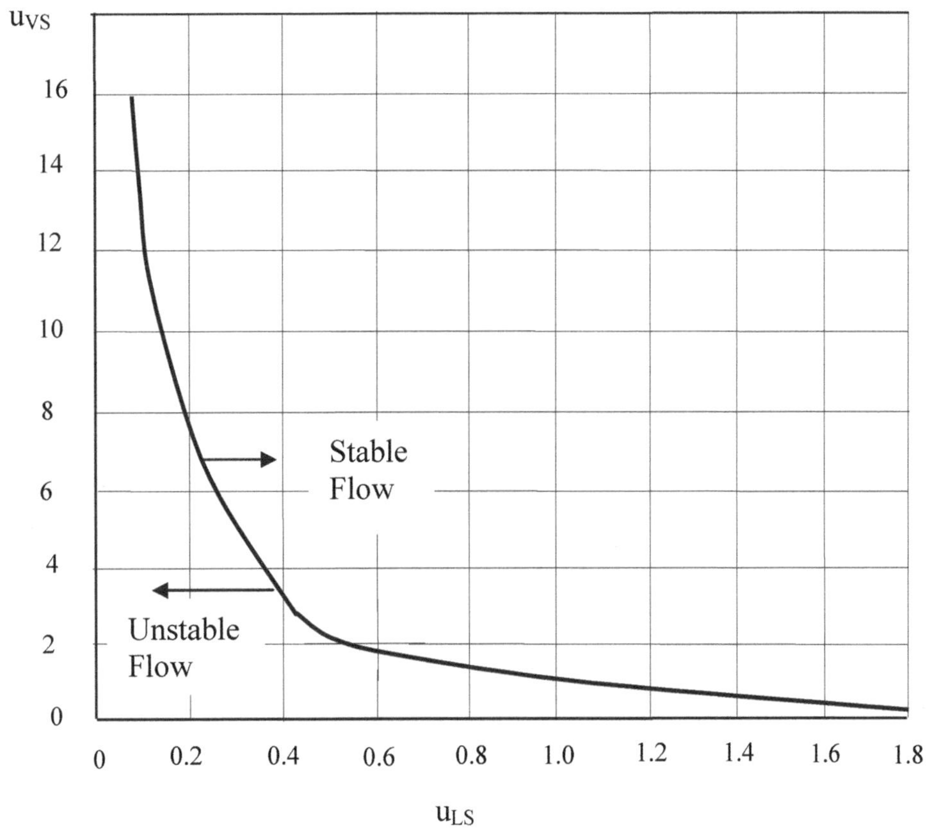

Figure 6.11

This is one of the simplest charts I have ever come across. It was used by the process department of one of the contractors I once worked for. The only variables are the superficial velocities of the vapour and liquid. The units of velocity are m/s.

This chart is the only one I know of with linear scales which make it very easy to use. Although it may appear to be oversimplified the 'working area' on the previous chart was usually around a single line which divided the annual flow area from the churn flow area.

MAPS BASED ON MODIFIED FROUDE NUMBERS

To use these charts the modified Froude numbers, Fr are defined as follows:

$$Fr_V = u_{VS}\,[\rho_V/(\{\rho_L - \rho_V\}gd_i)]^{0.5} \qquad [6.6]$$

And

$$Fr_L = u_{LS}\,[\rho_L/(\{\rho_L - \rho_V\}gd_i)]^{0.5} \qquad [6.7]$$

It should be noted that the velocities are once again superficial values. g is the gravitational acceleration, d_i is the inside diameter or the pipe or tube and the units are all SI.

UPWARDS FLOW IN VERTICAL PIPES

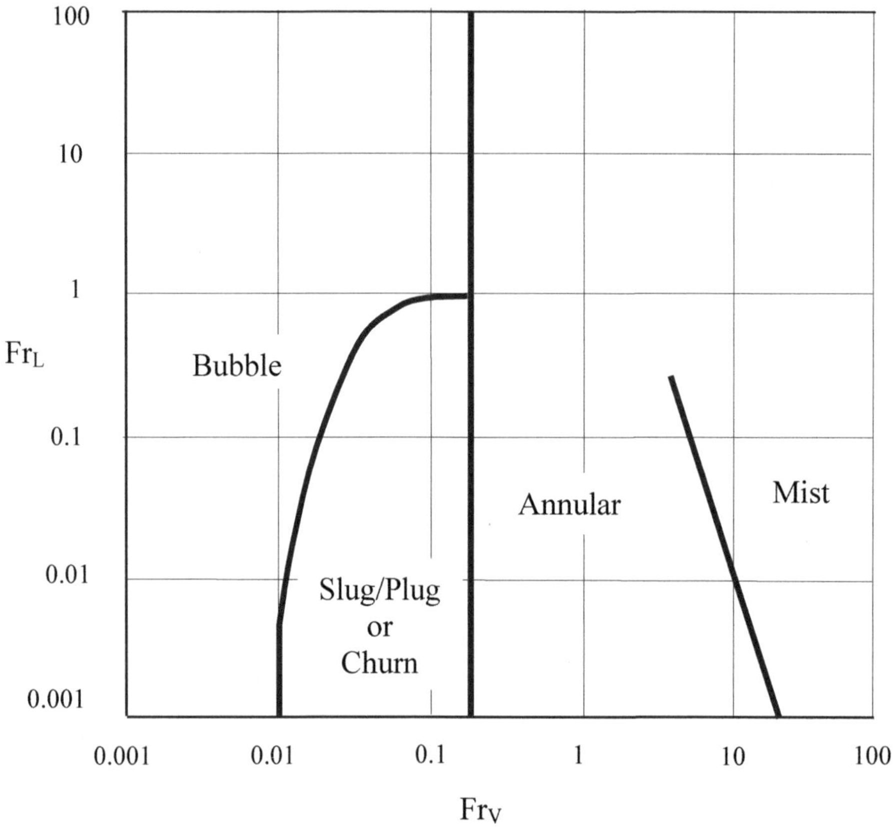

Figure 6.12

Once again the main 'working area' of the chart is around the vertical line which divides the annular flow region from the unstable flow area where the flow could be in the slug/plug or churn regime.

HORIZONTAL PIPES

This chart in figure 6.13 for horizontal pipes/tubes shows all the regions associated with all the flow regimes described in the earlier part of this chapter.

A common feature of the charts for the horizontal and upward vertical flow is that they do not attempt to distinguish between the slug, plug and churn regimes. They form a common unstable region which should be avoided.

The charts for both the vertical and horizontal flows are based on reading from plants rather than laboratory data. The data was limited but it was for hydrocarbons.

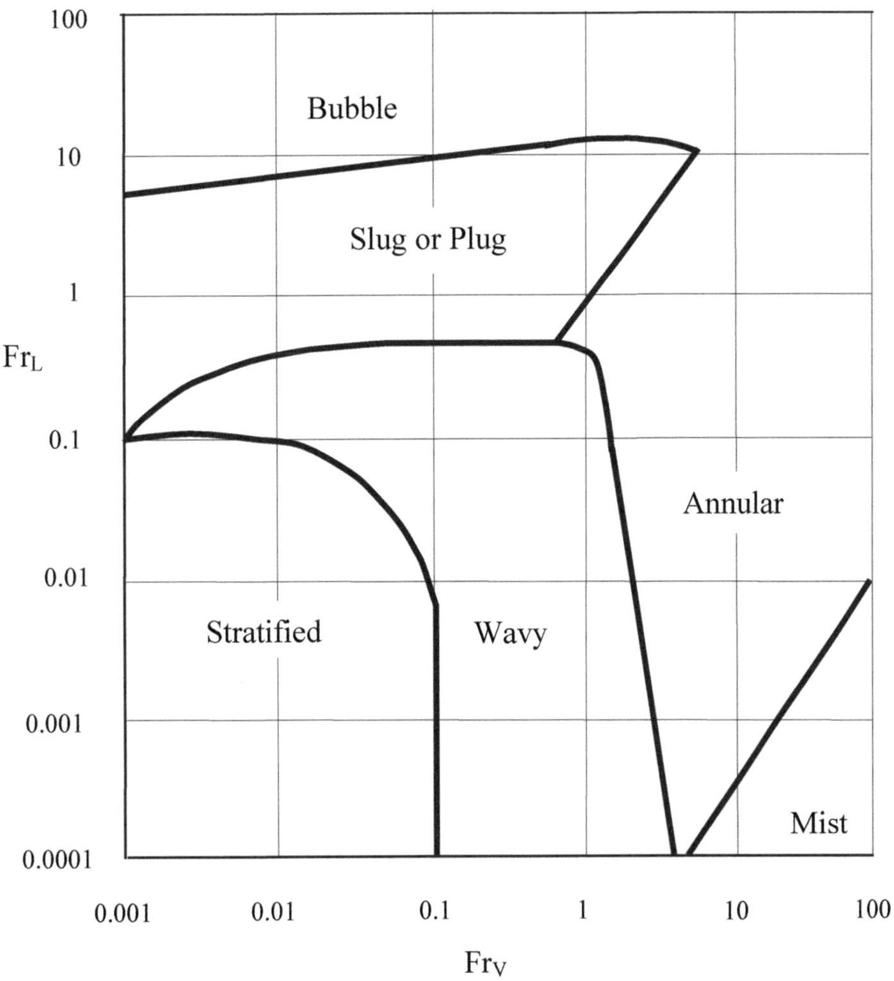

Figure 6.13

FAIR MAP FOR UPWARD VERTICAL FLOW

The coordinates for the Fair map are the total bulk/mass velocity, G_T in kg/m^2s on the vertical axis, (no superficial values this time), and the reciprocal of the Martinelli parameter, ϕ_{Ltt} on the horizontal axis.

$$G_T = \frac{\dot{m}_T}{A_X} \qquad [6.8]$$

$$\phi_{Ltt} = \left(\frac{1-y}{y}\right)^{0.9} \left(\frac{\rho_V}{\rho_L}\right)^{0.5} \left(\frac{\mu_L}{\mu_V}\right)^{0.1} \qquad [6.9]$$

All the lines dividing the different flow regimes are straight on a log/log scale and may be expressed as follows:

Mist/Annular line $\quad G_T = 2440\ \phi_{Ltt}$ [6.10]

Annular/unstable line $\quad G_T = 445\ \phi_{Ltt}^{1.23}$ [6.11]

Unstable/bubble line $G_T = 72\ \phi_{Ltt}^{1.382}$ [6.12]

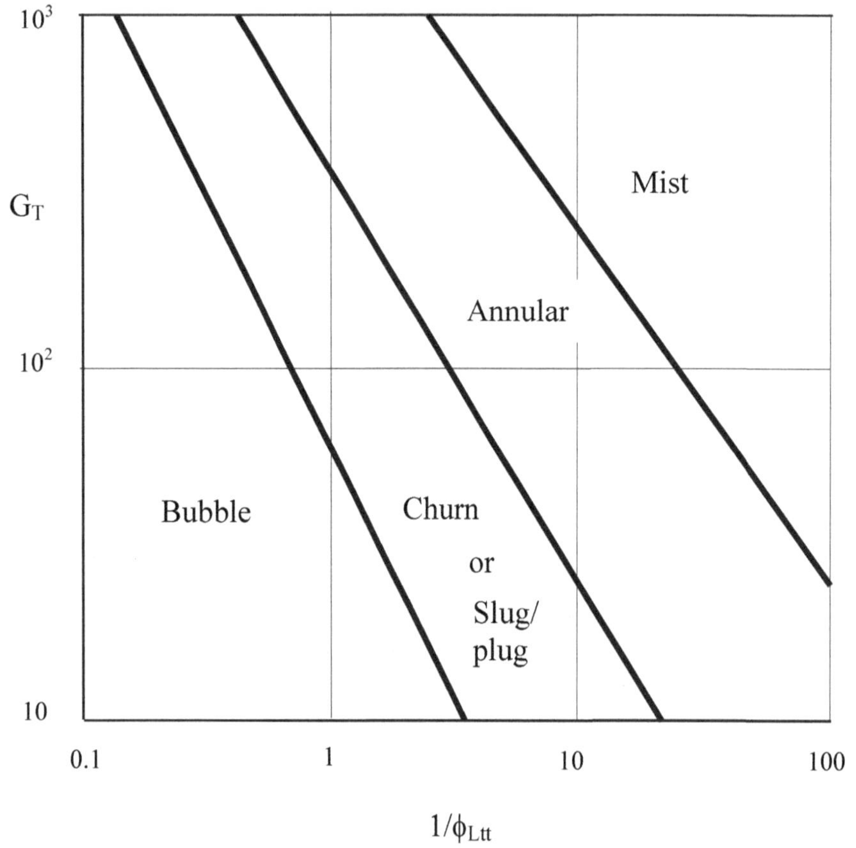

Figure 6.14

This flow map has been used in a least one modern computer program. The output of course may just indicate the flow regime and will not give any indication of how close to the dividing line a point is.

BAKER MAP FOR HORIZONTAL FLOW

The original Baker presentation of a horizontal flow map was first modified by Scott and then Schicht. I still think of it as a Baker map – partly because I have no idea how to pronounce Schicht! The novelty of this form of presentation is that two new dimensionless groups λ and ψ are introduced which compare the properties of the vapour phase with air and the liquid phase with water at ambient temperature and pressure. The logic of this of course is that a lot of the data on which the correlations were based used air/water mixtures at ambient conditions. These dimensionless groups are defined as follows:

$$\lambda = \left(\frac{\rho_V}{\rho_{air}}\ \frac{\rho_L}{\rho_{water}}\right)^{0.5} \qquad [6.13]$$

$$\psi = \frac{\sigma_{water}}{\sigma}\left(\frac{\mu_L}{\mu_{water}}\left(\frac{\rho_{water}}{\rho_L}\right)^2\right)^{1/3} \qquad [6.14]$$

The others terms in the coordinates are G_T which is the total bulk/mass velocity in kg/m^2s and y the mass fraction. For air/water mixtures at ambient temperature and pressure λ and ψ both equal 1.

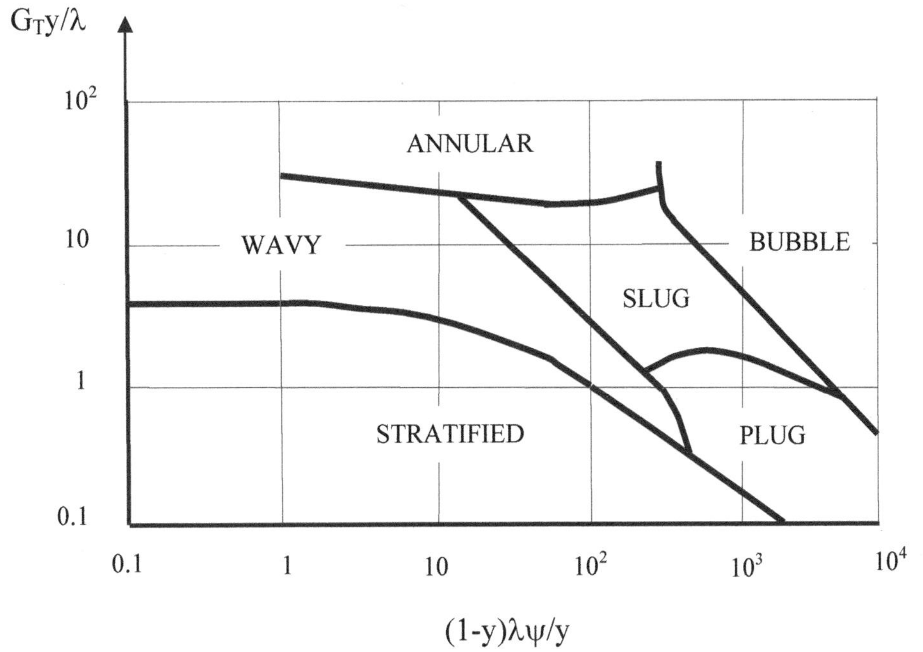

Figure 6.15

One very interesting feature of the Baker flow map is the introduction of the surface tension, σ. Surface tension must have some effect on flow patterns in two phase flow in a similar way that it affects entrainment in kettle reboilers and yet it is seldom taken into account.

Example 6.1

A vertical 8 inch schedule 40 pipe has 30 kg/s of a hydrocarbon flowing upwards. The densities of the liquid and vapour are 800 and 20 kg/m^3 respectively. The viscosities of the liquid and vapour are 0.5 and 0.02 cps respectively. The mass fraction of vapour is 0.05 and the surface tension of the liquid is 15 dynes/cm. Determine the probable flow regime.

Solution

Although it is not required for any of the calculations I always start by working out a volume or void fraction to get a picture of what is taking place.

$$\text{Volume of 1 kg of fluid } V = m_L/\rho_L + m_V/\rho_V = 0.95/800 + 0.05/20$$
$$= 0.00119 + 0.0025 = 0.00369 \text{ m}^3$$

$$\text{Void fraction } = V_V/V = 0.0025/0.00271 = 0.68$$

Although the mass fraction is quite small the vapour is occupying more space. This should help to associate this particular case with the flow regime illustrations at the beginning of this chapter.

We can investigate the flow regime predicted by the different methods available.

HEWITT & ROBERTS MAP

The inside diameter of the pipe can be found in useful tables 3 or elsewhere to be 203 mm, hence:

$$A_X = \pi/4\ d_i^2 = \pi/4\ 0.203^2 = 0.03237\ m^2$$
$$\dot{m}_V = y\ \dot{m} = 0.95 \times 30 = 1.5\ kg/s$$
$$\dot{m}_L = (1\text{-}y)\ \dot{m} = 0.05 \times 30 = 28.5\ kg/s$$

Momentum fluxes

$$\rho_L u_{LS}^2 = \frac{\dot{m}_L^2}{\rho_L A_X^2} = \frac{28.5^2}{800 \times 0.03237^2} = 969\ kg/ms^2$$

$$\rho_V u_{VS}^2 = \frac{\dot{m}_V^2}{\rho_V A_X^2} = \frac{1.5}{20 \times 0.03237^2} = 107\ kg/ms^2$$

Using these values with figure 6.10 shows that this case is predicted to be just annular flow but it is on the borderline between annular flow and churn flow. The annular flow is also close to becoming wispy annular.

SUPERFICIAL VELOCITIES

Superficial vapour velocity $u_{VS} = \dot{m}_V / \rho_V A_X = 1.5/20 \times 0.03237 = 2.3\ m/s$
Superficial liquid velocity $u_{LS} = \dot{m}_L / \rho_L A_X = 28.5/800 \times 0.03237 = 1.1\ m/s$

When these values are used with figure 6.11 stable flow is predicted.

FROUDE NUMBERS MAP

$$Fr_V = u_{VS}\ [\rho_V /(\{\rho_L - \rho_V\} g d_i)\ ]^{0.5}$$
$$= 2.3\ [20/(\{800\text{-}20\} 9.81 \times 0.203)]^{0.5}$$
$$= 0.26$$
$$Fr_L = u_{LS}\ [\rho_L /(\{\rho_L - \rho_V\} g d_i)]^{0.5}$$
$$= 1.1\ [800/\ (\{800\text{-}20\} 9.81 \times 0.203)]^{0.5}$$
$$= 0.79$$

Using these values with figure 6.12 shows that the flow is annular but it is close to the boarder of the unstable area.

FAIR MAP

$$G_T = \frac{\dot{m}_T}{A_x} = \frac{30}{0.03237} = 927 \text{ kg/m}^2\text{s}$$

$$\phi_{Ltt} = \left(\frac{1-y}{y}\right)^{0.9} \left(\frac{\rho_V}{\rho_L}\right)^{0.5} \left(\frac{\mu_L}{\mu_V}\right)^{0.1}$$
$$= (0.95/0.05)^{0.9} \times (20/800)^{0.5} \times (0.5/0.02)^{0.1}$$
$$= 3.1$$
$$1/\phi_{Ltt} = 0.32$$

Figure 6.14 shows that this case is in the unstable area.

Summary

Hewitt & Roberts	– Annular but borderline
Superficial velocities	– Clearly annular
Froude numbers	– Annular but borderline
Fair map	– Unstable

This is typical of the results you might expect for a particular case. I would conclude that the flow is just annular. The point on the fair map is very close to the edge of the map. It is also the only map with straight lines dividing the flow regimes. Straight lines are of course easy to incorporate into computer programs but this may compromise their accuracy.

Example 6.2

Repeat example 6.1 for the same pipe when it is horizontal.

Solution

FROUDE NUMBERS MAP

These are the same as for the vertical flow case.

$$Fr_V = 0.34 \quad \& \quad Fr_L = 1.74$$

Figure 3.13 clearly predicts that the low regime is slug or plug. It is probably slug flow because there is more vapour than liquid by volume.

BAKER MAP

If ambient temperature is taken to be 20°C the approximate relevant properties for water are; viscosity 1 cps, surface tension 67 dynes/cm and density 1000 kg/m^3. For air at 20°C and ambient pressure the density is 1.2 kg/m^3.

$$\lambda = \left(\left\{\frac{\rho_V}{\rho_{air}}\right\} \left\{\frac{\rho_L}{\rho_{water}}\right\}\right)^{0.5} = (20/1.2 \times 800/1000)^{0.5} = 3.65$$

$$\psi = \frac{\{\sigma_{water}\}}{\{\sigma\}}\left(\frac{\{\mu_L\}}{\{\mu_{water}\}}\left\{\frac{\rho_{water}}{\rho_L}\right\}^2\right)^{1/3}$$

$$= 67/15 \times (0.5/1 \times \{1000/800\}^2)^{1/3} = 4.11$$

For the vertical axis $G_T y/\lambda$ = 927 x 0.05/3.65 = 12.7
For the horizontal axis $(1\text{-}y)\lambda\psi/y$ = 0.95 x 3.65 x 4.11 / 0.05 = 285

Figure 6.15 shows that this case is clearly in the slug region.

Summary

Froude numbers	– Slug or plug
Baker map	– Slug

This case is clearly in an unstable region which is almost certainly slug flow.

CONCLUSIONS

The flow maps considered have had the following vertical and horizontal coordinates:

(1) Vapour superficial momentum flux – liquid superficial momentum flux.
(2) Vapour superficial velocity – liquid superficial velocity.
(3) Liquid modified Froude number – vapour modified Froude number.
(4) Total mass/bulk velocity – reciprocal of the Martinelli parameter.
(5) Total mass/bulk velocity, mass fraction and one dimensionless group – mass fraction and two dimensionless groups.

The reader will no doubt come across other maps with different coordinates. This demonstrates some of the problems of dealing with two phase flow. Maps which may give very accurate predictions for one type of fluid may be far less accurate for another type of fluid. Maps for a single fluid are naturally the most accurate and a map for water/steam was produced by Bennett, Hewitt, Kearsey, Keeys and Lacey as long ago as 1966. (This had the coordinates of mass velocity and vapour mass fraction expressed as a percentage.)

The maps I have drawn for this book are only approximate representations of the originals and are not intended for use on real work. They are there simply to illustrate the scope of what is available. Some of the originals had intermediate values on the scales, others did not. If you want to use a chart which has no intermediate values it is essential to put some in, particularly with a logarithmic scale. The reader should build up their own collection of flow maps. It is often useful to enlarge the original. For particularly dodgy pipes I would recommend using several maps as I did in example 6.1. Then of course there is your computer program. This may be regularly updated to take account of modifications to old correlations or the introduction of new ones. Read the manual and try and understand what is inside the program. If the output from your program is simply a statement of the flow regime then check it on a flow map to see how close it is to a boundary. Remember that heat exchangers do not always operate at their design conditions.

The data for the example for the vertical pipe was exactly the same as for the data on the horizontal pipe and yet the predicted flow regimes were quite different. This is quite normal. Another point to bear in mind is that the data used for producing flow maps is normally from laboratories where there will be a reservoir at the each end of the pipe. This is not always the case in a real plant and where the fluid has just been can affect the flow in the pipe. For example if the fluid in a horizontal pipe is in stratified or worse still wavy flow and it comes to a bend before rising vertically it will almost certainly be in an unstable flow regime until it settles down to the regime predicted for the vertical pipe. This phenomenon is not too difficult to visualise. Just imagine being in that boat or on a surf board progressing nicely along the pipe at the liquid/vapour interface and then confronting a right angle turn! The same thing can occur with two horizontal pipes but the effect is not as severe. One way to reduce this effect is to always use long radius bends rather than short radius bends for two phase flow. For reboiler pipework have the vertical pipe upstream of any horizontal pipe whenever possible.

This page has been left blank so that the next chapter starts on a right hand page.

BOILING – CHAPTER 7

INTRODUCTION

The general term for changing a fluid from the liquid state to the vapour state is **vaporisation**. There are two distinct mechanisms by which this can take place.

The fist of these is **evaporation**. This takes place at the surface of a liquid whenever the saturation vapour pressure corresponding to the temperature of the liquid is higher than the partial pressure of the vapour of the same fluid above the surface of the liquid. A common demonstration of this is after a downpour on a hot day. During the heavy rain the air will be close to being saturated and pools of liquid will be built up. When the rain stops the air temperature will rise increasing the difference between the saturation pressure of the pools of water and the partial pressure of the water vapour in the air and the pools of water will quickly evaporate. It is a very important process if you happen to be watching or better still playing in a cricket match! The study of air/water vapour mixtures is known as psychrometry and is covered in detail in my other book – ref 4. The same thing happens with hydrocarbons where the difference between the saturation pressure of liquid and the partial pressure of the hydrocarbon vapour controls the rate of evaporation

The second method of changing the phase of a liquid to a vapour is by **boiling**. This could involve a very complex definition – I prefer a very simple one – it is a heat transfer process characterised by the formation of bubbles.

The two methods of vaporisation may occur in series to change a liquid to a vapour. I saw a very spectacular example in a program on the National Geographic channel recently. At the bottom of an ocean along the dividing line between two tectonic plates there were some significant hot spots. There were lots of bubbles where the sea was clearly boiling. The bubbles were soon cooled and collapsed and never got anywhere near the surface. They were of course clearly heating up the sea in a small way and would eventually contribute to the evaporation from the surface. A more mundane example is when a pan of hot water is heated.

Boiling heat transfer is a very complex topic with sufficient scope to fill a book. I have one such book close by as I am writing this. It is Vaporisers by R A Smith – ref 7. It is a very readable book which I would recommend to the reader. It includes descriptions of a lot of items of equipment and processes other than shell and tube heat exchangers. This chapter is limited to what a thermal design engineer should understand about boiling to design the shell and tube units he/she may be faced with. Each type of unit involving boiling will be dealt with in separate chapters.

POOL BOILING

Pool boiling is defined as boiling in a relatively stagnant liquid where the only movement of the molecules is caused by changes in density of the liquid or by bubbles of vapour passing through the liquid. It can only take place within a force field such as the earth's gravitational field.

POOL BOILING CURVE FOR WATER

As in many other branches of heat transfer the most important and the most studied compound is water. The first experimental work relevant to understanding the behaviour of boiling water was undertaken by Nukiyama in1934. He used a 0.14 mm diameter platinum wire, submerged in the water which he slowly heated by passing an electrical current through the wire. This classical experiment is still carried out today in educational laboratories worldwide and has a nice spectacular finale when the glowing wire melts and the light is extinguished.

Figure 7.1 illustrates the different stages of heat transfer which takes place when a tube or wire is heated in water at atmospheric pressure.

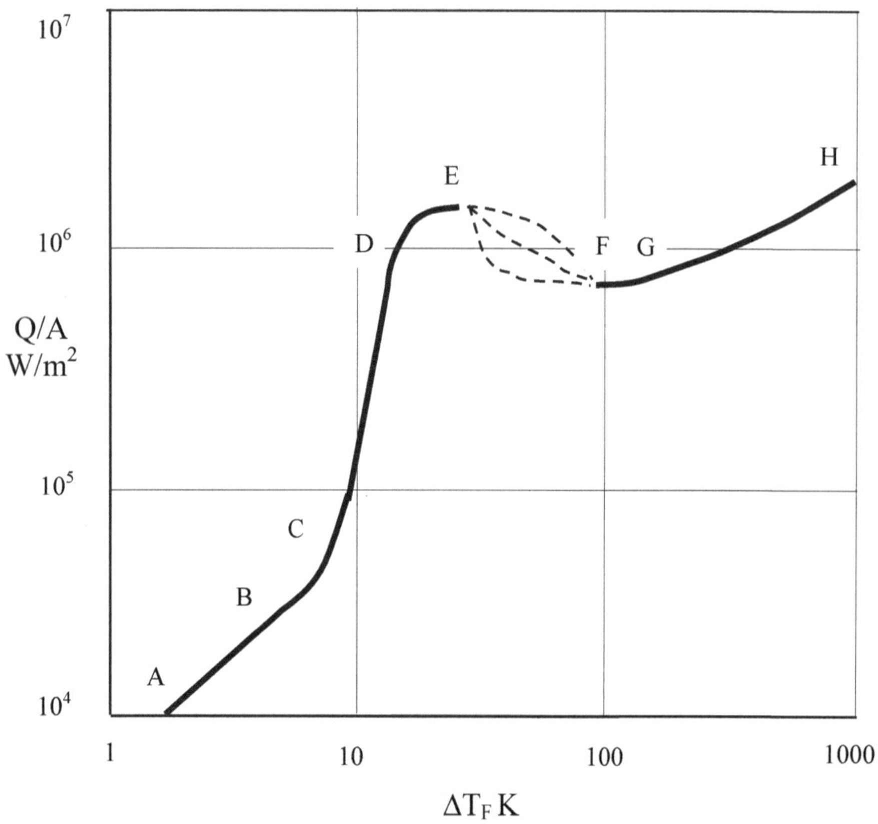

Figure 7.1

A-B Natural convection

If the water is initially well below the boiling point and the temperature difference between the surface of the tube and the bulk of the liquid, ΔT_F is increased gradually there will be only heating by natural convection for this part of the curve. Convective heat transfer produces very low heat transfer coefficients which is reflected in the slow rate of increase of the heat flux, Q/A. The rate of evaporation from the surface increases as the heat flux increases.

B-C Mixed boiling and convection

After point B bubbles will be observed from isolated points on the surface to the tube. Boiling has started but the main heat transfer is still by convection. As the temperature is further increased more points will be observed where bubbles are being produced. The heat transfer coefficients for boiling are far higher than for natural convection and as the number of sites producing bubbles increases the heat transfer coefficient increases. This is reflected by more significant increases of the heat flux and the slope of the heating curve becomes steeper. This process is sometimes referred to as **incipient boiling**. Evaporation increases from the surface but there is now some direct transfer of liquid to vapour as bubbles reach the surface.

C-D Nucleate boiling

This is the nucleate boiling region where the bulk of the liquid has reached its saturation temperature and there is no longer any liquid heating. Heat is transferred directly from the surface of the tube to the surface of the liquid in bubbles of vapour. The slope of the curve is very steep as the heat flux increases rapidly with quite small changes in temperature difference. The number of sites producing bubbles increases rapidly along this part of the curve.

D-E Nucleate boiling

Nucleate boiling is still taking place but the rate of increase is rapidly falling and the size of the bubbles is increasing. The change in temperature difference across this small part of the curve is higher than the change through the main nucleate boiling region C-D.

Point E

This is where the heat flux has reached its maximum value. It is therefore known as the **peak heat flux**. This is a very important point in the study of boiling heat transfer.

E-F Unstable boiling

If the temperature difference is increased beyond the temperature difference required to produce the peak heat flux the system becomes unstable until a much higher temperature difference if reached at point F when the system will once again become stable. Progressing from E to F involves bubbles increasing in size until some of them cover sections of the tube. These large bubbles may then burst leaving part of the tube free to resume nucleate boiling. The system may therefore oscillate between film and nucleate boiling. The heat flux falls off rapidly in this region. It is an area which should be avoided if possible during thermal design work.

Point F

At point F the curve has reached a minimum value. The system is again stable and the tube is completely covered by a film of superheated vapour. This point is known as the **Leidenfrost point**.

E-F Stable film boiling

Any further increase in temperature difference will once again start to increase the heat flux slightly. In this region the water is boiling on the surface of the film of superheated vapour. The heat flux is well below the maximum value and is much smaller than could have been achieved with a far lower temperature difference. This is the region which the thermal design engineer could encounter and may even have to design a unit operating in this region. The occasions when he/she may have to do this are as rare as a British spring bank holiday without any rain.

F-G Stable film boiling

Any further increase in temperature difference will once again start to increase the heat flux more significantly and eventually it could rise to even higher values than the peak heat flux. In this section of the curve radiation effects across the film become increasingly important but this region is of more interest to the physicist than the thermal design engineer.

Some practical demonstrations of film boiling which the reader may have witnessed is the quenching of steel in a bucket of water or the effect of spilling hot water on a cooking hob turned up to maximum. The latter can be quite explosive and very unstable. Of greater interest to the heat transfer engineer are more details of what is taking place during the more practical boiling processes.

BUBBLE FORMATION AND NUCLEATION SITES

Your average member of the public probably regards the surfaces of many of the metallic objects that they handle each day as being 'flat'. Any engineer knows better and it is all about 'relative roughness'. Even the most highly polished surface will have irregularities which can be clearly seen under a microscope. Figure 7.2 shows a very small notch in the surface of a tube. The notch has a cross section which is approximately square.

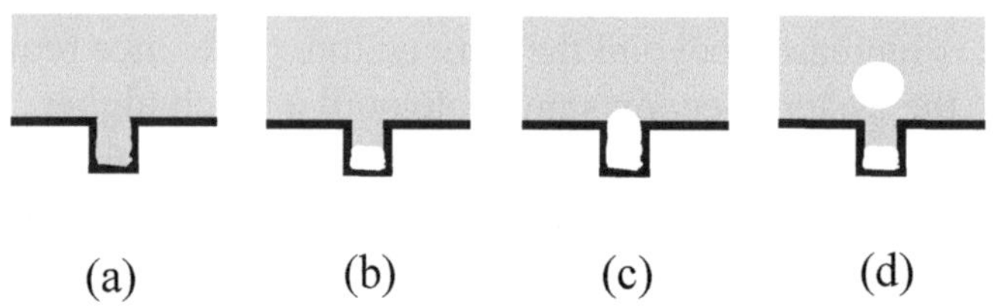

Figure 7.2

At (a) the notch is initially filled will liquid below its saturation temperature. Natural convection will be taking place on the surface but the natural convection in the notch will be restricted. If the liquid in the notch is compared with a similar volume of liquid in contact with the surface of the tube it is exposed to the warmer tube metal on three sides while the same volume of liquid on the surface is only exposed to the warm metal on one side. If there was no gravitational force and heat transfer was by conduction only, the liquid protruding into the metal would be acting as a fin and fin theory would apply.

Unlike evaporation a liquid must be heated to above its saturation temperature before bubbles will form and it starts to boil. A liquid above its saturation temperature is termed superheated liquid which is analogous to a vapour above its saturation temperature being termed a superheated vapour. As the liquid is heated a point will be reached, when the liquid at the bottom of the notch will become superheated while the bulk of the liquid above the notch is not. In fact the bulk of the liquid may not have even reached its saturation temperature. The degrees of superheat required are quite small and will vary depending on the fluid and the pressure. As soon as the liquid has sufficient superheat a bubble will start to form at the bottom of the notch, (b). This bubble will grow until it fills the notch and then starts to protrude from the notch, (c). Finally it will break free from the notch and rise into the liquid whilst a new bubble is starting to form at the bottom of the notch, (d).

The square section which I have used to illustrate the formation of a bubble was easy to draw and discuss but is not very realistic. Notches in the surface of a tube will come in many shapes and sizes so perhaps it worthwhile considering what is the ideal shape and size? The circumference of the notch compared with a unit of length along the flat surface should be a maximum. This will maximise the heat flow to the fluid in the notch and speed up the process of superheating the liquid in the notch. A deep rectangular notch would therefore be a lot better than a notch with a square section. Even better is a notch with a small opening and a large surface area below the opening. The ideal notch could therefore look something like the one illustrated in figure 7.3.

Figure 7.3

This is an unlikely shape for a notch in a normally produced tube. However tubes can be specially produced with surfaces which approach this ideal. Their use is normally restricted to low flux services with boiling refrigerants and will be discussed in more detail in chapter 13 on enhanced heat transfer.

A bubble has been produced from a nucleation site, so what happens to it then? Figure 7.4 shows two possibilities.

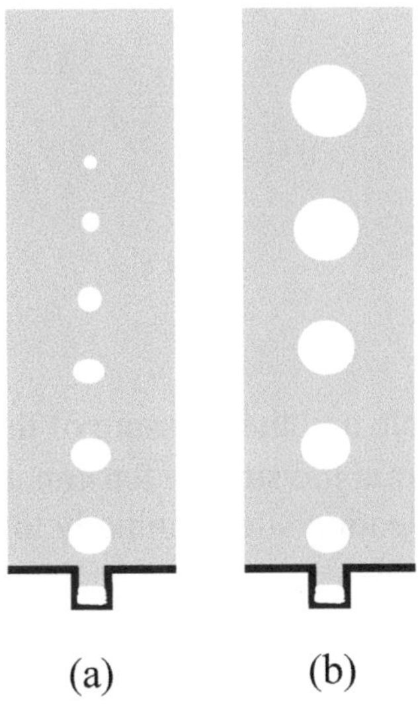

Figure 7.4

On the left (a) shows a bubble which has been produced when the bulk of the fluid is still well below its saturation temperature. It starts to rise through the subcooled liquid and is losing heat to the liquid. The heat it is losing is the enthalpy of phase change which it received in the notch so that some of the vapour must be condensing on the liquid vapour interface. The bubble therefore decreases in size until it eventually collapses and becomes part of the liquid. This is termed subcooled nucleate boiling. It will only happen in pool boiling if there is a large head of liquid, such as at the bottom of the sea in the example given in the introduction. As far as practical thermal designs are concerned subcooled boiling is most likely to occur in flow boiling inside tubes when the bubbles which form on the surface of the tube will be swept from the surface into the bulk of the liquid where they are cooled and collapse. Bubbles must of course decrease in size whenever they pass through a subcooled liquid. Water heated in a pan will show bubbles being formed before the temperature reaches 50°C. As these are detached from the bottom of the pan they will rise to the surface. They must decrease in size as they pass through the subcooled liquid but you would need a very sensitive measuring device to detect the change. (An interesting exercise to calculate the change in volume of the bubble?)

On the right in (b) the bubble just leaving the notch is surrounded by saturated liquid. There is only a very small temperature difference between the vapour in the bubble and the saturated liquid and so the heat transfer is negligible. The pressure of the vapour in the bubble is the sum of the pressure above the liquid surface and the pressure due to the static head of liquid above the bubble. In the diagram the pressure above the liquid is quite low and the static head will therefore be a significant part of the total pressure in the bubble. As the bubble rises its pressure will therefore fall and its volume must therefore increase. When the bubble reaches the surface it will be much larger than it was when it was released from the notch. At medium pressures there will be very little change in the size of the bubbles because the effect of the static head of liquid will be very small.

BUBBLE SIZE

We have seen how the size of bubbles may change but what size are they to start with? A bubble of vapour surrounded by liquid can be considered as a spherical shell in tension and the following equation will apply:

$$p_V - p_L = \frac{2\sigma}{R}$$

Rearranging gives:

$$R = \frac{2\sigma}{(p_V - p_L)} \qquad [7.1]$$

Where p_V and p_L are the vapour pressure in the bubble and the pressure of the surrounding liquid respectively with units of N/m^2. σ is the surface tension in N/m and R is the radius of the bubble in meters. Our nucleation site is unlikely to be a perfect sphere but the same form of equation will apply.

The difference in pressure is fixed by the degree of superheat the liquid requires to form a bubble. The degree of superheat required to form a bubble can vary with the nature of the site. The radius of the bubble is proportional to the fluid surface tension. Surface tension values fall as the temperature increases (see physical properties – in chapter 1). This will cause the size of the bubbles to decrease as the saturation temperatures/pressures increases. As the bubble size decreases more surface imperfections can become nucleation sites. There is an optimum size for a nucleation site for a particular fluid at a specified saturation temperature but the thermal design engineer has no control over that and it is well beyond the scope of this book.

I have often thought what would happen if there were no nucleation sites. Whilst researching the subject before writing this chapter I came across a very interesting answer in reference 25 by Fraas. He refers to laboratory experiments carried out with very pure water in meticulously cleaned beakers. The water was heated to over 25 K above the saturation temperature with no sign of boiling – that is a hell of a lot of superheat for water. When boiling did start it was so violent it was described as "explosive boiling". Had the whole beaker effectively become a single nucleation site?

ENTHALPY OF PHASE CHANGE h_{fg}

As the saturation temperature and pressure increase the enthalpy of phase change decreases until it eventually becomes zero at the critical point. The enthalpy of phase change determines how much heat must be added to a bubble of a fixed size. Therefore the higher the pressure the smaller the amount of heat required becomes. For a fixed rate of heating it is therefore possible to generate more bubbles from any nucleation site.

SUPERHEAT REQUIRED FOR NUCLEATE BOILING

A number of equations have been proposed for calculating the degree of superheat required for nucleate boiling. The following equation is by Frost and Dzakowic:

$$\Delta T_F = \left[\frac{8\sigma(Q/A)T_{sat}}{k_L h_{fg} \rho_V}\right]^{0.5} Pr_L \qquad [7.2]$$

Where ΔT_F is the temperature difference between the surface of the metal and the liquid saturation temperature in a nucleation site and is therefore the degree of superheat required to produce nucleate boiling. Pr_L is the liquid Prandtl number and T_{sat} is the absolute saturation temperature. The best way to demonstrate the use of this equation is by means of a simple example from everyday life.

Example 7.1

An old fashioned type of electric kettle has a coiled heating element 8 mm in diameter with a total effective length of 0.7 m. The power supplied to the kettle is 2 kW. How much does the liquid need to be superheated to produce nucleate boiling?

Solution

All the necessary data can be found from property tables.

At atmospheric pressure T_{sat} = 100°C =373.1 K
Surface tension s = 0.059 N/m
Prandtl number Pr_L = 1.73

Thermal conductivity $k_L = 0.68$ W/mK
Vapour density = 1/1.673 = 0.6 kg/m^3
Enthalpy of phase change h_{fg} = $(h_g - h_f)$ = 2676 -419 = 2257 kJ/kg

The heat flux Q/A = 2000/π 0.008 x 0.7 = 114000 W/m2

Using equation 7.2, $\Delta T_F = \left[\frac{8 \times 0.059 \times 114000 \times 373.1}{0.68 \times 2257000 \times 0.6}\right]^{0.5} 1.73 = \mathbf{8\ K}$

Example 7.2

If an experiment had been carried out in laboratory using the same heating element as in example 7.1 and the pressure was maintained at 10 bar how much liquid superheat would now be required to produce nucleate boiling?

Solution

The new saturation temperature is now 179.9°C (453 K) and the approximate physical properties are as follows:

Surface tension s = 0.038 N/m
Prandtl number $Pr_L = 0.97$
Thermal conductivity $k_L = 0.68$ W/mK (very small change)
Vapour density = 1/0.1944 = 5.14 kg/m^3
Enthalpy of phase change h_{fg} = 2015 kJ/kg

Using equation 7.2, $\Delta T_F = \left[\frac{8 \times 0.038 \times 114000 \times 453}{0.68 \times 2015000 \times 5.14}\right]^{0.5} 0.97 = \mathbf{1.45\ K}$

Example 7.3

If an experiment had been carried out in laboratory using the same heating element as in example 7.1 with hexane as the fluid instead of water and the pressure back to atmospheric pressure how much liquid superheat would now be required to produce nucleate boiling? The saturation temperature is 333.1 K and the physical properties are as follows:

Surface tension s = 0.015 N/m
Prandtl number $Pr_L = 4$
Thermal conductivity $k_L = 0.12$ W/mK
Vapour density = 3.1 kg/m^3
Enthalpy of phase change h_{fg} = 335 kJ/kg

Solution

Using equation 7.2, $\Delta T_F = \left[\frac{8 \times 0.015 \times 114000 \times 335}{0.12 \times 335000 \times 3.1}\right]^{0.5} 4 = \mathbf{24\ K}$

For the water the degree of superheat of the liquid required for nucleate boiling fell substantially at the higher pressure. The property dominating this change was the vapour density which is directly affected by the pressure.

The degree of superheat required for the hydrocarbon was much higher than that required by water at the same pressure.

The heat flux used in these examples is a modest value for the water but quite a high one for the hydrocarbon and is not too far off its peak value.

OVERALL EFFECT OF PRESSURE.

We have looked at some of the detailed effects of saturation pressure and temperature on boiling it is now time to look at the overall effect of pressure. Figure 7.5 is for n-Pentane pool boiling on a flat plate. If each line was to be extended it would have a similar shape to the pool boiling line in figure 7.1 for water. The lines are limited to the nucleate boiling region which was section C-D in the more general pool boiling curve for water.

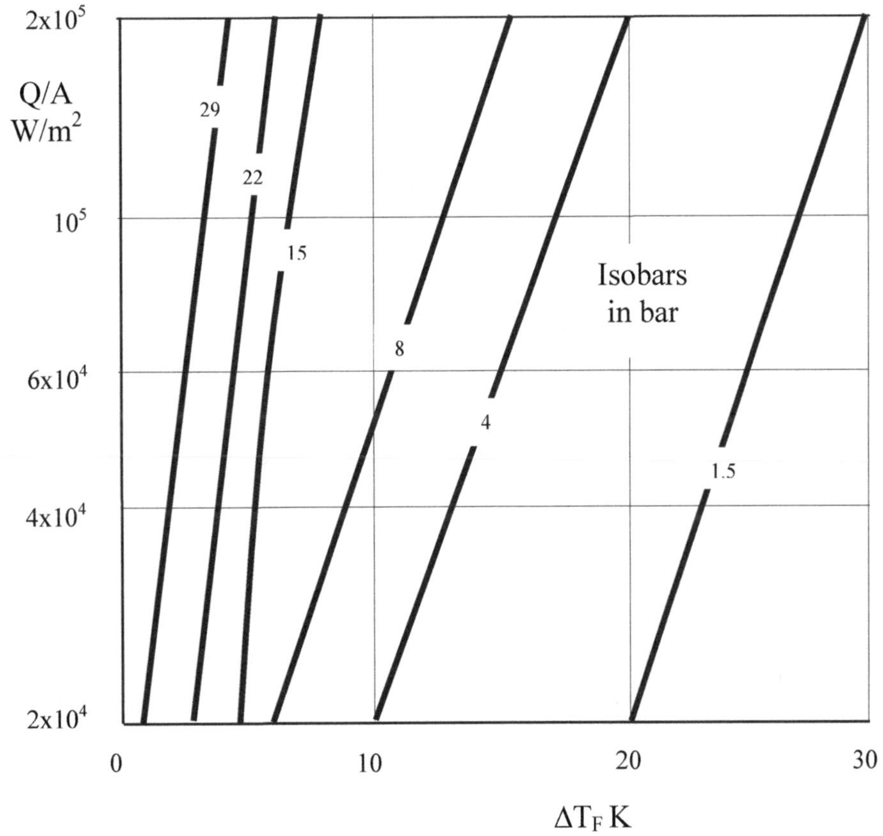

Figure 7.5

A boiling heat flux of 20000 W/m^2K requires a temperature difference of 20 K at a pressure of 1.5 bar. If the pressure is increased to 4 bar only half this temperature difference is required. If the pressure is increased to 29 bar a temperature difference of just over 1 K is required.

The change in the slope of the lines is also interesting. At the lowest pressures the temperature difference needs to be increased by 10 K to increase the heat flux from 20000 W/m^2K to 200000 W/m^2K. At the highest pressures only about 3 K is required to achieve the same increase in heat flux. It must of course be noted that the percentage increase in temperature difference is much higher for the high pressure lines than for the lower pressure lines – 300% compared with 100%.

It is now necessary to look at the peak heat flux in more detail.

PEAK HEAT FLUX $(Q/A)_P$

The prediction of the peak heat flux is a very important aspect of the study of boiling heat transfer. For practical design purposes we are usually trying to avoid it. Some of the methods used to predict the peak heat flux follow.

ADDOMS CHART

The chart shown in figure 7.6 is part of Addoms MIT thesis and was published in 1948. A more detailed chart with intermediate values for the two scales can be found in McAdams, (ref 5) on page 385 of the second edition.

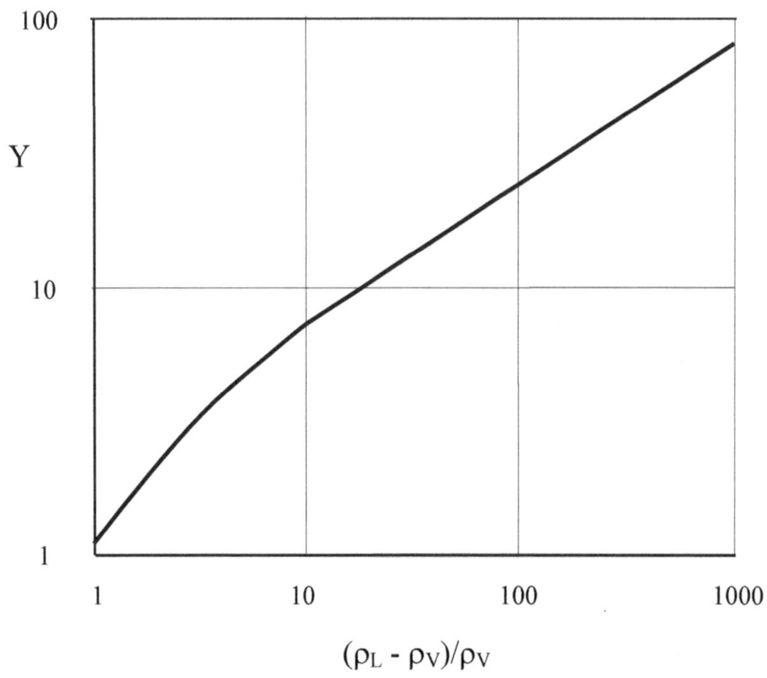

Figure 7.6

This is the chart we used in the days of hand calculations. On the horizontal axis the term $(\rho_L - \rho_V)/\rho_V$ is termed the buoyancy effect. The value for Y from the vertical axis was used in equation 7.3 to predict the peak heat flux.

$$(Q/A)_P = Y\, h_{fg}\, \rho_V \left(\frac{g\, k_L}{(\rho_L C_{pL})}\right)^{1/3} \qquad [7.3]$$

The term $k_L/(\rho_L C_{pL})$ is the thermal diffusivity which was multiplied by the acceleration due to gravity before being raised to the power of one third. The reader should note that the peak flux is directly proportional to both the enthalpy of phase change and the density of the vapour which are both dependant on the pressure. It is interesting that g is not included in the buoyancy term although it is taken account of in the equation. The physical property surface tension is conspicuous by its absence.

The data used to draw the chart was based on two different experiments with water, one using a platinum wire and the other a chrome plated tube together with data for hydrocarbons. The hydrocarbon data was by Cichelli and Bonilla from tests on a chrome plated copper disc. The fluids used were n-propane, n-pentane, n-heptane, benzene and ethanol. There were no doubt good reasons for using a disc rather than a tube but I can not give any explanation.

The chart in McAadams shows all the experimental points and there is a good fit with the line of the graph. However it is a big jump to take from a single disc with a pure fluid to a heat exchanger bundle with a mixture of fluids and we applied some very drastic safety factors to the values calculated from equation 7.3. These were as follows.

$(Q/A)_{max}$ For design

(a) 0.15 $(Q/A)_P$ or 77850 W/m^2 max (25000 BTU/hft^2) for natural circulation (Kettles and horizontal thermosyphons).
(b) 0.25 $(Q/A)_P$ or 110390 W/m^2 max (35000 BTU/hft^2) for forced circulation.
(c) 0.35 $(Q/A)_P$ or 141930 W/m^2 max (45000 BTU/hft^2) for low strength aqueous solutions.
(d) 0.45 $(Q/A)_P$ or 220780 W/m^2 max (70000 BTU/hft^2) for water.

The first two of these were used for hydrocarbons and we used only 15% of the predicted peak heat flux for natural circulation systems. Even then there was a cut off value of 77850 W/m^2 which we tried not to exceed. These systems are of course the most vulnerable to vapour binding towards the top and centre of the bundle.

PHYSICAL PROPERTY METHOD

This method was developed from test data on pure fluids the same as the Addoms method. The original equation by Zuber was later modified by Lienhard and Dhir and is shown below as equation 7.4.

$$(Q/A)_P = 0.2093\, h_{fg}\, \rho_V \left(\frac{\sigma \{\rho_L - \rho_V\}}{\rho_V^2} \right)^{1/4} \qquad [7.4]$$

The format is very similar to the Addoms format but there is no longer any need for a chart which makes it much simpler to use in a computer program. The peak heat flux is again directly proportional to both the enthalpy of phase change and the density of the vapour which are both dependant on the pressure. The other significant feature is that the property surface tension is now taken into account.

The constant in the equation is dimensional because the acceleration due to gravity has been incorporated into the constant. The constant in equation 7.4 is therefore only applicable to the SI system of units. In the US customary system the constant is 2420.

REDUCED PRESSURE METHOD

It was first suggested by Cichelli and Bonilla that the peak heat flux of a boiling fluid could be correlated using only the pressure of the fluid and its critical pressure. The constants have been refined over the years and equation 7.5 below is one of the more recent ones.

$$(Q/A)_P = 30400\, P_r^{0.28} (1 - P_r)^{0.73}\, p_C \qquad [7.5]$$

p_C is the critical pressure in bar and P_r is the reduced pressure, p/p_C which is dimensionless. The constant 30400 is dimensional and only applies to the SI system of units. (664 in the US customary system)

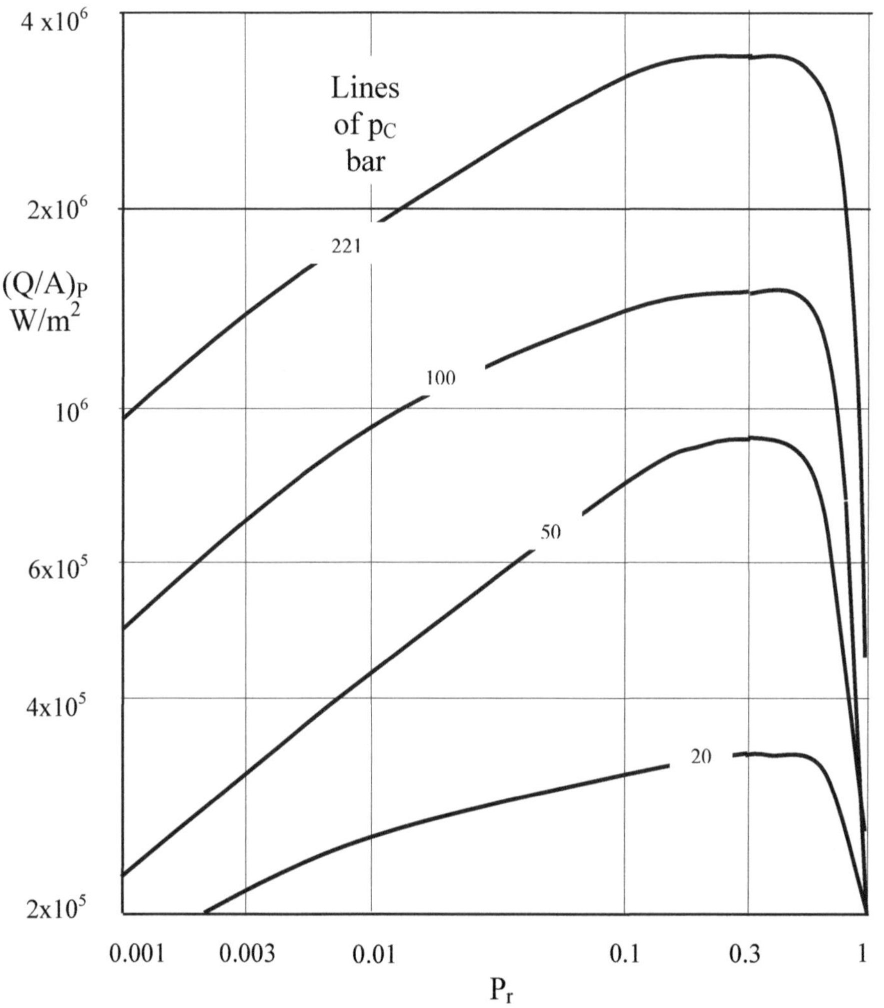

Figure 7.7

Figure 7.7 is a plot of equation 7.5 for four different values of critical pressure. The 221 bar curve is the critical pressure for water; the two lower pressure curves are typical values for hydrocarbons. The curves all show a steep rise in the peak heat flux as the pressure and hence the reduced pressure increases from the lowest value. The curves then flatten out above a reduced pressure ratio of about 0.1 reaching a maximum peak heat flux at about 0.3. Above a reduced pressure of 0.6 the peak heat flux rapidly falls towards zero at the critical point where the reduced pressure is 1.

This graph highlights two potential problem areas in boiling heat transfer – very low pressures and very high pressures (compared with the critical pressure for the fluid.) The equation is very easy to use with a calculator or spreadsheet and has been used in several computer programs.

Note that it is convenient to use the bar as the unit of pressure in equation 7.5 rather than N/m^2; this simply modifies the constant; other units must be basic ones. (The constant in equation 7.4 could have been modified to use the enthalpy of phase change in kJ/kg.) The best way to compare these methods is by means of numerical examples.

Example 7.4

What is the peak heat flux for water boiling at 10 bar Pressure?

Solution

The relevant data can be found in property tables and is as follows:

Critical pressure = 221 bar
Reduced pressure = 10/221 = 0.045
Surface tension s = 0.038 N/m
Vapour density = 1/0.1944 = 5.14 kg/m^3
Liquid density = 886 kg/m^3
Enthalpy of phase change h_{fg} = 2015 kJ/kg

Physical property method.

$$(Q/A)_P = 0.2093\, h_{fg}\, \rho_V \left(\frac{\sigma \{\rho_L - \rho_V\}}{\rho_V^2}\right)^{1/4}$$

$$= 0.2093 \times 2015000 \times 5.14 \left(\frac{0.038 \times 881}{5.14^2}\right)^{0.25}$$

$$= 2300000 \text{ W/m}^2\text{K}$$

Reduced pressure method.

$$(Q/A)_P = 30400\, P_r^{0.28}\, (1 - P_r)^{0.73}\, p_C$$

$$= 30400 \times 0.045^{0.28} \times (1 - 0.045)^{0.73} \times 211$$

$$= 2700000 \text{ W/m}^2\text{K}$$

Example 7.5

What is the peak heat flux for hexane boiling at 10 bar Pressure? The relevant physical properties are as follows:

Critical pressure = 30 bar
Reduced pressure = 10/30 = 0.333
Surface tension s = 0.005 N/m
Vapour density = 23.3 kg/m^3

Liquid density = 480 kg/m^3
Enthalpy of phase change h_{fg} = 230 kJ/kg

Solution

Physical property method.

$$(Q/A)_P = 0.2093\ h_{fg}\ \rho_V \left(\frac{\sigma \{\rho_L - \rho_V\}}{\rho_V^2} \right)^{1/4}$$
$$= 0.2093 \times 230000 \times 23.3 \left(\frac{0.005 \times 480}{23.3^2} \right)^{0.25}$$
$$= 290000\ W/m^2K$$

Reduced pressure method.

$$(Q/A)_P = 30400\ P_r^{0.28}\ (1 - P_r)^{0.73}\ p_C$$
$$= 30400 \times 0.333^{0.28} \times (1-0.333)^{0.73} \times 30$$
$$= 500000\ W/m^2K$$

DISCUSSION OF RESULTS

The peak heat flux values calculated by the two methods were in very good agreement for the boiling water. This is not too surprising given that so much of the data used to develop the correlations was based on water.

For the hexane the reduced pressure method predicted a peak heat flux value about 70% higher than the property method. I was surprised at this since I was expecting a more conservative value from the reduced pressure method. I hope I have not just dropped a clanger and I have no doubt that selecting a different fluid at a different pressure might reverse these results. This section of the book is not intended to be an in depth analysis of boiling heat transfer, I simply aim to give the reader the level of understanding that I needed to design shell and tube units. My own limited knowledge has increased while writing this chapter.

The first time I came across the reduced pressure method I was very sceptical of it. It did not seem right to me that an equation based only on the operating pressure and the critical pressure alone would produce sensible results for such a complex situation. However it does and I have come to accept this although it is not something I can conjure up an image of.

The difference in magnitude between the two values for the hydrocarbon did not surprise me even though I deliberately selected a pressure near the flatter part of the chart in figure 7.7. I am sure that if I had selected a very low or a very high pressure to move into the more dodgy areas the discrepancies might have been much greater. Differences of over 100% are not unusual in boiling heat transfer.

Everything that has been covered so far applies only to a single tube. It is now necessary to consider tubes that are part of a tube bundle.

BUNDLE CORRECTION FACTORS

Figure 7.8 compares the heat flux from a single tube to the heat flux it would have if it were an average tube in a tube bundle. The term average is important because it would perform almost as a single tube if it was in of the bottom tube row.

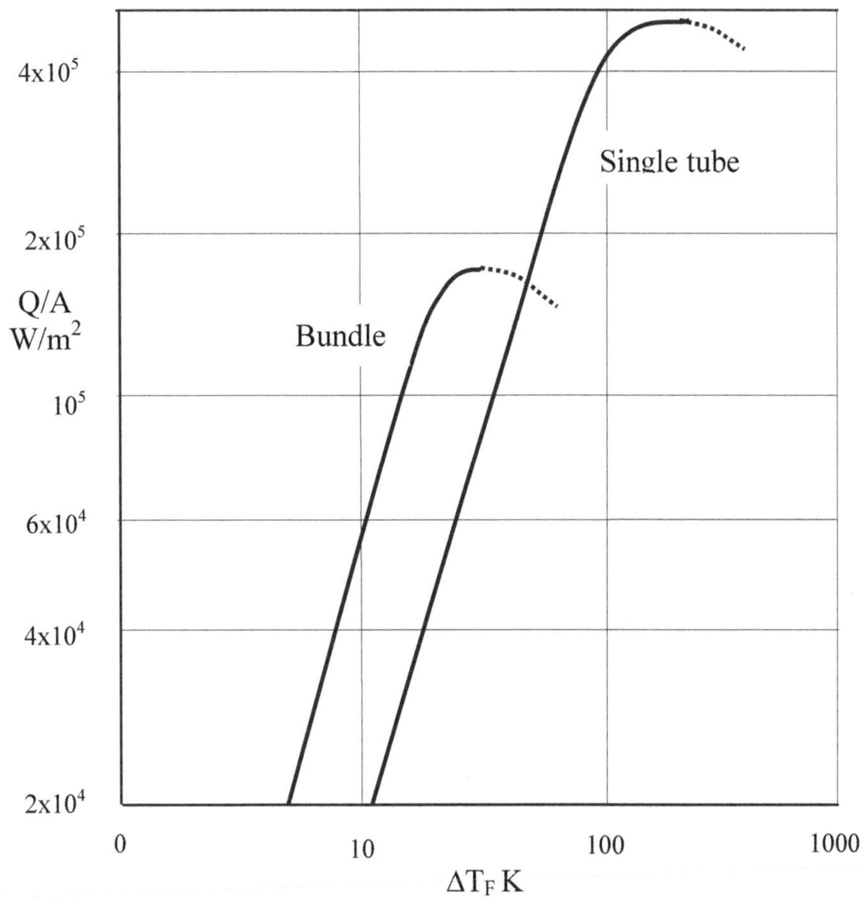

Figure 7.8

The data is for n-pentane boiling at 8 bar pressure. The critical pressure for pentane is about 34 bar which gives a reduced pressure of 0.28. This positions it along the flatter part of the curves on figure 7.7.

The first point to notice is the change in the magnitude of the peak heat flux. The value for the tube in the bundle has been reduced by a factor of more than two. This is easy to explain if the reader recalls the growth of the bubble size as the peak heat flux was approached on a single tube until the bubbles became so large that the system was tipped into the unstable region. A tube in the centre of a bundle not only has to discharge the bubbles it is generating but has to cope with the extra bubbles which have been generated by the tubes below. It will therefore be covered in vapour and enter the unstable region much sooner.

The second point to notice is that a much lower temperature difference is required to establish a fixed value of heat flux for the tube in the bundle compared with the single tube. The tube in the bundle only requires between about 40% and 50% of the temperature difference required by the isolated tube for the same heat flux. This is because of the turbulence created by the bubbles released from the tubes below. As explained in the sections on nucleation sites and bubble formation, without any disturbance a bubble will cling to the surface of the tube for some time before becoming detached and allowing another bubble to form in the same site. In a turbulent situation these bubbles are quickly wiped away allowing the nucleation sites to become more active.

The data used to plot figure 7.8 was for a single bundle. Clearly the effect of the heat flux on the average tube depends on the size of the bundle. For the peak heat flux a bundle correction factor, ϕ_b can be used in the following equation:

$$(Q/A)_{P\ bundle} = (Q/A)_{P\ tube}\ \phi_b \qquad [7.6]$$

One equation which has been used for ϕ_b is:

$$(Q/A)_{P\ bundle} = (Q/A)_{P\ tube}\ 3.1\ \frac{\pi D_b L}{A} \qquad [7.7]$$

The term $\pi D_b L/A$ is a dimensionless bundle geometry parameter, normally defined as the perimeter of the bundle divided by the heat transfer surface area of the bundle. I have read several papers which correctly note that for a single tube this parameter becomes 1, What they do not point out is that this leads to the nonsensical conclusion that a single tube which 'thinks it is part of a bundle' has over three times the peak heat flux of a single tube! The equation clearly has no meaning for bundles with a geometric parameter of more than 1/3.1 and the tubes in very small bundles will behave as if they were single tubes.

I am also unhappy with the basic definition used for the geometry parameter. To me, (and the dictionary I have just checked it in), perimeter is a linear dimension. I believe the definition should be as follows:

$$\text{Bundle geometry parameter, } \psi_b = \frac{\text{Bundle outer surface area}}{\text{Heat transfer surface area}}$$

Which for a tube bundle with a mean diameter D_b becomes:

$$\text{Bundle geometry parameter, } \psi_b = \frac{\pi D_b L}{N_t \pi d_o L} = \frac{D_b}{N_t d_o} \qquad [7.8]$$

This is a nice simple equation and the best test of the use of the equation is to apply it to a reasonable sized bundle. The most satisfactory way to do this is by means of an example.

Example 7.6

Calculate the bundle correction factor for a bundle 1 meter in diameter assuming 10% of the cross sectional area is unavailable for tubes due to pass partition lanes and nozzle space losses.

(a) For 19 mm tubes on a 25.4 mm square pitch.
(b) For 19 mm tubes on a 25.4 mm triangular pitch.

Solution.

Equation 2.16 can be used to find the number of tubes.

$$N_t = \frac{\text{Area available for tubes}}{FP_t^2}$$

For the square pitch $$Nt = \frac{\pi/4 \times 1^2 \times 0.9}{1 \times 0.0254^2} = 1095$$

For the triangular pitch $$Nt = \frac{\pi/4 \times 1^2 \times 0.9}{0.866 \times 0.0254^2} = 1265$$

The bundle correction factor, ϕ_b is 3.1 ψ_b therefore:

For the square pitch $$\phi_b = 3.1\frac{Db}{N_t d_o} = \frac{3.1 \times 1}{1095 \times 0.019} = 0.15$$

For the triangular pitch $$\phi_b = 3.1\frac{Db}{N_t d_o} = \frac{3.1 \times 1}{1265 \times 0.019} = 0.13$$

Example 7.7

What would the bundle correction factors calculated for example 7.6 have been if the bundle diameter had been?

(a) 0.5 m
(b) 1.5 m

Solution

Looking at the above equations shows that the number of tubes is directly proportional to the square of the bundle diameter. The bundle correction factor is directly proportional to the bundle diameter. Therefore overall the bundle correction factor is inversely proportional to the bundle diameter i.e. $\phi_b \alpha 1/D_b$.

For the 0.5 m bundle $\phi_b = 1/0.5 \times 0.15 = 0.3$ for the square pitch
$\phi_b = 1/0.5 \times 0.13 = 0.26$ for the triangular pitch
For the 1.5 m bundle $\phi_b = 1/1.5 \times 0.15 = 0.1$ for the square pitch
$\phi_b = 1/1.5 \times 0.13 = 0.09$ for the triangular pitch

SUMMARY AND DISCUSSION OF RESULTS

Bundle diameter D_b m	Tube pitch P_t mm	Correction factor ϕ_b
0.5	25.4 Square	0.3
0.5	25.4 Triangular	0.26
1.0	25.4 Square	0.15
1.0	25.4 Triangular	0.13
1.5	25.4 Square	0.1
1.5	25.4 Triangular	0.09

Table 7.1

For reasonably sized bundles the correction factors are sensible and significant. As expected they decrease as the size of the bundle increases. The smaller the bundle correction factor is the more susceptible the bundle will be to suffering from **vapour**

blanketing and **liquid starvation**. These are interchangeable expressions for the same phenomena. If a tube is covered with vapour it will be starved of liquid and no boiling can take place. Palen in ref 26 recommends that if the design heat flux is greater than 50% of the peak value for the bundle and the bundle correction factor is less than 0.1 then vapour release channels should be introduced to allow the vapour to escape. An alternative to vapour release passages is to use a larger tube pitch or a longer bundle to maintain the same surface area.

It should be noted that there is nothing in the equation for the bundle correction factor to distinguish between an inline and a rotated tube pitch arrangement. Clearly a vapour can escape more easily through the vertical passages characteristic of an inline square pitched bundle. I can not recall ever seeing a rotated square pitch arrangement being used in a pool boiling situation. Triangular pitches are only ever used with very clean fluids and very low design heat fluxes. The most common application of the triangular pitch is where the boiling fluid is a refrigerant and the MTD, and hence the heat flux is very low.

This has only been a brief introduction to some of the basics involved in predicting peak heat fluxes. There is a multitude of different equations and charts in the literature. Some equations simply use different constants; others may take account of more variables than others. Equations are regularly updated in modern computer programs and the accuracy is improving but for many situations there is probably still a long way to go.

The next topic to look at is the equally complex subject of the heat transfer coefficients in a pool boiling situation.

HEAT TRANSFER COEFICIENTS FOR PURE HYROCARBONS.

The following chart was published in the transactions of the AIChE in 1956. It was originally in imperial units which I have converted to SI. It is part of the very substantial contribution to the study of boiling heat transfer by Cichelli and Bonilla. The chart is applicable to pure hydrocarbons with a molecular mass between 50 and 100

The strong dependence of the heat transfer coefficient on the temperature difference is very clear. What the reader should also take note of are the very high values of the heat transfer coefficient compared with forced convection coefficients. They are of the same order of magnitude as condensing coefficients for pure fluids.

This is one of the charts we used to design heat exchangers in the days of hand calculations. There was very little to do in terms of arithmetic manipulation. It was a simple process to estimate a temperature difference from the MTD and knowledge of the other thermal resistances. This would need to be refined as an optimum solution was approached.

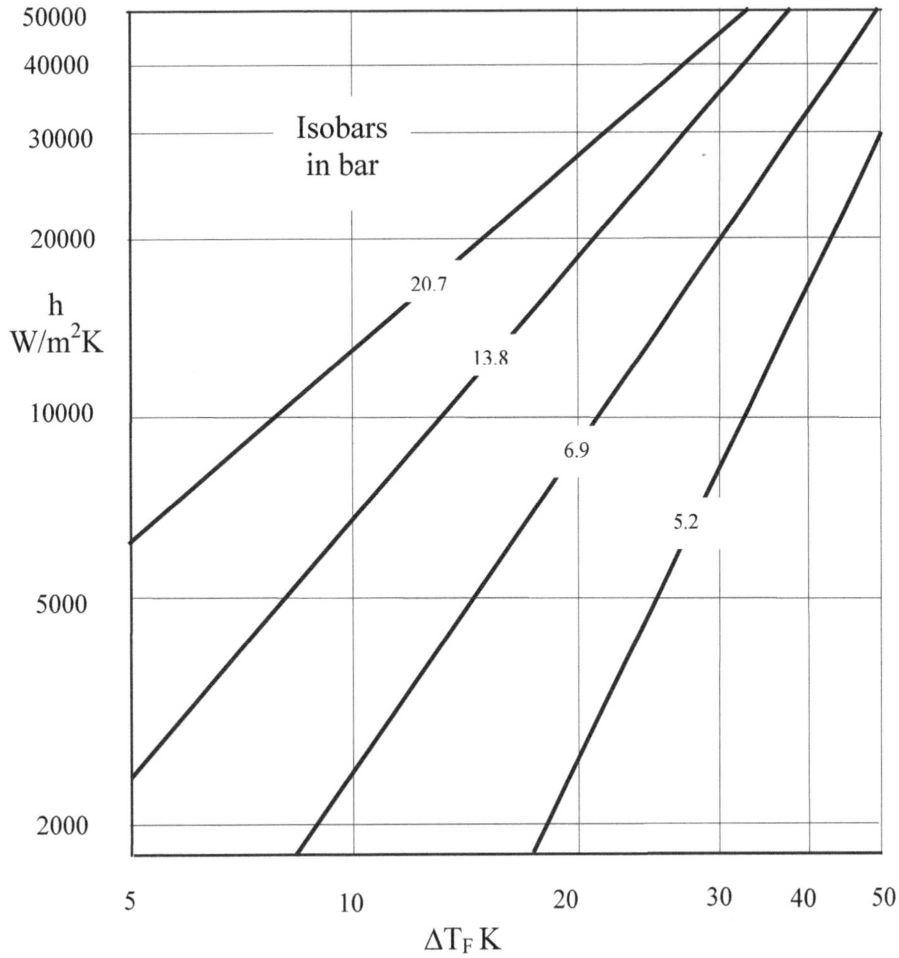

Figure 7.9

HYDROCARBON MIXTURES

Cichelli and Bonilla first noticed that if two pure hydrocarbons are mixed together to form a binary mixture then the heat transfer coefficients of the boiling mixture are always lower than the coefficients resulting from either fluid boiling alone.

It is frequently necessary to have to deal with mixtures of far more than two components. Sternling and Tichacek studied boiling in multi component mixtures and published some of their results in Chemical Engineering Science in 1961. For the most extreme range of boiling mixtures the coefficients were less than 4% of the values for the lightest compound boiling alone. Sternling called this the '**depletion effect**' and suggested that it might be due to the heavier components clogging up the nucleation sites once the lighter components had been boiled off.

The following chart – figure 7.10 also by Cichelli and Bonilla is for mixtures of hydrocarbons. The first thing the reader should look at is the magnitude of the heat transfer coefficients in figure 7.10 compared with those in figure 7.9 which is for pure hydrocarbons. The values of the heat transfer coefficients are only about 10% of those of the pure hydrocarbons.

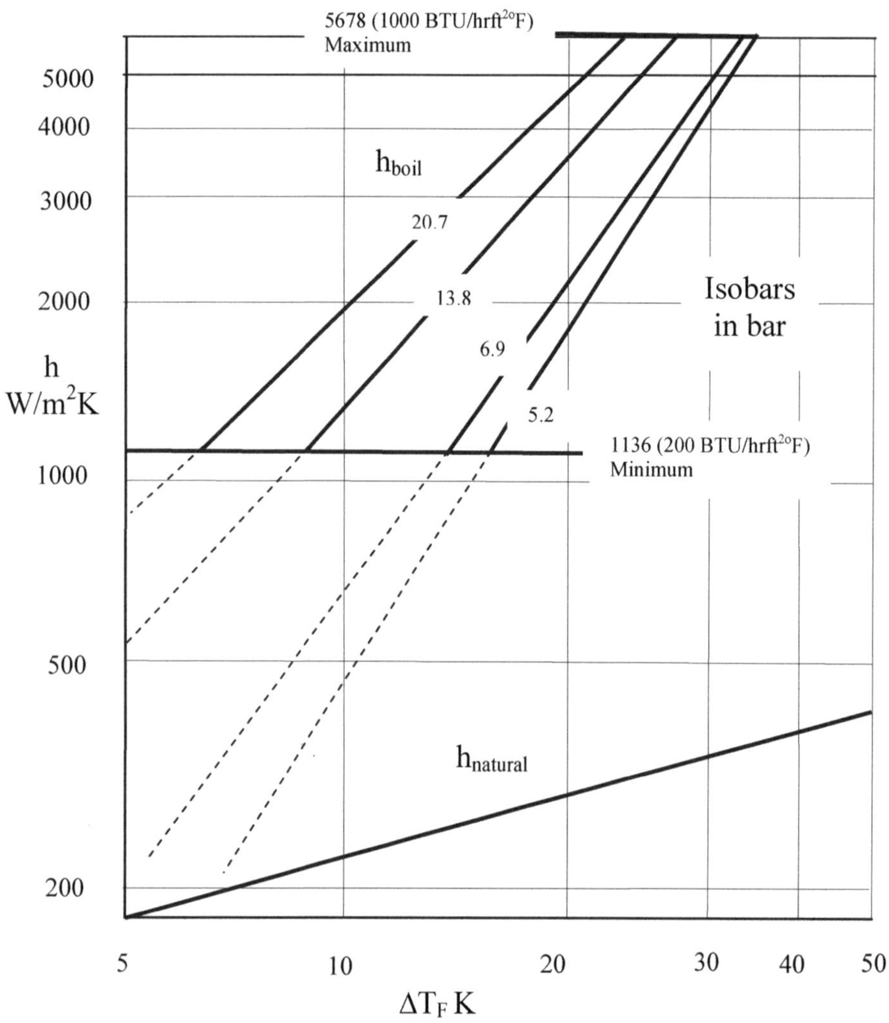

Figure 7.10

This chart is the actual one we used for hand calculations which has been converted from imperial units. As well as lines for boiling coefficients there is a line for natural convection coefficients. For the boiling coefficients there are two cut off lines. We never used a value below 1136 W/m2K (200 BTU/hrft^{2o}F) nor one greater than 5678 (1000 BTU/hrft^{2o}F).

OTHER CHARTS

Even back in the 1960's we had lots of charts for specific fluids or groups of fluids. For example we had one for Ammonia and another for some of the hydrocarbons used as refrigerants and of course water.

ACCURACY OF OLD CHARTS

The charts illustrated in figures 9 and 10 were practical design charts. Although the data on which they were based would have been for a single tube they had to they had to include sufficient safety factors to allow for bundle effects etc. For hydrocarbon mixtures there are big differences between the coefficients of binary mixtures and wide range boiling mixtures which was not adequately taken into account.

The chart we used for mixed hydrocarbons has been found to be very conservative in some instances. There is one case I recall very well. It was a steam heated vertical thermosyphon reboiler on a revamp job. The operators had fitted a level gauge to the shell to check the back flooding during the normal design operation. The liquid level had never fallen below the half way mark. The thermal calculations had been carried out during the days of hand calculations and I recognised the signature on the thermal data sheet. I checked the design using the old methods and it was spot on. The computer program I was using at the time was much closer to predicting the observed performance but it was still about 20% out. It was easy to decide that the existing shell would handle the new increased duty without any problems. It is widely accepted that boiling coefficients can be as much as 100% out but this one was a lot more than 100% in error. This is not too surprising knowing that the difference between pure hydrocarbons boiling and mixtures of hydrocarbons boiling can be a factor of ten.

EQUATIONS FOR PREDICTING h_{boil}

The equations used to predict nucleate boiling coefficients invariably use heat flux rather than the temperature difference used in charts to predict the heat transfer coefficients. Since heat flux is directly proportional to temperature difference ΔT_F is still a dominant parameter which may not be obvious to a reader who is new to the subject. One such equation is shown below.

$$h_{boil} = 0.0042\ p_C^{0.69}\ (Q/A)^{0.7} \times F_b F_p F_{BR} \qquad [7.9]$$

In this equation the critical pressure p_C is in kN/m2. F_b is a bundle correction factor which is calculated in a similar way to the bundle correction factor which was used in equation 7.7 to calculate the peak heat flux. F_p is a pressure correction factor and F_{BR} is a boiling range correction factor.

PRESSURE CORRECTION FACTOR F_p

In equation 7.9 the pressure correction factor can be simplified to $1.8P_r^{0.17}$ for some situations. For other situations it needs to be established from a log/log graph with the heat flux and the critical pressure as the variables. For very low pressures and vacuum operation the following equation is more applicable.

$$\text{Pressure correction factor} = 2.1\ P_r + \left(9 + \frac{1}{\{1 - P_r^2\}}\right) P_r^2 \qquad [7.10]$$

BOILING RANGE CORRECTION FACTOR F_{BR}

The boiling range of a mixture is defined as the difference between the dew point T_D and the bubble point T_B, hence:

$$BR = T_D - T_B \qquad [7.11]$$

The bubble point is the temperature at which the boiling starts and the lightest component in the mixture is starting to be changed into a vapour. The dew point is the temperature when the last drop of fluid of the heaviest component has just been evaporated.

The more components there are in a mixture the wider the boiling range will be and the lower the boiling heat transfer coefficient will be. F_{BR} can be found from an equation such as the one below.

$$F_{BR} = \frac{1}{1 + 0.023(Q/A)^{0.15}\,BR^{0.75}} \qquad [7.12]$$

There should be an optional input for the boiling range in a computer program. If this optional input is not used the program will just extrapolate to calculate the boiling range. This extrapolation is satisfactory for most situations but there will be cases when it is not which is why there is this option. Ideally the boiling range should be on the process data sheet but I have never known this happen. An experienced designer will put a value when considered necessary, if in doubt consult the process engineer.

.

Frankly I find it impossible to 'get my head round' equations such as 7.9, 7.10 and 7.12. With convective heat transfer and condensation I have a mental picture of what is taking place and how each variable is making its contribution to the heat transfer coefficient. If I try to do this with some of the equations used for boiling heat transfer coefficients I just start to get a headache. (No such problems with taking a value from a graph for a particular fluid when the only variable was the temperature difference!)

Modern computer programs of course do not get headaches and can handle a multitude of equations. There will be many equations in the program you are using and the program will be selecting the most appropriate one for each case. (Not necessarily exactly the same equations shown above. These are only intended to be indicative of what is likely to be in your program.) No matter how many equations there may be in your program boiling is still one of the dodgiest areas of heat transfer.

COMBINED BOILING AND CONVECTION.

The one situation when a boiling coefficient may be the only component of a shellside heat transfer coefficient is when a pure fluid is boiling at its saturation temperature. At the other extreme if a kettle reboiler is stripping off some light fractions and most of the flow of fluid is being discharged over a weir to a liquid outlet nozzle then liquid heating by natural convection and circulation through the bundle may dominate the shellside heat transfer coefficient.

For natural convection outside a single tube the heat transfer coefficient may be calculated from the following equation which the reader if probably familiar with.

$$Nu = 0.53\,(Gr\,Pr)^{1/4} \qquad [7.13]$$

Where Gr is the Grashof number

$$Gr = \frac{\beta g \rho^2 d_o^3}{\mu^2} \qquad [7.14]$$

Where β is the coefficient of cubical expansion.

The convective heat transfer coefficients during pool boiling outside tubes in a bundle are higher than a natural convection coefficient for a single tube. This is because of the circulation of fluid through the bundle. The amount of circulation through the bundle depends on the static head effects and the hydraulic resistance of the bundle. The pressure drop driving the circulation can be found from the following equation:

$$\Delta p = (\rho_L - \rho_{mix})gh \qquad [7.15]$$

Where ρ_{mix} is the average density of a column of fluid passing through the bundle and h is the average height of the bundle. This is balanced against the frictional pressure drop through the bundle. Calculating recirculation rates is not a problem for modern computer programs but was unrealistic for hand calculations.

If a single tube is considered in the middle of a bundle there is a two phase mixture of fluid flowing over it. The tube has no idea if this flow has been caused by a pump or a static head but the fact that fluid is flowing over the tube makes it a forced convection situation. If the velocity is very low this forced convective coefficient may be lower than the natural convective coefficient. Higher velocities will produce higher forced convection coefficients. Even the bottom row of tubes will be experiencing some forced convection and the ratio of the natural convective coefficient to the forced convective coefficient will vary throughout the bundle. Circulation boiling is a very complex situation and the best way to visualise what is taking place is to think of a convective coefficient. h_{con}, which may have natural and forced convection components or simply be the higher of the two values.

For any combined boiling and liquid heating situation the following energy balance applies:

$$Q_{total} = Q_{boil} + Q_{liquid} \qquad [7.16]$$

If the ratio of the boiling duty to the total duty, Q_{boil}/Q_{total} is defined as R_Q then the ratio of the liquid heating duty to the total duty becomes $(1 - R_Q)$.

The shellside heat transfer coefficient at the outside of the tube, h_o, can then be calculated by weighting the boiling and convective coefficients as follows:

$$h_o = \frac{1}{\dfrac{R_Q}{h_{boil}} + \dfrac{(1 - R_Q)}{h_{con}}} \qquad [7.17]$$

This is the equation we used when we took values from the chart in figure 7.7 to use in hand calculations. The convective coefficient was just a natural convection coefficient which in some situations might be very conservative compared the convective coefficients calculated by modern computer programs which can take full account of the circulation through the bundle.

The weighting of the two coefficients is fundamental and will still be part of any computer program. Another fundamental is the importance of the value of the ratio of the boiling duty to the total duty. Whenever I am faced with a design involving boiling one of the first things I still do, before I start to fill in any input sheets, is to note the value of R_Q. (I tend to multiply it by 100 and write it down as a %.)

A high value for R_Q indicates that the heat transfer is mainly from boiling and I am anticipating that this will result in a high shellside heat transfer coefficient. A low value of R_Q and I will be expecting a low value for the coefficient.

FLOW BOILING ON THE SHELLSIDE

Flow boiling on the shellside has received a lot less attention than flow boiling inside tubes. This is not surprising since it occurs far less frequently.

Once a bubble of vapour reaches the surface in a pool boiling situation it can not undergo any more heating and will leave as a saturated vapour. This is not the case in a forced convective situation where a fluid is being boiled in say an E type shell. Any vapour generated near the inlet will be in contact with tubes and liquid at a higher temperature until it reaches the outlet nozzle. During this time it will become superheated. Part of the heat load will therefore be vapour heating.

The shellside heat transfer coefficient is therefore a weighted combination of a vapour heating coefficient, a liquid heating coefficient and a boiling coefficient. Between the end baffles the convective coefficients will be from forced convection but in any dead spaces particularly in the end spaces natural convection may dominate.

The boiling heat transfer will still be governed by an equation similar to equation 7.9 for pool boiling but the bundle correction factor will be higher because of the higher velocities in forced convection compared with recirculation convection. When we used charts with hand calculations we used the same values as we would have for pool boiling knowing that these would be a bit conservative.

As always I would recommend imagining a picture of what is going on in a particular unit. One possible idealised picture is shown in figure 7.11 for a unit with vertically cut baffles.

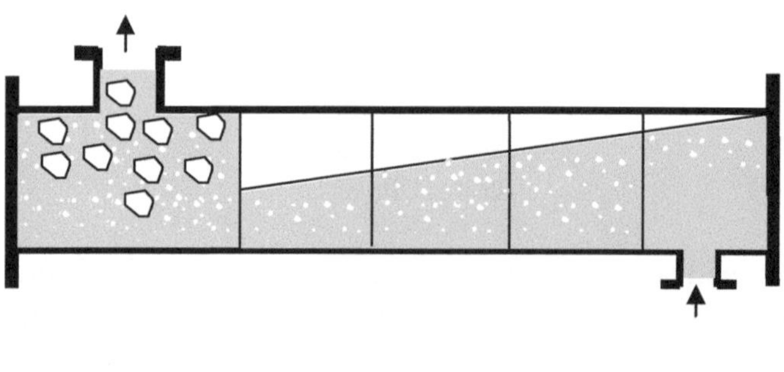

Figure 7.11

In order to form that picture it is essential to work out and jot down the void fractions at inlet, an approximate mid point and at the outlet. In figure 7.8 these fractions would be about 0, 0.3 and 0.5. In this case the liquid is subcooled at inlet and there is only a small quantity of vapour generated in the inlet space. The baffles are vertically cut and as the fluid moves through the exchanger and more vapour is generated there will be a tendency for the vapour to move to the top. The vapour must finally mix with the remaining liquid in the end space before it leaves through the outlet nozzle.

When we used hand calculations vapour and liquid heating coefficients were calculated in the same way as we did for condensers before we could establish an overall weighted shellside coefficient. Modern computer programs can carry out calculations for each baffle space. They will of course still be making assumptions about the nature of the flow in general and how much phase separation is taking place in particular.

Almost every spec I have ever read has contained a statement that vertically cut baffles should be used when there is two phase flow on the shellside. This is clearly necessary for most condensers although even there I demonstrated in chapter 5 on condensers that there should be exceptions when the volume flow rate of liquid is very low. I offer an interesting question to the reader, why should vertically cut baffles always be used for a fluid which is vaporising on the shellside?

I have put this question to many people over the years. The donkeys will look puzzled or just reply something like "because the spec says so", or "we always do it that way". The more thoughtful designer is more likely to say, "That's an interesting one", and we will start a conversation which never comes to any definite conclusion. I am still asking myself this question as I am writing this.

First let us consider an extreme case where there is a lot of vapour at a low pressure at inlet resulting in very high void fractions, (say 0.3, 0.6 and 0.9) throughout the unit as illustrated in figure 7.12.

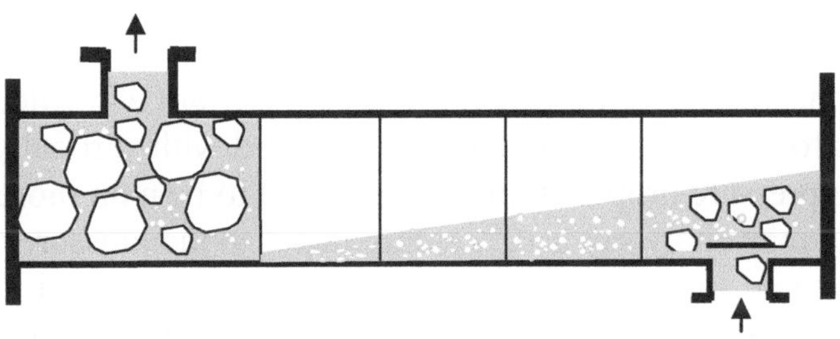

Figure 7.12

If the vapour and liquid phases do separate as shown in the sketch most of the heat transfer surface is only in contact with the vapour. This tendency is encouraged by the fact that with two phase flow at inlet, the inlet nozzle will be larger, and impingement protection will be required. (Rods might be better than the plate shown.) This will leave a gap at the bottom of the bundle along which some of the liquid may flow without touching the tubes. (Ears could be used at the bottom of the baffles to offset this but would your computer program understand what you want to try?) Particularly if the vapour pressure is low, the vapour heat transfer coefficient will be very low and although vapour heating may account for less than 10% of the duty it will require a high percentage of the surface area. This will result in a very uneconomic design.

If horizontally cut baffles are used for the same service all the tubes will be wet and therefore able to undertake boiling heat transfer. Any potential dead pockets in the end spaces are eliminated just as they are for single phase fluids. The liquid by passing the bundle will be reduced.

At the other extreme a service may be characterised by very low values of void fraction, (say 0, 0.05 and 0.1). For such a service I would run designs with both baffle orientations and if there was an advantage in using horizontal baffles I would not hesitate to use them. I would of course approach whoever it was necessary to approach and point out the advantage and cost saving whilst arguing that it was almost a single phase unit anyway and it would only be a minor deviation from the spec.

Returning to the case with the very high void fractions.It could be argued that it is almost a vapour heater, (by area) but the most persuasive argument concerns the problem with the liquid contact with the tubes. The saving in cost is potentially a lot higher but of course this depends on the thermal resistances of the fouling factors and the tubeside coefficient.

Some computer programs will give warning messages whenever horizontally cut baffles are used in a vaporising service. The warning is normally "there is limited data for horizontal baffles in this service". If there are so many specs saying that vertically cut baffles should be used this is not surprising. However whenever there is a warning message a thermal designer should be cautious. My main concern is that flow pattern may be nothing like the regular growth of vapour bubbles and the phase separation which I have used in the 'idealised' diagrams above. Chapter 6 on flow maps identified a wide range of flow regimes for the very simple geometry of the inside of a tube or pipe. The shellside of a heat exchanger with its combination of long flow, cross flow and untubed areas for pass partition lanes has a much more complex geometry. However this must also apply to vertically cut baffles.

I do not have any concerns about the way a computer program will calculate the heat transfer coefficients for horizontally cut baffles compared with vertically cut baffles. That only leaves the pressure drop and vibrations. I tend to be a bit more conservative on pressure drop and may apply say a 10% safety factor to the pressure drop between the end baffles. The same applies to avoiding vibration problems. I will often add a few extra supports to vulnerable tubes which I would not bother with for a design with no warning messages.

It is interesting that when I was training as a thermal design engineer vertically cut baffles were the norm with one important exception. The exception was when there was hydrogen in the boiling stream and the reason for this was the concern about phase separation. Phase separation is much more probable with hydrogen than with any other gas because of its low density.

I just accepted this at the time without asking the obvious question: if we use horizontal baffles for a 'potential problem' case why not use them for more straightforward cases? I have increasingly used horizontal baffles over the years pushing the high and low void fraction extremes but I still wonder why they are not more generally used.

Perhaps in the future using horizontally cut baffles for boiling services will become the norm and vertically cut baffles the exception?

FLOW BOILING INSIDE TUBES.

The possible flow regimes in two phase flow were covered in chapter 6. The flow regimes covered in chapter 6 did not make any distinction between condensing and vaporising they were simply for flows through an unheated tube. There is a very important difference between condensing and vaporising a pure fluid.

For condensing the ideal process is isothermal which maximises the MTD and is therefore isobaric. A practical process will try to approach this ideal and a very low pressure drop requires a very low velocity. The stratified flow for a horizontal tube shown in figure 6.3 is therefore a common flow regime for condensing inside tubes.

For vaporisation a force is required to keep the fluid moving. The velocities are therefore higher and there will be a significant pressure drop which may be provided by a pump or a static head. Consequently the stratified flow regime is never present when a fluid is vaporising inside horizontal tubes.

The other flow regimes may occur with the slight modification that there will be bubbles forming and being released from the tube wall whenever the wall temperature is high enough.

A lot of research has been done and a lot of papers have been published on boiling inside tubes particularly for steam in vertical tubes. This is because of the power industry rather than the petrochemical industry.

There are several situations where hydrocarbons are boiled inside the tubes of shell and tube heat exchangers. The overall heat transfer coefficient is a weighting of liquid and vapour convective coefficients and boiling coefficients just as they were for boiling on the shellside. The equations used are not necessarily the same as those used on the shellside and I did consider going through some of the interesting work by Chen and Shah. A sophisticated computer program can of course use different equations for different flow regimes; you can read about these in your computer manual. Some of the more important flow boiling applications will be dealt with in some detail in the appropriate sections such the chapter on vertical thermosyphon reboilers.

WATER TUBED BOILERS

The water tubed boiler has been a key component of a vapour power plant for many years. In chapter 5 on condensers I briefly described the power plant surface condenser because there are some features which can be applied to designing condensers for petrochemical plant applications. By the same logic it is useful to have a brief look at some of the salient features of a water tubed boiler.

Water enters the steam drum from the economiser at a temperature which is sufficiently below the saturation temperature to avoid flashing in the pipe. This inlet is below the liquid level in the drum. Saturated steam leaves the top of the steam drum and then enters the superheater.

A series of downcomer pipes are connected to the steam drum below the liquid level. The water entering the downcomers will be close to its saturation temperature but as it passes down the tubes the static head increases offsetting the loss in pressure due to friction which might otherwise cause flashing. To ensure that the downcomers deliver only liquid to the water drum they should not undergo any significant heating and are therefore situated in a cooler part of the furnace.

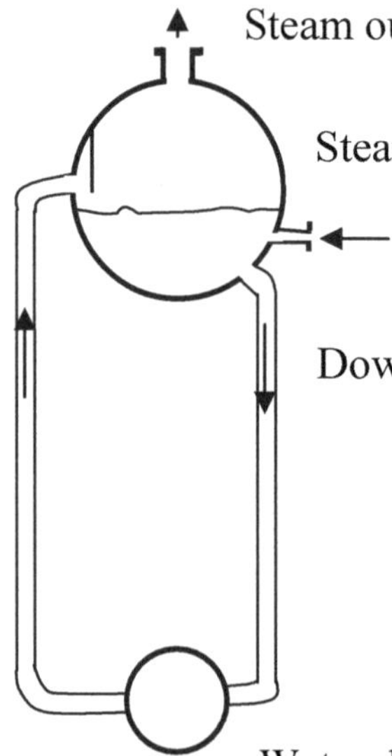

Figure 7.13

The risers are the steam generating section of the boiler and are exposed to the highest temperatures. The heat transfer to these tubes is mainly by radiation from flames and combustion gases. The steam leaving the risers is a mixture of saturated liquid and saturated vapour. These are separated in the steam drum and the saturated liquid is mixed with the feed water from the economiser before being recirculated via the downcomers.

The economiser is situated in the ductwork leading to the exhaust stack. The water inside the tubes has a very high heat transfer coefficient and the exhaust gas passing over the tubes has a very low heat transfer coefficient. This is a classic case for using extended surfaces and there is always some type of fin on the outside of the tubes. With moderate metal temperatures the tubes and fins are made of basic carbon steel.

The steam generating tubes, the risers, are very interesting as far as heat transfer is concerned. These plain tubes are exposed to the highest temperatures in the combustion area of over 1500°C and are normally made of carbon or very low alloy steel. Provided there is water or boiling water in the tubes the metal temperatures will remain close to the water temperature. (If the steam is being generated at say 40 bar the saturation temperature is 250.3°C). It is essential to ensure that the inside of the tubes are always wet. If a section of a tube dries out there will be a relatively low vapour heating coefficient inside the tube which will allow the metal temperature to rise and the tube will burst. This is where the term **burnout** originated. The most likely reason for burnout is a partial blockage which will restrict the flow through the tube. Dry wall conditions may be met during the design of shell and tube exchangers but although there will be a severe fall off in performance the consequences may not be as drastic because of the much lower temperatures.

The design of the steam drums which separate the saturated steam from the saturated vapour is very similar to the space above the tubes in a kettle reboiler and many of the correlations and standards that have been used for entrainment calculations originate from work on power plant steam drums. They will also incorporate intermittent and/or continuous blowdown systems. These have applications in kettle reboilers to prevent the concentration and eventual deposition of dissolved solids.

After the combustion gases have passed over the steam generating tubes their temperature will have fallen to perhaps 1000°C they are then directed over the superheater tubes. The heat transfer from the gases is now mainly by convection. Inside the tubes the vapour is being superheated from say 250°C to 350°C before it is passed to the turbine. The superheater has relatively low heat transfer coefficients inside and out side the tubes. This results in much higher metal temperatures than the tubes in the steam generating section even though the gases providing the heat are at a much lower temperature. The superheater is therefore constructed of plain tubes made out of a metal capable of operating at high temperatures such as stainless steel.

The exhaust gases leaving the superheater then pass over the economiser which was described above. Finally they are directed over an air preheater where the air for combustion is heated before being discharged to the stack. The thermal design of an air preheater is quite simple. There is a gas on both sides resulting in low heat transfer coefficients and plain tubes. The metallurgy may not be so simple. Particularly if solid or liquid fuels are used for combustion there will be acid dew points to take into account. Various materials have been used in different plants including glass!

SUPERCRITICAL STEAM GENERATION

An alternative boiler used for steam generation in the power industry is shown in figure 7.14

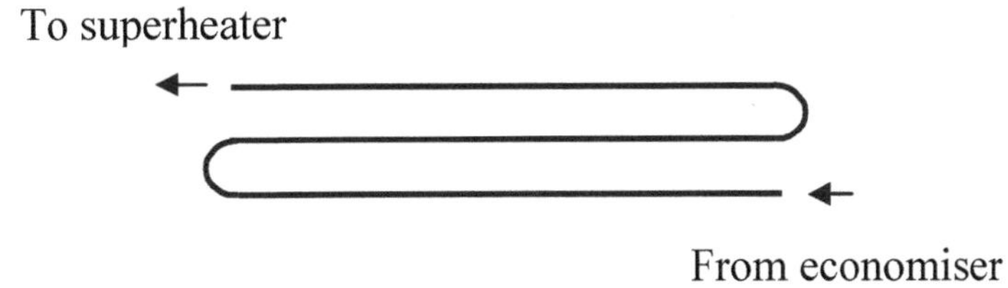

Figure 7.14

The steam drum, water drum, downcomers and risers are all replaced by a number of single tubes. These are most likely to be coiled around the sides of a cylindrical furnace rather coiled as shown in the diagram for convenience.

This is known as a '**once through boiler**'. A clear advantage of such an arrangement is that it is so simple. The key disadvantage is that any dissolved solids are likely be deposited inside the tube at some point. In any power plant the make up water is highly treated but it is still necessary to have blowdowns to remove concentrations of dissolved solids. This is not possible with a once through boiler; the make up water must therefore be 'super pure'.

The application of the once through boiler is in a supercritical steam plant when the term 'boiler' could be questioned because no boiling takes place. There is just a steady heating of high pressure hot water until it becomes high pressure steam. There is no observable change of phase and no enthalpy of phase change. The operating pressure must be at least 230 bar to keep it above the water critical pressure of 221.2 bar. The maximum temperature of the steam after superheating will be at least 540oC.

The reader should recall from early studies in thermodynamics that for a high cycle thermal efficiency the heat should be supplied at as high a temperature as possible and rejected at as low a temperature as possible. (Carnot efficiency = $\{T_{max} - T_{min}\}/T_{max}$.) The minimum temperature is fixed by the temperature of the coolant and therefore there has been continuous development to increase the average maximum temperature since the first plants were commissioned. When conventional plants had reached their temperature limits the next step was the supercritical vapour power plant cycle.

Supercritical cycles are not new. I presented a brief paper on them when I was a student in the early sixties. The first one in the UK was Margam B. It was built for the Steel Company of South Wales and not the Central Electricity Generating Board which one might have expected to have been the innovators in new power plant developments. It did impress on me the enormous power consumptions of metal processing plants.

The early supercritical plants did not last long before they were revamped as more traditional systems. One of the more interesting problems they experienced was due to the very high purity of the water. The water dissolved some of the copper in the brass condenser tubes which was then deposited on the turbine blades.

There then followed several decades when vapour power plants remained sub-critical. Today supercritical vapour power plants are once again being used to improve the efficiency of producing electricity. Detailed descriptions of vapour power plants can be found if ref 10 by Kakac.

CONCLUSIONS

This chapter has been limited to the basic theory of boiling and the development of some of the methods used to calculate heat transfer coefficients and pressure drops.

Most of the practical applications to thermal design are dealt with in the three separate chapters which follow this one.

POOL BOILING EXCHANGERS – CHAPTER 8

INTRODUCTION

My original title for this chapter was Kettle Reboilers but I soon realised that a lot of the topics I wanted to include had nothing to do with reboilers. What they did have in common is that the heat transfer was by pool boiling alone.

The design of the internal column reboiler is the first item covered followed by the design of the kettle type reboiler. The basic function of any type of reboiler is to provide a heat load to a fractionating tower/column. A fractionating tower is a vertical, tall relatively slim pressure vessel which is one of the most important and expensive parts of any hydrocarbon processing plant. It is supplied with a feed of mixed hydrocarbons and some of this feed is vaporised. The lighter fractions which become the vapour are discharged from the top of the tower to an overhead condenser whilst the heavier fractions which remain as a liquid are extracted from the bottom of the tower. Pressure and temperature gradients are established from the bottom to top and the tower which is filled with a series of trays. The design of the tower and the selection of the type of trays used is one of the most interesting and challenging aspects of the process engineer's thermal design work.

There are two alternatives to using pool boiling exchangers to provide a heat load to a tower – the horizontal and the vertical thermosyphon reboilers. These are covered in the chapters which follow this one.

The other main type of pool boiling exchanger is the evaporator. These are often part of a refrigeration system and are covered in the later part of this chapter.

STAB-IN/INTERNAL COLUMN REBOILERS

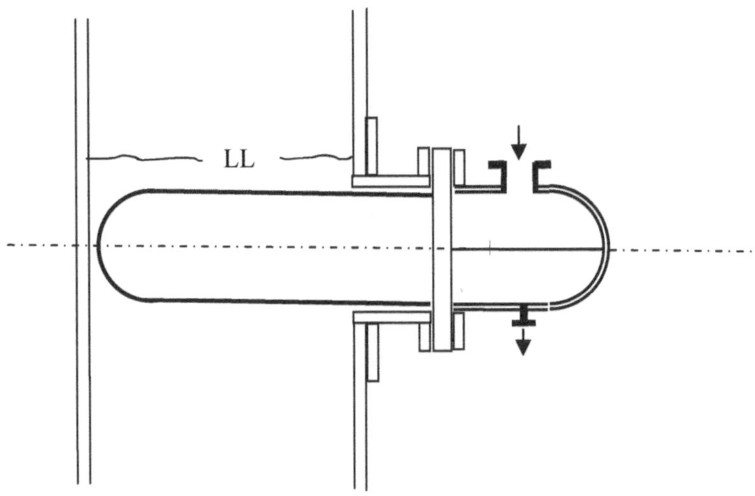

Figure 8.1

Figure 8.1 shows a section through a typical stab-in bundle. The tubes are invariably U tubes and the heating fluid is normally steam. The bundle is inserted through a nozzle in the side of the column. To maximise the length, the end of the bundle will be very close to the opposite side of the tower – typically no more than about 50 mm away. The tower and tube materials will normally be the same but if they are not the differential expansion must be taken into account when fixing this clearance. I have deliberately shown some typical nozzle details because these are crucial to fixing the depth of the port. The mechanical details of nozzles on towers are exactly the same as for shell and tube exchangers and have been covered in chapter 2. The internal space inside the nozzle becomes the equivalent of the port in a kettle.

The length of the tubes can be calculated from the following equation:

$$L = ID_{tower} + \text{Nozzle standout} + TS - \text{Tower/tube clearance} - R_U - d_o/2 \quad [8.1]$$

Where TS is the tubesheet thickness and R_U is the radius of the bend of the longest U tube. This is certainly a situation when the designer should never even consider using a standard length for the U tubes and the U bends should be fully effective for heat transfer.

It is very unlikely that your computer program will be adapted to design stab-in units so you just have to feed in normal kettle data and try not to confuse it. There is no shell and there are no shellside nozzles but the program may need an input. Use a very large diameter for the shell – at least the diameter of the column. Input a large number of large diameter nozzles at the top and the bottom – as big and as many as the program will take before complaining. The aim is to maximise the areas to get as close as possible to the actual cross sectional area of the tower.

The liquid level above the tubes may be specified by process otherwise the level is established in the same way as for a normal kettle which is covered later. Since these bundles are normally quite small there are unlikely to be any thermal problems due to static head effect or extremes of heat fluxes. However there is one potential problem which may occur with even a moderate flux. With no shellside nozzles it may be tempting to fill up the area within the SCD, (safe circle diameter) with tubes. (This is implied in figure 8.1). This will create a pocket of vapour within the port area which can only escape in an axial direction. If this takes place a length of the tubes adjacent to the tubesheet will become blanketed in vapour and become ineffective. Since the tube length within the port is normally over 25% of the total tube effective length this can lead to a serious underperformance. It is therefore good practice to leave a space at the top of the bundle as shown in figure 8.2.

To establish how large a space needs to be left requires some hand calculations. With an isothermal heating medium such as steam the vapour generated in the port area will be equal to the length of tube in the port divided by the tube effective length multiplied by the total vapour generated.

The simplest calculation is just to compare the flow area out of the top of the bundle in the port with the flow area out of the segmental section at the top of the port. The flow area out of the top of the bundle can be accurately calculated by taking an effective diameter across the centre line of the top row of tubes and multiplying this by the length of the port. The height of the segment can be safely calculated by taking the distance from the top of the top tubes to the top of the port.

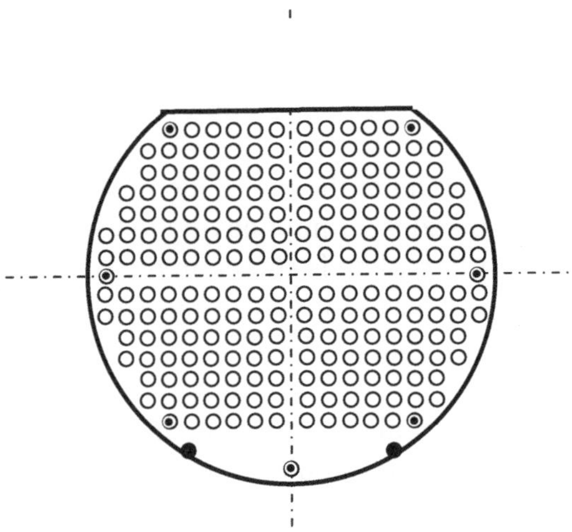

Figure 8.2

The other calculation is to compare pressure drops. The pressure drop through the bundle can be calculated using equation 7.15 for the static head driving force:

$$\Delta p = (\rho_L - \rho_{mix})gh$$

The pressure drop out of the segment and into the tower can be assumed to be one velocity head. The mixed density at this point should be used. This can be calculated from the vapour fraction leaving the bundle from a preliminary computer run. This should also be transposed into a void fraction to give an idea of how wet the top tubes are.

I will not propose any fixed guidelines pertaining to the relative values from the above calculations. The purpose of the calculations is to build up a picture and get a 'feel' for what will be taking place. If in doubt the safe option is to increase the height of the segment but increasing the height of the segment decreases the area available for tubes.

For a relatively low flux design a conventional U tube bundle as illustrated in figure 8.2 may be satisfactory. However as the height above the bundle increases it becomes more worthwhile to consider fitting more tubes in the vacant space at the bottom of the bundle. This can be accomplished by using horizontal U tubes rather than the more normal vertical U tubes as illustrated in figure 8.3.

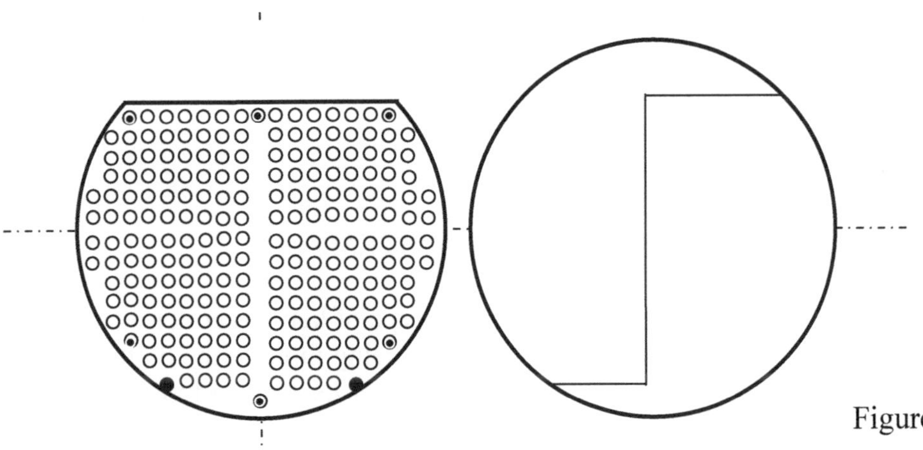

Figure 8.3

Rotating the U tubes also has a distinct thermal advantage. With the large pass partition lane being vertical it provides a vapour passage for the vapour to escape from the bundle area in the port in both a vertical and horizontal direction.

For designs with high fluxes requiring a large segment at the top of the bundle it is usually possible to avoid offsetting the channel nozzles without having to extend the length of the channel. The shape of the pass partition plate is shown in the right hand sketch of figure 8.3. Details of the design and the checks for this arrangement can be found in the section 'U tube pass partition lanes' in chapter 2.

The reader should note that I have clipped the top off all the support baffles. They serve no purpose and just add to the bundle weight. They could even be a menace if one of them were too close to the port opening. The support for the bundle inside the tower can be left to the mechanical engineer. He will probably use two angle or I sections to support the bundle under the bundle runners. The bundle runners are important even for small stab-in bundles. Without them it would be easy to get the bundle in but the baffles could catch on the edge of the port when withdrawing the bundle. The thermal designer should add a note to the thermal data sheet "The supports must be flush with the inside diameter of the port and not interfere with the flow into the bundle."

The advantages of the stab-in reboiler are the low cost because there is no shell and no process pipework. It is also less likely to foul than a conventional kettle undertaking the same service. This is because there is less accumulation of heavy constituents in the boiling fluid.

The main disadvantage is the size which limits it to relatively low duties. The maximum port which can be used is limited to 50% of the tower inside diameter and the diameter of the port and the inside diameter of the tower determine the maximum tube length. I can remember one occasion when process agreed to increase the inside diameter of the tower by 50 mm which was just sufficient to provide a bundle with sufficient surface area to satisfy the duty. This of course lost a lot of the cost advantage although it was only a relatively short section of the tower which needed to be increased in diameter.

Another disadvantage is that the bundle can only be removed for cleaning when the tower is shut down. This of course is a problem for the process engineer to worry about as is the loss of vertical space which would otherwise have been available for plates.

KETTLE REBOILERS

FRACTIONATING TOWER REBOILER SYSTEMS

Figure 8.4 below shows the basic features of the type of kettle reboiler which is connected to the bottom of a tower. A heat load is being input to the bottom of the tower by the vapour which is leaving the top of the kettle. The feed to the kettle is from the bottom of the tower and is entering the bottom of the kettle close to the tubesheet of the bundle. This maximises the flow path of the liquid across the bundle.

The liquid level is maintained above the tubes by a weir. In the bundle the lighter fractions of the hydrocarbon are boiled off and the heavier fractions are discharged over the weir. These heavy fractions are the bottoms product.

The diagram shows the liquid level beyond the weir being controlled by a valve which is actuated by a level control device. With this arrangement there is likely to be a 'hold up' requirement which is an aid to evening out fluctuations in the process flows. This is defined as the time it takes for the liquid level to fall from the top of the weir to the outlet nozzle when discharging at the design flow rate, (including any overcapacity). When a hold up is required it is necessary to increase the length of the shell. This is dealt with later in this chapter. Many kettles do not require any hold up: process may make sufficient provision for flow fluctuations the tower. If no hold up is required the length of the end compartment is controlled by the size of the outlet nozzle and any vortex breaker which may be required.

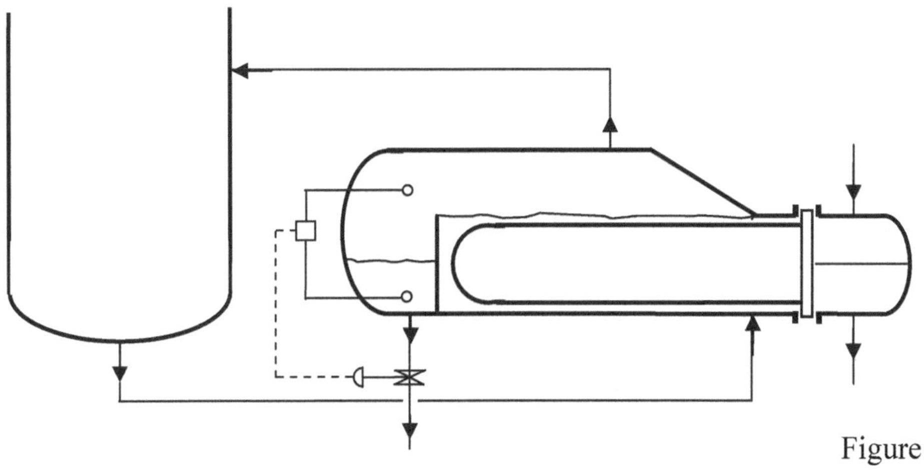

Figure 8.4

A U tube bundle is shown in the diagram so the heating fluid is likely to be steam. If a fouling stream is used as the heating medium a pull through, TEMA T type, floating head bundle would be used.

One of the advantages of the kettle type of reboiler, (particularly when compared with the stab-in bundle), is that the heat transfer surface area is unlimited therefore it can be designed for any heat load to the tower. The surface area of an individual shell is limited by the maximum weight of the bundle, (typically 10-12 tonne). However there is no theoretical limit to the number of shells in parallel which could be used. In most cases a single shell is sufficient

A process advantage of this type of kettle is that it can handle a wide range of percentage vaporisation rates; from as little as 10% vaporisation to over 80% vaporisation. (Kettles with 100% vaporisation are covered later.) Another process advantage is that it is insensitive to process changes such as part load operation.

The disadvantages are that they are relatively expensive compared with alternative types of reboiler and they are more likely to foul because of the recirculation around the bundle.

SHELL CONSTRUCTION

The near end of the shell is constructed in the same way as any 'normal' shell and tube unit. There is a flange which bolts on to the channel flange, holding the tubesheet in between them. There is then a short parallel section known as the port. This length should be as short as possible because there is no disengaging space above it. For some designs when the flange has a long neck it may simply be the length of the flange. In other cases the length of the port may need to be extended. This may need to be discussed with a mechanical engineer for critical cases.

There is then a conical section which connects the port section with the cylindrical section of the main part of the shell. From a thermal viewpoint the cone should have as steep an angle as possible to maximise the disengaging space. However there are mechanical problems if the angle is above 30°. The angle is therefore invariably 30° from the horizontal. The height of the cone is the difference between the shell inside diameter and the port inside diameter, D_P. The length is therefore $(D_S - D_P)/\text{Tan } 30$.

The far end of the shell is sealed by a dished head. This is invariably an ellipsoidal/elliptical head. The cross sectional shape is an ellipse with the diameter of the major axis twice the diameter of the minor axis. The length is therefore half the minor axis which is $D_S/4$.

LIQUID HOLD UP TIME

The calculation of the extra shell length required to provide a specified hold up time is best illustrated by means of a calculation.

Example 8.1

A kettle has a 2000 mm shell and a weir which is 300 mm above the centre line of the shell. The liquid has a flow rate of 15 kg/s and a density of 850 kg/m3. Calculate the required length of shell beyond the weir for a hold up time of 4 minutes.

Solution

Figure 8.5 shows a side elevation and a section through the end compartment of the kettle. The required length L_{req} is the distance from the inside of the weir plate to the tan line of the dished end.

The volume of liquid for a 1 m length of shell is calculated as follows:

$$\theta = \text{Cos}^{-1}(300/1000) = 72.5^\circ$$

The half chord length C/2 $= (1000^2 + 300^2)^{1/2} = 1044$ mm

Area of the two triangles $= 1044 \times 300 = 313209$ mm^2

Area of circle sector @ $2\theta = (2 \times 72.5)/360 \times \pi 1000^2 = 1265364$ mm^2

Segment area above weir $= 1265364 - 313209 = 952155$ mm^2

Area occupied by liquid $= \pi 1000^2 - 952155 = 2189438$ mm^2

The volume of liquid is therefore 2.19 m3/m length of shell

Volume of liquid required = 15/850 x 4 x 60 = 4.24 m^3

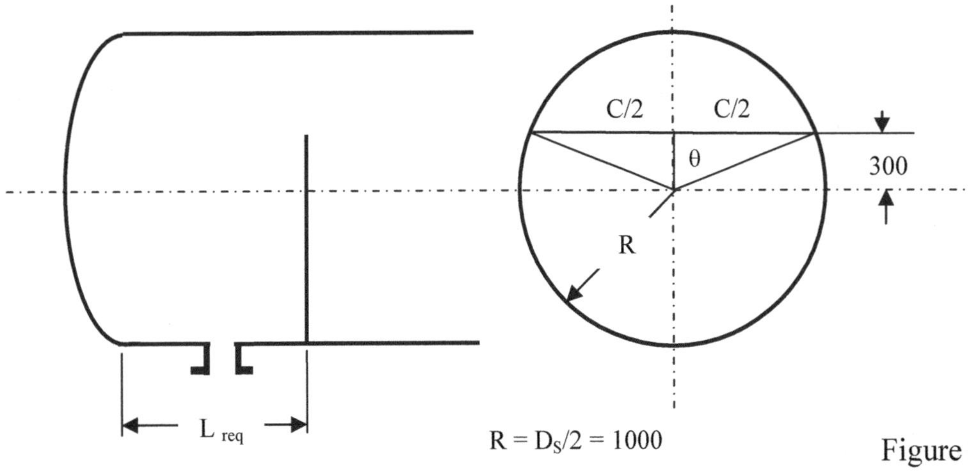

Figure 8.5

The length of shell for this volume is (4.23/2.19) = 1.934 m. However this ignores the volume in the dished head. The volume in a $D_S/4$ dished head is approximately equal to $\pi/24 \times D_S^3$. The length of shell for the same volume can be found as follows:

$$\pi/24\ D_S^3 = \pi/4\ D_S^2 \times L$$
$$L = D_S/6$$

This of course is for the whole of the dished head and there will be no liquid above the top of the weir. Since the volume in the head is a maximum at the centre and a minimum at the top and bottom it is safe to apply a correction factor of the height of the liquid level divided by the shell diameter. The correction factor in this example becomes (1000 + 300)/2000 = 0.65.The required length then becomes:

$$L_{req} = 1934 - 2000/6 \times 0.65 = 1717 \text{ mm}$$

Knowing that the correction factor is very safe I would round this down to 1700 mm.

LIQUID LEVELS AND WEIRS

There must always be a liquid level above the top of the tubes to ensure that the tubes in the top row remain wet and effective. In the days of hand calculations we used a minimum of 50 mm, (2 inches). However this would be increased if aqueous solutions were involved to allow for foaming. Foaming is a froth which can form of the top of boiling liquids and water is particularly susceptible to this because of the large difference between the densities between the liquid and the vapour particularly at low pressures. Most of our knowledge of foaming comes from the power industry. Of course hydrocarbon mixtures can foam but are less likely to do so. Some specs now require a minimum of 50 mm foaming allowance in addition to the minimum of 50 mm above the tubes giving a total of 100 mm. I think that this is often unnecessary and should be left to the discretion of the thermal design engineer after discussing it with a process engineer.

Having established a level above the tubes it is more convenient to work with dimensions from the centre line of the top row of tubes which is half a tube diameter below this. Eventually when a preliminary setting plan is drawn a dimension will be specified from the centre line of the shell rather than the centre line of the bundle.

It is important to look at the tube layout after a preliminary sizing computer run. (Count the U bends as being fully effective.) There will always be a loss of area available for tubes at the bottom of the bundle to provide space for an impingement device. When vertical U tubes are used there will also be an untubed space at the top of the bundle. This is not totally wasted space since it will provide a passage for vapour to escape from the port section of the shell. This is not as important as it was for the stab-in bundle because it will be a relatively small percentage of the effective tube length but it should still be watched.

For horizontal U tubes or pull through floating head bundles a computer program is likely to come up with some silly tube layouts at the top of the bundle. I have seen tube layouts with as few as two tubes on the top row. These two tubes are likely to have a negligible effect on the surface area. They will however contribute to increasing the cost of the unit well above their own costs. If they are used to establish the liquid level they will increase the diameter and possibly even the thickness of the shell. I always look carefully at the top row and sometimes the second from the top row of tubes and work out their percentage contributions to the total surface area. If the bundle surface area is still satisfactory without them then leave them out.

A drain notch may be fitted to the bottom of the weir plate. This is often a 50 mm diameter half circular notch. Process should decide if one is required.

RECTAGULAR WEIRS

Figure 8.6 shows a liquid flowing over a rectangular weir. Remote from the weir the liquid has a height H above the bottom of the weir which has a breadth b. As the liquid approaches the weir the level will fall as the fluid accelerates. When the liquid reaches the weir the level will have fallen to about 2/3 H.

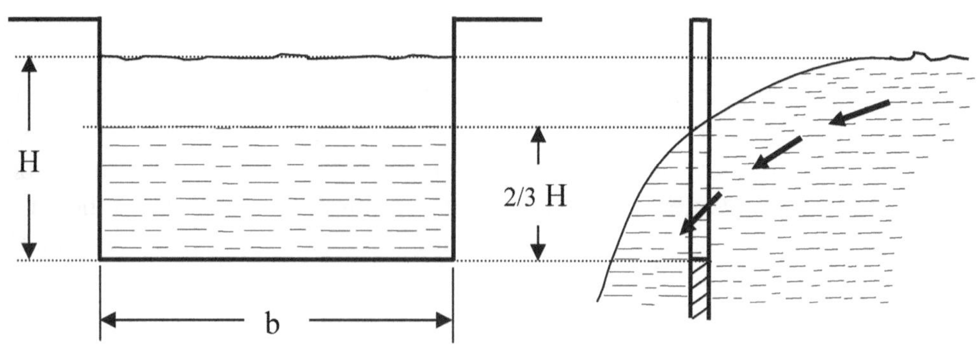

Figure 8.6

The height H above the bottom of the weir can be found from the following equation:

$$H = 1.5 \left(\frac{V}{b\ g^{0.5}} \right)^{2/3} \qquad [8.2]$$

V is the volume flow rate and g is the gravitational constant. I first found this equation in Vaporisers by Smith, ref 7 and the derivation can be found in many standard textbooks on fluid mechanics. It is useful to thermal design engineers not only for the type of weir used in kettles but for the small weirs at the tops of the castellated tubes used in falling film evaporators. Falling film evaporators are covered in detail on pages 1-7 of chapter 12 on special units.

An example is again useful to get a 'feel' for the magnitude of the elevation of the liquid level above the weir in a kettle reboiler.

Example 8.2

What will be the height of the liquid level above the weir in example 8.1?

Solution

$$H = 1.5 \left(\frac{15/850}{2.088 \times 9.81^{0.5}} \right)^{2/3} = 0.029 \text{ m}$$

The calculation does not take long and demonstrates that if a minimum of 50 mm above the height of the tubes had been used it would have already had a built in safety factor of 29/50 x 100 = 58%. Is there any need for another 50 mm in addition to this? For some cases my answer is no; and there may be cases where the 50 mm could justifiably be reduced.

I had designed many kettles without bothering about weir theory until one day I was presented with a particularly challenging case. It had a very large flow rate of liquid and only a small percentage of the lightest ends were being stripped off. The duty was mainly liquid heating. The large flow rate of liquid required two inlet nozzles and a distributor channel along the bottom of the shell. I came up with a preliminary design for the bundle and shell and looked at it, thought about it, and then went outside for a cigar to think about it some more.

My concern was that I could visualise some of the liquid by passing the bundle completely to pass straight over the weir without touching the tubes. The solution I came up with was to minimise this effect by restricting the length of the weir to the central part of the shell as illustrated in figure 8.7.

I fixed the breadth of the weir by measuring the width of the bundle between the centres lines of the outer tubes and then rounded this down to the nearest 10 mm. This resulted in a weir breadth which was about half the dimension it would have been with a conventional weir stretching across the full length of the chord above the 'liquid level'. This avoided liquid passing over the weir without at least coming near to the bundle and mixing with hot liquid which had passed through or been recirculated round the tubes.

This arrangement raised the liquid level and I knew it was time for me to investigate flow over weirs. Fortunately I had a copy of Mr Smith's book to hand and soon found what I needed. (Even when working on a short contract my first problem was trying to find enough shelf space to accommodate all the books and personal files that I took with me. By contrast I have known some people who arrive with a briefcase containing a calculator and some sandwiches!)

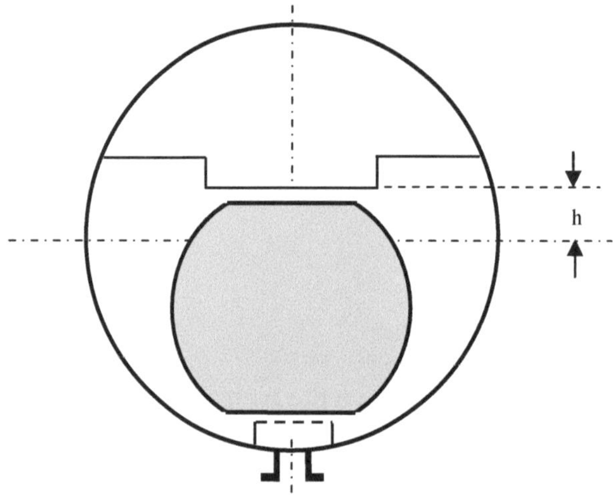

Figure 8.7

I calculated the liquid height above the weir using equation 8.2 and compared the value with what it would have been with a normal weir. I then fixed the dimension h, from the bottom of the weir to the centre line of the shell to give the same level of liquid above the tubes as there would have been with a conventional weir. I added about 50 mm to the value of H to fix the dimension of the vertical sides of the weir. Since the liquid level was unchanged there was no need to re-run the entrainment calculations or adjust the shell diameter.

There is no way of knowing if this measure did improve the design and if it did by how much. I know that it would not have affected the price of the unit. A vendor might include the cost of the weir in his estimate but he would not be concerned about a small change in the shape. The only way to establish if rectangular weirs can be advantageous is from a site test. If a unit with a large liquid heating duty is underperforming a retrofit would be a simple operation. It would not even be necessary to carry out any welding. The bottom section of the weir could be cut from the centre section and two side plates bolted on. If any reader tries this I would appreciate any feedback.

Since then I have used rectangular weirs several times for less extreme cases and I have often wondered if they should be the standard deign for all kettle weirs. This is something that one of the big clients such as BP or Shell, or one of the software suppliers such as HTRI or HTFS might like to take up.

NOZZLES AND PIPEWORK

For thermosyphon reboilers the pipework for the process fluid is an integral part of the thermal design. For kettle reboilers it is the responsibility of the process engineer. The thermal design engineer should have preliminary line sizes before starting a design.

The velocity in the liquid outlet nozzle may vary significantly depending on whether gravity or an extract pump is providing the necessary pressure drop. To avoid vapour entrainment a vortex breaker may be specified by process.

For a simple single inlet nozzle impingement protection needs to be provided if the velocity head is high enough to require one, (see nozzle momentum fluxes in chapter 2.) The nozzle should be near the channel end of the shell but there is no need to minimise the distance between the nozzle and the shell flange. Minimum distances between nozzles and shell flanges have been covered in chapter 2. I would use a rounded dimension for the preliminary setting plan which was between say 100 mm and 200 mm above the minimum distance. This is satisfactory for most kettles.

Sometimes an inlet distribution channel may be required as part of the specification for a particular kettle, (as illustrated in figure 8.7). A typical distribution channel is illustrated in figure 8.8.

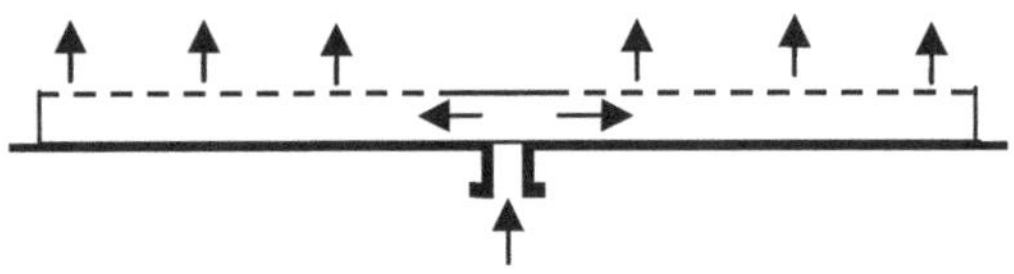

Figure 8.8

A distribution channel has vertical solid sides with a plate at the top which is perforated with a series of holes except for the section immediately above the nozzle which is at the centre of the plate. This channel runs the full length of the bundle and is fed by a central inlet nozzle.

One standard I have used for inlet distributors is to ensure that the cross sectional area of the channel is at least equal to the area of the nozzle which ensures that the maximum velocity in the channel is less than half the nozzle velocity. Use holes that are at least 6 mm in diameter and fix the number of holes by working out the cross sectional area required to give a velocity of 0.92 m/s and dividing this by the cross sectional area of one hole. The holes are then evenly distributed along the length excluding the section above the nozzle.

For perfect distribution the pressure drop along the channel should be negligible compared with the pressure drop through a hole. This implies either a small number of holes or a very large channel section neither of which is practical. The above standard is a compromise and will not give perfect distribution. For bundles with tube lengths of 6m and above distribution is much more even if two nozzles are used as illustrated in figure 8.9.

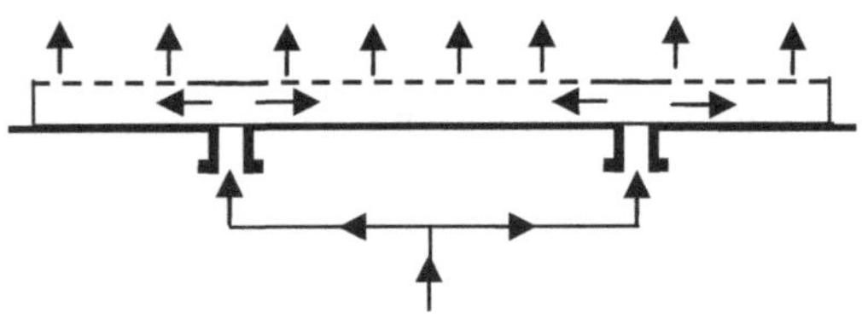

Figure 8.9

If two nozzles are used it is essential to specify that the inlet pipework should be symmetrical as shown.

The vapour outlet nozzle should match the line back to the tower. If two nozzles are required the cross sectional area of each nozzle should be at least half the cross sectional area of the line back to the tower and the connecting pipework should be symmetrical.

DISENGAGING SPACE

This is defined as the space above the liquid level where the vapour can separate from some of the liquid particles produced by the boiling. If the kettle has a weir the space beyond the weir should not be counted since this is a stagnant space that does not have a flow of vapour through it.

ENTRAINMENT, E

This is defined as the mass flow rate of liquid leaving a reboiler per unit mass of vapour.

$$E = \dot{m}_L/\dot{m}_V \qquad [8.3]$$

The reader should note that E is not the same as (1-x) where x is the dryness fraction which is defined as the mass of vapour divided by the mass of liquid plus vapour. I have often thought that it is a pity that we could not have stayed with a single definition since for practical values the numerical difference is very small. (If E=0.01 then x = 0.0099). There is a similar problem with the distinction between specific and absolute humidity in psychrometry.

It is worthwhile considering just how entrainment occurs. Two quite different mechanisms are possible. Whenever there is a vapour or gas velocity across an uneven liquid surface there is a tendency to detach some of the liquid from the crests and carry it along in the vapour stream. This is the mechanism which we have all experienced by a sea or lake or more dramatically on a sea or lake in a small boat. This is entrainment due to a horizontal velocity.

Entrainment can also take pace due to the vertical velocity of a vapour leaving the surface of a boiling liquid. This is illustrated in figure 8.10.

In (a) a bubble has just reached the surface of the liquid. The volume of vapour in the bubble is surrounded by a thin layer of liquid at the top and is in contact with the bulk of the liquid at the bottom.

In (b) the bubble has just burst, the thin layer of liquid which had been at the top of the bubble is broken up into a number of very small droplets of liquid. The bursting of the bubble leaves a depression in the surface of the liquid.

In (c) liquid rushes in to fill the depression which forces up a plume of liquid at the centre of where the bubble had been.

In (d) the top of the plume becomes detached and is carried upwards by the vertical velocity of the vapour. The drop of liquid formed in this way is much larger than the droplets from the top of the bubble. Typically 1000 μm compared with 40μm.

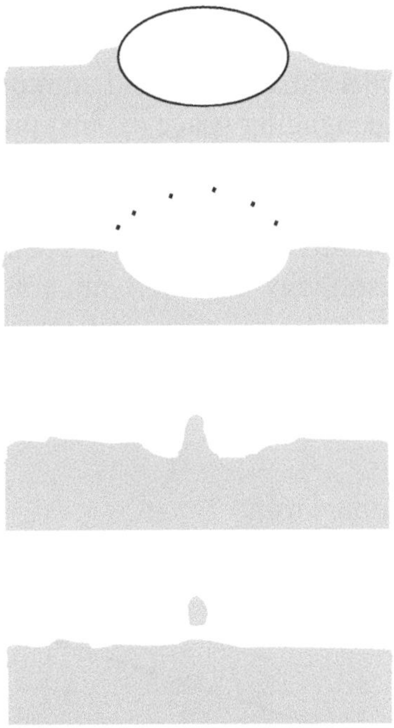

(a) (b) (c) (d)

Figure 8.10

Although there are a lot more small droplets than there are larger drops of liquid most of the entrainment is due to the larger drops. This is described in more detail by Newitt et al in reference 27.

It is not possible to produce a dry vapour from a boiling liquid without either subsequent heating, until there is a small degree of superheat, or throttling it where the reduction in pressure at constant enthalpy can have the same effect. The entrainment which can be tolerated depends on where the vapour is discharged to after leaving the kettle. This is a process responsibility and ideally an entrainment value should be specified on every process data sheet involving pool boiling equipment. This is not always done and a thermal design engineer should check the process flow diagrams to get a feel for what is required before discussing it with a process engineer.

The maximum entrainment which is likely to be tolerable is about 0.015 if the vapour is returned to a fractionating tower or passes to a knock out drum in a refrigeration system. At the other extreme 0.001 may be required in a refrigeration system with no knock out drum.

SHELL SIZING

Although the term entrainment was in common usage it was never part of the shell sizing process when I first started to design heat exchangers in Kellogg's London office. Disengaging space fulfilled the same function. We had a series of charts which we used to look up the allowable disengaging space for a particular fluid at the exchanger operating pressure. I have reproduced the chart for hydrocarbons in figure

8.11. The vertical axis is the number of pounds per hour of vapour allowed for one cubic foot of disengaging space. The horizontal axis, M is the average molecular weight of the hydrocarbons in the vapour. There were a total of 14 lines of constant pressure which ranged from low vacuum operation to 300 psia. Only half of these lines are shown in the diagram.

We used a series of other charts for specific fluid such as steam and ammonia. These charts were just single line plots of allowable disengaging space against pressure. The chart for steam had an in built safety factor of five!

The source of most of these charts was Tema Tentative Standards. I have never seen the original standards nor been given a good reason why they were not continued since they clearly contained a lot of useful data.

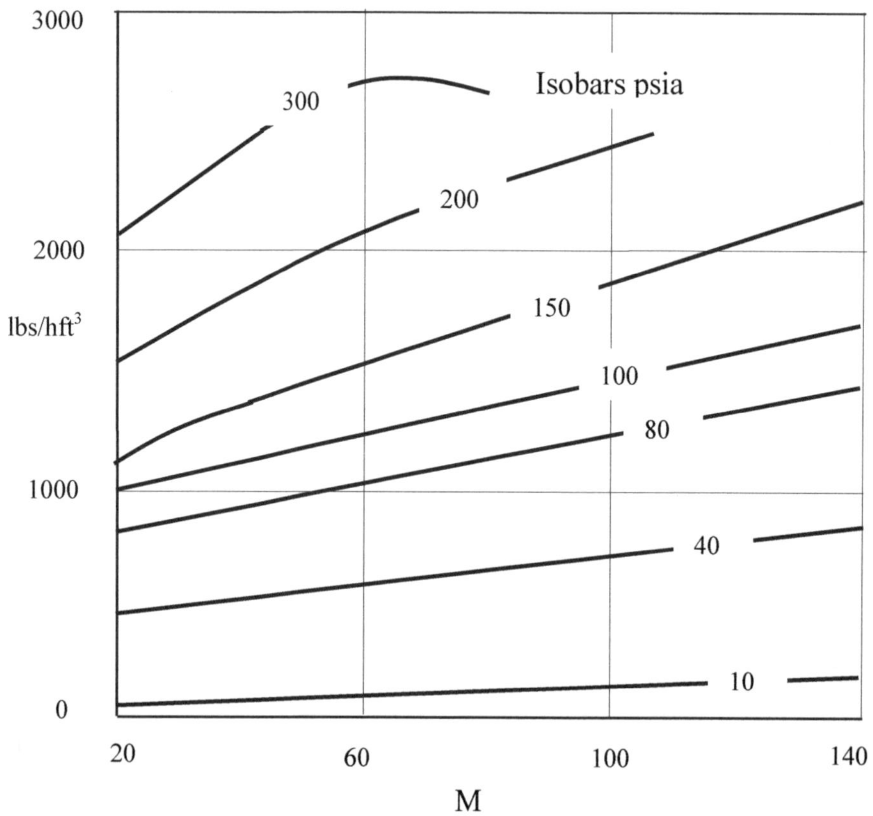

Figure 8.11

The mass flow rate of vapour was then divided by the allowable disengaging space per unit volume to give the total volume required above the surface of the liquid. This volume was then divided by the effective length of the bundle to fix the required cross sectional area above the liquid level.

Figure 8.12 shows a section through the shell and bundle of a kettle. The vertical length, A from the bottom of the shell up to the liquid level was normally fixed as being two inches above the top of the top tube row in the bundle as shown.

One way to establish a shell size would then be to use tables of chord lengths and areas of circular segments such as table D-14 in TEMA. This would have been tedious since a trial and error solution would have been necessary

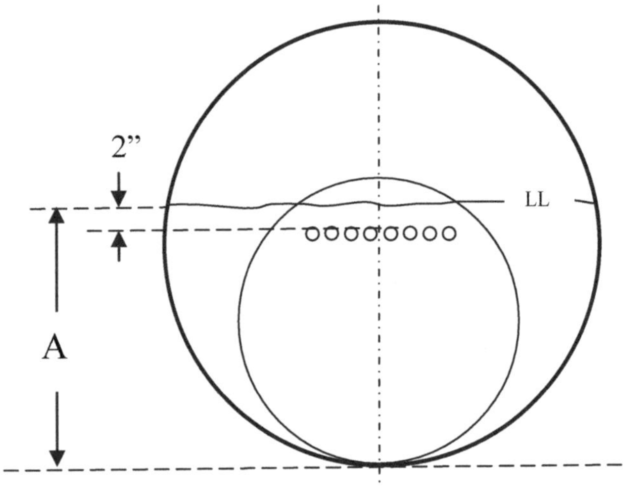

Figure 8.12

Figure 8.13 is also from Tema Tentative Standards and avoids a lot of repetitive calculations. My reproduction here is only an approximate one showing a sample of the lines of constant dimension A. Also the lines are not quite straight in the original chart. The chart is very simple to use. For example if a segmental area of 500 square inches is required in a kettle with a dimension A of 30inches a shell of about 45 inches is required.

There was a cut off point to give a minimum height of 12 inches above the liquid level, (shell ID –A). This minimum height is still in most specifications today as 300 mm, although I have sometimes seen this reduced to 250 mm. Without any specification limits I would use the 250 mm for small shells with a minimum risk of entrainment problems.

I have not attempted to convert the charts in this section to SI units because I know that they are unlikely to ever be used again. They are of historical interest and demonstrate the state of the art at some of the more progressive companies in the 1960's.

Despite the availability of the Tema Tentative Standards there were many other companies which still had a very crude and unsatisfactory approach to sizing kettle shells. They simply used a fixed shell to bundle ratio. The ratio would be between 1.3 and 1.5 which might change for ranges of shell sizes i.e. 1.5 for the smallest shells and 1.3 for the largest shells. This carried on in some companies for many years.
In 1967 I left Kellogg to join Stone and Webster where I was put in charge of a new heat exchanger section within the process department. Until then they had been using a single shell/bundle ratio to size their kettles. One of the first things I did was to review past kettle designs and compare the shell sizes with designs using the more scientific method outlined above. Some designs which used medium pressure fluids were in reasonable agreement but the designs with either low or high pressure fluids had significantly different shell sizes.

The high pressure units had shell sizes that were larger than necessary and the low pressure units had shells which were too small. I selected the worst case of the low pressure units and asked about its history. There had been problems with it due to high entrainment which had only been corrected by fitting a knock out drum in the vapour outlet line. Ironically with the high pressure units with their thick expensive shells being too large and the low pressure units with relatively thinner shells being too small I estimated that there would have been a slight cost saving in sizing the shells on a more scientific basis.

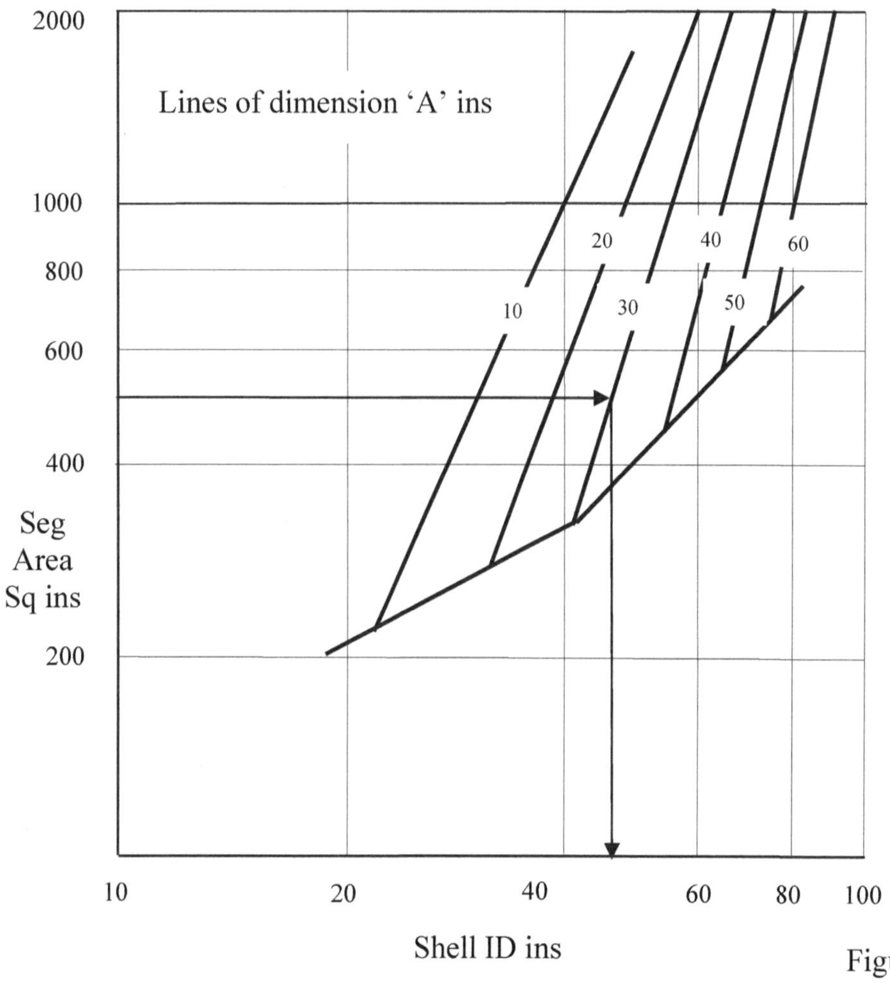

Figure 8.13

It may seem to the reader that this all happened a long time ago but I was involved in a very interesting case involving fixed shell/bundle ratios long after that in 1993. I was working on a contract for Bechtel when I was asked to check a kettle design by the head of the heat exchanger section. I came to the conclusion that the bundle size was OK but the shell size was not and told him that I could not find any calculations to show how he had arrived at his dimension. (I had checked the size using hand calculations based on TM9 which is fully covered later.) He was quite a spiky character with a sound mechanical but limited thermal background. It was the first of several technical clashes which were always left in the office and he was the life and soul of the party in the pub. He indignantly pointed to the computer output stating that there were no hand calculations because the computer had calculated the shell size.

The program we were using was the HTRI program RKH-1. I knew the program and its limitations. I pointed out that the shell size on the output sheet was based on fixed shell/bundle ratios and was only intended as a rough guide and use for very approximate estimate work. He clearly did not believe me so I set up the case on the computer and proceeded to make changes to the flows, number of outlet nozzles, etc and the shell size remained unchanged. He left me with a puzzled look on his face mumbling "alright then use your size". (It was 'safe', i.e. my size was larger than the size from the computer output.)

I was of course only a freelance worker and there were several jobs ongoing at Bechtel at the time so I never had chance to review how many kettles had been designed and installed using the shell sizes from RKH-1. I have often wondered how many other companies around the world might have been making the same mistake. **Know your program and understand its limitations!**

ENTRAINMENT CALULATIONS

In 1983 HTFS published a confidential paper TM9. It is effectively a joint paper by HTRI and HTFS so there was no need for thermal design engineers to make judgements as to whose approach to entrainment should be used. By now I assume that this paper is available to anyone working in the petrochemical contracting industry.

The basic equation in this paper is as follows:

$$u_{Vmax} = 1.77E^{1/3}\left(\frac{g\,\sigma\,\{\rho_L - \rho_V\}}{\rho_V^{\,2}}\right)^{1/4} \qquad [8.4]$$

Where u_{Vmax} is the maximum allowable vapour velocity which will ensure that the required entrainment E is not exceeded. The other symbols are the gravitational constant, the surface tension and the densities of the liquid and vapour. The constant 1.77 is dimensionless.

The maximum vapour velocity is the higher of the vertical velocity and the actual vapour velocity which is the resultant of the horizontal and vertical components.

When I first saw this equation I realized that it could be an improvement on the Tema Tentative Standards method using disengaging volumes that I had been using up until then. What impressed me was that horizontal and vertical vapour velocities were taken into account and the horizontal velocity could be changed by changing the number of outlet nozzles which clearly must have an effect on entrainment. With the old method we used one nozzle for every 3 m of effective tube length but the number of nozzles was never a variable in the calculations. (This 'rule of thumb' of one nozzle every 3 m is still worth bearing in mind.) The other innovation in the equation is that it takes into account the surface tension which was shown to be an important physical property in the chapter on boiling and must clearly affect the formation of droplets as discussed earlier in this section.

When I came to use the equation I found it was not as satisfactory as it appeared. Vapour velocities could vary enormously depending on the operating pressure and the molecular mass of the vapour so there was no way I could anticipate typical numerical values and of course I did not want a velocity value I wanted an entrainment value. I therefore decided to transpose the equation into the form I wanted. The result is as follows:

$$E = \frac{u_{Vmax}^{3}}{5.55} \left(\frac{1000\, \rho_V^{2}}{9.81\, \sigma\, \{\rho_L - \rho_V\}} \right)^{3/4} \qquad [8.5]$$

It is now clear to see how each variable affects the entrainment. The surface tension is now in dynes/cm. It is the unit used in most property tables and I have come across far too many serious errors being made getting the magnitude of units wrong. The 1000 in the top line converts this to the fundamental unit of N/m. The equation has been checked by a number of experienced engineers who subsequently started to use it.

For a particular design case:

$$E = u_{Vmax}^{3} \times A \text{ constant} \qquad [8.6]$$

It is still an iterative procedure to establish a value for E. The height of the liquid level above the bottom of the shell, dimension A is established as shown in the previous figure 8.12. A shell size is then estimated – at least (A + 300 mm) for the first estimate. The vertical height above the liquid level, h is then (D_S – A). The length of the chord, C and the cross sectional area of the segment, Ax is then taken from tables such as D-14 in TEMA.

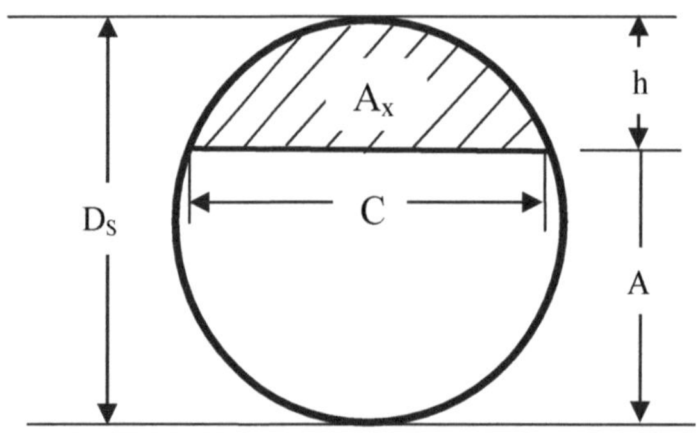

Figure 8.14

Calculate the vertical velocity u_{vert} from:

$$u_{vert} = \frac{V}{L_{eff} \times C} \qquad [8.7]$$

Where V is the volume flow rate in m^3/s and L_{eff} is the effective length of the tubes.

Calculate the horizontal velocity u_{horiz} from:

$$u_{horiz} = \frac{V}{A_x \times 2N} \qquad [8.8]$$

Where N is the number of vapour outlet nozzles.

Calculate the maximum velocity u_{Vmax} from:

$$u_{Vmax} = (u_{horiz}\, u_{vert})^{0.5} \qquad [8.9]$$

Or

$$u_{Vmax} = u_{vert}$$

Whichever is the greater. (Note that u_{Vmax} from equation 9 is simply the hypotenuse of a right angled triangle whose other sides are u_{vert} and u_{horiz}.)

I have usually found that equation 9 controls the maximum allowable vapour velocity. This is then used in equation 6 to calculate the entrainment. A new shell size is then estimated taking into account how far the first entrainment value calculated was from the target value. Two or three iterations are normally sufficient. (I could manage with 2 by converting back to the old units and using a faded copy of figure 8.13 for my first estimate of the shell size.)

I was on the same contract at Bechtel when I was asked to explain the above procedure to two junior heat transfer engineers. They were both keen bright young lads but neither of them had been deigning heat exchanger for more than a year or so. I gave them a calculation sheet which was basically the above equations. I went through this with them and as soon as I got to the end and mentioned iteration they looked at each other and in unison said “I could write a computer program for that”.

I asked them how long that would take and they replied, “only a few hours”, so I told them to carry on. I checked how they were doing an hour or so later and found they were busy writing out the entire D-14 table from TEMA. I stopped them and pointed out that this table is just a quick way of avoiding a series of basic geometric calculations which can be time consuming. They looked bemused as I wrote down the basic equations for calculating the area of a segment and the length of the chord. (See example 8.1 on liquid hold up times.) They discarded their table and in a very short time they proudly presented me with an Excel spreadsheet.

I was suitably impressed but a bit sceptical until I had tried it out on some past hand calculations. It worked and I could not find any fault in it. The iteration process was now just a simple matter of changing one number and a lot of potential arithmetical errors had been eliminated. The time taken to write the spreadsheet was more than recovered on the job we were working on at the time. I know that this spreadsheet was subsequently adopted by quite a few companies and is still used. I could probably just about manage to write it myself now and I am sure that younger readers will have no difficulty in reproducing it.

KETTLE NEAR ENDS

It is not always practical to use a ‘conventional’ kettle shell as illustrated earlier in figure 8.4. There are many occasions when the length of the conical section becomes excessive which would limit the positioning of the vapour outlet nozzles and reduce the disengaging space.

Figure 8.15 below shows a possible kettle design with an excessive conical section. The outlet nozzle is at the far end of the bundle when it should be above the centre of the bundle. If an entrainment calculation were to be carried out for such a design the number of outlet nozzles should be 0.5 in equation 8 since the vapour is only flowing in one direction towards the nozzle. It is also debatable as to what horizontal cross sectional area should be used since it is continually changing until it reaches the short parallel section near the nozzle. Such a design is clearly unsatisfactory

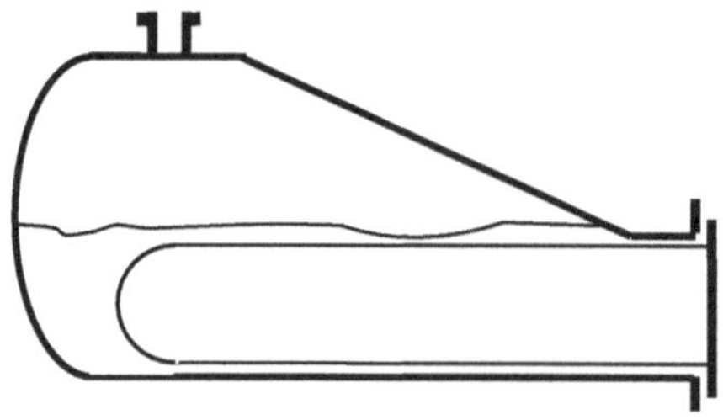

Figure 8.15

The need to review the near end construction of a kettle may arise in the following situations:

(a) Very large kettles
(b) Kettles which are limited to using a very short tube length.
(c) Low pressure units were a large shell/port ratio is required.

The length of the conical section in a conventional kettle is $(D_S\text{-}D_P)/\tan 30^\circ$. To establish the minimum distance of the centre line of the first outlet nozzle to the face of the tubesheet the following lengths must be added to this:

(a) The diameter of the nozzle (to allow for a possible reinforcing pad unless a long forged flange nozzle is used which increases the cost).
(b) 50 mm or twice the shell thickness.
(c) The length of the port flange.

Just calculating the length of the conical section is normally sufficient to establish if there is likely to be a problem. If in doubt carry on and calculate the full length. If this is excessive it is necessary to change the mechanical design and attach the port to a dished end as illustrated in figure 8.16.

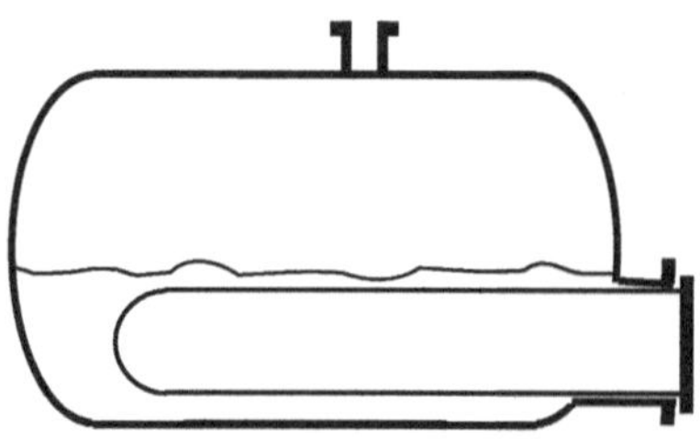

Figure 8.16

One thermal advantage of this arrangement is that the nozzle can be situated in its ideal position at the mid point of the effective length of the bundle. The other is that the port is no longer flush with the bottom of the shell. Raising the bundle above bottom of the shell helps the recirculation of the liquid and is particularly useful if any form of inlet distributor is required.

The downside is that the height of the liquid level above the bottom of the shell increases by the difference in height of the bottom of the port and the bottom of the shell. This increases the shell diameter and hence the cost.

The dished end at the far end of a kettle is invariably ellipsoidal. For a small shell with a single nozzle as shown in figure 8.16 it is probably not worth considering anything else for the near head. Kettles with multi tubeside passes generate vapour fairly evenly along the tube length. This is why the ideal position for a single vapour nozzle is above the centre of the effective length and at 0.25 L_{eff} and 0.75 L_{eff} for two outlet nozzles. However with single pass kettles vapour generation is only evenly distributed when both fluids undergo isothermal processes. If the temperature of the tubeside fluid falls significantly the vapour generation rate will be much higher at the hot end than the cold end. The ideal position for the vapour outlet nozzles is now not only where they will receive equal volumes of vapour but where the volume of vapour flowing to each nozzle is the same in both directions.

Single pass kettles are dealt with later in this chapter but they often require the vapour nozzle at the hot end to be as close to the tubesheet as is mechanically practical. This section will only briefly look at the basic mechanical details involved with kettles with dished ends.

Figure 8.17 shows details of a Klopper type of torispherical head which can be used as an alternative to an ellipsoidal head. Most of the face of the head has a radius equal to the outside diameter of the shell. There is a short parallel section which has a length equal to 2t where t is the thickness of the shell. These two sections are joined by another short section which has a radius equal to one tenth of the shell outside diameter hence they are also called decimal heads. The sections with the short parallel and small radius are called the knuckle. They are of course produced as a single forging.

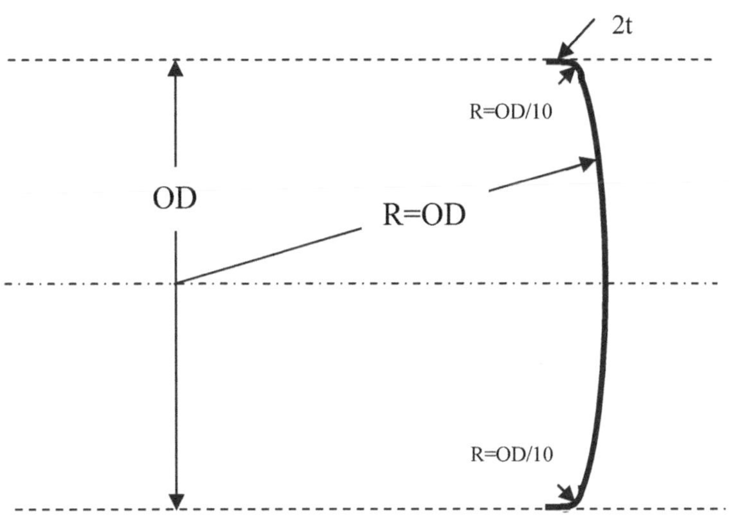

Figure 8.17

The height of the head can be calculated as follows:

$$h = OD\,(1^2 - 0.4^2)^{0.5} + 0.1OD + 2t$$
$$= 0.1835\ OD + 2t \qquad [8.10]$$

This compares with 0.25 OD for an ellipsoidal head. For say a 3 m shell this works out as about 600 mm compared with 750 mm. It is therefore possible to have the first shell outlet nozzle 150 mm closer to the tubesheet.

The other advantage of the torispherical head is that it may be possible to reduce the distance between the bottom of the bundle and the bottom of the shell which decreases the diameter of the shell.

Fixing the distance between the bottom of the bundle and the bottom of the shell is not a simple matter whichever type of dished end is used. The port may need to be increased in thickness to compensate for the loss of metal in the head. This is illustrated in figure 8.18 below.

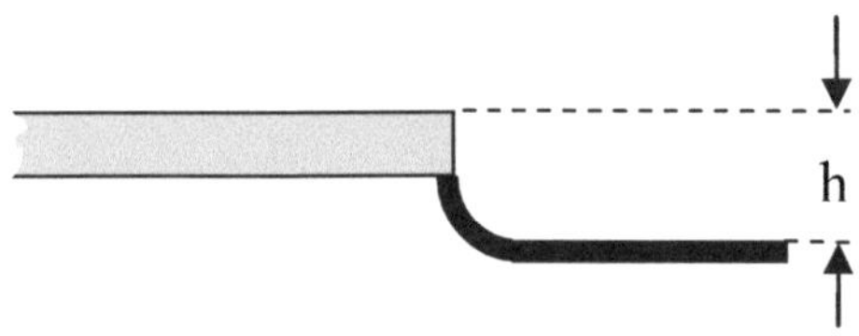

Figure 8.18

Establishing a value for the distance h should therefore be left to a mechanical engineer. I always consult with a mechanical engineer at an early stage whenever I consider using two dished ends in a kettle.

There is another type of torispherical head called the Korbbogen head but this has a knuckle radius of 0.154 OD. There is no thermal advantage in using this and it is more likely to be found in architecture than in engineering.

In most plants there are a lot of pressure vessels apart from the heat exchangers which are all likely to use ellipsoidal heads. The torispherical head is therefore non standard and will be more expensive and more difficult to source. I only consider using them when there is a significant thermal advantage to be gained by positioning the first vapour outlet nozzle as close to the tubesheet as possible.

EVAPORATORS

These exchangers are specified as TEMA K type exchangers but they are not reboilers. They are normally part of a refrigeration system.

TWO PHASE INLET NOZZLES

Positioning a liquid inlet nozzle is a relatively simple procedure. There are a lot more options with a two phase fluid at inlet. The vapour is the flash vapour produced during the expansion process across the throttle valve in the refrigeration cycle. This is important to remember because although a relatively low pressure drop may be specified on the process data sheet there should be no problem for process to make more available if required. An increase in pressure drop across any inlet arrangement into the kettle just reduces the pressure drop across the throttle valve.

Two phase inlets are normally above the liquid level. This arrangement reduces the number of bubbles reaching the surface of the liquid which reduces the number of drops of liquid carried upwards from the liquid surface. Some specifications foolishly state that all two phase inlets should always be above the liquid level. There are some situations where it is advantageous to have two phase inlets at the bottom of the shell below the bundle. These cases will be covered later. The first thing to consider is what type of entry to use to separate the liquid from the vapour for inlets above the liquid level. Some of the methods which can be used follow.

IMPINGEMENT PLATE/BAFFLE

The inlet nozzle is at the side of the shell parallel with the centre line of the shell. The discharge from the nozzle impinges on a plate where some of the liquid will be separated from the vapour immediately and the remaining flow will be directed along the side of the shell in two directions. The plate may be termed an impingement plate, an impingement baffle or a splash baffle/plate. The total length should be at least three times the nozzle diameter and the height should extend to the normal liquid level. It should not be placed so close to the nozzle that there is any restriction to flow. The bottom ends are normally attached to the shell by rectangular bars to provide mechanical strength. Only the basic plate is illustrated in figure 8.19.

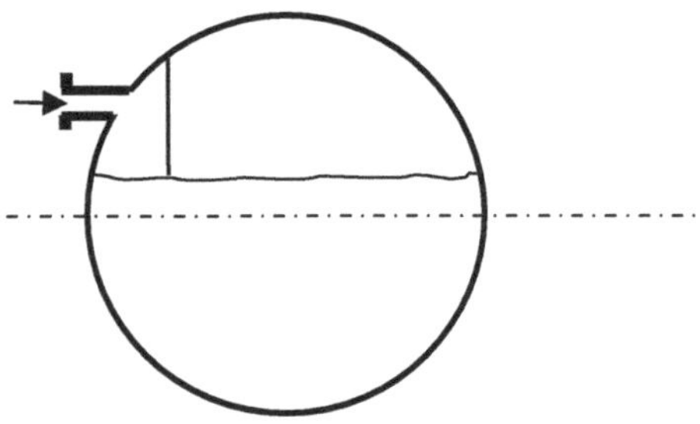

Figure 8.19

There are two schools of thought as to where two phase inlet nozzles should be placed in relation to the outlet nozzles. These are illustrated in figure 8.20 below.

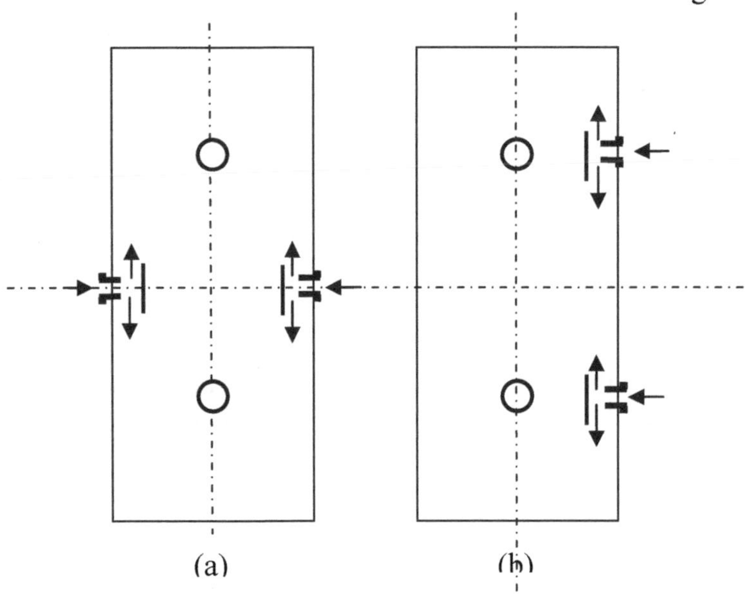

Figure 8.20

This shows two possible arrangements for a shell with two outlet and two inlet nozzles. In figure (a) the inlet nozzles are in between the two outlet nozzles. This is probably the most popular arrangement.

The argument against arrangement (a) is that the flows from the inlet nozzle are directed towards the outlet nozzles. To counter this, arrangement (b) directs the flows away from outlet nozzles.

I have discussed this with other experienced engineers without coming to any firm conclusions as to which arrangement is the most satisfactory. If there were only one inlet nozzle then clearly arrangement (a) would be used otherwise one outlet nozzle would be trying to handle the flash vapour plus its share of the vapour generated in the bundle. One nozzle may be sufficient when there is a relatively small amount of flash vapour.

The above arrangements assume that the vapour generation is fairly even throughout the length of the bundle. This is not always the case. Suppose one half of a bundle is generating twice as much vapour as the other half in a kettle which is supplied with a two phase flow with 25% vapour.

Only 75% of the mass flow rate of vapour leaving the kettle is produced by boiling in the bundle and 50% of this is produced in one half of the bundle and the remaining 25% in the other half. The ideal nozzle arrangement is now as illustrated in figure 8.21 where the lower half of the bundle is producing the most vapour.

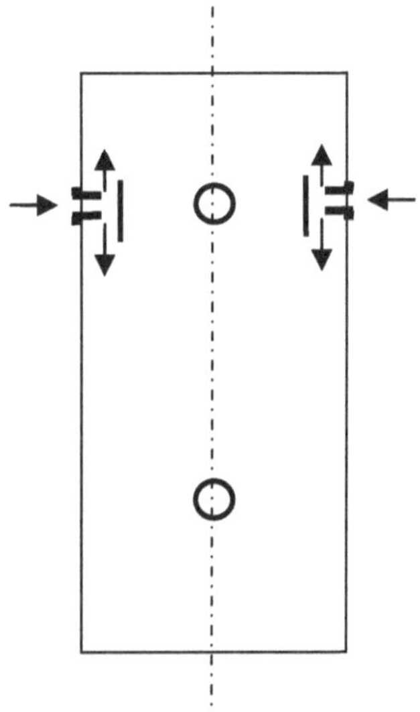

Figure 8.21

The positioning of the two phase inlet nozzles has now been used to even out the flow of vapour to each outlet nozzle and the entrainment calculation is straightforward.

Alternative shapes of plate

Although simple vertical plates are the most commonly used there may sometimes be advantages of using an alternative shape. One possible alternative is shown in figure 8.22 below.

Figure 8.22

The aim of this plate is to divert the liquid closer to the shell than would be the case with a simple vertical plate. I think it is an improvement on a simple vertical plate although it is difficult to assess how much of an improvement.

SPARGE PIPE DISTRIBUTORS

This arrangement is a bit more sophisticated and may have more flexibility than a simple impingement plate. The inlet nozzles are connected to pipes inside the shell which run in an axial direction. This is illustrated in figure 8.23. On the left hand side is a section through the nozzle and on the right hand side is a section through the pipe.

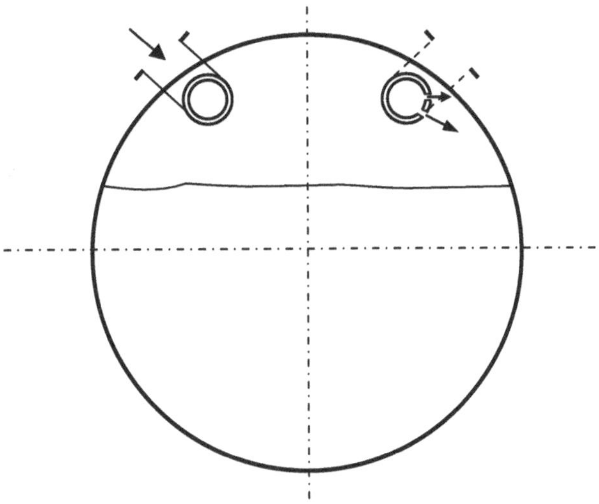

Figure 8.23

The distributor pipe has a series of holes which direct the fluid towards the side of the shell. As with any distributor a high pressure drop through the holes compared with the pressure drop through the pipe is required for good distribution. (About 10 to 1 to give to ensure no more than about 5% maldistribution.)For most two phase inlet nozzles extra pressure drop can usually be made available by offloading the expansion valve in the system. However a high pressure drop through the holes means high velocities which make it more difficult for the liquid to separate from the vapour so a compromise has to be accepted. The guidelines used for the spider distributor can be used.

SPIDER INLET DISTRIBUTORS

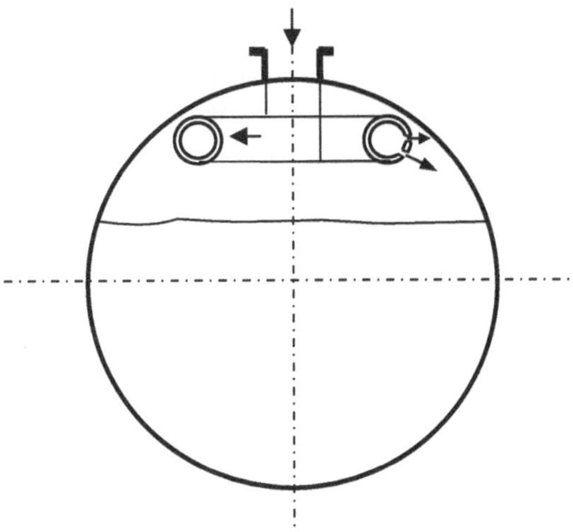

Figure 8.24

A section through a spider inlet distributor is shown in figure 8.24. It consists of a single top inlet nozzle which is connected to two distributor pipes. The left hand side shows the flow through the connecting horizontal pipe to a distributor pipe and there must be no holes opposite the connecting pipe. The left hand side shows a section through the distributor. Two rows of holes are shown but one row may be sufficient

Like any other two phase inlet they may be located at the end of the shell where there is the lowest vapour generation or they may run the length of the shell where the vapour generation does not vary. Some guidelines for sizing the pipes and holes based on momentum flux values are as follows:

(a) Inlet nozzle 4000 kg/ms^2 maximum.
(b) Inlet to the distributor sections 1000 kg/ms^2 maximum.
(c) Average through the holes 4000 maximum.

Although the momentum flux through the holes is a maximum it should be close to this value to give reasonable distribution provided the overall pressure drop is OK.

PIPE DISTRIBUTORS

A very simple pipe distributor is illustrated in figure 8.25. The nozzle is at the top and the fluid is diverted into a hard tee junction where it divides and is directed towards the sides of the shell.

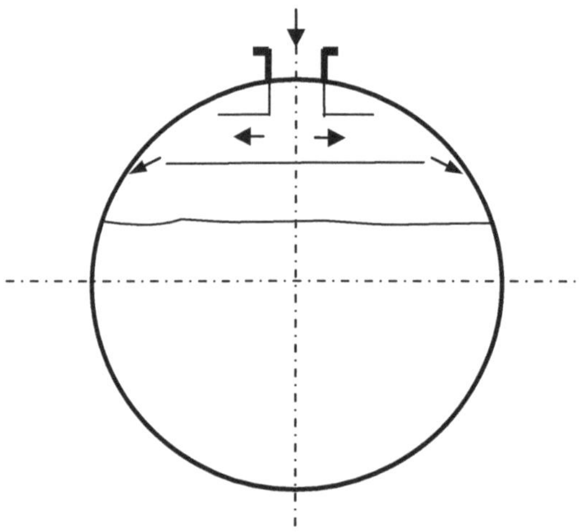

Figure 8.25

The aim of a pipe distributor is to start the separation of the liquid from the vapour in the pipe before it enters the shell as illustrated in figure 8.26 below.

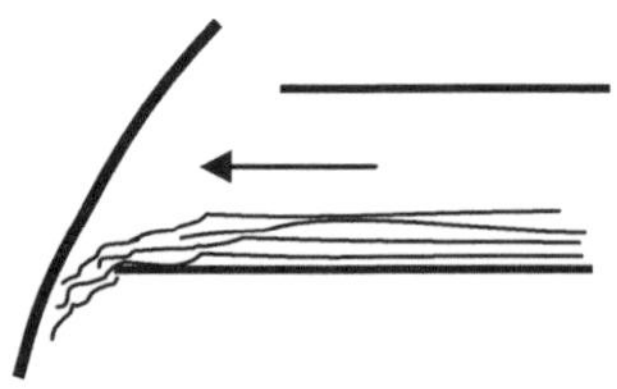

Figure 8.26

To aid the separation of liquid from vapour the ends of the pipe are tapered at an angle of no more that 45° to the horizontal axis. A lot of work has been done on tapers at the bottom of vertical tubes to prevent flooding but I am not aware of any work on the discharge from horizontal pipes with tapered ends in two phase flow. I am sure it must help and of course the removal of metal from the top of the tube provides an easy escape route for the vapour.

The effectiveness of this arrangement is significantly increased when four outlet passages are used as shown in plan view in figure 8.27. The velocity of the fluid entering the shell is only one quarter of the velocity in the nozzle and the momentum flux of the fluid entering the shell is only one sixteenth of the nozzle value. Although I would not place too much reliance on any flow map prediction for a fluid leaving such a short length of pipe after rounding a hard tee there is a good chance the flow will be stratified. (A 3200 kg/ms^2 nozzle momentum flux is reduced to only 200 kg/ms^2 before it enters the shell.)

I have deliberately drawn the inlet nozzle offset from the centre of axial pipe section. This would allow for a vapour outlet nozzle to be positioned at the mid point of the top half of the bundle. (Assuming the vapour generation pattern was the same as that assumed for figure 8.21.)

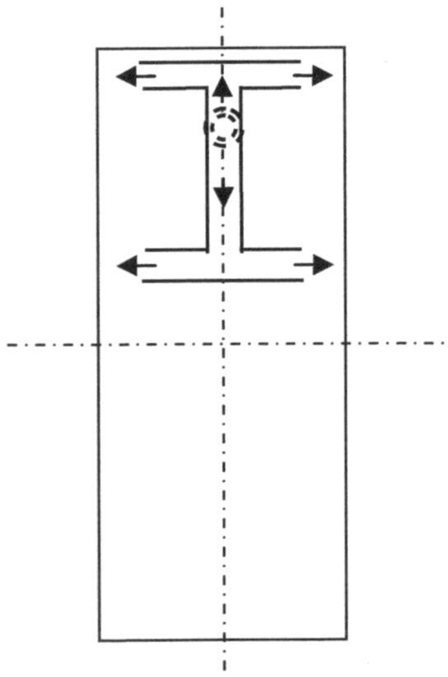

Figure 8.27

This deviation from a symmetrical arrangement will have a negligible effect on the distribution to the ends of the horizontal pipes because the frictional pressure drops will be very small compared with the pressure drops through the hard T's.

I first developed this arrangement when I was working on a low pressure unit with a specification which required spider type distributors. The diameter of the axial pipes was so large that about 40% of the disengaging space was taken up with the spider distributor. My arrangement took up about half this space and it was not too difficult to persuade the clients engineer to let me use it.

As well as saving space I like this arrangement because of the phase separation which will take place before the mixture enters the shell space. It also positions nozzles along the centre line instead of being offset which is better for the external piping. (The same as the spider in this respect but different from the simple impingement plate.) Since then I have used this arrangement many times. No doubt other people have independently come up with the same system but I have not met them.

BOTTOM INLETS

The vast majority of two phase inlets are correctly positioned above the liquid level and many specifications will state that this arrangement should be used. But like so many of the more interesting features of thermal design work there are exception to the 'rule'.

In some low flux kettles, (think of low temperature difference between the tube metal and the boiling liquid if that is easier), the ideal position for the inlet nozzle may be at the bottom of the shell. This situation is most likely to occur at the cold end of single tube pass kettles. The lower rows of tubes may be undergoing mainly natural convective heat transfer with a slow rate of bubble formation from a limited number of nucleation sites, (see incipient nucleate boiling in the description of figure 7.1 in chapter 7.) An inlet nozzle below these tubes will create turbulence and break up the laminar natural convection. It will also sweep bubbles off the tubes which will increase the effectiveness of the nucleation sites. These two effects will significantly increase the local shell side heat transfer coefficient which may otherwise be the controlling the heat transfer in part of the exchanger.

There has been some research which shows how the heat transfer coefficient can vary through a bundle. One of the more interesting pieces of work was carried out by Leong and Cornwell. Like most research they could only use a relatively small bundle which was 17 rows deep. The tubes were a practical 19 mm in diameter on an in line square pitch. I have approximately reproduced a diagram from their work in figure 8.28 below.

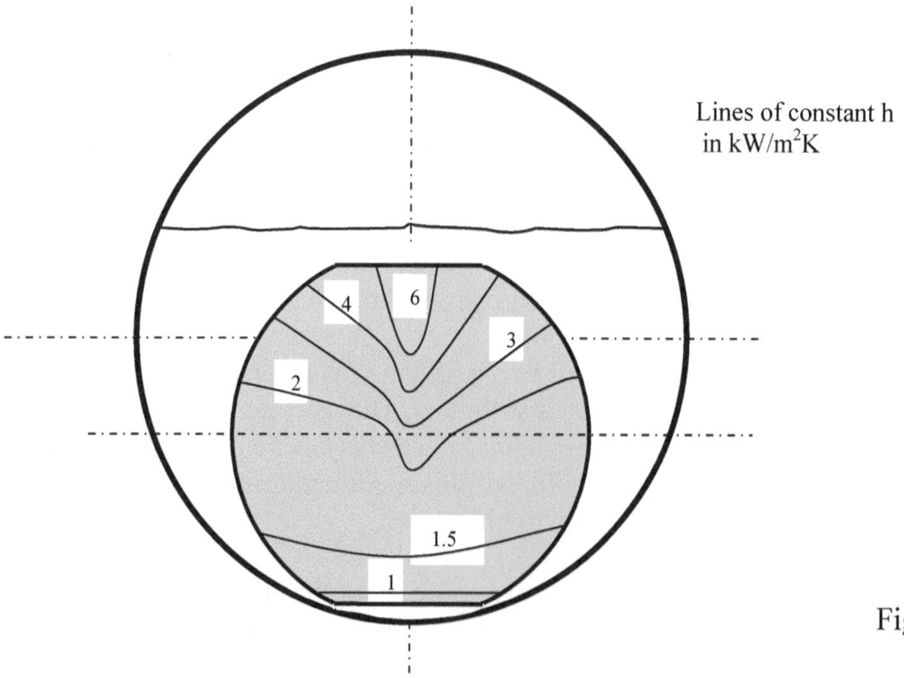

Figure 8.28

I have taken a bit of licence in drawing the constant heat transfer coefficient lines. Except for the line across the bottom row of tubes the other lines do not follow such smooth curves and I have also omitted two short lines of 1.5 kW/m^{2}K from either side of the bundle. (These only cover the outer tube rows.)

The important point this drawing demonstrates is the magnitude of the changes in the heat transfer coefficient through such a small bundle. Clearly the magnitude of the change may be even higher across a larger bundle. The actual values would of course be much lower for a particularly low flux bundle. Although the main aim of a bottom inlet nozzle is to try and increase the magnitude of the coefficients across the lower rows of tubes it will have some effect throughout the bundle.

Some programs may now allow bottom inlet nozzles to be input so that the benefits are immediately apparent. If they do not it is no excuse for not using them. If you have plenty of time you could try some hand calculations to justify increasing the coefficient but what I usually do is to simply make every other component of the design calculation as tight as possible, (minimum oversurface or slight undersurface) knowing that I have 'something in hand'.

To be as effective as possible an inlet distributor is advisable. This should cover a short length at the cold end and will even out the vapour loading in the disengaging space for the entrainment calculations.

VAPOUR ESCAPE LANES

In very high flux kettles there may be a danger of some of the tubes being starved of liquid which will result in a fall in heat transfer coefficient. Peak heat fluxes and the fall of the heat flux in the unstable region were discussed in chapter 7 on boiling. This is the opposite situation to the one just covered where low heat fluxes created design problems. Figure 8.29 shows a possible distribution of heat transfer coefficients in a high flux bundle.

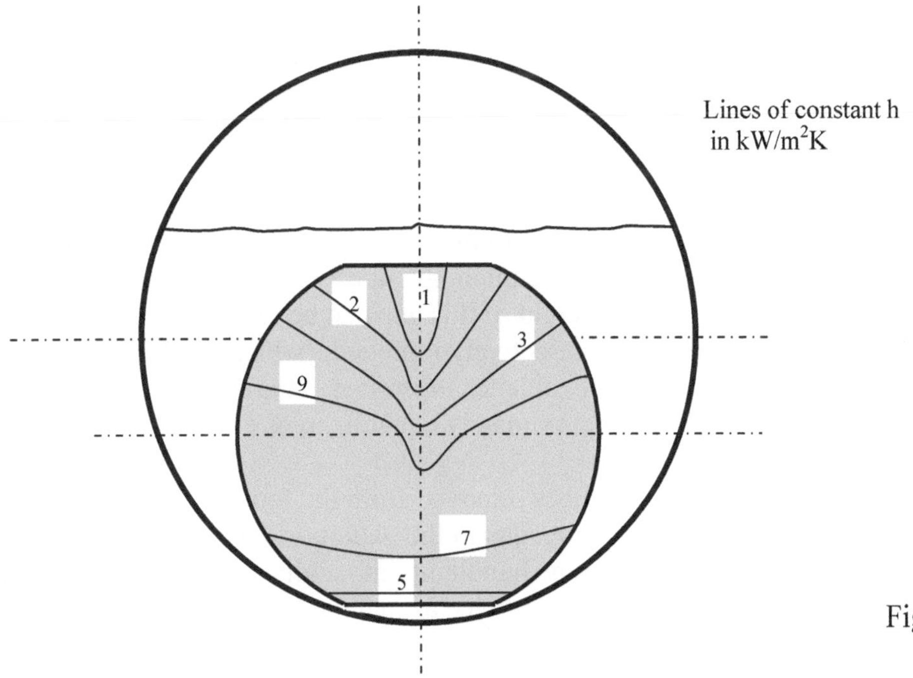

Figure 8.29

This diagram is not based on any research program it is just my lazy way of simplifying a very complex situation to help the reader understand what is likely to be taking place and find another use for a diagram which took me a long time to produce. This situation is likely to apply to a large bundle with several thousand tubes rather than the small experimental bundle with only a few hundred tubes which was used to produce the data in figure 8.28.

In this diagram the heat transfer coefficients at the bottom of the bundle are very reasonable and these will steadily increase through the bundle until the peak heat flux is reached, which in this case corresponds to a heat transfer coefficient of about 9 kW/m^2K. Above this line film boiling will start on some of the tubes. The proportion of the tubes involved in film boiling will increase towards the top and centre of the bundle creating an area with low heat transfer coefficients.

This raises an interesting philosophical question, "is the key to solving the problem getting liquid to the top tubes or getting vapour away from these tubes". I am quoting a very experienced engineer who I discussed this with recently. My view is that you can not have one without the other. If there is no vapour blanketing in an area the liquid will find its way into the bundle and if liquid is present there will be nucleate boiling on the tubes.

The traditional way to overcome this situation is to introduce vapour escape lanes. Figure 8.30 (a) shows a bundle with straight tubes and two passes on the tubeside. The heating fluid is single phase and problems with film boiling have been predicted in the top section of the bundle.

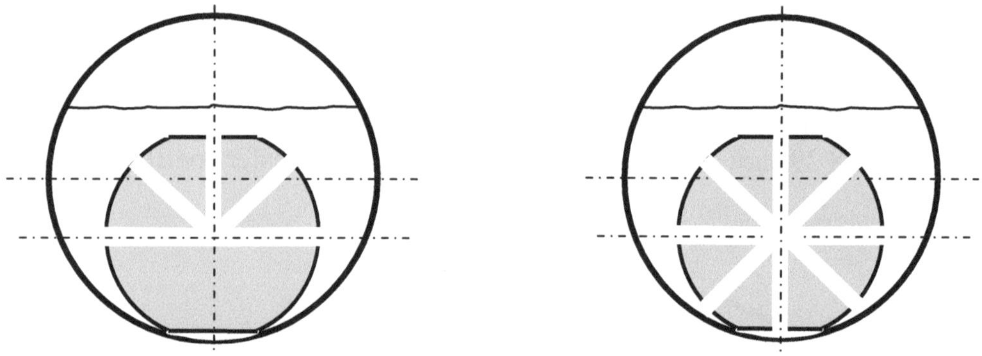

(a) Figure 8.30 (b)

There is already one vapour escape lane due to the tube dislocation required for the pass partition plate. With the normal square pitch layout this will be quite small, (see pass partition lanes in chapter 2). The height of this gap can be increased to make this lane more effective. Tubes can then be selectively removed, (which maintains the cleaning lanes) to form the three additional lanes illustrated. This splits the top of the bundle into four sections which are approximately equal in area.

If the bundle has vertical U tubes any tubes removed from the top of the bundle must also be removed from the bottom which results in the pattern of lanes illustrated in figure 8.31 (b). This results in breaking the bundle up into eight sections which may be even better for releasing vapour but loses twice as many tubes which may not be necessary if it is only the top of the bundle where problems have been predicted.

If vapour lanes are required it is necessary to carry out some calculations to decide how many tubes need to be removed. If the tubes are 19 mm in diameter on a 25.4 mm square pitch the vapour lanes will be as shown in figure 8.31 where the shaded tubes are the ones to be removed.

The two pass lanes at 45° will have a minimum width of $\sqrt{2}\,P_t - d_o = 16.9$ mm. The vertical lane will have a width of $2P_t - d_o = 31.8$ mm. If two lines of tubes are removed the gaps become 35.9 mm and 57.2 mm respectively.

How many lines of tubes need to be removed depends on the size of the bundle and how severe the film boiling warning messages are. I have always resorted to designing lanes which 'look right' and 'feel right' – not a very scientific approach. I would probably compare the combined width of the lanes with and effective diameter across the tubed section but nothing more. I have seen calculation procedures for establishing the number and width of vapour escape lanes but they are quite complex and rely on a lot of assumptions. The assumption plus the general uncertainty of the boiling heat transfer has never tempted me to try and use them and I have not come across any other experienced engineer who has.

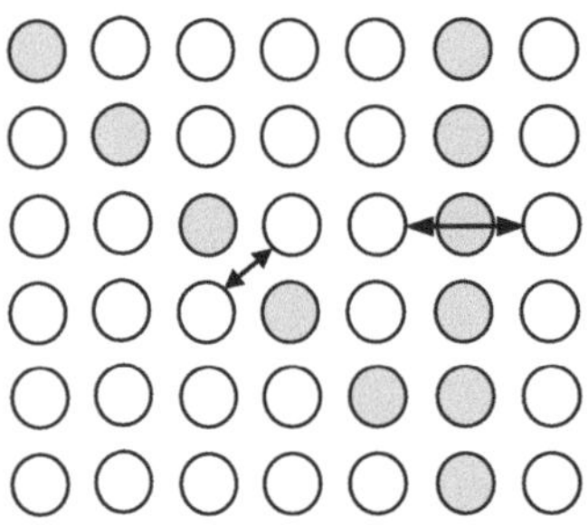

Figure 8.31

Whenever vapour lanes are incorporated into the design of a bundle the bundle size must increase to fit in the required number of tubes. This of course will predict that the problem is slightly worse because of the bundle correction factor and the extra rows of tubes. This puts pressure on minimising the tube losses for the vapour lanes.

As far as I can recall all the units I have come across with vapour lanes have had the 45° passages. During practical design work there is always pressure to get designs out and I have never questioned this approach before. While writing this it has struck me that the number of tubes removed for a 45° lane is about the same as for a vertical lane. However the width of the lanes is quite different. It is also much easier for a vapour to flow vertically upwards than at an angle. Perhaps we should be using a design such as the one illustrated below in figure 8.32.

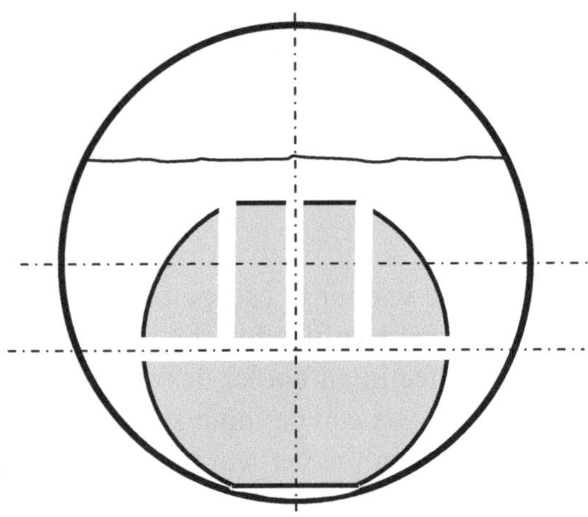

Figure 8.32

I will certainly consider this if I am ever faced with having to design another one. I would also consider extending the central lane to the bottom of the bundle which could be more effective at getting liquid into the centre than losing tubes opening up the horizontal lane across its full length. This would look something like figure 8.33 below.

Opening up the horizontal lane across the central section only seem to nicely balance this arrangement. The tube count for figure 8.33 would not be significantly different from the tube count of figure 8.30. (Removing one row in a vertical lane creates a passage similar in width to removing two rows in a 45° lane.) I think I might try this next time, if there is a next time.

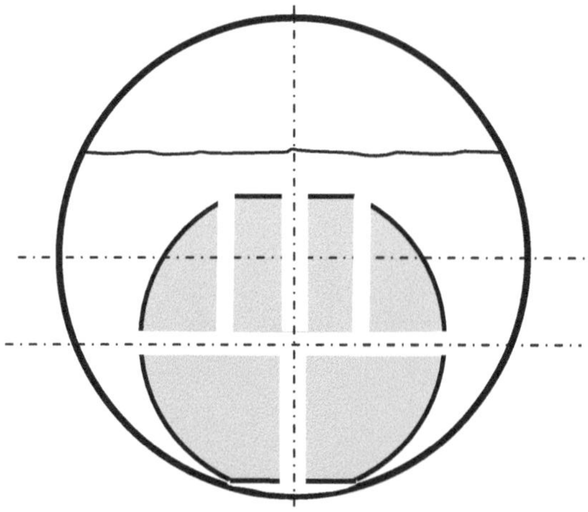

Figure 8.33

This final bundle arrangement has the advantage that it could be used with U tubes provided the bends were horizontal rather than vertical..

With any high flux kettle, even if you are just below the film boiling limit, it is advisable to incorporate a foaming height of say 100 mm in addition to the 50 mm minimum above the tubes when fixing the liquid level.

As with any other dodgy area of heat transfer one obvious solution is to avoid it. For example with a steam heated item reducing the pressure by throttling may be sufficient to avoid film boiling. This of course reduces the MTD and increases the surface area.

An alternative to using vapour escape lanes is to increase the tube pitch. This is a nice simple procedure which is easy to input into a computer program. However this significantly increases the size of the bundle and the bundle correction factor will be acting against you. The basic aim of increasing the tube pitch is to increase the width of the vertical passages in between the tubes to allow the vapour to escape. However this also increases the width of the horizontal passages in between the tubes which is not required for the fluid flow.

A compromise solution which would allow the vapour to escape but halve the extra area required for tubes would be to have **different values for the vertical and horizontal tube pitch.** (This is standard practice in air cooler design work where the horizontal and vertical distance between tube rows can be input separately.) For a kettle reboiler the minimum pitch could be used in the vertical plane and the pitch in the horizontal plane increased as required.

This could eliminate a lot of the guess work and hoping that 'if it looks right it will be alright' out of the design procedure. It would of course need to be incorporated into your computer program and before this could be done a lot of experimental work would be required. This could keep the boys and girls in the labs going for some time.

STATIC HEAD EFFECTS

In any pool of liquid the pressure at the bottom will be higher than the pressure at the surface. The steady increase in pressure from the top to the bottom can be calculated at any point from the equation $\Delta p = \rho gh$ where h is the depth below the surface. If the surface of the liquid is at its saturation temperature, (which it must be with a two phase inlet) then the liquid at the bottom will be below its saturation temperature.

For medium and high pressure units this effect is negligible and will have no effect on the design. For low pressure units it becomes increasingly important as the pressure falls and for vacuum units it becomes a critical aspect of the design. Ignoring static head effects can result in a severely undersurfaced unit.

It was quite a problem taking account of this effect in the days of hand calculation but today a modern computer program will do this for you provided the necessary information is input. It is essential to input two heating curves, one at the pressure above the liquid surface and the other at a higher pressure. The higher pressure curve should be the pressure at the surface plus the maximum pressure created by the static head effect.

If process have only provided one heating curve on the process data sheet you should immediately request another one if they are accessible. If you do not have access to process and the design is only an approximate one for a feed study you may have to fudge a second curve. I explained how to do this in the section on cooling and heating curves in chapter 1. If you do this, remember to put a note on the thermal data sheet so that if the design goes any further in the future it will be picked up and checked.

CALMING BAFFLES

The liquid level in a kettle may be very uneven and unstable particularly in a high flux kettle. If process is going to use the liquid level as a means of control they will require connections above the liquid level and below the minimum liquid level. Calming baffles can be used to isolate a small section of the exchanger where there are no bubbles rising through the liquid and the surface behind the baffle becomes relatively stable. This will normally be specified by process but the thermal design engineer should always check the process and instrumentation diagram, P & I D to see if one is likely to be required.

STEAM GENERATING KETTLES.

Steam generating kettles are similar to the old fire tubed boilers used to generate the low pressure steam which was supplied to reciprocating steam engines. In the old boilers the heating medium was the products of combustion from a firebox in our industry the heating medium could be from a variety of sources.

All heat exchanger generating steam have a number of features in common. One of the most important of these is that they are all very dependant on the quality of the feed water they are supplied with. All feed water supplies contain dissolved minerals and dissolved minerals will produce foaming. Even for relatively low flux designs that are well away from any film boiling limits a minimum of 100 mm should be allowed for foaming.

Whatever the initial amount of minerals in feed water the concentration value will increase in the kettle as the water is evaporated. If unchecked this would eventually completely foul up the exchanger. To counter this it is necessary to fit blowdowns. There are two basic types of blowdown.

Intermittent blowdowns

The main aim of the intermittent blowdown is to remove the sludge of solid particles which can accumulate at the bottom of the shell. A valve is opened at regular intervals or when considered necessary by the plant operators. When the sludge is removed a lot of liquid must also be carried away. For some kettles with a good quality of feed water this may be all that is considered to be necessary.

An intermittent blowdown connection must be fitted at the bottom of the shell as far away as possible from the inlet nozzle. Process should specify the size of this nozzle.

Continuous blowdowns

For lower quality feed waters a continuous blowdown may be required in addition to an intermittent blowdown. The take off point should ideally be placed in the hottest part of the shell where the concentration of dissolved minerals will be at its highest. This is of course is above the centre of the bundle. A sparge pipe arrangement can be used with a lot of holes in the top half of the pipe to minimise vapour entrainment. Any arrangement must not interfere with the bundle being withdrawn. It is advisable to have a liquid head of say 150 mm above the sparge pipe and this may mean fitting it to the side of the bundle rather than the ideal place at the top.

I will not attempt to offer and recommendations for the details of the design because I have only come across a couple of cases myself. If a continuous blowdown is required it will be specified by process and should be fully discussed with them particularly with regard to any vapour entrainment.

Dry pipes

As well as problems below the liquid level in steam generators there are also potential problems above the liquid level where entrainment is more likely than with any other fluid. As mentioned earlier in this chapter in the section in shell sizing there was a safety factor of five built into the standards we used to use for establishing the vapour disengaging space for steam in the days of hand calculations. In addition we also always specified a dry pipe.

A typical dry pipe is shown in figure 8.34. The sketch assumes a short bundle with a length of 3 m or less. The dry pipe always extended for the length of the bundle and we used an outlet nozzle at least every 3 m. For longer bundles there would be either a single dry pipe supplying two nozzles or two dry pipes.

The dry pipe was sloped from the horizontal plane. (The slope has been exaggerated in the sketch.) The idea of the slope is that any liquid particles which were separated from the vapour would run along the bottom of the dry pipe to be returned to the boiling liquid via a spout which would extend to just above the liquid level.

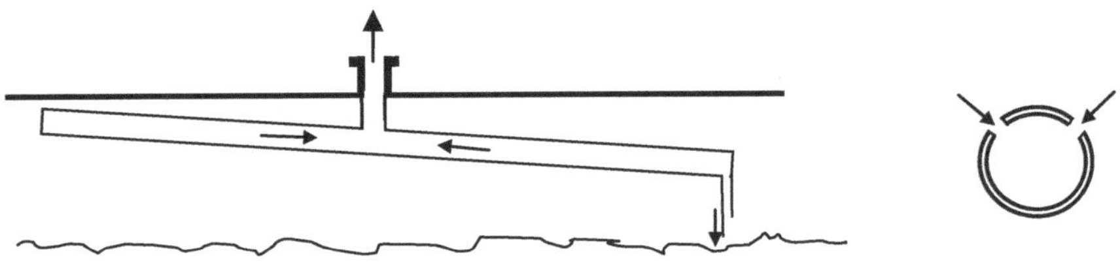

(a) (b)

Figure 8.34

The holes in the dry pipe were 12.7 mm in diameter and at 45° so that the vapour entered from the top as illustrated in (b). Well not quite as illustrated in the sketch, the holes on each side should be staggered so that the discharge from each hole is directed towards the inner wall of the pipe without meeting a jet from a hole on the opposite side. Any drops of liquid will run sown the wall to the bottom of the pipe. To achieve a reasonable distribution the average velocity head through the holes should be at least four times the maximum velocity head in the pipe. The dry pipe seems to have gone out of fashion in recent years in favour of the demister pad.

Demister pads

Demister pads consist of a blanket of knitted wire or plastic mesh held in a flat cylindrical case with cross bars across the bottom to hold the blanket in place.

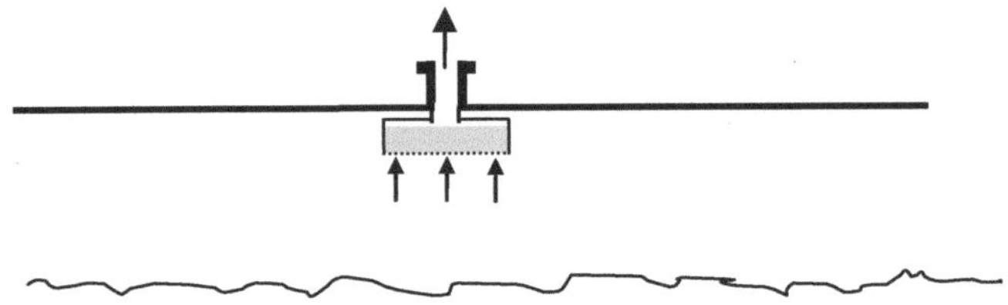

Figure 8.35

The vapour flow is normally upwards as shown in figure 8.35. As the vapour containing liquid droplets enters the mesh it must undergo a series of changes in direction which separates the liquid from the vapour. As droplets of liquid accumulate they increase in size. When they reach a sufficient size they will be able to fall back down the mesh to be returned to the surface of the liquid.

These are proprietary items and the reader should consult manufacturer's catalogues for guidance. It may be necessary to seek their advice on the most appropriate type, size and pressure drop for a specific application.

SINGLE PASS KETTLE DESIGN PROBLEMS

These units can present an interesting design challenge for a thermal design engineer. The procedure outlined below is based on data from an 18.2 MW feed chiller on an LNG plant I once worked on. The feed was condensing inside the tubes and the refrigerant supplied to the shellside had a vapour fraction of 0.4.

I have rounded the data to give numbers which will make it easier to follow and easier for me to present. For example the tubes were 12 m long and I have taken this to be the effective tube length between the inner faces of the tubesheets. I have also plotted the duty on a percentage basis rather than use the actual values from the design.

The MTD at the hot end was about 25 K and at the cold end about 3 K. This resulted in heat flux values of about 15000 W/m^2 at the hot end and less than 2000 W/m^2 at the cold end.

A preliminary computer run to establish the number of tubes required in the bundle was very straightforward.

The next step was to plot the heat release curve as shown in figure 8.36. If you are lucky your computer program will do this for you so it is only a matter of pressing the right buttons and walking over to the print machine to collect it.

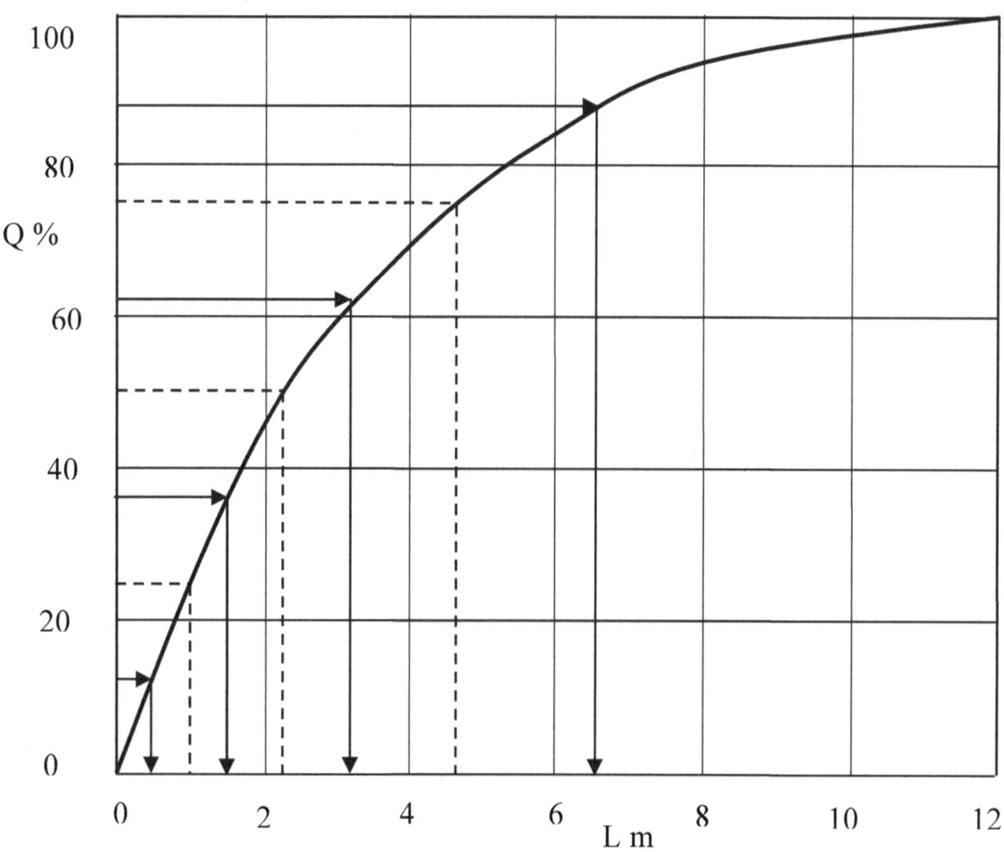

Figure 8.36

There are four vapour outlet nozzles used in the design and ideally each nozzle should be supplied with the same flow rate of vapour. Without taking the flash vapour into account the vapour produced will be proportional to the duty therefore each nozzle will be above a section of the bundle producing 25% of the duty. However each nozzle should receive equal flow rates of vapour from each direction so that the ideal nozzle positions are at 12.5%, 37.5%, 62.5% and 87.5% of the duty. The dividing lines between each section are shown as broken lines in figure 8.36. The lines with arrows indicate the mid point of each section.

In this case the ideal nozzle positions are at; 0.5, 1.55, 3.2 and 6.4 m from the hot end. This would have been the situation if the flash vapour had been separated from the saturated liquid prior to entering the kettle.

This diagram does not take long to draw and it gives an immediate picture of just how uneven the vapour production is along the length of the bundle. The positions of the first two vapour outlet nozzles are not possible with the normal type of kettle construction.

The situation is considerably improved when the flash vapour is taken into account. Only 60% of the total vapour delivered to the nozzles is produced from the boiling liquid the remaining 40% is the flash vapour. Since the vapour produced by boiling is proportional to the duty the vertical scale of figure 8.36 can be changed to percentage of the total vapour flow at outlet as shown in figure 8.37.

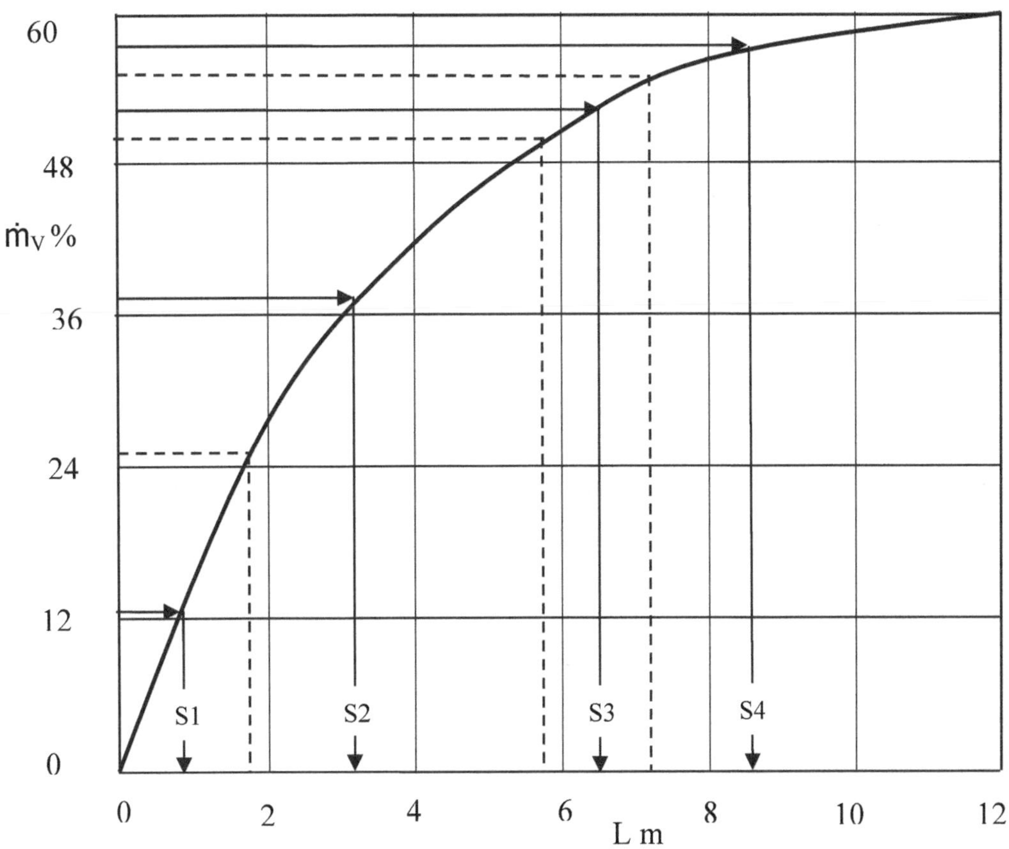

Figure 8.37

To keep the flow rates of vapour to each nozzle equal the two nozzles at the hot end must be supplied with all their vapour from the boiling liquid. The dividing lines identifying the first two sections are therefore at 25% and 50% of the total vapour generation rate. If the flash vapour is supplied equally to the two nozzles at the cold end each one will handle 20% flash vapour plus 5% of the vapour produced by boiling. The dividing line between these will be at 55% of the vapour produced by boiling.

It is convenient to label the vapour outlet nozzles as S1, S2, S3 and S4 starting from the hot end. The ideal positions for the nozzles are at 12.5%, 37.5%, 52.5% and 57.5% of the vapour generated as indicated by the arrowed lines.

It is interesting to note the difference in the lengths of the cold end sections served by nozzles S3 and S4, (1.5 m and 4.8 m). This is because of the sharp change in direction of the heat release curve in this region. It would not be practical to try and supply 50% of the flash vapour to one very long and one very short section. Since the vapour supplied to each nozzle is 80% flash vapour it is sensible to even out these two spaces and accept that there will be a small distortion of the entrainment calculations.

Taking account of the uneven distribution I extended section 3 to 2 m with the nozzle at the centre i.e. 5.3 m from the cold end. For section 4 I positioned the nozzle at 3 m from the cold end, (which are 9 m and 6.7 m from the hot end.)

Nozzle S2 is not a problem and can remain at its ideal location of 3.2 m from the hot end.

Nozzle S1 is a problem. Whilst carrying out this design it was necessary to have discussions with the mechanical engineers about the construction details. At an early stage it was clear that dished ends would be required and I was anticipating a problem with the location of S1. The ideal position of S1 was about 0.9 m from the hot end and the minimum distance was 1.6 m from the hot end. There is nothing that can be done about this and 1.6 m had to be used which would mean most of the vapour arriving at the nozzle from one direction only since this section was only 1.8 m long.

To allow for the minor distortions at the cold end and the major one at the hot end I added about 75 mm to the minimum shell ID required for even entrainment distribution. I then checked the entrainment in each section and compared them with the overall target of 0.005. (I checked the entrainment in each direction separately for sections 1 and 2.) Eventually I was reasonably satisfied with the nozzle positions which are summarised below in figure 8.38.

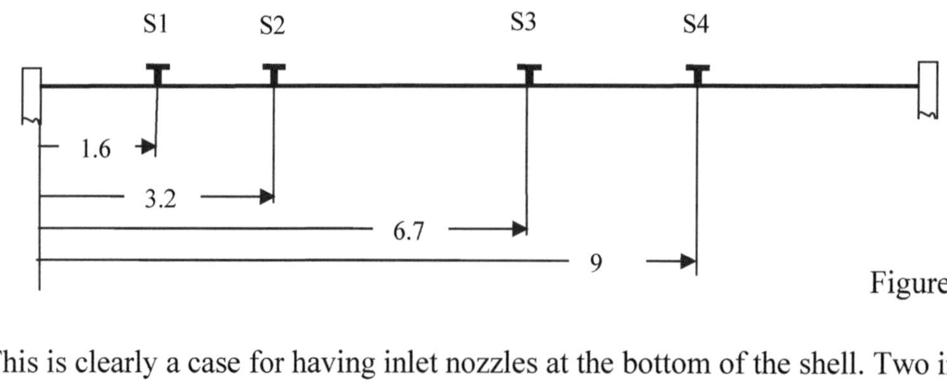

Figure 8.38

This is clearly a case for having inlet nozzles at the bottom of the shell. Two inlet nozzles were used and with a dished end construction there was plenty of room to fit the nozzles below the bundle together with distributors. The two inlet nozzles were directly below nozzles S3 and S4 and the distributors were 2 m long.

CHANNELS

It is worth a brief mention of the channels required for single pass kettles with condensing inside the tubes. The inlet channel has a central axial nozzle as illustrated in figure 8.39. This is constructed of a conical section with angles of 30^0 to the horizontal sandwiched between two parallel sections which join the two flanges.

The outlet channel has an offset nozzle so that the bottom of the channel is flush with the bottom of the connecting outlet pipe. This is illustrated in figure 8.39. The sloping section of the offset cone in at 30°. The outlet nozzle is also normally smaller than the inlet nozzles and the combined effect of the offset and the smaller nozzle to make the outlet channel considerably longer than the inlet channel. This construction is essential to prevent flooding of the lower tubes.

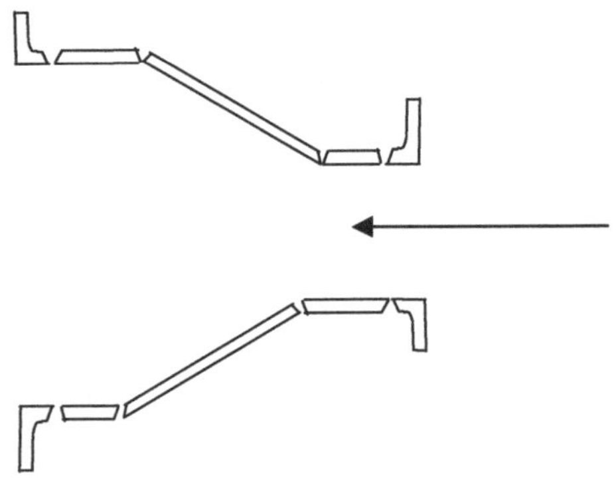

Figure 8.39

The methods of sizing individual sections which make up the nozzles have been covered in chapter 2. However I would always check with a mechanical engineer for the type of flanges that are appropriate and to make sure that my calculations for the parallel sections are reasonable.

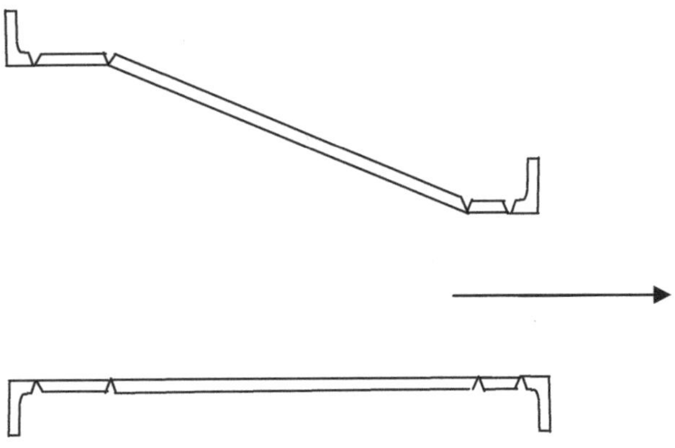

Figure 8.40

A thermal designer should be prepared to spend a lot of time designing this type of unit and at an early stage it is worth checking if this has been understood by others and that sufficient time has been allocated. They can be very expensive units. Doubly so in this case since there were two feed trains so that each exchanger designed would be duplicated.

CONCLUSIONS

Pool boiling exchangers offer more scope for innovation in the details of the mechanical construction than any other type of shell and tube unit. I have actually used or seen used all the methods of introducing a two phase fluid into a shell which I have described and I have no doubt that there are many others which I have never come across.

My venture into ways of avoiding vapour binding are not based a very extensive experience but I am quite pleased with my final offering in figure 8.33 which might be described as a tree shape vapour escape lane pattern and could be used with straight or horizontal U tubes. No doubt there will have been many other people somewhere who have used this. If they have I would like to hear from them.

Well I have had quite enough of large kettles for now and I think it is time to turn on the relatively small one in the kitchen.

HORIZONTAL THERMOSYPHONS – CHAPTER 9

Introduction

I started off the chapter on pool boiling exchangers by explaining that the basic function of any type of reboiler is to provide a heat load to a fractionating tower. The kettle exchangers also had other applications. The thermosyphon reboilers are mainly used in conjunction with fractionating towers but also have other applications.

In a horizontal thermosyphon the boiling normally takes place on the shellside of the exchanger and in a vertical thermosyphon the boiling normally takes place inside the tubes. I have guardedly used the expression 'normally' again because there may be the occasional exception.

The choice between using a vertical and a horizontal type of thermosyphon is down to process but the thermal design engineer may become involved in this decision. There may also be occasions when it is advantageous to change from one type to the other.

The key feature which distinguishes the thermal design of the thermosyphons from the thermal design of a kettle reboiler is the pipework connecting the exchanger to the tower. The kettle reboiler was designed by the thermal design engineer and the pipework by the process engineer. In any thermosyphon design the exchanger and the connecting pipework form a single hydraulic system and should be designed as such by the thermal design engineer.

TEMA H & G TYPES

Although the H and G type of exchangers have other uses their main application has traditionally been for horizontal thermosyphons. Figure 9.1 shows the basic flow patterns through a G type unit. The flow paths are established by a horizontal baffle which is fitted on or near to the centre line of bundle. If the tube layout precludes a central position it should be just below the centre line since the volume flow rate above the baffle is greater than the volume flow rate below the baffle. There is no need to provide any sealing device along this baffle.

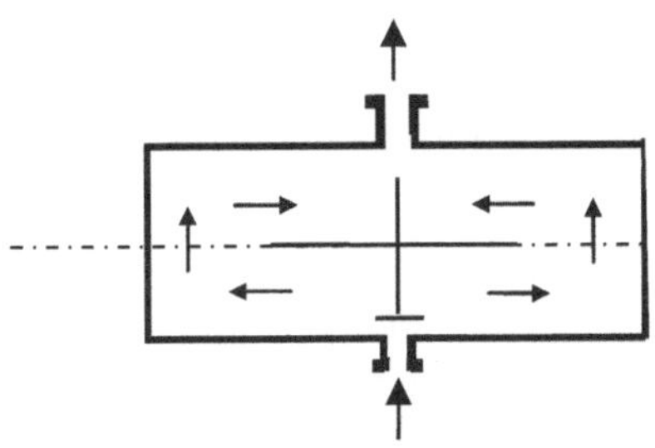

Figure 9.1

Figure 9.2 shows the basic flow patterns through an H type unit. As far as the shellside flow paths are concerned this is simply two G types joined together in the same shell. Thermodynamically they are not quite the same particularly if the hot fluid is single phase or condensing with a significant change in temperature. Any difference in the MTD and the tubeside heat transfer coefficients in the two halves will result in different rates of vapour generation in each section.

Assuming that there are two tubeside passes any differences are usually quite small and become even smaller if there are more than two tube side passes. However to cater for any slight differences in vapour generation it must help to clip the top off the central support baffle just above the top tube row as shown in the diagram. This may contribute to evening out the flows to each outlet nozzle.

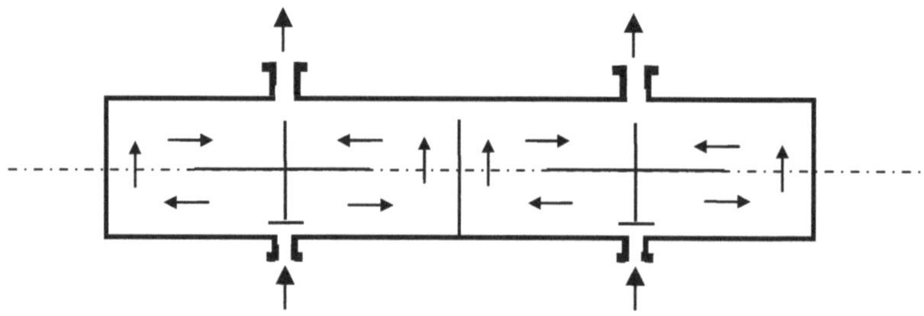

Figure 9.2

Horizontal thermosyphons are less likely to foul than kettle reboilers because of the presence of the circulating liquid. The surface area which can be provided is normally more than for a vertical thermosyphon unit and is unlimited when multiple shells are used. They do take up considerably more plot space than vertical units not only because of the length of the units themselves but they also need bundle withdrawal spaces for cleaning and maintenance. This may lead to them being sited some way from the tower which complicates the pipework.

HISTORICAL BACKGROUND

When I first started designing heat exchangers at Kellogg all horizontal thermosyphon units were either H or G types. The TEMA maximum unsupported length of about 1.5 m, (5 ft) for 19 mm tubes governed which type to select. Since we also kept to standard tube lengths it would invariably be 6 m tubes for H types and 3 m tubes for G types.

The horizontal baffles were always 1/3 of the effective tube length for the G type and 1/6 of the effective tube length for the H type. (Shorter than those illustrated in the diagrams which are about 1/2 and 1/4 of the tube length.) These were perforated with 6.4 mm, (0.25 inch) holes on a 32 mm, (1.25 inch) pitch. The idea of the holes was to ensure that there were no pockets of vapour in the corners under the baffle and to ensure that liquid would reach the corners above the baffle.

With such short horizontal baffles there was very little resistance to fluid flow and the heat transfer calculations were essentially the same as those we used to size the bundles in kettles. Boiling and natural convection coefficients were calculated and weighted to produce a shellside coefficient. We assumed a nominal value for the frictional pressure drop through the bundle of about 0.01 bar.

The inlet nozzles were sized for a maximum velocity of 0.9 m/s so that the momentum fluxes were very low and no impingement plates were used. The main pressure drop through the unit was therefore in the outlet nozzles and the resulting overall pressure drop was very low.

Because the frictional pressure drop through the exchanger was so low, it was possible to carry out the system hydraulic calculations independently. The thermal design was therefore by a heat exchanger engineer and the pipework by a process engineer.

When I moved on to set up a new thermal design section at Stone and Webster I decided that a thermal design engineer should do the hydraulic calculations for the horizontal thermosyphons as well as for the vertical thermosyphons. This was not too difficult a change since we were a group within the process department. I was accused of 'empire building' and there may have been some truth in that since I was very ambitious in those days.

SYSTEM HYDRAULICS

Horizontal thermosyphons are now always designed by a heat transfer engineer as an integral part of a single hydraulic system. Before starting any work on these units it is essential to get a copy of the tower drawing which shows the pipe connection points and the liquid level in the tower. The designer should be looking to identify the vertical distances between the liquid level/s they will be working with, the centre lines of any horizontal nozzles and the distance to the bottom of the tower where any vertical pipes are involved in the circuit. The basic systems are as follows:

ONCE - THROUGH REBOILER

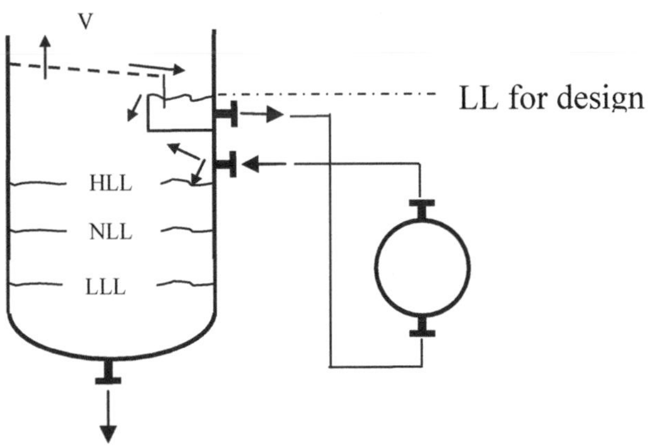

Figure 9.3

In this arrangement the feed to the reboiler is taken from a small open tank or bucket inside the tower. (It may be on one side as shown or in the centre of the column.) The liquid falling down the tower is directed into the tank. This maintains a liquid level at the top of the tank with some of the liquid spilling over into the bottom of the tower. A liquid level is maintained in the bottom of the tower to control the flow rate of the bottoms product. Normally three liquid levels will be shown; a high liquid level, HLL, a normal liquid level, NLL and a low liquid level, LLL.

The liquid feed supply to the reboiler is taken from the bottom of the overflow tank. The two phase return from the reboiler is discharged through the tower return nozzle to a space in the tower above the HLL. The liquid from the reboiler return falls to the bottom of the tower with the overspill. The vapour passes up the tower through a tray which is fitted with caps which allow the vapour to pass upwards while collecting the liquid on the surface of the plate. Figure 9.4 shows a simple cap arrangement.

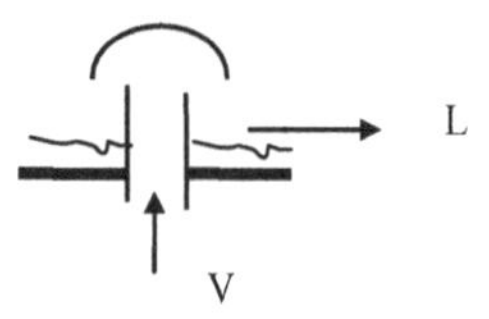

Figure 9.4

The liquid level for the reboiler design is the top of the overflow tank. **The liquid levels in the bottom of the tower are irrelevant to the thermal design of the reboiler**.

A variation on this arrangement is **when there is no overflow** from the collecting tank and all the liquid passes into the feed line to the exchanger. The liquid level for the thermal design may then be part way down the vertical feed pipe to the exchanger. This is a tricky situation and a design **using the maximum head may be unsafe**. It is therefore necessary to have runs with a range of heads below the maximum value and then make an assessment at to where the liquid level is likely to be in the pipe. As a rough guide this is normally about a third of the way down the pipe.

RECIRCULATING REBOILERS

Figure 9.5 below shows one type or recirculating reboiler. Instead of having a relatively small overspill tank there is large vertical plate which divides the bottom of the tower into two sections.

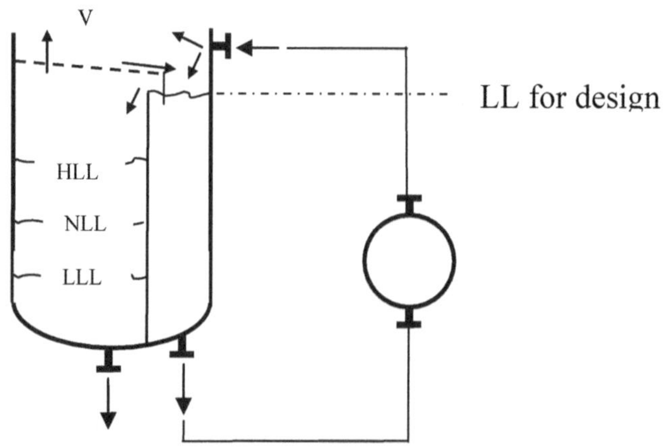

Figure 9.5

The diagram shows the section with the connection to the reboiler being the smaller of the two. This is not always the case. For this arrangement there is a large volume of liquid being recirculated and the liquid head can safely be taken as the top of the dividing plate, or the bottom of the weir if this is part of the dividing plate.

As with the once through reboiler the three liquid levels shown on the tower drawing are irrelevant.

Figure 9.6 below shows a fully recirculating reboiler where there is only one compartment at the bottom of the tower.

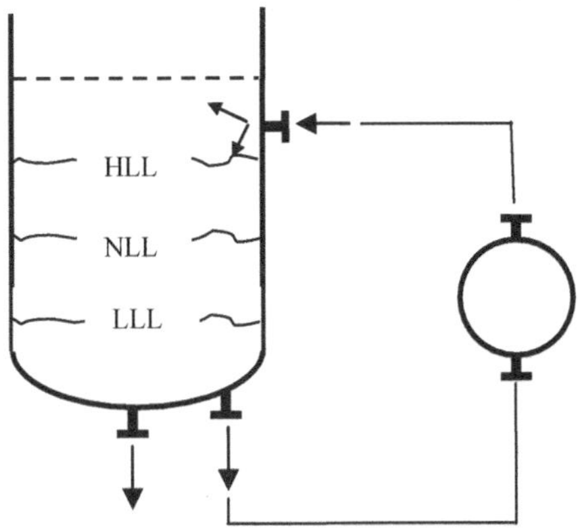

Figure 9.6

All three liquid levels must now be taken into account. The general approach is to design for the normal liquid level and then investigate the effect on the vaporisation rate and surface area when the reboiler is operating at the two extreme limits.

PUMP THROUGH REBOILER

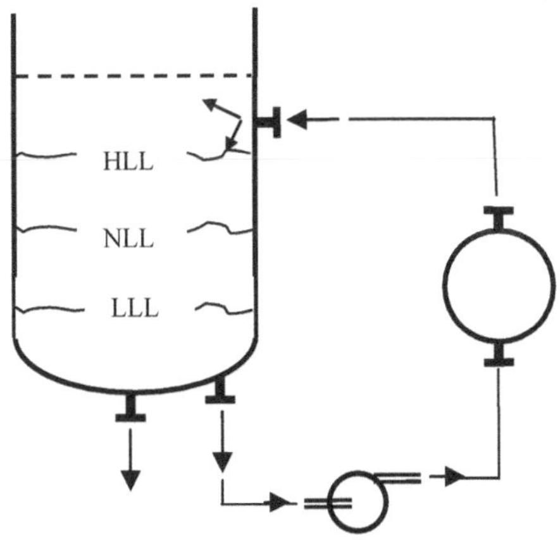

Figure 9.7

In this arrangement a pump is introduced into the system. The fluid flowing around the tower may be on either the shell side or the tube side of the exchanger. This type of design is used where the fluid is too viscous for a satisfactory circulation by gravity. The most likely TEMA type is an E which only requires a single inlet and out let nozzle. The cost of the pump and the power it consumes make this an unattractive but sometimes necessary option. Most viscous fluids require regular cleaning of the heat transfer surface. It is therefore more likely that it will be on the tubeside although a smaller heat exchanger will probably result from switching the fluids around.

For boiling on the shellside in this situation, (starting with all liquid), I would always use horizontally cut baffles, probably with variable pitched baffles since the pressurised liquid is unlikely to boil for a significant section of the exchanger. This avoids phase separation which could result in the top of the exchanger becoming ineffective. To minimise fouling all liquid heating may be used. The pressure from the pump is increased and a valve is fitted in the outlet pipe adjacent to the tower return nozzle. The compressed liquid is throttled through this valve producing a two phase flow to the tower. For this situation it is important to ensure that there is no boiling in the exchanger.

TOWER DRAWINGS

A heat transfer engineer should become familiar with tower drawings and recognise what he/she is looking for. There are lots of variations on the basic types illustrated above. Towers have lots of connections and details of internal components which are irrelevant to the design of the heat exchanger. It is often necessary to calculate the required dimensions from other dimensions. When there is any doubt about a dimension it should be discussed with the process engineer. Vertical dimensions on a tower drawing are normally given by reference to the tan line of the dished end at the bottom of the tower. These should be transposed to dimensions above grade.

As well as sorting out the vertical dimensions from sectional views a plan view of the tower will give the orientation of the nozzles on the side of the tower. In the above diagrams these have all been assumed to be in the same plane. They may be in quite different positions as illustrated in figure 9.8.

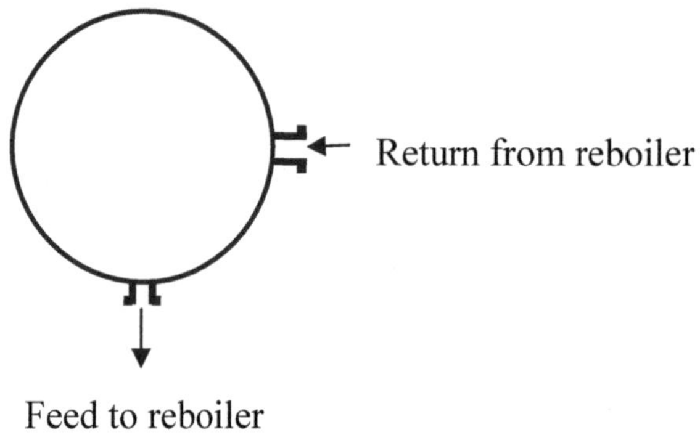

Figure 9.8

The orientation of the nozzles is essential to getting a picture of the probable lengths of pipes and more importantly the number of bends in each section. Piping runs can not be finalised until the heat exchanger is designed and the piping stress engineer has checked everything out. However it is sensible to get as close as possible to the final layout at an early stage.

ELEVATION DIMENSIONS

The dimensions taken from the tower drawings are required to calculate two crucial inputs to your computer program. These are the static head which provides the driving force for the circulation and the vertical height between the top of the exchanger and the centreline of the tower return nozzle. These are illustrated below for a simple recirculating system. Only the normal liquid level is shown.

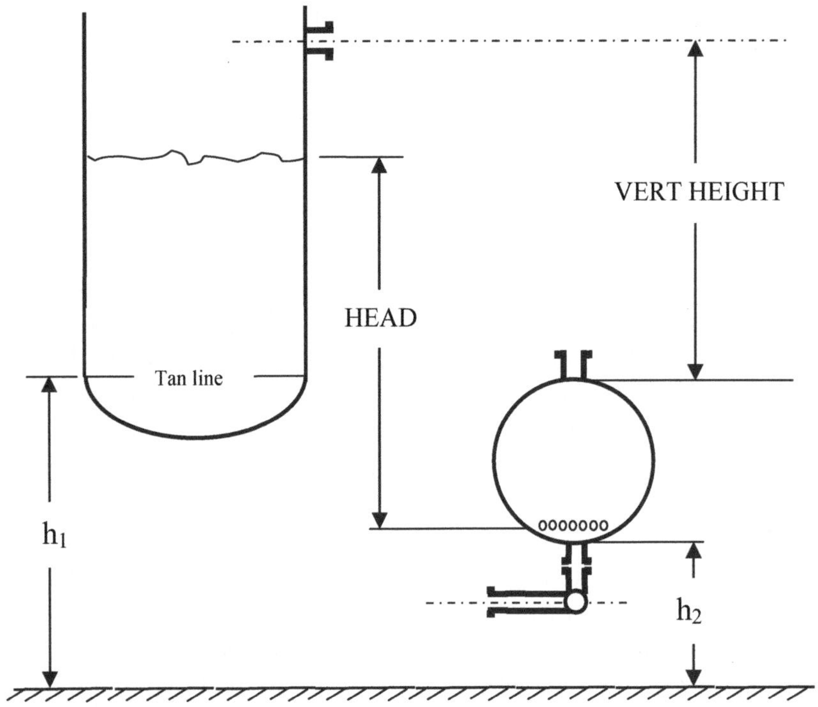

Figure 9.9

To maximise the static head the shell should be as close to grade as possible. The distance between the bottom of the shell and grade, h_2 is calculated as follows:

$$h_2 = \text{Shell thickness} + \text{Nozzle projection} + \text{Pipe flange length} + \text{Bend radius} + \text{Pipe flange radius} + \text{Clearance} \qquad [9.1]$$

The shell thickness need only be an approximation and the pipe data can be taken from manufacturers catalogues.

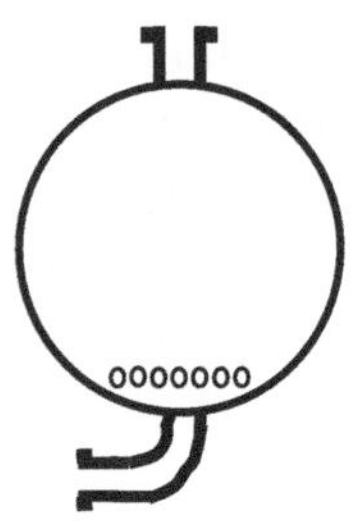

Figure 9.10

The height between the bottom of the shell and the centreline of the bottom pipe can be reduced by incorporating the bend into the feed inlet nozzles as illustrated in figure 9.10 but this is not standard practice.

Although long radius bends should always be used whenever possible for the outlet pipework short radius bends may be used in the inlet pipework.

The clearance between the pipe flange and grade should be at least 50 mm for a very small shell and would normally be about 100 mm. If in doubt consult a mechanical engineer.

The vertical height above the exchanger and the static head can now be calculated as follows:

$$\text{Vertical height} = \text{Return nozzle elevation} - h_2 - \text{shell ID} \qquad [9.2]$$

$$\text{Static head} = \text{LL Elevation} - h_2 - \text{height to lowest tubes} \qquad [9.3]$$

The height to the lowest tubes is the height from the bottom of the lowest tube row to the bottom of the shell. This height is often ignored but to omit it is an unsafe assumption. It takes very little time to take this dimension off the tube layout diagram. It is often quite small but can become more significant for a U tube bundle with vertical bends where the area unavailable for tubes at the bottom is the same as at the top which is controlled by the size of the outlet nozzle.

All the descriptions and diagrams so far have assumed that the reboiler is at grade as an alternative to a kettle. This is not always the case. Both horizontal and vertical thermosyphons can be used to supply a heat load part way up the tower. If the tower return nozzle from the reboiler it at an elevation of say 20 or 30 m above grade having the reboiler at grade would result in very severe problems in sizing the outlet lines. These exchangers are therefore often mounted on a structure above grade.

Normally these structures will be supporting lot of other items of equipment on a series of platforms at different heights above grade. If this is the case preliminary drawings of the structure must be accessed to see where it is proposed to fit in the reboiler. A preliminary computer run making very general assumptions about pipework etc will give an idea of the size of exchanger required.

There are far more potential problems associated with 'pulling a bundle in the air' to pulling a bundle at grade. The maximum length and weight limits could therefore be a lot lower. Your preliminary size for a single shell may clearly now require two shells (severely complicating the pipework or requiring an extra nozzle on the tower.). If only one shell is shown on the drawings it is time to start moving hardware or revising the bundle size limits and strengthening the structure. The thermal design engineer, the process engineer and whoever is responsible for the layout and structure should all be involved in determining the most economic solution to this complicated problem.

SYSTEM PRESSURE DROPS

The driving force for a thermosyphon system is provided by a static head of liquid. A stable flow rate of fluid around the system will be established when the pressure from this static head exactly balances the sum of all the pressure drops around the system. Starting at the point where the liquid feed leaves the tower and ending where the two phase mixture returns to the tower these resistances are:

(a) A contraction loss.
(b) Losses in the inlet pipework due to friction and discontinuities including bends and any T junctions or valves.
(c) A nozzle loss at the heat exchanger inlet.
(d) A friction loss through the bundle.
(e) An acceleration loss due to the change in density of the fluid through the bundle.
(f) A static head loss across the bundle.
(g) A nozzle loss at the exchanger outlet.
(h) Losses in the outlet pipework due to friction and discontinuities including bends and any T junctions or valves
(i) A static head loss in any vertical pipe
(j) An expansion loss.

Losses (b) and (h) will simply be listed as frictional losses in a computer output because part of the internal design procedure will be to convert discontinuities into equivalent lengths of pipe.

With a low pressure unit with significant flashing in the outlet line an extra acceleration term could be added to further complicate the situation.

MIXED DENSITIES FOR STATIC HEADS

The pressure drop due to a static head in two phase flow can be found from the following equation:

$$\Delta p = \rho_{mix}\, g\, h \qquad [9.4]$$

Some of the equations I have seen used to calculate ρ_{mix} from inlet and outlet densities ρ_1 and ρ_2 are listed below:

(1) $\rho_{mix} = (\rho_1\, \rho_2)^{0.5}$ [265]
(2) $\rho_{mix} = (\rho_1 - \rho_2) / \ln(\rho_1/\rho_2)$ [308]
(3) $\rho_{mix} = (\rho_1+\rho_2) / 2$ [400]
(4) $\rho_{mix} = 2 / (1/\rho_1+1/\rho_2)$ [175]
(5) $\rho_{mix} = \ln(\rho_1/\rho_2) / (1/\rho_2-1/\rho_1)$ [227]
(6) TEMA Tentative

The numbers in the square brackets are the result of using 700 kg/m^3 and 100 kg/m^3 for ρ_1 and ρ_2. The differences are interesting and highlight the potential errors in calculating a static head in two phase flow.

For hand calculations it was not unusual to work out values by more than one method and take the more conservative one but the simple arithmetic density (3) is clearly far too conservative.

I have not attempted to detail the TEMA Tentative method, (although I have used it) because it involved the use of a chart which took into account the temperature differences across the bundle.

Modern computer programs are likely to use mixed densities based on void fractions calculated from very complex empirical equations. The reader should check their manual if they wish to find out what is inside their program.

PIPING PRESSURE LOSSES

Only a small percentage of these loses are normally due to fluid friction in the pipes, most of them are discontinuity losses. There are two possible ways of combining frictional and discontinuity losses. Either the frictional losses must be converted into equivalent velocity heads or the discontinuity losses must be converted to equivalent lengths of pipe. The latter method is always used in this type of work. It has the advantage that it is easy to visualise a discontinuity such as a bend in terms of the length of pipe which would have the same pressure drop.

Table 13 in useful tables illustrates two approaches for finding equivalent lengths for fittings and other discontinuities. Table 13A gives values of equivalent lengths in meters for bends, valves and T junctions. I converted this table to SI units from a well established table which has been used for many years.

Table 13B is for changes in diameter of pipe sections. It can be used with any system of units because the table values are a ratio of the equivalent length of pipe to the smaller of the two diameters.

The only resistances not catered for in these tables are the inlet to the pipework from the tower and the exit from the pipework back into the tower. When we had to use hand calculations these were taken to be half a velocity head and one velocity head respectively. The inlet loss was a small part of the total circuit loss because the velocity of the liquid would be quite low. The exit loss however could be quite significant because the velocity of the two phase flow is a maximum at this point. However since the return flow is drastically reduced in velocity as soon as it enters the tower there will be some static regain so a judgement had to be made as to how much of the velocity head should be counted as a pressure loss. The reader who will be using a computer program does not have this problem even a relatively simple program should take account of the inlet and exit loss so that these items should not be included in the pipework. (Check your manual to be sure of this.)

HEADERS FOR H TYPES

When an H type shell is used for a none thermosyphon application the thermal design engineer just designs the exchanger and leaves it at that; someone else will design the pipework and take into account the pressure drops. When a single hydraulic system needs to be analysed life is not so easy.

The exit pipe from the outlet manifold is normally a larger diameter than the two outlets from the exchanger because it carries twice the flow rate. The T section in the manifold will have the same diameter to all connecting parts. For minimum hydraulic resistance between the exchanger nozzles and the T section the construction should be as illustrated in figure 9.11.

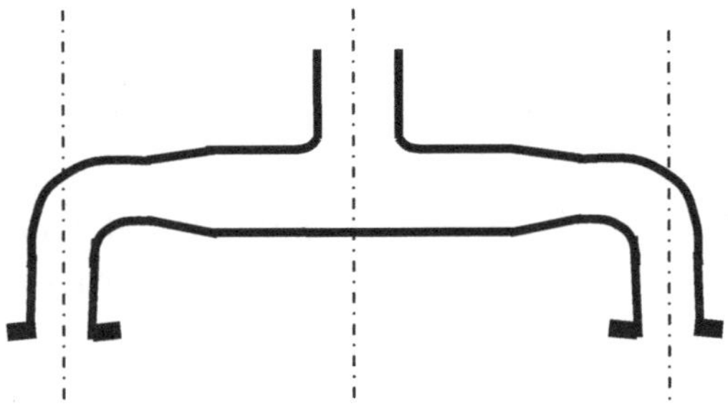

Figure 9.11

The components are a nozzle to match the exchanger nozzle, a long radius bend and an expander section. This is fine for most systems. However for low pressure systems with very high volume flow rates the size of the fittings increases and it may not physically possible to fit these sections in. If the reader does not already have their own collection of catalogues they need to borrow some from a mechanical engineer and check out the minimum distance between the centre line of the exchanger nozzle and the centre line of the outlet pipe. (This problem is now more likely to occur since so many clients have started to specify 80% of the TEMA value for the tube maximum unsupported length.) If the length required for the ideal fittings is too long the next thing to try are short radius bends instead of the long radius bends. If this is still not sufficient it is necessary to use another hard T junction as shown in figure 9.12.

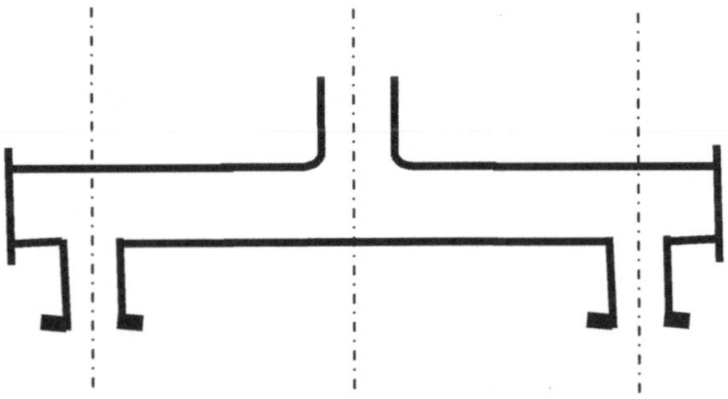

Figure 9.12

When the arrangement shown in figure 9.11 is used a drain should be specified in the header opposite the outlet from the T. This can be avoided if offset expanders are used but these require extra length.

OTHER TEMA TYPES

Since the advent of the more sophisticated modern computer programs there is a lot more flexibility for designing horizontal thermosyphon reboilers. For example the problem with length limitations controlled by the maximum unsupported length can be overcome by fitting support baffles which the program will take into account. The length of the horizontal baffle in the H and G type is now a variable and not a nominal fixed length. Increasing the length of the horizontal baffle will help to ensure that fluid ‘sweeps’ the face of the tubesheets. This should not be overdone and a guideline is to ensure that the crossover section has at least the same flow area as the long flow section underneath the baffle.

The H and G types are no longer the only TEMA types which can be used. Particularly if there is a substantial liquid heating component in the duty it is often advantageous to use up some of the system pressure drop in improving the shell side heat transfer coefficient. In this situation TEMA E and J types may now provide the optimum design. If a J type shell is used it is tempting to automatically use a 12J arrangement which minimises the loss of area unavailable for tubes at the top of the bundle and hence maximises the tube count for a particular size of shell. This of course complicates the outlet pipework and I have normally found that using a 21J, so that there is only one outlet pipe to deal with, is more satisfactory. The TEMA X type will give the lowest pressure drop and is most likely to be used where vibration problems have been predicted for other types.

TUBE SIDE VAPORISATION

I have only designed one horizontal thermosyphon reboiler which required the boiling fluid to be on the tubeside. When I was first given the process data sheet my immediate reaction was to go and see the process engineer and point out the mistake. I was assured that the process data sheet was OK and that a similar design had been used on a previous plant.

I am sure that I must have kept a copy of such an unusual design but it remains hidden in my chaotic filing system, it will probably emerge while I am working on the index. I remember it was only a very small unit and sited at grade and there were very long vertical pipes necessary to reach the tower connections. The heating fluid was a very viscous liquid with only a small allowable pressure drop. If the heating fluid had been inside the tubes a very much larger exchanger would have been required.

I had to design two exchangers and iterate between the two deigns. The first design was a 'normal' heat exchanger with the fluids as on the PDS. The second was a horizontal thermosyphon system with the recirculating fluid on the shell side and a fudged coefficient input for the tube side. After the final runs the sum of all the pressure drops through the shell side of the thermosyphon was approximately equal to the pressure drop through the tube side of the other design. Surprisingly there were no problems with the pipework. Often with small units an ideal size for the return pipe is not an option. Say a four inch vertical return pipe can be well over the top on pressure drop whilst severe instability problems are predicted for a six inch pipe.

VAPORISATION RATES %V

These are normally defined as the percentage of vapour produced by the thermosyphon system for the two phase stream returning to the tower. It is expressed on a mass basis therefore the following equation applies:

$$\text{Vaporisation rate} = \frac{\dot{m}_V}{\dot{m}_V + \dot{m}_L} \times 100 \qquad [9.5]$$

It is therefore simply the mass fraction/dryness fraction expressed as a percentage. There is another term which is often used for the same purpose:

RECIRCULATION RATIO R

This is defined as ratio of the mass of liquid compared with the mass of vapour of the two phase stream which is returned to the tower, therefore:

$$\text{Recirculation ratio R} = \frac{\dot{m}_L}{\dot{m}_V} \qquad [9.6]$$

The numerical values of these two quantities are compared in the table below:

R	% V
0.5	67
1	50
3	25
5	17
20	5

Table 9.1

This is a situation which often crops up in engineering. There are two different quantities which have been defined to serve the same purpose. At least in this case there should never be any confusion about the numbers; (unlike relative humidity and percentage saturation for example.)

Most people will tend to think in terms of one quantity or the other. Personally if I see a recirculation ratio of say 3 in a document I will immediately convert this to 25% vaporisation as I visualise what is required. Others may do the opposite. I can see the logic of recirculation ratios since carrying liquid around the system is what distinguishes the thermosyphons from the kettle but I will stick with percentage vaporisation in this book.

An important point I must emphasise it that the percentage vaporisation rate refers to the fluid returning to the tower. The value leaving the exchanger will always be lower. This is because the pressure loss in the outlet pipework will always produce some flash vapour. For most designs very little flash vapour is produced and the difference between the percentage vaporisation leaving the exchanger and entering the tower will be very small but for low pressure units with a high pressure drop in the outlet pipework it may be very significant.

LIMITING VALUES

The theoretical lower limit is zero when there is no recirculation. This will occur at start up for most units. The theoretical higher limit is 100% when something has gone badly wrong, such as a valve in the inlet line closing and the unit will temporarily act like a kettle.

Limiting values for design cases are difficult to establish and give guidelines for. There must be sufficient vapour generated to recirculate the fluid. However too much vapour will lead to the heat transfer surfaces drying out and boiling will cease. This will lead to an increase in the tube wall temperatures which in some cases could lead to tube failure.

Dry out of the heat transfer surface will take place with a lower heat flux inside tubes than outside the tubes in a bundle. It is therefore possible to use a higher vaporisation rate in a horizontal unit than a vertical unit for any specified fluid.

For horizontal thermosyphons I have read about maximum limits of between 25% and 50%. 50% is not unusual and I have designed units with over 50%. Unfortunately I have not kept a record of the highest one. Also whatever the design value may be there will be off design cases when it will increase. When reading papers or articles which recommend maximum limits it is important to scrutinise the background to the article. What may be a perfectly sensible limit for one type of tower may be too conservative for another type of tower. The key variables to look out for are the operating pressure and the average molecular mass of the vapour. These determine the density of the vapour component and the void fraction of the mixture.

As with most two phase flow problems I would always work out the vapour and liquid volumes assuming a homogeneous mixture at an early stage. It then helps to reinforce the 'picture' if you take a piece of normal graph paper, select one of the squares and fill in bits to represent the liquid component. If you do this it is easy to see why a low pressure steam generating reboiler may be limited to a maximum percentage vaporisation of only say 5%. At the other extreme I am sure that somewhere there will be a reboiler in a high pressure tower breaking down heavy hydrocarbons which is operating with a vaporisation rate of 70% without any problems, (other than needing very regular cleaning?)

Your computer program should of course do a lot of the thinking for you by giving warning messages if there is any danger of the tubes drying out. Also the process engineer will specify the vaporisation rate but there may be times when a small change will help with the design and it is important for the thermal design engineer to understand what is taking place before suggesting this.

EFFECT ON THE PRODUCT

Whatever the value of the percentage vaporisation rate specified on the process data sheet by the time the pipework has been finalised and the elevations fixed the actual value under design conditions will never be quite the same.

The value on the process data sheet is basically a 'target' value to be aimed for in the same way that we may target a design for say 5% oversurface but accept anything between 4% and 8% for a reasonably sized unit and perhaps over 20% for a very small unit.

The effect of changing the vaporisation rate by say 5% may have a negligible effect on the composition of the fluid returning to the tower for one design but a significant effect on another design. A key factor is the nature of the boiling fluid particularly the boiling range. I had thought – only very briefly, of delving into this aspect and offering some guidance. However there is such a wide range of possible process conditions and fluids that this must be left to the process engineer. Before I start work on any thermosyphon I like to discuss the operation with the process engineer. A key question is always how 'sensitive' is the tower operation to changes in the vaporisation rate in the exchanger. In most cases the value on the process data sheet is a maximum and a decrease of 5% or even more will not be a problem. This immediately gives the thermal design engineer a 'target range' rather than a target value to aim for. When this has been discussed with the process engineer I will occasionally state target and predicted values on the thermal data sheet. (Say target 40% and predicted 36%). The drawback to this is that it may not be 'politically correct' because data sheets are often scrutinised by people who do not understand anything about the design of thermosyphon systems.

Exceeding the process data sheet value by more than one or two percent is something which is most likely to crop up during a revamp design. All the hardware is fixed but the process has been changed and this might include the heating medium. There is no need for any assumptions about the pipework etc and your computer run may predict a vaporisation rate which is quite different from the one hoped for by process. If the composition has been changed significantly by the predicted vaporisation rate it may be necessary to ask for new heat release curves which fully cover the new range and rerun your program.

HEATING CURVES

None linear heat release curves were covered in detail in chapter 1. For most two phase flow situations it is desirable to have two curves which cover the full range of the operating pressure. For any thermosyphon this is essential. (The data should ideally be for temperature, vapour mass fraction and specific enthalpy.) The lowest pressure is the pressure at the space in the tower adjacent to the tower return nozzle. The highest possible pressure is the lowest pressure in the tower plus the pressure due to the static head of the liquid when the system is stationary. The static head is not known at the start of the calculations so that the higher pressure curve should be on the conservative side – for the maximum likely static head.

Some programs now strongly recommend three curves, the two discussed above and a third one at an intermediate value. I am not convinced that this is essential for all cases because reboilers have been designed for years with only two. However three curves can only increase the accuracy of the calculations and I now always ask for another curve when only two are provided.

Curves should also extend above the target vaporisation rate. Your program may well extrapolate into this region but it will not be as accurate as the data from a process simulation. This is particularly important when off design cases need to be investigated.

STATIC HEAD EFFECTS

Whatever the state of the liquid at its surface in the tower, it will certainly be pressurised by the time it reaches the inlet of the exchanger otherwise there would be no circulation. Another term for a pressurised liquid is a subcooled liquid which is the way we normally think of it in heat exchanger work. The degrees of subcooling depend on the temperature and the saturation temperature corresponding to the pressure at that point.

The degrees of subcooling might be quite low for a high or medium pressure system but will increase in significance as the pressure falls and can become quite a problem in vacuum systems.

Once the system is circulating the pressure at the inlet to the exchanger will never be the tower pressure plus the extra pressure due to the full static head due to the friction losses in the inlet pipework. The higher the friction losses are in the inlet pipework the lower the subcooling will be. This factor points to using up a significant proportion of the total system pressure drop in the inlet pipework particularly for low pressure systems.

Provided the input is satisfactory a modern computer program will take account of this effect by reducing the heat transfer coefficient and MTD in the exchanger. In general the effect of subcooling due to static head is relatively small in horizontal units because of the high heat transfer coefficients outside tubes. It can be much more of a problem in vertical units which is covered in the next chapter.

INSTABILITIES

For a fixed operating condition a horizontal reboiler can be considered in the same way as a 'normal' shell and tube exchanger which is supplied with a subcooled liquid and produces a two phase fluid on the shell side. It can therefore have the same problems as far as flux limitations; vapour blanketing and vibrations are concerned which are dealt with elsewhere in this book.

The key possible instability which distinguishes the horizontal thermosyphon system from most other systems is slug/plug flow in any vertical outlet pipe. The terms slug and plug are interchangeable in vertical flow and I think in terms of plugs of vapour sandwiched in between slugs of liquid. Slug/plug flow will result in disturbing noises but more importantly they may result in the pipe vibrating due to the slugs of liquid hitting the bend at the top. The possible types of flow regime in a vertical pipe have been covered in chapter 6 together with some of the procedures used to identify each regime. A modern computer program will include at least one method of predicting the flow regime in a vertical pipe. However the output usually simply states which flow regime is predicted for each method.

In my 'ideal bespoke' computer program for horizontal thermosyphons the output would include several flow maps with a cross in each indicating the predicted operating point on the map. This would show how close it was to the boundary of a change in flow regime. I know some companies have in house programs which will do this. In the absence of help from any program I have always plotted the 'progress' of the operating point by hand for each computer run varying say the diameter of the outlet pipe. This does not need to take up too much time and is well worthwhile.

The ultimate aim is to have an outlet pipe which is in annular flow for as wide a range of operating conditions as possible.

DESIGNING THE SYSTEM.

The first step is to find out as much as possible about the probable pipe runs and elevations from the tower drawing as explained earlier. An initial estimate of the static head will give a good idea of the available pressure drop.

Some of the available static head will be used up by the acceleration loss and the static head losses in the two phase stream. The computer program will take care of these and the designer should be thinking of how best to proportion the remaining pressure drop between the following:

(a) The inlet piping.
(b) The exchanger frictional pressure drop.
(c) The outlet piping.

With only liquid in the inlet piping it would be possible to design for a very low pressure drop. However it now seems to be generally agreed that it should provide a substantial resistance – up to 30%? One reason for this is the effect of the static head on the saturation temperature of the fluid entering the exchanger although this is not always significant. The other reason is a bit more subtle and not so easy to explain. In any thermosyphon system **upstream resistances seem to promote stability** and downstream resistances instability.

I have always thought of this in terms of flows of people. I have witnessed the chaos resulting from people moving from a large space, (low people flow resistance), to a narrow space, (high people flow resistance), and then back into a large space, (low people flow resistance). There are many situations where this can occur; the turnstile at the entrance to a football ground, the checking in desk at an airport etc. The solution is to force people to form an orderly queue in some way in the upstream space. This may involve physical barriers or firm stewarding; either way what is being implemented is introducing an upstream resistance to people flow. This results in a smoother flow of people and an increased throughput.

I am aware that I have now written too many pages without any diagrams to 'break up the text' so you can have a laugh at my attempt at an 'artistic drawing' in figure 9.13 – but it might help you to remember the importance of upstream resistances.

One aspect of the inlet pipework which has always been controversial is should it include a valve? I know that for many years some companies insisted on this in their specs. For the thermal design engineer this is a win, win situation. If the system works satisfactorily with the valve fully open this is fine if it operates better with an increase in the upstream resistance the valve can be partially closed. A valve is no longer a requirement in most of these specs.

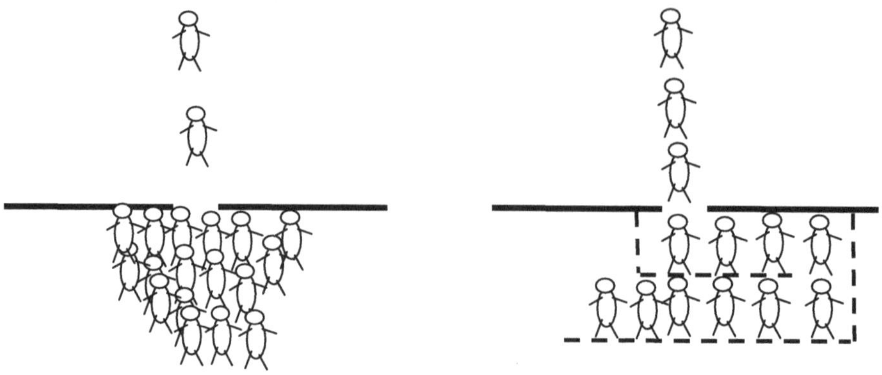

No upstream resistance With an upstream resistance

Figure 9.13

The valve that is invariably used is a butterfly valve. This has a very low resistance when fully open but very poor 'control characteristics' and is therefore not suitable for connecting to any system of instrumentation. I can understand any manually operated valve not being popular with site operators. With the advent of so much computer control the number of operating staff has been cut right back and the ones that are left will spend most of their time in a control room which can be remote from the valve which may need adjusting. In the UK this will probably just mean getting soaking wet but there are now parts of the world where venturing outside could mean catching a lot worse than a cold.

A valve can also be useful to provide a resistance during start up even when design operation is optimised when the valve is fully open.

If I have a horizontal reboiler to design and I know that there is another one on a similar plant which is operating with no problems and process do not want a valve I will not worry about designing a system with no valve. I will still make sure that there is some resistance in the inlet pipework – say a minimum of 15% but not as much as 30%. (Having too much resistance is very difficult to correct if there is a problem compared with adding an additional resistance.) If however I am faced with a new situation and I have any concerns about the operation I would always try and make out a case for including a valve.

It is not unusual to have to increase the inlet piping resistance to solve an operating problem or during revamp work. Although a valve is the ideal solution a much simpler and cheaper solution is to use a resistance orifice. This may not involve any change in the pipework. However a resistance orifice presents its own set of design problems. There are well documenting procedures for sizing the type of orifice plate which is used for measuring fluid flow but these devices use up a very small pressure drop, (only slightly more than a well designed venturi meter.) To produce a significant pressure drop through an orifice will require quite a small hole relative to the diameter of the pipe. A thin plate is ideal to fit into the pipework but then will the high velocity through the hole wear away the plate and increase the diameter? If a thick plate is used there are a variety of possible profiles through the orifice. There is a lot to take into account for such a small item of equipment.

One of the worst cases I can remember of a system being designed with too little resistance in the inlet line involved a unit where the target vaporisation rate was 30% and it was operating at far too low a value. When I checked the unit out the calculations were awash with conservatisms including a static head which was over twice the value that was required. My runs predicted a vaporisation rate of less than 15%. The solution was simple – fit a valve in the inlet line. It was not too difficult to persuade the others involved that the flexibility this arrangement offered was far more satisfactory than using an orifice plate.

Having taken up a lot of space making out a case for using a substantial resistance in the inlet piping I have come across units which have used up hardly any and yet they still worked OK.

THE EXCHANGER

As far as designing the exchanger is concerned a lot depends on the fouling factors and the type of heating fluid. Provided there is not too much liquid heating the boiling fluid will have a reasonably high heat transfer coefficient. For a particular fluid at a specified pressure, the boiling fluid heat transfer coefficient will depend on the temperature difference across the film. This is fixed by the temperature difference between the two fluids. In this situation the pressure drop used up over the exchanger will make very little difference to the surface area required and the traditional H and G types of unit are the first ones to try.

For wide boiling range fluids there will be a lot more liquid heating and a lower heat transfer coefficient which may well be the controlling thermal resistance. In this situation an E or J type shell should be used to try and optimise the shell side coefficient and minimise the surface area required. This will require using up a high proportion of the system pressure drop. Of course using H or G type shells with vertical baffles (or horizontal baffles if your program will accept them?), will also use up more pressure drop and improve the coefficient. I would try the E or J type first although with a liquid heating medium the H or G type shell must also be investigated. This is because the central horizontal baffle effectively divides the unit into two sections thermally i.e. it is similar to two shells in series as far as the MTD correction factor is concerned.

When H or G type shells are used for wide boiling range mixtures – particularly if the liquid is significantly subcooled, the ideal exchanger would have variable pitched baffles. A close pitched baffle arrangement below the horizontal baffle and wider pitched baffles above the horizontal baffle. I suspect it will be a long time before any computer program will accept this input and until then lot of 'fudging' is required for such a design.

As far as the overall system stability is concerned if a large proportion of the available system pressure drop has been consumed by frictional losses in the exchanger it is less necessary to use up pressure drop in the inlet pipework. Conversely if the exchanger has a low pressure drop it is more important to use up some pressure drop in the inlet pipework

THE OUTLET PIPING

In any circulating system upstream resistances will increase stability and downstream resistances are likely to cause instabilities. For a horizontal thermosyphon the downstream resistance is the outlet pipework which should ideally have a low pressure drop but there are very few cases where this is possible.

With a vertical pipe at some point in the system the design becomes a battle between trying to keep the pressure drop down and ensuring that there is annular flow in the vertical pipe which requires high velocities which result in high pressure drops.

Ideally the vertical pipe should always be immediately above the exchanger. If the two phase fluid leaves the exchanger through a bend and then subsequently enters a vertical pipe the predictions of the flow regime in the vertical pipe will be much less reliable.

Digressing to the sixties when thermosyphons were invariably low pressure drop H or G types with no baffles other than the central horizontal 'deflecting' baffles and the need for need for upstream resistances was not widely appreciated I saw some strange designs. The need for annular flow in the vertical pipe was understood and it was not unusual to see units designed using up most of the system pressure drop in the outlet pipework. I have seen units with outlet pipes smaller than the inlet pipes! Of course the methods available for predicting the flow regime in the outlet pipes were not as refined as they are now and if a valve was used in the inlet line this could be partially closed to give a more balanced system.

Although the main design problems are normally with the vertical pipe some horizontal reboiler systems may have quite long horizontal pipes. These should also be checked out for possible slug flow. There is no theoretical reason why a horizontal section downstream of the vertical section should not have a larger diameter to take it out of the slug flow region.

When there are more than one set of design conditions to satisfy it may not be possible to be confident of stable flow for each case. This should be discussed with process and piping. Piping can use extra supports to minimise the chance of any instability leading to damage.

For some systems it may be possible to avoid using any vertical pipe. This is often the case with a once through reboiler where the return nozzle to the tower is below the supply line from the tower. This is ideal and makes the system design a lot easier since there will be a relatively low pressure drop in the outlet pipe.

At the other extreme with a long vertical pipe it may be necessary to use up more than half the allowable pressure drop. By contrast, with a vertical thermosyphon it is normally possible to keep the pressure drop in the outlet pipe quite low.

SERIES & PARALLEL SHELLS

It is not unusual for a horizontal thermosyphon reboiler to have more than one shell. This may be because of bundle size limitations or a requirement to have one bundle on standby. Standby bundles are needed for some heavy fouling fluids so that one bundle can be cleaned whilst the other is in operation.

If each bundle has individual connections to the tower there are no complications to the thermal design – just appropriate notes on the thermal data sheet. If there are two shells with a common connection to the tower and one shell is on standby this is also not a problem but there will be extra resistances in both the inlet and outlet pipework.

With more than two shells the pipework can become much more complicated. Figure 9.14 shows part of the piping layout for a system I designed about 10 years ago.

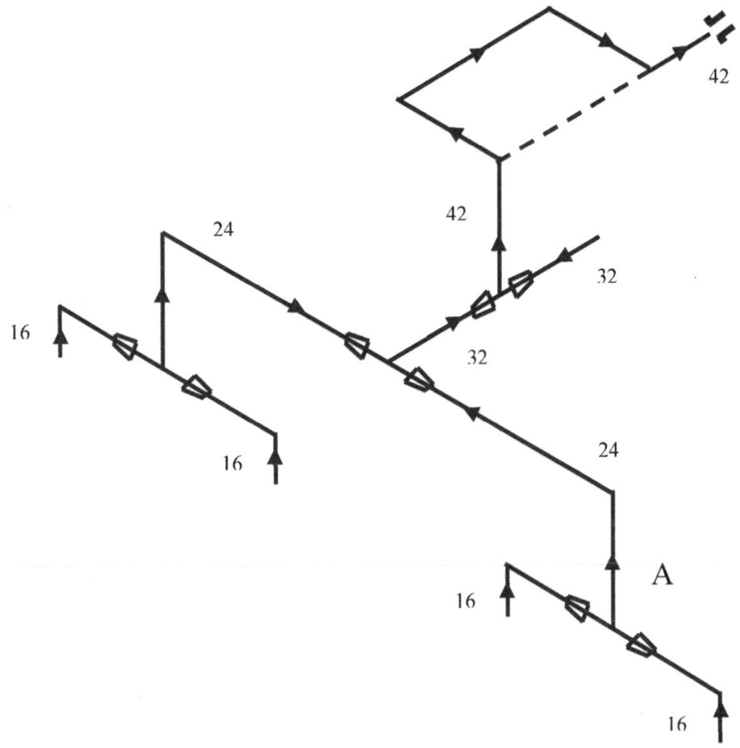

Figure 9.14

There are four H type shells with 16 inch outlet nozzles connecting to a single 42 inch tower return nozzle. At the top of the sketch the dashed line is the ideal final line to the tower nozzle as far as the system hydraulics are concerned. However this was quite a long line and resulted in stress problems. After consulting with the piping stress engineer I had to accept a system with two more bends as illustrated by the solid lines.

Some of the more sophisticated modern programs are now approaching the point where the may accept such a system by the time this book is published. Until then, and of course there will always be some designers using simpler programs, it is necessary to model the system. This involves using fictitious pipe sizes in the computer run which will result in the same frictional pressure drop as the actual system.

When I did this design the computer program would only handle a single H type exchanger. It was therefore necessary to model the outlet pipework from point A to the tower return nozzle.

The computer program will assume that all the outlet pipework in this system is 24 inches, (23.25 inch ID). The details for the 42 inch, (41.25 inch ID) pipe are set out in the following table:

Fitting	Equivalent lengths Le
Hard T	64
90° bends	14 x 4 = 56
Straight	30
Total	150

Table 9.2

I have taken the equivalent lengths from table 12 in useful tables. The bends are all long radius. In this section of the system the flow rate from the exchanger will have a much lower pressure drop due to this change alone. From useful tables 1 pressure drop varies inversely with the diameter to the power of 4.8. However the flow rate through this section is four times the flow rate from the exchanger. Since most of the pressure drops are due to discontinuity losses they will vary directly with the square of the mass flow rate. The change in equivalent length can therefore be found from the following equation:

$$\text{Change in Le} = \frac{(d_{ref})^{4.8}}{(d_i)^{4.8}} \times \frac{(\dot{m})^2}{(\dot{m}_{ref})^2} \qquad [9.7]$$

Where the subscript, 'ref', is for the reference quantities which in this case are 23.25 inches for the diameter and a mass flow rate which can be taken as unity. Inserting numbers in equation 9.7 gives:

$$\text{Change in Le} = \frac{(23.25)^{4.8}}{(41.25)^{4.8}} \times \frac{(4)^2}{(1)^2} = 1.02$$

This factor is multiplied by the equivalent length assuming a 24 inch pipe for this section. In this example the change is negligible, which is an indication that the pipe sizes are 'well balanced', this is not always the case. This calculation is repeated for the 32 inch pipe and the two modified equivalent lengths are added to the short section of actual 24 inch pipe equivalent length for the input to the program.

Having assumed that the pipework from the exchanger is all 24 inches any messages from the program concerning the flow regime will only apply to one short section of vertical pipe. Separate calculations must be undertaken to check the flow regimes in the 32 inch and the 42 inch pipes. If you do not have a program available to do this hand calculations were covered in chapter 6. With three separate vertical sections in this system each vertical pipe would be a lot shorter than it would have been for a single exchanger system. The predictions for the flow regimes will therefore be a lot less reliable. On the positive side instability in a short pipe is less likely to lead to damage than instability in a long pipe.

You may have access to an independent program for pipework, (check with process and see if they have one), which will give you the frictional pressure drop between the exchangers and the tower nozzle. Being a natural sceptic I would use it and then compare the results with the method outlined above to see how they compared.

The same procedure can of course be used for the inlet pipework

One simplification in the above system was its symmetry. You may have to deal with none symmetrical systems which are even more challenging. For example you may have an existing system like the one above but the plant is being revamped and process need to increase the heat load to the tower by adding two more shells in parallel to the system. There are various ways in which this could be achieved but whatever system is selected it will never be symmetrical.

In any none symmetrical system the vaporisation rates can never be the same from all the shells. Whenever there are two distinct paths through a combined circuit the only theoretical way to achieve a balance is by fitting a valve in the inlet piping of the path with the least resistance. The valve can then be manually controlled until the resistance is the same through both the possible flow paths.

Horizontal reboiler shells do not need to be in parallel; they can also be arranged in series or a series/parallel arrangement.

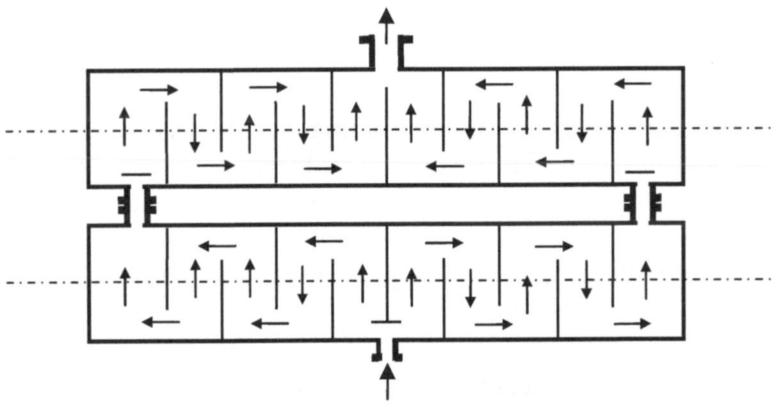

Figure 9.15

Figure 9.15 shows two shells in series where the bottom shell is a 12J and the top shell a 21J. There are several possible reasons why series shells may be required. One is on a revamp where there is only plot space for one shell when two shells are clearly required for the new duty.

One advantage of having shells in series is that the inlet and outlet pipework is simplified. Another is that the vertical height in the outlet pipework is reduced. In some cases this may not make too much difference but with a once through system it may be sufficient to eliminate the vertical leg altogether.

Although I have shown both bundles with the same baffle pitch in figure 9.13 it may be advantageous to use a wider pitch in the top bundle to optimise the heat transfer and pressure drop. The benefits of using more than one baffle pitch in flow boiling were discussed earlier in this chapter. If the internals of the two shells are not the same it will be necessary to carry out three computer runs. A run will be required for each individual shell and then a third run with a 'fudged' shell and heat transfer coefficients input from average values for the other runs. This must have the same frictional resistance as the total from the other two runs and can be used for the system hydraulics.

Another advantage of having two shells in series is that different heating fluids can be supplied to each shell. For example it is often desirable to recover heat from a process stream which can be supplied to the tower. If the heat recovered in this way is not sufficient to provide the full heat load to the tower then steam can be used to make up the balance. With different heating fluids the two shells are very unlikely to be the same. The only limitations to their design are that the lengths must be the same and the nozzles must match up. Figure 9.16 shows two H shells in series with different sized shells.

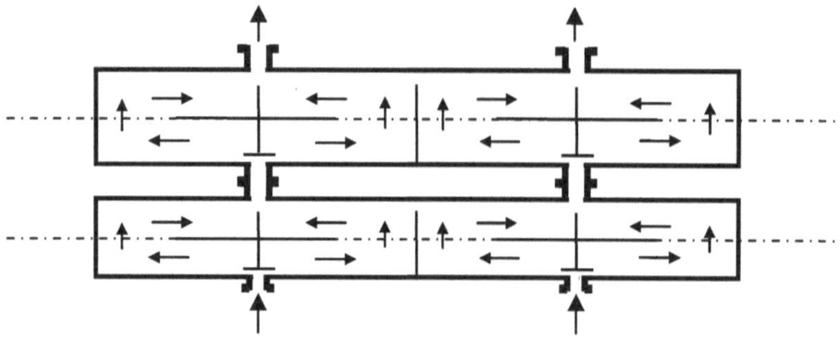

Figure 9.16

One plant where this arrangement was used in the original design was the Petronas plant at the Melaka Refinery in Indonesia. It worked OK for several years and then came the inevitable revamp. The saturated gas recovery unit had to undergo a debottlenecking process to increase the capacity and improve the plant thermal efficiency. The necessary extra heat load could be provided from a kerosene stream which was being lost in the original design.

It was decided that if two stripper reboilers in series were working OK then why not add a third. Foster Wheeler has an office in Singapore which was invited to quote for the work and the actual design of the reboilers was passed up to their design office in Reading in the UK.

With three units stacked in series there are mechanical as well as thermal problems. The two existing shells were at grade and these needed to be lifted up so that the new shell would be underneath and could be mechanically strengthened to support the other two. A side view of this arrangement is shown in figure 9.17 below.

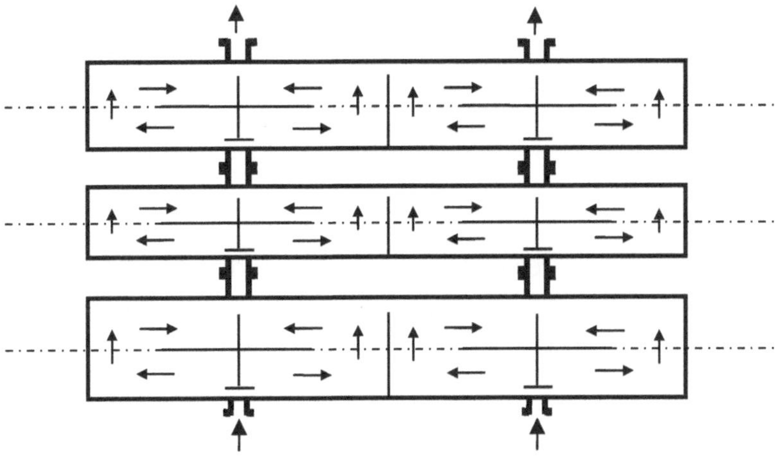

Figure 9.17

The problems involved with modelling two shells as a single shell for the design of the system hydraulics was discussed earlier. With three shells the problems were magnified. In this design the solution was to model the system by using a single vertical thermosyphon to represent the three horizontal shells as illustrated in figure 9.18 below.

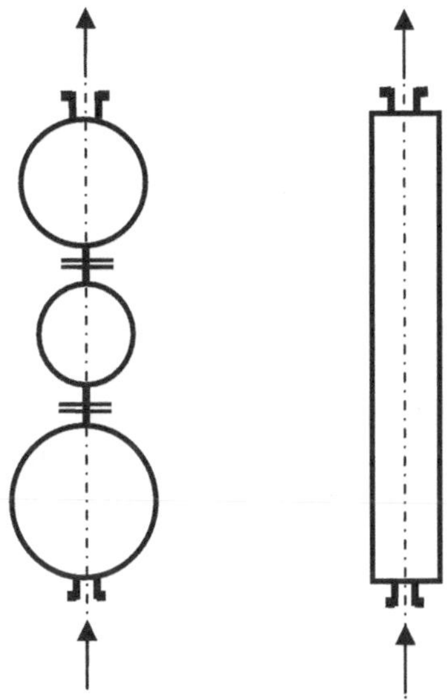

Figure 9.18

The vertical thermosyphon had the same vertical height as the three horizontal shells and was fudged until it had the same pressure drop as the sum of the actual pressure drops through the three horizontal shells. This run was then used to check out the system hydraulics and pipework.

The ultimate test for any innovation in heat transfer design is 'does it work'? In this case the modifications were made and when the plant was started up it performed as predicted. Not only was heat now being recovered from the kerosene stream but the steam consumption had been reduced producing a substantial increase in the plant thermal efficiency.

This design was named the 'Reboiler hat trick' and as far as I know it was the first time it had ever been used. Further details can be found on the Foster Wheeler website at 'exchangers@fwc.com'. Although I was at FW when this unit was designed I was working on another job at a site remote from the head office and my only contribution was suggestions for the overall approach over a pint at the pub. I do know the engineer who did the detailed design very well and at the time of writing this he is still at FW. If the reader is faced with a similar problem and does not have the confidence to carry out such a complicated design they could try contacting FW for help.

Three different reboiler shells using three different heating fluids may still be unique but like climbing Mount Everest or running the first four minute mile I do not expect it to remain that way for long. It is always a lot easier to follow than be the first. A glance at the typical congestion of pipes and equipment at the bottom of a tower show the advantage of having a single connection to and from three reboilers to the tower. This was a revamp job where innovation is often required. However innovations on revamps which work often become the basis for a new plant design. There is no theoretical limit to the number of shells. Will there ever be a four series shells arrangement?

STEAM GENERATORS

Steam generators are very much a 'special' for horizontal thermosyphon system designs. For tower reboilers the vertical height is limited by the tower connections. For steam generators a critical variable is often the height required to give adequate circulation. On some plants the required heights have to be seen to be believed. Whilst I was working at Kellogg in the sixties I was sent out to supervise some repair work on a plant in Holland. On the same site was an operating ammonia plant which I naturally went to investigate. Whilst I was looking up at the top platform one of the operating engineers said he had to go up and jokingly asked if I would like to join him. I jumped at the chance and when he could see that I was serious I was questioned about vertigo, fitness etc. I told him that I had been up a few rope ladders and a fixed ladder would be a piece of cake and I had had less to drink than normal the previous night.

At the bottom there was a bit of a breeze which was hardly noticeable. The climb to the top was not a 'piece of cake', it was dammed hard work and I felt like I had had a hard session in the gym by the time I got to the top but it was well worth the effort. The view was magnificent. With the Netherlands being so flat I could see for miles, sorry kilometres, around. The only thing I can compare it with is going up the Eifel tower in Paris and that is not as satisfying since is just involves getting into a lift which anyone can do. The wind was now a howling gale and I had to lean into it to stay upright without holding on to a rail. (The wind had not changed when I returned to grade.)

I do not suppose I would be able to do it now without filling in a wad of health and safety forms and having a special medical and of course it would be even harder work for a seventy year old and I might need a trial run up to a lower platform. I can recommend it to any reader who gets the chance and is up to it even if it involves filling in a lot of forms and disclaimers.

I regret not taking a note of the elevation but it was very impressive. I also regret never having designed a steam drum for an ammonia plant but I have designed systems for many other applications where less impressive heights have been sufficient.

The height required depends on the pressure of the steam being generated. As the pressure increases the difference between the density of the water and steam will decrease. The static head therefore needs to increase to compensate and maintain a good recirculation rate.

At the other end of the vertical scale from the ammonia plant steam systems some vendors will supply dual-drum packages where the drum is mounted piggyback style above the reboiler as illustrated in figure 9.19 below.

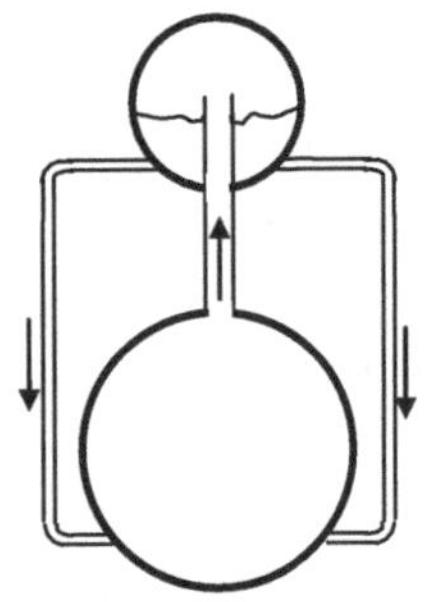

Figure 9.19

These are suitable for lower pressure systems which do not require too much static head. The steam drums can be quite complex with baffles, demister pads etc to control the quality of the steam leaving the drum. The central pipes carrying the two phase fluid from the reboiler to the steam drum are known as risers and the pipes carrying the water from the drum to the reboiler are termed downcomers just as they are in a water tubed boiler.

Horizontal thermosyphons can be used to recover heat from process streams and exhaust gas streams. The steam pressures can vary from a modest 3 bar up to 150 bar. They are all classified as 'waste heat boilers', a term which is also used for many other arrangements.

A feature that most of them have in common is that the hot gas pressure is quite low – often only just above ambient pressure. The allowable pressure drop is therefore very low which results in a single pass on the tubeside.

Another feature that they have in common is that the inlet gas temperature can be very high – up to 1400°C and there may be a substantial temperature drop between the inlet and outlet. The temperatures involved with most units are well above the design limits for the normal metals used in their construction. For a central section of a tube in contact with boiling water this not a problem since the metal temperature will only be slightly higher than the boiling water saturation temperature. The problems arise in the designs of the tubesheets and the method of fixing the tubes to the tubesheets. Fixed tubesheet designs for 'normal' exchangers are covered in chapter 11 on special mechanical features but these are not normally suitable for waste heat boilers.

FLEXIBLE TUBESHEETS

The potential problems which normal thick tubesheets would have due to the stresses caused by having a substantial temperature gradient along the horizontal axis are overcome by having a thin flexible tubesheet – usually no more than 25 mm thick. Figure 9.20 below shows the basic features of a flexible tubesheet.

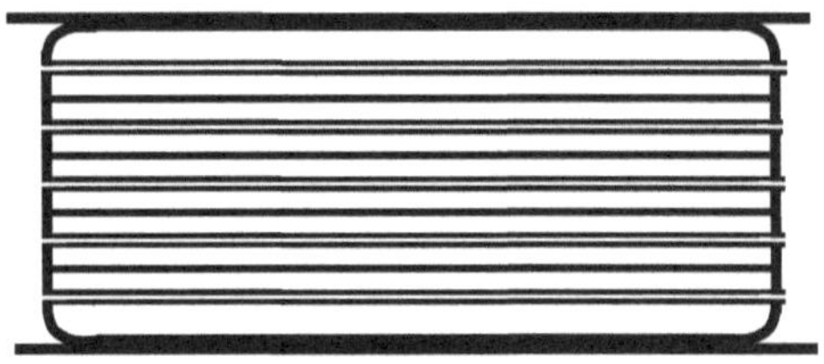

Figure 9.20

The face of the tubesheet at the hot gas inlet can be covered in insulation and protective ferrules inserted which extend through the tubesheet so that no part of the tubesheet is exposed to the hot gas. The shell side of the tubesheet is in contact with the boiling water and therefore has a temperature which is only slightly above that of the water saturation temperature.

The tubes are all at the same temperature and act as stays which keep the tubesheet flat. There are more tie rods to hold any support baffles than in a 'normal' exchanger and these are at the same temperature as the water as is the shell. The shell needs to withstand the normal hoop stress so it may be quite thick for high pressure units but there are negligible longitudinal stresses.

The tubesheets are connected to the shell by flexible knuckles. These are shown as simple bends in the above sketch but they may be more complex. These allow for any differential expansion between the shell and tubes.

Waste heat boilers are normally proprietary designed items with each manufacturer's design having some individual features but the overall design principles are the same. As with other types of proprietary equipment the fact that the vendor may give a thermal 'guarantee' does not absolve a thermal design engineer working for a contractor of all responsibility. Often a thermal guarantee is limited to corrective engineering which means that if a unit does not work as required they will design another one but there is no compensation for the cost of a new one or the loss of plant thermal efficiency. It is also often necessary to get a quick estimate of the size and static head required without waiting for a quote which may take some time.

DESIGN DETAILS

Figure 9.21 shows some of the details of a design I once worked on. The shaded sections are insulation. The hot gas enters the unit axially from the left hand side and leaves through a radial nozzle in the rear head. There were 620, 50.8 mm diameter tubes in a 2400 mm shell. The gas was a reactor furnace effluent which entered at 1133°C and left at 340°C. Ceramic ferrules were used at the entrance to the tubes. The saturation temperature of the boiling water was 256°C and the pressure 44 bar. The gas pressure was 1.5 bar and the pressure drop was only 0.02 bar.

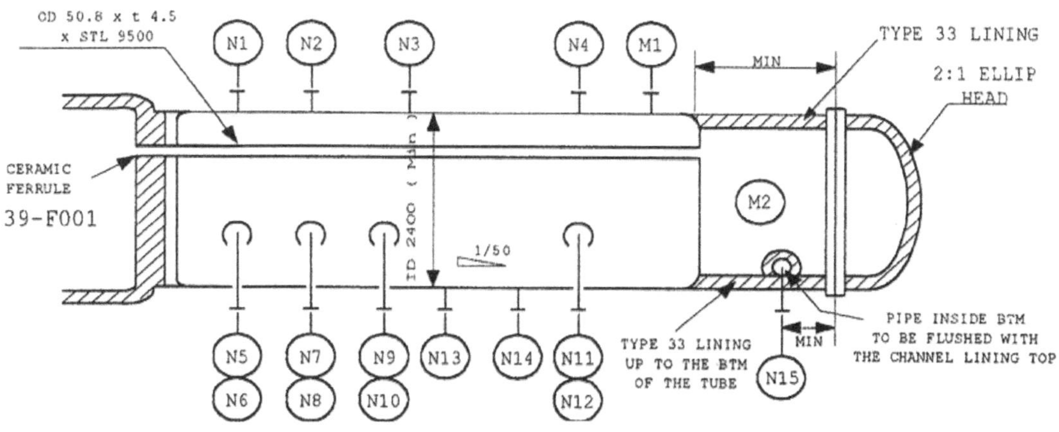

Figure 9.21

There are 4, 10 inch nozzles at the top of the shell numbered N1 to N4 which connect to the risers. There are two 6 inch downcomers for each riser and the connecting nozzles are numbered N5 to N12. N13 and N14 are 2 inch connection to a blowdown and a drain. M1 and M2 are 24 inch manholes.

An important feature of the thermal design is the axial positioning of the riser and downcomer nozzles. The steam generation rate along the shell is very uneven. The temperature difference at the hot end is 877 K and is only 84 K at the cold end. This is a similar situation to the single pass kettles which where covered in detail in chapter 8 and must be tackled in a similar way. The first riser nozzle, N1 should be as close to the hot end as mechanically possible. At the other end, nozzle N4 receives the steam from the cold end of the exchanger where about half of the length of the exchanger is required to generate about one quarter of the total steam generated.

CONCLUSIONS

The design of any thermosyphon reboiler is challenging compared with most other types of unit. The design of a single horizontal shell unit is relatively straightforward but as the number of shells increase the design problems escalate. The most complex situation I have come across is the Foster Wheeler 'hat trick' and I regret not having been personally involved with that one.

This page has been left blank so that the next chapter starts on a right hand page.

VERTICAL THERMOSYPHONS – CHAPTER 10

Introduction

The vertical and the horizontal thermosyphon reboilers fulfil exactly the same function as far as the overall process plant is concerned, namely to provide a heat load to a fractionating tower. However there are sufficient differences in the problems which face the thermal design engineer to justify a separate chapter for each of them.

As mentioned in the previous chapter, if the frictional pressure drop through a horizontal unit was sufficiently small the design of the heat exchanger and the system hydraulics could be separated, this has never been possible for vertical units. With a vertical unit the pressure drop through the exchanger is normally the dominant pressure drop in the hydraulic system.

When computer programs were first introduced I used a dedicated program for the design of the vertical thermosyphon. This was the HTRI program RTF. I found it to be an excellent program but I came across other people who found it to be too challenging and would spend a lot of time imitating the system hydraulics by 'going round in circles' without achieving any sensible output.

I was disappointed when RTF was integrated into one massive program, IST. I have always considered separate programs for different types of design to be the logical way forward. With separate programs there would have been far fewer bugs/mistakes made in subsequent developments and the input and output sheet would have fewer irrelevant items and be more user friendly.

Perhaps I am a bid old fashioned in my way of thinking but I will never forget an advert for technical staff at HTRI. The essential requirements were a PhD and a good knowledge of computing, some knowledge of heat exchangers was desirable but not essential. None of the experienced engineers I have worked with and learned from could have satisfied these requirements. This of course was at a time when computer programs were increasing in size and yet basic deficiencies like getting the number of nozzles right for some arrangements needed to be addressed.

Sometimes it seems like the computing challenges are dominating the engineering challenges rather than serving them. As computing capacities increase there is a compulsion to fill up all the available space as soon as possible. I am sure for some that the ultimate goal is to feed in process data and output a thermal data sheet and a setting plan – no need for a thermal design engineer? Well perhaps not quite that far, if anything goes wrong there is always a basic requirement to have someone to blame.

I am clearly trying to swim against the tide, which I often do, and it is time get back to the design of vertical thermosyphon exchangers.

THE SYSTEM HYDRAULICS

The tower connections to and from the reboiler may be from any of arrangements shown in figures 9.3, 9.5, 9.6 and 9.7 for the horizontal thermosyphons. The same system pressure drops are also present although the relative values of each component can be quite different. Once again before starting any thermal design work it is essential to get copies of drawings of all relevant features in the tower and the proposed position of the reboiler with respect to the tower connections. Figure 10.1 shows a simple arrangement for one possible type of unit.

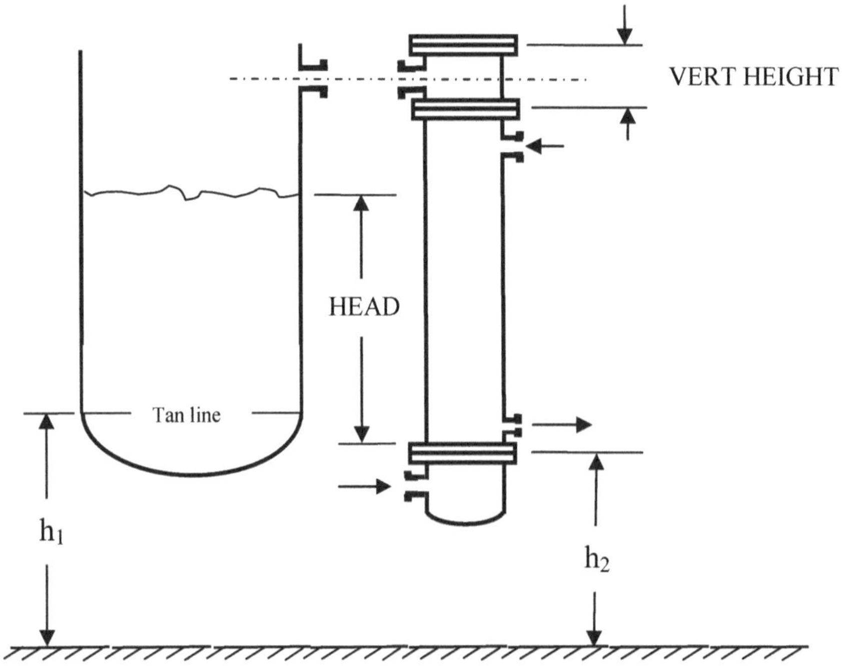

Figure 10.1

The two important vertical dimensions to input into the program are the static head driving the recirculation and the vertical height at outlet.

The liquid level with respect to grade is the same as it would be for a horizontal unit. The static head is from the design liquid level to the point where the liquid starts to be heated. This is the top of the bottom tubesheet for all vertical units.

For the exchanger construction illustrated the vertical height is the length of the outlet channel. If the tower nozzle is above the exchanger outlet nozzle then the distance from the top of the tubesheet to the top of the tower nozzle should be used. (Your program may require the distance to the centre line of the outlet nozzle.)

The elevation of the top of the bottom tubesheet clearly depends on the length of the tubes and the length of the tubes depends on dimension h_2 in the sketch. This dimension is fixed by the height required underneath the tubes when they are being cleaned. The mechanical construction of the exchanger will have a major impact on fixing this dimension. The TEMA type shown in figure 10.1 is a BEL but if the TEMA type is an AEL, the bottom section of which is shown in figure 10.2, then h_2 is taken from the bottom of the channel instead of the top of the channel.

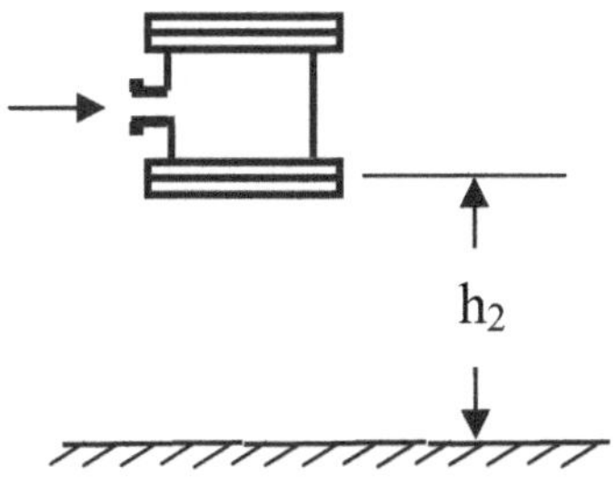

Figure 10.2

To clean the BEL only a cover needs to be removed at the top but at the bottom the channel needs to be removed which means disturbing the pipework. With an AEL design only two covers need to be removed. The A type head costs more than the B type head but is more convenient. The choice should be the client's who is going to operate the plant or a process engineer who is familiar with the client's requirements. A B type head may be acceptable for a small unit but not for the larger ones. Ideally of course it should be in a spec but the thermal design engineer needs to sort it out at an early stage.

The minimum height require for cleaning also needs to be sorted out. The cleaning process will be from the top and a receptacle for catching the crud will be positioned at grade underneath the tubes. A height of about 1 m is usually sufficient.

TOP/REAR HEAD TYPES

An alternative to the A/L type heads described above is the C/N type head which result in the TEMA CEN type of exchanger. This is illustrated in figure 10.3 (a) below.

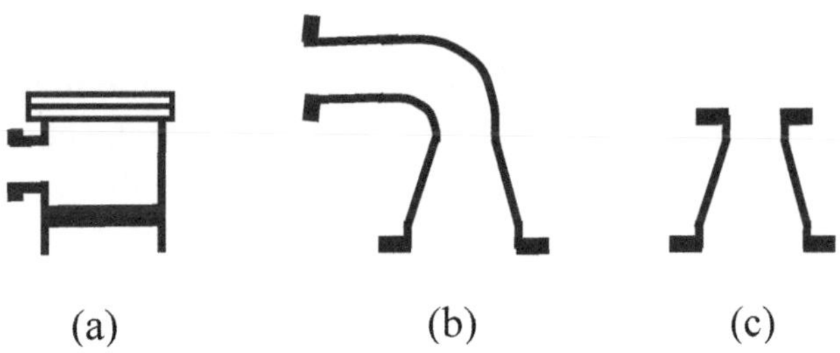

Figure 10.3

This construction reduces the cost and weight of the exchanger by eliminating two flanges. The length of the channel will also be reduced slightly. This will result in the same reduction in the vertical height required for the system hydraulics. As far as static head is concerned there will be a small reduction at the top which will be cancelled out by a small increase at the bottom.

One disadvantage of this type of construction is that maintenance of the tubes is not so easy. If tubes need to be plugged or replaced the outer ones are difficult to access through the channels.

Despite the maintenance problems most of the early vertical thermosyphon units I designed were CEN types. They now seem to have gone out of fashion and have not been used on the jobs I have worked on in recent years.

The ideal elevation for the reboiler exit nozzle is the same as the tower return nozzle as shown in figure 10.1 but this may not provide sufficient static head to achieve a satisfactory design. It is then necessary to lower the elevation of the exchanger with respect to the tower return nozzle. When extra head is required the first thing to try is an axial flow into a top channel which has a 90° bend as shown if figure 10.3 (b). The bend should have a long radius, (1.5 x NB) or 45° to minimise the hydraulic resistance. Short radius bends should not be used. The disadvantage of this construction is that the pipework must be disturbed to clean the tubes. The length of the channel section can be increased a bit by increasing the length of the parallel section between the flange and the conical section. The conical section will have an angle of 30°. As an alternative to using a cone and a long radius bend a mitred bend can be used. This is known as a 'lobster back'.

If the above constructions are still not adequate to provide sufficient static head then an axial nozzle as shown in figure 10.3 (c) which will connect to a vertical pipe may be required. This should be avoided whenever possible. However it is hydraulically better than having a side exit which then connects to a vertical pipe although this would have the advantage of being able to clean the tubes without disturbing the pipework – a difficult decision.

TOWER SKIRTS

The tower skirt height, h_l in figure 10.1 is calculated by a process or mechanical engineer. A minimum value will be established for the preliminary tower drawings which the thermal design engineer will be working to. For a horizontal reboiler this dimension is unlikely to change, this is not the case for vertical thermosyphons. Particularly if it means the difference between having to use a length of vertical pipe at outlet and not having one it is necessary to investigate the effect of increasing the skirt height. Often a small increase in height of say 1 m is sufficient to make a big difference to the design of the reboiler. This will of course increase the cost of the tower and will affect the elevation of every connection to and from the tower. It therefore needs to be discussed with the process engineer at an early stage.

TUBES

For horizontal thermosyphons the diameter of the tubes is determined by the pressure drop and cleaning requirements of the heating fluid. This will have only a relatively small effect on the hydraulics of the boiling liquid. For vertical thermosyphons the diameter of the tubes has a major impact on the system hydraulics.

Many specs will require a minimum tube diameter of 25.4 mm. This is a cleaning requirement since most streams tend to be relatively high fouling ones. Most vertical thermosyphons have 25.4 mm, (1 inch) tubes but it may be necessary to investigate the effect of using 31.8 mm, (1.25 inch) or even 38.1mm, (1.5 inch) tubes.
There is no theoretical reason why 19 mm, (0.75 inch) can not be used and some streams are far less fouling than others. Very occasionally a case will crop up when it is worth while discussing a deviation from the spec if there is a clear thermal advantage in using smaller tubes.

There is the same contrast between the design of horizontal and vertical units as far as tube length is concerned. It is not too difficult to keep to 'standard' tube lengths in horizontal units since the TEMA type and the baffle arrangement can be varied. Also when U tubes are used standard lengths are pointless. This is not the case with vertical units where a change of length from say 6.1 m (20 ft) to 4.877 m (16 ft) may not be ideal and a case may need to be made for using a none standard length.

6.1 m is usually the maximum tube length which is allowed. If this is based on the type of cleaning equipment used on the site there is no point in thinking about using anything longer.

NOZZLE ORIENTATIONS

I have often seen papers and articles on vertical thermosyphons where the shellside nozzles have been positioned on opposite sides of the shell as they would normally be on a horizontal exchanger. This is not necessary and having a shell nozzle facing the tower will complicate the pipework. I always show the nozzles on the same side facing away from the tower as illustrated in figure 10.1 since this is the most probable position to be convenient for the pipework. The shellside fluid will therefore have an even number of crosses. The orientation of the two nozzles must be the same when a single phase fluid is used for heating but their orientation with respect to the tubeside nozzles has no effect on the thermal design.

Although facing away from the tower is the most likely position there are three possible orientations facing away from the tower as shown below in figure 10.4

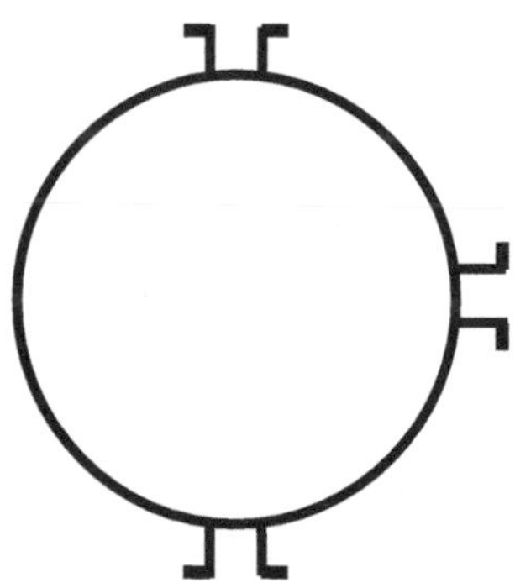

Figure 10.4

Since any of these might be the most convenient I always add a note to the thermal data sheet as follows: **'The orientation of the shellside nozzles could be changed to suit the pipework.'**

The same principle applies to the tubeside inlet nozzle where the orientation has no effect on the thermal design. Since this is more generally appreciated there may be no need for a note on the thermal data sheet. One arrangement may result in fewer bends but bends may be required to increase the flexibility of the pipework. A tubeside inlet nozzle may also be axial. This requires extra space underneath the exchanger so it is unlikely when the exchanger is at grade but it may be the most convenient arrangement when the exchanger is part way up a tower.

The orientation of the tubeside outlet nozzle is not so flexible since it is critical to the thermal design. Figure 10.1 shows a direct coupling or 'straight shot' arrangement. There may be a shut off valve in between the exchanger and tower nozzle but there are no bends. This is the ideal hydraulic arrangement. However for very large shells on a separate structure the combined expansion of the tower and the exchanger may make this arrangement unacceptable to the piping stress engineer. An arrangement such as the one shown in figure 10.5 may then be required.

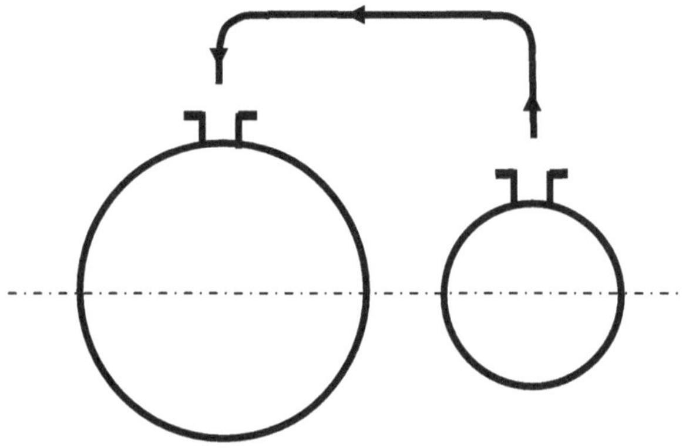

Figure 10.5

The thermal design engineer should check at an early stage if a 'straight shot' is possible. Unfortunately there will be times when he/she will be assured that it will be OK only to find a few weeks later when the piping stress engineer has done his calculations that it is not OK. This is very frustrating since it means that the unit must be checked through for the new configuration and may well need to be redesigned.

All vertical reboiler designs require regular coordination between the piping layout and piping stress engineers as well as the process engineer.

VENTS & DRAINS

When the heating fluid is condensing steam it is important to minimise the height of the condensate level at the bottom of the exchanger during full load operation. This means that the shell side outlet nozzle should be as close to the bottom tubesheet as possible. For small nozzles this will normally require a long forged flange type of nozzle to avoid using a reinforcing pad. For larger nozzles and short tube lengths it may be worthwhile to consider a siphoning arrangement. Whatever arrangement is used there will be some liquid level which will reduce the effective length of the tubes for heat transfer. This can be allowed for in the computer run by fictitiously increasing the thickness of the tubesheets. This should be allowed for when fixing the static head and a note should be added to the thermal data sheet explaining what you have done. (Otherwise someone down the line will think you have made a mistake in calculating the effective tube length.) Whether this added complication is necessary is a matter of judgement for each case.

The top of the shellside of all vertical units must have a vent opposite the inlet nozzle. This is essential for start up and may be required intermittently during normal operation. However it is debateable what happens to inerts which can accumulate during normal operation. Rather than accumulate at the top of the unit they may fall to the bottom of the unit as they are cooled and collect above the liquid level. A powerful argument in favour of the 'falling inerts theory' is that the molecular mass of air is close to double that of steam. To allow for this I always specify an extra vent opposite the outlet nozzle and about 150 mm above the expected liquid level.

Full details of venting and draining arrangements are covered towards the end of chapter 2.

DIFFERENTIAL EXPANSIONS OF THE TUBES/SHELL

Most vertical thermosyphon units are steam heated. Because steam has such a high heat transfer coefficient and low fouling factor compared with the likely heat transfer coefficient and fouling factor inside the tubes there will only be a small temperature difference between the shell and the tubes. This will inevitably result in a fixed tubesheet design without any need for any expansion bellows in the shell.

For other heating mediums, particularly liquids there will be a much higher fouling factor and a lower heat transfer coefficient on the shellside. If there is only a small temperature difference between the two streams it may still be possible to avoid an expansion bellows but when all possible operating conditions are taken into account this is unlikely. With any expansion bellows there is the choice of 'thin walled' or 'thick walled'. I am a 'thick man' for vertical thermosyphons. (Details of the two types are covered in the chapter on special mechanical features under fixed tubesheet designs.)

With any fixed tubesheet design where steam is not the heating fluid there is also the problem of cleaning. The only type of cleaning possible is chemical cleaning and process should have specified if this is acceptable on the process data sheet. If the shellside needs mechanical cleaning there is no option but to use a tubeside expansion bellows and there is no need to check out all the possible operating conditions. However fixing the elevation of the near/bottom head is much more difficult since it will depend on the type of bellows used.

SYSTEM PRESSURE DROPS

The system pressure drops are essentially the same as those for a horizontal thermosyphon which were listed on page 9 in chapter 9. The only difference is that the pressure drops through a bundle are replaced by pressure drops through the tubes and the two headers.

The relative values of the pressure drops can be quite different. A horizontal reboiler can be designed with quite a small pressure drop across the exchanger. A vertical thermosyphon should always be designed to use up most of the system pressure drop across the exchanger.

HEAT TRANSFER

A thermal design engineer should always have a mental picture of what is taking place. Figure 10.6 is my crude picture of what I imagine may be taking place inside a tube in a vertical thermosyphon reboiler.

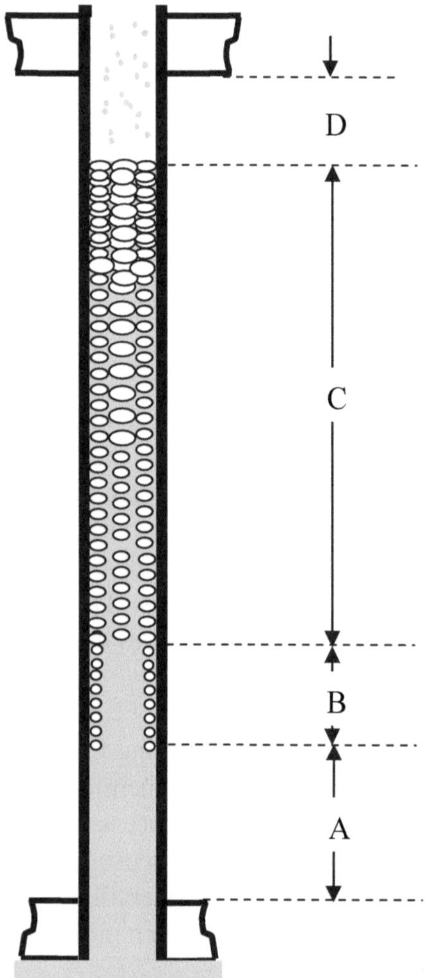

Figure 10.6

The five possible regions of fluid flow are as follows:

(a) **Liquid heating** – The liquid in contact with the tubes is sufficiently below its saturation temperature that no boiling is taking place. Heating is basically by forced convection although natural convection will be making a small contribution to the heat transfer coefficient. The heat transfer coefficients in this region are quite low.

(b) **Sub-cooled nucleate boiling** – The liquid in the boundary layer has reached a sufficiently high temperature that bubbles are starting to form. The bulk of the liquid however is still below the local saturation temperature. The bubbles therefore collapse as they are swept away from the tube wall. The heat transfer coefficients in this region are very high and may be as high as in the nucleate boiling region if there are sufficient nucleation sites.

(c) **Nucleate boiling** – The bulk of the fluid has reached its saturation temperature and bubbles no longer collapse as they move away from the tube wall. At the bottom of this region the volume fraction of vapour is significant and this increases up the tube until it reaches a value approaching unity at the top of this region. The heat transfer coefficients in this region are always very high..

(d) **Dry out/Mist flow** – The flow is now mainly vapour with entrained droplets of liquid. Very few of the liquid droplets are coming into contact with the tube wall and so there is little boiling taking place. The heat transfer is now mainly forced convection. Vapour heat transfer coefficients can be very low particularly at low pressures.

The nucleate boiling and the sub-cooled nucleate boiling regions both result in high heat transfer coefficients and may be lumped together. In an ideal vertical thermosyphon design they would be the only regions present. Both the liquid heating and the vapour heating regions result in relatively low coefficients and should therefore be minimised when it is not possible to eliminate them.

The vapour heating region is one that we can normally eliminate from the basic design case but it may be very significant during part load cases.

The liquid heating region will be present in all designs except those with a very high temperature difference at the inlet to the tubes when sub-cooled nucleate boiling may start at once. At the other extreme with very low temperature differences the liquid heating section may occupy up to about a third of the tube length. The liquid entering the tubes will always be sub-cooled. The degree of sub-cooling depends on the pressure at that point. The pressure at the entrance to the tubes depends on the static head and the pressure drop through the inlet pipework.

I will not attempt to delve into the many equations which your program may be using to calculate the heat transfer coefficients in each region. Part of your program output should include an incremental analysis which will identify each heating regime through the length of the tube. You should keep an eye on this as any design progresses to watch how any change you make impacts on the length of each section.

EFFECT OF STATIC HEAD

The higher the static head is the longer the liquid heating section of the tubes is likely to be. Increasing the resistance in the inlet pipework reduces the pressure at the inlet to the tubes and therefore reduces the length of the liquid heating section. However most of the system pressure drop is required for the pressure drop through the tubes so that a compromise must be reached.

The compromise is normally to use up to about 30% of the system pressure drop across the inlet pipework. There is another more important reason for using up a significant proportion of the total pressure drop across the inlet pipework and that is the effect it has on the system stability which will be dealt with later.

VAPORISATION RATES

Design vaporisation rates are normally much lower than for horizontal units – typically no more than about 30%. The main reason for this is that two phase flow in a tube will experience 'dry out' at a lower vaporisation rate than two phase flow across a bundle. This will not only reduce the heat transfer coefficient but accelerate the fouling and reduce the cleaning interval. Hence some computer programs will give a warning message if the vaporisation rate exceeds 50%.

Although a unit may be designed for a moderate vaporisation rate if the static head can vary it is very easy to exceed a value of 50% for low static head values. This is demonstrated in one of the examples later in this chapter.

INSTABILITIES

In a horizontal reboiler the potential instability problem was the nature of the flow regime in any vertical outlet pipe. If there is any vertical pipe in a vertical reboiler system there is the same potential problem from slug flow in the pipe. However it is normally possible to design a vertical thermosyphon system which has an outlet nozzle at the same elevation as the tower return nozzle or just below it when a bend is all that is required to match the two elevations.

Potential instabilities in vertical thermosyphons are generated within the tubes and affect the whole system. The instabilities take the form of flow oscillations where the fluid is moving up the tube as expected one instant and then reversing back down the tube against the overall pressure gradient provided by the static head a short time later. The cycle will then be repeated until the conditions which caused the oscillations are modified. These oscillations can be detected in the inlet line which is how they are monitored during experimental work.

Oscillations in fluid flow can lead to noise and vibrations. If an unstable state continued for any length of time the vibrations could lead to mechanical damage in the most severe cases. In most situations mechanical damage is unlikely but the effect on the thermal performance of the exchanger can still be very severe. Some experimental work has shown that the peak heat flux can be reduced to a quarter of the value it would have been without any oscillations and this may clearly have a drastic effect on the heat transfer. Several mechanisms have been identified which may lead to oscillations in a vertical thermosyphon. These include the following:

SLUG FLOW

Just as it is possible to have instabilities in a vertical pipe due to slug flow the same thing can occur in a single tube. If only a few tubes are involved this may not be a problem. With more tubes involved the slugs in each tube are unlikely to be in harmony which will minimise the effect on the overall system. Although it is possible to visualise instability due to slug flow it not generally thought of as being the most probable mechanism in most situations.

SUPERHEATED LIQUIDS

In the section on bubble size in chapter 7 I referred to a laboratory experiment where very pure water was heated in a clean glass beaker and became highly superheated before a very explosive form of boiling took place. The same phenomenon may be possible in vertical thermosyphon reboilers. However it would require a very smooth tube lacking in nucleation sites and a strange fluid which could become highly superheated before it started to boil. The strange fluids where this mechanism has been reported are liquid metals. I have never dealt with them and it is unlikely that the reader ever will either.

MIXED DENSITY CHANGES

A very important part of the pressure drop around any reboiler circuit is the static head of the two phase mixture. Different methods of calculating mixed densities were discussed in the previous chapter on horizontal thermosyphons. There it was shown that different calculation methods resulted in mixed densities values which could vary by over 100%. **Changes in the mixed density of the two phase flow are considered to be the most likely cause of instabilities**.

If there is a slight reduction in the mass flow rate into a tube the vapour fraction will increase and the mixed density decrease. This reduces the pressure drop of the two phase static head which is one of the pressure drops resisting flow around the circuit. This reduction of pressure drop will cause a sudden surge in the mass flow rate of liquid flowing into the tubes. This will reduce the vapour fraction, increase the mixed density and increase the pressure drop resisting flow. With an increase in system pressure drop the mass flow rate will fall again completing one cycle.

This cycle may be repeated indefinitely producing a pulsating flow. This pulsating flow has be investigated in laboratory experiments and backed up by field data.

There are many parameters which are known to affect the onset of unstable flow and the interaction between these parameters is very complex. However every thermal design engineer should be aware of how some key parameters can lead to either stability or instability. These parameters are:

(a) **Peak heat flux** – If peak heat flux values are exceeded an unstable region will be entered without any assistance from any other variables. However even approaching peak values will result in a large vapour fraction at the top of the tubes which will make the system more susceptible to instabilities.
(b) **Inlet velocity** – Only quite small changes in velocity can trigger off the oscillations described above. The lower the value of velocity the more likely this is to happen.
(c) **Subcooling** - Long lengths of subcooled liquid may instigate instabilities. As well as reducing the heat transfer coefficients liquid heating lengths should also be minimised to avoid instabilities.

STABILITY MAPS

Stability depends on many variables which may interact. The only approach to such complex situations is to use dimensionless groups and empirical data. The only stability prediction method that I am familiar with is the one used by HTRI in their program RTF. I assume the same approach is used in IST although it may have been developed and modified.

Three dimensionless groups are used which are plotted on a flow map which is divided into stable and unstable regions. This is illustrated in figure 10.7 below.

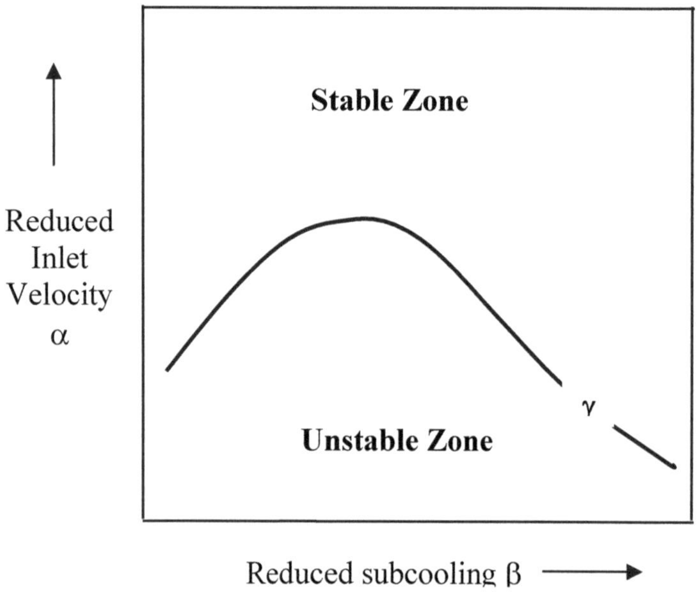

Figure 10.7

The dimensionless groups are as follows:

Reduced inlet velocity $\alpha = \frac{h\rho u d_i}{4Lq}$ [10.1]

Reduced subcooling $\beta = \frac{C_{pL}\Delta T_{sub}}{h}$ [10.2]

Reduced gravity $\gamma = \frac{(h\rho d_i)^2}{(4q)^2}\frac{g}{L}$ [10.3]

Where L and d_i are the tube dimensions, q is the heat flux, h is the liquid specific enthalpy, ρ the mixed density, ΔT_{sub} is the degree of subcooling at inlet, u is the liquid velocity at inlet and g is the gravitational constant.

I do not think that too detailed an understanding of these parameters is necessary for practical thermal design work so I will not attempt to delve into the two quantities which each dimensionless group is a ratio of. A brief discussion of each of them should be sufficient.

Reduced inlet velocity

The reduced velocity group features $\rho u d_i$ which the reader should recognise as being three of the quantities which appear in the Reynolds number. Comparing this group with the Reynolds number in convective heat transfer makes it easier to remember and understand its function. Low Reynolds numbers are associated with laminar flow and as the velocity is increased an uncertain region is passed through before fully developed turbulent flow is established.

In the same way low reduced inlet velocity values indicate an unstable zone and at higher values they indicate a stable zone. As with all maps involving fluid flow there is always an indeterminate region in between although a simple dividing line is shown on the diagram. It is easy to just think in terms of the inlet velocity being the main component of the vertical axis.

Reduced subcooling

The degree of subcooling is the quantity to concentrate on in this group. The stability threshold line is crossed at high values. Since the degree of subcooling controls the length of the subcooling region I tend to think of the horizontal axis as being the liquid heating length which is part of the output for each computer run.

Reduced gravity

To me the name of this group conjures up pictures of comparing the performance of a system on earth with the way it might behave in a space station. Perhaps reduced buoyancy might have been more appropriate. The product of ρg is a buoyancy force and a change in the buoyancy force is a key part of the mechanism which starts off oscillations.

There may well be stability maps which use different groups somewhere out there but I have not analysed any. Therefore I can not offer any alternative checks if RTF or IST is predicting possible problems. I do know that in the past when I have been working for a company which used TASC designs the one exception was that they used RTF for all vertical thermosyphon designs.

TUBE DIMENSIONS

As with any other type of single tube pass exchanger the pressure drop will be reduced by using either a shorter length or a larger diameter. For vertical reboilers increasing the diameter also increases the peak heat flux so that higher flux rates can be used before the onset of mist flow.

Because of the requirements for both satisfactory heat transfer and stability a far wider range of tube lengths and tube diameters need to be considered compared with other types of exchanger.

Tube lengths are usually limited to a maximum length of 20 ft, 6.1 m. There may be cases where longer tubes could be justified but they would be exceptional cases. For many designs it will be necessary to use lengths well below the maximum. Although there may be a list of preferred lengths in a spec there is often a good case for using none standard lengths.

Tube diameters can vary between ¾ inch, 19 mm and 2 inch, 50.8 mm giving a range of 6 different standard sizes. Some specs will specify a minimum of 1 inch, 25.4mm. If this is a client cleaning requirement it may necessary to keep to it but the optimum design for some units will be with using ¾ inch tubes.

LIQUID HEATING FLUIDS

The most common heating fluid for vertical thermosyphons is saturated steam or steam with a small degree of superheat which will behave in the same way as saturated steam. There are also many situations were a liquid is used. This may be to recover heat from a process stream or where steam at the right temperature is not available and a liquid heat transfer fluid is used.

For most other types of single pass exchanger it is normal to have the fluids in counter current flow which will maximise the MTD. Since there is unlikely to be any significant difference between the heat transfer coefficients in co-current and counter current flow in most exchangers this will minimise the surface area required. This is not the case for a vertical thermosyphon where there may be substantial differences between the heat transfer coefficients in co-current and counter current flow. There are two reasons for this:

(a) **Liquid heating region** – The length of the liquid heating section at the bottom of a tube is determined by the point at which sub-cooled nucleate boiling starts. This is fixed by the temperature difference in this region. The temperature difference is a maximum with co-current flow and will always result in a shorter liquid heating length and an improved heat transfer coefficient.

(b) **Maximum heat flux** – In counter current flow the temperature difference would be a maximum at the top of the tubes. In some designs this may be sufficient to approach the peak value which would reduce the heat transfer coefficient at the top of the tubes.

I am sure there must be situations when counter current flow could produce the optimum design and it only takes a very short time to try this out. The case investigated in example 10.1 may have been one of them but I doubt it and I would always start a design with a co-current flow arrangement.

Gases may also be used for heating. For hydrocarbons there is little difference between the liquid and vapour specific heat values. The heat released for a given temperature drop is therefore very similar. However more pressure drop is normally required for a gas stream to achieve a satisfactory heat transfer coefficient and there may be potential vibration problems.

SPARGING

Where it is not possible to achieve satisfactory circulation by other means the problem of lack of circulation may be overcome by sparging. This is the introduction of steam or an inert gas – normally nitrogen into the inlet stream. The sparging medium can be introduced either into the inlet pipe or the inlet channel.

The only times I have used sparging is where it has been specified by process as a result of experience on previous plants. It would therefore have been taken into account in the mass and energy balance for the tower. A reader thinking of using it where it has not been specified should consult process only after they are quite sure that there is no other solution.

DESIGNING THE SYSTEM

Proportioning the pressure drops resisting fluid flow is crucial to both the heat transfer and stability of any system design.

It is now well established that using up some of the total pressure drop in the inlet pipework can improve both heat transfer and stability. A figure of up to about 30% is all that can normally be spared. An exception might be in a vacuum system where a low static head is required and using up even more pressure drop in the inlet pipework could be an alternative to changing the liquid level in the tower. Whenever there is a wide range of possible operating conditions it is ideal to take up some of the pressure drop in the inlet pipework across a valve that can be manually adjusted to suit different operating conditions.

It is also well established that too much resistance in the outlet pipework can reduce heat transfer and promote system instabilities. This should therefore ideally be minimised. Some guidelines for a new design are as follows:

(a) Do not exceed 35% of the total system pressure drop in the outlet pipework. This is a value I have seen exceeded in many designs in the past. Often they have been designs for systems which have had problems and the solution has been to reduce this resistance. There are examples of this later in the chapter.
(b) The ratio of the exit pipe cross sectional area to the total tube cross sectional area should not exceed 0.4. HTRI recommends an ideal value of 1. This is a value I have never seen achieved in any practical design and I think it is unrealistic. However it is normally possible to keep above the 0.4 value.

PROCESS CHARACTERISTICS

The following is a list of some of the different types of processes which the reader may encounter with pointers to the likely design problems and solutions:

Narrow boiling range HC's – medium pressures.

These systems are normally one of the most straightforward to design. The heat transfer coefficients are good and the fouling relatively low. Peak heat flux limits are high and will only be approached with large temperature differences. The optimum design will often have ¾ inch tubes and the maximum tube length.

Wide boiling range HC's – medium pressures.

These systems will result in much lower heat transfer coefficients than narrow boiling range mixture because there will be much more convective heating at the bottom of the tubes. The peak heat flux limits are reduced and the fouling is higher. It will normally be necessary to use tubes with a minimum diameter of 25.4 mm and a length below the maximum. The tube diameter will increase and the length decrease as the boiling range increases.

Very wide boiling range HC's with high viscosities

Anything over 200 K can be thought of as a very high boiling range and if this is combined with a high liquid viscosity it will be very difficult to find a satisfactory design. Heat transfer coefficients will be very low and the fouling very high. Another potential problem with wide boiling range mixtures is that the heavier fractions which should be the last to vaporise may remain as liquid and start to concentrate at the bottom of the tubes. If this takes place the heat transfer coefficient becomes even lower.

Tube diameters will at or close to 50.8 mm and the tube lengths will be very short to reduce the system resistance. Everything should be done to maximise the static head driving the system.

This is the type of situation where sparging may be necessary.

Very high pressure HC's

The beat way to describe very high pressures is those where the operating pressures are approaching critical values. When any fluid approaches its critical point the difference between the values of the liquid and vapour density decreases. This will decrease the value of the mixed phase static head and make it more difficult to achieve a satisfactory circulation rate. The physical property data in general on the process data sheet will become more unreliable. Physical properties were covered in detail in chapter 1 on basics.

Heat transfer coefficients can be quite high provided there is no film boiling. However peak heat flux limits are very low and it is often difficult to avoid film boiling. To have confidence in a design it may be necessary to design it assuming all film boiling. Fouling is usually low and medium tube diameters and lengths are the norm. Everything should be done to maximise the static head driving the system.

High vacuum HC's

The heat transfer coefficients and the peak heat fluxes can be very low and the fouling quite high. There are usually potential instability problems to be avoided. The low heat transfer coefficients are mainly due to long liquid heating lengths although mist flow at the top of the tubes may also be present. The liquid heating section is very sensitive to the static head value.

Medium to high diameter tubes will normally be required together with an average/short tube length. Because of the large difference between the density of the liquid and vapour only a relatively small static head is required.

The combination of the requirement for a low static head and the sensitivity of the liquid heating length to the static head create a situation where I would always want to install a valve in the inlet line to these exchangers.

Polymerising HC's

The heat transfer coefficients and the peak heat fluxes can be reasonably high. There is normally a tube wall temperature value above which the tubes will rapidly foul. This often tends to dominate the design process.

It is normally possible to use smaller diameter tubes with an average tube length.

Aqueous solutions

The heat transfer coefficients and the peak heat fluxes can be very high. Fouling is very low for mixtures such as alcohol and water. For some salt solutions fouling may be low for normal operation but there is a danger of deposits at the top of the tubes and it may be very important to limit the surface temperature of the tubes.

It is usually possible to use smaller diameter tubes and utilise the maximum tube length.

Waste heat boilers

Boiling water always has the highest heat transfer coefficients and peak heat fluxes. If there is a large temperature difference at the hot end great care must be taken to avoid local patches of dry out.

Recirculation rates are the highest of any type of reboiler. The percentage vaporisation of higher pressure units is unlikely to be above 10% and may be less than 5% for low pressure units.

The mechanical construction of these items is very complex. The tubes often have very large diameters with lengths well above the normal maximum limit. Some details of an actual design are shown towards the end of this chapter..

CLEAN OPERATION

All heat exchangers must be clean at least once in their lives and many reboilers must be cleaned at regularly, sometimes at quite short intervals. The operation of a reboiler when it is clean may be significantly different than when it is fouled. Heat transfer coefficients will of course be higher but there is also the possibility that there may be instabilities when the unit is clean when no problems are predicted for the design case. The clean operation should always be investigated and process made aware of any potential problems.

For steam heated reboilers the ideal method of control for the clean case is by throttling the steam but this is not always possible. Control of steam heating systems and minimum condensing pressures was covered in detail towards the end of chapter 5 on condensers.

EXAMPLES

The following examples should give the reader an insight into the challenges faced when designing thermosyphon reboilers.

.

Example 10.1

This is a summary of a case which was reported in the Oil and Gas Journal in May 1989. It is an excellent example of just how bad designs for reboiler systems can be. The units were a primary demethanizer reboiler and a secondary demethanizer reboiler on an ethane extraction plant.

The original reboilers were brazed aluminium plate fin exchangers and these had failed. They were replaced by three vertical shell and tube units, two for the primary reboiler and one for the secondary reboiler. When these were installed and commissioned they were found to operate at about 30% of the required duty. The design should have been for 30% vaporisation but no vaporisation was taking place.

In the primary reboiler the heating fluid was a warm gas – the only case I have come across of a gas being used in a reboiler. The secondary reboiler used liquid propane as the heating medium. The author worked for the operating company and was presumably one of their process engineers. It was less than three years since he had graduated and was unlikely to have much experience in designing thermosyphon systems. There are two simple sketches in the article and both clearly show counter current flow. Changing this to co-current flow should have been the first and least expensive thing to try but there is no mention of it.

Eventually it was decided that the surface area was adequate and that the problem was due to a very low inlet velocity. Not surprising with all liquid heating inside the tubes. They investigated the options of using pumps to increase the circulation rate and of sparging with recycled liquid and recycled gas. Using a liquid as a sparge fluid would have been a very unusual solution. It was predicted that sparging would improve the reboiler performance but of course would upset the tower performance.

They decided that the most practical method of increasing the velocity was to reduce the cross sectional area through the tubes. With such a fraught history it is perhaps not too surprising that they opted to set up a pilot reboiler testing rig rather than rely on any more calculations. I have reproduced a sketch of the pilot reboiler instrumentation in figure 10.8 below. A sketch for the actual reboiler was similar except that there was not the same instrumentation. It also showed that the feed from the tower was under a constant static head from a bucket arrangement with an overflow.

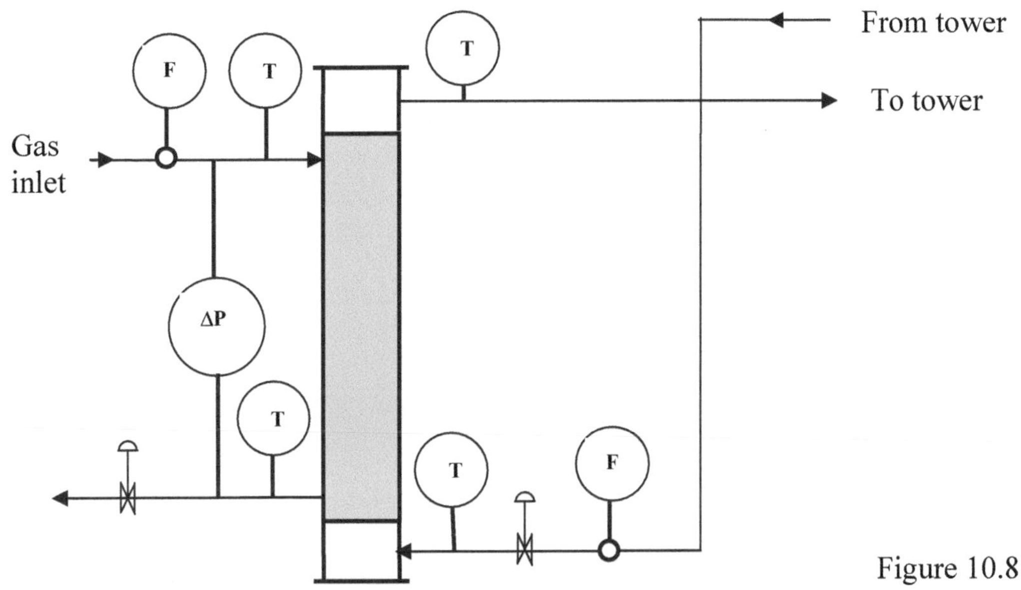

Figure 10.8

The instrumentation was clearly adequate to measure the performance of the test rig. The only reading missing is the reboiler operating pressure which was presumably taken from the tower instrumentation.

The first thing they tried was to use twisted aluminium strips as turbulators at the inlet to the tubes. Whilst this measure would not have significantly changed the cross sectional area it could have improved the heat transfer coefficient in the liquid heating length. In practice they had very little effect.

They then tried a number of tube inserts which would partially block the tubes and increase the velocity. The tubes used in the actual and test reboiler were ¾ inch diameter and 20ft long. The material and gauge was not mentioned. Three tube inserts were tried out with ½ inch, 7/16 inch and 3/8 inch diameters. The tube lengths were varied between the full 20ft and 16ft. The also tried 'fluted spheres' which is something I have not come across before and I am still not quite sure what they are.

The fluted spheres gave satisfactory heat transfer coefficients with no pressure drop problems as did the 7/16 inch tube inserts. The largest tube inserts gave pressure drops that were too high and the smallest tube inserts did not give a satisfactory increase in the heat transfer coefficient.

I have reproduced an interesting graph from this article – figure 10.9. It shows the effect of tubeside velocity on the overall heat transfer coefficient using the tube inserts.

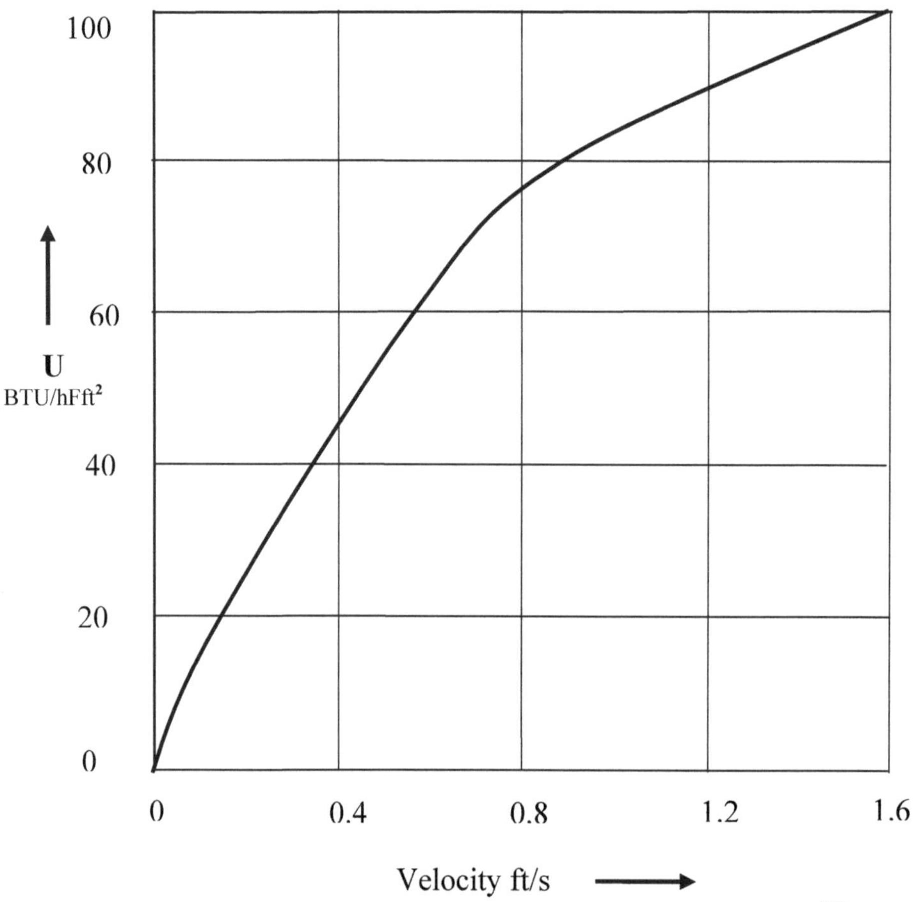

Figure 10.9

The original velocity in the tubes without the inserts was estimated to be 0.3 m/s. When the modifications had been made to the plant reboilers this had been increase to 1.08 m/s. There is no mention of the new vaporisation rate. However they did fit full length inserts after finding that there was a fall off in performance below an insert length of 17.5 ft. This suggests that even with the modifications made there was only boiling in a relatively short section at the top of the tubes.

They decided to make the inserts from aluminium to reduce the possibility of any vibrations damaging the tubes although a variety of metals had been used on the pilot rig. Each insert was a thin tube crimped at one end with three ridges equally spaced along the length to hold the insert in a central position. They needed about 30 km of inserts for the three shells!

Think of the money which might have been saved and how much less CO_2 would have been released into the atmosphere if the system had been well designed in the first place!

This example shows a solution for malfunctioning units with very low tube side velocities. If the reader ever comes across such a situation it is well worth considering this approach although I would hope that anyone considering it would have the confidence to carry the challenging calculations without the need to build a pilot plant.

It also raises the question of whether some form of inserts could be used in an initial design. I had often thought about this one before reading this article but always managed to find a more conventional solution. With a low fouling stream the wire mesh type of inserts for just a short length of tube might have a lot to offer.

Example 10.2 - Performance curves for a C4 Hydrogenation Reboiler.

This study is from a revamp design on a hydrogenation unit which I did in 1997. The program I used was RTF-2. I was fortunate to be working with a very experienced process engineer who I had known for many years. By the time I had finished I was on Rev 3 of the process data – unofficial issues of course – no point in confusing any paper circulating ‘engineers’ during the early stages of a design. I started by trying to keep the existing pipework sizes but they were not acceptable for the new duty and the outlet lines were increased from 20 inches to 24 inches and the tower nozzle was changed.

There were 2354, 25.4 mm diameter tubes inside a 1675 mm shell. The tube length was 5m.

The heating fluid was quench water which was in co-current flow to minimise the liquid heating at the bottom of the tubes. I ran the shell side on another program but the water heat transfer coefficient only accounted for less than 10% of the total thermal resistance so that this introduced hardly any error.

The process engineer wanted the performance at the design duty of 6.7 MW and at a turndown of 45% when the duty would be 3 MW. He also wanted to know how the unit would perform throughout the range of possible static heads. My final runs with the new pipework are summarised in the following tables.

The 50% in the stability column is a warning message that the normal recommended limit for the percentage vaporisation had been exceeded. I could not get a run with a static head of less than 1.2 m for the design case.

Full duty case

HEAD m	% VAP %	Q kW	STABILITY
6.14 HLL	16.9	6696	OK
3.59 NLL	30	6761	OK
2.99	35.6	6744	OK
2.49	44	6682	OK
1.99	56	6573	50%
1.49	75.3	6365	50%
1.2	93.7	6287	50%
1.1	99		No solution

Table 10.1

45% duty case

HEAD m	% VAP %	Q kW	STABILITY
6.14 HLL	7.0	3000	OK
3.59 NLL	12.9	3000	OK
2.99	15.8	3000	OK
1.99	29.5	3000	OK
1.49	45.5	3000	OK
0.99	73.5	3000	50%
0.84 LLL	86.7	3000	50%

Table 10.2

The values from the tables were then plotted on a graph as illustrated in figure 10.10.

I was surprised that I was able to get reasonable computer runs for such a wide range of static heads. All the points were very close to the curves drawn with hardly any spread. The two vaporisation curves clearly show a dramatic increase at the lower values of static head. There was not too much of a decrease in duty at low static heads for the design case but this was mainly due to fouling being 62% of the total thermal resistance. With a lower fouling exchanger the fall off would have been more severe for the very high vaporisation rates.

Predicting performance for extreme operating conditions is a bit like predicting the outcome of a football match. A top four team is playing a relegation candidate and the newspaper reporters, the television commentators and the bookmakers all point to one outcome and yet there is an upset and the minnow wins. Even with today's sophisticated programs if the reader is required to produce curves such as the ones shown in figure 10.10 they should make sure to point out the possible uncertainties for the limiting values.

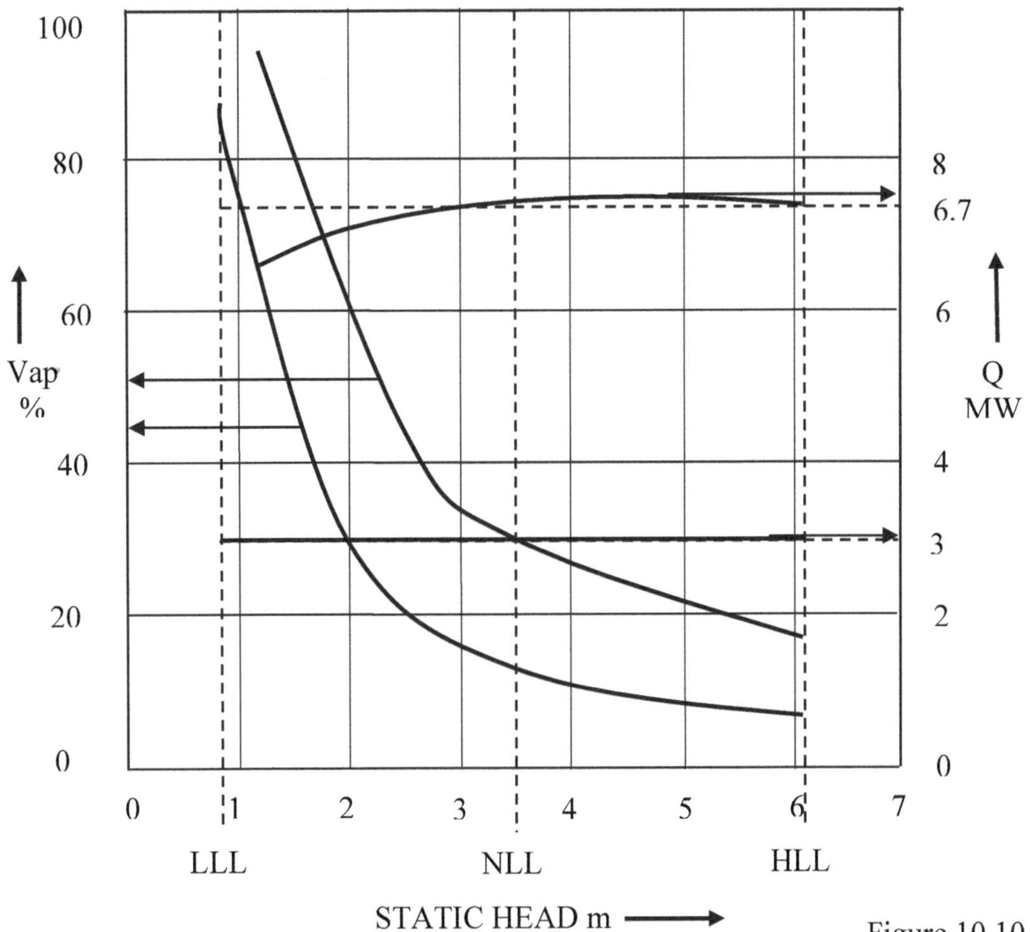

Figure 10.10

Example 10.3 - Performance of deoctanizer reboiler.

It is very unusual to get the chance to investigate the range of possible operating conditions demonstrated in example 10.2. However it does not take too much time to have extra runs to show how the duty and vaporisation will vary for a more limited range.

The following table was part of the thermal data sheet for a steam heated reboiler I once designed.

HEAD M	% VAP %	Q MW
3.2	35.7	4
3.9	28.4	4.15
4.3 NLL	25	4.2
4.8	21.5	4.1
5.3	18.4	3.9

Table 10.3

The target vaporisation rate was 25% which I just managed to achieve for the static head at the normal liquid level. However relatively small changes in static head were sufficient to give a range of between about 18% and 36% vaporisation and there was a reduction in the predicted duty at either side of the design case. There had been no request from process for a table or chart but I considered it to be necessary. It also justified the inclusion of the butterfly valve which I had specified in the inlet line.

Of course it is not only changes in static head which can change the vaporisation rate changes in the inlet and outlet pipework will also have an effect which is why it is essential to always have a note on the thermal data sheet stating that the unit should be rechecked if the pipework has not been finalised.

Example 10.4 - Performance curves for an acid recovery column reboiler..

This is an interesting design for a new plant. The client had another similar plant which had a much lower capacity. The reboilers for the existing plant had been well designed and were not causing any problems. The units for the new plant needed to be much larger and the increase in size could not be accommodated simply by increasing the number of tubes and the shell diameter. It would therefore be necessary to increase the tube length.

Whenever I face a challenging design and I know that there is a similar plant the first thing I do is ask for field data. This was quickly provided and I proceeded to run RTF to see how closely it would predict the field data. The results are summarised in table 10.4 below.

	Field data	RTF Predictions	
	U – W/m^{2}K	U – W/m^{2}K	% Vap
Clean case 1	1368	1393/1153	18/16
Clean case 2	1220	1393/1153	18/16
Fully fouled	910	801	15

Table10.4

They had recorded clean data on two occasions and fully fouled data once. I ran the RTF program for the clean case with no fouling on either side and with the tube side only clean which accounts for the two values in the table. It was a steam heated unit so that the shell side would never be cleaned and I thought that this would be the more realistic of the two runs.

I was pleased with the good agreement with the field data and the computer runs and noted that the computer runs were 'on the safe side'. This gave me confidence that provided I could find a design without any warning messages it would be OK to increase the tube length.

There were several operating cases to consider. Figure 10.11 are the curves drawn for the design case. I wanted to keep the % vaporisation as close as possible to the 15% predicted from the field data.

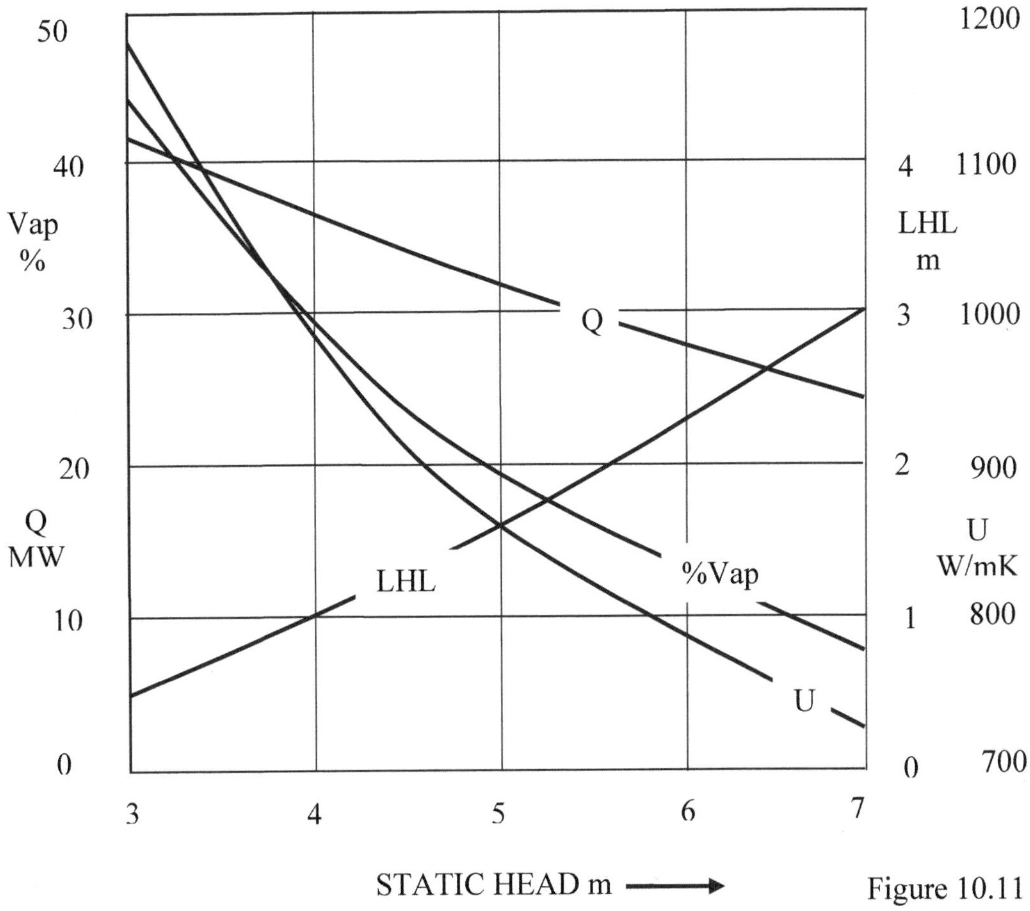

Figure 10.11

The curves are a plot of heat duty Q, %Vaporisation, Overall heat transfer coefficient and liquid heating length LHL against static head. The curves clearly show that the heat duty decreases as the length of the liquid heating section at the bottom of the tubes increases. This is a key feature of the design.

With a vaporisation rate of 15% the duty was only 90% of the required value. However the required value had a 10% safety factor specified for the duty and flows by process. A small increase to 20% was sufficient to achieve the overcapacity value. The only warning messages were at the extremes of the curves, in the anticipated operating area the runs were all OK. I decided that this solution was satisfactory.

There were two shells but I had already agreed with process that there could be two tower nozzles and the exchangers could be directly coupled to the tower without any bends in the outlet pipework. I managed to keep the frictional pressure drop in the outlet pipework to about 10% of the total. I used up over 20% in the inlet pipework most of which was taken up over a valve so that the system could be adjusted.

Examle10.5

This example is taken from some revamp work I once did. The original design had been a real cock up – not quite as bad as the one in example 10.1 but it is a good example of how not to design a vertical thermosyphon system.

The existing outlet pipework was all 14 inch. One computer run was sufficient to confirm where the problem lay. I therefore carried out a series of runs with different sizes of outlet pipe which are summarised in table 10.5 below.

NPS inches	Steam P bar	Steam T °C	% Vap %	Duty kW	ΔP_{out} % Total
14	2.6	130	90	5709	72
	2.0	120	65	4628	67
	1.9	118	60	4358	66
16	2.6	130	66	6158	63
	2.5	127	60	5750	60
	2.3	125	56	5550	59
18	2.6	130	51	6419	52
20	2.6	130	41	6571	40

Table 10.5

The steam temperatures and pressures have been rounded off for convenience of presentation so they will not all coincide exactly with steam table values. The last column is the pressure loss in the outlet pipework as a percentage of the total system pressure drop.

The service was known to be heavy fouling and there was a standby unit. The required duty of 6450 kW was only being achieved very briefly after cleaning before the performance fell off. It is not surprising that the unit quickly fouled up looking at the predicted vaporisation rate of 90% with the full steam pressure.

Increasing the outlet pipework to 18 inch might have been sufficient but I recommended using 20 inch. As well as new pipework this meant a new nozzle on the tower and a new top channel for the exchanger although the existing channel cover could be reused. Not a cheap a cheap modification.

The pressure drop in the outlet line is still a bit higher that I would have liked for a new design but 24 inch pipework with the existing exchanger was not a realistic option.

STEAM GENERATORS/WASTE HEAT BOILERS

The design principles for vertical steam generators are the same as for horizontal units but some of the construction details are quite different. The following three drawings are taken from the design of a unit generating steam at 49 bar from hot gases at 3 bar entering the unit at 320°C. With suitable protection of the metal surfaces it was possible to design the unit with SA 179 carbon steel tubes and low alloy materials for the other components.

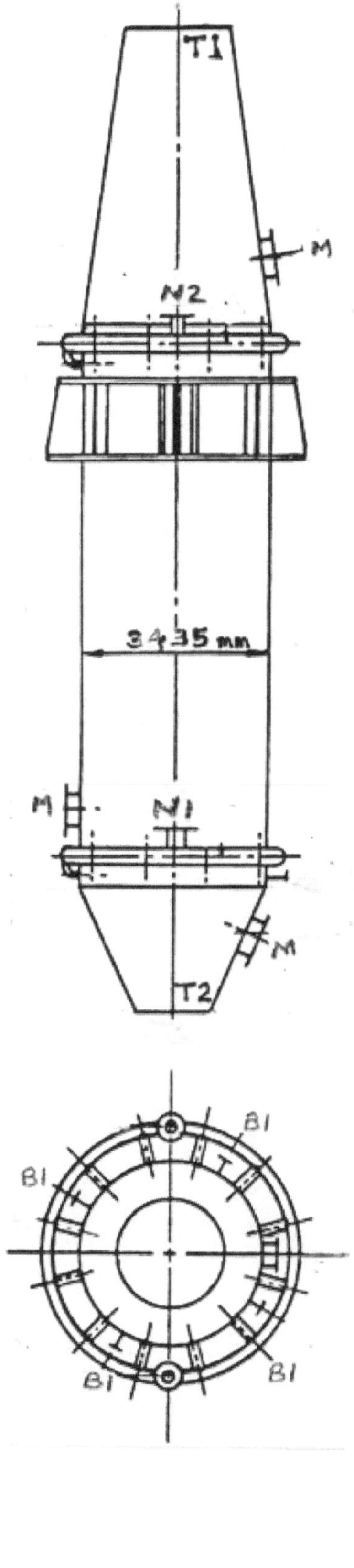

Figure 10.12

Figure 10.12 shows a side view of the unit and a plan view from the top. Figure 10.13 is a sectional view with details of the insulation.

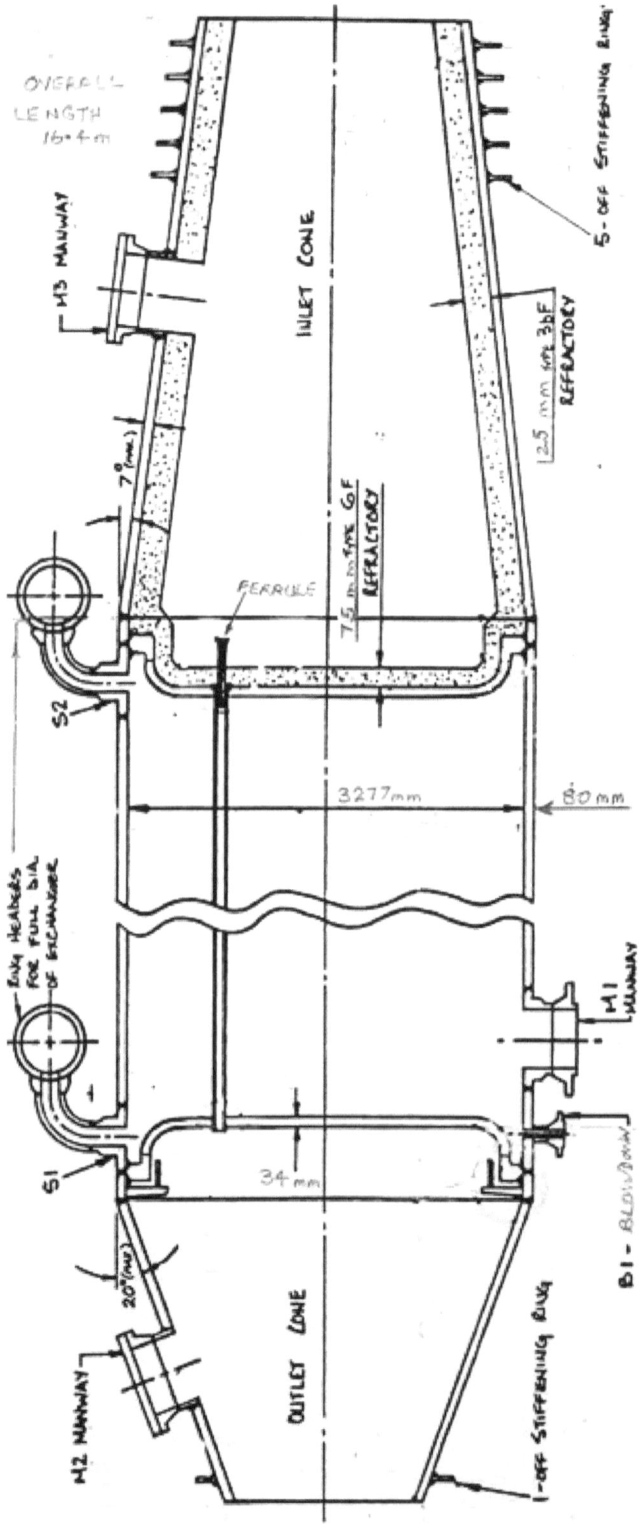

Figure 10.13

The design is over 20 years old and was originally in imperial units. The first two drawings I have updated to SI units so that a young reader can easily get a concept of the proportions without having to carry out any mental conversions. The last diagram showing the tube layout and shellside internals I have left in the old units.

There are clearly lots of interesting mechanical features in the design but I will only comment on these where they have an impact on the thermal design

The conical channels are tapered at different angles. The inlet is at 7° and the outlet 20° resulting in the inlet channel being much longer than the outlet channel. The long inlet channel allows for a gradual expansion of the hot gas which results in a good distribution of the gases into each tube.

The tubes are 6852 mm, (67.5 ft) long which is an odd length in either system of units. Once tubes are longer than the normal standard lengths they become specials and any length can be used. The overall length is 16.4 m which is longer than the swimming pool in the Gym I am currently using.

The ferrules used at the entrance to the tubes extended 50 mm beyond the inner face of the tubesheet. They relied on a small gap between the ferrule and tube to provide a thin insulating layer of static gas. This arrangement minimises the thickness of the ferrule which in turn minimises the pressure drop through the ferrule. They were manufactured to produce a slight interference fit and needed to be driven into the tubes.

Flexible stayed thin tubesheets have been used which were covered in the previous chapter on horizontal thermosyphons. These are attached to the shell in a way that makes it is possible to have the shellside inlet and outlet nozzles, S1 and S2 to the channel side of the tubesheets. This is a safeguard against forming a blanket of vapour under the top tubesheet which could insulate the top tubesheet from the boiling liquid and increase the tubesheet temperature. At the bottom this arrangement minimises the chance of deposits building up on top of the tubesheet because there are blowdown connections in the same annular space as the inlet nozzles.

Even distribution into and out of the shell is achieved by having 12 inlet and 12 outlet nozzles connected to ring headers. The ring headers have two nozzles which are connected to the risers and downcomers.

The final diagram shows the tube layout and baffle arrangement. I suppose the baffle arrangement could be described as quadruple segmental but they are normally simply referred to as strip baffles. This arrangement provides support for the tubes whilst minimising the resistance to fluid flow through the bundle. Note that the top baffle is type A which minimises the flow from the second space down to the annular section containing the outlet nozzles. This gives a better sweeping of the top tubesheet than would be the case with a type B baffle at the top. As far as a computer run on such an arrangement is concerned using triple segmental baffles would probably make little difference to the pressure drop. This would be safe for the frictional pressure drop and of course the designer should be conscious of what proportion of the total system pressure drop it accounts for.

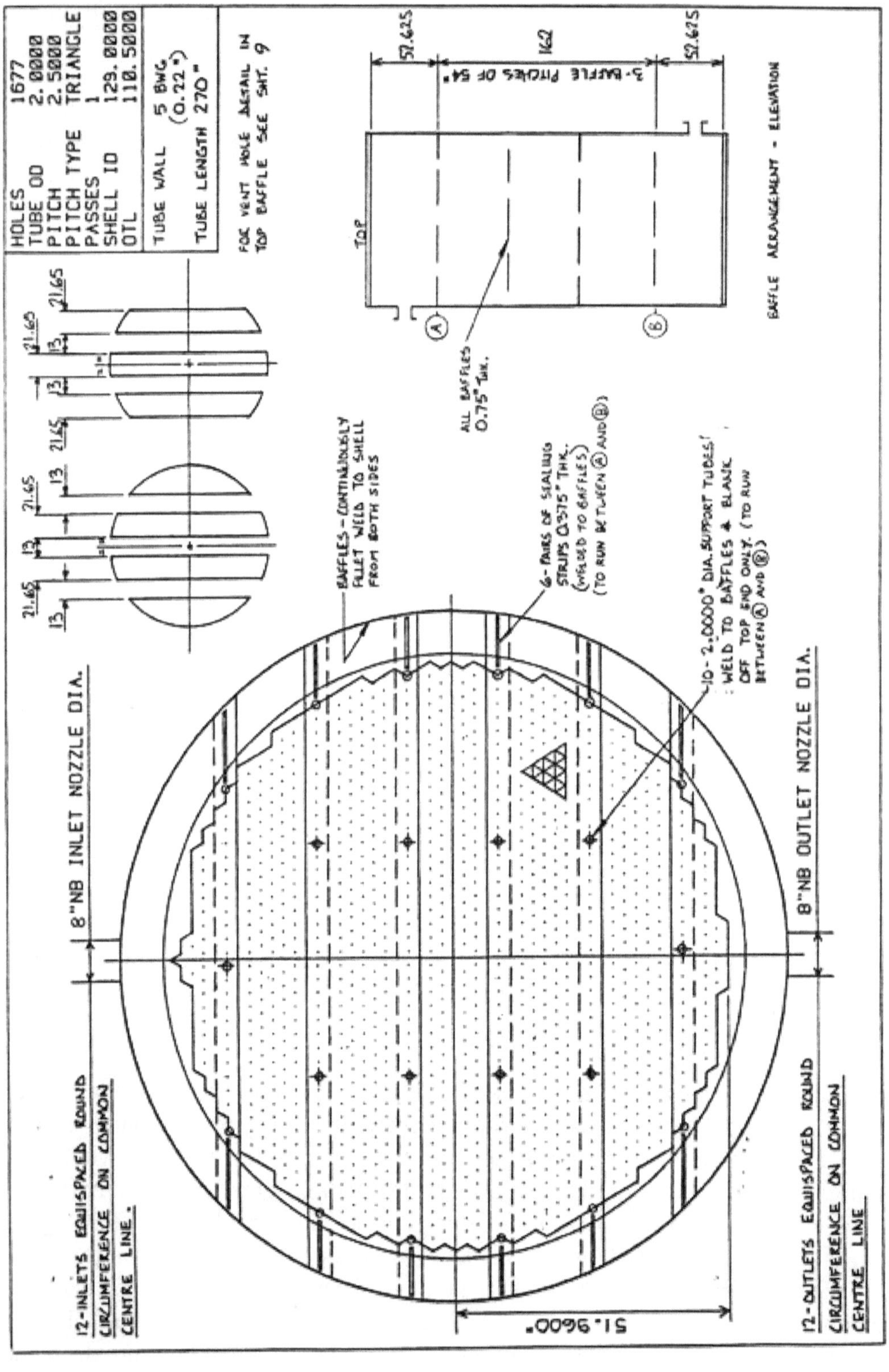
HOLES 1677
TUBE OD 2.0000
PITCH 2.5000
PITCH TYPE TRIANGLE
PASSES 1
SHELL ID 129.0000
OTL 110.5000
TUBE WALL 5 BWG (0.22")
TUBE LENGTH 270"
FOR VENT HOLE DETAIL IN TOP BAFFLE SEE SHT. 9
12-INLETS EQUISPACED ROUND CIRCUMFERENCE ON COMMON CENTRE LINE.
8"NB INLET NOZZLE DIA.
12-OUTLETS EQUISPACED ROUND CIRCUMFERENCE ON COMMON CENTRE LINE
8"NB OUTLET NOZZLE DIA.
51.9600"
BAFFLES - CONTINUOUSLY FILLET WELD TO SHELL FROM BOTH SIDES
6-PAIRS OF SEALING STRIPS 0.375" THK. (WELDED TO BAFFLES) (TO RUN BETWEEN A AND B)
10-2.0000" DIA. SUPPORT TUBES: WELD TO BAFFLES & BLANK OFF TOP END ONLY. (TO RUN BETWEEN A AND B)
ALL BAFFLES 0.75" THK.
TOP
3-BAFFLE PITCHES OF 54"
57.625
162
BAFFLE ARRANGEMENT - ELEVATION

Figure 10.14

LOW RESISTANCE T JUNCTIONS

I was recently faced with a novel problem in a reboiler system. There were two reboilers feeding the tower through a common tower nozzle. This situation frequently occurs when either one reboiler is on standby with only one working or there are two reboilers working with equal flows from each side.

In this case the normal operation required most of the flow to be from reboiler A with a trickle flow from reboiler B to keep this unit ticking over so that it could be quickly brought into service when needed. This is illustrated in figure 10.15 below.

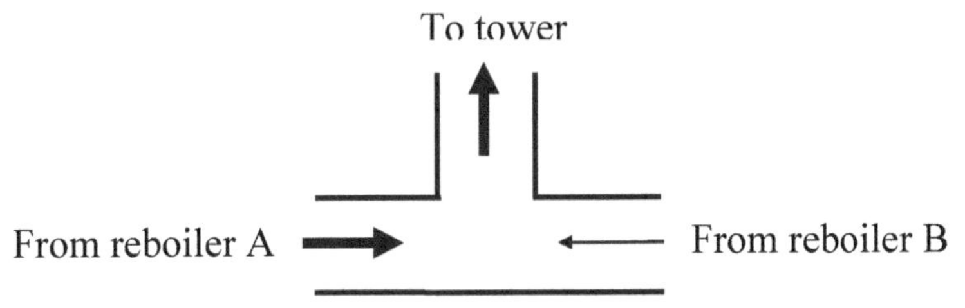

Figure 10.15

The danger of using the normal type of T junction illustrated is that a lot of the flow from A would take the easy path and finish up flooding reboiler A instead of rounding the hard T into the tower.

Being mindful of how important it is to minimise the hydraulic resistance in reboiler outlet pipework I thought that a novel problem warranted a novel solution. I started off by drawing two long radius bends on a sheet of graph paper as shown in figure 10.16 below.

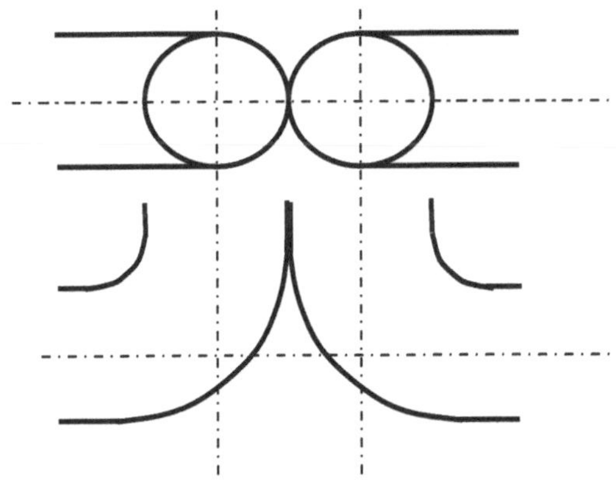

Figure 10.16

I then proposed to cut each bend through the vertical centre lines and discard the central section. The two outer sections could then be welded together as shown in figure 10.17 (a) below.

I realised that just welding the two sections together would still not be sufficient. I then added flow deflecting vanes and a short extension to the centre of the horizontal axis as shown in (b). The fluid flowing from either side could no longer 'see' the continuation of the horizontal pipe. I was therefore confident that hydraulically this design would work.

The next stage was to discuss this with the best mechanical engineers I knew. They could see no problems with fabrication but it would of course need a finite element analysis to investigate the stresses. They also suggested that stiffening webs along the centre lines might be required. Provided these were tapered at each end they would only make minor distortions to the fluid flow.

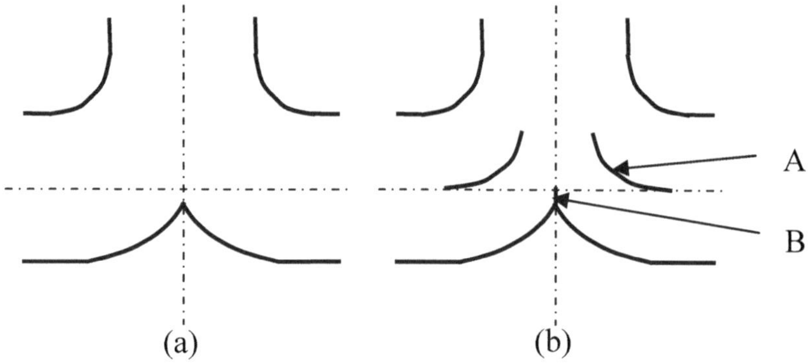

Figure 10.17

I offered my proposal to process and piping who were bemused and offered comments such as: "A bum solution", and "An interesting cheeky solution". In the end no one wanted to try anything new and I had to settle for a conventional 45° lateral T as shown in figure 10.18 below.

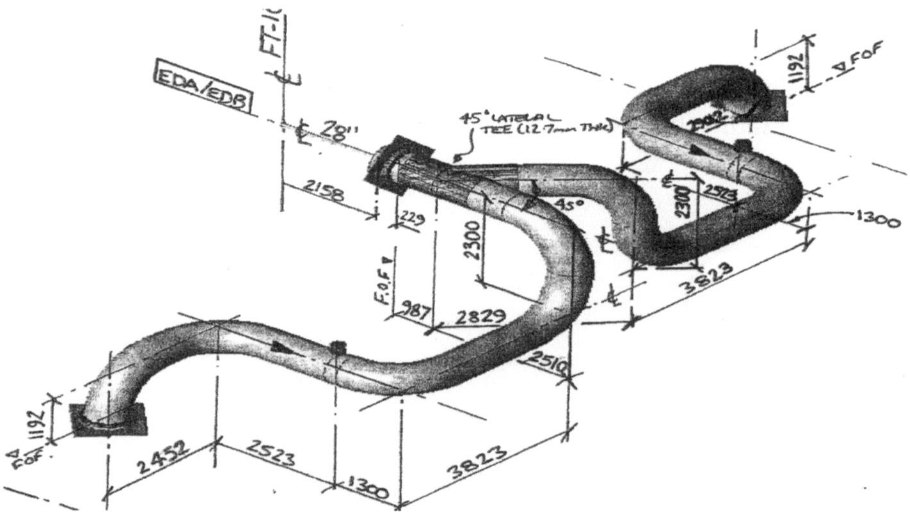

Figure 10.18

Although this proved to be yet another of my ideas which has never been tried out I still think that there is a place for a low resistance T in the industry. A lot of both horizontal and vertical reboiler systems require T junctions. They invariably dominate the frictional hydraulic resistance in the outlet pipework. They would of course cost a lot more as a single item but this would be more than recouped by having a more satisfactory system. It would make an interesting research project to establish the equivalent length of a low resistance T compared with a conventional T. If the results are as good as I suspect they could be then a manufacturer might take this up and offer them as a 'standard' fitting.

CONCLUSIONS & DISCUSSION

The design of vertical thermosyphons is one of the most challenging situations a thermal design engineer will face. They also provide an interesting criterion for assessing a thermal designer's ability. I can think of some designers who have been around as long as I have who will run a mile rather than pick up a vertical thermosyphon unit to design. At the other extreme the engineers I have come to respect would far sooner design one than a straightforward, boring, liquid/liquid unit.

One problem with designing vertical thermosyphons is that they can take up a lot of time. A single computer run with a target vaporisation rate and a design duty are all that there will be time for in a feed study to get an approximate price for a project. However if it is a serious design and is actually going to be built a lot more work is required looking at a wide range of possible operating conditions.

I once worked for Snamprogetti on the Dushanzi project which was to be built in the north of China. Politically it was a strange project. The full time staff at the UK office in Basingstoke did not seem to want to get involved. The task force was almost entirely contract labour except for the process engineers who had been sent over from the Italian office. One exception was a young graduate who joined four contract engineers in the thermal design group. He was very bright and as keen as mustard and would have made an excellent thermal designer if he had not seen the light and decided to widen his horizons later in his career.

My role was quite clear since I knew that I had been recruited because of my experience with reboilers. (I had negotiated a very good rate of pay – I never used agents if I could avoid them – somehow someone found out about the rates and I became known as Mr 10% which did not go down well with some of my 'supervisors' who were on less). Other roles were up for grabs and there was one individual who claimed to be experienced in both the thermal and mechanical aspects of heat exchanger designs and was given the job of supervising both the thermal and mechanical sections. I later found out that his thermal experience had been while he was working for a client. He had run some alternative cases on units that had already been finalised – I even recognised the units because I had recently designed them whilst working at Foster Wheeler.

As a supervisor he seldom did any technical work and managed to clash with most of the mechanical engineers who had realised his technical limitations in their field. The first of many clashes with me was when he produced a progress analysis for the thermal designs. He smugly pointed out that the new graduate had turned round more units than I had during the precious few weeks. He had also discovered that another thermal designer was checking designs much quicker than anyone else in our group.

The work I had been doing on the thermosyphons and kettles on the refrigeration plant had involved spending a lot of time with the piping and process engineers particularly the latter. Up until then I had a rather stereotyped view of Italians which was limited to jokes about armies and diving footballers. The Italian process engineers I worked with were very sound and they were excellent company in the pub.

Fortunately one of our thermal design team was a very good thermal design engineer who had once worked full time at Snamprogetti and I now know very well. Paul was the last to join the group and had to spend most of his time on the plate fin units but he was well aware of what was taking place in the group. He knew the hours that needed to be spent on each design. After the meeting he nicknamed the individual who was so quick with the checking as 'the lightning checker'. He had observed, as I had that he was only checking the input and output data without any analysis to determine if it was an optimum design. He also knew that I had been spending a lot of time with the new graduate closely supervising his work.

Both Paul and the process engineers had communication channels above our 'supervisor' which helped to straighten things out. Ironically my only concern had been that I was not spending sufficient time on the reboilers!

One last story on designing vertical thermosyphons may be of interest to the reader. I was working at Bechtel at the time and the head of the heat exchanger section asked me to check two vertical thermosyphon units he had designed. I was told that they only needed a quick check because the order had already been placed and they were in the process of being built – a strange situation!

They had been designed using RTF. I did not like the first one I checked. The output was full of warning messages and was clearly unsatisfactory. I quickly went on to look at the second one and there was no output at all –just an input and a no solution message. I then found that there were some hand calculations and the unit had been pro rated from the first design using the same coefficient and modifying the MTD!

When I reported back it was not a happy meeting. I made it clear that I could not sign off either unit but I would try and sort out the mess. I managed to get a run without any warning messages for the first unit by using a larger outlet nozzle and pipework but it was about 30% undersurface. The pipework for the second unit was OK and it was about 25% oversurface. Both units had similar heat loads and they were on the same tower. It was time to see the process engineer who had designed the tower.

The process engineer was quite sympathetic and could see the funny side of my dilemma. He spent a lot of time looking at the effect of repositioning the heat loads while I redesigned the exchangers for the heat loads they could deliver. Finally we came to a solution which we could both accept. There was still time to change the tower nozzle but there was an embarrassing extra cost for changing the nozzle on the exchanger which was already being assembled.

Good luck with designing your vertical thermosyphon reboilers.

SPECIAL MECHANICAL FEATURES – CHAPTER 11

Introduction

I found that chapter 2 on classification and construction was getting out of hand, particularly when I kept on finding omissions and improvements which meant shuffling around a lot of pages to keep the text and drawings in sensible positions. On reflection perhaps I should have found some way to split it up into at least two chapters. Anyway chapter 2 included most of the features common to most shell and tube exchangers. This chapter covers construction features which are not the norm but are nevertheless very important for some designs.

INLET & OUTLET DISTRIBUTORS

There are two basic arrangements for distributing a fluid into and out of the bundle of a heat exchanger.

ANNULAR DISTRIBUTORS

When shell and tube heat exchangers are required to handle large volume flow rates of gases or vapours then large nozzles are required. Using standard, single inlet/ outlet nozzles may create the following problems:

(a) Excessive loss of space available for tubes.
(b) The nozzle diameter may exceed 50% of the shell diameter.
(c) Excessive inlet/outlet lengths which may result in poor heat transfer rates in the spaces adjacent to the nozzles and possible vibration problems.
(d) Uneconomical channels in units with very high pressures on the tube side but low pressures on the shellside. A lot of the cost of high pressure units is in the channel. If a distributor is used there is no loss of space available for tubes which minimises the channel diameter.

Figure 11.1 below shows a side view through the centre of a shell with a distributor.

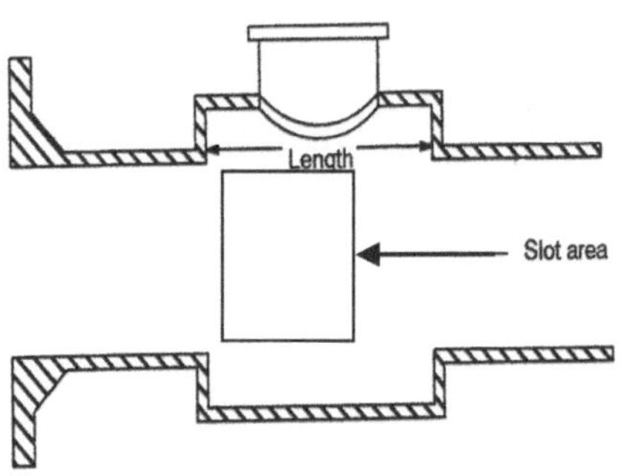

Figure 11.1

This is a box type of distributor where the sides of the distributor are at 90° to the shell. The sides may also be sloped when they are termed cone type distributors. The slot area is normally offset from the nozzle centre line towards the tubesheet to minimise any dead space.

Figure 11.2 below shows a sectional end view.

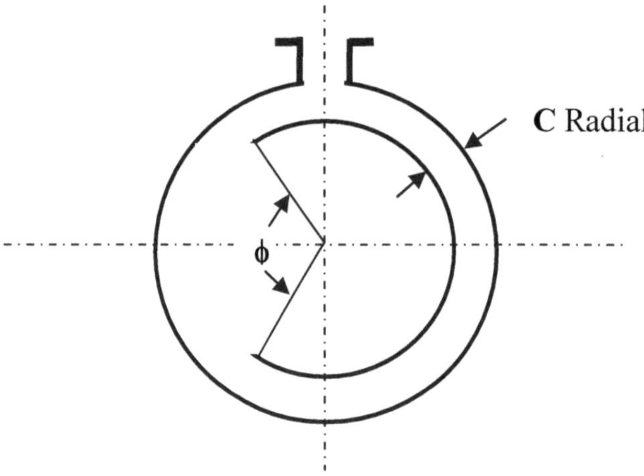

Figure 11.2

The radial clearance C is the inside diameter of the distributor minus the outside diameter of the shell. ϕ is the angle of the slot opening into the shell. If L is the length of the distributor and L_S the length of the slot the flow paths may be calculated as follows:

The nozzle

This can be sized using the normal criteria for a shell nozzle with an impingement plate. With an impingement plate the minimum distance from the nozzle to the plate is $D_N/4$. This fixes the **minimum** radial clearance C for the distributor. This should then be rounded up to the nearest 10 mm. For nozzles less than 10 inch NB where the above method predicts less than 50 mm clearance then 50 mm should be used. (This might occur on a small diameter shell with a high pressure on the tubeside.)

The distributor

The flow is in two directions away from the nozzle the flow area is therefore 2 x LC. The distributor area should never be less than the nozzle area the minimum value can therefore be found as follows:

$$\pi/4\ D_N^2 = 2LC$$

Hence

$$L_{Smin} = \frac{\pi\ D_N^2}{8C} \qquad [11.1]$$

There is also a mechanical consideration which may affect the minimum length of the distributor. This is illustrated in figure 11.3 below.

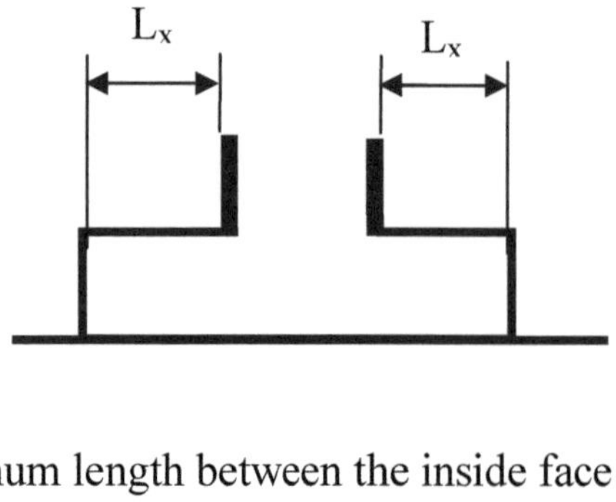

Figure 11.3

The minimum length between the inside face of the distributor and the outside diameter of the nozzle L_x should be 150 mm. The distributor length L would then become 300 mm plus the outside diameter of the nozzle. (Check with a mechanical engineer if in doubt.)

The slot area

With any distributor arrangement at the inlet to a bundle the fluid entering the bundle will impinge directly on to the tubes. It is therefore essential to keep below the TEMA momentum flux limits of 2232 kg/ms^2 (500 lb/fts^2) for none abrasive single phase fluids and 744 kg/ms^2 (1500 lb/fts^2) for all other gases and vapours. This establishes the minimum slot area.

Any slot will have a circumferential length proportional the angle ϕ in figure 11.2. The area can therefore be found from the following equation:

$$\text{Slot area } A_S = \frac{\phi}{360}\pi D_S L_S \qquad [11.2]$$

The angle and length of the slot can be adjusted to give sensible rounded values. The only limitation on the length of the slot is that it must not exceed the length of the inlet/outlet space for the first baffle.

Slot positions

The position of the slot/slots depends on the type of baffles used. Three possible positions are shown in figure 11.4 below.

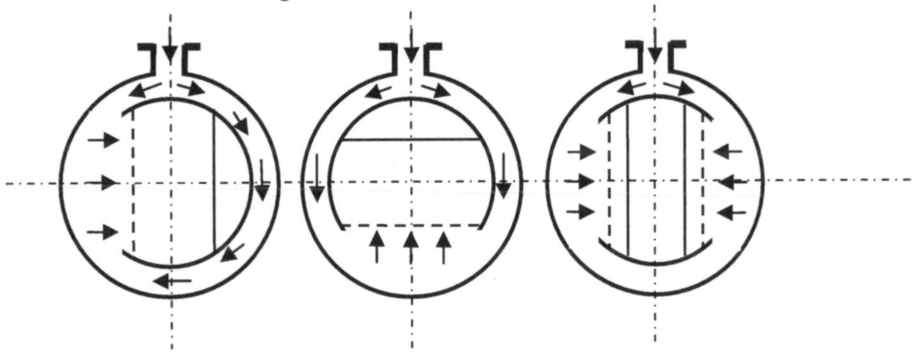

(a) (b) (c) Figure 11.4

Positions (a) and (b) are for single segmental baffles. The first baffle cut is remote from the opening to ensure that there is no dead space. For the horizontally cut baffles the distribution of the flow into the bundle will be quite good because all the fluid has taken equal flow path lengths around the distributor. For the vertically cut baffles it is clearly not as good because more fluid will take the shorter path around the distributor into the bundle. To take account of this distortion you can allow some safety factor on the momentum flux into the bundle.

Position (c) shows a two slot arrangement for use with double segmental baffles. The flow paths are the same on each side which improves distribution compared with a single segmental baffle. In this arrangement there are two openings which must be taken into account when sizing the slots. This is probably the most common arrangement because large volume flow rates often require double segmental baffles for both pressure drop and vibration reasons.

Tapered 360° opening

Figure 11.5 below shows a sketch of a distributor with a tapered slot opening around the full circumference of the shell.

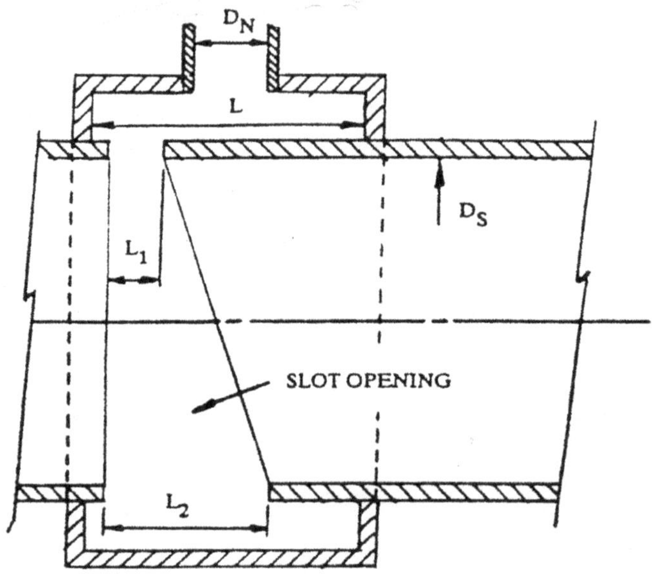

Figure 11.5

For this arrangement:

$$\text{Slot area } A_S = \pi D_S (L_1 + L_2)/2 \qquad [11.3]$$

With this design of slot it is theoretically possible to produce an even flow rate into the bundle around the circumference. I have never used this arrangement so I will leave it up to the reader to try and devise an equation relating L_1 and L_2 to an average value.

Example 11.1

This is taken from a combined feed exchanger I once designed. There were eight shells in series arranged in four banks of two stacked shells. The feed on the shell side started off as all liquid and finished up mainly liquid. The effluent on the tube side started off as all vapour and finished up mainly liquid. The client had experience in cleaning inside U tubes so BEU units were used to minimise the cost and reduce the chance of gasket leakages. I used double segmental baffles with cone distributors as shown if figure 11.6 and 11.7.

The opening into the bundle does not quite cover the full circumference because it was necessary to provide support for the far end under the U tube support plate. I left it to mechanical to decide what supports they need between the bottom of the shell and the bottom of the distributor. I have shown a single support in figure 11.7 but two could have been used since this would not affect the fluid flow path.

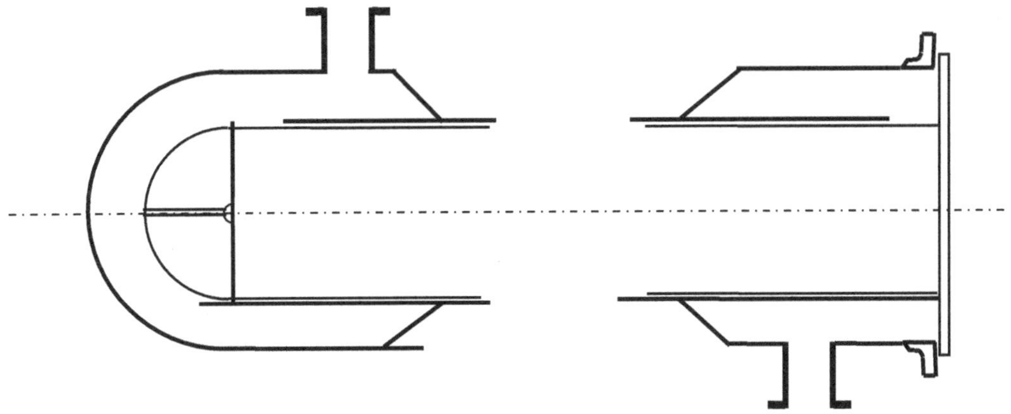

Figure 11.6

At the near end the shell flanges were on the distributor rather than on the shell which is the normal situation for most distributors. Mechanical supports could be added between the distributor and shell provided there was not too much interference with the flow around the distributor.

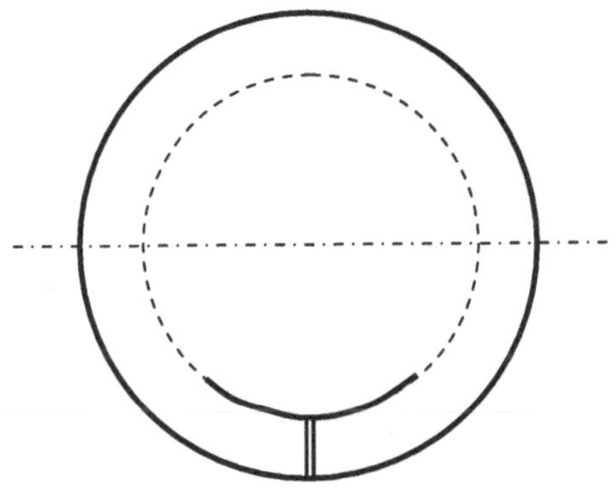

Figure 11.7

I would not use a design with a 360° opening as shown in figure 11.5 with any removable bundle because of the danger of the bundle snagging on the gap at the bottom.

RING HEADERS WITH MULTIPLE NOZZLES

These can be used where it is necessary to distribute a fluid around the entire periphery of the bundle. One such situation is for vertical waste heat boilers and there is a practical example of such a design at the end of chapter 10 on vertical thermosyphon reboilers.

FIXED TUBESHEET EXCHANGERS

The first procedure in analysing the mechanical requirements for a fixed tubesheet exchanger is down to the thermal design engineer, the process engineer and one of the clients engineers. I would recommend compiling a table such as the one below.

	Temperatures °C						Pressures barg		Cycles
	Shellside fluid		Tubeside fluid		Mean metal				
Case	In	Out	In	Out	Shell	Tube	Shell	Tube	
1									
2									
3									
4									
5									
6									
7									
8									
9									

Table 11.1

Cases which may be included in the list are as follows:

(1) **Normal operation**. Metal temperatures from the design run.
(2) **Start up** – After the tubes have been cleaned there will be no fouling factor on the shell side. A separate computer run is required.
(3) **No shell side pressure.** – If a pump fails there will be little or no pressure. The metal temperatures for this case should be the most severe of cases 1 and 2.
(4) **No tube side pressure**. – As for 3.
(5) **Shell side hydrotest.** – Shell side test pressure as determined by the mechanical design code, ambient temperature and pressure on the tube side.
(6) **Tube side hydrotest** – As for 5.
(7) **Emergency.** – Process should provide details of any possible emergency case which is not covered elsewhere.
(8) **Steam out –** This is unlikely to be a problem for the shell side unless the shell side and tube side materials are different. If the tube side only requires steam out then it will probably define the controlling case. One way to overcome this is to add an instruction plate to the exchanger and a section in the operating manual that makes it clear that steam should be applied to both sides when the tubes are cleaned.
(9) **Storage/shutdown**. – This will only apply to exchangers that have different materials for the shell and tubes and the climatic conditions for the site and fabricator are significantly different. For example I once worked on a job where the exchangers were likely to be built in India where the workshop could have been at say 30°C and the winter design temperature for the site was -40°C.

The tube wall temperatures should be for the worst pass if there is more than one tube side pass.

The pressures used for the operating cases are normally the operating pressures. Some clients may require the design pressures to be on the safe side.

The last column in table 11.1 is a relatively recent addition to tables that I have used for many years. The maximum number of thermal cycles that a particular case is likely to be taken through during the life of the exchanger is necessary for the mechanical engineer to carry out fatigue calculations. A few brief incursions above a particular stress limit may be acceptable when regular incursions would not. Figures for this column should be provided by the process engineer or client.

The list is not exhaustive it is up to the thermal design engineer and the process engineer to think up any other possible situations. For condensing fluids it may be necessary to consider full vacuum possibilities. I have worked with a list of 12 cases in the past.

It is normal practice to 'round' the metal temperature values from computer runs – either up or down to give safe values.

MECHANICAL CALCULATIONS

The table 11.1 is then passed to a mechanical engineer who will run one of the many computer programs now available to analyse FTS items. The aim of these calculations is to determine:

(a) The maximum stresses in the tubes.
(b) The maximum stresses in the shell.
(c) The thickness and stresses in the tubesheets.

If the stresses are within the limits of the code being used are acceptable the unit is satisfactory without any expansion bellows. If they are not an expansion bellows is necessary and a required movement for the bellows will be calculated.

It is beyond the scope of this book to go into the details of these calculations. I vaguely remember once have to do them using hand calculations and once was quite sufficient. However one aspect of these calculations still puzzles me to this day. A key component of the calculations is working out the stresses in the shell and tubes due to thermal expansion alone.

There are two possible cases. If the shell is hotter than the tubes then the tubes will be in tension. I have no problem with this case. When the tubes are hotter than the shell it becomes more interesting. The tubes are in compression and there is a limiting case when the tubes start to buckle. Euler's theory is used to analyse the point of buckling and I can still remember the problems I did as a student using his equations. These involved examples such as one leg of a high water tower becoming overstressed and the spectacular images of what happened next.

However I can not visualise anything very dramatic taking place if a tube 'buckles'. Any bending of the tube will immediately relieve any thermal stresses. Exchangers are assembled in a horizontal position and the tubes are never quite straight. The bending of tubes between supports is very obvious when tubes are stacked on a rack prior to use. Figure 11.8 shows an exaggeration of the shape of a tube supported by two baffles between the tubesheets.

Figure 11.8

The tubes are bent before the heat exchanger is used. Surely a compressive stress will just bend them a little bit more? I have put this question to many mechanical engineers over the years. Their answers have ranged from "do not worry about thermal stresses when the tubes are in compression" to the more pragmatic answerer "you could be right but we only have one design method." I have never had a convincing argument against the tubes just bending a bit more.

The attitude of not worrying about tubes in compression goes back to the days before vibration checks. We now know that if a tube is in compression it could make a potential vibration problem worse. Vibrations are covered in chapter 14.

The current methodology of designing FTS items is something we have to accept but perhaps it is time for someone to come up with a 'modified Euler theory' for application to heat exchanger designs? Meanwhile I always feel more confident when a design is near the limit for requiring a bellows if the tubes are in compression.

DIFFERENTIAL TEMPERATURE LIMITS

Many client specs will specify a maximum value for the difference in temperature between the tube wall and the shell before an expansion bellows is required. Typical values range from 28 K to 50 K. I am invariably against blanket limits in design specs and this is no exception. If the calculations clearly indicate that there will be no problems with a higher value I can see no reason for not using one. This is well illustrated in example 11.2.

Like many fixed limits and prohibitions in specs they usually originate from a past bad experience in the field. In the case of FTS designs I suspect that most of the field problems originate in either failure to cover all possible situations in table 11.1 or using expanded tubes. With the tubes in either compression or tension over a period of time the tubes may start to loosen and leak. I would always recommend a full strength weld for all FTS items.

SHELL EXPANSION BELLOWS

There are two distinct types of expansion bellows which can be used in FTS items. Figure 11.9 below is generally known as a thick-wall bellows.

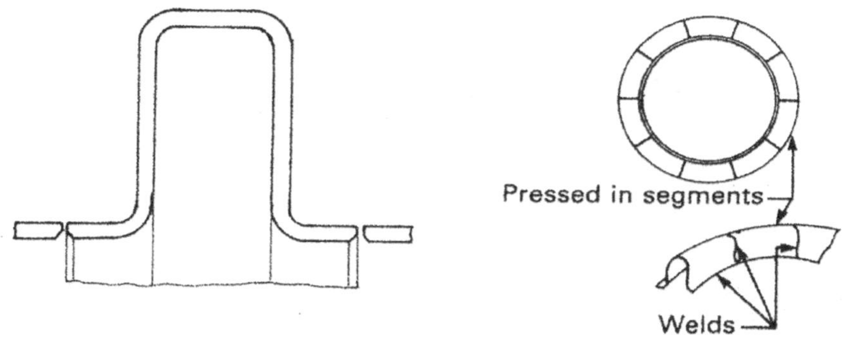

Figure 11.9

A number of pressed segments are welded together to form a single convolution. The material used is invariably the same as the shell and so is the thickness. The advantage of this construction is its strength and robustness. The disadvantage is that the movement is restricted to a maximum value of about 5 mm. In theory this restriction can be overcome by having more than one convolution and there may be an occasional situation where this might be justified. In practice if more movement is required then it is much more economical to use a thin-wall bellows as illustrated below in figure 11.10.

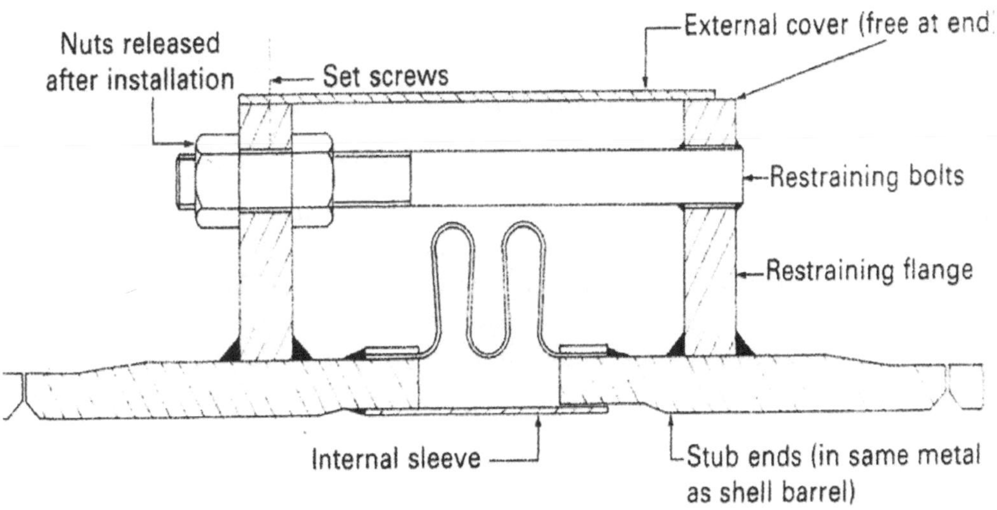

Figure 11.10

The convolutions are very thin compared with the thickness of the shell and although the sketch only shows two there is no limit to the number which may be used to satisfy any required movement.

The disadvantage of the thin-wall bellows is that it is quite fragile and special precautions have to be taken during construction, transportation and installation on site. The sketch shows a method of protecting the bellows. The details of construction can be left to the manufacturer of the bellows and the mechanical engineer on the project to sort out.

The materials used for thin-wall bellows are seldom the same as for the shell. The materials which may be used include; austenitic stainless steels, Incoloy, Inconel and Monel. The material needs to be specified on the thermal data sheet. The project metallurgist/technologist should always be consulted on the choice of material. It can be tempting but dangerous to save money by using a cheap material for a thin-wall bellows. The worst cock up I have heard of is when stainless steel was selected for one job. The site was a few miles from the sea and was not classed as coastal. However anyone who has lived a few miles from a coast will confirm that there are times during stormy weather when spray can carry a long way inland. Chloride ions in sea water are deadly for stainless steel particularly very thin stainless steel. The thin walled bellows on that plant did not last long before they all had to be replaced.

Both sketches in this section were originally drawn by Ted Saunders and have featured in a number of publications. He was the chief engineer at Whessoe Heavy Engineering Ltd in Darlington UK. I have very pleasant memories of my dealings with him in the sixties when I was head of the heat exchanger section at Stone and Webster. That was in the days when Whessoe could quote for all the shell and tube units for an ethylene plant. Sadly I am not sure we have any vendor in the UK who could do that now.

RETROFITTING AN EXPANSION BELLOWS

In recent years I seem to have spent far more time on revamp work than on grass root plants. A FTS item with no expansion bellows, which was satisfactory for the original service, may still have sufficient surface area to be useful for the new conditions but is not satisfactory for the new expansion problems.

It is not too difficult to visualise retrofitting a thin-wall bellows and it is not unusual to replace these during the life of the exchanger. However I was once working on a job which now required a bellows for an existing unit and only a thick-wall bellows was acceptable to the client. This was discussed among our group and no one was quite sure if it could be done. Motherwell Bridge was contacted. They could see no problems and by coincidence had just completed a retrofit of a thick-wall bellows for another customer.

THICK or THIN?

The thick-wall bellows have a limited movement but this is normally sufficient for most applications leaving a choice of types. Often a client will be decisive on this one in their specs. I have a little ditty “BP are thick, Shell are thin.” I have not read their current specs to check if this still holds.

If the thermal design engineer has to make a choice I always feel more comfortable with thick-wall bellows on vertical units but have no preference for horizontal units. With a thick-wall bellows there are no venting and draining problems. With a thin-wall bellows venting and draining is more difficult in either orientation.

DESIGN METHODS & STANDARDS

Appendix A in the ninth edition of TEMA has a section on the equations which are used to design fixed tube sheet exchangers. TEMA also has a computer program.

The main standard for the design of expansion bellows is the Standards of the Expansion Joint Manufacturers Association which is abbreviated to EJMA. This is written in the same style as TEMA and runs to over 180 pages. Any reader wishing to find out more on this topic should look at a copy.

Example 11.1

I once picked up a vent condenser to check. It was situated at the top of a structure above the main overhead condenser and there were very conservative restrictions on the length and weight of any removable bundle. There was cooling water inside the tubes which required mechanical cleaning and U tubes were not acceptable. The shell side fluid was steam with entrained air which was being vented at atmospheric pressure and the condensate recovered. The design temperature was about 110°C and the design pressure was very low since the operating pressure was only just above atmospheric.

I was looking at a design with two AES shells in parallel. They would have worked of course, but the cost and the maintenance seemed ridiculous for this service. Although he had now left the job I knew quite well the individual who had designed this unit and he would have kept to all the specs regarding cleaning and no expansion bellows etc without too much thought of the overall picture.

If ever there was a good ‘special case’ for using an expansion bellows this was surely it. If the bellows had leaked there would be some extra loss of condensate until it was repaired – hardly a disaster. After checking the process data in the input I went straight for a single shell fixed tubesheet design. I even managed to achieve a higher cooling water velocity which would help the case I was preparing.

I listed all the possible conditions for the fixed tubesheet design and about **90 K** was the maximum temperature differential between the shell and tubes. I left my new design on the desk of an excellent mechanical engineer together with a simple note – “John, what type of bellows?”

I called in to see him the next day and was amazed to hear “none”. He explained that with such a low design pressure the unit was OK without any bellows. That made my day and saved the client a lot of money. There would have been a lot more CO_2 in the atmosphere if the unit had been checked by someone with a similar outlook to the original designer.

Example 11.2

This example goes back to the days when Procon were still around and does not have a happy ending. It was one of my first short contracts during a ‘vacation’ from the academic world. The job I was working on was an oil refinery.

There were about four units which I checked that were AES types with very clean fluids on the shellside. I remember one was a gasoline product cooler. They all had very moderate temperature differentials.

I could see the potential for saving a lot of money by making them into fixed tubesheet exchangers. Not only saving money but designing safer units because a FTS unit with strength welded tubes is a lot less likely to leak than an AES unit with all the extra gaskets to worry about.

I discussed this with other members of the group and the supervisor of the heat exchanger section. He took it up with the engineering manager. They all agreed that it was a good idea and told me to carry on and change them.

A week or so later I was called in to see the America project manager. He was very blunt in stating "we do not use FTS exchangers on refineries". He listened to my explanation of the advantages then replied. "It does not matter how much better they might be, the savings are peanuts in terms of the overall cost of the plant and we never use FTS exchangers." That was it; I had to change them all back again.

The technological curtains drawn between different branches of the same industry can still be very frustrating.

FLOATING HEAD EXPANSION BELLOWS

TEMA types AES Spl or BES Spl with a single pass on the tubeside were covered in chapter 2 on classification and construction. The drawings were thumb nail sketches and did not give any details of the complexity of these arrangements.

Figure 11.11

Figure 11.12 below is a section showing the details of one possible type of tube side expansion bellows. Figure 11.11 is a photograph of a floating head expansion bellows during assembly in Whessoe's workshop. Restraining bars have been fitted prior to the tube side hydraulic test. The expansion bellows are not as fragile as those used for thin-wall bellows on the shell side but they still need handling with great care.

I will leave it to the reader to count how many gaskets there are in figure 11.12. Any gasket is a potential source of leaks. This together with the cost is why tube side expansion bellows are avoided whenever possible.

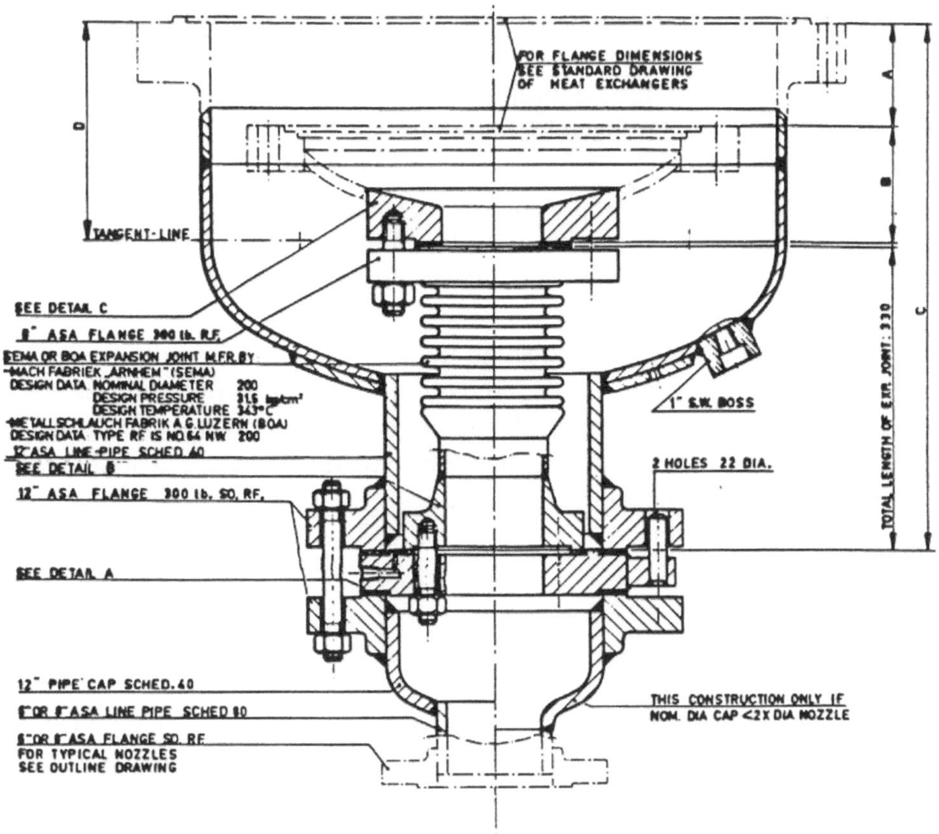

Figure 11.12

If a tube side expansion bellows is to be used I normally leave it to the mechanical engineer to decide what type is appropriate. All the thermal design engineer needs to know is what impact it has on the fluid entering or leaving the shell side and an approximate length for the preliminary setting plan.

DOUBLE PIPE & MULTI TUBE SPECIALS

Double pipe and multi tube exchangers are used when a small heat transfer surface area is required or when true counter flow is necessary. They also have an advantage when the shell side heat transfer coefficient is controlling the overall coefficient because longitudinal fins can be added to the outside of the tubes. **The economic break point between a small shell and tube and a double pipe/multi tube unit is normally about 20 m^2.**

For a job with lots of exchangers it is standard practice to isolate the small units at the start and send the process data sheets off for a quote to one of the specialist manufacturers. Borderline cases my later be designed as a shell and tube unit so that the prices can be compared. The specialist manufacturers have developed a wide range of near and far head arrangements to suit most services. They can then be classified in a similar way to the one that TEMA uses to classify shell and tube units. If the reader does not already have one, I suggest that they send off for a catalogue which will explain all the possible construction details.

What may be economic for a reasonable number of small units is not necessarily economic for a single item. A single item of any type of equipment must be accompanied by a plethora of documentation just the same as a batch of a dozen items. Over the years specs seem to have increased in size faster than banker's bonuses and the paperwork which has to be internally prepared and then externally digested will result in the price of a single small item rocketing. There are also not many manufacturers who are competent to provide a thermal and mechanical design service for double pipe/multi tube items which compounds the problem.

The best solution to purchasing one or two small items is often to thermally design them in house. They can then be added to a requisition for a batch of reasonably sized shell and tube units and the overall cost for the batch may hardly change. Since the complex head fitting which a specialist manufacturer would use are not available to the thermal designer he/she has to improvise. These are what I term **'double pipe/multi tube specials'.**

Example 11.3

Figure 11.14 on the next page is a double pipe special which I designed many years ago. It is a true double pipe with a single inner pipe surrounded by an annulus formed by an outer pipe for the 'shell side' fluid. Neither side required mechanical cleaning and the momentum fluxes through the shell side nozzles did not require impingement protection.

Short supports were welded to the outside of the tube to maintain it in a central position as illustrated in figure 11.13 below.

Figure 11.13

If the pressure drop on the annulus side is tight the supports could be tapered at the ends to reduce the drag. If there is plenty of pressure drop available this is not necessary.

The far end enclosure is a standard pipe fitting which governed the length of the U bend.

If the inside pipe had required mechanical cleaning flanged connections could have been used at the near end of each section. It should be noted that cleaning around a large diameter bend with a large inside diameter is a lot easier than cleaning around the bends of a typical BEU type exchanger.

This type of design could also be used with longitudinal fins if the annulus side heat transfer coefficient is strongly controlling.

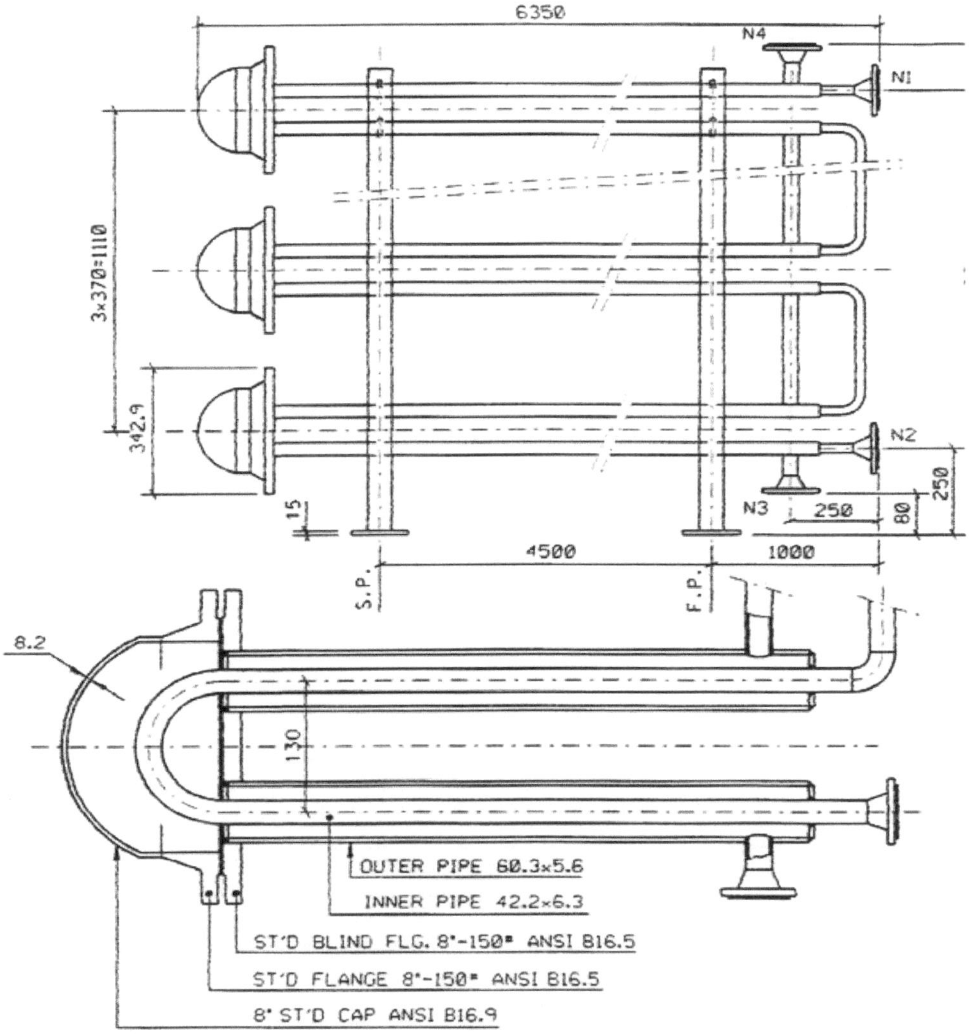

Figure 11.15

Example 11.5

Figure 11.15 below is a multi tube special. The inside of the tubes needed cleaning and required full access. However once again the bends of the U tubes would be very large compared with the inner rows of a normal U tube bundle and could be mechanically cleaned. (See the section on cleaning in chapter 14 on fouling.) There was condensing steam on the shell side and it was decided that access was not required so that flanges could be eliminated from the far end to reduce the cost.

I used standard pipe T junctions with reducer/expander sections to match up with the shell side nozzles.

It should be noted that by incorporating a T junction in the shell the connection is the same diameter as the shell. When pipe fittings are used there is no '50% of the diameter' restriction as there are with shell and tube units. This enlarged opening meant that I could stay under the TEMA maximum limit for the inlet momentum flux.

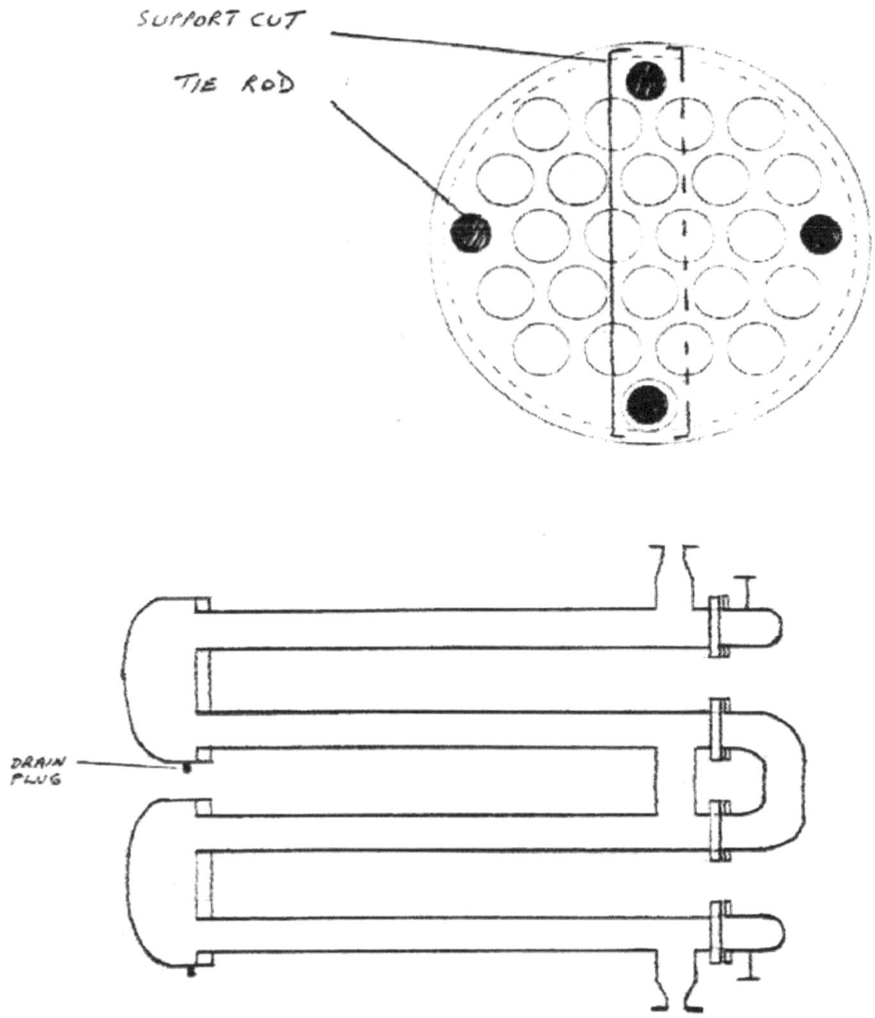

Figure 11.15

The support baffles in this design are the same type as would be used in a shell and tube unit. An astute reader may be critical of there being only one tube row overlap which might infringe some specs. With such a small unit only one row overlap is required – another exception to a normally sensible rule.

I have not shown the details of the U bends. I made an estimate of the surface area they would contribute to the heat transfer surface area to use in the thermal calculations and left the details up to the mechanical engineer.

Example 11.6

The final example shown in figure 11.16 is not taken from the petrochemical industry. The work was for a local manufacturer that I have been working with for many years. It is a heater for a sewage digestion tank. Standard pipe fittings have been used which are easy to identify from this view from the fabrication drawings.

The sludge from the digester was heavy fouling and the unit was cleaned after every batch cycle. The heating fluid through the annulus was clean warm water in a closed circuit.

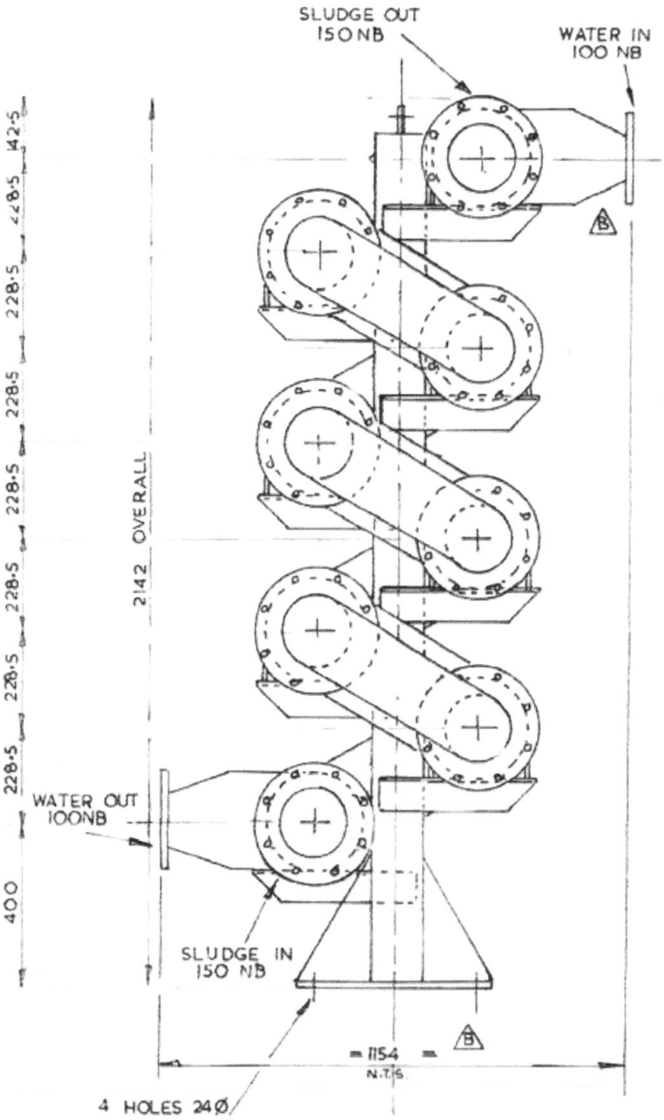

Figure 11.16

I had some flexibility in the sludge flow rate so that I managed to fully use up the allowable pressure drop with a velocity of 2.2 m/s. I never had any adverse feed back so I assume that the unit was satisfactory.

SUMMARY

These examples should be sufficient to give the reader a good idea of the possible arrangements that can be used with standard pipe fittings to design special double pipe and multi tube exchangers.

With some of the earlier computer programs a number of fudges were required to achieve a satisfactory run. Most modern programs should accept them without any problems except that you may need to pro rate the shell inlet momentum flux by hand.

HIGH PRESSURE UNITS

TEMA shows a single type of high pressure enclosure – the 'D' type head. There are many variations of the D type head to suit different applications.

An obvious first question to ask, but not too easy to answer is "what is high pressure?" Certainly anything above 140 bar would qualify but some units over 70 bar may need to be treated as high pressure items. The highest pressures I have ever had to deal with were about 250 bar.

The type of construction required depends not only on the pressure but also on the diameter. The main problem that high pressures present is how to avoid leaks. To seal the gasket on a TEMA A type head require a force from the bolting that is greater than the force from the channel fluid which is trying to move the head away from the channel. As either the diameter or the pressure increases a thicker cover and heavier bolting is required. A point is reached were it becomes either impractical or uneconomical to use a thicker cover and heavier bolts and a high pressure enclosure is required.

The following drawings have been taken from a Combustion Engineering catalogue.

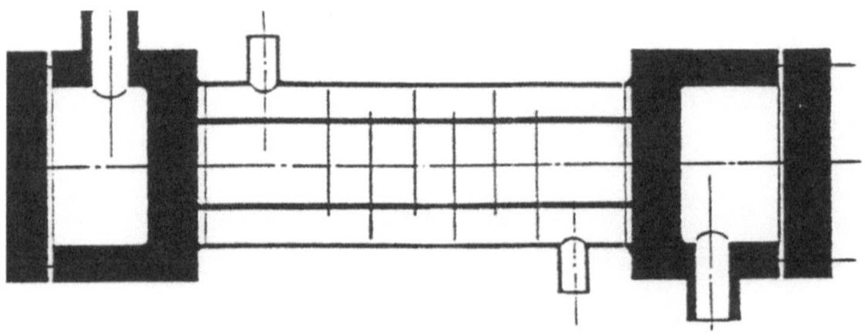

Figure 11.17

Figure 11.17 is a DED unit with high pressure enclosures at both ends of a low pressure shell. The channel is sufficiently thick to accommodate stud bolts without the need for any flanges.

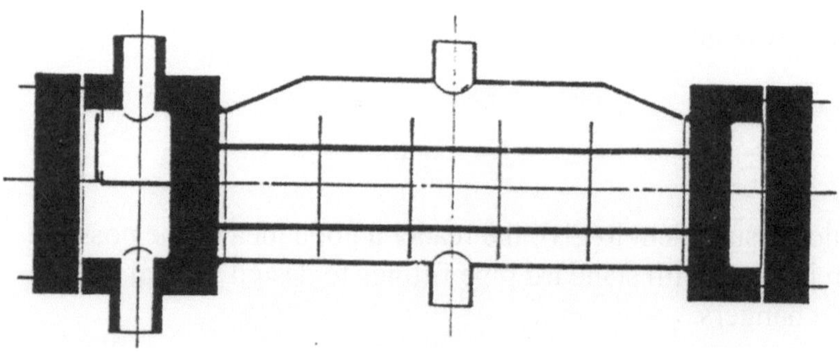

Figure 11.18

Figure 11.18 is a DKD unit with high pressure enclosures at both end of a low pressure shell. The main difference between the heads on this unit and the last one is to accommodate two tube side passes. A feature to note is the pass partition arrangement in the near head. With an A type head there would be a single horizontal pass partition which would have a gasket seal against the channel cover. In a high pressure enclosure the horizontal pass partition stops short of the cover and a vertical plate is bolted on to seal the turnaround compartment.

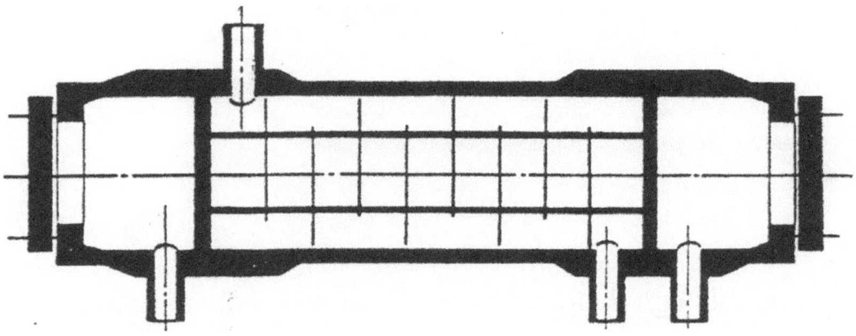

Figure 11.19

This is a DED unit with high pressures on both sides of the exchanger. The diameter of the channel cover plates has been reduced to reduce the thickness. It would probably be fixed at a standard manhole size. This gives access to all the tubes for cleaning and plugging but would not allow some of the outer tubes to be replaced.

The first three examples in this collection are the only ones with straight tubes, most high pressure shell and tube exchangers use U tubes. This means that only one expensive head is required instead of two.

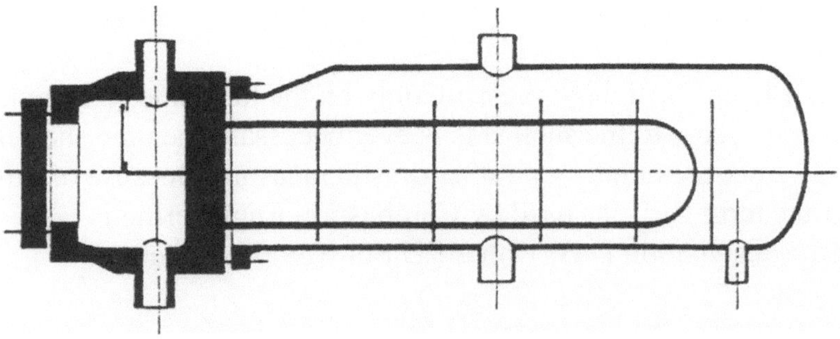

Figure 11.20

Figure 11.20 is a DKU unit. The head is a very similar construction to the one in the previous example and the low pressure shell is bolted to the high pressure head.

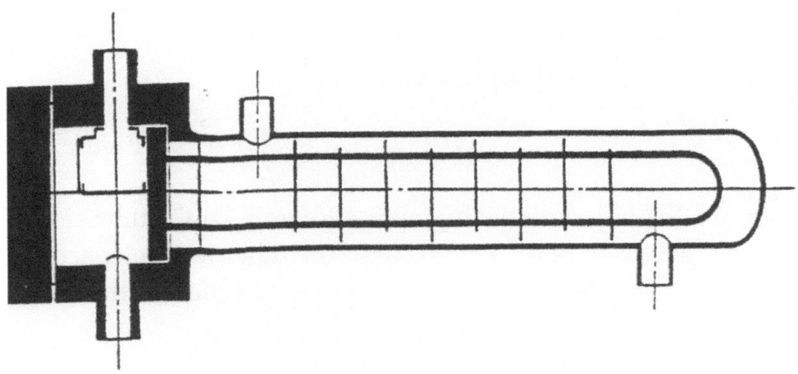

Figure 11.21

Figure 11.21 is a DEU with a large flat cover. The shell is welded to the channel. This arrangement allows the bundle to be removed through the channel for either maintenance or replacement.

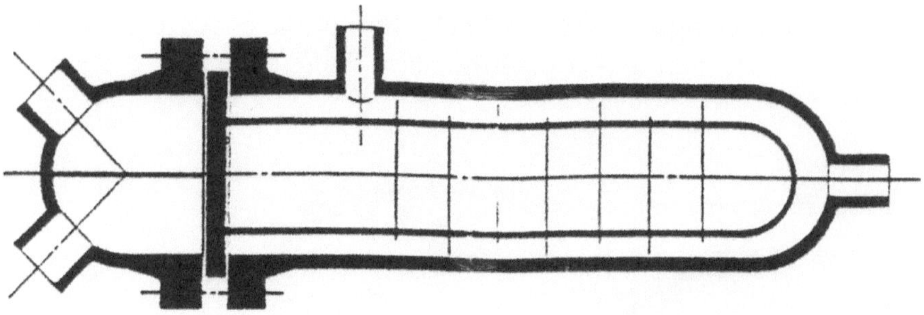

Figure 11.22

Figure 11.22 is a BEU unit although there are high pressures on both sides. This construction would only be suitable for smaller diameters and/or pressures at the bottom of the high pressure range.

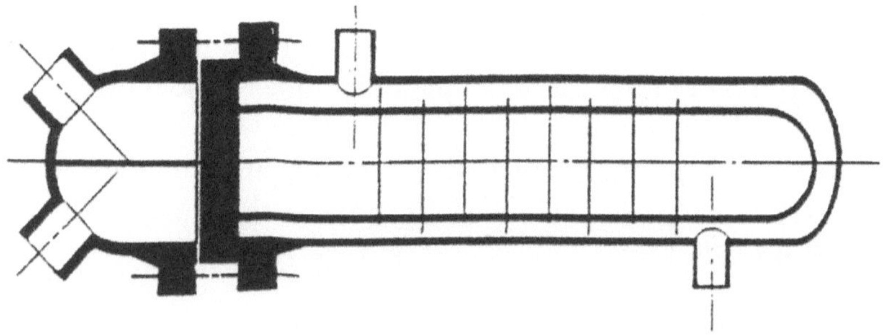

Figure 11.23

Figure 11.23 is also a BEU unit with high pressure only on the tube side. The bundle is no longer removable. If access to the shell side is ever necessary then the shell must be cut and moved away from the bundle – a rather drastic measure. Because of this it is debateable whether the term BEU, as used by Combustion Engineering is appropriate. Perhaps the designation D/BEU or BEU Spl would be a better description.

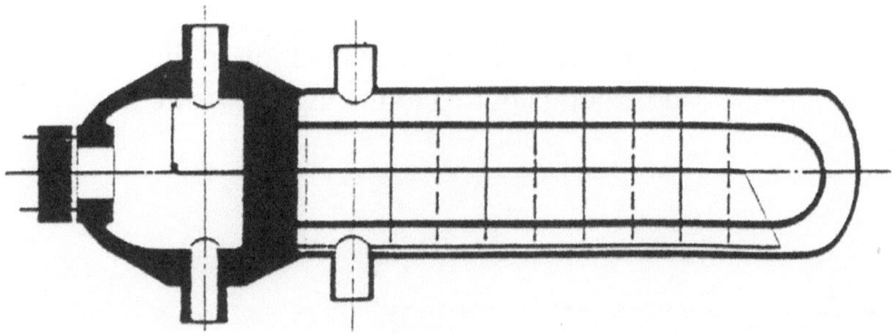

Figure 11.24

Figure 11.24 is certainly a DFU unit with the high pressure only on the tube side. The channel cover is a manhole access to allow the inside of the tubes to be cleaned or if necessary plugged. As with the previous example the only way to access the shell side is to cut the shell.

All the examples so far have relied on a 'conventional' bolting arrangement to seal the gaskets and the high pressure is acting to try and pull the seal apart. The remaining types of enclosure all use the high pressure to compress the gasket so that it is less likely to leak.

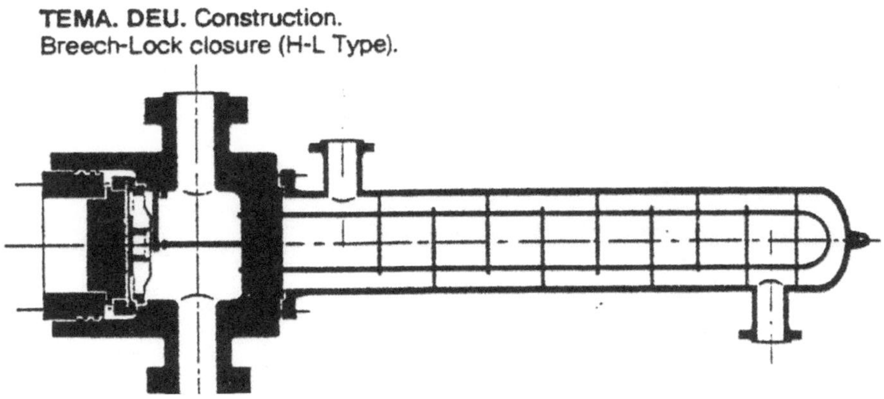

Figure 11.25

Figure 11.25 has the high pressure only on the tube side. The channel 'cover' in this design is held in place by a threaded cylindrical section. The gasket is then compressed between these two sections.

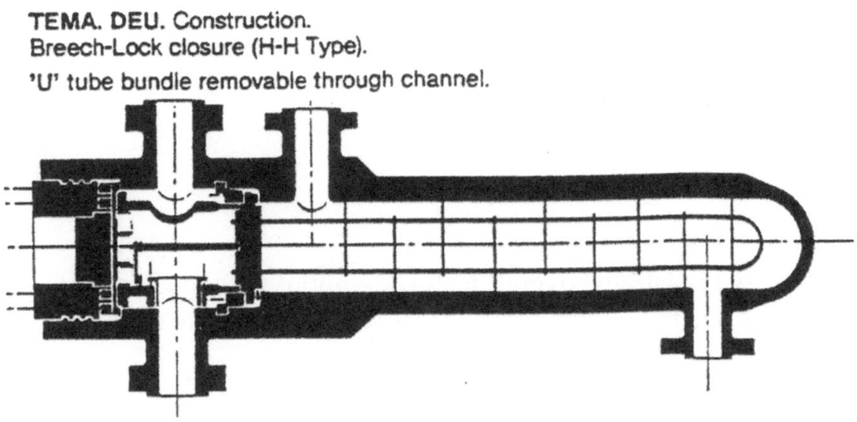

Figure 11.26

Figure 11.26 has high pressure on both sides. The 'Breach-Lock' method of sealing is the same as for the previous example.

To get a better feel for what is involved with the type of unit which require internally threaded channels I would recommend a visit to the Tema India website. They have a very impressive colour photograph of the channel of a unit designed for high pressures on both sides. The channel has an OD of 2.13 m and is 203 mm thick. The total weight of the unit is over 100 tonnes.

The next unit in figure 11.27 is a 'multilock' closure. The key features of this design are a split ring which is mounted in a groove in the channel and a tapered cylindrical section to hold it in place. The final seal is provided by a hemispherical cover which is welded to the channel and the channel cover. To access the channel and bundle it is necessary to cut through this section. The diameter of the cross section of the hemispherical welded cover is usually quite small compared with the diameter of the channel.

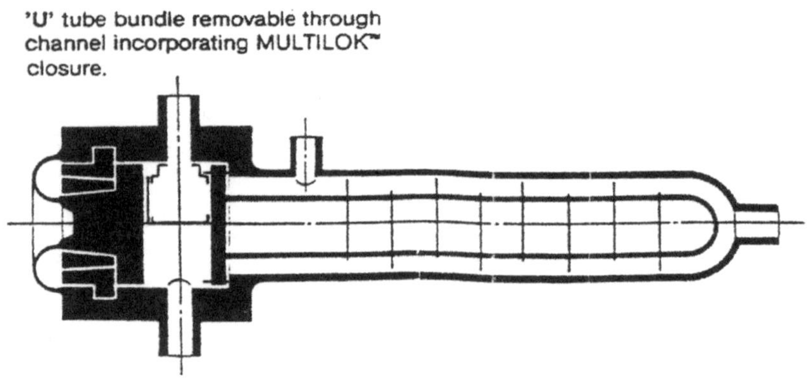

Figure 11.27

Figure 11.28 is the last variation from the Combustion Engineering catalogue and combines some of the features of the previous illustrations.

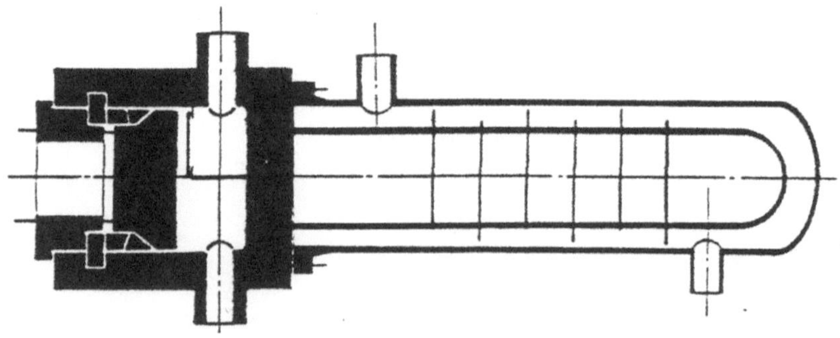

Figure 11.28

All the diagrams so far have illustrated overall features without very much detail. Figure 11.29 below shows a detailed picture of a Hemlock head from a Yuba catalogue.

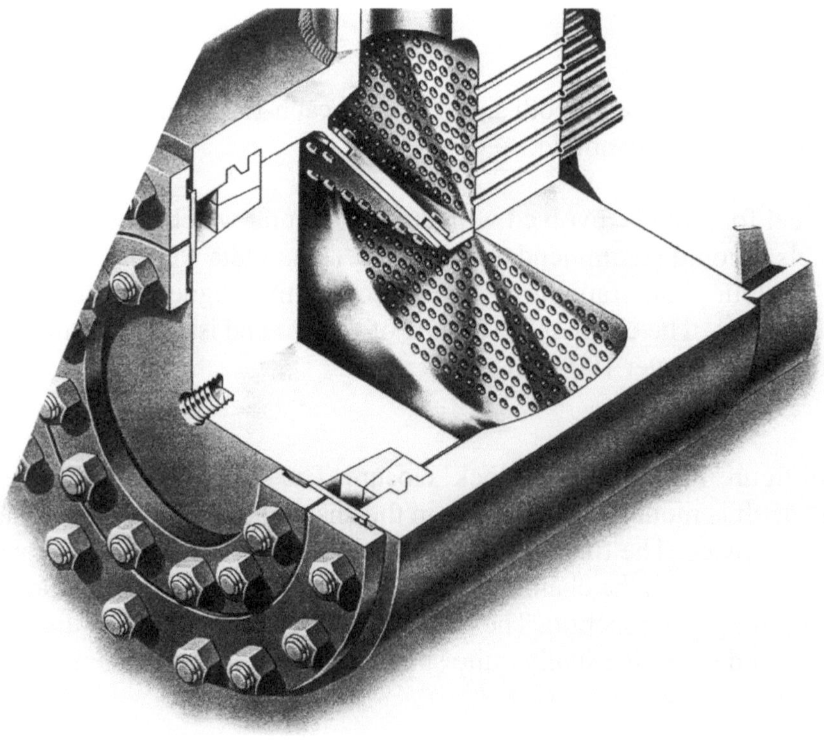

Figure 11.29

NOZZLES

Figures 11.25 and 11.26 show forged welding neck type of nozzles which connect to flanged pipework. Most of the other diagrams just show a general type of nozzle which could be anything. For the low pressure shells some form of flanged nozzles would be specified in the normal way. For the high pressure connections there are often no flanges. The nozzles are welded to the pipework to avoid any possibility of leaks. This has no impact on the thermal design but must be made clear on the thermal data sheet.

SUPPORTS

For horizontal units the ideal position for the two supports is where they each take half of the weight of the unit plus the weight of any fluid for the worst possible case. (This may be during a hydraulic test for a gas/gas exchanger.) The positioning of supports is not always easy with low pressure designs but there are extra problems to take into account with high pressure items particularly those with relatively short tube lengths compared with the diameter.

Figure 11.30 shows a high pressure unit with conventional supports. Support A has been positioned as close to the bottom shell nozzle as possible.

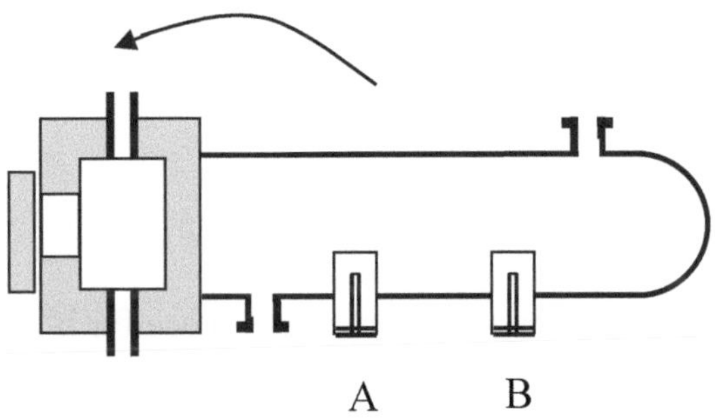

Figure 11.30

The mechanical calculations may show that support B is taking none of the load and there is a net anticlockwise moment which would tip the exchanger over with the support A simply acting as a pivot point. In this situation it is necessary to design support B as illustrated in figure 11.31 so that it is holding down the right hand half of the shell.

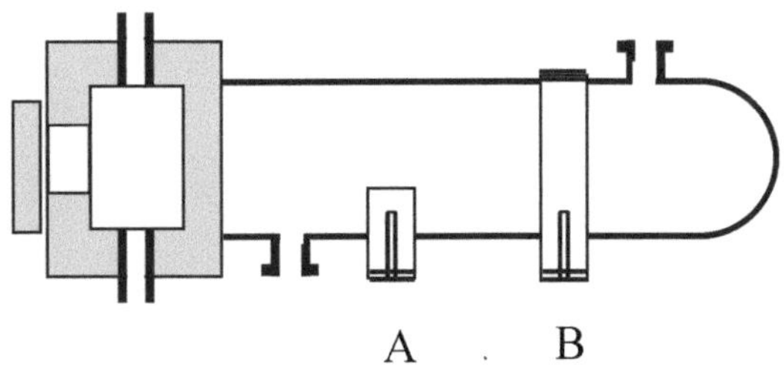

Figure 11.31

In this example the situation would have been considerably improved if the shell side nozzle nearest the channel had been at the top and support A had been much closer to the channel. This is not always possible – look back at the DFU unit illustrated in figure 11.26 for one such case. Even when the first support is close to the channel there may be cases when it is necessary to design for one of the supports to be holding down the exchanger. The alternative of course is to provide a support under the channel. Any channel support requires a special mechanical design to suit the type of channel being used.

One of the last high pressure units I worked on was a very interesting case where there was a number of possibilities but no clear cut solution to the support problem. I discussed this at length with the head of the mechanical section. In the end we opted out and I added a simple note to the thermal data sheet; "Vendor to propose and show details of the method of support in the quote". I do not normally like 'opting out' but in this case it was a mechanical not a thermal problem!

IMPACT ON THERMAL DESIGN

The construction of all high pressure exchangers has an impact on the thermal design. There is more pressure than normal to fully utilise the maximum allowable tube length and minimise the diameter – don't worry about standard tube lengths even with straight tubes – they will be a heavier gauge than the tubes on other units so that any spare tubes would only be for a single unit.

Work closely with a mechanical engineer. The construction of a high pressure channel may determine some features of a low pressure shell such as:

(a) The position of the nozzle nearest to the channel.
(b) The bundle/shell clearance.
(c) The minimum bend to use for U tubes and the tube gauges for each row.

The costs to be saved by optimising a high pressure unit design are far higher than for a low pressure exchanger.

SPECIAL THERMAL DESIGNS – CHAPTER 12

Introduction

This chapter on special designs is not an afterthought. Too many special features in the earlier chapters would have detracted from the points I was trying to make there. It would have also been difficult to decide which chapter to include some of them in since some of the designs rely on design features from more than one chapter.

FALLING FILM EVAPORATORS

Falling film evaporators are used to concentrate solutions, particularly those which are very sensitive to wall temperatures because they are prone to crystallisation and rapid fouling. Steam is the most common heating medium although liquids may be used. The evaporation which takes place inside the tubes is downwards and as vapour is produced the fluid will accelerate reaching a maximum value at the bottom of the tube. A separate separation vessel is an integral component of these designs as illustrate in figure 12.1 below.

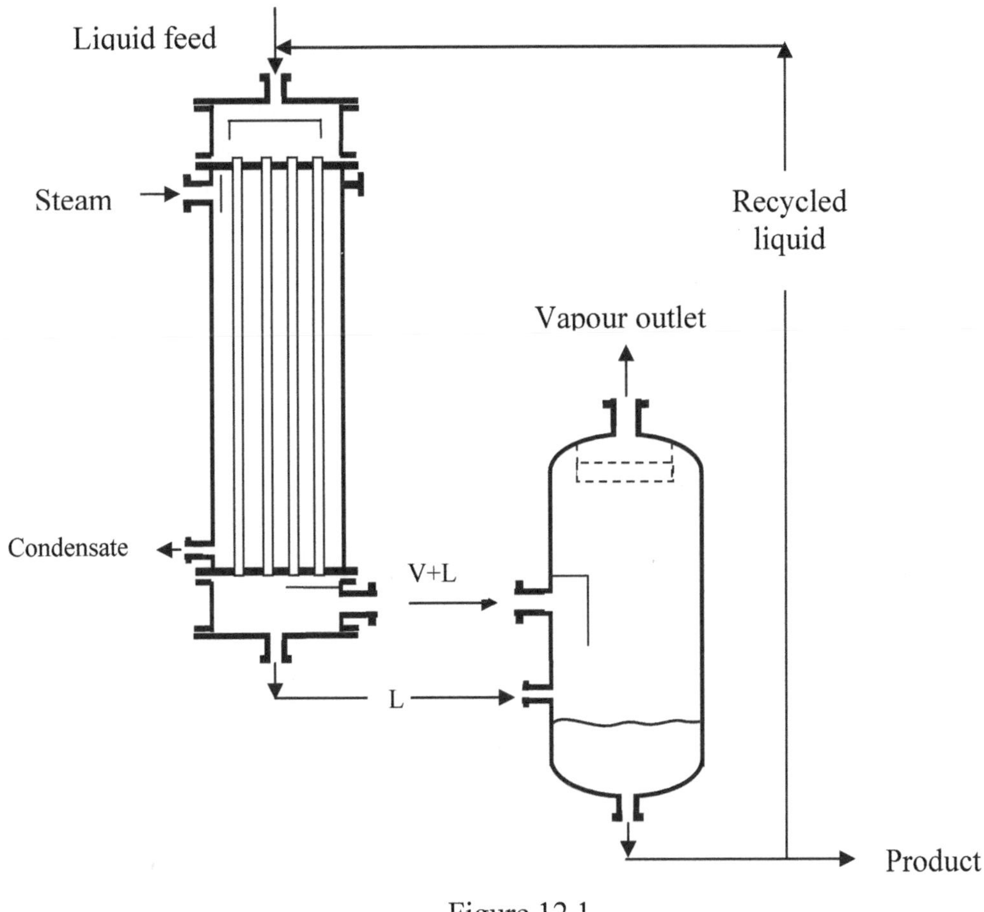

Figure 12.1

The liquid feed enters the top of the exchanger and special arrangements must be made to distribute the liquid evenly to each tube. The liquid and vapour leaving the bottom of the tubes undergoes a preliminary separation in the bottom channel of the exchanger. I have indicated a short deflector plate over the vapour outlet nozzle. This would be a semi circular or inverted V shaped plate to minimise the liquid entrainment. Most of the liquid will fall to the bottom of the channel from where it is passed directly to the bottom of the separation vessel.

The vapour with entrained liquid enters the separator behind some form of deflector plate which redirects the low downwards. This could be any of the types of deflector plates used to introduce two phase flow into kettle reboilers which was covered in chapter 8 or perhaps better still by a distributor pipe as illustrated in figure 12.2 below.

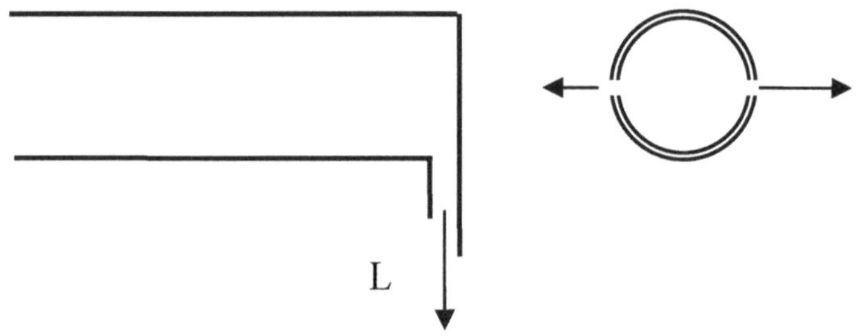

Figure 12.2

This will effectively give a second stage to the separation process with very little increase in pressure drop provided there are plenty of holes of sufficient diameter.

Above the vapour inlet there are a number of options. The first is to have a series of horizontal baffles. Each reversal of flow direction will knock out a bit more entrained liquid. With a central distributor pipe the ideal baffling would be double segmental to maintain symmetry and minimise ineffective spaces. This is shown in figure 12.3 below.

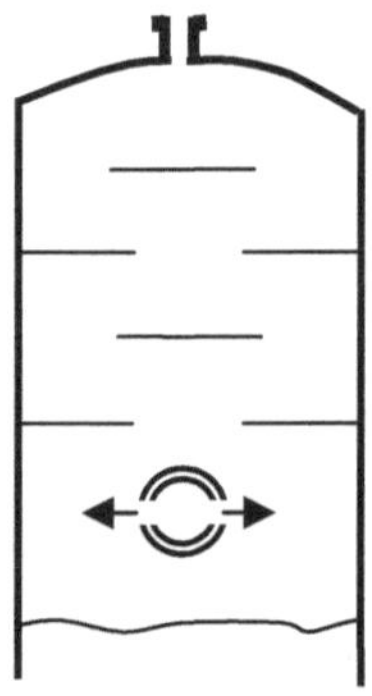

Figure 12.3

Baffles should be attached to the vessel in a way that leaves a typical baffle/shell clearance around the outer edges. This will allow any liquid droplets that are knocked out of the vapour stream to run down the side of the vessel. This is illustrated in figure 12.4 for a central baffle.

The second is to use a demister pad as shown in the broken lines at the top of the vessel/knock out drum. For a real belt and braces job a pair of baffles and a demister pad could be used.
The complexity of the vessel design will depend on the maximum acceptable entrainment value for the vapour being discharged from the vessel. Fixing this value

is down to process. Entrainment calculations are similar to those detailed in chapter 8 on kettle reboilers.

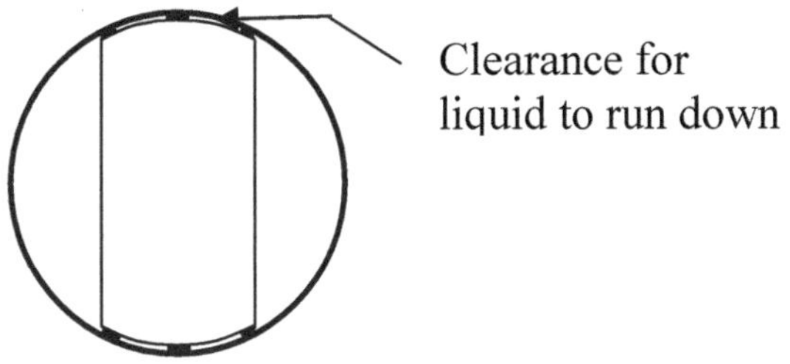

Figure 12.4

The design of the separation vessel/knock out drum may be carried out by a process engineer but I have often found that an inexperienced process engineer is very happy to leave it to the thermal design engineer and even a more experienced one may appreciate some input.

LIQUID DISTRIBUTION

Simple tube end weirs

The tubes protrude above the tubesheet. The liquid is directed from the inlet nozzle to form a reservoir around the periphery of the bundle. The level is then just allowed to build up until it overflows into each tube. For this arrangement to have any chance of success it is necessary to have all the tube ends almost perfectly level.

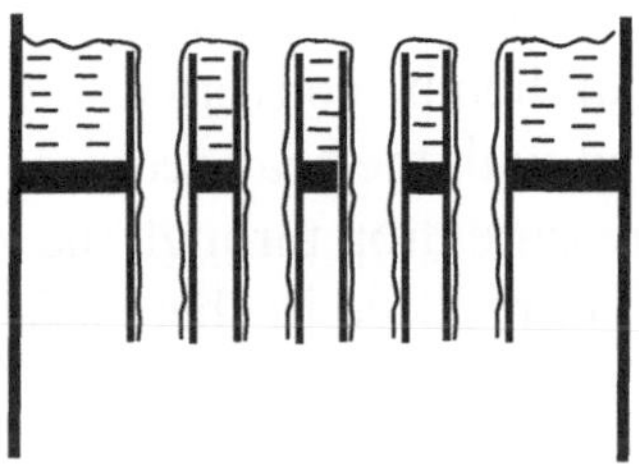

Figure 12.5

This is the cheapest but least effective way of distributing liquid into the tubes. I have read a research paper in which tests were carried out across 15 rows of tubes on a triangular pitch and the ratio of the flow rates to individual tubes was 3:1! I would not recommend ever using this simple arrangement.

Notched/slotted tube end weirs

Distribution is significantly increased by using tubes with slots cut into the ends. The slots provide a resistance to fluid flow which is greater than the resistance of a fluid just flowing over the top of a tube. As with any hydraulic system increasing the resistance at the point of delivery always improves the distribution.

A liquid level will build up above the bottom of slots as illustrated in figure 12.6. There will still be a higher liquid level at the periphery than at the centre of the bundle. To minimise the difference in liquid level the height of the tubes above the tubesheet can be increased whilst the length of the notch remains the same. Say a total tube protrusion of 300 mm and a slot length of 75 mm. These figures are only quoted to give some idea of the relative proportions.

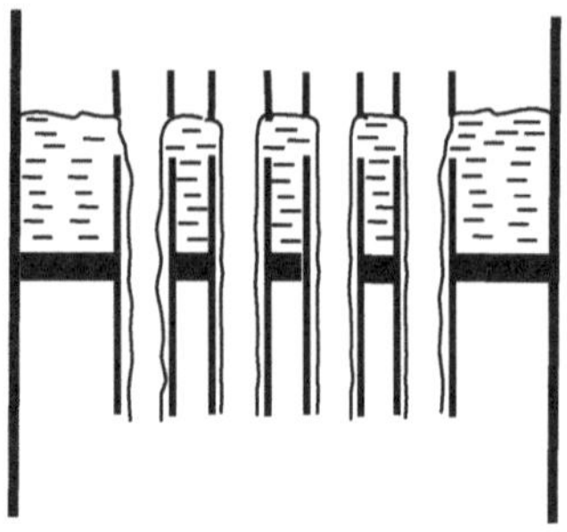

Figure 12.6

The length of the notch must be sufficient to ensure that the outer tubes do not also overflow over the top of the tubes.

The greater the total height of tube protrusion the smaller will be the variation in the liquid level. To understand this it necessary to consider the path taken by an element of fluid from the outer reservoir until if falls down an inner tube. As it crosses each row there will be a pressure drop determined by the velocity at that point. The velocity depends on the effective diameter and the height of the liquid. The smaller this pressure drop is compared with the pressure drop through the slots the less variation there will be in the liquid level.

The pressure drop across a few rows of tubes can be crudely estimated from a fudged computer run on a cross flow unit. A pressure drop can then be converted into mm head of the liquid involved. This can then be compared with the head required to achieve the design flow rate through the slots. The pressure drop through the slots can be found from the following equation which was first introduced in chapter 8:

$$H = 1.5 \left(\frac{V}{b\ g^{0.5}} \right)^{2/3} \qquad \text{[Eq.2 from Ch 8]}$$

H is the height of the liquid above the bottom of the slot, V is the volume flow rate to one tube, b is the breadth of a notch multiplied by the number of slots and g is the gravitational constant.

The pressure drops in mm of liquid across a couple of sample tube rows, (taking into account the change in mass flow rate and effective diameter) for different values of tube protrusion will just give the designer a good 'picture and feel' for the situation. (Varying the tube pitch might also be investigated.) If sensible values of these variables result in pressure drops that are significantly lower than the value of H then the distribution will be OK. This simple arrangement is likely be satisfactory for small bundles only. For a larger bundle the first calculation across the outer tube row may be sufficient to rule out this arrangement and it will be necessary to use a distributor to introduce the liquid more evenly across the bundle or use inserts which will have much higher hydraulic resistances than a slot arrangement.

Trough distributors

Figure 12.7 below shows what is generally known as a trough distributor. I am not sure of the origin of the term but it is basically just a pipe distribution system.

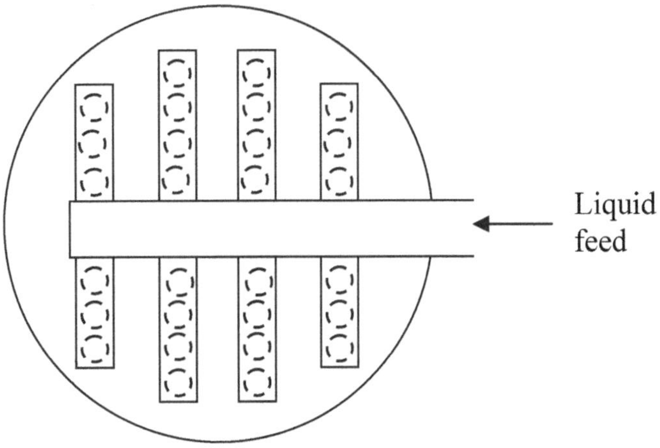

Figure 12.7

The feed enters a primary pipe and is then directed into a number of secondary pipes. From there it is directed into a number of vertical delivery pipes. For this arrangement a radial nozzle would be used in the top channel rather than the axial nozzle shown in figure 12.1. A large diameter should be used for the primary pipe – much larger than the channel inlet nozzle and most of the pressure drop should be used in the delivery pipes to give a good distribution. (I am sure that by now a program will be available to design pipe distribution systems.)

A decision now has to be made as to whether to use plane tube ends or slotted tubes. I would personally continue with slotted tubes. The very crude calculations which were outlined previously may be used to investigate the flow from a distributor pipe to the most remote tube. This will help to determine how many supply pipes are required.

It may be necessary to leave out selected tubes to accommodate the delivery pipes which should ideally be tapered and introduce the liquid above the face of the tubesheet. This is the only type of arrangement for distributing a liquid to the top of a falling film evaporator that I have any personal experience of. I have never had to design a unit that was small enough not to need one.

Notes on the tube end slots.

Problems have been reported with large diameter slotted tubes. A situation may arise when some of the flow from the slots pours into the centre of the tubes and then falls to the bottom of the tubes without being heated. This is clearly unsatisfactory and may be overcome by directing the flow from the slots towards the surface of the tubes.

Figure 12.8 shows; (a) a plan view of normal slots and (b) a plan view of slots at an angle which direct the flow onto the surface of the tubes.

For 19 mm or 25.4 mm tubes I doubt if there is ever likely to be a problem however very large diameter tubes may be used for some applications and it would be a sensible precaution to use angled slots. I have no experience in designs with large diameter tubes and can not offer any guidance to check for when angled slots should be introduced.

(a) (b) Figure 12.8

There is clearly an extra cost in producing angled slots and an alternative for larger tubes might be to use castellated tubes with a large number of slots. It might be worth contacting a manufacturer for a cost comparison.

Tube end inserts

There are many types of inserts figure 12.7 below from a paper by Bromley and Read.

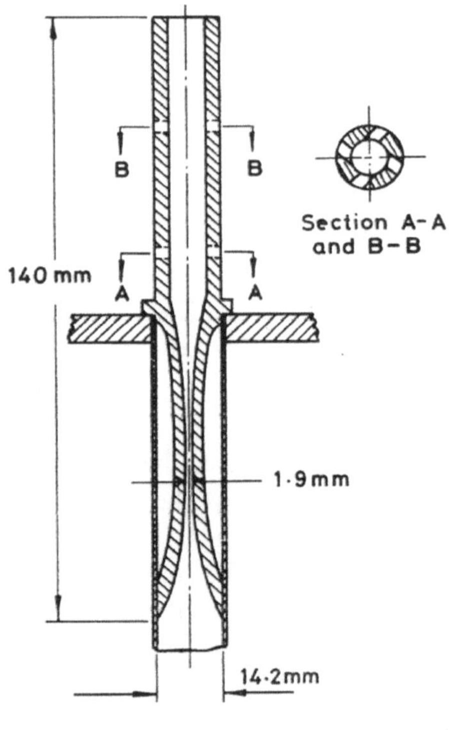

Figure 12.9

A lot of research seems to have been carried out on tube end inserts for falling film evaporators. I am not sure how many of these have found their way onto the market. The principles involved in their design are very sound but some of them would be quite costly to produce. They are therefore likely to be propriety items and guidance in using them should be available from the manufacturer. (A similar situation to the tube inserts available to protect the end of tubes in waste heat boilers.)

Angled holes rather than slots were used in the section of the insert which protruded above the tubesheet. A further resistance to flow was provided by a venturi shaped section which was inserted into the tubes. It was reported to give a very good distribution but I would not like to have tried to drill the angled holes through the tubes!

One big advantage I can see of using inserts is that the mechanical construction is more conventional and any type to tube/tubesheet joint can be used. The tubes are therefore easier to maintain and could be replaced if necessary. I will certainly consider using them if I ever design another falling film evaporator.

HEAT TRANSFER & PRESSURE DROP

The liquid film flowing over the surface of the tubes may be in either laminar or turbulent flow. As the fluid is heated evaporation takes place from the surface of the film. The process is therefore analogous to condensation with the heat flow in the opposite direction. As with condensation the heat transfer can be thought of as conduction through a layer of liquid and any physical property which affects the thickness of the film as well as the liquid conductivity will affect the heat transfer coefficient.

If the heat flux is high enough nucleate boiling may take place. Although this would temporarily increase the heat transfer coefficient it could dry out the surface of the tubes which would lead to rapid fouling. For any heat sensitive fluid it is much better to ensure that nucleate boiling does not take place.

The pressure drop through the tubes is normally very low and is mainly at the bottom of the tubes where the vapour velocity is a maximum. The acceleration loss may be significant and the frictional pressure drop is likely to be less than the discontinuity loss due to the sudden enlargement at the exit.

Although I have designed one of these units using hand calculation I remember feeling a bit uneasy about them because of all the assumptions that were necessary. There are now computer programs available and I doubt if anyone will ever use hand calculations again unless they are a part of a student training program.

RECYCLE

Figure 12.1 showed a line taking liquid from the bottom of the separator back to the liquid feed line. Recycling may have two distinct functions. One is controlling the concentration level of the product. The other is to ensure that there is always sufficient liquid flowing down the tubes to prevent the tubes from drying out.

TUBE SIZES

With a very low pressure drop through the tubes the diameter of the tubes is strongly controlled by the operating pressure. I have only ever had to deal with units operating at a reasonable positive pressure which suited relatively small tubes. However I have read that tube sizes of over 2 inches may be required for some vacuum systems.

INERT GASES

Some very high vacuum systems may utilise an inert gas to keep the total pressure just above ambient which ensures that there is no air leakage into the system. (Ambient air will always contain some water vapour.)

I have never dealt with such a system but there is a section which covers it in reference 7 by R A Smith.

SUMMARY

I have not been involved in the design of many falling film evaporators during my time in industry. Each one has therefore been a separate challenge and I have never developed on overall design strategy. Having said that, as I was writing this section I found myself trying to do just that. It is amazing what having a bit more time to spend on a topic without the pressure of working to a deadline can do. I have tried to cover most of the basic principles used in their design without being too specific. In some sections I have just been throwing a few ideas around. I hope the reader may find it useful in this context.

REFLUX CONDENSERS

Reflux condensers are also known as 'knock back' condensers or dephlegmators. They are partial condensers which condense some liquid from a stream of saturated vapour. The feature which distinguishes them from other partial condensers is that the flow of vapour is upwards inside vertical tubes. The liquid which condenses flows downwards in the opposite direction.

Figure 12.10 below shows a typical construction. The top head may contain a means of separating any entrained drops of liquid from the vapour stream. A demister pad is shown but baffles could also be used as described for the separator which was part of the falling film evaporator system. There could also be a normal head with a single circular deflector plate below the nozzle and a separate knock out drum.

The bottom head is normally elongated to provide a holding area for the liquid and a level control device. The liquid product is then extracted through a nozzle at the bottom.

Inside the tubes there is a core of vapour surrounded by liquid as shown in figure 12.11 (a). The condensing heat transfer at any point is determined by the conduction through the layer of liquid. For the more normal case of condensing with both the liquid and the vapour flowing downwards the shear force at the liquid vapour interface is acting in the same direction as the gravitational force which reduces the thickness of the liquid layer and increases the condensing heat transfer coefficient.

When this shear force is acting in the opposite direction it is holding up the liquid and increasing the thickness of the liquid layer and reducing the condensing coefficient. However it is also making the liquid film more wavy and irregular which increases the surface area of the liquid at the interface. The effect of vapour shear is therefore very difficult to predict.

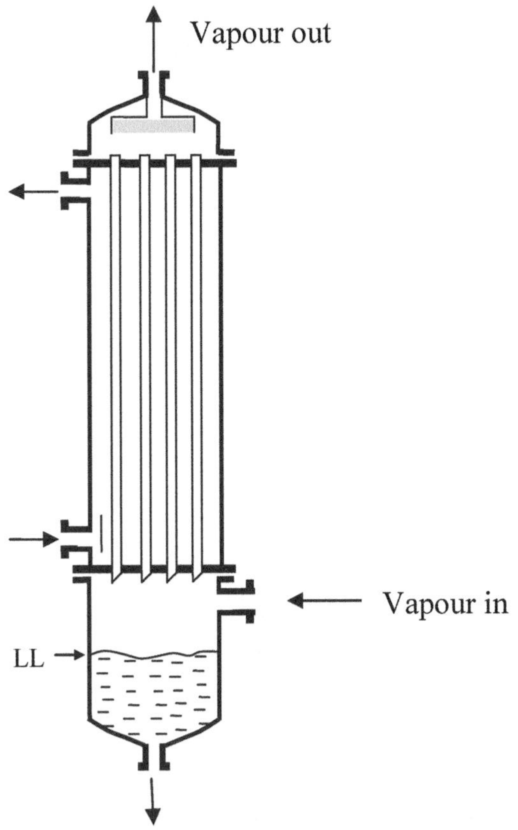

Figure 12.10

In addition to the thermal resistance of the condensate layer there is the resistance of the boundary layer of the gas stream. This can be calculated using the usual form of equation for evaluating a heat transfer coefficient for a gas flowing inside a tube with modified constants to suit the application.

The last hand calculations I used for designing one of these units followed the Silver method. This assumed a straight line condensation curve from which an overall temperature/specific enthalpy gradient could be calculated as follows:

$$\frac{\Delta T}{\Delta h} = \frac{(T_2 - T_1)\,\dot{m}}{Q} \qquad [12.1]$$

The gradient at any point Z, where the vapour mass fraction is y then becomes:

$$Z = y\,C_{pV}\frac{\Delta T}{\Delta h} \qquad [12.2]$$

This was then used to determine a combined heat transfer coefficient for the tubeside from the following equation:

$$\frac{1}{h_{tube}} = \frac{1}{h_{cond}} + \frac{Z}{h_{vapour}} \qquad [12.3]$$

A modern computer program will calculate the tubeside heat transfer coefficient at increments down the tube length. There will also be many more equations for calculating both the condensing and vapour coefficients for application to different compositions of the condensing stream.

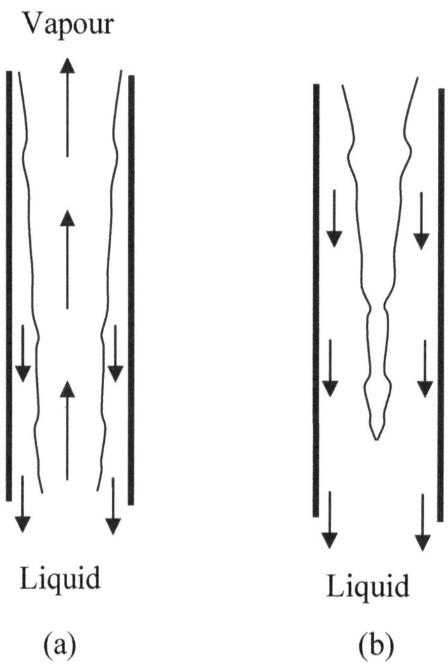

Figure 12.11

FLOODING

A focal point in the design process centres on the bottom of the tubes. The flow pattern shown in figure 12.11 (a) is what is required to take place. However if the tube is long enough and the condensate loading high enough a point may be reached when the liquid occupies the whole of the tube cross sectional area as shown in figure 12.11 (b). Vapour will no longer be flowing up the tubes so that the tube will become ineffective. This is of course an unstable situation because eventually the liquid must drain from the tube and condensation will be resumed.

The point at which the bottom of the tubes could become full of liquid is known as the flooding point and is defined in terms of a flooding critical velocity.

The critical velocity at which flooding is predicted is a superficial vapour velocity. That is the velocity the gas entering the bottom of the tubes would have if it alone occupied the full cross sectional area of the tubes. In the days of hand calculations the paper I used was "Flooding in a vertical updraft partial condenser" by K G English et Al – Chem. Eng. Progress1963 – Ref 28.

The empirical equation they developed is as follows:

$$G_F = \frac{1550\, d_i^{0.3}\, \rho_L^{0.46}\, \rho_V^{0.5}\, \sigma^{0.09}}{\mu_L^{0.11}\, (\text{Cos}\,\theta)^{0.32}\, (L/G)^{0.07}} \qquad [12.4]$$

Where

G_F = Superficial vapour mass velocity for flooding, lb/hr ft^2
G = Superficial vapour mass velocity for total entering tubes, lb/hr ft^2
L = Superficial liquid mass velocity for total leaving the tubes, lb/hr ft^2
d_i = Tube inside diameter in inches
ρ_L = Liquid density, lb/ft^3
ρ_V = Vapour density, lb/ft^3
σ = Surface tension, dynes/cm
θ = Angle of cut on the tube ends

The gravitational constant is conspicuous by its absence but is of course taken account of in the numerical constant.

The angle of cut of the tube ends is a very important variable for the designer. The angle is defined with reference to the horizontal plane as illustrated in figure 12.12 below.

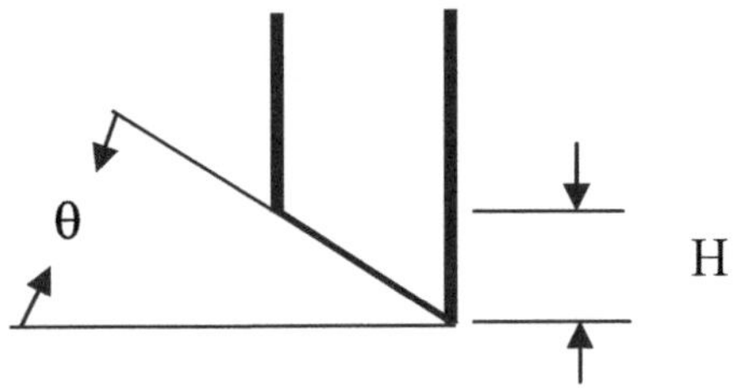

Figure 12.12

The effect of varying the angle θ in the English equation is shown in table 12.1 below.

Angle θ	Increase in G_F %	H in mm for tube diameter		
		19 mm	25.4 mm	31.75 mm
0	-	-	-	-
30	5	11.0	14.7	18.3
45	12	19	25.4	31.8
60	25	33.0	44.0	55.0
75	54	71.1	94.8	118.5
80	*75*	*108*	*144*	*180*
85	*118*	*218*	*290*	*363*

Table 12.1

I have added the height H which represents the length of tube lost for effective heat transfer. I have also added values for angles 80° and 85° because I like to investigate the limits of any correlation, (when I have the time!) I have shown these in italics.

It is easy to see why English et Al limited their research tests to an angle of 75° because above this value the loss in effective tube length becomes increasingly unacceptable. The table shows that there is little point in using an angle of 30° or less and the increase between 60° and 75° is very significant and could make the difference between the tubes flooding and not flooding. It is the percentage loss in effective tube length that is important so that a relatively small angle of say 45° may be the optimum for a very short tube length whilst an angle of 75° might be optimum for very long tubes. I have never read of any research to confirm that the percentage increase in flooding velocity continues to increase above 75° as the equation predicts.

Like so many dated equations you may not find exactly the same one in your computer program. For example in recent years HTRI have carried out their own tests and concluded that the power of Cos θ should be 0.38 rather than 0.32 not too drastic a change. When their predictions are compared with the English paper the percentage increase in critical flooding velocity was 14% compared with the English value of 12% for the 45° cut and 65% compared with the English value of 54% for the 75° cut. This is quite a tribute to English et Al who developed such a complex equation so long ago with less sophisticated measuring devices than are now available to research workers.

I will end this section on a lighter note. I remember once doing my own research on the effect of a taper on the discharge of a fluid from the end of a pipe. It must have been a rather boring session at the pub because when I returned I was still thinking about the English paper that I had recently read. I took out a garden hose and systematically pared off varying tapers with my Swiss army knife and observed the changes in the discharge from the end. The results of my research were never published!

COOLING MEDIUMS

Figure 12.10 shows a reflux condenser with a liquid cooling medium. For low temperature items a boiling fluid may be used. The direction of the shellside fluid flow now becomes upwards as illustrated in figure 12.13 below.

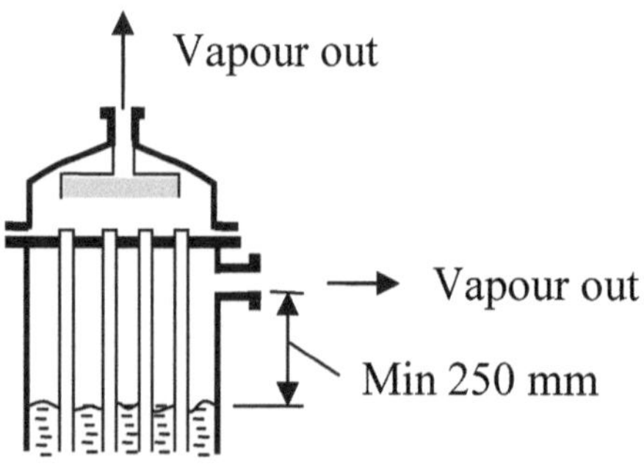

Figure 12.13

To minimise entrainment it is necessary to maintain a liquid level below the outlet nozzle. In the absence of any specific data for entrainment in this situation I have always used the same procure that I would for a kettle as described in chapter 8 using a mean cross flow value for the horizontal velocity above the top of the tubes. As for kettles there should be a minimum of 250 mm from the bottom of the nozzle to the liquid level.

The length of tube above the liquid level is relatively ineffective for heat transfer and should not be included in the effective length.

STAB-IN CONDENSERS

Stab-in reboiler bundles which are used to provide a heat load to a tower are not unusual and were covered in detail at the beginning of chapter 8. Stab-in condensers such as the one illustrated in figure 12.14 below are less common

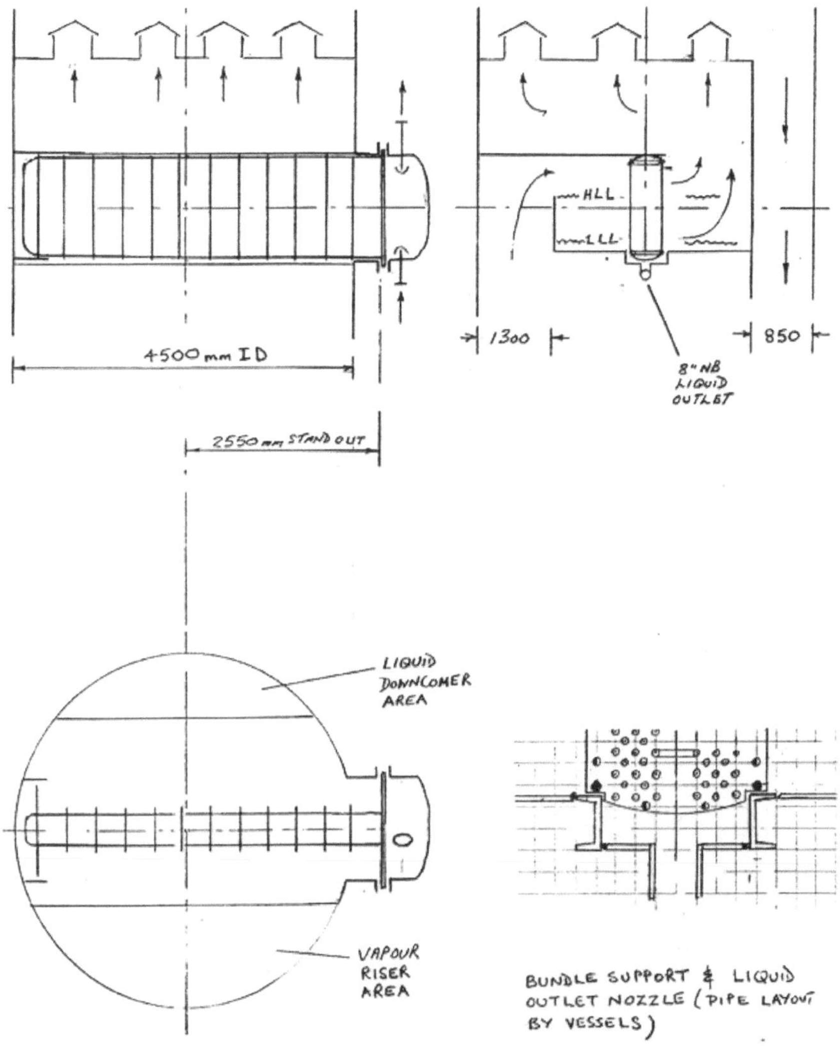

Figure 12.14

The sketches are my own from a real job and are not to scale. (I could produce a much better quality of sketch now that I am getting the hang of the drawing program in word.) Below the section of the tower where the condenser was fitted was a reservoir with an overflow to provide a constant head for the reboiler which was supplying the heat load to the bottom of the tower. The net heat load was to be varied by changing the liquid level in the condenser. The allowable pressure drop across the condenser was very low and varied as the liquid level varied - a maximum of about 1 kPa at high liquid level.

There is a liquid downcomer space shown in the side view of the top left right hand sketch and in the plan view of the bottom left hand sketch. The same sketches show the vapour riser space at the opposite side of the tower.

The requirements for this unit presented a number of challenges. It was clear at an early stage that a cross flow design was the only one which would satisfy the pressure drop requirements and to have confidence in the heat transfer calculations it would need to be a very flat bundle. Separate computer runs were needed for the tubes above the liquid level and those below the liquid level for each set of operating conditions.

The top and bottom of the bundle were fully tubed so that it was necessary to offset the channel nozzles which connected to the cooling water system. U tubes were used and I was concerned with by passing around the ends of the U tubes so I arranged for the far end of the bundle to have a full support plate which fitted in a short enclosure. This made the U bends ineffective for heat transfer but with the wide gaps in the tower port the length between the main support and the face of the tubesheet was fully effective.

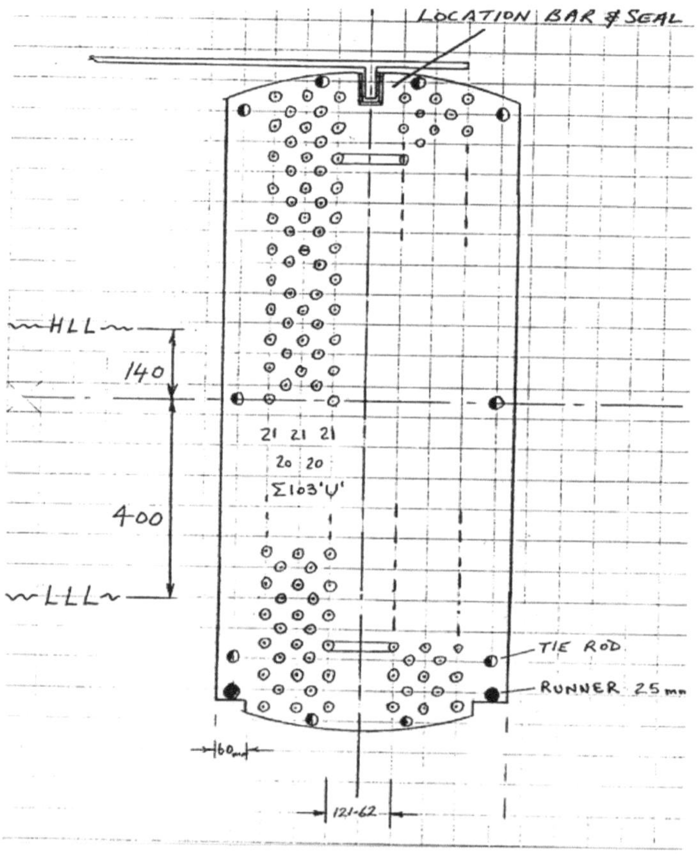

Figure 12.15

I had to draw a tube layout shown above by hand. I used a computer program with very large inlet and outlet nozzles to get an initial layout and then drew this on graph paper to get the scale right.

STAB-IN CONDENSER AND SUBCOOLER

The process requirements for this design were very similar to the ones for the previous design with an additional requirement for a minimum degree of subcooling of the condensed liquid. The layout of the system is shown in figure 12.16 below.

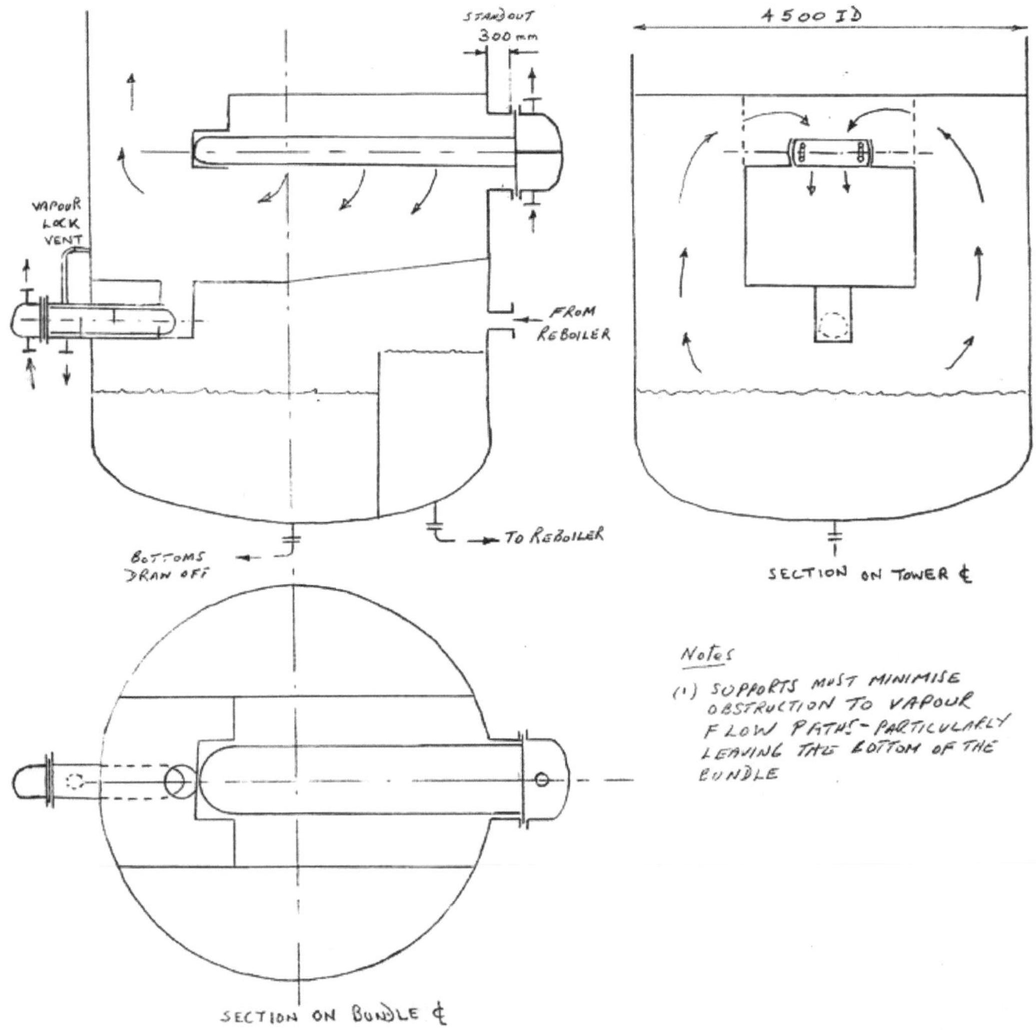

Figure 12.16

There would have been some subcooling in the previous example as there is with most condensers. With a minimum degree of subcooling it was necessary to introduce a separate subcooler. The only pressure drop available for the shell side was the small static head in the pipe leading down to the end of the shell which was an extension of the inside diameter of the port..

A flat cross flow bundle was again used for the condenser which was now in the horizontal plane. With this arrangement radial nozzles could be used in the channel for the cooling water connections. The U bends were again not effective but the area in the port was. I left the support arrangements to the vessel engineer; these were to be checked later by a thermal design engineer.

LNG Streams & Flow distribution

LNG production requires some very large expensive heat exchangers. To liquefy the natural gas it must cooled from around ambient temperature to a very low temperature so that it can be stored and transported at a drastically reduced volume. This is achieved by passing the gas through a series of refrigerant cooled heat exchangers which progressively cool and liquefy the gas. The refrigerants, starting from the hat end would typically be propane, ethylene and methane. Each refrigerant may operate at more than one temperature level. The natural gas therefore passes through the tube side of at least three exchangers.

I have worked on several of these plants but only at the stage where it was a matter of getting a design out by tomorrow and I had to work within preset restrictions. Some of these restrictions concerned an obsession with trying to maintain homogeneous flow through each tube.

Homogeneous flow through each tube is an understandable and natural instinct. The ideal situation to liquefy natural gas would be a single tapered tube which started with a large diameter and finished up with a small diameter while it was progressively cooled by each refrigerant. In the real world we need to design individual heat exchangers and the velocity in each one can be changed by either changing the tube diameter or the number of tubes.

I have come across some interesting measures used to achieve homogeneous flow. One of these is illustrated in figure 12.17 below.

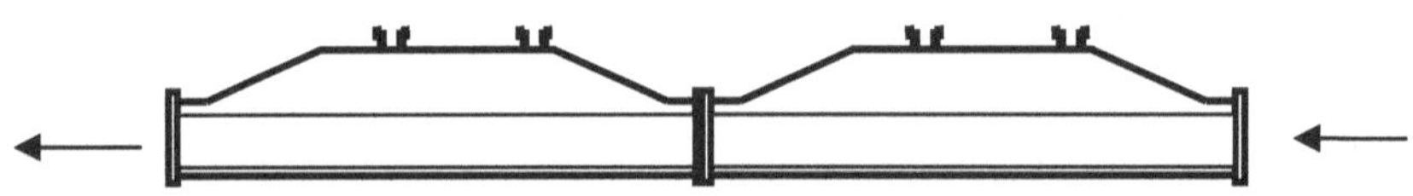

Figure 12.17

Two exchangers were butted together without any channel to ensure that the flow from an individual tube in one unit passed directly into a single tube in the next unit. Access to the tubes required the exchangers to be mounted on locomotive type tracks so that the heat exchangers could be moved along the tracks for inspection and maintenance.

Both exchangers therefore had to be designed with the same number of tubes which meant coordinating the designs and trying to achieve an optimum design for the two units in combination.

Restricted plot sites become even more interesting. To maintain homogeneous flow when the natural gas stream needs to be turned through 180^{0} requires an interesting bend as illustrated in figure 12.18 below. I was once in the position where I had to deal with this situation and it was clearly not practical to have a dividing plate for each tube row – I think I eventually compromised on one every three tube rows. Of course once you have more than one vertical row of tube there is still the possibility of phase separation.

Figure 12.18

Whilst I was working on these projects I kept on asking myself the obvious question is all this really necessary? Is phase separation such a bad thing? If this had been in the days of hand calculations I may have suggested that some hours should be allocated to check this out but it would have been very time consuming. With a modern computer program it took very little time.

My simple logic was that there are two extreme cases possible:

(a) Homogeneous flow – each tube gets its share of liquid and vapour.
(b) Stratified flow – all the liquid from the previous exchanger will flow through the lower tubes and all the uncondensed vapour will flow through the upper tubes.

In practice I would expect that there would be a gradual transition with the lowest tubes receiving all liquid and the highest tubes all vapour and the proportion changing for the intermediate tubes.

I took a heat exchanger that had been designed for homogeneous flow and investigated what would happen if the flow was stratified. I set up two computer runs; one with all vapour entering the tubes and the other with all liquid entering the tubes. I then varied the number of tubes in each run keeping the total the same as for the homogeneous flow design. When the pressure drops from the two runs were the same I had established a stratified flow case.

Comparing the two runs showed a very marginal advantage for the stratified flow case! The reason for this was that the liquid cooling coefficient for the lower tubes was surprisingly high. One reason for this is that the heat transfer properties for very light hydrocarbons are much better than for typical hydrocarbons – particularly the low viscosities. My conclusion was that it did not matter a dam whether the flow was stratified or not but both possibilities should be investigated to find the worst case scenario.

I only did this calculation for a single unit in the train but I would have expected similar results for the other units. If the results had been the same for all the units then each unit could have been individually optimised without any restrictions and there would have been a lot of cost savings in using normal bends and methods of access. I was not working in a very enlightened office on this project and my diversion was looked upon as 'wasting time'.

Some years later a proposal for a new LNG plant came up whilst I was working at Stone & Webster. Their American office would have engineered the project which was for Shell. A meeting was arranged in the Netherlands. The top brass – project and process managers etc. were flown over from the USA but they needed some technical personnel and it was cheaper to send these over from the UK office. I was a last minute replacement for the head of the heat exchanger section when he was taken ill. I had to give a couple of presentations with very little preparation. One of these was on the heat exchangers in the refrigeration train. I told them about the work I had done on one of these units and suggested an early study to check out my approach for the whole train. I was more than just a bit apprehensive in case someone came up with something I had not taken into account. No one did; and Shell made the unusual request that if S & W were awarded the contract I would be seconded to the US office to work on the project. They never knew that I was not a full time S & W employee and I heard no more about it so I assume S & W were not awarded the contract.

There are some situations in two phase flow when a good distribution is essential to the design. There are others where it may be costly and unnecessary. The reader should be aware of both possibilities.

GLYCOL INJECTION IN HEAT EXCHANGERS

If a gas stream consisting only of water vapour and inerts such as air is cooled it will eventually reach the dew point temperature where the mixture is saturated. Cooling below the dew point will result in liquid water being deposited on the cooling surfaces. This is a common process in air conditioning systems.

Natural gas streams invariably contain water vapour and if a natural gas stream is cooled it will it will be able to hold less and less water vapour as the temperature is lowered but instead of remaining as liquid water it may form hydrates by reacting with the hydrocarbons. Hydrates are solid crystals and they will be deposited on the cooling surfaces – usually the inside of tubes. Most of the main hydrocarbon constituents of natural gas; methane, ethane and butane may combine with water to form solid hydrates. The temperatures and pressures which determine which hydrates are most likely to be deposited is something we can leave to the process engineers.

The solution to avoiding hydrate deposits is to introduce a liquid which has a higher affinity for water than the hydrocarbons have. The liquid used is one of the ethylene glycols – usually just referred to as glycol. It is necessary for the glycol to absorb any water droplets before they have time to react with the hydrocarbons.

A very simple line diagram of a glycol drying system is illustrated in figure 12.19. The glycol is introduced into the warm gas before it enters the tubes of the heat exchanger. After leaving the tubes of the heat exchanger the glycol is removed and the cool dry gas proceeds to the next stage of processing. The glycol is circulated through a regenerator where the water is removed before being reintroduced back into the gas stream.

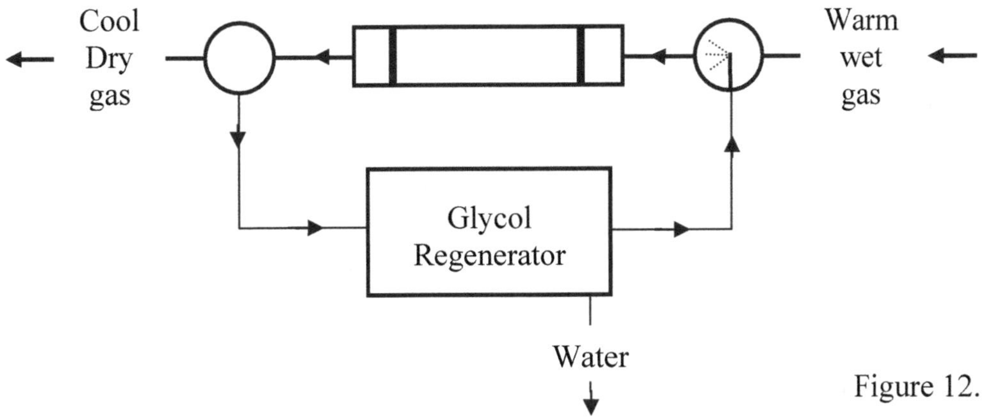

Figure 12.19

GLYCOL DISTRIBUTION

A critical part of any design is to ensure that each tube receives some glycol – ideally exactly the same amount in each tube. The glycol may be introduced in a separate item as shown in the diagram or in a specially designed exchanger inlet channel.

Where the injection takes place before the exchanger the channel should be fitted with an axial nozzle as illustrated in figure 12.20 below. A perforated plate with a diameter of about 150 mm more than the nozzle inside diameter will help to even out the distribution. Perforated plates with 10 mm diameter holes on a 20 mm pitch have been successfully used.

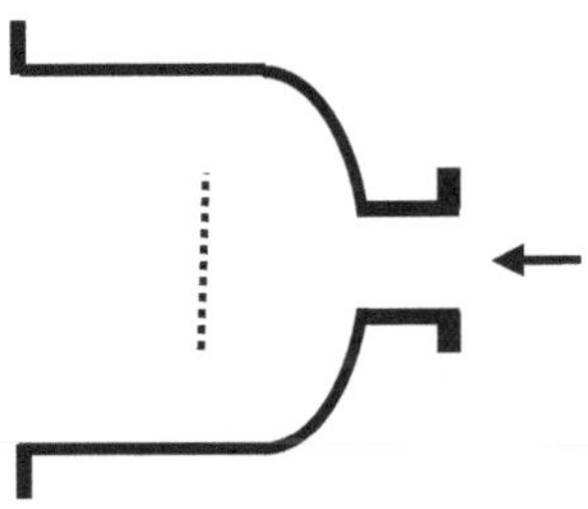

Figure 12.20

Probably a better solution is to use good quality spray nozzles in the channel as illustrated in figure 12.21 below. The minimum number of nozzles is four but more may be required for large units.

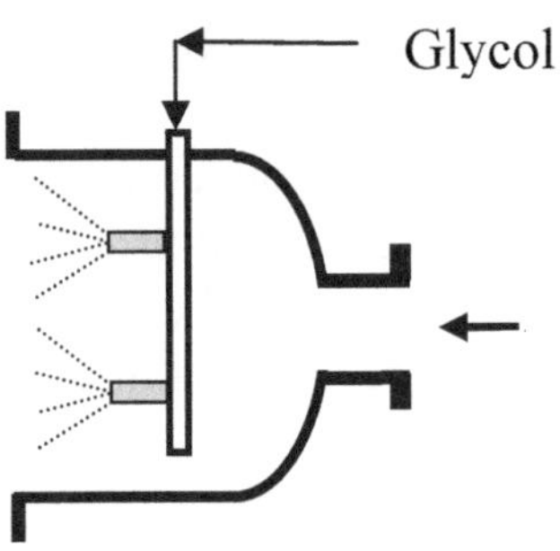

Figure 12.21

Whichever system is used the face of the tubesheet should be covered in glycol. A tube which protruded from the tubesheet more than an adjacent tube would receive less glycol. The only way to ensure that all tube ends are the same is to grind them all off so that they are flush with the tubesheet.

Whatever efforts are made to ensure good distribution perfect distribution is impossible and to ensure that each tube has sufficient glycol to avoid hydrates forming an excess of glycol is normally used. This may be two or three times the minimum theoretically required and is a process decision. Process must deal with potential hydrate problems in many areas other than heat exchangers. Line pressure losses may be a problem on cold nights on some sites and whenever a wet gas stream is throttled over a valve the consequent cooling is likely to cause problems. Gas compositions on different sites will all be different and the requirements to avoid hydrate deposits will vary.

PLANT TURN DOWN

During part load operation on any exchanger the velocity inside the tubes will fall and the oversurface will allow the unit to overcool the warm gas stream. Both of these conditions make hydrate formations more likely. One way of avoiding this is for two shells in parallel to be used in the original design so that one shell can be shut off during part load operation.

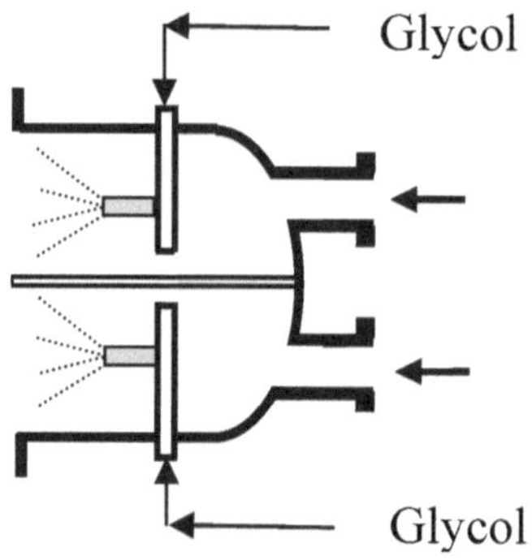

Figure 12.22

A less expensive alternative is to have two warm gas inlet nozzles and use a partition plate as shown in figure 12.22. This effectively splits the exchanger into two halves which may be operated in the same way as two shells in parallel.

GAS VELOCITY, HEAT TRANSFER & PRESSURE DROP

As far as the tube side thermal design is concerned the most important variable is the velocity. Knight – ref 40 has suggested an empirical method for checking the velocity which is based on operating data from a number of plants. He identified a 'critical minimum velocity', although he expressed this in terms of a momentum flux. His equation is as follows:

$$\rho_g u^2 = 670 \qquad \text{kg/ms}^2 \qquad [12.5]$$

Hydrate formations were only observed for cases where the momentum flux fell below this value.

Knight also suggested maximum velocity limits of 8.5 m/s at 6.9 bar and 6.0 m/s at 69 bar with linear interpolation in between these limits. Above his velocity limits he predicted that the formation of foam like emulsions was possible.

It is interesting that Knight found that the best correlation for his low limit was in terms of a momentum flux but that velocities were more satisfactory for predicting a high limit. The above data is only applicable to tube lengths up to 12 m.

Knight's work was published in 1975 and a lot of data has been accumulated since then. His recommendation for heat transfer coefficients and pressure drops did not rely on the usual complex equations for two phase flow. His method used multiplier factors which were used with values for gas phase only calculations.

The multipliers for his pressure drops were plotted against tubeside outlet temperature for different values of tube diameter as illustrated in figure 12.23 below.

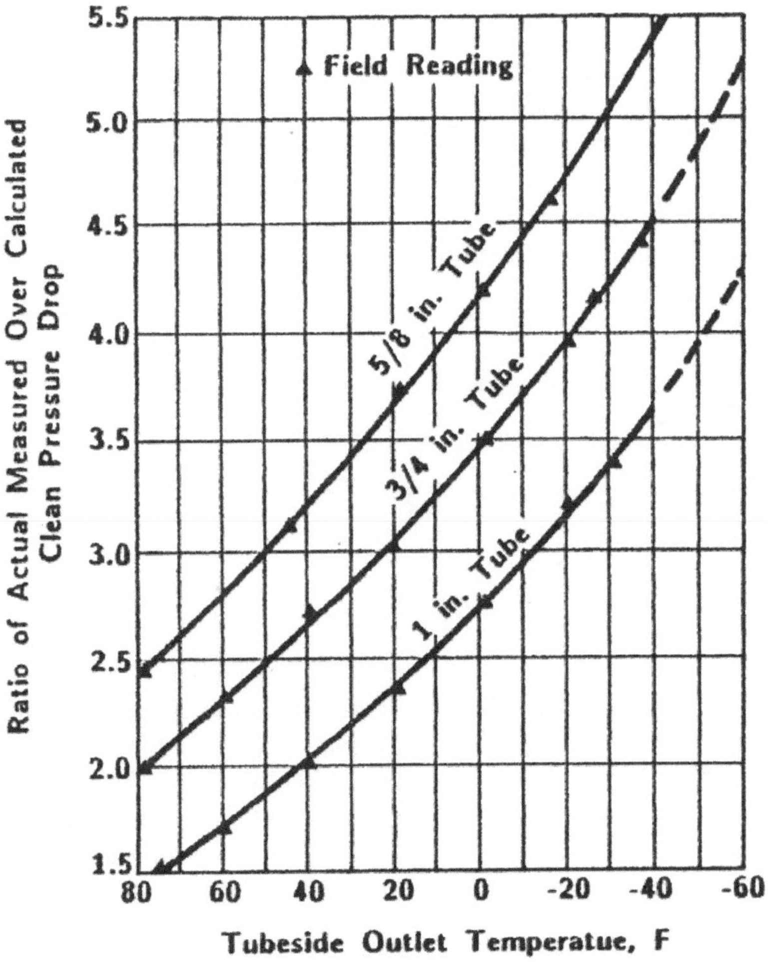

Figure 12.23

Unfortunately the units are the old US customary, the same as the original paper. It is interesting that the smallest diameter tubes are 5/8 inch as most specs require a minimum of ¾ inch.

For heat transfer his multipliers were plotted against tubeside outlet temperature for different values of tube side velocity as illustrated in figure 12.24 below.

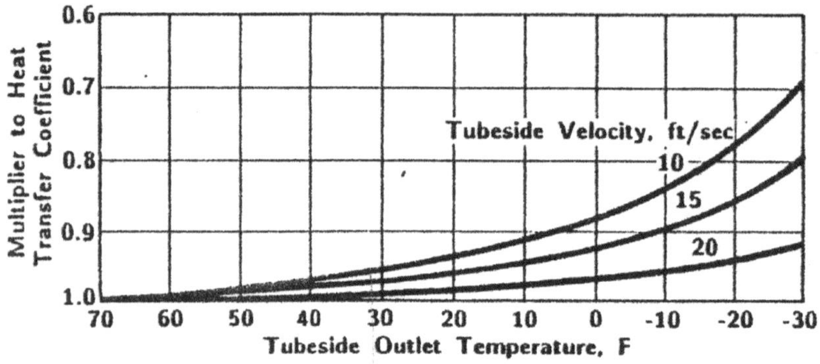

Figure 12.24

For most of the working ranges likely to be encountered the heat transfer multiplier is relatively small compared with the pressure drop multiplier.

Laminar Flow Annulus Analytical Method

One of the first analytical methods used in a computer program was based on the assumption of annular flow with a laminar layer of liquid around the inside of the tube as illustrated in figure 12.25 below.

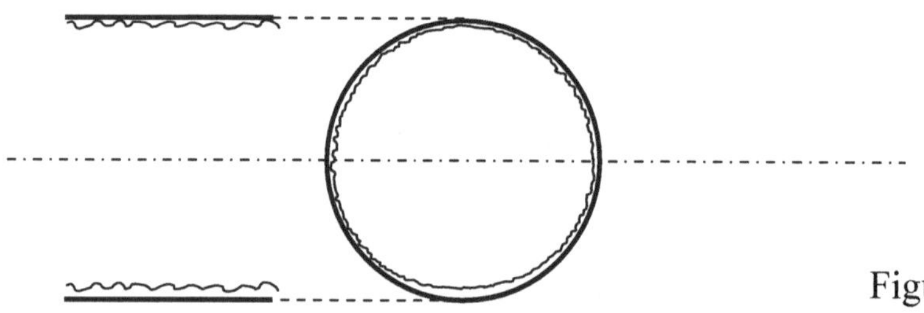

Figure 12.25

The mean thickness of the film was established by a pressure balance. An equivalent diameter for the annulus was used to determine a liquid heat transfer coefficient. A gas coefficient was calculated in the usual way using the inside diameter of the annulus as the characteristic dimension. The thermal resistance of the gas boundary layer and the liquid layer were then added together to give a combined tube side thermal resistance and hence a tube side heat transfer coefficient.

The uneven surface of the gas/liquid interface was allowed for by introducing a fixed multiplier of 1.5 to the friction factor used for normal metal surfaces. This was built into the pressure drop calculation.

Fortuine-Hamersma Analytical Method

This method is based on laboratory tests using a 50 mm diameter 1.4 m long glass tube. Tests were carried out through a range of velocities starting with very low values when the flow regime was stratified to sufficiently high values to establish annular flow. The liquid was observed to progressively climb up the wall of the tube as the velocity increased. Figure 12.26 shows a section through the tube at an intermediate velocity. For each value of velocity the proportion of the tube wall covered in liquid was estimated visually.

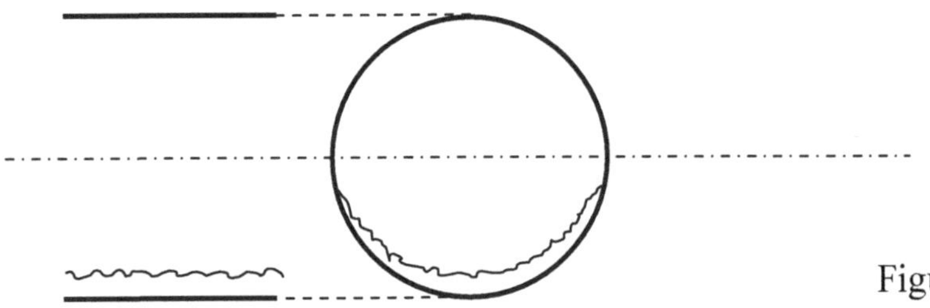

Figure 12.26

An equation was developed to predict the proportion of the tube covered in liquid. The equation uses the liquid Froude number and appropriate empirical constants.

The calculation methods developed for heat transfer and pressure drop therefore take into account the different flow regimes which can exist between very low and very high velocities.

These experiments were only part of Hamersma's PhD thesis and could not be expected to be too exhaustive. One limitation is the use of only one tube diameter which is larger than typical industrial tube diameters. The other is that the tests were only carried out with pure water and a 25% by mass mixture of glycol in water. Typical industrial mixtures are 60%-80% glycol in water. (This of course is the mixture entering the exchanger and the mass percent of glycol will fall from the inlet to the outlet as moisture is absorbed.)

CONCLUSIONS

As always the thermal design engineer must work with the program available and read the manual to establish which correlations are being used. The equations used will be updated as more information is added to the provider's data banks. The current HTRI program uses the laminar annular flow method and it is therefore essential to use a high enough velocity to make sure that laminar flow is established. Many two phase flow situations are a bit dodgy and ethylene glycol injection is no doubt one of the dodgiest.

The approach I would recommend for these systems is as follows:

(a) Take two squares on a sheet of graph paper and pencil in an area to represent the liquid fractions at inlet and outlet to get a picture of the proportions. I do this with any two phase flow problem.

(b) Check with process to find out if there is any operating data from a similar item on another plant.
(c) Establish a preliminary design which is in laminar flow. Then check the wall temperatures at outlet. These should ideally be a few degrees above any hydration temperature limit.
(d) Have another run on the design from (c) with vapour only without any glycol to get a feel for effect the glycol is having on the heat transfer coefficient and pressure drop.
(e) Check the momentum flux and the velocity against Knight's criteria. This does not take up much time and if the momentum flux is above 670 kg/ms^2 and the velocity is below his upper limit it adds to the confidence you can have in the design. If there is a severe discrepancy further investigation may be prudent. It may just be that your case is well outside the limits of the gas steams that Knight took reading from. If this is not the case it may influence your final design.
(f) Given the uncertainty of the correlations it is reasonable to add safety factors to the pressure drop and heat transfer. Say up to 20% on pressure drop and up to 10% oversurface unless process has already specified a figure above 10%.

These units are normally large expensive items and it is well worth spending extra time on them to be confident that they will work satisfactorily.

My own experience of these systems is very limited and I am indebted to John Round for giving me the benefit of his expertise. This has significantly broadened the scope of this section. It remains the only complete section of the whole book which has been proof read by another thermal design engineer.

ENHANCED HT SURFACES – CHAPTER 13

Introduction

Like the previous two chapters on special designs this chapter is not an afterthought. I decided at an early stage to make enhanced surfaces which result in enhanced heat transfer a separate topic. This makes it easier to categorise the alternatives available and discuss where they should be considered. I have limited this chapter to the type of tubes and heat transfer augmentation which can be used in shell and tube units.

HISTORICAL BACKGROUND

The early development of enhanced surfaces in heat exchangers took place in the refrigeration and air conditioning industry. The fluids they were dealing with were simple and relatively clean. The pressures they had to deal with were either atmospheric or moderate.

They were also designing for a mass market. A lot of their work is 'catalogue engineering' – the nearest standard size condenser is used to fulfil a specified duty. This allowed units to be tested before they were added to a range of products.

In petrochemical work each exchanger needs to be individually designed to satisfy a wide range of temperatures, pressures and fluids. Different fluids require different materials of construction. I can not remember any use of enhanced surfaces from when I first started to design heat exchangers other than the high fins used in air coolers and the longitudinal fins used in double pipe units.

LOW FINNED TUBES.

CONSTRUCTION

The first low fin tubes were made from copper or aluminium, soft metals which are easy to shape. As production methods improved more material were included and low finned tubes are now available in most materials including; carbon steel, stainless steel, nickel alloys, duplex, zirconium and titanium.

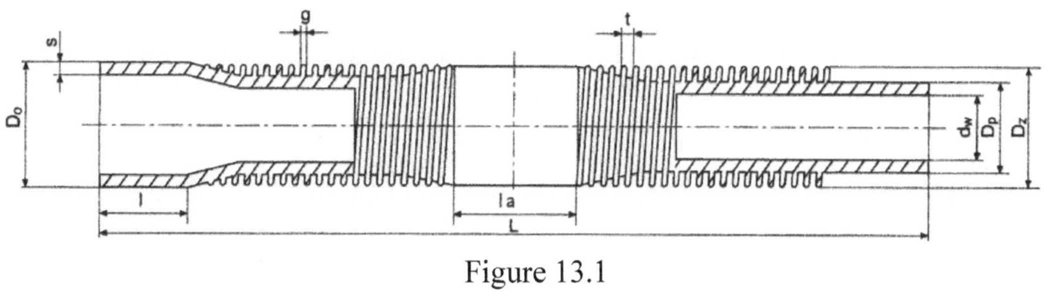

Figure 13.1

Figure 13.1 above is taken from the Aircofin website and shows all the salient features of a low finned tube. The range of outside diameters of the plane area at the ends of the tubes is the same as for bare tubes and these are attached to the tubesheets in the same way. Table 13.1 below gives the data which accompanies the diagram.

Do mm	Dw mm	Dp mm	Dz mm	S mm	g mm	la mm	t Fins/inch
10	5.5	7.1	9.8	1.0	0.3	6-25	19;25
11	6.5	8.1	10.8	1.0	0.3	6-25	19;25
12	7.4	9.1	11.8	1.0	0.3	6-25	19;25
16	11.5	13.1	15.8	1.5	0.3	10-25	19;25
19	14.5	16.1	18.8	1.5	0.3	10-25	16;19;25
22	17.5	19.1	20.8	1.5	0.3	10-25	16;19;25
25.4	20.5	22.5	25.2	1.5	0.3	10-25	16;19;25

Table 13.1

The length of the plane ends l needs to be only slightly longer than the thickness of the tubesheet. Although there is a column for this data there are no values in the table.

The main finned section of the tube has a substantially reduced internal diameter, dw and a slightly reduced 'over the fin' outside diameter, Dz – in this table only 0.2 mm less than Do. Dp is the outside diameter of the bare tube in the finned section.

The thicknesses of the bare ends of the tubes, S is interesting because they do not conform to BWG values. The 1.5 mm could be a rounded up 17 BWG but the 1.0 mm is in between a 19 and a 20 BWG. Clearly the BWG method of specifying tube thicknesses has not been used in this table.

The thickness of the fins, g is an average value. Although the fins are show as rectangular in section in the diagram they will always be thicker at the root than at the tip. This is actually a better shape than a rectangular section as far as the heat transfer through the fin is concerned.

The plane or bare section at the centre of the diagram, la is often termed a 'land'. It is a short section – which can be varied in length to suit the thicknesses of the baffles or support plates.

There is another column in the in the manufacturers table which I have omitted for convenience of presentation. It is the overall length of the tube, L. The values are specified as "up to" i.e. maximum values. For the four smallest tubes this is 10 m and for the other three 12 m. If lands are required it is clear that unlike a plane tube unit the tubes must be manufactured to suit a specified exchanger. This is likely to be the case even if lands are not required. It is therefore nonsensical to try to keep to 'standard tube lengths' when using this type of tube, (just as it is for any U tube exchanger.)

The last column is the number of fins/inch, t. In the diagram it is shown as the distance between two fins. It is interesting that this is the only column which is in old units. This is one situation where it is useful to use old units since it is a lot easier to visualise than the number of fins in one metre.

Computer input

The HTRI input for a low finned exchanger tube is as follows:

(a) Number of fins per unit length – 1/m
(b) Fin root diameter – mm (Dp)
(c) Fin height – mm ({Dz-Dp}/2)
(d) Average fin thickness – mm (g)
(e) Outside surface area per unit length – m^2/m
(f) Wall thickness of the finned section – mm ({Dp-dw}/2)
(g) Are the tubes finned in the baffle holes? Yes or No.

It can be annoying when manufacturers provide data in a different format to the one required for the computer program we are using. A wall under fin thickness and a fin height could have been used in the table instead of the two diameters.

The outside surface area required for the computer input is not in the table as it is with data from some other manufacturers. To use these tubes the designer would have to calculate a value, probably making the simple assumption of rectangular cross sections which would introduce a small error. No doubt this data would be provided if requested.

The thickness of the baffles will have been imputed elsewhere or defaulted to a TEMA value. This is used to determine the length of the lands. The number of lands will be determined from the baffle/support plate pitch. If the design gets to the point where it is going to be manufactured the tubes will need to be carefully specified unless the tube manufacturer is also going to construct the whole unit which is the ideal scenario.

To land or not to land?

This is an easy question to ask but not too easy to answer. Let us start with a simple diagram which shows how low finned tubes are produced – figure 13.2 below.

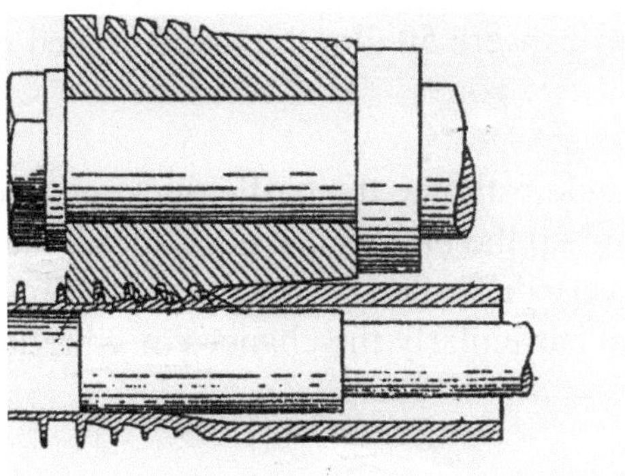

Figure 13.2

The diagram shows a single cutting tool. There are three cutter assemblies at 0°, 120° and 240° together with a cam arrangement to lift the cutters off the tubes when lands are required.

Producing tubes without lands requires a much simpler control system particularly if standard length are used for the tubes. Tubes with lands are therefore much more expensive.

If lands are not used with low finned tubes the contact area between the tube and the baffle is reduced. If a baffle with the same thickness as required by TEMA is used there is a danger that the fins will be flattened. This will increase the baffle/tube clearance which would increase 'A' stream leakage and also increase the chance of vibration problems. To avoid this potential problem the baffles ideally need to be thickened up so that the contact length remains the same.

Consider a 25.4 mm finned length of the 19 mm tube in table 13.1.

For 16 fins/inch
Contact length = 0.3 x 16 = 4.8 mm
Thickness increase factor = 25.4/4.8 = 5.3

For 25 fins/inch
Contact length = 0.3 x 25 = 7.5 mm
Thickness increase factor = 25.4/4.8 = 3.4

For 36 fins/inch
Contact length = 0.3 x 36 = 10.8 mm
Thickness increase factor = 25.4/10.8 = 2.35

I have added the 36 fins/inch case from a HPT catalogue. (Note the fin average thickness is the same for all cases.) Baffle thickness increase factors can now be defined as follows:

$$\text{Ideal baffle thickness increase factor} = \frac{\text{Bare tube contact length}}{\text{Finned tube contact length}} \quad [13.1]$$

$$\text{Actual baffle thickness increase factor} = \frac{\text{Bare tube contact length}}{\text{Tube contact length used}} \quad [13.2]$$

It is easy to see that the 'ideal' contact length can lead to some very thick baffles. I once worked on a job for a very big client who required the 'ideal' baffle thicknesses in their specs. They had a representative in the office and I showed him the consequences of keeping to their specs. I seemed to remember that we came to a bit of a compromise but only a small one. The baffles were 50 mm thick and looked more like tubesheets than baffles!

Using average fin thicknesses will always underestimate the baffle thickness increases required. When lands are not used thickening up the baffles is a good idea. What thickness increase factor should be used is very difficult to decide. A decision should be taken after considering the overall design particularly the chances of vibration problems.

Any program will assume TEMA baffle thicknesses if nothing else is input. Where lands are not used the HTRI program will make a small modification to the 'A' stream calculations. This modification slightly reduces the flow through the tube/baffle clearances. In theory this is not unreasonable because the fins will be acting in the same way as a labyrinth gland. Of course it will take no account of any flattening of the fins.

Now that most manufacturers seem to be able to offer finned tubes with lands the costs will have come down compared with tubes without lands. I suspect that in the future most designs will use lands.

RANGE OF LOW FINNED TUBES.

There are now many manufacturers who can supply low finned tubes. As far as I am aware there is no Low Finned Tube Manufacturers Association to provide any standardisation. The oldest catalogue I have is for Wolverine Tube which is dated 1968. They then offered tubes for a limited range of materials at either 16 or 19 fins/inch. The fin height for all tubes was 1.6 mm.

I have extracted the values in table 13.2 from an up to date HPT catalogue. There are also values for 30 fins/inch and a footnote saying that 19 fins/inch could be available.

	Fins/inch			
	26	28	36	43
Height mm	1.24	0.89	0.66	0.56
Thickness mm	0.33	0.3	0.3	0.25

Table13.2

As materials become harder and more difficult to work there is a trend for the number of fins/inch to increase. The table also illustrates that as the number of fins/inch increases the height of the fins decreases. This is no doubt partially explained by experience in forming fins in harder materials there is also an interesting thermal basis for this.

I decided against filling up pages with fin theory which the reader will no doubt have already covered in a basic heat transfer course. Two important features from this theory are:

(a) As the fin height increases the fin efficiency decreases.
(b) As the thermal conductivity of the fin metal decreases the fin efficiency decreases.

The harder materials all have low values of thermal conductivity. For example the different titanium grades can vary between 10-20 W/mK. This compares with 50 W/mK for carbon steel and 390 W/mK for copper. Hence more, shorter fins are required to optimise the heat transfer for these materials.

The term area ratio is often quoted in the literature. This is usually defined as the ratio of the external finned surface area compared with the plane inside surface area of the tube. These values range from about 2.5 to 4. A more useful and basic comparison would be the ratio of the finned outside surface area to the bare outside surface area of an unfinned tube. This would reduce these values because a bare tube has typical values of between 1.1 and 1.3 for the outside/inside surface area depending on the tube gauge used. The higher values are naturally quoted by manufacturers because it makes the tubes appear more attractive. They would also be used in any hand calculations. The HTRI input avoids the use of area ratio by requiring only the finned outside area per unit length.

Wolverine Tube us the term 'Trufin' for their product. HPT use the term 'Fine-Fin', there are no doubt other terms to describe propriety products which are all low finned tubes.

When designing a low finned exchanger it is necessary to take data from a particular catalogue. It is therefore necessary to check with purchasing if any product you are considering using is on any approved list of suppliers for the job your are working on.

HEAT TRANSFER

The basic equation for calculating the overall heat transfer coefficient was first introduced in chapter one and is repeated below.

$$\frac{1}{U_{calc}} = \frac{1}{h_o} + f_o + R_w + f_i \frac{d_o}{d_i} + \frac{1}{h_i}\frac{d_o}{d_i}$$

For finned tubes this needs to be modified as follows:

$$\frac{1}{U_{calc}} = \frac{1}{h_o} + f_o + R_w + f_i \frac{A_o}{A_i} + \frac{1}{h_i}\frac{A_o}{A_i} \qquad [13.3]$$

This is a simplification assuming a fin efficiency of 100% but it is all that is required to emphasise some important points. The area ratio A_o/A_i has typical values ranging from 2.5 to 4. The two thermal resistances inside the tube are multiplied this ratio, therefore it is not worthwhile considering the use of finned tubes if either:

(a) There is a high fouling resistance inside the tubes (say above 0.00025 m^2K/W), or
(b) A low heat transfer coefficient inside the tubes.

Manufacturers will recommend the use of finned tubes whenever the ratio of the inside heat transfer coefficient to the outside heat transfer coefficient is the same as the area ratio of the finned tubes. This recommendation is a bit on the low side and a ratio for the two coefficients of between 3 and 5 is normally required to produce an optimum design unless;

(a) A finned tube design results in a single shell compared with two shells using bare tubes. In this situation the extra cost of pipework and the extra pressure drop used up in the pipework using two shells must be taken into account.
(b) On a revamp where there are many cost savings to be made by replacing an existing bare tube bundle with a finned tube bundle to cater for a higher duty.

Single phase heat transfer.

Low pressure gas streams are the fluids most likely to result in low heat transfer coefficients and are therefore most likely to be suitable for finned tubes. With a normal segmental baffle arrangement, (either single or double segmental), some of the flow will be in long flow and some in cross flow. In the long flow sections of the shellside, finned tubes may actually decrease the heat transfer coefficient because pockets of gas in between the fins will have an insulating effect. In the long flow sections the fins should ideally be longitudinal fins. In a high fouling service Low finned tubes are therefore most likely to be worthwhile in cross flow and NTIW designs.

Viscous liquids also produce low heat transfer coefficients and in theory might be a situation where low finned tubes could be used. A catalogue I have in front of me lists oil heaters and coolers as typical applications for their tubes. Some times there may be a warning that they should not be used in laminar flow, (which results in the lowest coefficients). For these fluids we would most likely be using a 45° tube pitch arrangement which provides cleaning lanes and also creates a mechanical turbulence which makes laminar flow very difficult to define. I would never consider using finned tubes for this type of design without some field data from a similar item on another plant.

Condensing heat transfer

Tube manufacturers may quote the use of low finned tubes in a condensing service although they may qualify this by stating; "Provided the condensing is shear controlled rather than gravity controlled". Whatever the mechanism controlling condensation the use of finned tubes is always dodgy when there is any condensation. In any low velocity area there will be gravity controlled condensation and the condensate film will be thicker than it would be on a bare tube resulting in a lower heat transfer coefficient. Of course if the condensing load is a large proportion of the total load there is unlikely to be a low coefficient in a well designed bare tube unit.

This leaves units which are basically gas coolers with some condensation and the design of these units was covered in chapter 5.

Boiling heat transfer

In most cases boiling heat transfer produces high heat transfer coefficients and only bare tubes would be used. The exception is when there are very low temperature differences. For these cases it is difficult to establish sufficient superheat to activate the nucleation sites found in bare tubes. I went into the theory of bubble formation in some detail in chapter 7 on boiling.

For low MTD applications there may be a significant advantage in using low finned tubes. For any boiling situation there is also the effect of fouling to consider. This may be very low for some pure fluids. One of the main applications has therefore been in evaporators in the refrigeration industry. In any refrigeration cycle there is always pressure to minimise the MTD which maximises the overall COP, (Coefficient of Performance). In the petrochemical industry one situation where we regularly have to deal with boiling refrigerant fluids operating with low MTDs is in an LNG plant.

Although low finned tubes are much better than bare tubes for low temperature difference applications they do not provide the 'ideal shape' for a nucleation site. Surfaces which provide nucleation sites that are much closer to the ideal shape are now available and will be covered later in this chapter.

FOULING

A lot of fluids we have to deal with are mixtures rather than pure fluids and the fouling factors are much higher than for pure fluids. Some manufacturers will say that low finned tubes are less likely to foul than plane tubes; but to quote Mandy Rice-Davies "They would say that wouldn't they?"

However such claims are not made without any data to back them up. One of the most impressive examples is illustrated in figure 13.3. This is taken from Yokell ref 9.

Figure 13.3

The photograph shows a single HPT finned tube in a bare tubed bundle. It is clear that the finned tube has fouled less than the bare tubes. It should be noted that the experimental tube is situated in between the edges of the baffles. It would therefore be in cross flow. It would have been interesting to compare the thermal performance of this tube with an outer tube in the window which would have been mainly in long flow. There are no details of the shell side fluid.

This is a very simple experiment to set up and data from a field case is worth far more than simulated data from laboratory experiments. I have never worked directly for a client but if I ever did I would consider a series of such experiments on a new plant, (with more than one tube in each experiment). There might be a few years delay in the payback, but when the inevitable revamp came along replacement bundles utilising finned tubes in a high fouling service could be used with confidence.

One case I have read about was a revamp replacement bundle for a kettle reboiler. This is a situation where I would never have considered using finned tubes. Perhaps I should and with experimental evidence I would have the confidence to do so.
One theory behind the reduction in fouling on finned tubes is that the thermal gradient along the fins together with any longitudinal expansion and contraction of the tubes as process conditions change induces stresses in the fouling layer which cause it to crack and break away. This clearly only applies to very hard types of deposits and tar like or biological deposits would be sufficiently flexible to avoid cracking. The wide range of different types of fouling layers is covered in the next chapter. Experimental evidence of how a particular process or service stream will foul on a finned tube surface is invaluable.

The cleaning of low finned tubed bundles is also claimed to be easier and quicker than for bare tubed bundles. It would be interesting to have the views of plant operators on this one.

MATERIALS

It is interesting that manufacturers now offer low finned tubes in carbon steel. It does not need too much corrosion before a fine finned tube could become a bare tube with a reduced outside diameter. A lot can happen to a heat exchanger from the time it leaves the fabricators works to when it is commissioned. I once worked on a job where some of the exchangers had been on site for several years. Then of course there is the hydro testing. Someone, somewhere, must be using them and it would be interesting to know how well they performed and how long they lasted. I would need a lot of convincing to even consider using carbon steel.

The alternatives to carbon steel are all far more expensive. I will once again avoid any cost comparisons although the cost is often the most important consideration. This is because I have seen drastic changes in the cost of some metals over a very short period of time and anything I could come up with today might be out of date before the reader looks at it. Current prices are essential in any cost comparison and most companies have specialists who should be consulted for each job.

The most important point to remember about materials is that some process and service fluids may require exotic materials for bare tube designs. For these cases the use of low finned tubes becomes much more attractive.

DECISION DIAGRAM

The decision diagram in figure 13.4 is a useful way of rounding off this section. To avoid it becoming too complex I have only included clear cut decisions. For example I have omitted the size of the surfaces which are required when a finned tube design may be possible with a single shell but two would be needed using bare tubes.

In most cases it will be necessary to get cost estimates for a low finned tube unit compared a bare tubed unit. Some manufacturers will offer to do this for you but it would be unrealistic to expect an unbiased assessment and I am writing this book to help engineers who look for challenges rather than just looking for easy ways out.

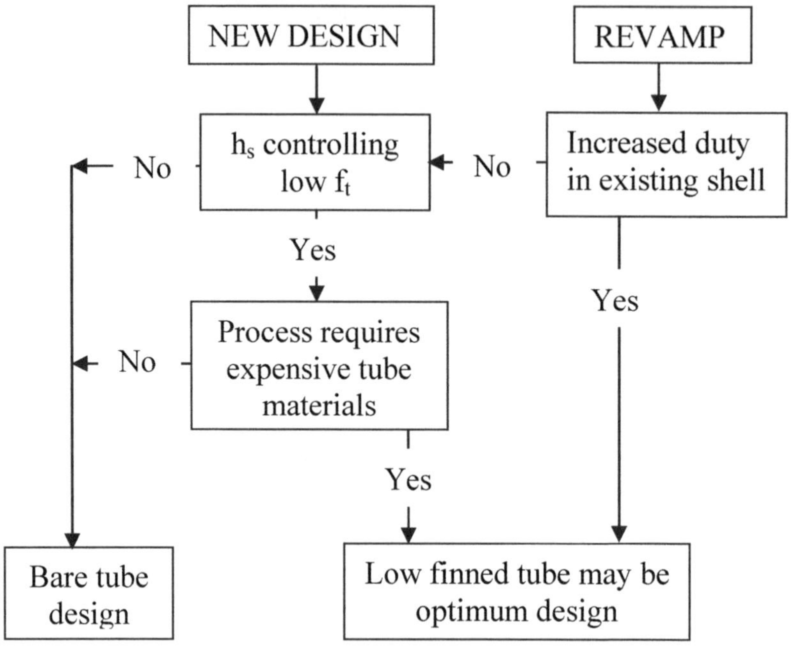

Figure 13.4

DOUBLE ENHANCED TUBES

Figure 13.5 below is taken from a current HPT catalogue. The outside of the tube has low fins and the inside has a surface enhancement in the form of helical grooves which are termed ribbed surfaces.

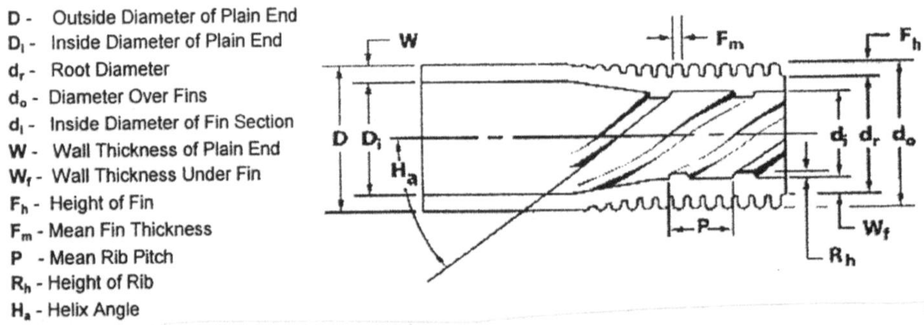

Figure 13.5

The height of the rib fin is only about 0.25 mm which only increases the surface area of the inside of the tube by about 35%. However I can easily see how the mechanical turbulence created by the fin will increase the heat transfer rate, at the expense of increased pressure drop of course. HPT claim an increase in heat transfer coefficient of about 100% at the expense of a 50% increase in pressure drop. It is interesting that this data is based on sea water inside the tubes. Any type of cooling water inside plane tubes will produce high heat transfer coefficients and I can not understand introducing any complication to improve these.

The effect on fouling is a big unknown. The manufacturer simply states that the effect of fouling may be reduced.This type of tube seems to be very much in the experimental stage and is not a serious option at the present time. If they ever do become a serious option in petrochemical work they will no doubt be initially introduced on revamp work where using normal low finned tubes is not quite sufficient to avoid replacing the shell as well as the bundle. It will need an adventurous client to try this out for the first time.

TUBE SIDE ENHANSMENT

Helical grooves

Helical grooves were illustrated earlier in figure 13.5 for double enhanced tubes. HPT are also producing tubes with an internal helix and bare exterior. One proposed application is tubes for vertical thermosyphon reboilers. There is certainly a strong incentive to improve the tubeside coefficients on some vertical thermosyphons – particularly those with long liquid heating lengths. However the improved heat transfer may only be required in the lower half of the tubes and there are other ways of achieving this. (See chapter 10 on vertical thermosyphon reboilers.) I am not convinced that this type of tubeside enhancement has a future in the industry.

Twisted tape inserts

Figure 13.6 shows a twisted tape insert. The diagram is also twisted a bit but I have spent too much time already trying to correct this.

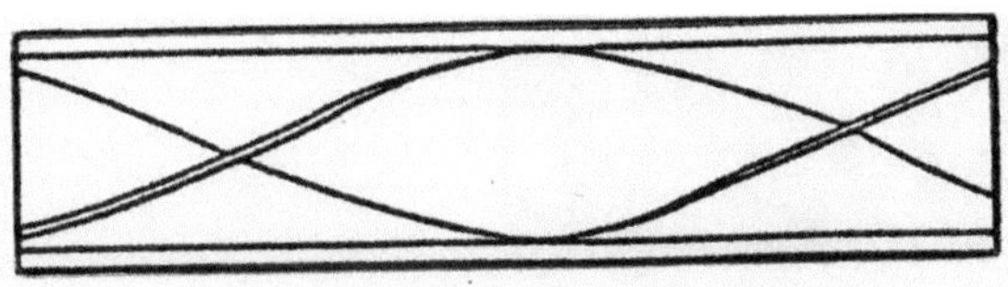

Figure 13.6

This is the simplest form of in tube turbulator and is illustrated in many books and articles. It must impart a helical path to the fluid which may improve the heat transfer coefficient at the expense of pressure drop. I could not find any data on either or anything concerning its effect on fouling.

I doubt if they are worth the added complication and I have never come across them being used.

Internal longitudinal low fins

Figure 13.7 below shows an internal longitudinal finned tube which can be used to enhance an internal heat transfer coefficient. I could not find a suitable catalogue sketch for this one and had to produce it myself nor could I find any details the heights and density of the fins. It is another form of surface enhancement which I have no experience of. One of its main uses seems to be with doubly enhanced tubes where different methods of enhancement can be used for the external surface.

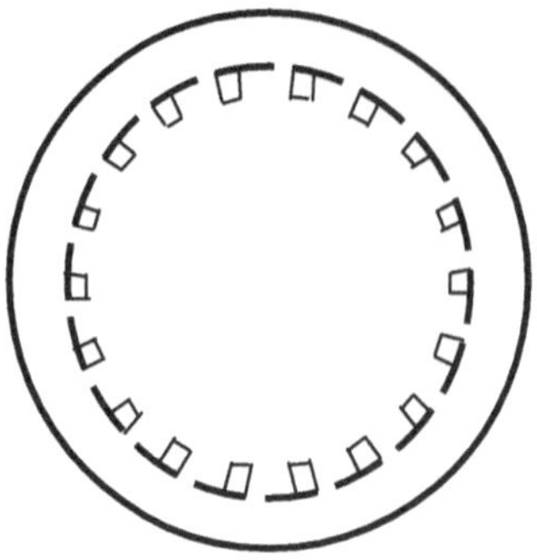

Figure 13.7

UOP produce tubes with longitudinally finned inner surfaces and a high flux coating on the outer surface.

Wire fin tube inserts

Figure 13.8 below is the type of wire fin tube insert produced by Cal Gavin

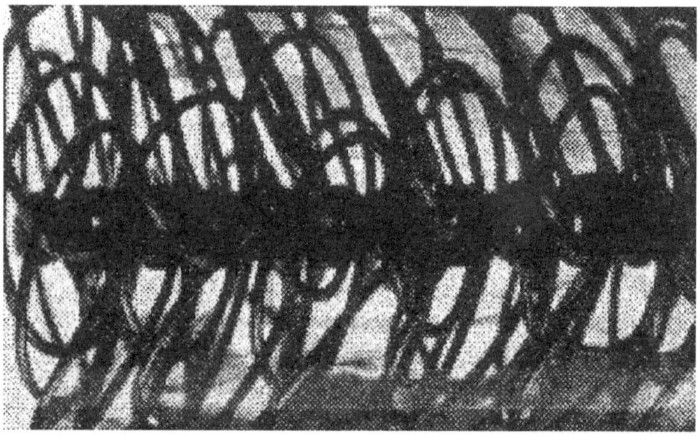

Figure 13.8

Figure 13.9 below is from a Midland Wire catalogue.

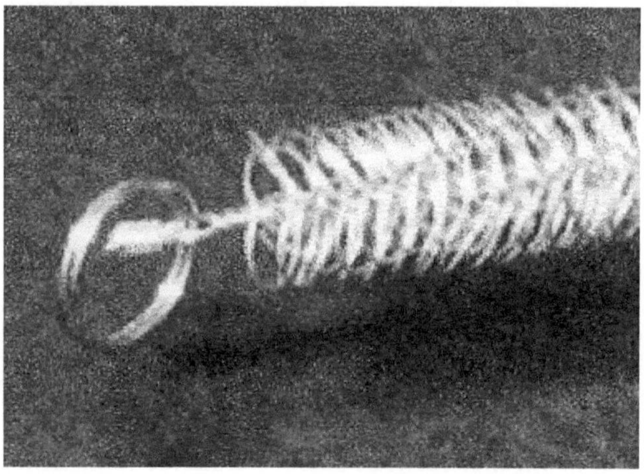

Figure 13.9

The Cal Gavin turbulators have been around for many years and they claim to have installed over 5000 units. I only came across Midland Wire when I started to research this topic. This is a sad admission on my part since their works at Redditch UK is only about 40 kilometres from where I am writing this book. I had always assumed that Cal Gavin was the only source of this type of insert – they still quote a "unique patented wire matrix element" on their web site. Whenever there is a single source of any product this must be reflected in the price and it is nice to know that there is now some competition.

A wire matrix inside a tube will certainly create a mechanical turbulence which will break up the boundary layer and increase heat transfer rates. This is well illustrated on a video clip on the Cal Gavin web site. This shows a dye being introduced into a clear liquid which is presumably water. (Classic Reynolds experiment.) Both Cal Gavin and Midland Wire have carried out testing to establish correlations for heat transfer and pressure drop for the different type of matrixes they offer. The variables in the different types are the gauge of the wire and the number of helixes involved. The diagram from Midland Wire clearly shows the method of maintaining the position of the insert inside the tubes and facilitating withdrawal.

Cal Gavin have produced a design and simulation software package called hiTRAN.SP which can be used as either a stand alone program or incorporated into HTRI or HTFS programs. The latest version is 4.5.0 which indicates that it has been around for some time. Midland Wire has manual methods for calculating both heat transfer and pressure drop on their web site. The format of these equations will be familiar to any heat transfer engineer. I find it odd that these have not been incorporated into some form of program. An interactive program with manual input for the heat transfer constant j_h and the friction factor f from their chart with Reynolds Number on the horizontal axis would be relatively simple and cut out the chance of a lot of numerical errors.

I can not see any significant difference in the product offered by the two companies so I presume that the Cal Gavin patent has expired. It would be interesting to compare the results for the heat transfer and pressure drop by the two methods for a range of specified cases.

As with all heat transfer enhancements there is the ubiquitous question of the effect on fouling resistance. Suppliers will have plenty of cases to quote showing how fouling resistances have been reduced and I have no reason to doubt this data. However I wonder if there have also been cases where a bit of corrosion combined with a lot of fouling has meant an insert snapping off and becoming stuck in a tube.

Clients and plant operators could find out a lot about the suitability of using turbulators for particular streams by carrying out low cost experiments similar to those covered earlier for the low finned tube. These could be set up for a new plant or an existing plant at a shutdown prior to any revamp. A single tube could be fitted with a wire matrix turbulator in selected units. At the next shutdown the state of the tube with the turbulator could be compared with the other tubes. This would be a very 'safe' experiment as far as fouling is concerned because the fluid flow through the tube/tubes with the turbulator would be less than the flow through the other tubes and therefore more likely to foul. A variation on this experiment would be to fit short lengths of turbulator in the tube which would reduce the difference in flow rates. The ideal experiment would have four turbulators; one full tube length, one half tube length, one a quarter tube length any one very short one. Although the assessment of the fouling in each one would be subjective without the complication of adding instrumentation it would be very useful.

My experience of wire matrix turbulators inside tubes has all been on revamp work and it has always involved Cal Gavin. Some of these were a long time ago when it was simply a matter of consulting them for a particular case. Now that design tools are available it is much easier for a thermal design engineer to investigate their use in the search for that elusive optimum design.

ENHANCED SURFACES FOR BOILING APPLICATIONS

In chapter 7 on boiling, nucleation sites were discussed in detail. The characteristics of the 'ideal' nucleation site shown in figure 13.10 are:

(a) Maximum metal contact surrounding the site.
(b) Relatively small hole to retain the bubble until it had sufficient superheat to move away from the surface as a bubble without collapsing.

Figure 13.10

There are now many products on the market which provide surfaces which approach the ideal nucleation site. Some of these are illustrated below in figure 13.11 below.

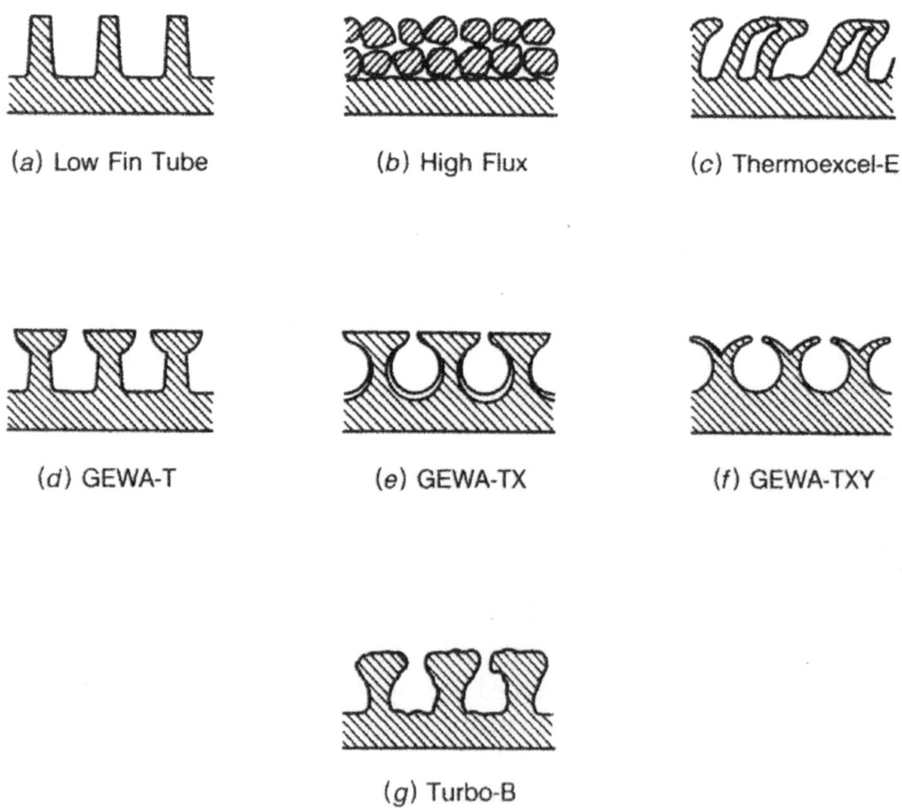

Figure 13.11

I am grateful to Kakac – ref 10 for assembling this sample collection. There are many manufacturers which offer the low finned tubes which have already been covered for general applications. The high flux tube is produced by UOP, the Thermoexcel-E by Hitachi, the three GEWA options by Wieland and the Turbo-B by Wolverine.
If the space between adjacent tubes is considered as being the nucleation site for the low finned tubes, type (a), then they are clearly not ideal as far as the small hole at the top is concerned.

Surfaces type (c) through to type (g) all have the shape of the ideal nucleation site. They all have the small opening at the top with different designs to maximise the contact area below the hole. They all have something else in common – they are all two dimensional nucleation sites. The grooves are all almost circumferential - on a fine helix the same as the low finned tubes. The bubble will therefore start off as elongated rather than near spherical in shape.

The high flux tube, type (b) is the only one where the nucleation sites receive heat from three directions. These tubes are produced by bonding a thin porous metal matrix to the surface of the tubes. This produces a large number of nucleation sites. The liquid is drawn into the matrix by capillary attraction. The manufacturers claim that heat transfer coefficients can be improved by factors of between 10 and 30 compared with bare surfaces. (Enhancement factors.)

ENHANCEMENT FACTOR

This term is defined as follows:

$$\text{Enhancement factor} = \frac{h_{\text{enhanced tube}}}{h_{\text{bare tube}}} \qquad [13.4]$$

Enhancement factors depend on:

(a) The boiling fluid
(b) The type of enhanced surface being used.
(c) The temperature difference between the metal and the saturation temperature of the boiling fluid.

Manufacturers will have sponsored research for their types of tube for a number of fluids through a range of temperature differences. When faced with a new fluid it is necessary to assume an enhancement factor for a similar fluid. This is always dodgy since physical properties such as surface tension and enthalpy of phase change can vary considerably. Unfortunately most of the early research in this field was on fluorinated hydrocarbons which are now banned because of their impact on the ozone layer. The current state of their data bank is something I would always enquire about when considering using a particular type of enhanced surface from a supplier.

TEMPERATURE DIFFERENCE

The temperature difference ΔT between the metal and the fluid saturation temperature – often termed the superheat temperature is a key factor in deciding the most appropriate type of enhanced surface to use.

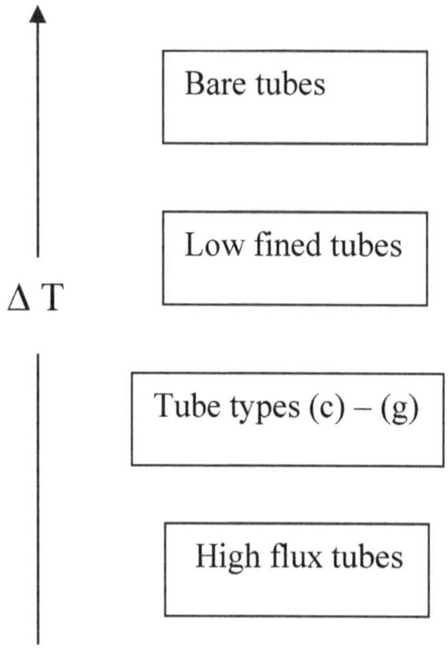

Figure 13.12

Figure 13.12 is a very rough guide to selecting the right type of tube. For very low temperature differences the high flux tubes have a definite advantage. Kakac – ref 10 quotes temperature differences of below 1.0 K as being sufficient to establish boiling in some cases. Probably 1 to 2 K is the limit for most situations. At other extreme if there is a sufficiently high temperature difference to establish nucleate boiling on bare tubes enhanced surfaces would only be considered on a revamp.

If the boiling fluid is subcooled, probably due to the static head effect there may be sufficient natural convection to give the low finned tubes an advantage.

FOULING

The main applications for enhanced surfaces are pure refrigerants where the fouling resistance for bare tubes is always relatively low. For high flux tubes UOP state that "conventional fouling resistances do not appear to be necessary". To make this quite clear it means that they use a value of zero for the fouling resistance. (Unlike plate and plate fin exchangers where they are normally just reduced.) I am still in a quandary over this one. The first time I came across it I was not convinced despite all the field data they had to back it up. On reflection, **assuming they are right, perhaps we should be looking in the opposite direction and using a zero fouling resistances for bare tubes also for pure refrigerants?**

With the exception of the brittle fouling layers on the low finned tubes, which have already been covered, I would not consider any of the surfaces in figure 13.11 to be suitable for a high fouling service. They have been used however with a lot of fluids other than pure refrigerants such as in the boiling of binary mixtures. Further examples can be found in the case study referred to later.

MIXED TUBE BUNDLES

Tube bundles do not need to consist of either bare tubes or enhanced surface tubes. They can be a mixture as illustrated for a kettle reboiler in figure 13.13 below.

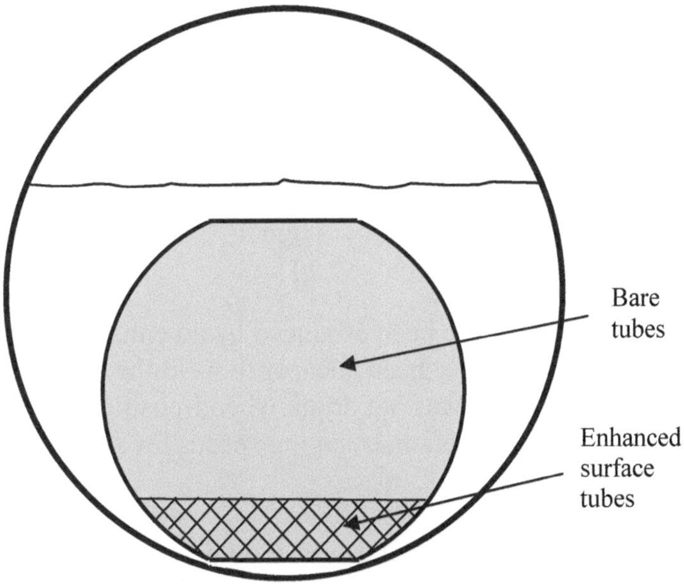

Figure 13.13

For large bundles, particularly those operating at low pressures the local states and hence the local heat transfer coefficients can vary significantly from the bottom to the top of the bundle. In a bundle with only bare tubes those at the bottom may be undergoing insipient nucleate boiling at best and only natural convection at the worst. Further up the bundle the static head will be lower and less superheat is required to form bubbles. Towards the top of the bundle there may be full nucleate boiling and a very satisfactory heat transfer coefficient. The output from a modern computer program will indicate when this is taking place.

One solution is to use enhanced tubes for the whole bundle which is the solution many manufacturers would recommend. The most economic solution however may be to use the enhanced tubes at the bottom of the bundle only. Even a few rows of tubes which are producing bubbles may be sufficient to increase the circulation rate and the subsequent increased turbulence will wipe off any bubbles formed on the bare tube surfaces which would otherwise take some time to become detached. This makes the nucleation sites on the bare tubes more effective. Together with a small reduction in static head effect, this may be sufficient to produce a satisfactory heat transfer coefficient throughout the bundle.

I have no doubt that one day it will be possible to input mixed tube data into a computer program. Until then any mixed tube bundle will require two runs and some fudging to simulate the conditions for the enhanced tube run which will then be used as the input for the bare tube run

I know of one case where using mixed tube bundles was forced on the designer rather than being an option. Process had set a very interesting challenge. The unit was a single pass kettle with a very high temperature difference at the hot end and a very low temperature difference at the cold end. Enhanced surface tubes were required for the cold end to achieve nucleate boiling. When the unit was run will all enhanced tubes there was a problem with excessive heat fluxes at the hot end where 'dry out' was predicted. The mixed tube solution is illustrated in figure 13.14 below.

.

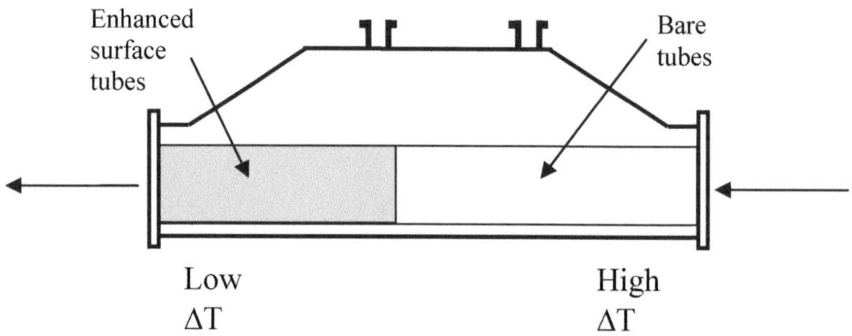

Figure 13.14

The combined surface tubes could have been produced by an enhanced tube manufacturer but I suspect it might have been cheaper to weld the enhanced tube section to the bare tube section. (No I was not drunk when I positioned the shell side outlet nozzles – remember the section on disengaging spaces for single pass kettles in chapter 8?)

This is not too bad a case to simulate since it can be thought of as two shells in series and no doubt a two shell option was offered to process who had good reasons for rejecting it.

Using enhanced surfaces for the lower tubes only is not new but studying this case got me thinking again. There must be many cases where the optimum design would combine the two types of tube as illustrated in figure 13.15 below.

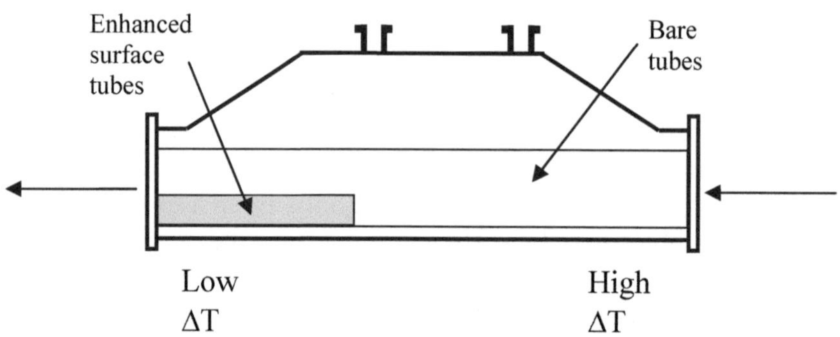

Figure 13.15

As far as I know this has never been tried but there is always a first time. Perhaps a reader will accept the challenge and find a small place in shell and tube heat exchanger design history?

CASE STUDY

This is an interesting case I found on the net concerning a succession of revamps for a C2 Splitter on a Chevron ethylene plant – ref 39. The paper was presented to the AIChE at a conference under the heading "Ethylene Revamps and Retrofit Technology." A lot of the paper concerns the trays in the tower and will only be of interest to process engineers, I only intend to highlight some of the work carried out on the shell and tube units.

Splitter Condenser

The original overhead condenser was a TEMA type CKN with three shells A, B and C size 1450/2440 x 12192. The refrigerant on the shell side was propylene. The overheads which passed through the tubes would be mainly condensables with some inerts.

During the first expansion the bare tubes in shell B were replaced with UOP high flux tubes with a normal bare surface inside the tubes. This produced an interesting situation which made it possible to compare the performance of the two types of tubes.

At the next expansion the tubes in shells A and C were replaced. This time the new tubes were the double enhanced variety with a high flux outer coating and low finned inner surface. This allowed the performance of the high flux tubes with and without the low finned inner surface to be compared.

Splitter Reboiler

The original splitter reboiler was a vertical TEMA type CEN with four shells, size 1245 x 6096. The photograph shows the shells are independently close coupled to the tower. The bare tubes used were 19.05 mm. The heating fluid was propylene refrigerant. This entered the top as a superheated vapour and left at the bottom as a subcooled liquid.

There are clean fluids at both sides of the unit. In the original design there would have been a reasonable heat transfer coefficient on the shell side provided wet wall condensing took place at the top. On the tube side there would have been a reasonable heat transfer coefficient at top but there may have been a problem with a liquid heating length at the bottom.

The revamp was certainly a challenge. It was decided to reuse the existing shell and channel covers but to increase the size of the outlet pipe and nozzle. The new bundle utilised 31.75 mm tubes which were 'fluted', (longitudinal low finned), on the outside with a high flux coating on the inside.

There are no details of the runs to show how each coefficient improved. For the shell side condensing the characteristic dimension would still be the vertical height, the same as for plane tubes, and the increased surface area, (presumably by a factor of about 2.5) would certainly increase the coefficient. On the tube side the boiling coefficient would increase and the use of a high flux surface could be sufficient to promote nucleate boiling at the bottom of the tubes and eliminate any liquid heating length.

Improved performances

The improvement in the individual heat transfer coefficients is difficult to sort out from the data given because the increased performances enabled the process conditions to be changed. The changes in process conditions together will the new trays in the tower and the heat exchanger upgrades resulted in an impressive increase in production of 70%.

CONCLUSIONS

Writing this chapter has opened my eyes to some applications of enhanced surfaces which I would never have thought of before. The case study is a classic example of how a revamp is the instigator of innovation. However where the application of a particular type of enhanced surface to a particular fluid has been well tried and tested, we should perhaps be using them more often in new designs.

FOULING – CHAPTER 14

Introduction

Fouling resistances were first introduced in chapter 1 in the basic equation for calculating the overall heat transfer coefficient. Fouling is another topic which could be the subject of a complete book. I have a copy of one such book - the Fouling Notebook by Bott – ref 16 which I would recommend to the reader.

Fouling invariably accounts for a significant part of the total thermal resistance across a heat exchanger and in some cases may be over 50%. It therefore represents a significant part of the capital cost. There are also running costs which are less obvious. Cleaning may involve interruption and sometimes shut down of a plant where the costs can be far higher than the capital costs of the exchangers.

The other hidden cost of fouling is the increase in pressure drop which increases pump/compressor running costs. Some of the ways of dealing with the effect of fouling on pressure drop were covered in chapter 4.

Fouling is one of the few topics in this book which is of vital interest to the general public. Arteriosclerosis is an example of fouling in the human body. Cholesterol deposits inside arteries will increase the resistance to the flow of blood which must either reduce the flow of blood or force the heart to work harder to provide the extra pressure drop required.

The allocation of fouling resistances to fluid streams is normally by process but for single units and very small plants it may be passed over to the thermal design engineer and I have had to do it on many occasions particularly when working for small manufacturers. Even if the reader never has to specify a fouling resistance some knowledge of fouling mechanisms can be essential to producing an optimum thermal design.

HISTORICAL BACKGROUND

When I first started to design heat exchangers I was given a copy of the current TEMA to read. When I came to the section on fouling resistances I approached it in my usual inquisitive way and looked down the list to identify the limiting values. The lowest was steam with a fouling resistance of 0.0005 hrft^{2}F/Btu (0.00009 m^{2}K/W). The highest was the Chicago sanitary canal with a fouling resistance of 0.01 hrft^{2}F/Btu $\leq$ 3 ft/sec > 125°F (0.00176 m^{2}K/W $\leq$ 0.9 m/sec > 52°C). This is a factor of 20 between the two values.

The value for steam is still commonly used although I have sometimes wondered if any is necessary since steam is often used to clean heat exchangers.

The Chicago sanitary canal is no longer listed in the current addition of TEMA. I am not sure whether this means it no longer exists or because it has been cleaned up a bit. Perhaps the decision was just political; it was not a record to be proud of. At the time I tried to visualise it by comparing it with some of the worst waterways I had seen in industrial Lancashire. If I had ever visited Chicago in those days it would have been top of my list of my places to visit just to check it out. It is still part of thermal design folklore although I have never met anyone who has designed an exchanger which used cooling water from it.

The other thing I remembered about my first introduction to fouling factors is having it explained to me that they served a dual purpose. One was to allow for the build up of a layer of dirt which would reduce the overall heat transfer coefficient and the other was to provide a hidden safety factor.

The safety factor aspect was taken advantage of by some companies in the sixties including Kellogg where I first started thermal design work. As I explained in chapter 1 the fouling resistance inside the tubes needs to be referred to the outside diameter of the tubes before being added to the other thermal resistances. This increases the value of the fouling resistance in the ratio of the tube outside/inside diameter. Ignoring this step in the thermal design calculation will result in a smaller, cheaper, more competitive exchanger. This practice was justified by the assumption that all fouling resistances were specified at the outside diameter. This was clearly not the case since we never changed them if the fluid streams were switched and when the same stream passed through the shellside of one unit and then the tubeside of another unit is series they never changed. I was never happy with this procedure and when I moved on and became responsible for thermal design methods at Stone and Webster I never used it. As far as I know all companies now use the correct procedure.

With the introduction of computer programs and the development of more sophisticated heat transfer correlations the spotlight for further improvements in thermal design has increasingly turned on fouling resistances.

DEFINITION

Undesirable deposits on a heat transfer surface which increase the resistance to both heat transfer and fluid flow.

They are often referred to as fouling factors and I am as guilty as anyone for loosely using this term. A factor is a ratio which is dimensionless; a fouling resistance has the same units as any other thermal resistance i.e. m^2K/W.

STANDARD VALUES

I could easily have filled up a few pages of this book with typical values of fouling resistances for both service and process streams as many books have but I can not see any point in that. The reader will have access to many such lists. The latest edition of TEMA has 5 pages; I have seen a Shell DEP with over 20 pages and there are many other sources.

The TEMA list is the one I first encountered and will be the one most readily available to the reader. I would like to highlight some aspects of the TEMA list. In order to do this I have reproduced two sections, with permission. One is the section on fouling factors for crude oil. Table 14.1 gives recommended fouling resistances for crude oil in two different states. The other is table 14.2 which gives recommended fouling resistances for cooling water. Both tables are in US customary units but since I only want to compare values there is no need to update them to the SI system of units.

	Crude oil fouling resistances in hrft²F/Btu					
	0 to 250°F			250 to 350°F		
u ft/s	< 2	2 - 4	> 4	< 2	2 - 4	> 4
Dry	0.003	0.002	0.002	0.003	0.002	0.002
Salt	0.003	0.002	0.002	0.005	0.004	0.004
	350 to 450°F			450 °F and above		
u ft/s	< 2	2 - 4	> 4	< 2	2 - 4	> 4
Dry	0.004	0.003	0.003	0.005	0.003	0.003
Salt	0.006	0.005	0.005	0.007	0.008	0.006

Table 14.1

Table 14.1 clearly demonstrates the fouling resistance dependency on temperature and velocity. It is the only table I can recall which gives three bands for the fluid velocity. This is no doubt because crude oil is one fluid where it is often necessary to design with very low velocities to stay within allowable pressure drop limits.

One interesting aspect is that there is not indication of the source of the crude oil. I know that there are very significant differences between the composition of crude oils from different fields and the methods of processing them which is reflected in the price.

	Cooling water fouling resistances in hrft²F/Btu			
Hot fluid temperature	≤240°F		240 to 400°F	
C Water temperature	≤125°F		≥125°F	
C Water Velocity	≤ 3 ft/s	> 3 ft/s	≤ 3 ft/s	> 3 ft/s
SOURCE				
Sea water	0.0005	0.0005	0.001	0.001
Brackish water	0.002	0.001	0.003	0.002
Cooling tower - treated	0.001	0.001	0.002	0.002
Cooling tower -untreated	0.003	0.003	0.005	0.004
River - minimum	0.002	0.001	0.003	0.002
River - average	0.003	0.002	0.004	0.003
Hard - > grains/gallon	0.003	0.003	0.005	0.005
Condensate	0.0005	0.0005	0.0005	0.0005

Table 14.2

The table for cooling water is the one a thermal design engineer is most likely to be familiar with. There is a note in TEMA that the table values should be modified if the hot fluid temperature exceeds 400°F (204°C). This is an increasingly unlikely scenario with the drive to maximise plant efficiency by recovering heat whenever possible.

The first thing to note about table 14.2 is that the fouling resistances vary by a factor of ten. The second is that there are a series of step changes. The most dramatic of these is for river water minimum value in the lower temperature range which changes by 100% at the velocity break value. This may seem unlikely but is theoretically possible if the fouling mechanism changes.

There have been a number of criticisms of the TEMA tables which are summarised in Ted Saunders book – ref 6. These are as follows with my own comments in brackets:

(a) **There is no stated operating period**. (This may vary from a few weeks to the lifetime of the exchanger. TEMA uses the term "reasonable" cleaning intervals. "Reasonable" is a term lawyers have made fortunes from defining and re-defining. My interpretation of the TEMA time scale is the typical time between major shutdowns – say one year.)
(b) **There is no distinction between the shell side and the tubeside.** (This is a very difficult one to assess.)
(c) **There is no mention of the tube diameter and pitch.** (For flow inside tubes the relationship between the tube inside diameter and fouling is covered in more detail later as is the effect of the clearance between tubes for shell side flow. Most TEMA exchangers have 19 or 25.4 mm tubes with a clearance of 6.35 mm. Since the effect of tube diameter and clearance can be taken account of in any thermal calculations this seems to be a pointless criticism.)
(d) **Bulk fluid temperatures are specified rather than surface temperatures for tables such as 14.1 and 14.2.** (Fouling factors are normally specified by process engineers and surface temperatures are not known until after some preliminary thermal calculations. Surface temperatures are the ones which are important in cooling water systems. This is a point which a thermal design engineer must always be conscious of. The TEMA values could be modified to reflect this. This would require the process engineer to specify a range for the fouling resistance and leave the final choice to the thermal design engineer.)
(e) **A relationship between fouling resistances and temperature and velocity is only provided for crude oil and cooling water.** (Clearly velocity and temperature must have an effect on fouling resistances for other fluids. Process will often specify limits for either or both of these variables for specific designs.)

There is one important omission from Ted's list which is understandable. He is an excellent mechanical engineer with a good knowledge of but limited experience in thermal design.

(f) **The heat transfer processes are limited to single phase fluids.** For the single phase fluids there may be subtle differences between the heating and cooling values. However more dramatic differences in fouling resistances are encountered when a fluid is changing phase compared with when it is being heated or cooled as a liquid or vapour.

It is easy to sit back and criticise the TEMA fouling resistances but for many years they have provided a standard for the petrochemical industry. A comprehensive coverage of fouling resistances would mean a significant enlargement and I would not like to see it expanded to two volumes.

Increasingly clients are likely to be compiling their own list of fouling resistances which are based on plant experience. An example is the Shell DEP mentioned earlier. This contains one entry which emphasises point (f) in the above list and is for a cumene plant where the fouling resistance for the cumene column bottoms reboiler is twice the value for the cumene column bottoms cooler.

An experienced designer will always have such lists to hand. If a fouling resistance on a process data sheet does not 'look right' the first step is to compare it with a value from at least one of these lists for a comparable stream. Process engineers do make mistakes and if a fouling resistance is unusual, but correct, it is important to understand why. This might have an impact on the thermal design.

EFFECT ON OVERALL HEAT TRANSFER

The effect of fouling resistances on overall heat transfer can vary enormously depending on the values of the other thermal resistances. Equation 1.2 from chapter 1 can be modified as follows:

$$\frac{1}{U_{calc}} = \frac{1}{U_{clean}} + \text{Total fouling resistance } R_{ft} \qquad [14.1]$$

Where the total fouling resistance, R_{ft} is:

$$R_{ft} = R_i \frac{d_o}{d_i} + R_o \qquad [14.2]$$

If the total fouling resistance is say 0.0004 m^2K/W and the clean overall heat transfer coefficient is 1000 W/m^2K the service heat transfer coefficient becomes 714 W/m^2K. The extra surface area that is required for fouling can then be expressed as a % as follows:

$$\% \text{ Extra surface area} = \frac{(U_{clean} - U_{calc})}{U_{clean}} \times 100 \qquad [14.3]$$

$$= \frac{(1000 - 714)}{1000} \times 100$$

$$= 29\%$$

To investigate the effect of the fouling on the overall heat transfer coefficient I have fixed the value of the fouling resistance at 0.0004 m^2K/W and then calculated the % extra surface area required for a range of clean coefficients. I have then repeated the calculations for a very low fouling resistance of half this value and a high fouling resistance of double this value. These calculations are summarised in table 14.3 below.

	$R_{ft} = 0.0002$		$R_{ft} = 0.0004$		$R_{ft} = 0.0008$	
U_{clean}	U_{serv}	Extra area %	U_{serv}	Extra area %	U_{serv}	Extra area %
50	50	1	49	2	48	4
100	98	2	96	4	93	7
200	192	4	185	7	172	14
500	455	9	417	17	357	29
1000	833	17	714	29	556	44
1500	1154	23	938	38	682	55
2000	1429	29	1111	44	769	62
2500	1667	33	1250	50	833	67
3000	1875	38	1364	55	882	71
3500	2059	41	1458	58	921	74
4000	2222	44	1538	62	952	76

Table 14.3

The lowest coefficients in the table are most likely to be encountered with a low pressure gas on both sides of the exchanger. Typical examples are to be found in waste heat recovery units in the building services industry. It is clear that the fouling resistance has very little impact on the design.

A common practice in such situations is to ignore fouling resistances altogether and just design for a fixed % oversurface – typically 10%. An example from the petrochemical industry is in the design of air coolers where a fouling resistance is seldom applied to the air side of a unit.

The highest coefficients in the table are often found in condensers using cooling water. For such units getting the fouling resistance right has a big impact. For the clean coefficient of 4000 W/m^2K halving the fouling resistance results in only 44% extra surface area being required – a drop of 18%. If the fouling resistance is doubled the extra surface are required is 76% - an increase of 14%. Even higher clean coefficients than the maximum values shown in this table are found in surface condensers. There the allocation of a fouling resistance; or the elimination of it by introducing a continuous cleaning system is a critical part of the design.

From a practical thermal design standpoint it is often possible to influence the fouling resistance by changing the velocity. This is clear from tables 14.1 and 14.2. This will be covered in more detail for cooling water later.

FOULING TIMESCALE

Figure 14.1 shows an idealised fouling curve for the build up of deposits on a heat exchanger tube surface. The surface is initially clean so that the fouling resistance is zero. As the unit is commissioned a period of time may elapse before there is any noticeable fouling. A layer starts to build up and the fouling resistance increases. Eventually it will reach a steady value and no further increase will take place. This is termed the asymptotic or plateau value.

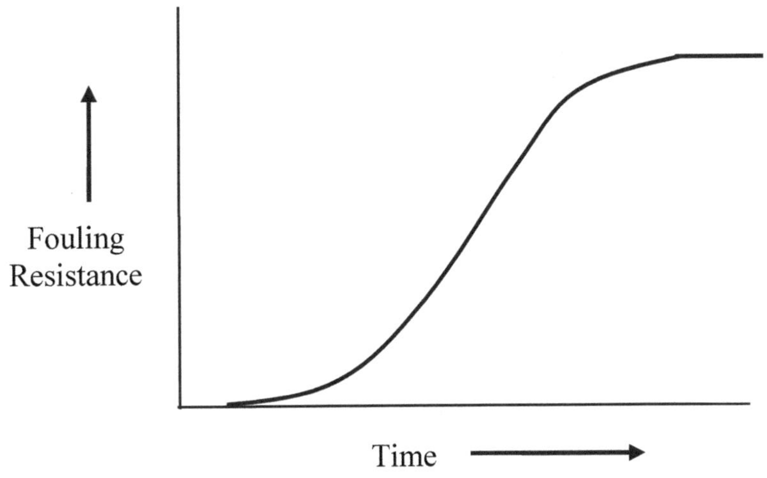

Figure 14.1

The time taken to reach the asymptotic value is very variable. With some types of fouling it may take only a day or two, in other case it might take many months. This type of curve is termed a fouling factor-time curve. Figure 14.1 is the classical one which is most frequently referred to.

Some other shapes of fouling resistance-time curves are shown in figure 14.2 below.

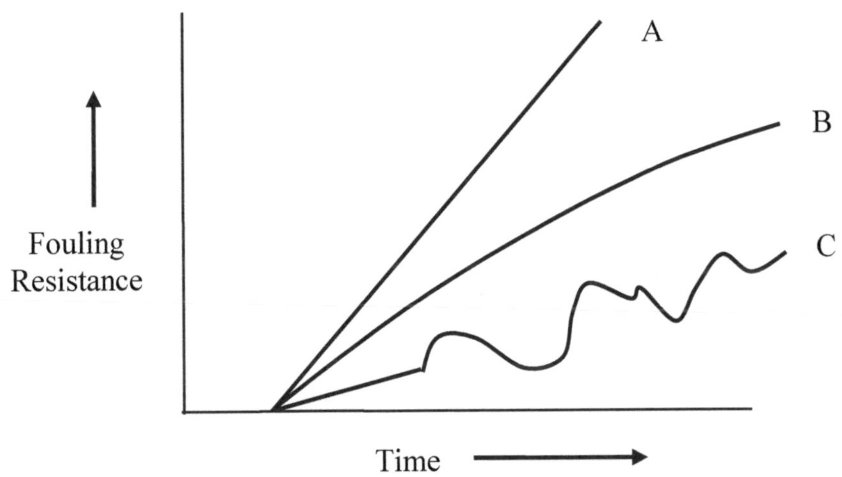

Figure 14.2

'Curve' A is linear. It represents a steady build up of deposits without ever reaching a maximum value. Curve B is similar to curve A because no maximum value is reached but the rate of build up decreases with time. This is a more realistic curve than A since the rate of build up of deposits is likely to change as the temperature at the surface of the deposits changes.

Curve C follows a mean path in an erratic manner with highs and lows at either side of a mean path and is therefore known as a saw tooth fouling resistance-time curve. This shape of curve can result from deposits building up and then breaking away to be carried off by the fluid stream. If there is more than one type of deposit the more adhesive ones will remain to keep the mean fouling resistance increasing. This build up and detachment of particles may carry on indefinitely or reach a mean asymptotic value.

To give some idea of a typical timescale I have reproduced figure 14.3 from Bott for a visibreaker which fouled heavily until a chemical antifouling compound was added to the process stream.

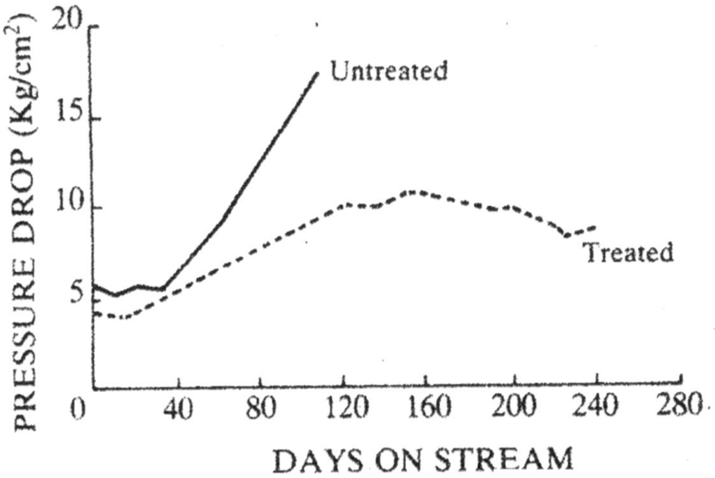

Figure 14.3

To measure a fouling resistance in an operating unit is very difficult since four other thermal resistances are involved in the heat transfer. However pressure drop is relatively easy to measure. The increase in pressure drop can be used to estimate the thickness of the fouling layer and provided the thermal conductivity of the fouling layer was known can be used to calculate the value of the fouling factor during the build up of the deposits.

This is a very easy calculation using a computer program – just increase the thickness of the fouling layer until the pressure drop matches the observed pressure drop. It is also a very simple hand calculation as the following example demonstrates.

Example 14.1

The tubes of a heat exchanger are 25.4 mm 16 BWG average wall. The pressure drop when the unit is clean is 0.6 bar. After a period of time in operation the pressure drop increases to 0.96 bar. Calculate the thickness of the fouling layer and the value of the fouling resistance assuming the thermal conductivity of the fouling layer is 2.6 W/mK.

Solution

A 16 BWG tube has a wall thickness of 1.65 mm so that the inside diameter is 25.4 – 2 x 1.65 = 22.1 mm. The inside diameter after fouling will be (22.1 – 2t) where t is the thickness of the fouling layer. The first section of useful tables at the back of this book gives the relationship between pressure drop and inside diameter; $\Delta p \propto d_i^{-4.8}$.

Therefore

$$\frac{\Delta p_{fouled}}{\Delta p_{clean}} = \left(\frac{d_i}{d_i - 2t}\right)^{4.8} \qquad [14.4]$$

$$\frac{0.96}{0.6} = \left(\frac{22.1}{22.1 - 2t}\right)^{4.8} = 1.6$$

Hence t = **1.03 mm**

As explained in chapter 1 it is not necessary to use the mathematically correct cylindrical boundary conditions for this type of calculation, plane boundary conditions are sufficiently accurate, therefore:

$$R_{ft} = \frac{t}{k} \qquad [14.5]$$

$$= \frac{0.00103}{2.6}$$

$$= \mathbf{0.0004\ m^2K/W}$$

This example illustrates the danger of not using a fouling layer thickness for designing heat exchangers. Only about one mm of fouling was sufficient to increase the pressure drop by 60%

Example 14.2

What would be the new pressure drop and the percentage increase if the fouling layer in example 14.1 built up to 2 mm?

Solution

Rearranging equation 14.4

$$\Delta p_{fouled} = \Delta p_{clean} \left(\frac{d_i}{d_i - 2t} \right)^{4.8}$$

$$= 0.6 \left(\frac{22.1}{22.1-2x2} \right)^{4.8}$$

$$= \mathbf{1.56\ bar}$$

Ratio of Δp's = 2.61

% increase = **161%**

This pressure drop would be well over any acceptable limit and was the result of a change in the inside diameter of only 22%.

Example 14.3

What would be the thickness of a 0.0004 m^2K/W fouling resistance composed of serpentine which has a thermal conductivity of 1.04 W/mK? Assuming it is inside the same tubes as used in the previous examples calculate also the percentage increase in pressure drop.

Solution

Rearranging equation 14.5

$$t = R_{ft}\, k = 0.0004 \text{ x } 1.04 = 0.00042 \text{ m} = \mathbf{0.42\ mm}$$

Using equation 14.4

$$\frac{\Delta p_{fouled}}{\Delta p_{clean}} = \left(\frac{d_i}{d_i - 2t} \right)^{4.8} = \left(\frac{22.1}{22.1-2x\ 0.42} \right)^{4.8} = 1.2 \text{ (\textbf{20\% increase})}$$

It is very informative to repeat the last calculation for a number of different materials which are known to be foulants. In the following table the value of the fouling resistance has been kept constant at 0.0004 m^2K/W.

Material	Thermal Conductivity W/mK	Fouling Thickness mm	Increase in Δp %
Iron oxide Fe_2O_3	0.61	0.24	12
Biofilm	0.71	0.28	14
Calcite	0.93	0.37	18
Serpentine	1.04	0.42	20
Gypsum	1.32	0.51	27
Magnesium phosphate	2.16	0.83	48
Calcium sulphate	2.34	0.9	53
Calcium phosphate	2.6	1.0	60
Iron oxide Fe_3O_4	2.87	1.09	68
Calcium carbonate	2.94	1.12	70

Table 14.4

I have taken the values for the thermal conductivities from Kakac and Liu – ref 2. It is the most comprehensive list I have seen. Their book covers other types of heat exchangers and their main target market is presumably undergraduates. I was therefore surprised and grateful to find so much useful information on fouling. There was however no reference to the source of the thermal conductivity data.

I would have to seek advice from a chemist or technologist to explain the difference in the values for the two oxides of iron. The ferrous oxide, Fe_3O_4 or rust is red and is the one most likely to be encountered. The ferric oxide, Fe_2O_3 is black and on its own can form a stable protective layer such as when it is used on gun barrels. In a fouling situation it is normally mixed up with other contaminants particularly ferrous oxide.

The range of thermal conductivity values, and hence the thickness of the fouling layers is a factor of about four. Apart from possibly the Biofilm, a practical fouling resistance is unlikely to be composed of a single constituent.

The increase in pressure drop may be just as important as the reduction in overall heat transfer when deciding if a unit needs to be cleaned. It is therefore often used a means of assessing the state of the fouling in a heat exchanger.

CLEANLINESS FACTOR CF

This is defined as the ratio of the fouled overall heat transfer coefficient to the clean overall heat transfer coefficient. There are two possible fouled overall coefficients; the service coefficient and the calculated coefficient. The service coefficient is $Q/A\ \Delta T_m$ and the calculated overall coefficient has already been used in this chapter as equation 14.1. There is normally only a small difference between the two values as explained in chapter 1. The calculated value must be used in defining cleanliness factor.

$$CF = \frac{U_{calc}}{U_{clean}} \qquad [14.6]$$

Cleanliness factor is a term I never use and I have included it here to try and cover any terms the reader may come across. I understand it is most likely to be used in the power industry. It is of course an alternative way of saying that an exchanger has been designed for a specified amount of extra surface rather than using a fouling factor

FOULING COEFFICIENTS

In recent years some companies have started to specify fouling coefficients, h_f instead of fouling resistances, R_f where $h_f = 1/R_f$. If a computer input is set up to only accept fouling resistances two simple hand calculations are all that is necessary although each will introduce the possibility of an arithmetic error.

I can sympathise with the logic of this development because it does not make too much sense to have some terms in the equation for calculating an overall heat transfer coefficient so numerically different from other terms. Expressing them both in the same units makes it easier to compare values. However any computer output will list all the thermal resistances as either a ratio or percentage of the total thermal resistance and this is what I always look at to compare values.

If we could start again from scratch I would prefer to go the other way and calculate film heat transfer resistances rather than heat transfer coefficients. Sadly this would make the term U value obsolete.

FOULING MECHANISMS

An important starting point in assessing a fouling situation is to understand the basic mechanisms of the fouling processes. For a molecule of foulant to become attached to a tube surface it needs to pass through the fluid boundary layer. At the point where the boundary layer is in contact with the surface the velocity is zero. This is where fouling must take place. The temperature at this interface may be higher or lower than the bulk fluid temperature depending on whether the fluid is being heated or cooled.

Epstein – ref 36 was one of the first to develop a classification for the mechanisms involved in fouling. These are as follows:

PARTICULATE DEPOSITION

An example of pure particle deposition is the accumulation of soot in chimneys from a coal fire. A wood fire will produce liquid particles which solidify on contact with cool surfaces to form tar like deposits. A modern incineration plant will produce both types of deposit together with many others which is a problem for the surfaces of any heat recovery equipment. In any power plant a boiler using a solid or liquid fuel, particles may be deposited on any of the heat transfer surfaces from the steam generating tubes to the air preheater.

If a river is the source of cooling water it will contain particles of clay and other suspended mineral and organic particles. The concentration of particles will be very variable depending on the rainfall in the river catchment area.

Particles in fluid streams which flow through tubes may not be too much of a problem if they are the only source of fouling. However combined with other sources of fouling they become more serious. For example biofouling may stick particles to a surface and particles may fill in any gaps in other types of fouling.

Particles in a fluid stream on the shell side of a heat exchanger may simply be deposited in any dead areas due to gravity. Larger particles may lodge in the gaps of the baffle/shell and baffle/tube clearances resulting in a large increase in pressure drop.

CRYSTALLISATION

Crystallisation is the deposition of salts on heat transfer surfaces and is often termed scaling. The most usual salts involved are calcium and magnesium carbonates, silicates and phosphates. This is one type of fouling which most members of the public will have experience of. The build up of deposits on the heating surfaces of domestic kettles is something which must be regularly dealt with. How often attention is required depends on the 'hardness' of the water supply. Hardness is a measure of the concentration of salts in the supply.

There are two distinct ways a solute may behave in a solvent. For example sugar will dissolve in water and if the temperature of the water is increased it is possible to dissolve more sugar. If the solution is then cooled down it will solidify. This principle is used in making jam or marmalade where only quite small quantities of other substances are involved in the cheapest varieties. Solidification when a fluid is cooled can be described as 'normal solubility behaviour'.

There is another form of solubility behaviour where heating a solution can result in solidification. This is termed **inverse solubility** behaviour and this is what takes place in a domestic kettle or on the surface of a heat exchanger tube from aqueous solutions. For crystallisation to occur, supersaturation of the salt solution involved must take place where the liquid is in contact with the solid surface. It is therefore highly temperature dependant.

A fouling layer deposited by crystallisation may be very variable depending on the mix of salts involved. The deposits may be hard or soft and can vary from white to dark grey in colour. Very hard deposits may build up to a significant thickness and have even been reported as completely blocking some tubes. They may also lead to an asymptotic fouling resistance value since the fouling layer will reduce the temperature of the solution at the solid/liquid interface.

BIOFOULING

Biological growths on heat transfer surfaces can be roughly classified into two groups depending on the size of the organisms. Macrofouling is due to the growth of animals such as barnacles, mussels, serpulid worms, hydroids and plants such as seaweed and pond weed. I once read a report on the fouling of the condenser tubes at Blyth power station in the UK. The steady circulation of sea water which was warmed up as it passed through the tubes provided ideal conditions for mussels which were growing to an eatable size.

During a wreck dive in the red sea I once picked up a bottle which was covered in coral. Back in the boat I was disappointed to find that it was just a modern beer bottle. I asked the local dive leader how long it was likely to have been in the water and he replied “a couple of weeks”. Given the right conditions macro biological fouling can be very rapid and severe.

Microbiological fouling is due to much smaller plants and animals such as bacteria, moulds and algae. The colour is a guide to the constituents and can vary from almost transparent to a very dark green. Examples can be seen in any water garden. Even in fast flowing water falls a green slimy filamented layer is usually present. It is very easy to remove the bulk of it but the layer next to the surface can be very difficult to remove. Microbiological fouling will always be found where macro biological fouling is present. It may also aid other forms of fouling. For example pitting corrosion may often start under layers of slime.

CHEMICAL REACTION

Chemical reactions can take place within the fluid stream which distinguishes this category from the next one of corrosion which involves the metal of the heat transfer surface. Fouling resulting from chemical reactions is usually due to organic materials containing inorganic elements or compounds such as a liquid stream of organics which contains molecules of oxygen. The oxygen reacts with selected organic material to form compounds which adhere to the surface. It can also be the mechanism causing polymerisation on heat transfer surfaces.

Deposits derived from chemical reactions can have a wide range of properties from a loose powdery deposit of carbon to a much more stable layer resulting from polymerisation. Chemical reaction fouling can occur over a wide range of temperatures from ambient to those found in combustion systems.

CORROSION

This involves a chemical reaction between constituents of the fluid stream and the metal of the heat transfer surface. The most common of these is the reaction between ferrous tubes and dissolved oxygen in cooling water which leads to a slow oxidation process and the familiar rusting.

Corrosion technology is a subject in its own right and is well beyond the scope of this book. The practical solutions to corrosion problems are:

(a) Select a material for the tubes which will not corrode.
(b) Treat the fluid with a corrosion inhibitor.
(c) Accept that corrosion will take place and use thicker tubes which will have a limited life.

FREEZING

Freezing is a physical process which may occur whenever a liquid is in contact with a heat transfer surface below the change of phase temperature of the liquid. It is always likely to take place when a heat exchanger is used to produce chilled water. It is not always a problem since in some case it may cover only part of the tubes near the outlet and may reach a stable asymptotic state as the ice provides an insulating layer. Wax deposits from some hydrocarbon streams can also be classed as freezing although there is some debate as to whether this should be classed as crystallisation.

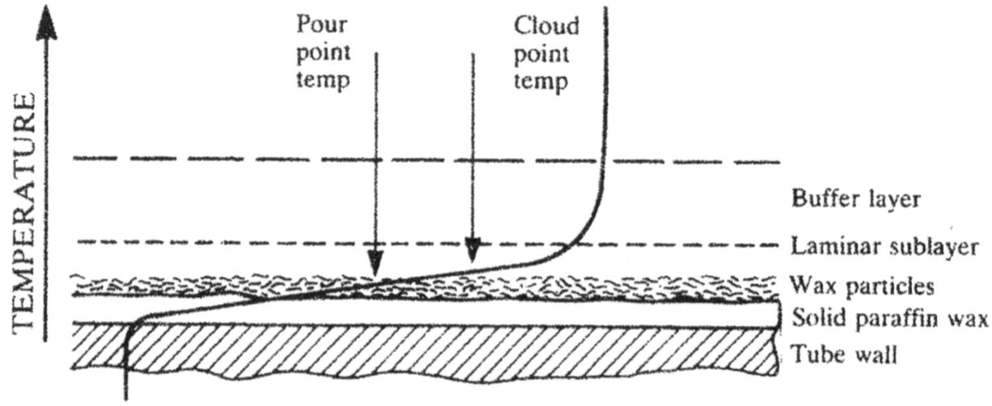

The deposition of paraffin wax

Figure 14.4

Figure 14.4 above from work by Gudmundsson - ref 37 shows the temperature profile through the boundary layer of a stream of paraffin wax as it is being cooled below the temperature where it solidifies.

SUMMARY

The freezing of water is one of the few practical cases where there is only one fouling mechanism taking place. In most situations more than one mechanism will be involved. These may interact and one may be dependant on another.

FOULING & TEMPERATURES

Bulk fluid temperatures are specified by process. They should also specify any surface temperature limits which would lead to rapid fouling. These may be high limits for process or service streams which are being heated such as the process side of reboilers or the service side of cooling water units. They may also be low limits to avoid pour point problems.

One possible solution to either problem is to use co-current flow rather than counter current flow. This will even out the surface temperatures throughout the unit. This is often necessary in air coolers to avoid pour point problems in winter. This will always produce a reduction in MTD leading to an increase in the surface area required.

In shell and tube units the reduction in MTD is very dependant on the terminal temperature differences as discussed in chapter 1. Assuming there is no temperature cross and both arrangements are possible the effect on the MTD is very dependant on the terminal temperature differences.

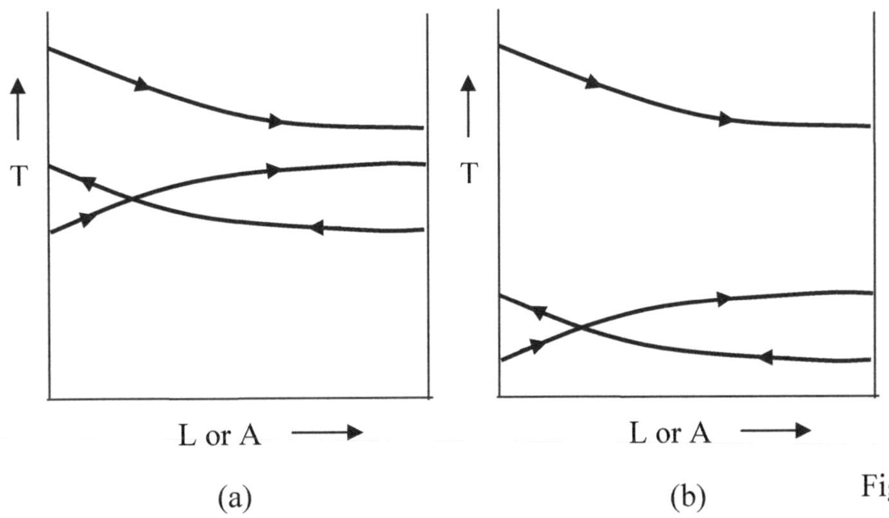

Figure 14.5

In figure (a) there are relatively small terminal temperature differences in counter current flow and changing to co-current flow would result in a drastic reduction in the MTD. In figure (b) there are relatively large terminal temperature differences in counter current flow and changing to co-current flow would result in a much smaller reduction in the MTD.

Of course there are not many shell and tube units that are in true counter current or co-current flow. A typical unit is a mixture of co-current, counter current and cross flow. However tube wall temperature problems are often localised and may be solved by quite simple modifications. Figure 14.6 below shows two possible arrangements for the flows through an E type shell.

The hot fluid is on the shell side and the cold fluid inside the tubes. In arrangement (a) the hot fluid entering the unit is exchanging heat with the cold fluid which is just about to leave the unit and is therefore at its maximum temperature. The tube wall temperature will therefore be at a maximum possible value for the two fluids involved.

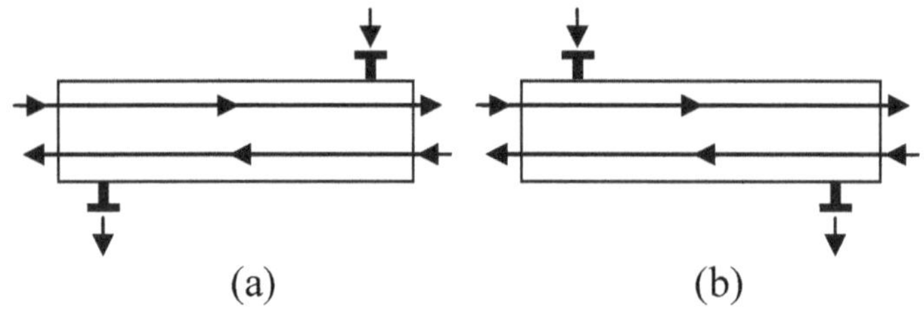

Figure 14.6

In arrangement (b) the hot fluid entering the unit starts to exchange heat with cold fluid at a much lower temperature. The tube wall temperature will therefore be much lower. If the potential tube wall temperature problem had been for a maximum value of the cold fluid, say cooling water, then arrangement (b) should be used.

Conversely if the potential problem had been a pour point problem in the hot fluid then arrangement (a) should be used to avoid the coldest section of the hot stream exchanging heat with the cooling stream at its minimum temperature.

The difference in the MTDs of the two arrangements is negligible, (Zero if an overall MTD correction factor is used - which is necessary for hand calculations). A modern computer program will enable a designer to check the tube wall temperature throughout the unit.

FOULING & VELOCITIES

Velocity is normally the most important variable which affects fouling. It is also a variable which the thermal design engineer has some control over. In general the higher the velocity the lower will be the degree of fouling and this applies to any fluid. The most important of these is cooling water. Figure 14.7 below shows the variation of fouling resistance with cooling water velocity.

This is not just an idealised curve such as the ones I have seen in other books and papers it has been taken from the client's spec from a job I once worked on. The shape however follows the form of an 'ideal' curve and I do not know how much of it was based on test data. The cooling water was treated with zinc phosphonate and acid inhibitors.

The first thing to take note of is the range of values of the fouling resistance. There is a factor of ten difference between the values for the highest and lowest velocities. The most substantial change in values is for velocities below 1.5 m/s and there is only a relatively small change above about 2 m/s. The preferred range is in between these values.

Even within the preferred range there is a variation of about 60% and knowing that it is not always possible to achieve a minimum value of 1.5 m/s this very soon rises to over 100%. When the cooling water fouling resistance is a large proportion of the total thermal resistance, which it invariably is, it is worth the trouble to get it right. I therefore took the central part of the chart between 1 and 2 m/s and magnified it several times on a printing machine and then added intermediate values to both scales. I passed this around so that other people working on the same job could use it.

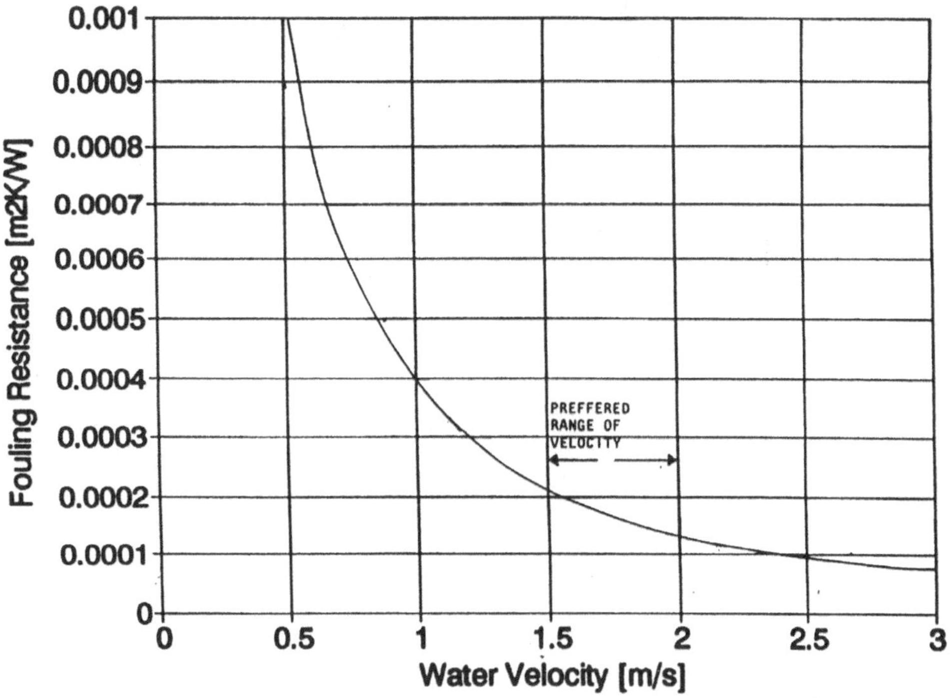

Figure 14.7

Some people might regard it as a 'pain in the neck' to change a fouling resistance each time a design is modified. I always find it very refreshing to be able to work in a way that I have always advocated.

Unfortunately I have found it very rare to work for such an enlightened client. All too often there are just some step values, as in TEMA, or worst till just a single value of fouling factor. In these situations I have no hesitation in 'bending the rules'. Suppose I have a design with a high fouling resistance, (as a percentage of the total thermal resistance), which is say 4% undersurface but I have maximised the cooling water velocity by using up the entire allowable pressure drop. I know that increasing the shell size to add more surface area will reduce the velocity. If I add 4% extra surface area the unit may still be 2.5% undersurface because the heat transfer coefficients will also have fallen. So I add a bit more surface area and I have still not quite reached the target zero percentage oversurface. This is known as 'chasing ones tail' and you may find that you are getting close to the minimum allowable cooling water velocity before you reach the target.

I have been conscious of fouling resistances following curves such as the one in figure 14.7 since my early days as a heat transfer engineer and I know that the design with 4% undersurface is the optimum one for performance as well as cost. If I am working with enlightened engineers and clients I prefer to leave it at that with an explanatory note on the thermal data sheet. There have been other times when I have taken the 'politically correct' way out and just fudged it to show a value of say 0.1% oversurface. The easiest way to do this is to add an enhancement factor to the lowest heat transfer coefficient; it is very unlikely anyone will spot it. I will leave it to reader to think up any other ways to fool those who 'think like accountants'.

CONTROL

The control of cooling water items is very important. A glance at the P&ID is sufficient to establish how process intends to do this. It is especially important if the cooling water is to be supplied at different temperatures in summer and winter.

The worst case scenario is if the temperature of the process stream is controlled by throttling the cooling water. If the unit is started up in late winter a very low flow rate of cooling water may be required, especially if it is operated at part load. The very low velocities will result in very heavy fouling throughout the unit particularly if the main source of fouling is biological. Then spring arrives and the cooling water temperature increases and the throttle valve is fully open but the unit still has no chance of achieving its design duty without taking it off line for cleaning.

There are many types of service which do not require any form of control in which case none should be provided to prevent anyone from misguidedly using it. If a product is going to be delivered to a storage vessel and there are no potential overcooling problems the cooling water velocity should always be at its maximum value. ('Let it rip' is an expression often used in this situation.) Another example of this is in a vapour power plant where lowering the cooling water temperature results in a lower condenser pressure and improved plant thermal efficiency. This was very noticeable when I was at sea in a steam ship. In there was a very noticeable increase in maximum speed, (weather permitting) in the north Atlantic compared with when we were in the tropics.

Where overcooling is a problem the best solution is to maintain a constant cooling water temperature throughout the year by means of a recirculation system. In theory this can be used on a single unit as illustrated in figure 14.8 below.

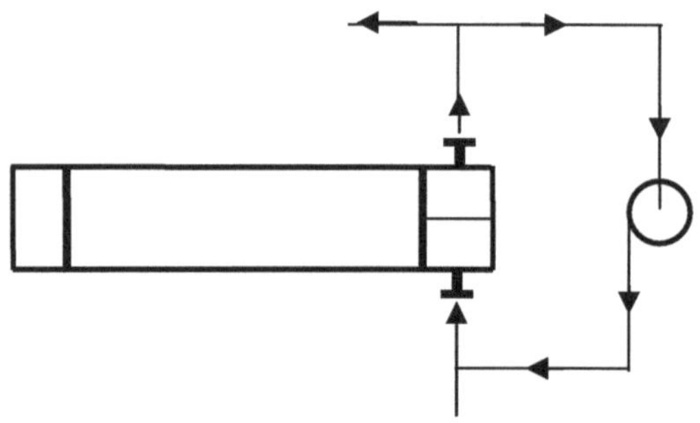

Figure 14.8

I have omitted the valves and control system which together with the pump make this a very expensive option. It is much less costly if such a system is designed to serve more than one exchanger.

One of the most imaginative recirculation systems I have worked with was for a plant in Norway. There was a constant temperature reservoir constructed by isolating a section of a fjord. The constant temperature was maintained by mixing the return water from the exchangers with water from the fjord in winter. For the warmest part of the summer water was taken directly from the fjord.

Another interesting cooling water exchanger I remember was when process had proposed using the dreaded throttling method of control. It was not a particularly bad case since there was only a very modest difference between the summer and winter temperatures which might have been just acceptable for a 'tightly designed' unit utilising all the allowable pressure drop. However there was also a requirement for a 30% overcapacity which would have resulted in a very low cooling water velocity in winter and I suspected that the fouling resistance would drastically increase above the value on the PDS.

I discussed this with process and they were soon convinced about the potential fouling problem. I asked them the obvious question, "Why do you need 30% overcapacity?" They replied that it was to cater for an increase in capacity at a later date and would not be required for a couple of years. I soon suggested putting temporary plugs in selected tubes to keep up the velocity for the immediate service and then removing them when the higher duty was required. This proposal was accepted and a design that had started off looking very straightforward became more challenging. I have no feedback as to whether they remembered to take the plugs out again when the plant capacity was expanded!

Another of my daft ideas?

Negative feedback is well known to travel much faster than positive feedback and there was none from the previous example. Perhaps it is worthwhile to investigate the possibility of selective temporary tube plugging as a special design tool.

Shutting down and restarting heat exchangers to modify the number of active tubes is clearly a none starter but what about a system that did not involve interrupting the operation? U tubes are increasingly being used in cooling water systems and most of them are only two pass units. This means that it is only necessary to consider plugging a selection from half of the tube holes. Bars with conical plastic end pieces could be grouped together and connected to the outside via shafts through gland seals. These could then be inserted or retracted as required to keep the cooling water velocity within acceptable limits. One or two of these arrangements should be sufficient and would only require the occasional attention.

I will leave the details and cost to a mechanical engineer. I can see no thermal or hydraulic problems. I can foresee objections from site engineer who will say that they have more than sufficient adjustment to make to equipment. It would certainly make an interesting student design project or a topic for discussion in the pub.

ALLOWABLE PRESSURE DROPS

The allowable pressure drop for the cooling water supply is specified by process. On most jobs exactly the same pressure drop will be allocated to each exchanger. A cooling water system is designed using the same principles as those used to design a hot water radiator system for heating a building. Details of this design methodology and many examples can be found in the book I wrote with Dr Eastop – ref 4. A broad overview of how both systems are designed is as follows:

(a) Draw a line diagram of the system showing the flows required in each pipe.
(b) Identify the index unit. This is the unit which is most remote hydraulically from the pump.
(c) Select preliminary pipe sizes.
(d) Calculate the pressure drops in each section of pipe.
(e) Add resistances – usually valves to balance up the pressure drops. The pressure drops around each individual hydraulic path should be equal.

A very simple system is illustrated below in figure 14.9.

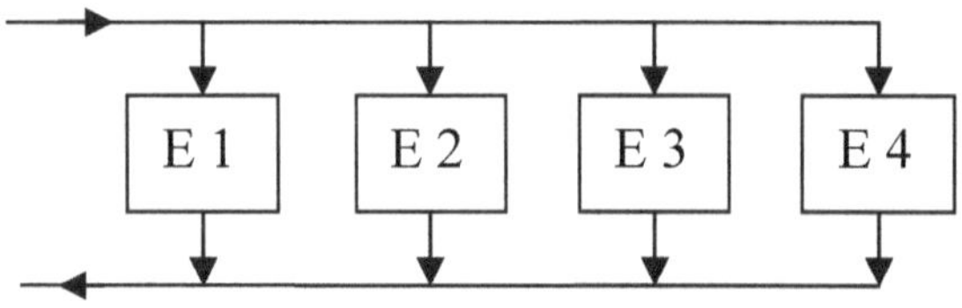

Figure 14.9

A thermal design engineer working on a unit supplied with cooling water should check the cooling water flow diagram to see where the exchanger fits into the system. In this example unit E4 is clearly the index unit. If this is the unit you are working on there is no point in considering asking process for more pressure drop. However if you are working on unit E1 the potential for more pressure drop is clearly there. This may mean process resizing some pipes but this is a small price to pay if a bit more allowable pressure drop results in an exchanger with a high cooling water velocity rather than a marginal dodgy one.

TUBE MATERIALS

A Client specification will invariable contains a list or table of cooling water velocity limits for different materials. Table 14.5 below has been taken from the spec of a well known international company. It is typical and other specs may have slightly different values.

	Velocity m/s		
MATERIAL	Preferred	Minimum	Maximum
Aluminium brass	1.5	1.0	2.2
Copper nickel (90 Cu-10 Ni)	1.8	1.0	2.5
Copper nickel (70 Cu-30 Ni)	2.1	1.0	3.0
Carbon steel	1.5	1.0	2.2
Stainless steel	2.5	2.0	4.5
Titanium	3.5	2.5	5.0

Table 14.5

There are several interesting features of such tables. The first of these is the minimum velocity of 1 m/s for the top four materials. When I first started to design heat exchanger this would have been 0.9 m/s, (3 ft/s) and still is in TEMA. I am not sure if the logic for this change was a new rounded number to suit a new system of units or an increasing awareness of how the minimum velocity affects fouling.

The minimum value for the other two materials has always puzzled me and I have never been given a satisfactory answer to the question why? They are regularly impossible to achieve within the allowable pressure drop.

The maximum velocity limits are normally only encountered in exchangers with a single tube side pass or two pass units with relatively short tubes. They are intended to avoid erosion problems hence the very high limits for stainless steel and titanium which are less susceptible to erosion than the other materials.

The maximum limit of 2.2 m/s for carbon steel is the one most likely to be encountered. Erosion is most likely to occur at the ends of the tubes. In chapter 3 I discussed the TEMA momentum flux limits for flow of any fluid inside tubes. There I gave an example of a case where the TEMA limit of 8900 kg/ms2 was grossly exceeded. The liquid being heated in the tubes was a fairly heavy hydrocarbon with a density of about 900 kg/m^3. The momentum flux in the tubes was about 20000 kg/ms^2 and the velocity 4.7 m/s. The velocity of water to give the same momentum flux is 4.2 m/s. The tubes were made of carbon steel and Incoloy 825 ferrules were fitted to the ends of the tubes.

I can do no know how the erosion characteristics of cooling water might compare with those of the hydrocarbon used in this case but the principles involved are the same. If I ever come across a case where a cooling water velocity of say 3m/s was attainable I would consider using it with ferrules or a protection plate to avoid erosion. There would certainly be no problems with fouling.

The high velocity limits for the stainless steel and titanium tubes are unlikely to be encountered but there are exceptions. I once designed some very high pressure units with thick walled stainless steel tubes. They were single pass units with typical tube lengths of 2-3 m and the maximum velocity limit for the job fixed the number of tubes.

The nature of the metal surface can affect the fouling resistance. It is generally accepted that the smoother the tube wall is the less prone it will be to fouling. It might therefore make sense to use different fouling resistances for different tube materials. This is not something that the thermal design engineer is likely to be involved in and is not normally considered. I have worked on too many jobs where the tube material for the cooling water units has been suddenly changed in mid job and there was never any change in the fouling resistances.

SLURRIES

Slurries are an interesting case as far as velocity limitations are concerned. Slurries contain solid particles which cause erosion of the tubes. It is therefore necessary to apply a much closer tolerance to the range of permitted values.

	Velocity m/s	
	Minimum	Maximum
Straight tubes	1.0	2.1
U Tubes	1.0	1.8

Table 14.6

Table 14.6 is an example from a spec I once worked to. U tubes are more likely to erode than straight ones. There will also be a slight thinning of the tubes during bending so that the inner tubes where erosion is most likely will have less metal to spare before failure.

CASE STUDIES

There are a lot of interesting case studies in Bott – ref 16. I thought about reproducing a few but found it difficult to make a selection. Like the rest of this book I have tried to limit examples to those I have had some personal experience of or have not been published in other books.

CLEANING PROCESSES

One of the early decisions to be made in the design of any heat exchanger is if either stream requires mechanical/hydraulic or chemical cleaning. The way that this decision affects the choice of the construction of the design has been covered in several chapters. I have a limited practical experience of cleaning heat exchangers. At sea it was one on the more unpleasant tasks we had to carry out. It is not only interesting but important for a thermal design engineer to have some knowledge of the processes involved in cleaning.

WHERE AND BY WHOME?

Cleaning may take place in situ or remotely in a workshop or cleaning area. For remote sites it may be necessary to train maintenance staff to carry out cleaning operations. For other sites it is increasingly likely that specialist contractors will be used. It is a dirty and often potentially dangerous job. Techniques and the equipment used have become increasingly sophisticated over the years. A quick search on the web will identify many such companies. Their sites have lots of coloured diagrams of equipment and even videos of cleaning processes. The diagrams I have used are relatively simple and all in black and white in keeping with the rest of this book.

HYDROCLEANING

Figure 14.10 shows the tube side of an exchanger being cleaned in the cleaning area of a workshop.

The application of high pressure water jets to clean fouled surfaces is now in everyday use. I don't actually have one but I know a friend who does. I borrow it to clean garden furniture once every year or two. It is much more effective than a scrubbing brush. The only drawback is that you do not get as much exercise and that feeling that you not only need but deserve a pint. Using one is sufficient to appreciate some basic safety requirements. I would never use one without some form of eye protection and it is essential to ensure that there is nothing downstream which may be damaged.

Figure 14.10

If there are safety issues with a simple domestic tool these are magnified in an industrial application. I have a page from an equipment supplier in front of me which offers three models for tube side cleaning. The water pressures vary from 700 bar to 2800 bar. For shell side cleaning there are two models operating at 700 bar and 1400 bar.

For tube side cleaning a nozzle is fitted to the end of a lance. The lance is then traversed from one end of the tube to the other. The traditional way of doing this with relatively low pressure systems was for one operator to manually insert the lance while another operator held the other end. The second operator would then walk the length of the tube until the nozzle reached the far end. This became known as 'lance walking'. This method is still in use for appropriate applications.

Any jet of fluid produces a reactive force which can be felt in the simple domestic appliance. For the highest pressures used to clean heat exchanger tubes this is not realistic and automatic and semiautomatic machines are used which need to be bolted to the exchanger. Figure 14.9 is of a semiautomatic hydraulic system.

For cleaning the shell side of a tube bank multiple thin lances are used which are passed through the cleaning lanes.

The advantages of water jet cleaning are that there is little chance of damaging the tubes and the waste produced is no more noxious than the original foulants.

MECHANICAL CLEANING

Mechanical cleaning is necessary when the fouling deposits are too hard to remove by using a water jet. Figure 14.11 below is a picture of a very simple hand tool which is basically just a pneumatic drill with a special expandable extension piece which allows the cutting head or brush to be passed into the tubes.

Although this simple tool may still be used for small jobs mechanical cleaners can be automated and semiautomated just like any other drilling machine.

Figure 14.11

For really serious deposits and fully blocked tubes a cutting head may be required. Figure 14.12 below shows some of the types of head that can be used for cutting. From left to right these are a scraper head, a conical toothed cutter and a drill head.

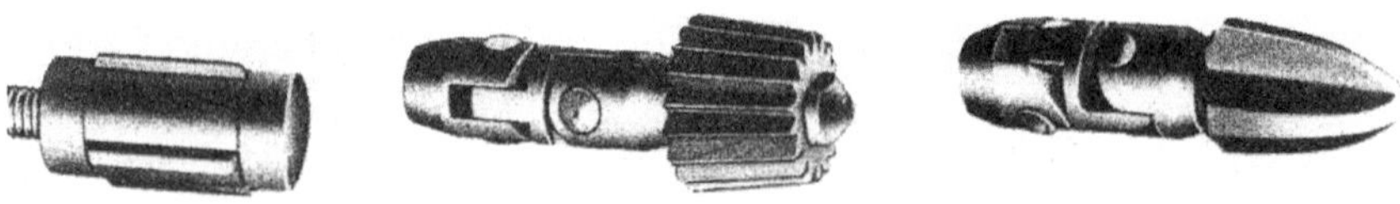

Figure 14.12

For other types of fouling layers the head is likely to be a brush head. There is an incredible variety of this type of head some of which are illustrated in figure 14.13 below. There is also a wide variety of materials used for the brushes ranging from metals to plastics to suit both the nature of the fouling and the type of metal the tube is made from.

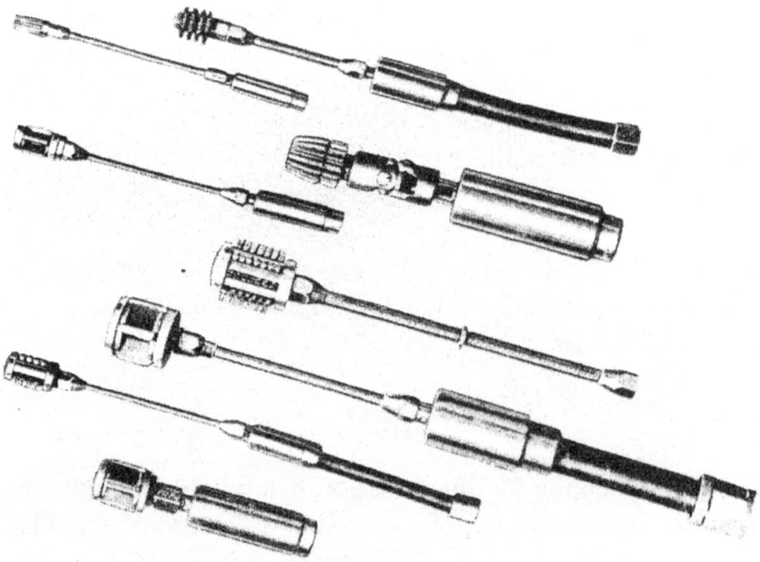

Figure 14.13

With mechanical cleaning more than one pass through the tubes may be required with a different head fitted for each pass. It is well beyond the scope of this book to look into when and how each type of head should be used I am quite content to leave this to the specialists.

One problem with mechanical cleaning is the removal of the fouling material which has been detached from the tubes. The solution is to combine the mechanical cleaning with hydraulic particle removal.

The machine is therefore supplied with water as well as air. Since only particle removal is required a much smaller flow rate of water is used and the supply pressure is modest compared a water jet system – typically about 20 bar. Figure 14.14 below is a picture of a heavy duty air driven tube cleaning machine with a water flushing system.

Figure 14.14

The main disadvantages of mechanical cleaning is that some types of head may damage the tubes and it takes longer than water jet cleaning which increases the costs.

The diagrams in the last two sections have been taken from Yokell – ref 9 who goes into cleaning technology in a lot more detail than I have.

U TUBES

When I first started designing heat exchangers using U tubes for high fouling services such as cooling water was out of the question. The argument against was very simple – it was not possible to clean around the bends, particularly those in the inner rows. If any tube was not cleaned around the full bend any deposits remaining would result in a reduction in velocity leading to rapid fouling and further blockage until eventually the tube may become inactive. The pressure drop through the remaining active tubes would of course increase.

In more recent years I have noticed an increasing number of clients not only allowing but in some cases specifying that U tubes should be used. A U tube bundle not only has the obvious advantage of reduced capital cost it is also much quicker to extract for remote cleaning since the far end of the unit is not touched. This results in a much lower downtime for a cleaning operation.

When U tube bundles are to be cleaned it is carried out in two stages. The initial stage is the cleaning of the straight lengths using the optimum cleaning heads for this operation. A flexible drive shaft is then added to the end of the rigid drive shaft and the most appropriate head used to clean around the bends.

In 2001 I had an interesting very personal experience which demonstrated just how flexible some pieces of equipment can be. In this case it was just an inspection job – no rotating shaft, just a retractable sampling grab and camera inside a flexible tube. The technical term for the procedure is a colonoscopy. The patient is normally heavily sedated without being out for the count. I had already omitted to take the preliminary pills and told the surgeon I did not want any sedation because I did not want to miss anything. He had never had such a request before but eventually agreed and the nurse had to put her syringe away.

It was by far the longest time I have spent on a TV screen and I was very grateful not to have missed it. The bowel follows a very tortuous path including one very tricky hairpin bend but we made it all the way to where I used to have an appendix.

We were chatting all the way through the procedure starting with the engineering of the probe. One topic he brought up was the difficulty and time spent in training doctors to use the probe. The simulators they had were not up to much so they were almost straight on to practicing on the real thing and had to complete over 100 closely supervised cases before they could work alone. I suggested that they should practice on animals which were about to be slaughtered anyway so that puncturing the bowel would not be such a disaster. He replied that pigs would be ideal since they had such a similar physiology and they were used in Germany but could not be used in the UK.

As I drove back home, (another good reason for not having an anaesthetic), I was already mentally designing training rigs using dead pigs colons inside slightly agitated solutions to simulate live body conditions. I kept in touch with the surgeon for several weeks until I found out that the University of Tel Aviv had already thought of my best designs.

Back to cleaning U bends. After seeing what that body probe could do I am not surprised that cleaning devices have now been developed that can clean right round U bends and I would expect the use of U tubes in fouling services to increase.

One question I would like to ask a tube cleaning company is do they take into account the different tube gauges when thicker gauges are specified for the inner tube rows of U tube bundles?

TUBE SIZES

One of the reasons I have been given in the past for fixing maximum allowable tube lengths is limitations in the length of cleaning equipment.

Often a minimum tube diameter of 25.4 mm is specified for heavy fouling services. This is usually a blanket requirement for a fixed value of fouling resistance and applies to all exchangers irrespective of the nature of the fouling.

While researching the cleaning of heat exchangers I looked at the Tubetech website. They give some details of a very interesting case. It is a vertical 'Texas tower' feed/effluent exchanger on a plant in the Philippines. It had not been cleaned for eight years! The 3186 tubes were 14 mm in diameter and 18 m long. (Or were they? They mentioned an access point which means they could have been only 9 m but they talked about drilling 18 m?) They tackled the fouling in both the upward and downward direction which might have limited the lance length to just over 4.5 m, whatever they did it was a rather impressive undertaking.

This example reinforces a view I have had for some time – perhaps we are too conservative in placing limitations on tube sizes?

CHEMICAL CLEANING

Chemical cleaning may take place in situ or at a remote location. It may involve soaking or continuous circulation of the cleaning fluid. It is a very specialised topic and any decisions taken are in the fields of the chemist and the technologist. For example inhibited hydrofluoric acid is the most effective agent for removing iron oxides but can not be used if the deposits contain more than a very small amount of calcium.

It is normally a last resort method of cleaning because of the problems of dealing with the subsequent waste. Examples of typical cleaning agents and their application can be found in references 9 and 19. It is very unlikely that a thermal design engineer will be involved with any decisions concerning chemical cleaning. All he/she has to do is make sure that nozzles are specified on the TDS if the PDS states that chemical cleaning is required.

SOOT BLOWING

This method is used in boilers which are outside the scope of this book. It is however an interesting example of intermittent cleaning. Steam jets are directed onto the outside of the tubes to remove carbon deposits. This is carried out at intervals, perhaps once per shift.

COMBUSTION

McCarthy and Smith in ref 38 report a novel cleaning method for a mixed C_4 fractionating column kettle type reboiler. The standard method was to pull the bundle and place it in an incinerator to burn off all the rubber like deposits. They were recalling a situation from their early years in the industry in the late fifties. It would be interesting to know if this practice is still in use anywhere today.

CONTINUOUS CLEANING

Continuous on line cleaning systems have been used for many years. I have only ever heard of them being used in cooling waters systems although in theory they may be applicable to other fluids. The introduction of a continuous cooling system justifies the use of a zero fouling resistance. They are costly to install and run and are therefore only considered for exchangers where the fouling resistance would otherwise be a very significant proportion of the total thermal resistance to heat transfer and/or cleaning downtime is very expensive. The main applications are for surface condensers and the condenser of large refrigeration plants. There are two distinct types of continuous cleaning systems namely:

Brush Type System

These systems consist of:

(a) A brush and two catching baskets for each tube as illustrated in figure 14.15.
(b) A four-way flow diverting valve which is shown diagrammatically in figure 14.16.
(c) A control panel.

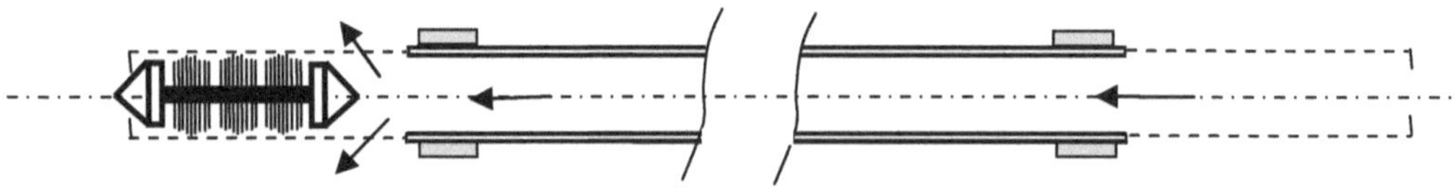

Figure 14.15

There are several sections of brush mounted on a plug with cylindrical ends. The ends have a good clearance in the tubes but the brushes have a small interference fit with the tubes. The brush materials are selected to suit the type of fouling. The diagram show the water flowing from right to left along the tube. When the direction of the water flow is reversed the plug is drawn into the tube and then forced along the tube until it reaches the catching basket at the other end.

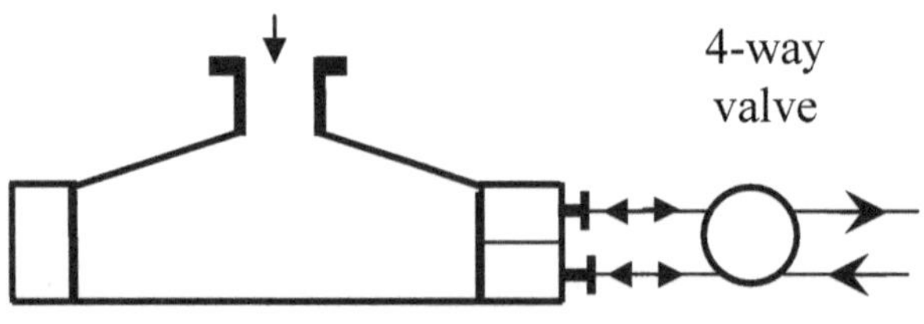

Figure 14.16

If the reader is curious about the four-way diverting valve there are some nice pictures on the WSA Engineering Systems website but there are no details of the internals to show exactly how it operates.

Cleaning intervals are selected to just keep the tubes clean. The maximum interval recommended by WSA is 8 hours. This is ideal for once per shift so that the system only needs to be semiautomatic. I could not find any details of how long the brushes last before they need replacing which would require a shutdown.

This type of system needs to be installed as part of the original design. It also raises an interesting thermodynamic dilemma. There will be an optimum water flow direction to maximise the MTD and this should be the mode of operation for most of the time. It will therefore be necessary to have two flow reversals during a cleaning operation. The longer this takes the better as far as cleaning is concerned to ensure that all brushes make two passes of the tubes. As far as the plant thermal efficiency is concerned the cleaning interval should be as short as possible.

I have never worked on one of these systems to check out the effect on the plant performance. I suspect the effect on the main banks of condensing tubes would be quite small. However if an air cooler section was part of the design the effect of supplying it with the warmest cooling water instead of the coolest cooling water would be quite dramatic. I covered air cooling sections in surface condensers in chapter 5 which the reader may wish to refer back to. The purpose of an air cooling section is to minimise the loss of steam with any entrained air which is passed to the air ejectors.

The maximum interval between cleaning cycles was stated to be eight hours but the recommendation for a particular case might be say 2 hours. The lower the time interval between cleaning the greater will be the proportion of the total operating time spent with the cooling water going the wrong way. The loss of steam any loss in thermal efficiency during this period should be taken into account in any cost comparisons with alternative systems.

Ball Type System

A simple outline sketch of this type of system is shown in figure 14.17 below.

These systems consist of:

(d) A large number of very flexible rubber balls. Each ball has a small interference fit inside the tubes. When they are deaerated they should have the same density as the cooling water to reduce the chance of them sinking to the bottom or rising to the top of the inlet header.
(e) A collection tank (not shown in the diagram) for storing the balls, rejecting old balls and introducing new balls.
(f) A device for injecting the balls into the cooling water supply.
(g) A ball trap/strainer to recover the balls from the return water.
(h) A recirculation pump which recirculates some of the cooling water together with the balls.
(i) A control panel.

This system involves no interruptions to the thermal operation of the plant. Its thermal disadvantage is that by recirculating some of the cooling water the mean supply temperature is slightly increased which should be taken into account.

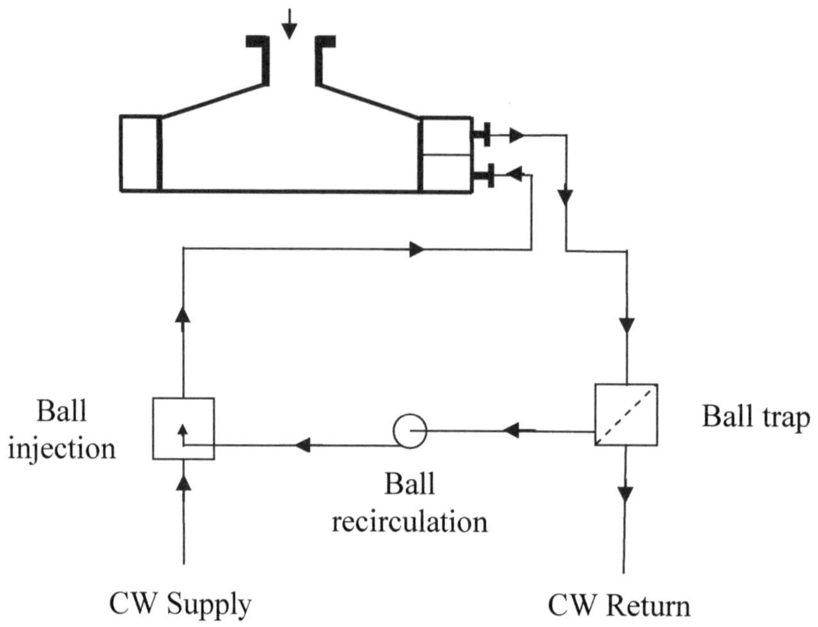

Figure 14.17

One potential problem is that if one tube on the outside of the bank did start to foul the extra hydraulic resistance through that tube would increase which would reduce the flow through it. It would therefore be less likely to capture a ball to clean itself. I have no operating experience to know if this takes place. Provided there are no such problems there is no need for any shutdown to replace any parts.

A big advantage of this system is that it is easier to retrofit. I once worked on a plant which had a surface condenser that was underperforming at its design conditions and yet more duty was required from it during a revamp. Extra cooling water and pressure drop were available but it still had no chance of being able to deliver the new duty. I ran the unit with a zero fouling resistance and it was just acceptable so a recirculating ball system was installed. I had some feedback on this one. It worked fine and the client was very satisfied.

STEAM CLEANING

One method of descaling makeup water evaporators is by steam heating the shell side and then rapidly cooling the tubes. This induces thermal stresses in the fouling layer which causes it to crack and detach from the surface of the tubes.

A hydro-steam cleaner is a mobile machine which needs a supply of low pressure steam and a solution of a cleaning agent. The temperature and flow rate of the solution are controlled by the steam injection. The mixture is delivered to the fouled surfaces via a hose with a spray nozzle.

VIBRATIONS – CHAPTER 15

Introduction

If I had ever managed to find a co-author for this book my first action would have been to take him/her out for a good session at the pub to cement our new relationship. When I considered that the time was right I would have paid compliments to his depth of knowledge in the field of vibrations compared with my own and suggested that this chapter should be one of his sections of work. Now I will just have to get on with it.

My reluctance to write this chapter is because I still feel that I have a lot more to learn than I have to impart to others. It is a topic where the 'pictures' that I try to conjure up can become very blurred when I am trying to visualise what is taking place. Perhaps I am not alone – TEMA now offers methods of predicting potential vibration problems with the proviso "Due to the complexity of the problem, the TEMA guarantee does not cover vibration damage." At least I have some experience of real cases which may be of interest to the reader.

HISTORICAL BACKGROUND

Vibration problems have been around long before I started to design heat exchangers in the mid sixties. I can not recall any problems which led to damage during my time at Kellogg. They had an excellent system from the initial thermal design through to a thorough thermal check of the vendor drawings. After I had left I was told that they had an interesting 'noise problem' on one of the units on an Ammonia plant. There was a very audible sound but after carrying out all the checks they could they found no evidence of vibrations which might lead to tube failure. The sound was not particularly annoying compared with many of the other noises on a plant of that era and they became known as 'the singing exchangers.'

The first vibration problem I was involved in was after I had taken charge of the thermal design group at Stone & Webster. A unit had failed on a plant in Grangemouth so I had to pack my bags go up there to sort it out.

It was typical bleak Scottish weather, either raining or looking like rain was about to start at any time. The exchanger I found myself looking at was equally depressing. A couple of tubes had already failed, some were leaking and a number of others were clearly about to join them.

The unit was designed to cool a gas stream on the shell side with cooling water passing through brass tubes which were supported by carbon steel baffles. The tube layout was horrific. There was no impingement protection and the tubes at the top of the bundle were actually just inside the nozzle end projection as shown on the next page in figure 15.1.

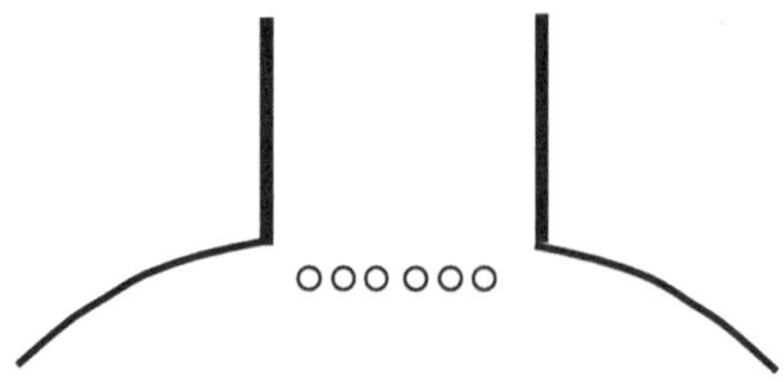

Figure 15.1

The tubes that had failed were in the top two rows. They had just snapped off at the face of the tubesheet due to work hardening. There were signs of severe fretting of some of the more vulnerable tubes where they passed through the baffles but away from the nozzle area they were not too bad. My immediate patch up recommendations to get the unit back on stream was as follows:

(a) Remove sections from all the tubes in the top three rows between the tubesheet and the first baffle. Leave rest of the tubes in place to avoid excessive bypassing. Plug the ends.
(b) Perforate the next row of tubes but leave them in place with the ends plugged to act as impingement rods.

Back at the office I designed a replacement bundle with baffles from the same material as the tubes and tight tube/baffle clearances. It had impingement protection and an extra support for the first five rows of tubes under the nozzle. I managed to fit the same number of tubes into the shell with a flatter bundle – the existing bundle had the tubes spread out to form as round a bundle as possible!

I came to the conclusion that if the correct material had been used for the baffles and a competent thermal design engineer had checked the vendor drawings there would have been no problems. This should have been standard good practice even then.

From that era the only failures that I came across or heard of could all be put down to unsatisfactory designs or checking procedures. This seemed to justify a complete lack of vibration checks for new designs which was then the norm.

Then came the advent of the computer program and automatic vibration checks for most units. I was back at Stone & Webster on one of my many return visits as a consultant. I was just starting to work on a cracked gas aftercooler. The last one I had designed had been several years before using hand calculations so my first move was to discuss the design approach with Dave Edge who was then the head of the section. I remember well his comments: "We always use cross flow designs now – CST gives so many warning messages for the type of designs that you did that they should have fallen to pieces years ago." I was taken aback before I managed to ask; "So how many have failed then?" "Well actually none; they are all still working fine but CST predicts severe problems so we never use them now." The colour returned to my cheeks and I immediately felt much better but my somewhat cynical attitude to some warning messages had already started.

The designs we had used when I had been working there full time were H types to maintain a low pressure drop but with double segmental baffles to keep the tubes well supported and maintain relatively short unsupported tube lengths. I have no recriminations about the designs we used to produce but I realised that the age of the computer vibration checks had arrived and thermal design work would never quite be the same again.

Even before vibration checks became standard practice another parallel change had been taking place in the design of shell and tube units. They were getting bigger! For many years the maximum shell diameter for units with removable bundles had been 60 inches. For units with none removable bundles the limit was 80 inches. Over the years I have designed floating head units with 85 inch shells and some of the fixed tubesheet items now being used could swallow up the house I am in whilst writing this book.

As the size of shell and tube units increased the probability of vibration problems increased and the need for vibration checks grew in importance. Increases in shell sizes to take larger volume flow rates of fluid meant that nozzles increased in size and with them the length of the inlet and outlet spaces which are the most vulnerable part of any unit.

FLOW INDUCED VIBRATIONS

All structures have a natural frequency which has the potential to vibrate when subjected to an appropriate excitation. I am sure that by now the reader will have seen some of the old films of the infamous Tacoma Bridge being destroyed by a steady but not unexpectedly high wind. More recently there was the Millennium Bridge in London which was never damaged but could only be crossed with restrictions on its pedestrian traffic. I assume it was designed by an architect rather than an engineer and it was certainly by someone who knew nothing about the way armies had never crossed bridges in a coordinated marching order. Still it looked pretty!

If this type of phenomena can happen to a relatively solid structure like a bridge it is not surprising that there are always potential problems with a very flimsy component such as a heat exchanger tube.

NATURAL FREQUENCY f_N

The natural frequency of a tube is governed by exactly the same variables that determine the way it will bend. All readers should have covered a basic course in strength of materials which involved how a beam will bend when it is uniformly loaded. **Any variable which resists bending will resist vibrations by increasing the natural frequency of the tube.** The variables which cause and resist bending are as follows:

Uniform loading

The uniform loading for a tube is its own weight plus the weight of any fluid inside the tube. (I have used average wall tubes to simplify the calculations throughout this section – for CS tubes they would be minimum wall.) The weight providing the uniform loading is the mass multiplied by the gravitational constant. It is worthwhile comparing the mass of the fluid and the mass of the tube for a 19 mm diameter tube made out of carbon steel and titanium. I have assumed that the fluid inside the tubes is water with a density of 1000 kg/m^3 in the following table:

Tube metal & gauge	Mass of metal kg	Mass of water kg	Total mass kg
CS 14 BWG	0.881	0.1728	1.054
Ti 20 BWG	0.231	0.2343	0.465

Table 15.1

For the carbon steel tube the increase in mass due to the water inside the tube is about 20% which is not insignificant. For the titanium tube the mass of the water is greater than the mass of the metal!

Unfortunately the mass of the tube plus the fluid inside the tube have been found to under predict the loading on the tube because of the effect of the fluid outside the tubes and the proximity of the tube to any other solid surface – normally adjacent tubes. This correction is termed the **hydrodynamic mass**. Figure 15.2 has been reproduced from TEMA, with permission.

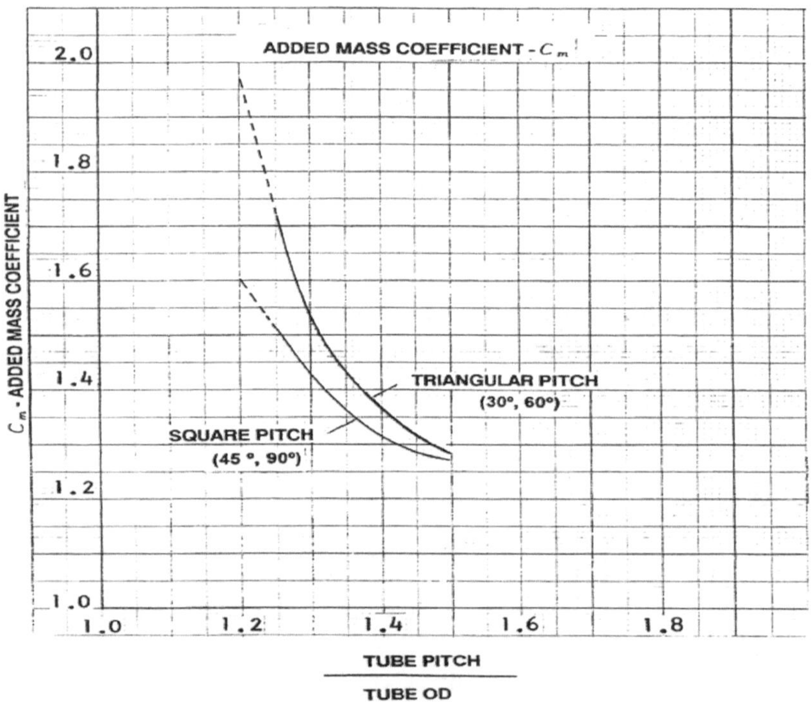

Figure 15.2

The added mass coefficient on the vertical axis is multiplied by the mass of the shell side fluid displaced, (the external volume of the tube multiplied by the density of the shell side fluid) to give the hydrodynamic mass. The values of the added mass coefficient in this diagram are for calculations in US Customary units. I have reproducing it here only to comment on the form of the curves.

For any tube arrangement the hydrodynamic mass increases as the tube pitch decreases. As the tube pitch decreases the fraction of the cross sectional area of the bundle occupied by the shell side fluid decreases. I can offer no explanation as to why this should increase the effective mass of the tubes.

The orientation of the tube pitch has no effect on the hydrodynamic mass. For any triangular pitch orientation the hydrodynamic mass is higher than for any square pitch orientation. Once again I can not think of any explanation for these phenomena.

The dotted lines on both curves are extrapolations. Experimental testing stopped at the lowest tube pitch to diameter ratio we normally use of 1.25. It is academically interesting that tests were carried out well beyond the pitch to diameter ratios which are normally used. The most practical section between 1.25 and 1.33 could hardly be described as a curve!

Since a correction for hydrodynamic mass is based on experimental data I have no reason to doubt it but I still can not 'get my head round it'.

The type of support

The three possible types of support are illustrated in figure 15.3 below. The first case (a) is simply supported at both ends. The second case (b) is fixed rigidly at both ends. The third case (c) is fixed at one end and simply supported at the other end.

(a) (b) (c)

Figure 15.3

There is a very easy way to demonstrate the difference in bending between beams with simple point supports and built in end supports. Just ask a friend or colleague to hold a long ruler balanced on their index fingers say 50 mm from each end. Then carefully place a weight at the centre and observe the deflection – any reasonably sized weight will do, the more adventurous could try a cup of coffee! Then repeat the experiment with the person holding the ruler grasping it firmly at both ends. This will give a feel for the difference in the effect of the two types of support on the bending of the ruler.

Figure 15.3 (c) has one built in end support and one simple support. If the above experiment is repeated with this arrangement the deflection will lie somewhere in between the previous two values. This is a very important arrangement for heat exchangers since it is the way the tubes are supported in the all important inlet and outlet spaces.

Our equation for determining the natural frequency of a tube must therefore take into account the length between supports, **L** and the type of support. The latter is taken into account by a constant C_N which depends on the type of support. The values of C_N for the three arrangements in figure 15.3 are shown in table 15.2 below.

Support type	Constant C_N
(a)	31.7(9.9)
(b)	72.4(22.4)
(c)	49.6(15.4)

Table 15.2

The type (b) support which is built in at both ends is unlikely to be encountered in heat exchanger design work. The figures in brackets are the values the reader will find in TEMA for the equation in the US Customary units.

Second moment of area I

The second moment of area of any cross section defines the way the area is distributed about a specified axis. It is the sum of all increments of area multiplied by the distance squared from each increment to the specified axis.

Unfortunately this variable has been wrongly defined as moment of inertia in the eighth edition of TEMA and was not corrected in the ninth edition. HTFS define it correctly. Moment of inertia is defined in terms of the distribution of mass about an axis. It is used in engineering dynamics and would be used to calculate the energy stored in a flywheel or the increase in rotational speed of a ballerina as she changes the position of her arms. In the situation we are looking at nothing moves – unless the tube starts to vibrate!

To investigate the effect of the second moment of area on the bending of a heat exchanger tube is relatively simple since bending is assumed to take place about the neutral axis which is on the centre line of the tube as illustrated in figure 15.4 below.

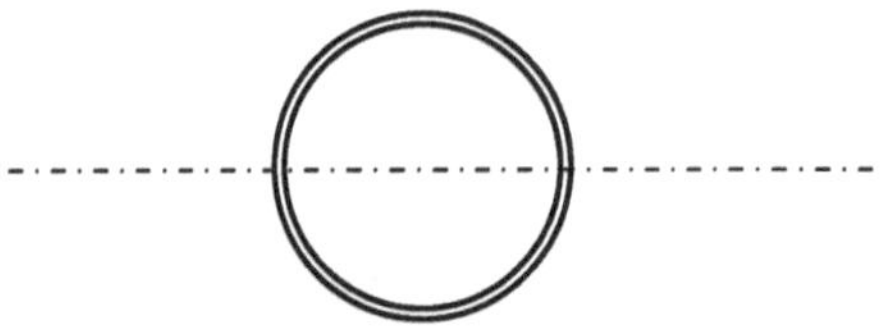

Figure 15.4

The equation for the second moment of area of a tube about its neutral axis is as follows:

$$\text{Second moment of area } \mathbf{I} = \frac{\pi}{64}(d_o^4 - d_i^4) \qquad [15.1]$$

It is worth a brief look at how tube diameters and gauges affect the second moment of area. Table 15.3 below shows two tubes diameters and gauges commonly used for shell and tube units.

Tube d_o mm	Tube d_i mm	Second moment of area **I** m^4
19.05	14.83 (14 BWG)	0.4079×10^{-8}
25.4	21.18 (14 BWG)	1.0531×10^{-8}
19.05	13.51 (12 BWG)	0.4828×10^{-8}
25.4	19.86 (12 BWG)	1.2778×10^{-8}

Table 15.3

Tables D7 and D7M in TEMA give values for all the diameters and tube gauges we are likely to ever use. The table shows that changing the tube gauge only makes a difference of about 18% to the second moment of area whereas changing the tube diameter changes the second moment of area by about 160%.

Modulus of elasticity E

This is a property of the material and varies with temperature. Values of E for most of the materials we are likely to use can be found in table D-10 in TEMA. Unfortunately the values in TEMA are only in US Customary units and need to be converted to SI. The units for the modulus of elasticity in the SI system are N/m^2.

Axial tube stress multiplier A_S

Although I have absolutely no practical experience of any musical instrument I know that changing the tension of the wires in any stringed instrument changes the note by changing the way the wire will vibrate. I have some experience of applying tension to a string when I am trying to keep a straight line between posts or building a brick wall.

Tension reduces the bending deflection of a string or a heat exchanger tube and increases the natural frequency. Unlike a string a tube may also be in compression and this will reduce the natural frequency.

Axial stresses are dealt with by means of a stress multiplier. The equation for calculating the stress multiplier may take a number of forms. I like the way Ted Saunders presents it in ref (6) although I have changed it a bit to suit the nomenclature I am using:

$$A_S = 1 \pm \frac{(W_A L^2)^{0.5}}{(\pi^2 EI)^{0.5}} \qquad [15.2]$$

Where W_A is the axial load in Newtons. For a fixed exchanger geometry and tube material the multiplier depends on the square root of the axial load. The sign is positive for a tensile load and negative for a compressive load. In theory there will always be a very small axial load in any exchanger but it is normally only worth taking into account for fixed tubesheet items.

I eventually managed to locate a diagram which will give the reader some idea of the effect an axial stress may have. This is shown below in figure 15.5.

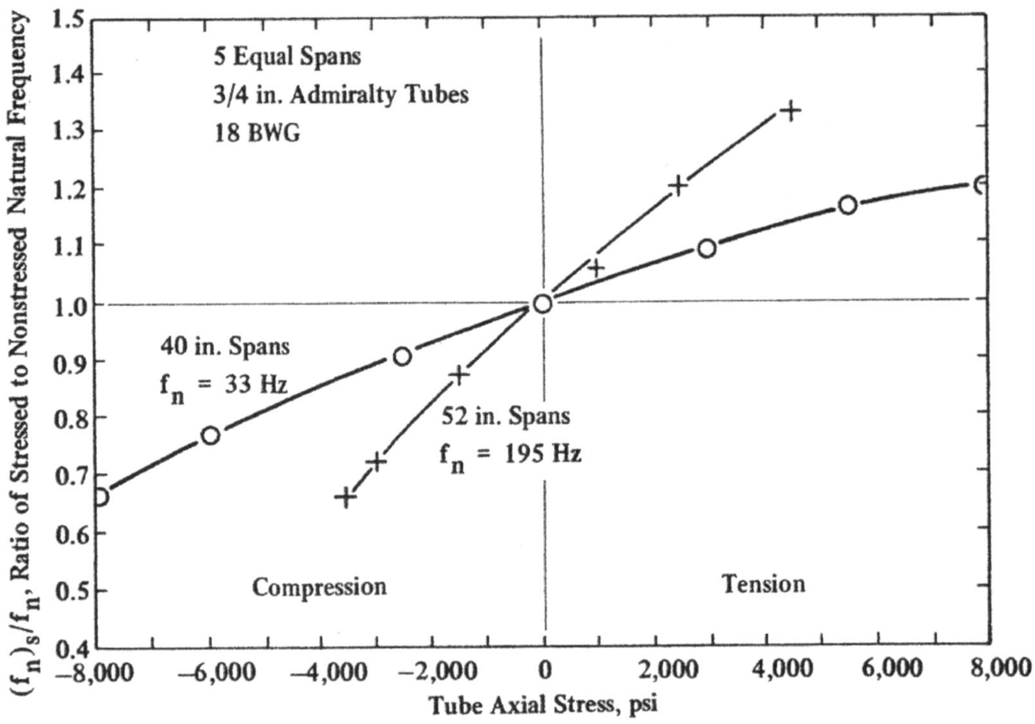

Figure 15.5

This diagram is taken from some experimental work by Kissel and there is a good fit between the points and the curves. Once again the units are US Customary but the axial stress factor is dimensionless and it is an easy matter for a reader to convert the stresses to MN/m^2 by multiplying by 6.89×10^{-3} if required.

The axial stress multiplier is not an easy correction to handle during a thermal design because stress calculations are not normally carried out until the thermal design is finished. Personally I have ignored it when the tubes are in tension – this is a safe omission. When the tubes are in compression it is not safe to leave it out. Despite this I have still been known to leave it out if I am working on a design which is predicted to be well clear of any potential vibration problems. If in doubt I would try and get a mechanical engineer to run one of their programs. As a very last resort I would do some simple hand calculations making safe assumptions. For a feed design I have often just made a note somewhere and then forgot about it.

On reflection and after studying figure 15.5 I think I might have been too lax in the past. Quite a modest compressive stress is sufficient to result in a correction factor of 0.7 for a wide span which is very significant and unsafe. At the other end of the scale a modest tensile stress is sufficient to take a dodgy design into a safe region.

Although I am impressed by this piece of research I am critical of some key aspects. Firstly I have never considered using an 18 BWG brass tube – 16 BWG is the norm for brass. I can not think of any good reason for the choice of tube gauge. Secondly the tube spans are larger than we would try and use in an exchanger with potential vibration problems. It is a pity that the work was not extended to include a third, smaller span. Lastly it would be very interesting to see these experiments repeated for the more usual CS tubes. I suspect that brass tubes were used because they are more likely to fail than CS tubes subjected to the same vibrations.

Equation for natural frequency f_N

The equation for calculating the natural frequency of a tube in SI units is as follows:

$$f_N = 0.04944 \frac{A_S C_N}{L^2} \frac{(EI)^{1/2}}{(m_E)^{1/2}} \quad H_Z \qquad [15.3]$$

Where m_E is the equivalent mass of metal plus the tube side fluid plus the hydrodynamic mass of the shell side fluid in kg i.e

$$m_E = m_{Tube} + m_{TubeFluid} + m_H \quad kg/m \qquad [15.4]$$

A_S is the axial stress multiplier, (1 for no stress) and C_N is the frequency constant which depends on the number of spans and the end conditions.

For any specified tube and tube arrangement the only variable at the disposal of a thermal design engineer is the unsupported length and the natural frequency of the tube is inversely proportional to this quantity, therefore:

$$f_N = \text{A constant} \times \frac{1}{L^2} \qquad [15.5]$$

This is probably the most important equation to remember about tube vibrations. It makes it very easy to predict the effect of adding an extra support for the tubes in a vulnerable area.

PRACTICAL EXCHANGERS

Considering single spans for the three different methods of fixing is a logical place to start predicting the natural frequency but I have never come across an exchanger with built in ends and no intermediate supports and the other two cases are physically impossible. A real heat exchanger tube is supported as illustrated in figure 15.6 below.

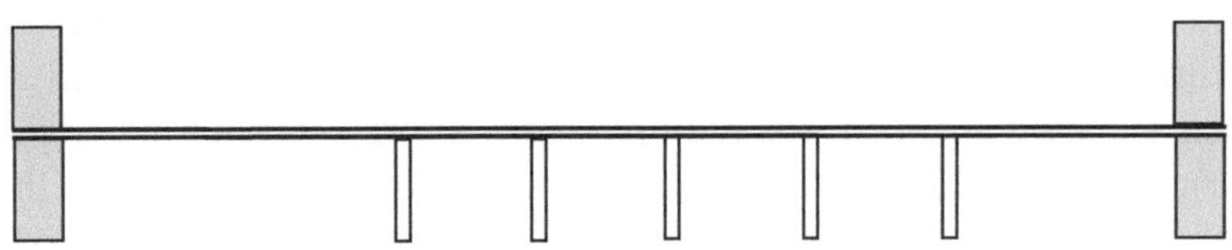

Figure 15.6

Tubes which are supported by all the baffles will have lengths equal to the inlet and outlet lengths which are built in at one end and a number of intermediate lengths equal to the baffle pitch which are simply supported at both ends. Tubes which pass through the baffle window have end lengths equal to the length of the end space plus one baffle pitch and intermediate lengths equal to twice the baffle pitch.

A lot of the experimental work which has been used to establish the empirical constants was using ideal tube banks with equal spacing along the full length of the tubes. This is the same as the approach used to establish heat transfer coefficients and pressure drops across tubes. Modern computer programs attempt to take into account the complexity of the support arrangement in real exchangers.

EXCITATION MECHANISMS

A tube will not vibrate unless it is excited. I am tempted to offer an analogy but I will leave that to the reader's imagination. The source of any exciting mechanism is the fluid flowing across the tubes. Any exciting mechanism will have its own frequency. It is likely that whenever a fluid is flowing across a tube there will be some movement of the tube which may be classed as a vibration but is harmless. Serious problems can arise when an exciting frequency coincides with the natural frequency of the tube. This condition is known as **resonance**.

If resonance occurs the magnitude of the vibration will suddenly increase and may lead to tube failure. The possible exciting mechanisms are as follows:

FLUID ELASTIC WHIRLING

The term whirling comes from an analogous situation with rotating shafts. The reader may have seen a laboratory experiment which demonstrates this in quite a spectacular manner. Problems with rotating shafts were encountered long before our problems with heat exchangers.

My first experience of resonance in rotating shafts was when I was in Marine Engineering. The system arrangement there is even more complicated than for a heat exchanger tube. At one end of the arrangement there is a propeller which cuts through the water at a regular frequency which depends on the number of blades and at the other end an engine. The engine room is in the parallel section of the vessel so that a long shaft is required to connect the propeller to the engine.

The ships I sailed in invariably had 'barred speed ranges'. These were resonance speeds which could produce vibrations which could be felt throughout the ship. They were clearly marked in red bands on the control panel and had to be quickly passed through when a change of speed was required. Just before I left marine engineering to join the petrochemical field I went on the trials run for a new ship. One of the many objectives of a trial run is to check the barred speed ranges. We found an extra one which had not been anticipated. My first lesson in vibration problems!

On a lighter note I have experienced a barred speed range in an old banger I used to drive. It was most uncomfortable between 55 and 60 mph so I had to drive either above or below these limits. Eventually I fixed it by getting the wheels rebalanced.

In heat exchangers we can not afford to have any barred fluid velocities because any fluid velocity below the design value could be maintained indefinitely at part load. It is therefore necessary to ensure that the natural frequencies of the tubes are always above those of any exciting frequencies and do not forget the likely revamp when the fluid velocity may be increased by 20%!

The mechanism of fluid elastic instability is best explained with reference to a diagram such as the one shown below in figure 15.7.

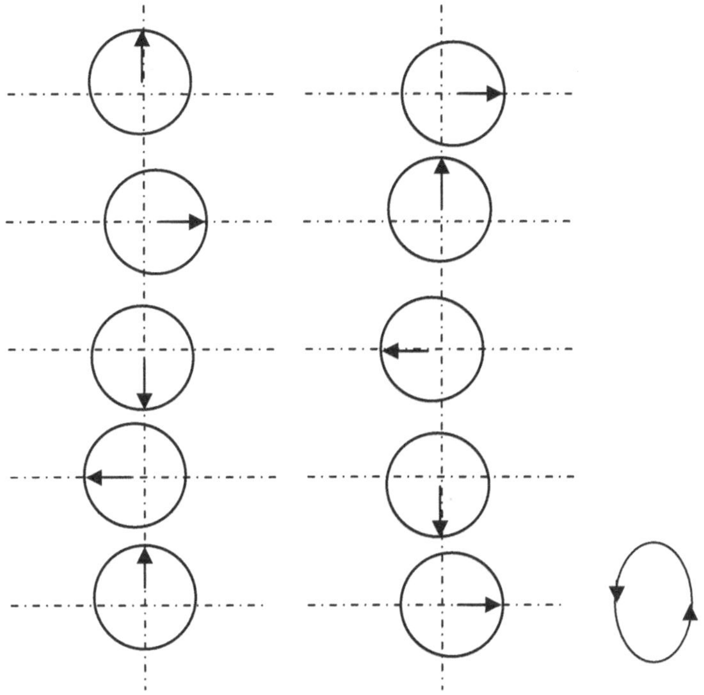

(a) (b) (c) Figure 15.7

Alternate tubes move 90° out of phase with each other. The diagram (a) shows a point in time when the odd numbered tubes are being displaced in the vertical plane and the even numbered tubes are being displaced in the horizontal plane. This is followed by a change in the planes the tubes are being displaced from as shown in (b). There is a smooth transition from one plane to the other therefore any point on the tube surface is moving through a curved path. This path has been shown to be ellipsoidal as illustrated in (c).

The nearest comparison I can think of is the motion of a skipping rope where the ends are held and the rope rotates in a curve. It is now easy to see why this form of motion is termed whirling.

Critical velocity u_{Crit}

The critical velocity is defined as the velocity of the fluid across the tubes which will result in resonance between the natural frequency of the tube and the exciting frequency.

A single equation for predicting the critical velocity has been used by Ted Saunders in reference (6). This equation was proposed by Connors and is as follows:

$$u_{Crit} = \beta\, f_N\, d_o \frac{(m_E\, \delta_o)^{0.5}}{(\rho\, d_o^2)^{0.5}} \qquad [15.6]$$

Where β is the instability constant which can be found from table 15.4 below, ρ is the density of the shell side fluid and δ_o is the log decrement of the tube.

Pitch angle	Tube pitch/diameter ratio		
	1.25	1.33	1.5
30°	4.6	5.2	6.8
60°	3.8	4.0	4.1
90°	3.9	3.9	4.1
45°	3.9	4.5	5.7

Table 15.4

The instability constants are empirical values from experimental work. The log decrement of the tube is the dodgy variable.

The log decrement for a sinusoidal wave is calculated from the ratio of the peak values of successive waves as the magnitude of the waves decrease due to damping. For a full mathematical analysis of this phenomenon I would recommend that the reader consults a good book on Engineering Dynamics such as the one I still have from my student days by Swanson – ref 29. Before writing this section I reread the relevant sections in his book. There are lots of diagrams and equations which are relatively easy to follow but well outside the scope of this book. Figure 15.8 below illustrates the effect of resonance as the exciting frequency approaches the natural frequency in both directions.

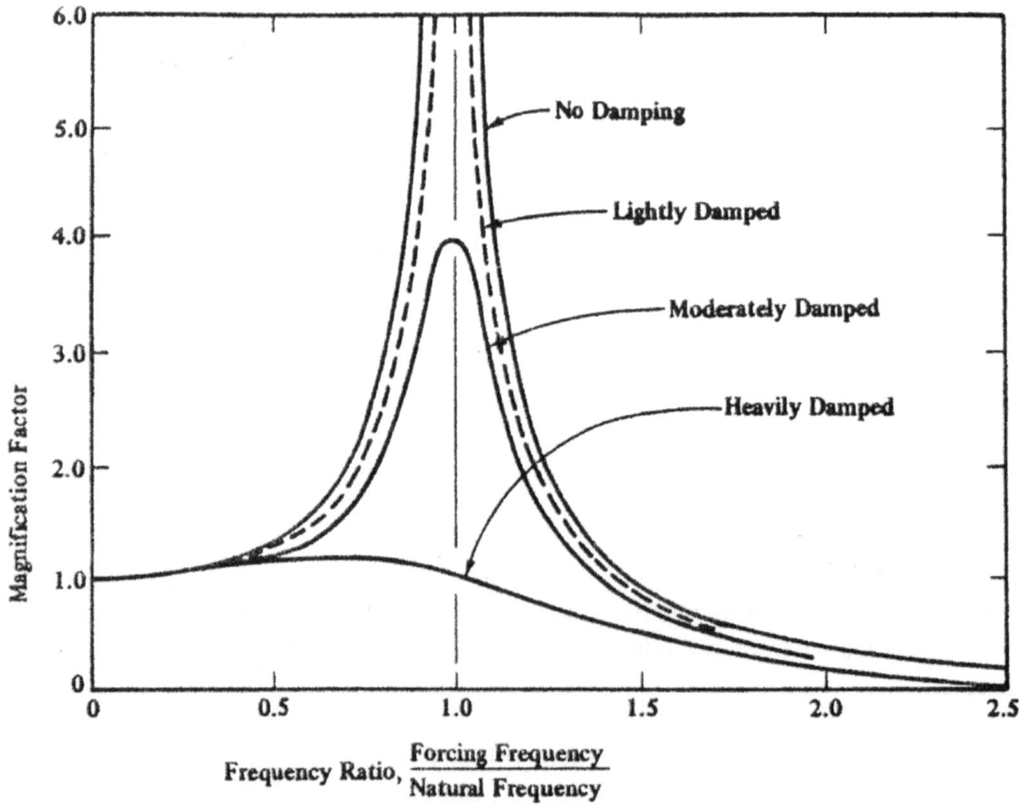

Figure 15.8

When a resonance frequency is reached there is an equation which predicts that the magnitude of a vibration will be infinitely large for a system with no damping. Long before infinity could be approached any tube would be destroyed. The system damping for a heat exchanger tube depends on many variables. These include the variables which describe the mechanical arrangement of the system together with characteristics of the shell side fluid. TEMA provides a method for calculating a damping factor in section 6V-8. There are different equations for liquids, vapours and two phase flow. The latter even includes a surface tension function. Some heat exchanger programs take the conservative approach and assume no damping which I am thankful for because it gives me an excuse for not trying to explain the details of the TEMA equations and the logic behind them. I suggest the reader should have a look through this section and make up their own minds.

In recent years the equations for predicting the critical velocity have been refined and their accuracy improved. The procedure in TEMA is very easy to follow in a step by step manner but involves using multiple equations so that the effect of a single variable is unclear and it is only in old units.

In practice thermal design engineers have to rely on the predictions from the program they are using. Some programs will calculate a critical velocity and then alert a designer if the velocity at any point in a tube bundle exceeds 80% of this value. Knowing some of the uncertainties in the correlation they are using I have often wondered if this margin for error is sufficient. Then when I think about all the steps involved I know that each of them may contain a safe assumption and knowing how accumulative safety factors can build up I feel better about the 80% guideline.

This is the one vibration message I have never even considered ignoring because it is so easy to visualise the consequences of getting it wrong.

VORTEX SHEDDING

Vortex shedding is easy to visualise because everyone will have observed it at some time. My own immediate picture is of the eddies which form downstream of rocks in a fast flowing trout stream. Areas are created where the direction of flow is against the direction of the main flow of the stream. A lure cast close to the rock may also be observed to be very quickly pulled under the surface without any fish being involved. This clearly demonstrates that there are some eddies which are acting at right angles to the surface of the stream as well as the ones that are easy to see on the surface. An angler on a stretch of water he knows well will have observed how different rocks and areas of protruding banks produce different eddy patterns and know how to use them to his/her advantage.

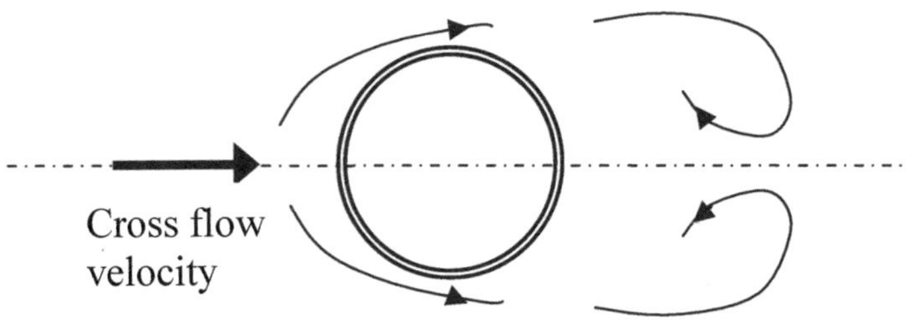

Figure 15.9

In fluid dynamics we use the term vortex rather than eddy. Figure 15.9 shows the vortexes formed as a fluid passes over a single tube this is the same flow pattern as a single rock in midstream. I am not going to attempt a three dimensional view to show the vortexes at right angles to the direction of flow!

Figure 15.10 shows the vortexes formed as a fluid passes through a section of a tube bank.

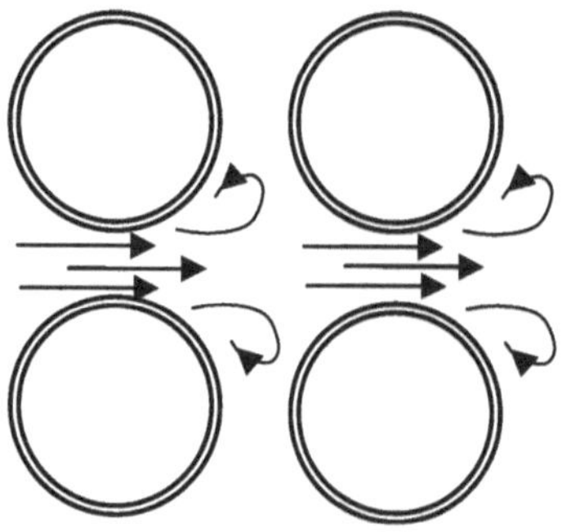

Figure 15.10

Under steady flow condition vortexes are shed in the wake of the tube at regular intervals. This produces alternating forces on the tubes in two directions:

(a) At 90° to both the direction of flow and the tube axis there is a lift force.
(b) In the direction of flow there is drag force. The frequency of the drag force is twice that of the lift force but it has a lower amplitude.

To investigate the effect these forces can have on vibrations it is necessary to be able to predict the frequency of vortex shedding. As with so many situations in thermofluid dynamics a dimensionless number is introduced.

Strouhal Number S

The Strouhal Number is defined as follows:

$$S = \frac{f_V d_o}{u} \qquad [15.7]$$

Where f_V is the vortex shedding frequency and u is the maximum cross flow velocity at the minimum flow area across a bank of tubes. Like all dimensionless groups the Strouhal Number must be a ratio of two quantities. The two quantities in this case are both velocities so that the Strouhal Number may be thought of as the ratio of the velocity of the tube as it is being displaced to the velocity of the fluid stream. Rearranging equation 15.7 gives:

$$f_V = \frac{Su}{d_o} \quad \text{Hz} \qquad [15.8]$$

The frequency of vortex shedding across a bank of tubes is highly dependant on the tube arrangement. Values of the Strouhal Number for different tube arrangements may be presented as charts or in tables. The following table is based on the work of Chen and Fitz-Hugh.

Pitch angle	Tube pitch/diameter ratio		
	1.25	1.33	1.5
30°	0.18	0.22	0.29
60°	0.8	0.75	0.61
90°	0.42	0.35	0.29
45°	0.58	0.5	0.57

Table 15.5

It is clear from equation 15.8 that the higher the value of the Strouhal Number the higher the vortex shedding frequency will be. Looking at a table of numbers is normally difficult to digest but comparing numbers is easier to handle. In table 15.5 the comparison between two sets of numbers sticks out like a sore thumb. The Strouhal Numbers for the 60° tube arrangement are about four times higher than those for the 30° arrangement for the two most commonly used pitch/diameter ratios!

I would have expected there to be some difference but not one of that magnitude. This is another of those situations in vibration analysis that I can not visualise and offer any sensible explanation for but it has been clearly demonstrated during experimental work and I have no reason to doubt it.

In table 15.4 there is also a 21% decrease in the instability constant in the equation used to predict the critical velocity for fluid elastic whirling for the 60° arrangement compared with the 30° arrangement.

The effect of a greatly increased vortex shedding frequency and a more modestly reduced critical velocity has made the 60° arrangement very unpopular in thermal design work. Many specs apply a compete ban on its use. This is no doubt due to the fact that many units which have had tube failures had a 60° tube arrangement. If they were designed before the advent of rigorous vibration calculations this is not too surprising. This is one of the many issues which I do have strong views on. Now that we can carry out vibration checks we can identify which units may be vulnerable and which ones will not have any problems. As I have explained elsewhere in this book there are cases where the 60° pitch results in the optimum design and if there are no vibration problems predicted there is no sensible reason why a 60° tube pitch arrangement should not be used. (It may also be a cure for an acoustic problem!)

TURBULENT BUFFETING f_{TB}

This is a phenomena associated with highly turbulent streams which tends to limit it to gases. Energy is selectively extracted from the gas stream which may result in vibrations. Forming a 'picture' of this is once again beyond me. The equation used to predict the frequency produced in this way, f_{TB} was proposed by Owen and is as follows:

$$f_{TB} = \frac{u\,d_o}{P_{Lt}\,P_{tt}} \left(3.05\,\{1 - d_o/P_{tt}\}^2 + 0.28\right) \qquad [15.9]$$

Where P_{tt} and P_{Lt} are the transverse and longitudinal tube pitches defined in figure 15.11 below.

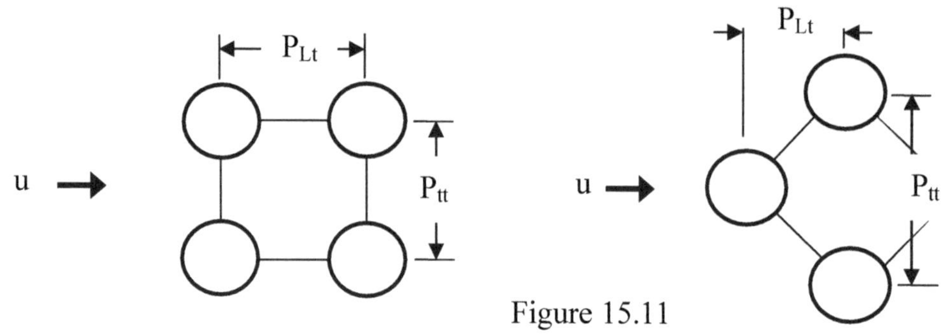

Figure 15.11

It is recommended that turbulent buffeting frequency should be kept to less than half the tube natural frequency therefore:

$$\frac{f_{TB}}{f_N} < 0.5 \qquad [15.10]$$

Comparison between the amplitudes of exciting mechanisms

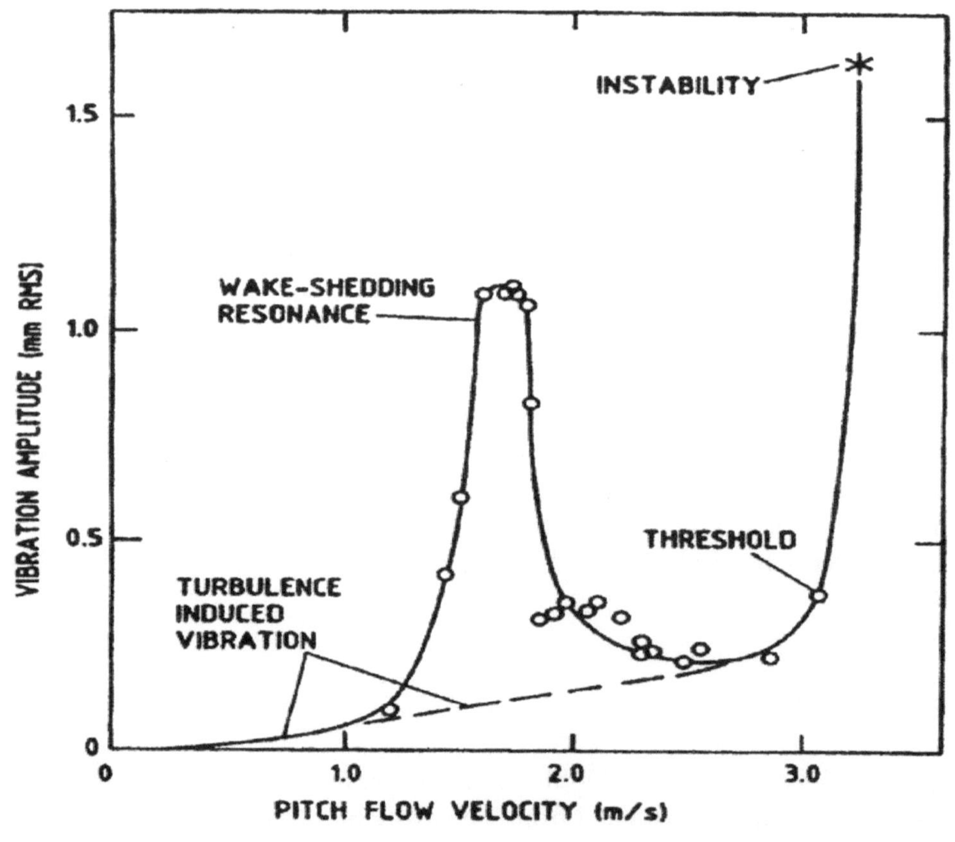

Figure 15.12

Figure 15.12 above is a comparison between the amplitudes of the three exciting mechanisms discussed so far. The vertical axis is the amplitude in mm. These are 'root mean square' values so that the maximum values will be somewhat higher. The curves show the relative importance of the three mechanisms. The diagram has been drawn for one particular case but the trend will be similar for any case.

PARALLEL FLOW

Vibrations can be started by fluid flow in the passages in between tubes in the axial direction – long flow. This is only likely to occur with very long unsupported tube lengths. Reports of problems seem to have been limited to the nuclear industry.

I have worked twice in the nuclear industry and I can vouch for the very long unsupported lengths. I did the heat transfer and pressure drop calculations for a novel grab designed to recover short graphite samples from the reactor without cooling it down. The grab had a camera on the end to check that the sample was firmly held and could not fall off during retraction. It had a temperature limit so that it needed to be kept cool with a stream of gas. I have just glanced at some of the calculations I did and the lengths involved were over 30 m.

As far as the design of shell and tube units are concerned this is one that the reader can forget about.

VIBRATION TRANSFERENCE

I have highlighted this phenomenon by giving it a separate section rather than just mentioning is elsewhere. When dealing with say a long unsupported span in the inlet space it is easy to forget the other end of the tube in the outlet space. A fluid velocity across one end of a tube may excite a vibration at the other end of the tube if that span has the lowest natural frequency.

Whether a vibration excited remotely is as severe as it would be if excited locally is subject to some debate. That it may take place is not.

U BENDS

Up until this point I have not mentioned U tubes. This was intentional to try and 'keep things simple'. A U tube is a cantilever simply supported at both ends. This reduces their natural frequency and makes them particularly susceptible to vibrations.

They can also be **a good example of vibration transference**. A U tube may have a full support plate and no fluid flowing across the bends of the tubes but the bends can still vibrate.

Calculating the natural frequency of a U tube is more complicated than for a straight tube as illustrated in figure 15.13 below.

The variables are L_b, the baffle pitch (P_b in my nomenclature), r the radius of the U bend to the centre line of the tube and s which is the distance from the tan line to the centre of the support plate. The nomenclature is from TEMA and the only significant change I have made to one of the diagrams in TEMA is to have the support plate at the bottom which is normal practice.

The distance from the tan line to the support plate was discussed in detail in chapter 2. It is typically about 50 mm, half the thickness of the support plate is added to this to give the dimension s. There are an infinite number of possibilities for the ratio of s/r between the theoretical limits of 0 and 1.0. This also applies to the ratio L_b/r between a lower limit of 0 and a limit determined by the maximum allowable baffle pitch for a given tube diameter. An upper limit of 6 more than covers any practical geometry.

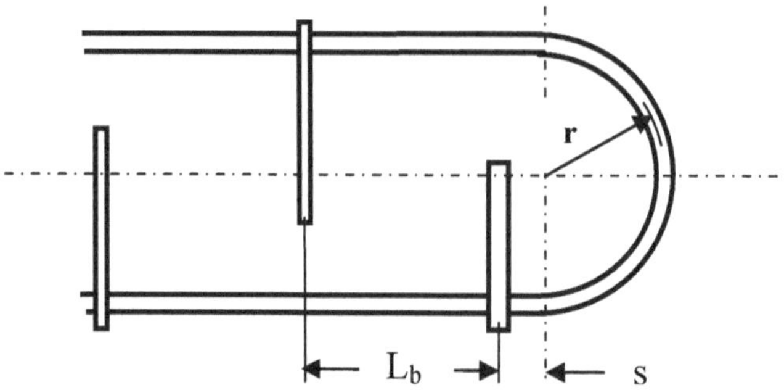

Figure 15.13

The data for finding the appropriate constant for the natural frequency equation has therefore been presented in graphical form rather than in tables. Even then over 4.5 pages are required in TEMA compared with less than half a page for the straight tube possibilities.

An axial tube stress multiplier for U tubes is mentioned but not covered in TEMA – just a reference to a book by Singh and Soler. I have not read the book but I am aware that axial stresses may occur in U tubes.

A condensing vapour in a vertical U tube may have significantly subcooled liquid in the outermost tubes at the bottom. This may result in a large temperature difference between the top of the tube and the bottom of the tube. The top of the tube will be in compression and the bottom of the tube in tension. It is easy to understand why TEMA left it out!

There is an interesting and very readable paper by Nieh and Zhang – ref 30. It proposes some modifications to the TEMA data and it includes a sketch of the test rig used which will give the reader some idea of how the constants were measured.

Horizontal U tubes

Towards the end of writing this section on U tubes I suddenly became conscious of something I should have spotted years ago. The entire section in TEMA is devoted to U tubes in the vertical plane. What about U tubes in the horizontal plane?

At this point I was ready for another coffee and while the kettle was boiling I went into the workshop, bent a piece of thick wire and balanced a weight on it with the wire in both planes. The difference was quite dramatic and as I was carrying out my experiment I was thinking "silly old sod – you may not be an expert on vibrations but at least you do understand second moments of area (and the efinition of moment of inertia)!

The two possible tube orientations are illustrated in figure 15.14.

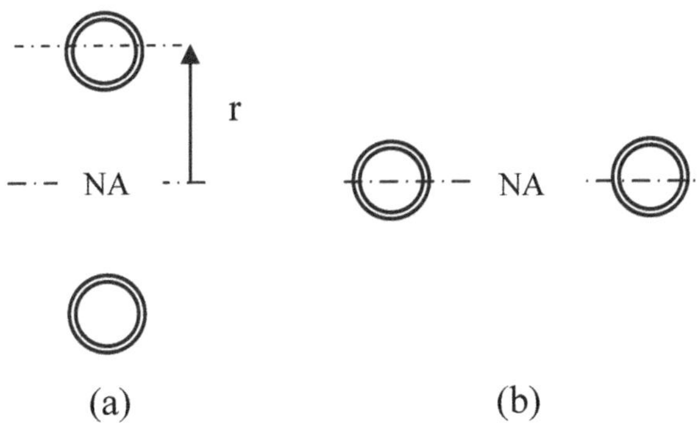

Figure 15.14

Consider a horizontal U tube with a diameter of 19.05 mm and gauge of 14 BWG; for case (b), equation 15.1 can be used as follows:

$$\text{Second moment of area } \mathbf{I} = \frac{\pi}{64}(d_o^4 - d_i^4) \times 2 = \mathbf{0.8158 \times 10^{-8}\ m^4}$$

Using the theory of parallel axes for the vertical tube, case (a) becomes:

$$\text{Second moment of area } \mathbf{I} = \left\{\frac{\pi}{64}(d_o^4 - d_i^4) + r^2 \frac{\pi}{4}(d_o^2 - d_i^2)\right\} \times 2$$

Using a very modest value for r of 10 to simplify the arithmetic results in:

$$\text{Second moment of area } \mathbf{I} = 0.8158 \times 10^{-8} + 0.01123 \times 2$$
$$= \mathbf{0.02246\ m^4}$$

The difference between the two values is a factor of 2.75×10^6! Since F_N is directly proportional to I to the power of 0.5 the natural frequency of the horizontal tube is less than the natural frequency of the vertical tube by a factor of about 1660!

I went to bed thinking that I must have missed something but today I still can not see anything wrong with the above analysis.

After all that analysis for vertical U tubes the ones that would seem to most vulnerable to vibration problems must be horizontal U tubes. Most U tube bundles are in the vertical plane but when more than two tube side passes are used it is normal to use the bends in the horizontal plane. If TEMA does not take this into account it is unlikely that your computer program will either?

I am puzzled as to why this has not been noticed in the field in the same way that problems with the flow at 60° to a triangular pitch arrangement have been picked up. Perhaps problem have been noticed and solved because it is a lot easier to strap up U bends than work on the main section of the exchanger which has straight tube lengths.

Of course just a single support on a horizontal U tube bundle changes the whole system as far a bending is concerned as illustrated in figure 15.15 below.

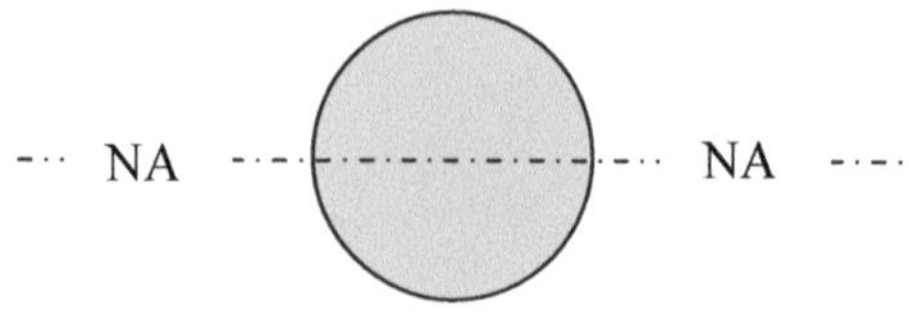

Figure 15.15

The second moment of area is no longer relevant for a single tube because each tube row is connected to the adjacent rows. Any bending calculation would therefore be for the bundle rather than for a single tube. The second moment of area for the bundle would be far higher than it would be for a single vertical tube.

The only potential problem is therefore for a bundle with horizontal tubes and no U bend support. Next time I design such a unit I must remember to specify a U bend support even if the computer program indicates that none is required.

EFFECT ON HEAT TRANSFER AND PRESSURE DROP

There have been reports both from the field and the laboratory of significant increases in shell side pressure drop when tubes vibrate. This is not too surprising.

What I have never read is any reports about is the effect of vibrations on heat transfer coefficients. I have some personal experience of this from my time in the academic world which may be worthwhile passing on.

I supervised a final year student project to investigate the effect of vibrations on heat transfer coefficients. The test rig had to be cheap and simple. A double pipe exchanger was used with steam at atmospheric pressure in the annulus and low pressure air passing through the tubes. A vibrator was attached to the centre of the tube via a rod which passed through a simple gland seal. The tube wall was assumed to be at 100°C so that air temperatures and pressures at inlet and outlet together with the flow rate were all that was necessary to calculate the heat transfer coefficient of the air inside the tubes.

Tests were carried out without any vibrations which correlated quite well with accepted equations. These tests were then repeated with the tube vibrating.

Before I had suggested this work as being suitable for a student project I had decided that a dimensional analysis with the usual variables used for convection plus the additional variables of the vibration frequency and amplitude offered the most promising approach. This resulted in two new dimensionless numbers. One was the frequency amplitude ratio and the other had exactly the same form as the normal Reynolds number with the characteristic length being the amplitude of the vibration.

The course the student was on was a sandwich degree which limited the time spent on the project to about five months. The analysis from the few tests carried out was very interesting but not too surprising. Where there would have been laminar flow without any vibrations the heat transfer coefficient increased by over 30% for specific vibration inputs.

The external examiners from other universities expressed surprise that I had only marked the project in the 2-1 bracket. They both agreed that it was the best one they had reviewed at Wolverhampton that year. One went so far as to say he had seen less academic content is some of the masters' projects he had recently reviewed. I explained that I had found it necessary to lead the student through the project at each stage with insufficient innovation from him and that was accepted. I was working with some lecturers who would have jumped at the chance for 'their' student to be awarded the project prize for the year. Bragging rights is the staff room were not my forte. When I awarded a first class mark it was because the student merited it.

I would have liked to have carried on with this line of research with further student projects and to have extended it to investigate the effect on very viscous liquids. However the storm clouds of 'educational progress' had been gathering for some time. The government had decided that there would be a massive expansion of higher education. For the educational administrators this meant 'bums on seats'. They thought that it was ridiculous that a few square meters of valuable space could be occupied by a single student. Our laboratories had to be 'rationalised'. The first battle had been lost and a few years latter so was the war. Engineering as we understood it closed down. The 'happy ending' of course is that there are now many more students leaving Wolverhampton University with degrees and higher degrees and some of them are eminently qualified to stack the shelves at Sainsbury. The down side is that real engineering is just part of its history.

TWO PHASE FLOW

Figure 15.16 below shows two extreme pictures of two phase flow on the shell side of a heat exchanger.

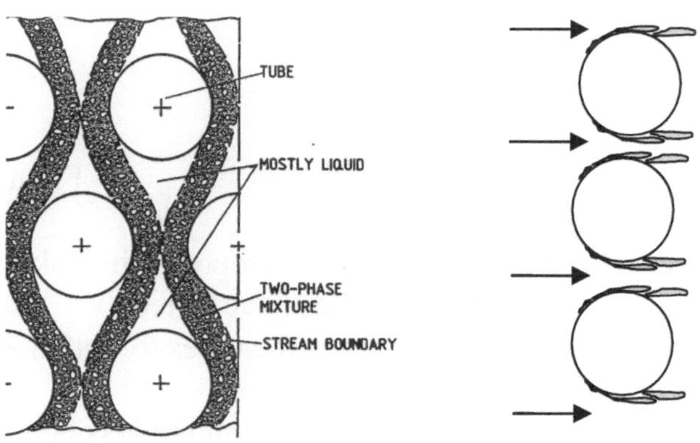

(a) (b) Figure 15.16

On the LHS, figure (a) there is very little vapour – low void fraction. On the RHS, figure (b) there is very little liquid –high void fraction.

In between these extremes there are many different possible flow patterns just as there were for flow inside tubes which were detailed in chapter 6 on flow maps. If predicting vibrations in single phase flow is a bit dodgy doing the same for two phase flow is even more difficult.

TUBE DAMAGE

Many tubes vibrate without causing any damage. When damage occurs one or more of the following mechanisms can normally be identified:

(a)**Fatigue** – This is due to repeated bending and is most likely to lead to failure at the face of the tubesheet. Brass and copper tubes are particularly susceptible.
(b) **Tube loosening** – Locked in tube end and tubesheet stresses may be relieved. This can weaken the tube/tubesheet joint causing the tubes to leak.
(c) **Tube weld cracking** – Cracks propagate from the root of the weld as illustrated in figure 15.17. This drawing is from Yokell – ref 9 which has an excellent section on vibration damage. This type of failure is more likely if the tubes have not been lightly rolled after welding.

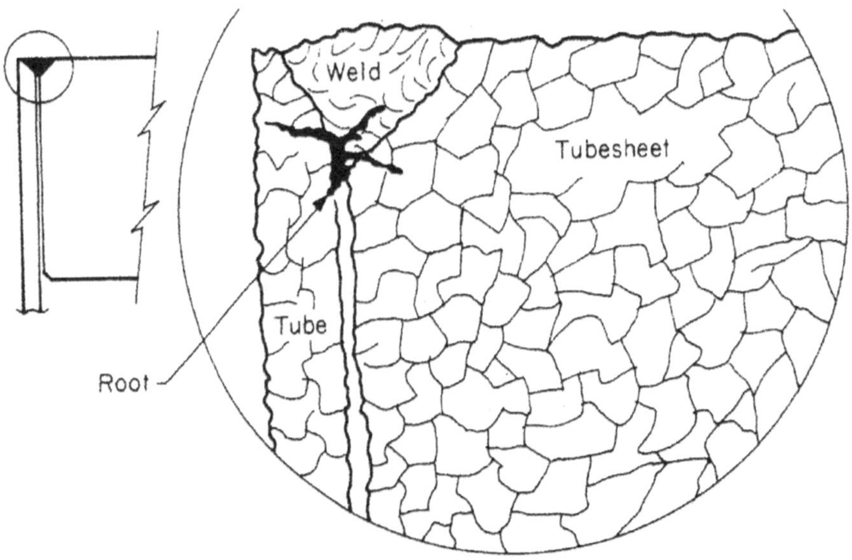

Figure 15.17

(d) **Tubesheet ligament cracking** – Cases have been reported in Yokell ref 9 where the tubesheets have cracked before the tubes have failed.
(e) **Collision damage** – Impact at mid-span causes the tubes to flatten and thin before failing.
(f) **Baffle damage** – Wear can take place due to vibration impacts between the tube and the baffle. The tube is much thinner than the baffle and will suffer the most damage. This form of failure can be very rapid if the baffles are made out of a harder material than the tubes. It is also more likely with high tube/baffle clearances.

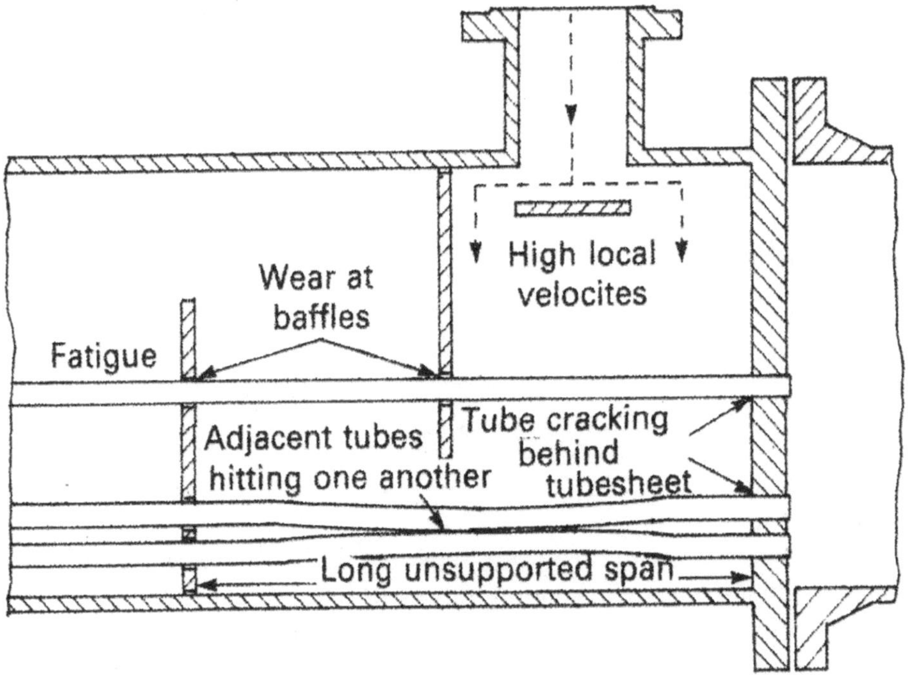

Figure 15.18

Figure 15.18 identifies the areas in a typical shell and tube unit which are susceptible to tube damage. It is taken from Ted Saunders book – ref 6. The reader should not forget that the high velocities around the impingement plate may be exciting vibrations at another section along the tube length.

COMPUTER PROGRAM OUTPUTS.

Thermal design engineers have little option but to trust the program they are using. Some of the early versions of the programs were very conservative in some respects. Over the years conservatism has been reduced as more data has become available. Some early programs had correlations which are no longer used such as the Thorngen Damage Numbers.

The current program I am most familiar with is the one by HTRI. I will try and highlight the more important features of their vibration output. The output has data for three positions in the bundle; inlet, centre and outlet.

Flow velocities.

There are four values of fluid velocity tabulated:

(a)The first of these is the window parallel velocity which is used to calculate the critical damage number limit and is unlikely to be the cause of problems.
(b)The next is the bundle cross flow velocity calculated from the B stream flow fraction between the tips of the baffles i.e. the B stream.
(c)Then follows the bundle/shell clearance velocity i.e. the C stream

(d) The last is the bundle cross flow velocity calculated from the cross flow area at the tip of the baffle i.e. the point at which the B stream has a maximum value.

Fluidelastic instability check.

The highest value of (b), (c) and (d) above is repeated and compared with the critical velocity. A * against the fluid velocity is shown if the ratio of the velocity/critical velocity is greater than 0.8. If the flow velocity is equal to or higher than the critical value vibrations problems are likely. If the ratio of the velocity/critical velocity is greater than 0.8 but less than 1 vibration problems are possible.

Tube vibration check

This section of the output starts with the vortex shedding ratio. Figure 15.12 earlier in this chapter illustrated that the vortex shedding ratio will reach a resonance state at a much lower fluid velocity than the velocity required to cause Fluidelastic problems. It is therefore not unusual for this item to be highlighted by a *. A * will be present whenever the vortex shedding ratio is above 0.5.

A vortex shedding ratio of above 0.5 indicates that there will be some vibrations but the magnitude may be insufficient to cause any problems. Below the vortex shedding ratio is a figure for the cross flow amplitude and another line for the gap between the tubes. If the amplitude due to cross flow is over half the gap between the tubes then the tubes will clash at the centre of the span as illustrated in figure 15.18 and there will be collision damage. Amplitudes below half the gap may be sufficient to cause fatigue damage over a period of time and it is necessary to keep well clear of half the gap between the tubes. The program states that vibrations are still probable for values between 0.25 and 0.5. For vibrations problems to be unlikely the value of the cross flow amplitude should be less than 0.1 of the gap between the tubes.

There is also a line for the parallel flow amplitude with the same restrictions and limitations but this is unlikely to be the source of any problems.

Other programs.

It has been a long time since I used TASC as a main design program and I am not familiar with the latest vibration data in the current ASPEN-TASK program. There are also a number of specialised vibration programs such as X-Vib. I have used these on a number of occasions but not sufficiently often to become familiar with them so that once again I have nothing to offer the reader.

Despite all the lengths we may go to in avoiding potential vibration problems it is still an area where there is still a lot of uncertainty and it is sensible to have a guarded note on the thermal data sheet such as “This unit has been checked for vibration problems and none are predicted.” Never use the term “vibration free”.

AVOIDING VIBRATION PROBLEMS

An experienced thermal design engineer will pick up a process data sheet and know within a couple of minutes if vibrations are likely to present problems and in extreme cases dominate the thermal design. At the other extreme a design may appear to be straightforward with no likelihood of vibration problems and after the first computer run a problem is identified which presents a challenge.

Some of the ways that can be used to avoid vibration problems are the same as the ones discussed in chapter 4 for utilising the shell side pressure drop, these variables are summarised below:

(a) **Baffle pitch** – change from single segmental to double segmental baffle pitch. Even though the pressure drop may be OK with single segmental baffles changing to double segmental could avoid vibration problems by reducing the maximum unsupported length.
(b) **Type of shell** – change from an E type shell to a J type shell. The benefits are the same as for (a). Changing to a type X shell or using a NTIW design can eliminate the chance of vibration problems but is often a high cost last resort solution.
(c) **Tube pitch** – increasing the tube pitch results in a reduction in unsupported length for a given pressure drop.

To these we can now add a further variable:

(d) **Tube pitch orientation** – this changes the cross flow velocity and the Strouhal number as discussed earlier in this chapter.

These measures alone are unlikely to produce that elusive optimum design. The rest of this section will be devoted to extra measures which can be used with them.

ADD EXTRA SUPPORTS

Adding extra supports to reduce the unsupported tube length can have a very significant effect on potential vibration problems since $f_N \alpha 1/L^2$. The trick is to add supports which do not have too much negative impact on pressure drop or heat transfer. Figure 15.19 shows a gas cooler with horizontal baffles. The tubes are well held between the end baffles but there is a very long inlet space, (S type head) and the outlet space is about twice the baffle pitch.

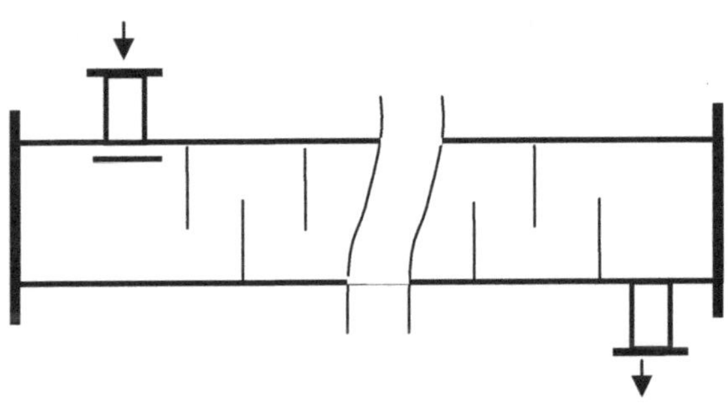

Figure15.19

The vulnerable areas are around the inlet and outlet nozzles and the long unsupported tube lengths. Figure 15.20 shows where supports could be added to reduce the risk of vibration problems.

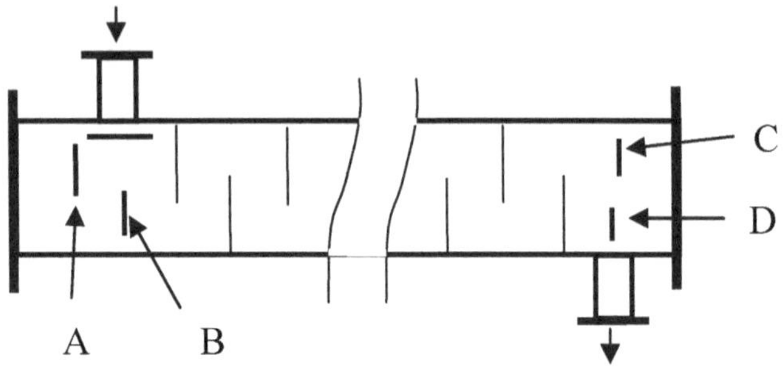

Figure15.20

I have added two supports in the inlet space. Support A would start about 3 mm above the tubes in the top row and at the bottom it would be at the same level as the cut of the first baffle. Support B would pick up the first tube in the baffle cut and end 3 mm below the bottom tube row. This arrangement should not have too much effect on the pressure drop and it may even be argued that it may improve the heat transfer by increasing the flow across the face of the tubesheet.

In the outlet space I have added a short support D over the centre line of the outlet nozzle. This may only pick up say 5 or 6 rows of tubes where the fluid is accelerating towards the outlet nozzle. Support C picks up the top row of tubes and extends to pick up the first row of tubes supported by the first baffle. C is not necessarily in the same vertical plane as D. There is a flow path above C and between C and D.

There is no need to have two rows overlap for support baffles. This is a sensible requirement for main baffles to keep the bundle together but for additional supports it is not necessary.

The reader should not get carried away putting supports in every design. It is a matter of engineering judgement when and where supports are required. It might be argued that when extra supports have been added a vibration program should be run to check out the revised support system and that is one of the reasons for developing these programs. For most cases it is sufficient to identify the length that the computer has used to predict a potential problem and then modify this on the output together with the new natural frequency using $f_N \alpha 1/L^2$.

Figure 15.21 below is the 12J unit – figure 2.41 which was discussed in chapter 2. The tubes are well held at inlet but there are long tube spans in the outlet spaces where vibration problems are predicted.

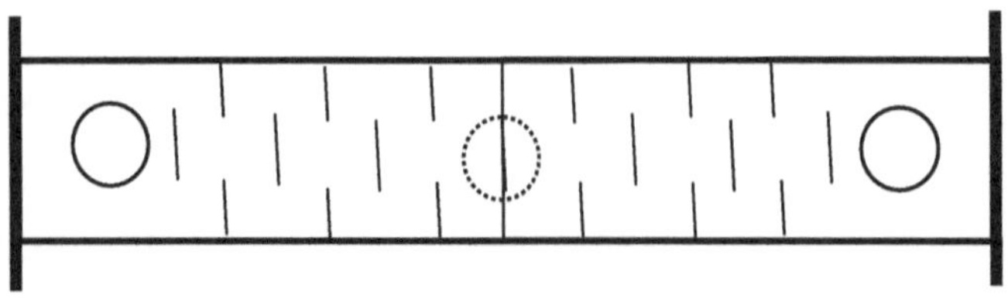

Figure 15.21

The problems can be overcome by fitting an extra support baffles in the exit spaces as illustrated below.

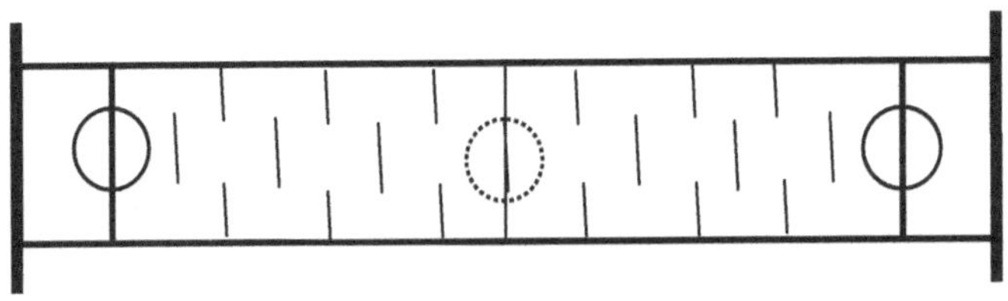

Figure 15.22

The shape of the support baffle is shown in figure 15.23 below. It is basically two outer baffles joined together at the top.

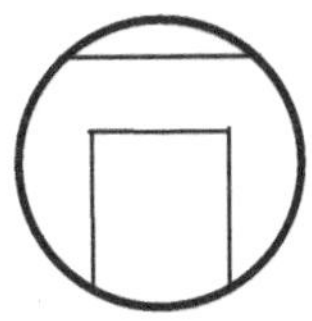

Figure 15.23

I have seen this arrangement used without joining the two outer baffles. Supporting the tubes under the outlet nozzle where there are high velocities can only be of benefit and in this case will also help to divert some of the fluid towards the face of the tubesheet.

The situation is the same for a 21J unit. I now hardly ever specify a double segmental design without one of these supports.

Unlike horizontally cut baffles for standard vertically cut baffles there is no easy way to add extra supports without unacceptable interference in the fluid flow paths. I would however recommend one simple support under inlet nozzles which is illustrated in figure 15.24 below.

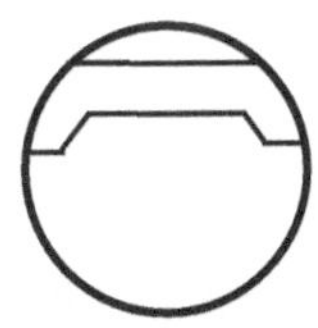

Figure 15.24

This design of this support should pick up the top few rows of tubes together with the outer tubes of some of the rows below it. It should be positioned at or near the centre line of the nozzle and when sensibly proportioned should not have too much effect on the pressure drop.

Analysing reports of tube failures has made me very aware of the vulnerability of the outer one or two tubes in the rows at the top of the bundle. Problems have arisen from high C stream velocities at the top of the bundle which programs may not pick up. This is particularly important for tubes close to impingement plates. (Which is a good argument in favour of using rods rather than plates?) Tubes close to sealing strips may also be vulnerable.

A short support 'without the ears' over the outlet nozzle as shown in figure 15.19 can also be beneficial. Of course ears could be required to pick up tie rods and will do no harm.

I would always use these measures for units that are only just OK as far as vibrations are concerned. (All the stars have disappeared but I know they will return for the revamp!) On reflection I think it might be a good idea to use these simple supports as a standard good practice.

REDUCE THE BUNDLE ENTRANCE VELOCITY

Some condensers and partial condensers may have a very large volume flow rate at in inlet and low volume flow rate at outlet. The potential for vibration problems are all in the inlet space. For such cases opening up the inlet space may provide the optimum solution. This has already been discussed with reference to condensing steam in chapter 5 but can be used with other condensers. Figures 15.25 and 15.26 below show a unit with an enlarged inlet space and the type of baffles which can be used.

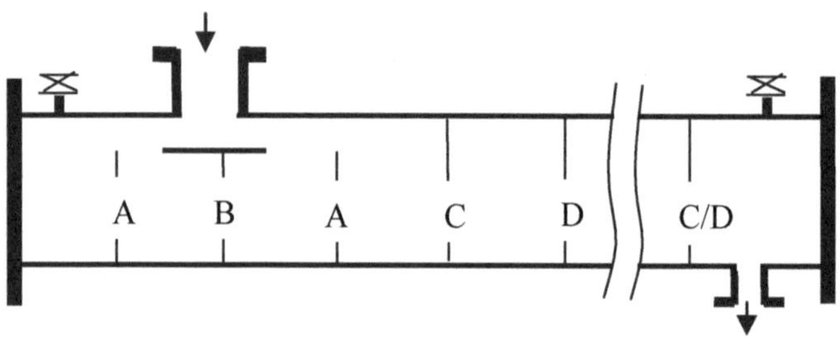

Figure 15.25

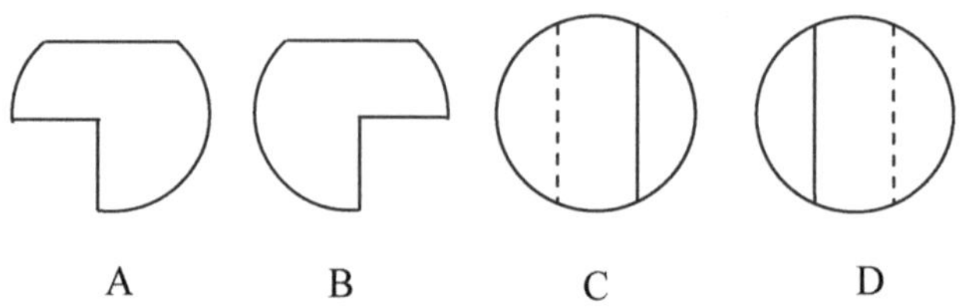

Figure 15.26

This arrangement is only suitable for units where there will be wet wall condensation as soon at the vapour comes into contact with any of the tubes and the overall heat load is mainly due to condensation.

The length of the inlet space becomes a variable together with the number of type A and B baffles in this space and the number of tube rows fully covered by them. The first tubes the vapour crosses are well held and the velocity kept well below the values which could cause vibration problems. The rest of the exchanger has conventional vertically cut baffles and most of the allowable pressure drop can be used up in this section where there will be a higher percentage of vapour cooling.

The inlet space can be run as a cross flow unit and the load matched up with the remaining duty in the rest of the unit. A bit of fudging will be required to achieve a balance. The error in doing this will be quite small if the shell side coefficient in this region results in a thermal resistance that is small compared with the total thermal resistance.

INLET & OUTLET DISTRIBUTORS

These are normally a relatively expensive way of getting fluids into and out of tube bundles to avoid vibration problems. They have been covered in the first section of special mechanical features in chapter 11.

U TUBE SUPPORTS

Potential problems from U tubes vibrating are normally only of concern with large diameter bundles except in cases where the U tubes are horizontal and the natural frequency will be very low. For these cases there should always be a central support for any size of bundle.

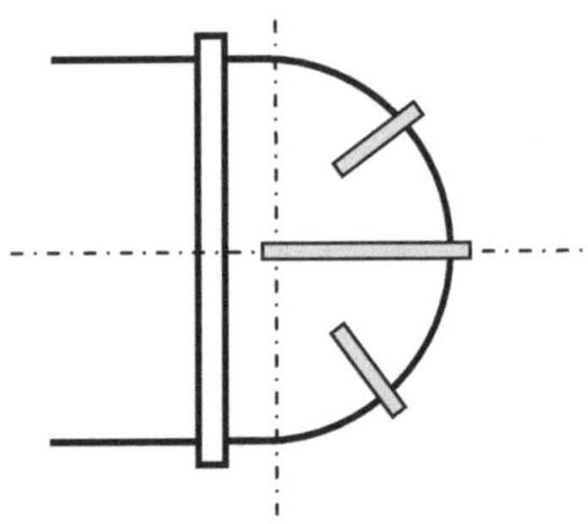

Figure 15.27

For large diameter U tube bundles with normal vertical bends vibration problems can be avoided by simply increasing the number of supports. Figure 15.26 shows a support at the centre line of the bundle. This is normally sufficient for most bundles and although I have shown the support picking up all the tubes it is not really necessary to pick up the short inner tubes. For very large bundles a central support alone may not be sufficient and additional supports at 45° may be required. These will normally only pick up the long outer rows. The detailed construction of U bend supports will depend on the tube pitch and orientation and can be left to the mechanical engineers.

If a design is only for a feed estimate it saves time to pass the buck on and have a simple note on the thermal data sheet such as "The U tubes should be well supported and the maximum length of any bend should not exceed 80% of twice the baffle pitch."

'ROD BAFFLE TYPE' SUPPORTS

The rod baffle exchanger was described in chapter 2 and figure 15.28 is repeated from that section. This type of exchanger was developed specifically to overcome vibration problems and rods instead of segmental baffles are used throughout the unit.

Rods can also be used in a conventionally baffled exchanger as an anti-vibration measure. To ensure that the tubes are well held four sets of rods are required. Two sets of rods are arranged in the horizontal plane and two in the vertical plane.

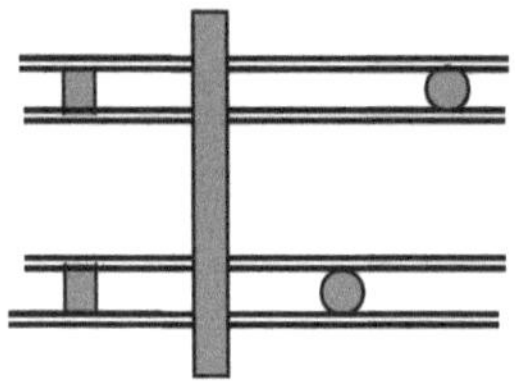

Figure15.28

A single set of four rods can then be specified as a support in the inlet and/or outlet space. All the rods are attached to a cylindrical ring with an inner diameter equal to the bundle OTL. This ring can then be attached to adjacent supports/baffles to ensure no longitudinal movement. A 'rod baffle type' support will pick up all the tubes as illustrated in figure 15.29 below.

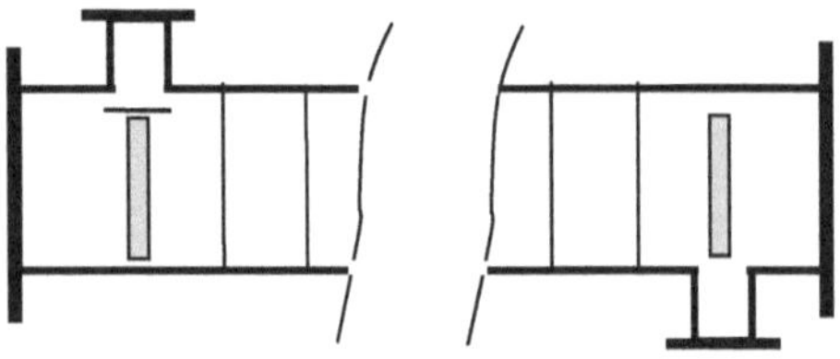

Figure 15.29

Provided there is a good space left between each row of rods this type of support will use up very little pressure drop. If in doubt the reader can do as I once did when a rod baffle option became available as a computer program and run a unit as a rod baffled exchanger and then work out the pressure drop across a set of four rods. (I was using this arrangement well before then – justifying the low pressure drop with some simple hand calculations.)

Although I have used this arrangement with new designs the main application has been on revamp work. It has of course been used for many years as a retrofit measure for bundles with vibration problems.

An obvious limitation of course is that this type of support can only be used with a square pitch arrangement – 90° or 45° with clear cleaning lanes.

EMbaffle SUPPORTS

This type of support can be fitted in between any type of conventional baffles. Chapter 2 has more details. They are a relatively recent Shell Global proprietary design. They do a similar job to the 'rod baffle type' of support providing a full support at the expense of a small pressure drop. I can see their use increasing in the future if current programs adapt to using them and of course the very big **but** – the price is right!

Whilst researching this chapter I came across a very interesting section on Holtec baffles in Yokell – ref 9. There is a photograph which clearly shows their construction. This immediately struck me as showing baffle/supports which are very similar to the EMbaffle. They are also a proprietary item and predate the EMbaffle. The lawyers and patent officers must have had a field day with this one! Or maybe Shell bought them out? In which case the EMbaffle is not so new?

NTIW & CROSS FLOW UNITS

When all else fails we always finish up with one of these types of design. By adding intermediate supports they can be vibration free – except of course for acoustic problems which will be covered next.

I have commented on these designs many times elsewhere but I will repeat one adage that I would recommend to the reader: "When you have added just sufficient intermediate supports to turn out all the stars – add one more for the revamp."

ACOUSTIC VIBRATIONS

Introduction

I have deliberately kept this as a separate topic although it is included in the same vibration output from a computer program. Although any type of vibration is instigated by the fluid flow it is the natural frequency of the shell rather than the tubes which needs to be excited. Acoustic vibrations are also very unlikely to lead to any tube failures but tolerance to noise is decreasing all the time and there are now very few sites which would allow 'singing exchangers'.

The topic is just as 'hairy' and unpredictable as tube vibrations. I can recall sitting with the head of the KTI exchanger section during on of my contracts in the Netherlands when he told me this story. He had designed a unit which the computer program predicted would have severe acoustic problems. He had decided to fit two acoustic baffles with the recommended spacing. When the unit was installed and commissioned there was an unacceptable noise!

The only field modification he could think of was to remove the two acoustic baffles since this measure was certain to change something! The acoustic baffles were removed and when the unit was restarted there was no noise!

SONIC VELOCITY IN GASES u_{sonic}

The sonic velocity through a perfect gas can be calculated from either of the equations below:

$$u_{sonic} = (\gamma R T_A)^{0.5} \quad [15.11]$$

$$u_{sonic} = (\gamma p/\rho)^{0.5} \quad [15.12]$$

T_A is the absolute temperature in K, ρ is the density in kg/m^3, p is the pressure in N/m^2 and R is the specific gas constant for the gas in J/kgK. (To transpose from one equation to the other use $\rho = p/RT$). γ is the ratio of the specific heats at constant pressure and constant volume and is therefore dimensionless. The reader should note the simplicity of the equations compared with the equivalent ones in US Customary units which confusingly contain the gravitational constant which has nothing to do with speed of sound.

For a perfect gas

$$R = \frac{Ro}{M} = \frac{8314}{M} \quad [15.13]$$

$$R = C_p - C_v \quad [15.14]$$

And

$$\gamma = \frac{C_p}{C_v} \quad [15.15]$$

Ro is the universal gas constant in J/kmol K and M is the molar mass in kg/kmol. (M is basically a relative mass. When I was a student the mass of any molecule was compared with an atom of hydrogen having a value of exactly 1. This presented chemists with some problems which are beyond the scope of this book. At the 1971 General conference of Weights and Measures it was decided that Carbon having a value of exactly 12 would be used as a reference. The value for the hydrogen atom had to be adjusted to its present value of 1.008.)

Example 15.1

Calculate the value of γ and the sonic velocity at 450 K (177°C) for (a) Hydrogen and (b) Air.

Solution

The specific heats at constant pressure and the molecular masses can be found from tables such as ref 15. The specific heats are 14.5 kJ/kgK for hydrogen and 1.005 kJ/kgK for air. The molecular masses are approximately 2 for hydrogen and 29 for air.

For hydrogen

$$R = \frac{Ro}{M} = \frac{8314}{2} = 4157 \text{ J/kgK} = 4.157 \text{ kJ/kgK}$$

$$C_v = C_p - R = 14.5 - 4.157 = 10.34 \text{ kJ/kgK}$$

$$\gamma = \frac{C_p}{C_v} = \frac{14.5}{10.34} = \mathbf{1.4}$$

$$u_{sonic} = (\gamma R T_A)^{0.5} = (1.4 \times 4157 \times 450)^{0.5} = \mathbf{1617\ m/s}$$

For air

$$R = \frac{Ro}{M} = \frac{8314}{29} = 287 \text{ J/kgK} = 0.287 \text{ kJ/kgK}$$

$$C_v = C_p - R = 1.005 - 287 = 0.718 \text{ kJ/kgK}$$

$$\gamma = \frac{C_p}{C_v} = \frac{1.005}{0.718} = \mathbf{1.4}$$

$$u_{sonic} = (\gamma R T_A)^{0.5} = (1.4 \times 287 \times 450)^{0.5} = \mathbf{425\ m/s}$$

It is convenient to work in kJ/kgK when calculating γ but J/kgK must be used in the equation for calculating the velocity of sound.

It is important to note that although the specific heat values for the two gases considered are drastically different the values of γ are the same. This is because the main constituent gases of air – N_2 and O_2 are diatomic gases the same as H_2.

The sonic velocities differ by a factor of about 4 because of the different values of the specific gas constant. A value for R is not something I can visualise so that to explain the difference in sonic velocity it is necessary to switch to thinking in terms of density knowing that $R \alpha 1/\rho$. The density of the air in this example is 4157/287 = 14.5 times greater than the density of the hydrogen. Changes in the density of atmospheric air are sufficient to change the Mach number for aircraft flying at high altitude compared with sea level.

There as always been the option to input a value of γ in the HTRI programs but most people I worked with did not take this option and accepted the default value of 1.4 for air assuming it to be 'safe' – **it was not**!

I was working on an exchanger with a potential acoustic vibration problem and carefully read through the relevant sections of the manual. The default to the value for air seemed strange since cases with air on the shell side of a shell and tube unit are very rare in this industry. Experimental work using air is of course very common and I suspect that this is the reason that it was chosen as the default gas. The only other comment I remember is that consideration needed to be given to unusual gases such as hydrogen! My interest had been aroused so I carried out the above calculation and found that they were exactly the same.

From my time as a lecturer the value γ for air was imprinted on my mind as well as the number of the house I am living in. The only other value I remembered was 1.67 for the monatomic gases helium and argon. This was because the value of γ together with the pressure ratio determined the thermal efficiency of the Joule cycle which is the basic cycle for gas turbines. Closed gas turbine cycles invariably use helium as the working fluid mainly because of the high value of γ.

In petrochemicals I was of course working mainly with hydrocarbons – so just what was a typical value of γ for a hydrocarbon? I realised that I had no idea so I repeated the above calculations for a number of hydrocarbons which are summarised in table 15.6 below.

Gas	M	C_p kJ/kgK	R kJ/kgK	C_v kJ/kgK	γ	u_{sonic} m/s
H_2	2	14.5	4.2	10.3	1.4	1617
Air	29	1.01	0.29	0.72	1.4	425
Steam	18	1.93	0.46	1.46	1.32	523
CH_4	16	2.7	0.52	2.18	1.24	572
C_2H_4	28	2.06	0.297	1.77	1.17	395
C_6H_6	78	1.6	0.11	1.49	1.07	227
C_8H_{18}	114	2.4	0.073	2.33	1.03	184

Table 15.6

It is clear that for hydrocarbons the value of γ decreases as the molecular mass increases. I have just briefly looked at trying to find an equation relating M and γ – without success.

Whilst writing this book I have added a column to include sonic velocities. These are all for a temperature of 450 K.

What was clear is that for a heavy hydrocarbon gas using a default value for γ of 1.4 can result in an unsafe error in the calculation of the sonic velocity of over 15%.

As far as designing heat exchangers is concerned it is worth remembering that **the higher the density of the gas involved the lower the sonic velocity will be and hence the possibility of acoustic vibration problems increase.**

I have included steam in the above table because it is a common fluid in thermal design work. The reader should not take too much notice of the value I have calculated. It might be OK for high vacuum work such as a surface condenser when the behaviour of steam approaches that of a perfect gas. For the pressures we are likely to use it does not. If it did we would not need steam tables. For most steam applications the steam is condensing and there are unlikely to be any acoustic problems. A steam desuperheater is the most likely case where there could be a problem. I would recommend using a safer value of 1.25 for desuperheaters.

I was working at Foster Wheeler at the time I did the above work on γ. I passed the results of my findings around the thermal group, (still a department in those days!) for comment. No one could find any fault with it so it became generally used by other members of the group. Some of them contacted HTRI on occasions between 2001 and 2006 and received a variety of replies. One included the sentence "1.4 is a common default value often used throughout the literature and we do not recommend changing it". A later one included a thank you for reporting the error in HTRI's "common understanding" of which gases will (and will not) violate our current default assumption of $\gamma = 1.4$ and suggesting improvements.

The default was eventually changed to 1.2 which to me still seems strange. It only needs the molar masses to be added to information already available on a typical input to get the calculation correct and allow for the differences in the size of the molecules in the gas. The changes required to the program must be quite simple – no iteration is required. I would not go so far as to say I could do it but I am sure most young readers could. Table P3.3 – section 8-20 in TEMA has values of γ for a selection of gases.

TUBE BANKS

Parker in ref 35 found that the velocity of sound across a tube bank is reduced by a factor of about 20% for tubes on a triangular pitch and 18% for tubes on a square pitch. I will assume that any necessary correction has been carried out rather than complicating the issue by introducing two values for the sonic velocity.

SONIC VELOCITY IN LIQUIDS

The velocity of sound in liquids can also be calculated from a very simple equation:

$$u_{sonic} = (\kappa/\rho)^{0.5} \qquad [15.16]$$

Where κ is the compressibility in N/m^2 and ρ the density in kg/m3. The only problem with this equation is that it is not too easy to find values for the compressibility. Fortunately the sonic velocity does not vary too much for hydrocarbon liquids and a typical value of about 1300 m/s is sufficient to get a feel for relative values.

For the lightest molecule in the table the ratio of the sonic velocity of the liquid to the gas is about 2.3 at the other extreme for the heaviest molecule the ratio is about 7. A typical ratio for water is about 4.

For liquids the velocity of sound is normally too high to cause acoustic problems in heat exchangers. For two phase flow at low void fractions with plenty of liquid around there should also be no problems. As the void fraction increases and the mixture approaches becoming all vapour there may be problems. Just how much liquid is required to avoid acoustic problems is difficult to assess. The equations used to calculate the velocity of sound in two phase flow require a sonic velocity value for the liquid as well as for the vapour.

There are a number of lengthy complex equations to calculate the sonic velocity in two phase flow. They include the mass or void fractions together with the density and sonic velocity of each phase. I have no comments to make on these equations so I can not see any point in reproducing one here. The reader can check up what is in the program they are using by reading the manual.

NATURAL FREQUENCY f_N

The natural frequency in acoustics must have been well researched long before calculations were needed for heat exchangers. Any musical instrument which is based on 'wind' rather than 'string' is based on acoustic vibrations in some form of chamber. The nearest comparison with a heat exchanger is an organ pipe. The following equation applies to such situations:

$$f_N = \frac{m\ u_{sonic}}{2L} \qquad [15.17]$$

Where f_N is the natural frequency of the system with units of cycles/second - Hz; m is the mode number and L is a characteristic dimension.

In chapter 3 on single phase heat transfer outside tubes a simple starting point and the way a lot of basic research is carried out is to examine what takes place in a rectangular duct before progressing to the more complex geometry of a cylindrical shell. This is also the case for studying acoustics.

Figure 15.30 below is a section across a rectangular duct – a wind tunnel.

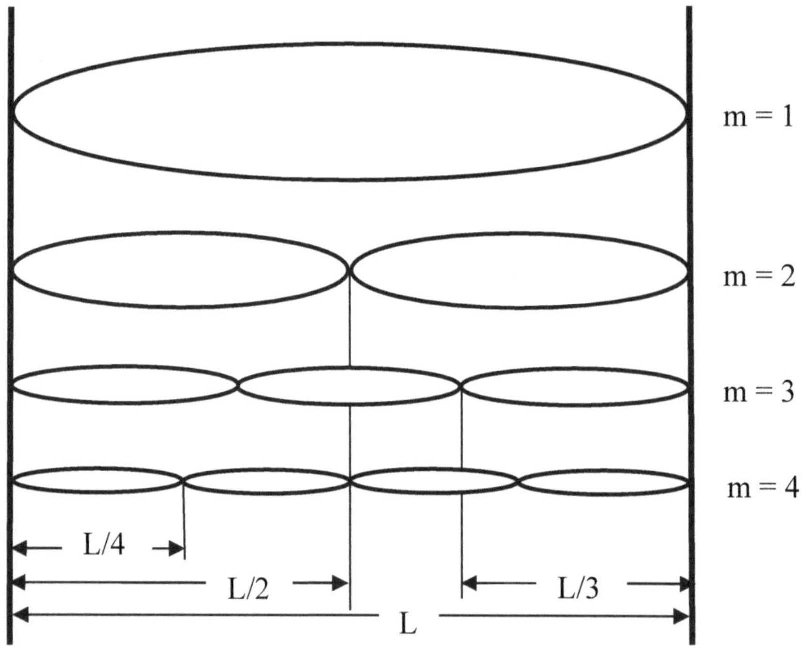

Figure 15.30

Experimental tests have shown that a noise can be produced by passing a flow of air through a wind tunnel. At low air velocities there is no noise. As the velocity of the air is increased a critical value will be reached where there is a very audible sound. As the velocity is increased above this critical value the noise will die away. A further increase in velocity will locate a second critical velocity where a second sound will be heard which is less audible than the first one. Increasing the velocity again will silence this noise. In theory this could carry on indefinitely with successive critical velocities producing less and less sound.

In a wind tunnel the characteristic dimension is the width of the tunnel. The first acoustic mode is half a complete sine wave. This is termed the fundamental natural frequency. The wavelength is the distance between the two minimum values and the amplitude is the maximum deviation from the minimum value. The higher modes occur at exact fractions of the characteristic length as shown on the diagram.

In an actual test the results could be a bit more complicated unless the duct had a square cross section because there would also be a series of critical values and noises in a plane at 90° (into the paper) where the characteristic dimension would be the depth of the duct. For a none square duct it would be easy to identify the first mode but for the higher modes it could get confusing.

CIRCULAR DUCTS

For a circular duct such as the shell of a heat exchanger the tangents at either end of a diagonal can be thought of as a rectangular duct with very short sides. Figure 15.31 below shows a section through a heat exchanger shell and the acoustic modes across the horizontal centre line. The direction of the fluid flow is downwards in the vertical plane which is the normal flow direction for a cross flow unit.

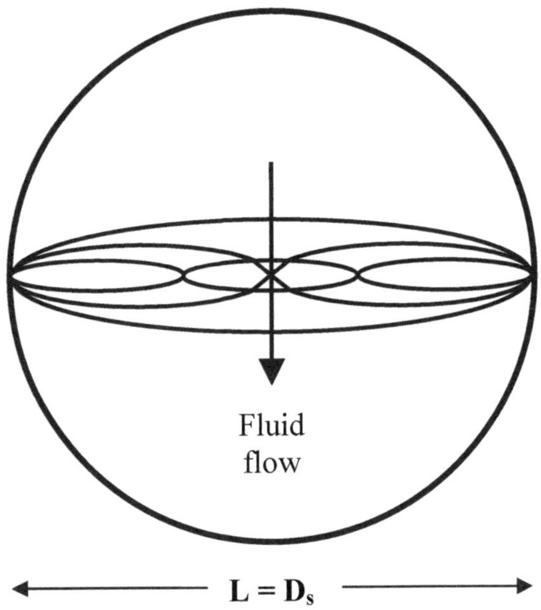

Figure 15.31

I have drawn this diagram by superimposing the waves from figure 15.30. I have omitted the fourth mode to avoid confusion and in reality we are unlikely to be concerned with any modes beyond the first three. It is clear that the amplitude of the mode decreases as the number of the mode increases.

DERESONATING/ACOUSTIC BAFFLES

Acoustic or deresonating baffles are installed in tube banks to 'detune' the acoustic system. The object is to change the natural frequencies of the system by changing the characteristic length in equation 15.17. Figure 15.32 shows the position of two acoustic baffles.

There is one basic rule to establishing the position of an acoustic baffle and that is to avoid any of the node points which would result in a new system where the length between the baffle and the shell would be a simple fraction of the shell diameter. For example if only one baffle was required and it was fitted on the vertical centre line it would be in the ideal place to avoid the first mode vibration but in exactly the right place to produce a second mode vibration.

The dimensions on the diagram have been put forward as the ideal positions to avoid the nodes of any higher modes by HTRI. In a real tube layout it is not practical to achieve these exact dimensions. I would recommend that the 0.45 D_s should be taken to be a maximum value. Just see where this is on the layout you are looking at and select the nearest row of tubes to eliminate and replace with a baffle that gives a value lower than 0.45 D_s.

If a second acoustic baffle is required 0.18 D_s should be taken to be a minimum value. Just see where this is on the layout you are looking at and select the nearest row of tubes to eliminate and replace with a baffle that gives a higher value than 0.18 D_s.

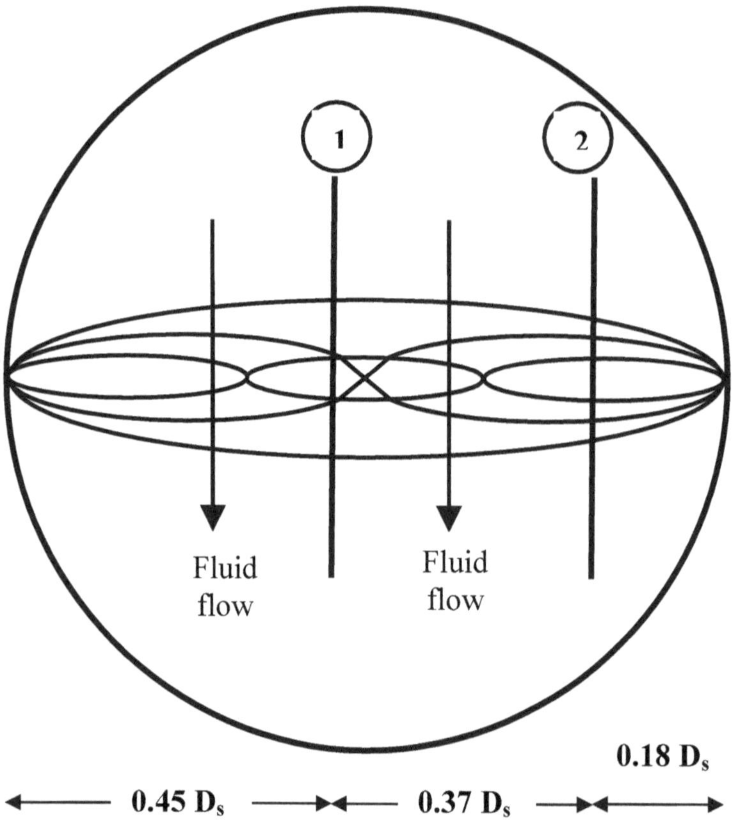

Figure 15.32

I do not know how the HTRI figures of 0.45 and 0.18 were determined. I do know that at one time HTFS recommended using 1/3 and 2/3 of the shell diameter. This was presumably based on the third mode not being considered to be important and maximised the chances of avoiding the second mode. Assuming the third mode may be important my simple logic would have aimed to position the baffle at the midpoint between 1/3 and 1/2 resulting in 5/12 or 0.42 when only one baffle was required.

Deresonating baffle height.

All the above analysis has assumed a standing wave across a diameter. Barrington in ref 31 and 32 has postulated that first mode standing waves can be established around the circumference of the shell. These depend on the tube pitch and figure 15.32 shows the waves for a square and a rotated square pitch.

The length of the standing waves in this case is 0.707 D_s. A similar diagram can be drawn for a triangular and a rotated triangular pitch which produce a standing wave of 0.866 Ds.

If this is possible is it not also possible to have a standing wave across a chord of the shell outside of the bundle? This is where it all gets very hairy and I do not pretend to fully understand it. I take the very pragmatic view that if there is any possibility of problems which can be avoided by a simple measure at a negligible cost then they must be taken.

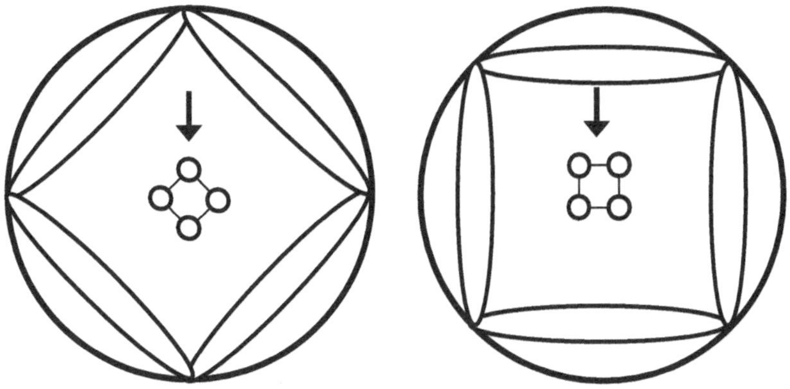

Figure 15.33

The simple measure in this situation is to **extend the acoustic baffles beyond the bundle as far as possible without interfering with the fluid flow**. To ensure that there is minimal interference with the fluid flow the minimum gap between the top and bottom of the baffle and the shell should be compared with the effective diameter of the section of the bundle that the fluid will flow across. Calculating effective diameters was fully explained towards the end of chapter 4. Particular care must be taken when there are two baffles and one of them is at about 0.18 D_s as illustrated in figure 15.32. There is of course no reason why the two baffles need to be the same height.

Example 15.1 - a practical design

Figure 15.34 is taken from a job I did in 1995. It is a NTIW design, the shell has an inside diameter of 1270 mm and the tubes are 25.4 mm on a 36 mm, 30^0 triangular pitch. I was working at KTI and they were using the HTFS program hence the positioning of the acoustic baffles at approximately 1/3 and 2/3 of the shell inside diameter. The layout has been drawn on OPTU with hand modifications. The crosses indicate the positions of two rows of impingement rods. These not only protect the tubes but help to distribute the fluid evenly across the bundle.

The baffles have been extended as far as possible beyond the bundle without interfering with the fluid flow. The bundle is a U tube so there was no loss of tubes for positioning some of the tie rods. In this design I managed to support the acoustic baffles without losing any tubes.

I have just glanced through several pages of hand calculations. They include the calculation for γ – fifteen years ago! There was also a sufficient change in temperature to justify calculating different values for the sonic velocity at inlet and outlet.

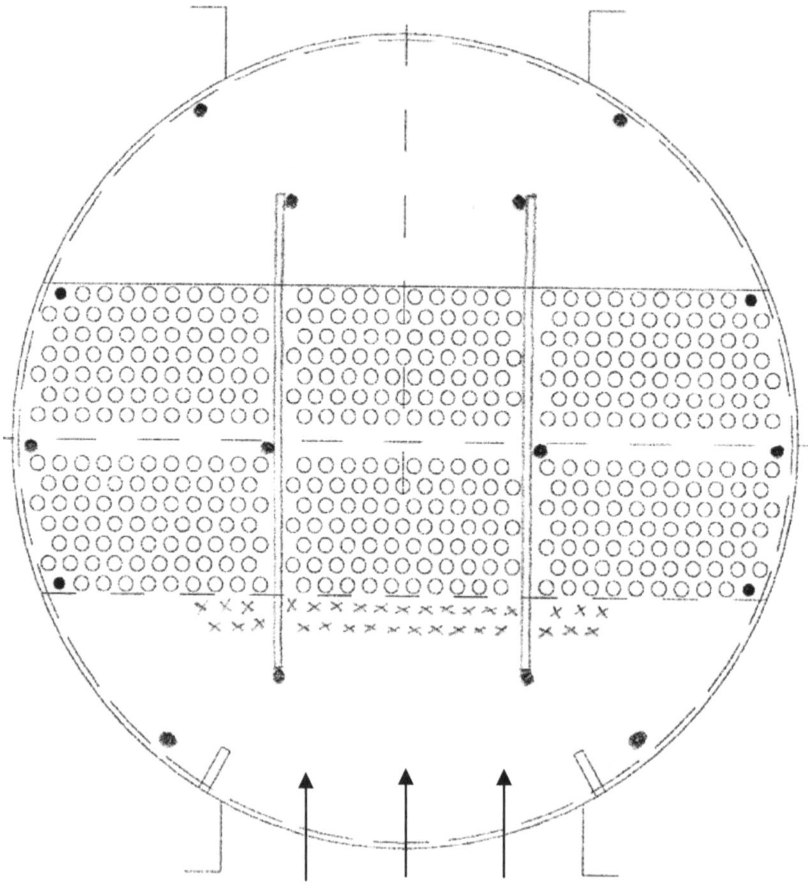

Figure 15.34

Thickness Length & Support

The only equation I can find for calculating the thickness of a deresonating baffle, t_b is suggested by HTRI and is as follows:

$$t_b >> \frac{\rho_F h_b}{\rho_M} \qquad [15.18]$$

Where ρ_F and ρ_M are the densities of the fluid and baffle metal respectively and h_b is the height of the baffle.

The use of the symbol >> is interesting yet quite clear, it means very much greater than – but how much is very much? Blevins in ref 33 pointed out that for a deresonating baffle to be effective it must be capable of reflecting a sound wave rather than just passing it on. This implies that it should be significantly more substantial that the fluid. (Unlike the skin of a drum which is designed to vibrate.) I will explore this further by means of an example.

Example 15.2

The fluid on the shell side of an exchanger is a gas at 10 bar pressure and 450 K temperature. To cure a potential acoustic vibration problem a deresonating baffle with a height of 1.2 m is to be fitted. The baffle will be made from steel with a density of 7850 kg/m^3. Estimate a minimum thickness of the baffle if the mean molecular mass of the fluid is:

(a) 28
(b) 78

Solution

Taking values of R from table 15.6.

For fluid (a)

$$\rho_F = \frac{p}{RT} = \frac{10^6}{297 \times 450} = 7.48 \text{ kg/m}^3$$

$$t_b >> \frac{\rho_F h_b}{\rho_M} = \frac{7.48 \times 1.2}{7850} = \mathbf{1.14 \text{ mm}}$$

For fluid (b)

$$\rho_F = \frac{p}{RT} = \frac{10^6}{110 \times 450} = 20.2 \text{ kg/m}^3$$

$$t_b >> \frac{\rho_F h_b}{\rho_M} = \frac{7.48 \times 1.2}{7850} = \mathbf{3.09 \text{ mm}}$$

Decision time – how much >> do we make the baffle? Fortunately in this example TEMA comes to the rescue for part of the problem. **The minimum thickness of any deresonating baffle must be 5 mm (3/16 inch).** This should be plenty for fluid (a), but is it >> for fluid (b)? I would probably increase it to 10 mm on a Monday and 12.7 mm on a Friday it is that type of decision.

The thickness of the baffle is directly proportional to the density of the shell side fluid. I selected the mean values of the relative molecular masses to cover a reasonable range and the gases/vapours are not necessarily pure ethane or benzene. For two phase flow the density would be the homogeneous mixed density which could result in much higher values and hence a thicker baffle.

Inserting an acoustic baffle can change the effective diameter for cross flow and hence the heat transfer and pressure drop. The easy way out of positioning a deresonating baffle is simply to remove a vertical row of tubes and just allow for this in the tube count. With some tube arrangements this is not sufficient. Figure 15.35 shows an inline square pitch arrangement before and after inserting a deresonating baffle by just removing a row of tubes.

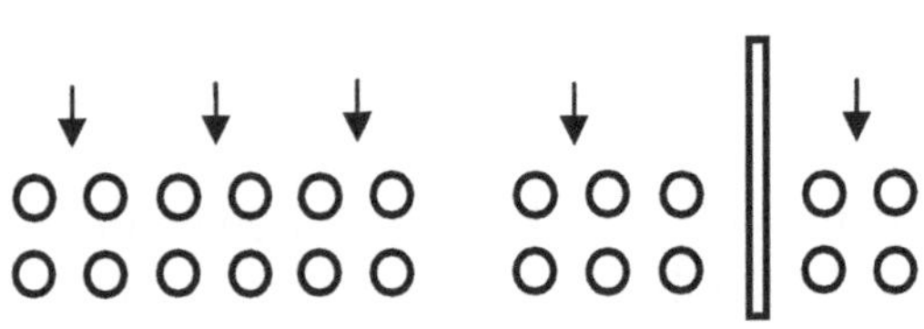

(a) (b

Figure 15.35

Consider a common arrangement of 19 mm tubes on a 25.4 mm pitch. The width of the flow path between three vertical rows of tubes is twice the gap between the tubes – 12.8 mm. After fitting a 5 mm acoustic baffle the gap between the same tubes becomes: 2 x 25.4 – 19 – 5 = 22.6 mm. This is sufficient to create a short ‘F’ stream by pass the same as a vertical pass partition lane would. The ideal way to overcome this would be to add dummy tube part sections as illustrated in figure 15.36.

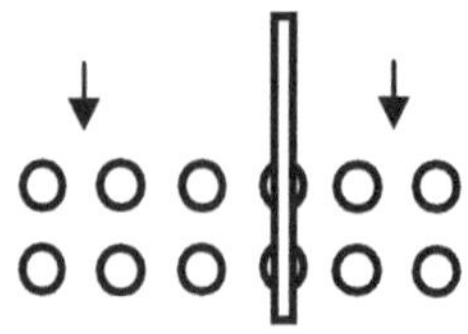

Figure 15.36

I have never used this arrangement – to be honest I have only just thought of it! It would be an interesting one to throw at some of my mechanical friends with the usual questions: Any problems? And how much would it cost?

When faced with this problem it is easy to go back on the earlier dodgy decision about the thickness of the baffle and make it thicker. Once again the easy decision I have known many people take is to make the baffle the same thickness as the tube outside diameter. Not too bad with 19 mm tubes but 25.4 mm tubes?

Even making the deresonating baffle the same thickness as the tube outside diameter does not leave a gap that would have the same effect on pressure drop and heat transfer as across the rest of the tube bank. As explained in chapter 2 on blocking pass partition lanes there is a lower resistance to fluid flow between a flat surface and a row of tubes than between two rows of tubes due to the lack of turbulence created by the flat plate.

The ideal gap between a row of tubes and a flat plate lies somewhere between the tube/tube clearance and half this value. I would recommend specifying a distance between the tube rows adjacent to the deresonating baffle to reflect this as shown in figure 15.37 below,

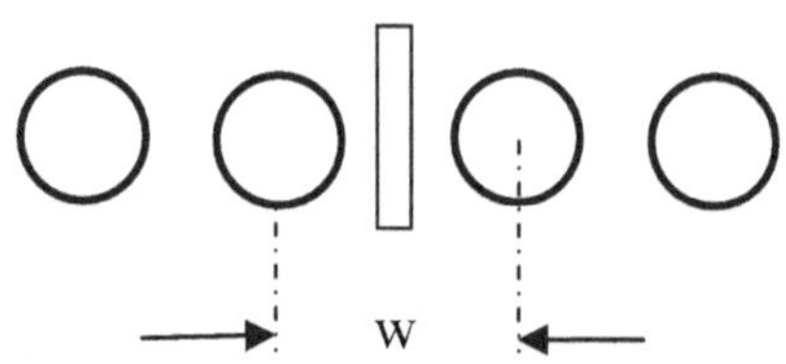

Figure 15.37

For this example I would use a distance w of 1.5 x 6.4 + 19 + 5 = 33.6 rounded up to say 35 rather than 34 mm because I know some people like to work in 5 mm steps. (I have never been given a good reason for this in modern designs!)

In most of these types of design sealing strips will have been specified so that there will be no increase in C stream bypassing if the tube count is left as it is. However it does not take too much time to look at the tube layout and see where it might be possible to fit in a few more tubes to compensate for the loss of a complete vertical row. A note can then be added to the tube layout diagram specifying the distance w and adding "1 (or 2) extra tubes to be added to rows x, y and z". Your computer program is unlikely to be able to handle this but it will not present any problems to a good vendor.

An acoustic baffle needs to be supported, particularly with a relatively thin baffle and the usual way is with tie rods although with a NTIW design it will also be supported by attaching it to the normal baffles. This often means losing a few more tubes. It is essentially a mechanical problem and if in any doubt about how many tubes to omit ask a mechanical engineer.

Lastly the length of the acoustic baffle, TEMA and others recommend that it should stop 50-75 mm from the tubesheet face or end support plate face. I can think of no acoustic reason for this. Thermally it will create two small bypass areas. The reason must be mechanical to ensure that acoustic baffles are not welded to tubesheets and there are no potential expansion problems.

ANOTHER WAVE?

In addition to the standing waves discussed so far Fitz-Hugh in ref 34 proposed another possible type of acoustic wave which may occur in cylindrical shells which have a baffle spacing of less than 85% of the shell diameter. (This would rule out most cross flow and many NTIW units.) Equation 15.17 for the natural frequency is of a 'normal standing wave' in a shell is:

$$f_N = \frac{m\ u_{sonic}}{2D_s}$$

For the first mode this can be transposed to:

$$f_N = \frac{0.5\ u_{sonic}}{D_s} \qquad [15.19]$$

This second series of waves can be found by multiplying the value in this equation by a constant which can be found from the abbreviated table below.

Mode m	1	2	4	8
Constant	1.172	0.972	.0.846	0.768

Table 15. 7

For the first mode equation 15.17 can therefore be expressed as:

$$f_{Nfh} = \frac{m\ u_{sonic}}{2D_s} \text{ x constant} = \frac{1 \text{ x } u_{sonic} \text{ x}1.172}{2D_s}$$

$$f_{Nfh} = \frac{0.59\ u_{sonic}}{D_s} \qquad [15.20]$$

Where f_{Nfh} is the Fitz-Hugh variant of the first mode natural frequency for a heat exchanger shell.

The possibility of the two series of natural frequencies is taken from HTRI sources. They then proceed to use equation 15.19. Equation 15.20 is used by HTFS and their explanation is simply that equation 15.19 is only correct for rectangular ducts and needs to be modified for heat exchanger shells.

I find the concept of there being two series of natural frequencies so close together a bit difficult to swallow but I am in no position to suggest which one is the more accurate. Since there are other constants involved in predicting the possible exciting mechanisms they might both lead to the same conclusion. (This is comparable with the way they each define the equivalent diameter for heat transfer through an annulus – see the section on equivalent diameters in chapter 3.)

Personally I would prefer to keep the 2 in any equation as a reminder that the waves we are considering are half a full sine wave.

I originally decided to dodge the issue of the two equations and leave this section out altogether but it is something that the reader is likely to come across in the literature. I hope I have managed to explain it rather than add to the confusion.

EXCITATION MECHANISMS

In the same way that an exciting frequency was necessary to produce a mechanical vibration an exciting frequency is necessary to produce a noise. A noise will be produced whenever a natural frequency of the system coincides with an exiting frequency. This state is known as resonance and the term 'lock in' is also used in acoustics. The exciting mechanisms for acoustic vibrations are the same as for mechanical vibrations, namely:

(a) Vortex shedding frequency f_V
(b) Turbulent buffeting frequency f_{TB}

These are both predicted from complex empirical equations which have already been covered sufficiently for the reader to appreciate the variables involved.

To allow for some tolerance a barred band is introduced for each of these mechanisms which is as follows:

$$0.8\ f_V < f_N < 1.2\ f_V \qquad [15.21]$$

$$0.8\ f_{TB} < f_N < 1.2\ f_{TB} \qquad [15.22]$$

Although these limits apply to any frequency mode, their importance decreases as the mode number increases.

CHEN NUMBER Ψ

Just matching an exciting frequency with a natural frequency does not always result in a problem. There may be a noise that a bat or a young dog could detect but they are unlikely to be on a petrochemical site. For acoustic resonance to become a problem it also requires an adequate energy input. A means of measuring the energy input required was proposed by Chen and the number that was named after him has become an industrial standard. To understand the Chen number the reader might find the following simplified definition useful:

$$\text{Chen number } \Psi = \frac{Re}{S} \times \text{A tube bank constant} \qquad [15.23]$$

Where $Re = \frac{\rho u d_o}{\mu}$ & $S = \frac{f_V d_o}{u}$

$$\text{Tube bank constant} = \left(\frac{P_{tL} - d_o/n_L}{P_{tL}}\right)^2 \frac{d_o}{n_t P_{tt}} \qquad [15.24]$$

P_{tL} is the longitudinal tube pitch and P_{tt} is the transverse tube pitch. Values for n_L and n_t are taken from the following table:

Tube pitch orientation	n_L	n_t
30°	2	1
60°	2	2
90°	1	1
45° (Pt/do ≤ 1.7)	2	2
45° (Pt/do > 1.7)	2	1

Table 15.8

The tube bank constant is fixed for any fully specified tube bank. The variables are the tube longitudinal and transverse pitches, the tube pitch orientation and the tube outside diameter.

Equation 15.23 is worth committing to memory. Equation 15.24 and table 15.8 have been included for completion only unless you have a photographic memory.

The tube bank constant is a dimensionless constant and since the Reynolds number and the Strouhal number are dimensionless the Chen number is also dimensionless. (It is a nice change to have no confusion with units!)

To emphasise the energy component of the Chen number; dividing Re by S produces a ρu^2 term on the top line of the equation – momentum flux. Most thermal design engineers tend to think in terms of momentum flux values as a measure of the energy in a flowing fluid.

In the same way that Reynolds number was used as a guide to identifying fluid flow regimes the Chen number is used as a guide to identifying if there is sufficient energy in a fluid stream to sustain an acoustic wave that may become a problem.

Threshold limits of 2000 and 4000 were put forward by Chen who suggested they should be used as follows:

$\Psi < 2000$ - No acoustic problems
$2000 > \Psi < 4000$ - Possible acoustic problems
$\Psi > 4000$ - Probable acoustic problems

It is easy to remember the lower limit because it is about the same as the Reynolds number when laminar flow inside a tube may become turbulent. Our main interest however is for values of the Chen number equal to or greater than 4000. Figure 15.38

is plotted for a Chen number of 4000 with the Reynolds number on the vertical axis and the tube pitch ratio on the horizontal axis. There are curves for each of the normal tube pitch arrangements.

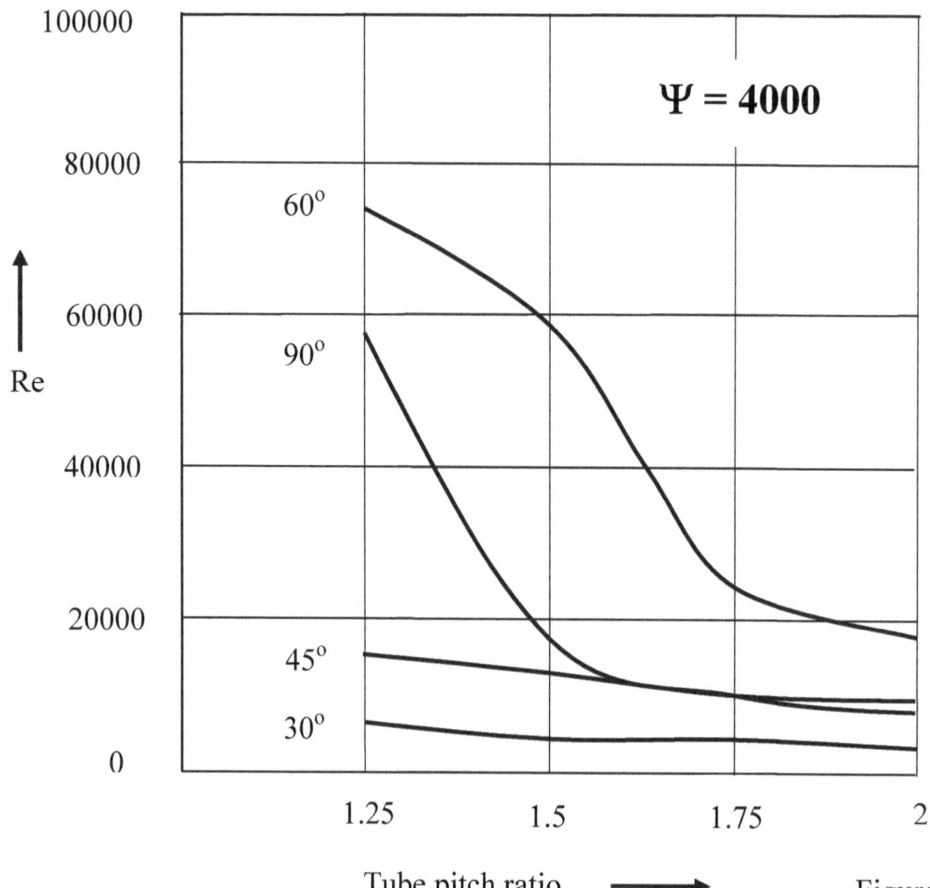

Figure 15.38

Like so many research based graphs these extend into tube pitch ratios that are seldom used. Over 90% of practical designs will utilise a ratio of 1.25 or 1.33. This is quite a small proportion of the graphs.

It is clear it that there is a substantial difference between the Chen numbers for the 60° and 90° layouts and the 45° and 30° layouts. The distinguishing feature between these two groups of layouts is that the first group has gaps that would enable an observer to see right across the bundle in the direction of fluid flow. The second group does not.

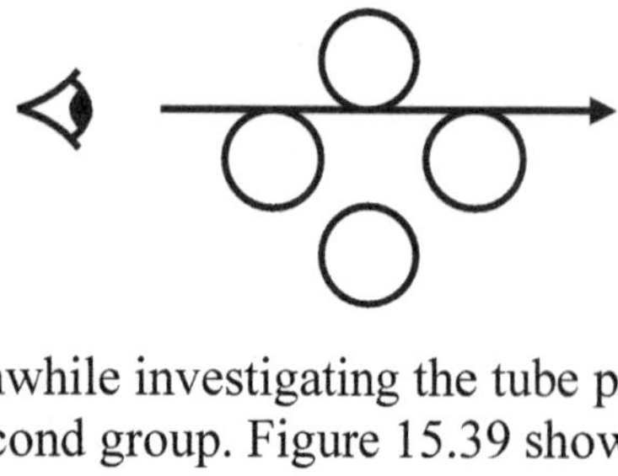

Figure 15.39

It is worthwhile investigating the tube pitch/diameter ratio required to open up a gap for the second group. Figure 15.39 shows the eye of an observer looking into a bundle with a 45° tube pitch at the point where any increase in the tube pitch would open up a gap. The pitch/diameter ratio at this point is 1.414.

Table 15.8 below shows the minimum P_t/d_o values before a gap will be opened up for the four arrangements together with actual gaps for commonly used P_t/d_o ratios.

		Gap as a % of P_t	
Pitch arrangement	P_t/d_o min	$P_t/d_o = 1.25$	$P_t/d_o = 1.33$
60°	1.15	8.3	15.4
90°	1	25	33.3
45°	1.414	No gap	No gap
30°	2	No gap	No gap

Table 15.9

Looking at the gaps or absence of gaps in this table emphasises the difference between the two groups of tube arrangements. It is clearly easier to transmit energy through a fluid stream when there are no intervening obstacles. The difference in the P_t/d_o ratio required to create a gap could also explain the relative positions of the 45° and 30° arrangements. However if this logic is followed then the 90° pitch arrangement should be above the 60° arrangement. This has puzzled people with a lot more understanding of acoustics than I have.

The only aspect of the high pitch to diameter ratios on the graph which is interesting is that as the tube pitch increases above about 1.4 they become much closer together and would presumably tend to merge at very high values where the gaps would occupy an increasing proportion of the flow area.

From a practical design point of view it is clear that 90° and 60° arrangements should be used in exchangers whenever there are acoustic problems rather than the 45° and 30° arrangements. On several occasions I have started a design with a 30° pitch, run into acoustic vibration problems, and then eliminated the problem by switching to a 60° pitch. I have just looked at a couple of cases. In the first one the Chen number was over 13000 for the 30° layout compared with less than 500 for the 60° layout. The design was satisfactory with a 60° layout without any acoustic baffle. In the other case the Chen number of was over 200000 for a 30° layout compared with less than 13000 for the 60° layout. This was still a problem but a manageable one by fitting deresonating baffles.

Acoustic characteristics are the most powerful argument against some client specs which may disapprove of and even bar the 60° pitch. For really stubborn ones it may be necessary to get an E mail from your computer provider to convince them that your diagnosis is correct.

Changing from 30° to 60° produces a dramatic improvement in acoustic characteristics. The downside is that mechanical vibrations in general and the dreaded fluidelastic instability in particular are more likely. However I have found that the two sets of possible problems do not normally arise together in the same exchanger. Most units with acoustic vibration problems are NTIW or cross flow designs where potential mechanical vibration problems are easily dealt with by adding extra intermediate supports.

HTRI say that the vast majority of the exchangers they have in their records with acoustic vibration problems have been ones with a 45° pitch. I suspect that this is because far more units have been designed with a 45° pitch than with a 30° pitch which is even more susceptible to potential problems.

DECISION CHART

The following decision chart/flowchart is a useful aid in analysing acoustic vibrations.

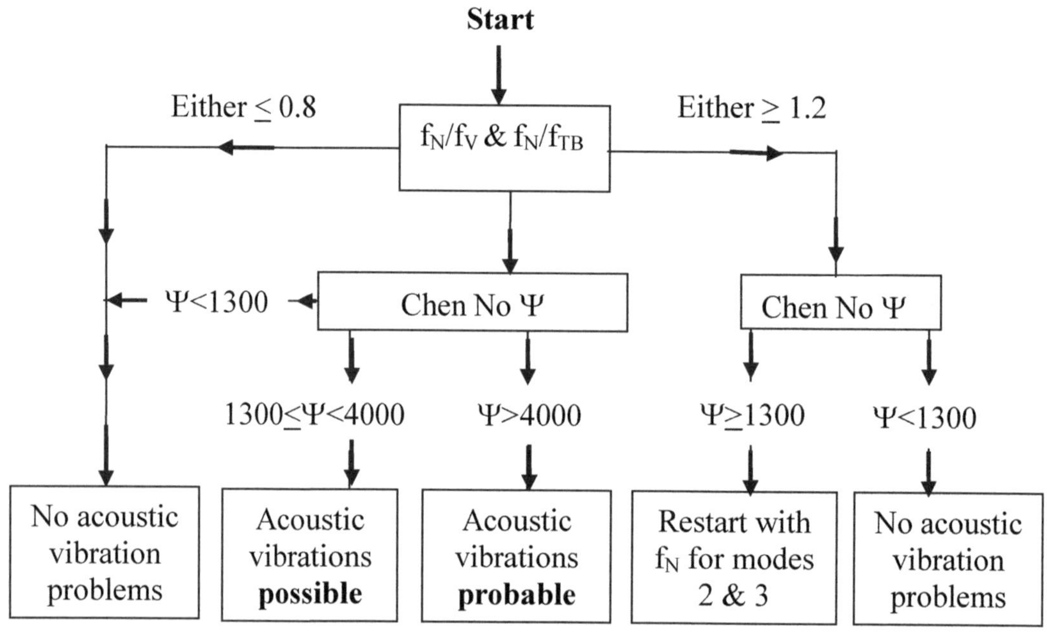

Figure 15.40

The lower limit of 1300 for the Chen number is more conservative than the 2000 proposed by Chen and is recommended by HTRI. Values of the frequency ratios, f_N/f_V and f_N/f_{TB} which are higher than 1.2, are the ratios which have missed the first mode resonance but may excite resonance at one of the higher modes.

EFFECT OF FOULING

There have been reports of exchangers that produced significant noise when they were first commissioned. The noise then reduced to acceptable levels as the unit started to foul up. Clearly the fouling was damping out the acoustic vibrations in some way. This is not a phenomenon which has not been quantified in any way to enable it to be taken into account. However if you have a unit which is only just OK for acoustic vibrations and it is also a heavy fouling unit like come cracked gas aftercoolers it is nice to know that you may have an extra safety factor up your sleeve!

DISC AND DONUT BAFFLES

Disc and donut baffle arrangements are an interesting case as far as acoustic vibrations are concerned because no acoustic problems have ever been reported as far as I know. Along a diagonal the flow is radial for two paths in opposite directions. There is also no fixed tube pitch orientation as far as fluid flow is concerned. A mechanically viewed square pitch is also a rotated square pitch along a diagonal at 45^0 to the horizontal and many 'odd angles' in between these limits as explained in chapter 2. I doubt if anyone will bother to try and analyse this arrangement until the first problems emerge.

ACOUSTIC MATERIAL COVER

Some field problems have been 'solved' by covering the shell with material which reduces the sound transmission from the exchanger to an acceptable level. I have no experience of what materials have been used but a series of air gaps would have a similar effect to double glazing.

If there were sufficient demand I have no doubt that someone would come up with a design for a curved vacuum tube which might really do a good job. I suspect that in all these cases an acoustic expert has been called in rather than a thermal design engineer.

OUTSIDE SOURSES OF EXCITATIONS

Cases have been reported where acoustic vibration problems with heat exchanger shells have been caused by vibrations transmitted along pipes from nearby machinery. Compressors are the most likely culprit. If this is identified as a cause the solution is to do something to the compressor or prevent the transmission along the pipe. I wonder if simply inserting a flexible bellows might work. It could do in the short term but I would leave it to a mechanical engineer to calculate how long it would last!

NOISE LEVELS

Predicting and measuring noise levels is now standard practice for many item of equipment on a plant as well as for shell and tube units. The area that a thermal designer is most likely to have been involved in is the design of air coolers. A long time ago I decided that air coolers would justify another book rather than just extending this one to the point where I would still be a long way from the end. A co-author who had spent a lot of time working for a vendor would also be highly desirable to cover some of the more subtle details. I always try never to say never so this might be a project for the future.

As far as this book is concerned I thought it might be useful to have a means of comparing the sound levels from other situations which can be used as a comparison with any values that may be reported in an article or a field problem.

Source	SPL in dB
Rocket engine	180
Turbojet engine	160
Siren	150
Loudspeaker at a rock concert	140
Jackhammer	120
Trumpet	115
Bellicose politician	90
Air cooler design limit	85
Normal conversation	70
Open plan offices	50
Offices & conference rooms	40

Table 15.10

The worst case that HTRI have records of is one which registered 165 dB. This would certainly have required some attention. The air cooler limit of 85 would probably not be a problem and we have all become accustomed to the loud politicians so that anything below 90 would not justify remedial action on most plants.

CONCLUSIONS

Fluid elastic whirling is the vibration problem which is most likely to result in very rapid tube failures. Vortex shedding and turbulent buffeting problems are more likely to result in slower long term damage.

Acoustic vibration problems do not result in tube damage. The priority for avoiding them may depend on where the plant is situated. There may be quite different attitudes to noise for a plant in Europe which may have residential property just across the road from the site to a plant in the middle of a desert.

I started this chapter by expressing my reluctance to write it at all because of lack of confidence in my own understanding of the subject. I have finished writing far more pages than I originally estimated and I have learned a lot from having the time to research the subject from different sources. My images of vibration situations are less blurred than they were but there are still quite a few blank spaces.

LIST OF USEFUL TABLES

USEFUL TABLE 1

Effect of changing a single variable in single phase flow

HEAT TRANSFER COEFFICIENT

TUBE SIDE

	TURBLENT	LAMINAR
Velocity	$u^{0.8}$	$u^{0.33}$
Specific heat	$Cp^{0.33}$	$Cp^{0.33}$
Viscosity	$\mu^{-0.47}$	no change
Thermal conductivity	$k^{0.67}$	$k^{0.67}$
Density	no change	no change
Tube inside diameter	$d_i^{-0.2}$	$d_i^{-0.67}$

SHELL SIDE

	TURBLENT
Velocity	$u^{0.6}$
Specific heat	$Cp^{0.33}$
Viscosity	$\mu^{-0.27}$
Thermal conductivity	$k^{0.67}$
Density	no change
Tube outside diameter	$d_o^{-0.4}$

Note

(1) A change in mass flow rate is the same as a change in velocity.
(2) A change in density produces a corresponding change in velocity so the product ρu in the Reynolds number remains the same.
(3) The product $\rho u = G$, the bulk or mass velocity.

PRESSURE DROP

	TUBE SIDE	SHELL SIDE
Velocity (mass flow rate)	$u^{1.85}$	$u^{1.85}$
Viscosity	$\mu^{0.24}$	$\mu^{0.2}$
Density	$\rho^{-1.0}$	$\rho^{-1.0}$
Tube diameter	$d_i^{-4.8}$	
Baffle pitch		$P_b^{2.85}$

Note

(1) The exponent for velocity (mass flow rate) for friction losses only inside a tube is 1.8. Using 1.85 allows about 20% of the pressure drop for discontinuity losses (Nozzles, tube inlet and outlet etc) which have an exponent of 2.

(2) The exponent of velocity (mass flow rate) for cross flow outside a tube is 1.8. The exponent of velocity for long flow friction losses is also 1.8. However the turn around losses has an exponent of 2, the same as the nozzle losses, hence I have used the same compromise exponent as for the tube side.

CONDENSING HEAT TRANSFER

For either side	Liquid thermal conductivity	k^1
	Liquid density	$\rho^{1/2}$
	Liquid viscosity	$\mu^{-1/3}$

Note – the exponent of density assumes that the density of the vapour is small compared with the density of the liquid.

BOILING HEAT TRANSFER

The heat transfer coefficients are strongly dependant on temperature difference, the operating pressure and the critical pressure but there are no simple relationships which are worthwhile committing to memory.

USEFUL TABLE 2

To convert a quantity in the units in the left hand column to those of another system multiply by the number in the column headed by the required change. For example to change a heat transfer coefficient of 1000 W/m^2K in the SI system to US customary units 1000 x 1/5.678 = 176.1 $BTU/hft^{2o}F$

HEAT TRANSFER COEFFICIENT

		SI	MKS	US
SI	W/m^2K	1	1/1.163	1/5.678
MKS	$kcal/m^2h^oC$	1.163	1	1/4.882
US	$BTU/hft^{2o}F$	5.678	4.882	1

FOULING FACTOR

		SI	MKS	US
SI	m^2K/W	1	1.163	5.678
MKS	$m^2h^oC/kcal$	1/1.163	1	4.882
US	$hft^{2o}F/BTU$	1/5.678	1/4.882	1

HEAT FLOW RATE

		SI	MKS	US
SI	W	1	1/1.163	3.412
MKS	kcal/h	1.163	1	3.968
US	BTU/h	1/3.412	1/3.968	1

HEAT

		SI	MKS	US
SI	kJ	1	1/4.187	1/1.055
MKS	kcal	4.187	1	3.968
US	BTU	1.055	1/3.968	1

HEAT FLUX

		SI	MKS	US
SI	W/m^2	1	1/1.163	1/3.154
MKS	$kcal/h\ m^2$	1.163	1	1/2.712
US	$BTU/h\ ft^2$	3.154	2.712	1

THERMAL CONDUCTIVITY

	SI	MKS	US
SI W/mK	1	1/1.163	1/1.731
MKS kcal/mh°C	1.163	1	1/4.882
US BTU/hft°F	1.731	1.4882	1

SPECIFIC HEAT CAPACITY

	SI	MKS	US
SI kJ/kgK	1	1/4.187	1/4.187
MKS kcal/kg°C	4.187	1	1
US BTU/lb °F	4.187	1	1

SPECIFIC ENTHALPY & LATENT ENTHALPY (See note 1)

	SI	MKS	US
SI kJ/kg	1	1/4.187	1/2.326
MKS kcal/kg	4.187	1	1.8
US BTU/lb	2.326	1/1.8	1

DYNAMIC (ABSOLUTE) VISCOSITY

	SI	MKS	US
SI kg/ms (= Ns/ m^2)	1	1000	2420
MKS cP	0.001	1	2.42
US lb/fth	1/2420	1/2.42	1

KINEMATIC VISCOSITY

	SI	MKS	US
SI m^2/s	1	1000	38750
MKS cPm^3/kg	0.001	1	38.75
US ft^2/h	1/38.75	1/38.75	1

SURFACE TENSION

	SI	MKS	US
SI N/m	1	1000	1/14.6
MKS dynes/cm	0.001	1	1/14600
US lb/ft	14.6	14600	1

MASS FLOW RATE

		SI	MKS	US
SI	kg./s	1	3600	7938
MKS	kg/h	1/3600	1	2.205
US	lb/h	1/7938	1/2.205	1

PRESSURE

		SI	MKS	US
SI	bar ($10^5 N/m^2$)	1	1.02	14.5
MKS	kgf/cm^2	1/1.02	1	14.22
US	lbf/in^2 (psi)	1/14.5	1/14.22	1

PRESSURE HEADS

	kN/m^2	mm Hg	in Hg	mm H_2O	in H_2O
kN/m^2	1	7.5	0.2953	102	4.02
mm Hg	1/7.5	1	1/25.4	13.6	1/1868
in Hg	1/0.2953	25.4	1	345.4	13.6
mm H_2O	1/102	1/13.6	1/345.4	1	1/25.4
in H_2O	1/4.02	1.868	1/13.6	25.4	1

Quantities where the SI units and the MKS units are the same

MOMENTUM FLUX (ρu^2)	kg/ms^2	x	0.672 (almost 2/3)	=	lb/fts^2
DENSITY	kg/m^3	x	0.0624	=	lb/ft^3
VELOCITY	m/s	x	3.281	=	ft/s
MASS	kg	x	2.205	=	lb
LENGTH	m	x	3.281	=	ft
AREA	m^2	x	10.76	=	ft2
VOLUME	m^3	x	35.31	=	ft^3

TEMPERTURE $^oC = (^oF - 32)/1.8$ & $K = {}^oR/1.8$

NOTES

(1) Only one number needs to be remembered for each conversion

(2) The term latent heat is still widely used in the industry. In the SI system of units heat is a unit of energy not a unit of a physical property. We should be using the term specific enthalpy of phase change. That is a bit of a mouth full so I have used the term latent enthalpy.

USEFUL TABLE 3

Maximum internal pressures for pipes and pipe sized shells.

The tables which follow enable you to quickly select the correct schedule for a nominal bore pipe. There are three different allowable stress values to suit the material at the design temperature being used for each nominal bore size. For any other values of allowable stress the maximum pressure is directly proportion to the stress. For most carbon steel applications using the middle table with for a stress of 15000 psi will be on the safe side.

Pipe sizes are still in US customary units and look likely to remain that way for some time to come. All the data needed for a thermal design program is in both US and SI units.

As far as heat exchanger shells are concerned pipe sizes are normally only used up to about 18 inch nominal bore. You will of course require larger diameter pipes for thermosyphon reboiler pipework.

2 INCH NOMINAL BORE PIPE

STRESS 12000 psi 82.7 MN/m² *MAXIMUM WORKING PRESSURES*

					Corrosion allowance (mm)					
					0		1.5		3	
Sch.	d_o ins	t ins	d_i ins	d_i mm	psi	bar	psi	bar	psi	bar
Std.	2.375	0.154	2.067	52.5	1451	100.1	805	55.5	112	7.7
40	2.375	0.154	2.067	52.5	1451	100.1	805	55.5	112	7.7
X-S	2.375	0.218	1.939	49.3	2111	145.6	1470	101.4	782	53.9
80	2.375	0.218	1.939	49.3	2111	145.6	1470	101.4	782	53.9
120	2.375	0.218	1.939	49.3	2111	145.6	1470	101.4	782	53.9
160	2.375	0.344	1.687	42.8	3526	243.2	2900	200.0	2225	153.4

STRESS 15000 psi 103.4 MN/m² *MAXIMUM WORKING PRESSURES*

					Corrosion allowance (mm)					
					0		1.5		3	
Sch.	d_o ins	t ins	d_i ins	d_i mm	psi	bar	psi	bar	psi	bar
Std.	2.375	0.154	2.067	52.5	1813	125.0	1006	69.4	140	9.7
40	2.375	0.154	2.067	52.5	1813	125.0	1006	69.4	140	9.7
X-S	2.375	0.218	1.939	49.3	2639	182.0	1838	126.8	977	67.4
80	2.375	0.218	1.939	49.3	2639	182.0	1838	126.8	977	67.4
120	2.375	0.218	1.939	49.3	2639	182.0	1838	126.8	977	67.4
160	2.375	0.344	1.687	42.8	4408	304.0	3626	250.1	2781	191.8

STRESS 17500 psi 120.7 MN/m² *MAXIMUM WORKING PRESSURES*

					Corrosion allowance (mm)					
					0		1.5		3	
Sch.	d_o ins	t ins	d_i ins	d_i mm	psi	bar	psi	bar	psi	bar
Std.	2.375	0.154	2.067	52.5	2116	145.9	1174	81.0	164	11.3
40	2.375	0.154	2.067	52.5	2116	145.9	1174	81.0	164	11.3
X-S	2.375	0.218	1.939	49.3	3079	212.3	2144	147.9	1140	78.6
80	2.375	0.218	1.939	49.3	3079	212.3	2144	147.9	1140	78.6
120	2.375	0.218	1.939	49.3	3079	212.3	2144	147.9	1140	78.6
160	2.375	0.344	1.687	42.8	5143	354.7	4230	291.7	3245	223.8

3 INCH NOMINAL BORE PIPE

STRESS 12000 psi 82.7 MN/m² *MAXIMUM WORKING PRESSURES*

					Corrosion allowance (mm)					
					0		1.5		3	
Sch.	d_o ins	t ins	d_i ins	d_i mm	psi	bar	psi	bar	psi	bar
Std.	3.500	0.216	3.068	77.9	1376	94.9	942	65.0	488	33.7
40	3.500	0.216	3.068	77.9	1376	94.9	942	65.0	488	33.7
X-S	3.500	0.300	2.900	73.7	1959	135.1	1528	105.4	1076	74.2
80	3.500	0.300	2.900	73.7	1959	135.1	1528	105.4	1076	74.2
120	3.500	0.300	2.900	73.7	1959	135.1	1528	105.4	1076	74.2
160	3.500	0.438	2.624	66.6	2982	205.7	2558	176.4	2112	145.7

STRESS 15000 psi 103.4 MN/m² *MAXIMUM WORKING PRESSURES*

					Corrosion allowance (mm)					
					0		1.5		3	
Sch.	d_o ins	t ins	d_i ins	d_i mm	psi	bar	psi	bar	psi	bar
Std.	3.500	0.216	3.068	77.9	1720	118.6	1178	81.2	610	42.1
40	3.500	0.216	3.068	77.9	1720	118.6	1178	81.2	610	42.1
X-S	3.500	0.300	2.900	73.7	2449	168.9	1910	131.7	1345	92.8
80	3.500	0.300	2.900	73.7	2449	168.9	1910	131.7	1345	92.8
120	3.500	0.300	2.900	73.7	2449	168.9	1910	131.7	1345	92.8
160	3.500	0.438	2.624	66.6	3728	257.1	3198	220.6	2640	182.1

STRESS 17500 psi 120.7 MN/m² *MAXIMUM WORKING PRESSURES*

					Corrosion allowance (mm)					
					0		1.5		3	
Sch.	d_o ins	t ins	d_i ins	d_i mm	psi	bar	psi	bar	psi	bar
Std.	3.500	0.216	3.068	77.9	2007	138.4	1375	94.8	712	49.1
40	3.500	0.216	3.068	77.9	2007	138.4	1375	94.8	712	49.1
X-S	3.500	0.300	2.900	73.7	2857	197.0	2229	153.7	1570	108.3
80	3.500	0.300	2.900	73.7	2857	197.0	2229	153.7	1570	108.3
120	3.500	0.300	2.900	73.7	2857	197.0	2229	153.7	1570	108.3
160	3.500	0.438	2.624	66.6	4349	299.9	3731	257.3	3080	212.4

4 INCH NOMINAL BORE PIPE

STRESS 12000 psi 82.7 MN/m² *MAXIMUM WORKING PRESSURES*

					Corrosion allowance (mm)					
					0		1.5		3	
Sch.	d_o ins	t ins	d_i ins	d_i mm	psi	bar	psi	bar	psi	bar
Std.	4.500	0.237	4.026	102.3	1164	80.3	827	57.0	479	33.0
40	4.500	0.237	4.026	102.3	1164	80.3	827	57.0	479	33.0
X-S	4.500	0.337	3.826	97.2	1693	116.8	1358	93.7	1011	69.7
80	4.500	0.337	3.826	97.2	1693	116.8	1358	93.7	1011	69.7
120	4.500	0.438	3.624	92.0	2252	155.3	1920	132.4	1575	108.6
160	4.500	0.531	3.438	87.3	2790	192.4	2461	169.7	2119	146.1

STRESS 15000 psi 103.4 MN/m² *MAXIMUM WORKING PRESSURES*

					Corrosion allowance (mm)					
					0		1.5		3	
Sch.	d_o ins	t ins	d_i ins	d_i mm	psi	bar	psi	bar	psi	bar
Std.	4.500	0.237	4.026	102.3	1455	100.3	1034	71.3	599	41.3
40	4.500	0.237	4.026	102.3	1455	100.3	1034	71.3	599	41.3
X-S	4.500	0.337	3.826	97.2	2116	145.9	1698	117.1	1264	87.2
80	4.500	0.337	3.826	97.2	2116	145.9	1698	117.1	1264	87.2
120	4.500	0.438	3.624	92.0	2815	194.1	2400	165.5	1969	135.8
160	4.500	0.531	3.438	87.3	3488	240.6	3077	212.2	2649	182.7

STRESS 17500 psi 120.7 MN/m² *MAXIMUM WORKING PRESSURES*

					Corrosion allowance (mm)					
					0		1.5		3	
Sch.	d_o ins	t ins	d_i ins	d_i mm	psi	bar	psi	bar	psi	bar
Std.	4.500	0.237	4.026	102.3	1697	117.0	1207	83.2	698	48.1
40	4.500	0.237	4.026	102.3	1697	117.0	1207	83.2	698	48.1
X-S	4.500	0.337	3.826	97.2	2469	170.3	1981	136.6	1475	101.7
80	4.500	0.337	3.826	97.2	2469	170.3	1981	136.6	1475	101.7
120	4.500	0.438	3.624	92.0	3284	226.5	2800	193.1	2297	158.4
160	4.500	0.531	3.438	87.3	4069	280.6	3589	247.5	3091	213.2

6 INCH NOMINAL BORE PIPE

STRESS 12000 psi 82.7 MN/m² *MAXIMUM WORKING PRESSURES*

					Corrosion allowance (mm)					
					0		1.5		3	
Sch.	d_o ins	t ins	d_i ins	d_i mm	psi	bar	psi	bar	psi	bar
Std.	6.625	0.280	6.065	154.1	924	63.7	697	48.1	463	31.9
40	6.625	0.280	6.065	154.1	924	63.7	697	48.1	463	31.9
X-S	6.625	0.432	5.761	146.3	1459	100.6	1233	85.0	1001	69.0
80	6.625	0.432	5.761	146.3	1459	100.6	1233	85.0	1001	69.0
120	6.625	0.562	5.501	139.7	1937	133.6	1712	118.1	1481	102.1
160	6.625	0.719	5.187	131.7	2541	175.2	2317	159.8	2088	144.0

STRESS 15000 psi 103.4 MN/m² *MAXIMUM WORKING PRESSURES*

					Corrosion allowance (mm)					
					0		1.5		3	
Sch.	d_o ins	t ins	d_i ins	d_i mm	psi	bar	psi	bar	psi	bar
Std.	6.625	0.280	6.065	154.1	1155	79.7	871	60.1	579	39.9
40	6.625	0.280	6.065	154.1	1155	79.7	871	60.1	579	39.9
X-S	6.625	0.432	5.761	146.3	1824	125.8	1541	106.3	1251	86.3
80	6.625	0.432	5.761	146.3	1824	125.8	1541	106.3	1251	86.3
120	6.625	0.562	5.501	139.7	2421	167.0	2140	147.6	1851	127.7
160	6.625	0.719	5.187	131.7	3176	219.0	2897	199.8	2611	180.1

STRESS 17500 psi 120.7 MN/m² *MAXIMUM WORKING PRESSURES*

					Corrosion allowance (mm)					
					0		1.5		3	
Sch.	d_o ins	t ins	d_i ins	d_i mm	psi	bar	psi	bar	psi	bar
Std.	6.625	0.280	6.065	154.1	1348	93.0	1016	70.1	676	46.6
40	6.625	0.280	6.065	154.1	1348	93.0	1016	70.1	676	46.6
X-S	6.625	0.432	5.761	146.3	2128	146.8	1798	124.0	1460	100.7
80	6.625	0.432	5.761	146.3	2128	146.8	1798	124.0	1460	100.7
120	6.625	0.562	5.501	139.7	2825	194.8	2497	172.2	2160	149.0
160	6.625	0.719	5.187	131.7	3705	255.5	3380	233.1	3046	210.1

8 INCH NOMINAL BORE PIPE

STRESS 12000 psi 82.7 MN/m² — *MAXIMUM WORKING PRESSURES*

					Corrosion allowance (mm)					
					0		1.5		3	
Sch.	d_o ins	t ins	d_i ins	d_i mm	psi	bar	psi	bar	psi	bar
Std.	8.625	0.322	7.981	202.7	812	56.0	638	44.0	460	31.7
40	8.625	0.322	7.981	202.7	812	56.0	638	44.0	460	31.7
X-S	8.625	0.500	7.625	193.7	1288	88.8	1114	76.8	937	64.6
80	8.625	0.500	7.625	193.7	1288	88.8	1114	76.8	937	64.6
120	8.625	0.719	7.187	182.5	1901	131.1	1728	119.2	1552	107.0
160	8.625	0.906	6.813	173.1	2450	169.0	2279	157.2	2104	145.1

STRESS 15000 psi 103.4 MN/m² — *MAXIMUM WORKING PRESSURES*

					Corrosion allowance (mm)					
					0		1.5		3	
Sch.	d_o ins	t ins	d_i ins	d_i mm	psi	bar	psi	bar	psi	bar
Std.	8.625	0.322	7.981	202.7	1016	70.1	797	55.0	575	39.7
40	8.625	0.322	7.981	202.7	1016	70.1	797	55.0	575	39.7
X-S	8.625	0.500	7.625	193.7	1610	111.0	1393	96.1	1171	80.8
80	8.625	0.500	7.625	193.7	1610	111.0	1393	96.1	1171	80.8
120	8.625	0.719	7.187	182.5	2376	163.9	2160	149.0	1940	133.8
160	8.625	0.906	6.813	173.1	3063	211.2	2849	196.5	2630	181.4

STRESS 17500 psi 120.7 MN/m² — *MAXIMUM WORKING PRESSURES*

					Corrosion allowance (mm)					
					0		1.5		3	
Sch.	d_o ins	t ins	d_i ins	d_i mm	psi	bar	psi	bar	psi	bar
Std.	8.625	0.322	7.981	202.7	1185	81.7	930	64.1	671	46.3
40	8.625	0.322	7.981	202.7	1185	81.7	930	64.1	671	46.3
X-S	8.625	0.500	7.625	193.7	1878	129.5	1625	112.1	1367	94.3
80	8.625	0.500	7.625	193.7	1878	129.5	1625	112.1	1367	94.3
120	8.625	0.719	7.187	182.5	2772	191.2	2520	173.8	2264	156.1
160	8.625	0.906	6.813	173.1	3573	246.4	3323	229.2	3069	211.7

10 INCH NOMINAL BORE PIPE

STRESS 12000 psi 82.7 MN/m² — *MAXIMUM WORKING PRESSURES*

					Corrosion allowance (mm)					
					0		1.5		3	
Sch.	d_o ins	t ins	d_i ins	d_i mm	psi	bar	psi	bar	psi	bar
Std.	10.750	0.365	10.020	254.5	736	50.8	596	41.1	454	31.3
40	10.750	0.365	10.020	254.5	736	50.8	596	41.1	454	31.3
X-S	10.750	0.500	9.750	247.7	1021	70.4	882	60.8	740	51.0
80	10.750	0.594	9.562	242.9	1224	84.4	1085	74.8	944	65.1
120	10.750	0.844	9.062	230.2	1781	122.8	1643	113.3	1502	103.6
160	10.750	1.125	8.500	215.9	2440	168.3	2303	158.8	2163	149.2

STRESS 15000 psi 103.4 MN/m² — *MAXIMUM WORKING PRESSURES*

					Corrosion allowance (mm)					
					0		1.5		3	
Sch.	d_o ins	t ins	d_i ins	d_i mm	psi	bar	psi	bar	psi	bar
Std.	10.750	0.365	10.020	254.5	920	63.4	746	51.4	568	39.2
40	10.750	0.365	10.020	254.5	920	63.4	746	51.4	568	39.2
X-S	10.750	0.500	9.750	247.7	1277	88.1	1102	76.0	925	63.8
80	10.750	0.594	9.562	242.9	1530	105.5	1356	93.5	1180	81.4
120	10.750	0.844	9.062	230.2	2227	153.6	2054	141.7	1878	129.5
160	10.750	1.125	8.500	215.9	3050	210.3	2878	198.5	2704	186.5

STRESS 17500 psi 120.7 MN/m² — *MAXIMUM WORKING PRESSURES*

					Corrosion allowance (mm)					
					0		1.5		3	
Sch.	d_o ins	t ins	d_i ins	d_i mm	psi	bar	psi	bar	psi	bar
Std.	10.750	0.365	10.020	254.5	1074	74.1	870	60.0	663	45.7
40	10.750	0.365	10.020	254.5	1074	74.1	870	60.0	663	45.7
X-S	10.750	0.500	9.750	247.7	1490	102.8	1286	88.7	1080	74.5
80	10.750	0.594	9.562	242.9	1785	123.1	1582	109.1	1376	94.9
120	10.750	0.844	9.062	230.2	2598	179.2	2396	165.2	2191	151.1
160	10.750	1.125	8.500	215.9	3558	245.4	3358	231.6	3155	217.6

12 INCH NOMINAL BORE PIPE

STRESS 12000 psi 82.7 MN/m² *MAXIMUM WORKING PRESSURES*

					Corrosion allowance (mm)					
					0		1.5		3	
Sch.	d_o ins	t ins	d_i ins	d_i mm	psi	bar	psi	bar	psi	bar
Std.	12.750	0.375	12.000	304.8	635	43.8	517	35.7	398	27.4
40	12.750	0.406	11.938	303.2	689	47.5	571	39.4	452	31.2
X-S	12.750	0.500	11.750	298.5	855	59.0	737	50.8	618	42.6
80	12.750	0.688	11.374	288.9	1194	82.3	1077	74.3	958	66.1
120	12.750	1.000	10.750	273.1	1779	122.7	1663	114.7	1545	106.6
160	12.750	1.312	10.126	257.2	2395	165.2	2279	157.2	2162	149.1

STRESS 15000 psi 103.4 MN/m² *MAXIMUM WORKING PRESSURES*

					Corrosion allowance (mm)					
					0		1.5		3	
Sch.	d_o ins	t ins	d_i ins	d_i mm	psi	bar	psi	bar	psi	bar
Std.	12.750	0.375	12.000	304.8	794	54.8	646	44.6	497	34.3
40	12.750	0.406	11.938	303.2	861	59.4	714	49.2	565	39.0
X-S	12.750	0.500	11.750	298.5	1069	73.7	922	63.6	773	53.3
80	12.750	0.688	11.374	288.9	1493	103.0	1346	92.8	1197	82.6
120	12.750	1.000	10.750	273.1	2224	153.4	2078	143.3	1931	133.2
160	12.750	1.312	10.126	257.2	2993	206.4	2849	196.5	2703	186.4

STRESS 17500 psi 120.7 MN/m² *MAXIMUM WORKING PRESSURES*

					Corrosion allowance (mm)					
					0		1.5		3	
Sch.	d_o ins	t ins	d_i ins	d_i mm	psi	bar	psi	bar	psi	bar
Std.	12.750	0.375	12.000	304.8	926	63.9	754	52.0	580	40.0
40	12.750	0.406	11.938	303.2	1005	69.3	833	57.4	659	45.4
X-S	12.750	0.500	11.750	298.5	1247	86.0	1075	74.1	902	62.2
80	12.750	0.688	11.374	288.9	1741	120.1	1570	108.3	1397	96.3
120	12.750	1.000	10.750	273.1	2595	179.0	2425	167.2	2253	155.4
160	12.750	1.312	10.126	257.2	3492	240.8	3324	229.2	3153	217.4

14 INCH NOMINAL BORE PIPE

STRESS 12000 psi 82.7 MN/m² *MAXIMUM WORKING PRESSURES*

					Corrosion allowance (mm)					
					0		1.5		3	
Sch.	d_o ins	t ins	d_i ins	d_i mm	psi	bar	psi	bar	psi	bar
Std.	14.000	0.375	13.250	336.6	577	39.8	469	32.3	361	24.9
40	14.000	0.438	13.124	333.3	677	46.7	569	39.2	461	31.8
X-S	14.000	0.500	13.000	330.2	776	53.5	669	46.1	560	38.6
80	14.000	0.075	12.500	317.5	1185	81.7	1078	74.3	970	66.9
120	14.000	1.094	11.812	300.0	1772	122.2	1666	114.9	1559	107.5
160	14.000	1.406	11.188	284.2	2331	160.8	2226	153.5	2119	146.1

STRESS 15000 psi 103.4 MN/m² *MAXIMUM WORKING PRESSURES*

					Corrosion allowance (mm)					
					0		1.5		3	
Sch.	d_o ins	t ins	d_i ins	d_i mm	psi	bar	psi	bar	psi	bar
Std.	14.000	0.375	13.250	336.6	721	49.7	587	40.5	451	31.1
40	14.000	0.438	13.124	333.3	846	58.3	712	49.1	576	39.7
X-S	14.000	0.500	13.000	330.2	970	66.9	836	57.7	700	48.3
80	14.000	0.075	12.500	317.5	1481	102.1	1348	93.0	1213	83.7
120	14.000	1.094	11.812	300.0	2215	152.8	2083	143.7	1948	134.3
160	14.000	1.406	11.188	284.2	2914	201.0	2782	191.9	2649	182.7

STRESS 17500 psi 120.7 MN/m² *MAXIMUM WORKING PRESSURES*

					Corrosion allowance (mm)					
					0		1.5		3	
Sch.	d_o ins	t ins	d_i ins	d_i mm	psi	bar	psi	bar	psi	bar
Std.	14.000	0.375	13.250	336.6	841	58.0	685	47.2	526	36.3
40	14.000	0.438	13.124	333.3	987	68.1	831	57.3	672	46.3
X-S	14.000	0.500	13.000	330.2	1132	78.1	975	67.2	817	56.3
80	14.000	0.075	12.500	317.5	1728	119.2	1572	108.4	1415	97.6
120	14.000	1.094	11.812	300.0	2585	178.3	2430	167.6	2273	156.8
160	14.000	1.406	11.188	284.2	3400	234.5	3246	223.9	3091	213.2

16 INCH NOMINAL BORE PIPE

STRESS 12000 psi 82.7 MN/m² *MAXIMUM WORKING PRESSURES*

					Corrosion allowance (mm)					
					0		1.5		3	
Sch.	d_o ins	t ins	d_i ins	d_i mm	psi	bar	psi	bar	psi	bar
Std.	16.000	0.375	15.250	387.4	503	34.7	409	28.2	314	21.7
40	16.000	0.500	15.000	381.0	676	46.6	582	40.1	487	33.6
X-S	16.000	0.500	15.000	381.0	676	46.6	582	40.1	487	33.6
80	16.000	0.844	14.312	363.5	1166	80.4	1072	73.9	978	67.4
120	16.000	1.219	13.562	344.5	1724	118.9	1631	112.5	1538	106.1
160	16.000	1.594	12.812	325.4	2310	159.3	2218	153.0	2125	146.6

STRESS 15000 psi 103.4 MN/m² *MAXIMUM WORKING PRESSURES*

					Corrosion allowance (mm)					
					0		1.5		3	
Sch.	d_o ins	t ins	d_i ins	d_i mm	psi	bar	psi	bar	psi	bar
Std.	16.000	0.375	15.250	387.4	629	43.4	511	35.2	393	27.1
40	16.000	0.500	15.000	381.0	845	58.3	728	50.2	609	42.0
X-S	16.000	0.500	15.000	381.0	845	58.3	728	50.2	609	42.0
80	16.000	0.844	14.312	363.5	1457	100.5	1340	92.4	1223	84.3
120	16.000	1.219	13.562	344.5	2155	148.6	2039	140.6	1922	132.6
160	16.000	1.594	12.812	325.4	2888	199.2	2773	191.2	2657	183.2

STRESS 17500 psi 120.7 MN/m² *MAXIMUM WORKING PRESSURES*

					Corrosion allowance (mm)					
					0		1.5		3	
Sch.	d_o ins	t ins	d_i ins	d_i mm	psi	bar	psi	bar	psi	bar
Std.	16.000	0.375	15.250	387.4	734	50.6	597	41.2	458	31.6
40	16.000	0.500	15.000	381.0	986	68.0	849	58.6	711	49.0
X-S	16.000	0.500	15.000	381.0	986	68.0	849	58.6	711	49.0
80	16.000	0.844	14.312	363.5	1700	117.2	1564	107.9	1426	98.3
120	16.000	1.219	13.562	344.5	2515	173.4	2379	164.1	2243	154.7
160	16.000	1.594	12.812	325.4	3369	232.3	3235	223.1	3100	213.8

18 INCH NOMINAL BORE PIPE

STRESS 12000 psi 82.7 MN/m² *MAXIMUM WORKING PRESSURES*

					Corrosion allowance (mm)					
					0		1.5		3	
Sch.	d_o ins	t ins	d_i ins	d_i mm	psi	bar	psi	bar	psi	bar
Std.	18.000	0.375	17.250	438.2	446	30.8	362	25.0	278	19.2
40	18.000	0.562	16.876	428.7	675	46.6	592	40.8	508	35.0
X-S	18.000	0.500	17.000	431.8	599	41.3	515	35.5	431	29.7
80	18.000	0.938	16.124	409.5	1151	79.4	1068	73.7	984	67.9
120	18.000	1.375	15.250	387.4	1729	119.2	1647	113.6	1564	107.9
160	18.000	1.781	14.438	366.7	2293	158.1	2211	152.5	2129	146.8

STRESS 15000 psi 103.4 MN/m² *MAXIMUM WORKING PRESSURES*

					Corrosion allowance (mm)					
					0		1.5		3	
Sch.	d_o ins	t ins	d_i ins	d_i mm	psi	bar	psi	bar	psi	bar
Std.	18.000	0.375	17.250	438.2	557	38.4	453	31.2	348	24.0
40	18.000	0.562	16.876	428.7	844	58.2	740	51.0	635	43.8
X-S	18.000	0.500	17.000	431.8	748	51.6	644	44.4	539	37.2
80	18.000	0.938	16.124	409.5	1439	99.2	1335	92.1	1230	84.8
120	18.000	1.375	15.250	387.4	2162	149.1	2059	142.0	1955	134.8
160	18.000	1.781	14.438	366.7	2866	197.7	2764	190.6	2661	183.5

STRESS 17500 psi 120.7 MN/m² *MAXIMUM WORKING PRESSURES*

					Corrosion allowance (mm)					
					0		1.5		3	
Sch.	d_o ins	t ins	d_i ins	d_i mm	psi	bar	psi	bar	psi	bar
Std.	18.000	0.375	17.250	438.2	650	44.8	529	36.5	406	28.0
40	18.000	0.562	16.876	428.7	985	67.9	863	59.5	741	51.1
X-S	18.000	0.500	17.000	431.8	873	60.2	752	51.9	629	43.4
80	18.000	0.938	16.124	409.5	1679	115.8	1557	107.4	1435	99.0
120	18.000	1.375	15.250	387.4	2522	173.9	2402	165.7	2280	157.2
160	18.000	1.781	14.438	366.7	3344	230.6	3225	222.4	3104	214.1

20 INCH NOMINAL BORE PIPE

STRESS 12000 psi 82.7 MN/m^2 *MAXIMUM WORKING PRESSURES*

					Corrosion allowance (mm)					
					0		1.5		3	
Sch.	d_o ins	t ins	d_i ins	d_i mm	psi	bar	psi	bar	psi	bar
Std.	20.000	0.375	19.250	489.0	400	27.6	325	22.4	250	17.2
40	20.000	0.594	18.812	477.8	641	44.2	566	39.0	491	33.9
X-S	20.000	0.500	19.000	482.6	537	37.0	462	31.9	387	26.7
80	20.000	1.031	17.938	455.6	1138	78.5	1063	73.3	988	68.1
120	20.000	1.500	17.000	431.8	1695	116.9	1621	111.8	1546	106.6
160	20.000	1.969	16.062	408.0	2280	157.2	2207	152.2	2132	147.0

STRESS 15000 psi 103.4 MN/m^2 *MAXIMUM WORKING PRESSURES*

					Corrosion allowance (mm)					
					0		1.5		3	
Sch.	d_o ins	t ins	d_i ins	d_i mm	psi	bar	psi	bar	psi	bar
Std.	20.000	0.375	19.250	489.0	501	34.6	407	28.1	312	21.5
40	20.000	0.594	18.812	477.8	802	55.3	708	48.8	614	42.3
X-S	20.000	0.500	19.000	482.6	672	46.3	578	39.9	483	33.3
80	20.000	1.031	17.938	455.6	1422	98.1	1329	91.7	1235	85.2
120	20.000	1.500	17.000	431.8	2119	146.1	2027	139.8	1933	133.3
160	20.000	1.969	16.062	408.0	2850	196.6	2758	190.2	2666	183.9

STRESS 17500 psi 120.7 MN/m^2 *MAXIMUM WORKING PRESSURES*

					Corrosion allowance (mm)					
					0		1.5		3	
Sch.	d_o ins	t ins	d_i ins	d_i mm	psi	bar	psi	bar	psi	bar
Std.	20.000	0.375	19.250	489.0	584	40.3	475	32.8	364	25.1
40	20.000	0.594	18.812	477.8	935	64.5	826	57.0	716	49.4
X-S	20.000	0.500	19.000	482.6	784	54.1	674	46.5	564	38.9
80	20.000	1.031	17.938	455.6	1660	114.5	1551	107.0	1441	99.4
120	20.000	1.500	17.000	431.8	2473	170.6	2364	163.0	2255	155.5
160	20.000	1.969	16.062	408.0	3326	229.4	3218	221.9	3110	214.5

24 INCH NOMINAL BORE PIPE

STRESS 12000 psi 82.7 MN/m^2 *MAXIMUM WORKING PRESSURES*

					Corrosion allowance (mm)					
					0		1.5		3	
Sch.	d_o ins	t ins	d_i ins	d_i mm	psi	bar	psi	bar	psi	bar
Std.	24.000	0.375	23.250	590.6	333	23.0	270	18.6	207	14.3
40	24.000	0.688	22.624	574.6	618	42.6	556	38.3	493	34.0
X-S	24.000	0.500	23.000	584.2	446	30.8	383	26.4	320	22.1
80	24.000	1.218	21.564	547.7	1119	77.2	1057	72.9	994	68.6
120	24.000	1.812	20.376	517.6	1708	117.8	1646	113.5	1584	109.2
160	24.000	2.344	19.312	490.5	2260	155.9	2199	151.7	2137	147.4

STRESS 15000 psi 103.4 MN/m^2 *MAXIMUM WORKING PRESSURES*

					Corrosion allowance (mm)					
					0		1.5		3	
Sch.	d_o ins	t ins	d_i ins	d_i mm	psi	bar	psi	bar	psi	bar
Std.	24.000	0.375	23.250	590.6	416	28.7	338	23.3	259	17.9
40	24.000	0.688	22.624	574.6	773	53.3	695	47.9	616	42.5
X-S	24.000	0.500	23.000	584.2	557	38.4	479	33.0	401	27.7
80	24.000	1.218	21.564	547.7	1399	96.5	1321	91.1	1243	85.7
120	24.000	1.812	20.376	517.6	2135	147.2	2057	141.9	1980	136.6
160	24.000	2.344	19.312	490.5	2825	194.8	2749	189.6	2672	184.3

STRESS 17500 psi 120.7 MN/m^2 *MAXIMUM WORKING PRESSURES*

					Corrosion allowance (mm)					
					0		1.5		3	
Sch.	d_o ins	t ins	d_i ins	d_i mm	psi	bar	psi	bar	psi	bar
Std.	24.000	0.375	23.250	590.6	485	33.4	394	27.2	302	20.8
40	24.000	0.688	22.624	574.6	902	62.2	811	55.9	719	49.6
X-S	24.000	0.500	23.000	584.2	650	44.8	559	38.6	467	32.2
80	24.000	1.218	21.564	547.7	1632	112.6	1542	106.3	1450	100.0
120	24.000	1.812	20.376	517.6	2490	171.7	2400	165.5	2310	159.3
160	24.000	2.344	19.312	490.5	3296	227.3	3207	221.2	3117	215.0

26 INCH NOMINAL BORE PIPE

STRESS 15000 psi 1,034.5 bar *MAXIMUM WORKING PRESSURES*

					Corrosion allowance (mm)					
					0		1.5		3	
Sch.	d_o ins	t ins	d_i ins	d_i mm	psi	bar	psi	bar	psi	bar
Std.	26.000	0.375	25.250	641.4	307	21.2	249	17.2	191	13.2
40	26.000	0.688	24.624	625.4	570	39.3	512	35.3	454	31.3
X-S	26.000	0.500	25.000	635.0	411	28.3	353	24.3	295	20.3
80	26.000	1.218	23.564	598.5	1029	71.0	972	67.0	914	63.0
120	26.000	1.812	22.376	568.4	1567	108.1	1510	104.1	1452	100.1
160	26.000	2.344	21.312	541.3	2070	142.8	2013	138.8	1956	134.9

STRESS 15000 psi 103.4 MN/m² *MAXIMUM WORKING PRESSURES*

					Corrosion allowance (mm)					
					0		1.5		3	
Sch.	d_o ins	t ins	d_i ins	d_i mm	psi	bar	psi	bar	psi	bar
Std.	26.000	0.375	25.250	641.4	383	26.4	311	21.4	239	16.5
40	26.000	0.688	24.624	625.4	712	49.1	640	44.1	567	39.1
X-S	26.000	0.500	25.000	635.0	514	35.4	442	30.5	369	25.4
80	26.000	1.218	23.564	598.5	1286	88.7	1215	83.8	1142	78.8
120	26.000	1.812	22.376	568.4	1959	135.1	1887	130.1	1815	125.2
160	26.000	2.344	21.312	541.3	2588	178.5	2517	173.6	2445	168.6

STRESS 17500 psi 120.7 MN/m² *MAXIMUM WORKING PRESSURES*

					Corrosion allowance (mm)					
					0		1.5		3	
Sch.	d_o ins	t ins	d_i ins	d_i mm	psi	bar	psi	bar	psi	bar
Std.	26.000	0.375	25.250	641.4	447	30.8	363	25.0	278	19.2
40	26.000	0.688	24.624	625.4	831	57.3	747	51.5	662	45.7
X-S	26.000	0.500	25.000	635.0	599	41.3	515	35.5	431	29.7
80	26.000	1.218	23.564	598.5	1501	103.5	1417	97.7	1333	91.9
120	26.000	1.812	22.376	568.4	2285	157.6	2202	151.9	2118	146.1
160	26.000	2.344	21.312	541.3	3019	208.2	2936	202.5	2853	196.8

28 INCH NOMINAL BORE PIPE

STRESS 12000 psi 82.7 MN/m² *MAXIMUM WORKING PRESSURES*

					Corrosion allowance (mm)					
					0		1.5		3	
Sch.	d_o ins	t ins	d_i ins	d_i mm	psi	bar	psi	bar	psi	bar
Std.	28.000	0.375	27.250	692.2	284	19.6	231	15.9	177	12.2
40	28.000	0.688	26.624	676.2	528	36.4	474	32.7	420	29.0
X-S	28.000	0.500	27.000	685.8	381	26.3	327	22.6	273	18.8
80	28.000	1.218	25.564	649.3	952	65.7	899	62.0	845	58.3
120	28.000	1.812	24.376	619.2	1448	99.9	1394	96.1	1341	92.5
160	28.000	2.344	23.312	592.1	1909	131.7	1857	128.1	1803	124.3

STRESS 15000 psi 103.4 MN/m² *MAXIMUM WORKING PRESSURES*

					Corrosion allowance (mm)					
					0		1.5		3	
Sch.	d_o ins	t ins	d_i ins	d_i mm	psi	bar	psi	bar	psi	bar
Std.	28.000	0.375	27.250	692.2	356	24.6	289	19.9	221	15.2
40	28.000	0.688	26.624	676.2	660	45.5	593	40.9	526	36.3
X-S	28.000	0.500	27.000	685.8	476	32.8	409	28.2	342	23.6
80	28.000	1.218	25.564	649.3	1191	82.1	1124	77.5	1057	72.9
120	28.000	1.812	24.376	619.2	1810	124.8	1743	120.2	1676	115.6
160	28.000	2.344	23.312	592.1	2387	164.6	2321	160.1	2254	155.4

STRESS 17500 psi 120.7 MN/m² *MAXIMUM WORKING PRESSURES*

					Corrosion allowance (mm)					
					0		1.5		3	
Sch.	d_o ins	t ins	d_i ins	d_i mm	psi	bar	psi	bar	psi	bar
Std.	28.000	0.375	27.250	692.2	415	28.6	337	23.2	258	17.8
40	28.000	0.688	26.624	676.2	770	53.1	692	47.7	613	42.3
X-S	28.000	0.500	27.000	685.8	556	38.3	478	33.0	399	27.5
80	28.000	1.218	25.564	649.3	1389	95.8	1311	90.4	1233	85.0
120	28.000	1.812	24.376	619.2	2111	145.6	2034	140.3	1956	134.9
160	28.000	2.344	23.312	592.1	2785	192.1	2708	186.8	2630	181.4

30 INCH NOMINAL BORE PIPE

STRESS 12000 psi 82.7 MN/m² *MAXIMUM WORKING PRESSURES*

					Corrosion allowance (mm)					
					0		1.5		3	
Sch.	d_o ins	t ins	d_i ins	d_i mm	psi	bar	psi	bar	psi	bar
Std.	30.000	0.375	29.250	743.0	265	18.3	215	14.8	165	11.4
40	30.000	0.688	28.624	727.0	492	33.9	442	30.5	392	27.0
X-S	30.000	0.500	29.000	736.6	355	24.5	305	21.0	255	17.6
80	30.000	1.218	27.564	700.1	886	61.1	836	57.7	786	54.2
120	30.000	1.812	26.376	670.0	1345	92.8	1296	89.4	1246	85.9
160	30.000	2.344	25.312	642.9	1772	122.2	1723	118.8	1673	115.4

STRESS 15000 psi 103.4 MN/m² *MAXIMUM WORKING PRESSURES*

					Corrosion allowance (mm)					
					0		1.5		3	
Sch.	d_o ins	t ins	d_i ins	d_i mm	psi	bar	psi	bar	psi	bar
Std.	30.000	0.375	29.250	743.0	332	22.9	269	18.6	206	14.2
40	30.000	0.688	28.624	727.0	615	42.4	552	38.1	490	33.8
X-S	30.000	0.500	29.000	736.6	444	30.6	382	26.3	319	22.0
80	30.000	1.218	27.564	700.1	1108	76.4	1046	72.1	983	67.8
120	30.000	1.812	26.376	670.0	1682	116.0	1620	111.7	1557	107.4
160	30.000	2.344	25.312	642.9	2215	152.8	2153	148.5	2091	144.2

STRESS 17500 psi 120.7 MN/m² *MAXIMUM WORKING PRESSURES*

					Corrosion allowance (mm)					
					0		1.5		3	
Sch.	d_o ins	t ins	d_i ins	d_i mm	psi	bar	psi	bar	psi	bar
Std.	30.000	0.375	29.250	743.0	387	26.7	314	21.7	241	16.6
40	30.000	0.688	28.624	727.0	717	49.4	645	44.5	571	39.4
X-S	30.000	0.500	29.000	736.6	518	35.7	445	30.7	372	25.7
80	30.000	1.218	27.564	700.1	1293	89.2	1220	84.1	1147	79.1
120	30.000	1.812	26.376	670.0	1962	135.3	1890	130.3	1817	125.3
160	30.000	2.344	25.312	642.9	2584	178.2	2512	173.2	2440	168.3

32 INCH NOMINAL BORE PIPE

STRESS 12000 psi 82.7 MN/m² *MAXIMUM WORKING PRESSURES*

					Corrosion allowance (mm)					
					0		1.5		3	
Sch.	d_o ins	t ins	d_i ins	d_i mm	psi	bar	psi	bar	psi	bar
Std.	32.000	0.375	31.250	793.8	248	17.1	201	13.9	154	10.6
40	32.000	0.688	30.624	777.8	460	31.7	414	28.6	366	25.2
X-S	32.000	0.500	31.000	787.4	333	23.0	286	19.7	239	16.5
80	32.000	1.218	29.564	750.9	829	57.2	782	53.9	735	50.7
120	32.000	1.812	28.376	720.8	1256	86.6	1210	83.4	1163	80.2
160	32.000	2.344	27.312	693.7	1653	114.0	1606	110.8	1560	107.6

STRESS 15000 psi 103.4 MN/m² *MAXIMUM WORKING PRESSURES*

					Corrosion allowance (mm)					
					0		1.5		3	
Sch.	d_o ins	t ins	d_i ins	d_i mm	psi	bar	psi	bar	psi	bar
Std.	32.000	0.375	31.250	793.8	311	21.4	252	17.4	193	13.3
40	32.000	0.688	30.624	777.8	576	39.7	517	35.7	458	31.6
X-S	32.000	0.500	31.000	787.4	416	28.7	357	24.6	298	20.6
80	32.000	1.218	29.564	750.9	1036	71.4	978	67.4	919	63.4
120	32.000	1.812	28.376	720.8	1570	108.3	1512	104.3	1454	100.3
160	32.000	2.344	27.312	693.7	2066	142.5	2008	138.5	1950	134.5

STRESS 17500 psi 120.7 MN/m² *MAXIMUM WORKING PRESSURES*

					Corrosion allowance (mm)					
					0		1.5		3	
Sch.	d_o ins	t ins	d_i ins	d_i mm	psi	bar	psi	bar	psi	bar
Std.	32.000	0.375	31.250	793.8	362	25.0	294	20.3	225	15.5
40	32.000	0.688	30.624	777.8	672	46.3	603	41.6	535	36.9
X-S	32.000	0.500	31.000	787.4	458	31.6	417	28.8	348	24.0
80	32.000	1.218	29.564	750.9	1209	83.4	1141	78.7	1072	73.9
120	32.000	1.812	28.376	720.8	1832	126.3	1764	121.7	1696	117.0
160	32.000	2.344	27.312	693.7	2411	166.3	2343	161.6	2275	156.9

34 INCH NOMINAL BORE PIPE

STRESS 12000 psi 82.7 MN/m² *MAXIMUM WORKING PRESSURES*

					Corrosion allowance (mm)					
					0		1.5		3	
Sch.	d_o ins	t ins	d_i ins	d_i mm	psi	bar	psi	bar	psi	bar
Std.	34.000	0.375	33.250	844.6	234	16.1	189	13.0	145	10.0
40	34.000	0.688	32.624	828.6	433	29.9	389	26.8	344	23.7
X-S	34.000	0.500	33.000	838.2	313	21.6	269	18.6	224	15.4
80	34.000	1.218	31.564	801.7	778	53.7	734	50.6	690	47.6
120	34.000	1.812	30.376	771.6	1178	81.2	1135	78.3	1090	75.2
160	34.000	2.344	29.312	744.5	1549	106.8	1505	103.8	1461	100.8

STRESS 15000 psi 103.4 MN/m² *MAXIMUM WORKING PRESSURES*

					Corrosion allowance (mm)					
					0		1.5		3	
Sch.	d_o ins	t ins	d_i ins	d_i mm	psi	bar	psi	bar	psi	bar
Std.	34.000	0.375	33.250	844.6	292	20.1	237	16.3	181	12.5
40	34.000	0.688	32.624	828.6	541	37.3	486	33.5	431	29.7
X-S	34.000	0.500	33.000	838.2	391	27.0	336	23.2	280	19.3
80	34.000	1.218	31.564	801.7	973	67.1	918	63.3	863	59.5
120	34.000	1.812	30.376	771.6	1473	101.6	1418	97.8	1363	94.0
160	34.000	2.344	29.312	744.5	1936	133.5	1881	129.7	1827	126.0

STRESS 17500 psi 120.7 MN/m² *MAXIMUM WORKING PRESSURES*

					Corrosion allowance (mm)					
					0		1.5		3	
Sch.	d_o ins	t ins	d_i ins	d_i mm	psi	bar	psi	bar	psi	bar
Std.	34.000	0.375	33.250	844.6	341	23.5	276	19.0	212	14.6
40	34.000	0.688	32.624	828.6	631	43.5	567	39.1	502	34.6
X-S	34.000	0.500	33.000	838.2	456	31.4	392	27.0	327	22.6
80	34.000	1.218	31.564	801.7	1135	78.3	1071	73.9	1007	69.4
120	34.000	1.812	30.376	771.6	1719	118.6	1655	114.1	1591	109.7
160	34.000	2.344	29.312	744.5	2259	155.8	2195	151.4	2131	147.0

36 INCH NOMINAL BORE PIPE

STRESS 12000 psi 82.7 MN/m² *MAXIMUM WORKING PRESSURES*

					Corrosion allowance (mm)					
					0		1.5		3	
Sch.	d_o ins	t ins	d_i ins	d_i mm	psi	bar	psi	bar	psi	bar
Std.	36.000	0.375	35.250	895.4	220	15.2	179	12.3	137	9.4
40	36.000	0.688	34.624	879.4	408	28.1	367	25.3	325	22.4
X-S	36.000	0.500	35.000	889.0	295	20.3	253	17.4	212	14.6
80	36.000	1.218	33.564	852.5	734	50.6	692	47.7	650	44.8
120	36.000	1.812	32.376	822.4	1110	76.6	1068	73.7	1027	70.8
160	36.000	2.344	31.312	795.3	1457	100.5	1416	97.7	1374	94.8

STRESS 15000 psi 103.4 MN/m² *MAXIMUM WORKING PRESSURES*

					Corrosion allowance (mm)					
					0		1.5		3	
Sch.	d_o ins	t ins	d_i ins	d_i mm	psi	bar	psi	bar	psi	bar
Std.	36.000	0.375	35.250	895.4	276	19.0	224	15.4	171	11.8
40	36.000	0.688	34.624	879.4	510	35.2	458	31.6	406	28.0
X-S	36.000	0.500	35.000	889.0	369	25.4	317	21.9	265	18.3
80	36.000	1.218	33.564	852.5	917	63.2	865	59.7	813	56.1
120	36.000	1.812	32.376	822.4	1387	95.7	1335	92.1	1283	88.5
160	36.000	2.344	31.312	795.3	1821	125.6	1770	122.1	1718	118.5

STRESS 17500 psi 120.7 MN/m² *MAXIMUM WORKING PRESSURES*

					Corrosion allowance (mm)					
					0		1.5		3	
Sch.	d_o ins	t ins	d_i ins	d_i mm	psi	bar	psi	bar	psi	bar
Std.	36.000	0.375	35.250	895.4	322	22.2	261	18.0	200	13.8
40	36.000	0.688	34.624	879.4	596	41.1	535	36.9	474	32.7
X-S	36.000	0.500	35.000	889.0	431	29.7	370	25.5	309	21.3
80	36.000	1.218	33.564	852.5	1070	73.8	1009	69.6	949	65.4
120	36.000	1.812	32.376	822.4	1618	111.6	1558	107.4	1497	103.2
160	36.000	2.344	31.312	795.3	2125	146.6	2065	142.4	2004	138.2

USEFUL TABLE 4

Minimum nozzle thicknesses and actual inside diameters.

SI UNITS

		0.0 mm CA			1.5 mm CA			3.0 mm CA			6.0 mm CA		
PS ins	D_o mm	min mm	Sch No	D_i mm	min mm	Sch No	D_i mm	min mm	Sch No	D_i mm	min mm	Sch No	D_i mm
3/4	26.7	2.6		20.9	4.1	**L**	19	5.6	**L**	19	8.6		19
1	33.4	3.0		26.6	4.5	**W**	25	6	**W**	25	9	**L**	25
3/2	48.3	3.3		40.9	4.8	**N**	38	6.3	**N**	38	9.3	**W**	38
2	60.3	3.5		52.5	5		51	6.5		51	9.5	**N**	51
3	88.9	4.8		77.9	6.4		74	6.9	160	67	11		76
4	114	5.3		102	6.8		97	8.3		87	11		87
6	168	6.3	40	154	7.8	80	146	9.3		146	12	160	132
8	219	7.2		203	8.7		194	11.4		194	14.4		173
10	273	8.2		254	9.7		243	11.4	80	243	14.4		243
12	234	8.4		303	9.9		303	11.4		289	14.4		289
14	556	8.4		333	9.9		333	11.4		317	14.4	80	317
16	406	8.4		381	9.9	40	381	11.4		381	14.4		363
18	457	8.4		429	9.9		429	11.4	40	429	14.4		409
20	508	8.4		489	9.9		489	11.4		478	14.4	40	455
24	610	8.4		575	9.9		575	11.4		575	14.4		575

NOTES

(1) The minimum thicknesses have been calculated allowing for a pipe tolerance of minus 12.5% and then adding on the corrosion allowance.
(2) The applicable pipe schedule is the one which results in a wall thickness equal to or greater than the minimum value.
(3) The inside diameters are the ones which should be used in thermal design calculations
(4) Tables only apply to 150 # and 300 # ratings.

Table 4 in US customary units is on the next page.

US CUSTOMARY UNITS

		0.0 ins CA			1/16 ins CA			1/8 ins CA			1/4 ins CA		
PS ins	D_o ins	min ins	Sch No	D_i ins	min ins	Sch No	D_i ins	min ins	Sch No	D_i ins	min ins	Sch No	D_i ins
3/4	1.05	.1		0.92	.161	**L**	0.75	.22	**L**	0.75	.34		0.75
1	1.32	.12		1.05	.177	**W**	1.0	.24	**W**	1.0	.35	**L**	1.0
3/2	1.9	.13		1.61	.189	**N**	1.5	.25	**N**	1.5	.37	**W**	1.5
2	2.37	.14		2.07	.197		2.0	.26		2.0	.38	**N**	2.0
3	3.5	.19		3.07	.252		2.9	.27	160	2.62	.43		3.0
4	4.5	.21		4.03	.268		3.83	.33		3.44	.45		3.44
6	6.62	.25	40	6.06	.307	80	5.76	.37		5.76	.48	160	5.19
8	8.62	.28		7.98	.343		7.63	.4		7.62	.52		6.81
10	10.7	.32		10	.382		9.56	.44	80	9.56	.56		9.56
12	12.7	.33		11.9	.39		11.9	.45		11.37	.57		11.4
14	14	.33		13.1	.39		13.1	.45		12.5	.57	80	12.5
16	16	.33		15.0	.39	40	15.0	.45		15.0	.57		14.3
18	18	.33		16.9	.39		16.9	.45	40	16.9	.57		16.1
20	20	.33		18.8	.39		18.8	.45		18.8	.57	40	18.8
24	24	.33		22.6	.39		22.6	.45		22.6	.57		22.6

The values for the internal pressures are from the current TEMA and can be safely used. The values for the external pressure have been taken from the red TEMA of 1959 and are not present in additions since then. They must therefore only be used as a very rough guide and the working pressure must be calculated for any tube close to the limits in this table.

TABLE 5 - INTERNAL & EXTERNAL WORKING PRESSURES OF TUBES AT VARIOUS VALUES OF ALLOWABLE STRESS

ALLOWABLE STRESS psi/*bar*			6000	*413.8*	8000	*551.7*	10000	*689.7*	12000	*827.6*	14000	*965.5*	16000	*1103.4*	18000	*1241.4*	20000	*1379.3*
Tube d 3/4 inch 19.05mm	20 BWG	INTERNAL	581	*40.1*	775	*53.4*	969	*66.8*	1163	*80.2*	1357	*93.6*	1551	*107.0*	1745	*120.3*	1939	*133.7*
		EXTERNAL	260	*18.0*	340	*23.0*	420	*29.0*	498	*34.0*	570	*39.0*	650	*44.0*	720	*50.0*	800	*55.0*
	18 BWG	INTERNAL	827	*57.0*	1102	*76.0*	1378	*95.0*	1654	*114.1*	1930	*133.1*	2205	*152.1*	2481	*171.1*	2757	*190.1*
		EXTERNAL	367	*25.3*	479	*33.0*	590	*40.7*	698	*48.1*	806	*55.6*	912	*62.9*	1018	*70.2*	1123	*77.4*
	16 BWG	INTERNAL	1117	*77.0*	1489	*102.7*	1862	*128.4*	2234	*154.1*	2607	*179.8*	2979	*205.4*	3352	*231.2*	3724	*256.8*
		EXTERNAL	514	*35.4*	676	*46.6*	837	*57.7*	995	*68.6*	1152	*79.4*	1308	*90.2*	1464	*101.0*	1618	*111.6*
	14 BWG	INTERNAL	1456	*100.4*	1942	*133.9*	2428	*167.4*	2913	*200.9*	3399	*234.4*	3885	*267.9*	4370	*301.4*	4856	*334.9*
		EXTERNAL	684	*47.2*	905	*62.4*	1125	*77.6*	1343	*92.6*	1560	*107.6*	1776	*122.5*	1991	*137.3*	2206	*152.1*
	12 BWG	INTERNAL	1973	*136.1*	2631	*181.4*	3289	*226.8*	3946	*272.1*	4604	*317.5*	5262	*362.9*	5920	*408.3*	6578	*453.7*
		EXTERNAL	939	*64.8*	1243	*85.7*	1545	*106.6*	1845	*127.2*	2144	*147.9*	2442	*168.4*	2739	*188.9*	3035	*209.3*
	10 BWG	INTERNAL	2501	*172.5*	3335	*230.0*	4169	*287.5*	5003	*345.0*	5836	*402.5*	6670	*460.0*	7504	*517.5*	8338	*575.0*
		EXTERNAL	1200	*82.8*	1583	*109.2*	1963	*135.4*	2339	*161.3*	2713	*187.1*	3086	*212.8*	3456	*238.3*	3825	*263.8*
	8 BWG	INTERNAL	3203	*220.9*	4271	*294.6*	5339	*368.2*	6407	*441.9*	7475	*515.5*	8543	*589.2*	9611	*662.8*	10679	*736.5*
		EXTERNAL	1682	*116.0*	2229	*153.7*	2773	*191.2*	3314	*228.6*	3845	*265.2*	4393	*303.0*	4930	*340.0*	5465	*376.9*
Tube d 1 inch 25.4 mm	20 BWG	INTERNAL	432	*29.8*	576	*39.7*	720	*49.7*	864	*59.6*	1008	*69.5*	1152	*79.4*	1296	*89.4*	1440	*99.3*
		EXTERNAL	180	*12.0*	235	*16.0*	280	*19.0*	330	*23.0*	385	*26.0*	435	*30.0*	480	*33.0*	530	*36.0*
	18 BWG	INTERNAL	611	*42.1*	815	*56.2*	1019	*70.3*	1223	*84.3*	1427	*98.4*	1631	*112.5*	1835	*126.6*	2039	*140.6*
		EXTERNAL	254	*17.5*	329	*22.7*	402	*27.7*	473	*32.6*	544	*37.5*	613	*42.3*	682	*47.0*	749	*51.7*
	16 BWG	INTERNAL	822	*56.7*	1097	*75.7*	1371	*94.6*	1645	*113.4*	1919	*132.3*	2194	*151.3*	2468	*170.2*	2742	*189.1*
		EXTERNAL	365	*25.2*	476	*32.8*	586	*40.4*	694	*47.9*	801	*55.2*	906	*62.5*	1011	*69.7*	1115	*76.9*
	14 BWG	INTERNAL	1066	*73.5*	1422	*98.1*	1778	*122.6*	2133	*147.1*	2489	*171.7*	2844	*196.1*	3200	*220.7*	3556	*245.2*
		EXTERNAL	489	*33.7*	642	*44.3*	793	*54.7*	942	*65.0*	1091	*75.2*	1238	*85.4*	1384	*95.4*	1530	*105.5*
	12 BWG	INTERNAL	1432	*98.8*	1910	*131.7*	2388	*164.7*	2865	*197.6*	3343	*230.6*	3821	*263.5*	4298	*296.4*	4776	*329.4*
		EXTERNAL	673	*46.4*	890	*61.4*	1105	*76.2*	1319	*91.0*	1532	*105.7*	1744	*120.3*	1955	*134.8*	2166	*149.4*
	10 BWG	INTERNAL	1801	*124.2*	2401	*165.6*	3001	*207.0*	3602	*248.4*	4202	*289.8*	4802	*331.2*	5403	*372.6*	6003	*414.0*
		EXTERNAL	853	*58.8*	1129	*77.9*	1404	*96.8*	1678	*115.7*	1951	*134.6*	2222	*153.2*	2493	*171.9*	2763	*190.6*
	8 BWG	INTERNAL	2281	*157.3*	3041	*209.7*	3801	*262.1*	4562	*314.6*	5322	*367.0*	6082	*419.4*	6843	*471.9*	7603	*524.3*
		EXTERNAL	1095	*75.5*	1446	*99.7*	1794	*123.7*	2141	*147.7*	2485	*171.4*	2827	*195.0*	3168	*218.5*	3508	*241.9*

TABLE 6 - TUBE SPECIFICATIONS

ASME SPECIFICATION		PRODUCTION	TUBE WALL	TOLERANCE	
A179	LOW CARBON STEEL	SEAMLESS	MINIMUM	+20	-0
A214	CARBON STEEL	WELDED	MINIMUM	+20	-0
A210	MED CARBON STEEL	SEAMLESS	MINIMUM	+20	-0
A192	C.STEEL FOR HP SERVICE	SEAMLESS	MINIMUM	+20	-0
A334	CS & LOW ALLOY FOR LT SERVICE	SEAMLESS	MINIMUM	+20	-0
A334	CS & LOW ALLOY FOR LT SERVICE	WELDED	MINIMUM	+20	-0
A268	11.5-13.5 Cr TYPE 410	SEAMLESS	AVERAGE	+10	-10
A268	11.5-13.5 Cr TYPE 410	WELDED	AVERAGE	+10	-10
A213	T5/T11 LOW ALLOY (FERRITIC)	SEAMLESS	MINIMUM	+20	-0
	SS 300 SERIES (AUSTENITIC)	SEAMLESS	MINIMUM	+20	-0
A249	SS 300 SERIES (AUSTENITIC)	WELDED	AVERAGE	+10	-10
A789	DUPLEX 2205 UNS31803	SEAMLESS	AVERAGE	+10	-10
A789	DUPLEX 2205 UNS31803	WELDED	AVERAGE	+10	-10
B111	NON-FERROUS	SEAMLESS	MINIMUM	+16	-0
B111	90/10, 70/30, ADM, AL-BRONZE	SEAMLESS	AVERAGE	+8	-8
B165	MONEL (70/30 Ni/Cu)	SEAMLESS	MINIMUM	+22	-0
B163	MONEL N04400 (U Tubes only)	SEAMLESS	AVERAGE	+10	-10
B704	INCONEL 625 UNS N06625	WELDED	MINIMUM	+28	-0
B704	INCOLOY 825 UNS N08825	WELDED	AVERAGE	+12.5	-12.5
B622	HASTELLOY	SEAMLESS	MINIMUM	+20	-0
B622	HASTELLOY	SEAMLESS	AVERAGE	+10	-10
B163	INCOLOY 825 UNS N08825	SEAMLESS	MINIMUM	+(20-25)	-0
B163	INCOLOY 825 UNS N08825	SEAMLESS	AVERAGE	+(10-12.5)	-(10-12.5)
B338	TITANIUM UP TO 19.1 mm d_o	SEAMLESS	MINIMUM	+20	-0
B338	TITANIUM UP TO 19.1 mm d_o	WELDED	AVERAGE	+10	-10
B338	TITANIUM ABOVE 19.1 mm d_o	SEAMLESS	MINIMUM	+20	-0

Blank rows have been left in the above table to enable the reader to add additional tube specifications.

TABLE 7 – PARTITION LANE DUMMY TUBES

The tables in this section are for triangular or rotated triangular pitch tubes which are either symmetrical or displaced about the lane axis. All dimensions are in mm. For symmetrical arrangements the clearances between the dummy tube and adjacent active tubes have the same value. For non symmetrical arrangements there are two clearances which ideally should not have the same value.

Values for the gap can be found in table 2.3 in chapter 2.

For U tubes when dummy tubes are used together with active inclined or crossed tubes the best approach is to expand a section around the partition lane of the tube layout from your output and find suitable places 'by eye'. When there are no crossed or inclined tubes involved the most satisfactory solution is to use sealing strips (which may be any width) and just specify the clearances.

TABLE 7A

Triangular pitch tubes (as viewed), symmetrical about the pass partition lane.

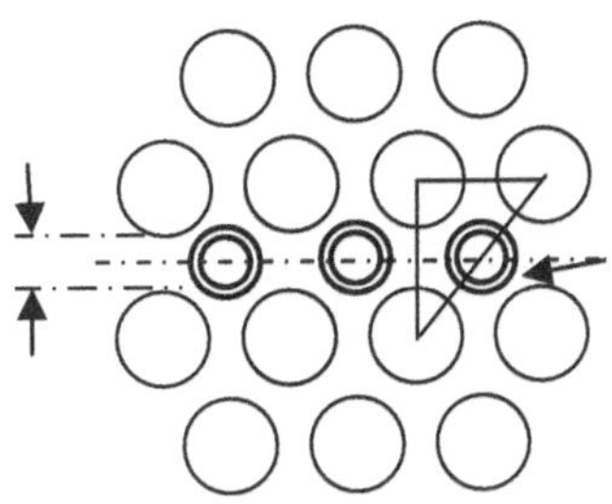

Figure 2.70

Tube	Tube	GAP	B	Diagonal	Dummy	Clearance
d_o	P_t		d_o+GAP	$(B^2+P_t^2)^{0.5}$	d_d	0.5 (Diag-d_o-d_d)
19.05	23.81	13	32.05	39.93	16	2.4
19.05	25.4	13	32.05	40.89	16	2.9
19.05	23.81	16	35.05	42.37	16	3.7
19.05	25.4	16	35.05	43.29	16	4.1
19.05	23.81	19	38.05	44.89	19	3.4
19.05	25.4	19	38.05	45.75	19	3.8
19.05	23.81	22	41.05	47.46	22	3.2
19.05	25.4	22	41.05	48.27	22	3.6
25.4	31.75	13	38.4	49.83	16	4.2
25.4	31.75	16	41.4	52.17	19	3.9
25.4	31.75	19	44.4	54.58	22	3.6
25.4	31.75	22	47.4	57.05	25.4	3.1

The dummy tubes must be on the centre line – specify only the diameter of the dummy tubes.

TABLE 7B

Triangular pitch tubes (as viewed), displaced by half a tube pitch about the pass partition lane.

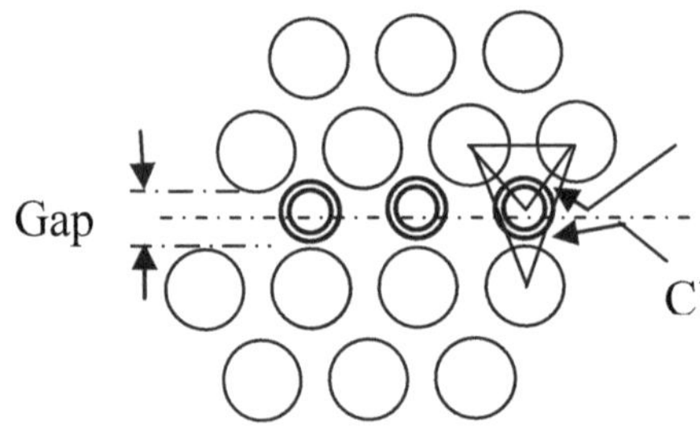

Figure 2.71

Tube	Tube	GAP	B	C1	B_1	Diagonal	Dummy	C2	C Mean
d_o	Pt		d_o+GAP	(Pt-d_o)/2			d_d		(C1+C2)/2
19.05	23.8	13	32.05	2.4	15.4	19.46	9.5	5.2	3.8
19.05	25.4	13	32.05	3.2	14.6	19.35	9.5	5.1	4.1
19.05	23.8	16	35.05	2.4	16.6	20.46	13	4.4	3.4
19.05	25.4	16	35.05	3.2	15.9	20.31	13	4.3	3.7
19.05	23.8	19	38.05	2.4	18.1	21.70	16	4.2	3.3
19.05	25.4	19	38.05	3.2	17.4	21.50	16	4.0	3.6
19.05	23.8	22	41.05	2.4	19.6	22.97	19	3.9	3.2
19.05	25.4	22	41.05	3.2	18.9	22.73	19	3.7	3.4
25.4	31.8	13	38.4	3.2	16.0	22.56	13	3.4	3.3
25.4	31.8	16	41.4	3.2	19.0	24.78	13	5.6	4.4
25.4	31.8	19	44.4	3.2	20.5	25.95	16	5.2	4.2
25.4	31.8	22	47.4	3.2	22.0	27.15	19	4.9	4.1

From the details in chapter 2.

$$B_1 = (B - C1 - d_o/2 - d_d/2)$$

$$\text{Diagonal} = [B_1^2 + (P_t/2)^2]^{0.5}$$

$$\text{Clearance C2} = (\text{Diagonal} - d_o/2 - d_d/2)$$

The dummy tubes are not on the centre line – specify the diameter of the dummy tubes and the clearance C1.

TABLE 7C

Rotated triangular pitch tubes (as viewed) symmetrical about the pass partition lane.

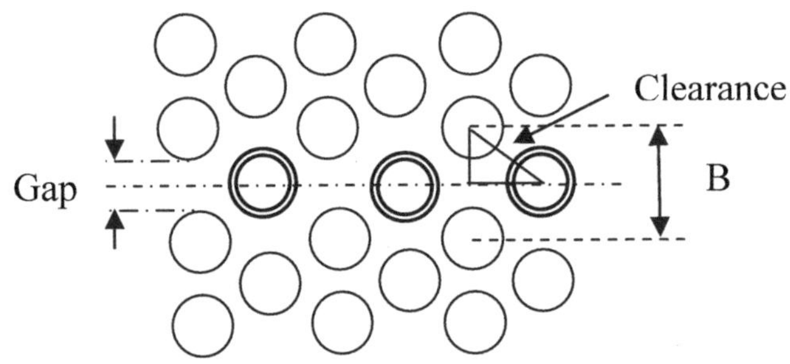

Figure 2.74

Tube d_o	Tube Pt	GAP	B/2 $(d_o+GAP)/2$	Diagonal $((B/2)^2+(0.866P_t)^2)^{0.5}$	Dummy d_d	Clearance Diag-$d_o/2$-$d_d/2$
19.05	23.81	13	16.0	26.11	22	3.2
19.05	25.4	13	16.0	27.21	22	3.5
19.05	23.81	16	17.5	27.06	22	4.2
19.05	25.4	16	17.5	28.12	22	4.4
19.05	23.81	19	19.0	28.06	25.4	3.5
19.05	25.4	19	19.0	29.08	25.4	3.7
19.05	23.81	22	20.5	29.09	25.4	4.5
19.05	25.4	22	20.5	30.09	25.4	4.7
25.4	31.75	13	19.2	33.54	25.4	5.0
25.4	31.75	16	20.7	34.42	25.4	5.8
25.4	31.75	19	22.2	35.34	32	3.5
25.4	31.75	22	23.7	36.30	32	4.4

From the details in chapter 2.

$$\text{The diagonal} = ([0.866P_t]^2 + [B/2]^2)^{0.5}$$

$$\text{Clearance} = (\text{Diagonal} - d_o/2 - d_d/2)$$

The dummy tubes must be on the centre line – specify only the diameter of the dummy tubes.

TABLE 7D

Rotated triangular pitch tubes (as viewed) with the tubes displaced by the height of a tube triangle about the pass partition lane.

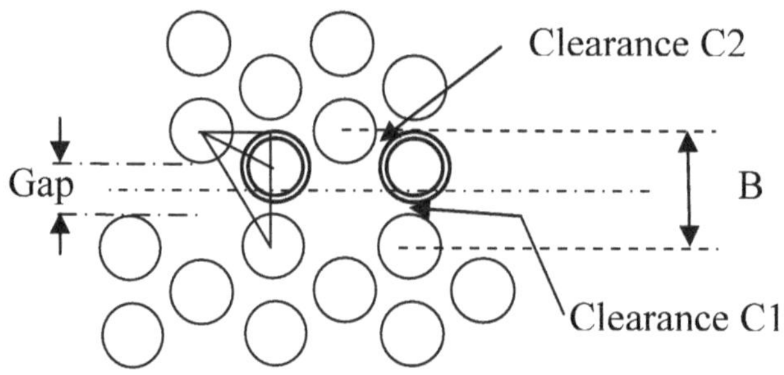

Figure 2.76

Tube	Tube	GAP	B	C1	B_1	Diagonal	Dummy	C2	C Mean
do	Pt		d_o+GAP	(Pt-d_o)/2			d_d		(C1+C2)/2
19.05	23.81	13	32.05	2.4	10.65	23.21	19	4.2	3.3
19.05	25.4	13	32.05	3.2	9.85	24.10	19	5.1	4.1
19.05	23.81	16	35.05	2.4	13.65	24.73	19	5.7	4.0
19.05	25.4	16	35.05	3.2	12.85	25.47	19	6.4	4.8
19.05	23.81	19	38.05	2.4	15.15	25.58	22	5.1	3.7
19.05	25.4	19	38.05	3.2	14.35	26.26	22	5.7	4.5
19.05	23.81	22	41.05	2.4	18.15	27.47	22	6.9	4.7
19.05	25.4	22	41.05	3.2	15.65	27.00	25.4	4.8	4.0
25.4	31.75	13	38.4	3.2	9.83	29.20	25.4	3.8	3.5
25.4	31.75	16	41.4	3.2	12.83	30.34	25.4	4.9	4.1
25.4	31.75	19	44.4	3.2	15.83	31.72	25.4	6.3	4.7
25.4	31.75	22	47.4	3.2	18.83	33.32	25.4	7.9	5.5

From the details in chapter 2.

$$B_1 = (B - C1 - d_o/2 - d_d/2)$$

$$\text{Diagonal} = [B_1^2 + (0.866P_t)^2]^{0.5}$$

$$\text{Clearance C2} = (\text{Diagonal} - d_o/2 - d_d/2)$$

The dummy tubes are not on the centre line – specify the diameter of the dummy tubes and the clearance C1.

TABLE 8 – OTL's FOR FLOATING HEAD EXCHANGERS

TABLE 8A SPLIT RING FLOATING HEADS – TEMA S TYPE

SHELL ID ins	SHELL ID mm	Clearance ins	Clearance mm	OTL ins	OTL mm
12	304.8	1.25	32	10.75	273
13.25	336.55	1.25	32	12	305
15.25	387.35	1.25	32	14	356
17.25	438.15	1.25	32	16	406
19.25	488.95	1.25	32	18	457
23.25	590.55	1.375	35	21.875	556
27	685.8	1.625	41	25.375	645
29	736.6	1.625	41	27.375	695
31	787.4	1.625	41	29.375	746
33	838.2	1.75	44	31.25	794
35	889	1.75	44	33.25	845
37	939.8	1.75	44	35.25	895
39	990.6	1.75	44	37.25	946
42	1066.8	1.75	44	40.25	1022
44	1117.6	1.875	48	42.125	1070
47	1193.8	1.875	48	45.125	1146
49	1244.6	1.875	48	47.125	1197
52	1320.8	1.875	48	50.125	1273
54	1371.6	1.875	48	52.125	1324
57	1447.8	1.875	48	55.125	1400
59	1498.6	1.875	48	57.125	1451
62	1574.8	2.125	54	59.875	1521
64	1625.6	2.25	57	61.75	1568
67	1701.8	2.25	57	64.75	1645
69	1752.6	2.25	57	66.75	1695
72	1828.8	2.5	64	69.5	1765
74	1879.6	2.5	64	71.5	1816
77	1955.8	2.5	64	74.5	1892
79	2006.6	2.5	64	76.5	1943
81	2057.4	2.75	70	78.25	1988

For welded tubes 2 x 3 = 6 mm (0.25 ins) should be added to the bundle/shell clearances in the above table. This decreases the OTL by the same amount.

TABLE 8B PULL THROUGH FLOATING HEADS – TEMA T TYPE

Shell ins	Shell mm	Bolts ins	Pressure psig	Pressure barg	OTL ins	OTL mm	Clearance ins	Clearance mm
12	305	0.75	1030	71	7.875	200	4.125	105
12	305	1	1640	113	7.125	181	4.875	124
13.25	337	0.75	1035	71	9.125	232	4.125	105
13.25	337	1	1290	89	8.375	213	4.875	124
15.25	387	0.75	910	63	11.125	283	4.125	105
15.25	387	1	1235	85	10.375	264	4.875	124
17.25	438	0.75	695	48	13.125	333	4.125	105
17.25	438	1	1165	80	12.375	314	4.875	124
19.25	489	0.75	640	44	15.125	384	4.125	105
19.25	489	1	910	63	14.375	365	4.875	124
23.25	591	0.75	560	39	19	483	4.25	108
23.25	591	1	837	58	18.25	464	5	127
23.25	591	1.25	1090	75	17.625	448	5.625	143
27	686	0.75	480	33	22.5	572	4.5	114
27	686	1	680	47	21.75	552	5.25	133
27	686	1.25	900	62	21.125	537	5.875	149
27	686	1.5	1210	83	20.375	518	6.625	168
29	737	0.75	455	31	24.5	622	4.5	114
29	737	1	655	45	23.75	603	5.25	133
29	737	1.25	900	62	23.125	587	5.875	149
29	737	1.5	1235	85	22.375	568	6.625	168
31	787	0.75	425	29	26.5	673	4.5	114
31	787	1	570	39	25.75	654	5.25	133
31	787	1.25	780	54	25.125	638	5.875	149
31	787	1.5	1065	73	24.375	619	6.625	168
33	838	0.75	375	26	28.5	724	4.5	114
33	838	1	555	38	27.5	699	5.5	140
33	838	1.25	780	54	27.125	689	5.875	149
33	838	1.5	1085	75	26.375	670	6.625	168
35	889	0.75	360	25	30.5	775	4.5	114
35	889	1	540	37	29.75	756	5.25	133
35	889	1.25	685	47	29.125	740	5.875	149
35	889	1.5	955	66	28.375	721	6.625	168
35	889	1.75	1210	83	27.625	702	7.375	187
37	940	0.75	345	24	32.375	822	4.625	117
37	940	1	530	37	31.625	803	5.375	137
37	940	1.25	695	48	31	787	6	152
37	940	1.5	975	67	30.25	768	6.75	171
37	940	1.75	1075	74	29.5	749	7.5	191
39	991	0.75	330	23	34.375	873	4.625	117
39	991	1	475	33	33.625	854	5.375	137
39	991	1.25	690	48	33	838	6	152
39	991	1.5	870	60	32.25	819	6.75	171
39	991	1.75	1000	69	31.5	800	7.5	191

Continued on the next page.

Table 8B continued.

Shell ins	Shell mm	Bolts ins	Pressure psig	Pressure barg	OTL ins	OTL mm	Clearance ins	Clearance mm
42	1067	0.75	305	21	37.375	949	4.625	117
42	1067	1	440	30	36.625	930	5.375	137
42	1067	1.25	590	41	36	914	6	152
42	1067	1.5	830	57	35.25	895	6.75	171
42	1067	1.75	950	66	34.5	876	7.5	191
44	1118	0.75	295	20	39.25	997	4.75	121
44	1118	1	435	28	38.5	978	5.5	140
44	1118	1.25	590	39	37.875	962	6.125	156
44	1118	1.5	760	50	37.125	943	6.875	175
44	1118	1.75	985	59	36.375	924	7.625	194
44	1118	2	1200	71	35.625	905	8.375	213
47	1194	0.75	285	20	42.25	1073	4.75	121
47	1194	1	405	28	41.5	1054	5.5	140
47	1194	1.25	560	39	40.875	1038	6.125	156
47	1194	1.5	730	50	40.125	1019	6.875	175
47	1194	1.75	855	59	39.375	1000	7.625	194
47	1194	2	1035	71	38.625	981	8.375	213
49	1245	0.75	260	18	44.25	1124	4.75	121
49	1245	1	395	27	43.5	1105	5.5	140
49	1245	1.25	515	36	42.875	1089	6.125	156
49	1245	1.5	670	46	42.125	1070	6.875	175
49	1245	1.75	880	61	41.375	1051	7.625	194
49	1245	2	1000	69	40.625	1032	8.375	213
52	1321	0.75	255	18	47.25	1200	4.75	121
52	1321	1	375	26	46.5	1181	5.5	140
52	1321	1.25	490	34	45.875	1165	6.125	156
52	1321	1.5	650	45	45.125	1146	6.875	175
52	1321	1.75	860	59	44.375	1127	7.625	194
52	1321	2	950	66	43.625	1108	8.375	213
54	1372	0.75	250	17	49.25	1251	4.75	121
54	1372	1	345	24	48.5	1232	5.5	140
54	1372	1.25	490	34	47.875	1216	6.125	156
54	1372	1.5	655	45	47.125	1197	6.875	175
54	1372	1.75	790	54	46.375	1178	7.625	194
54	1372	2	925	64	45.625	1159	8.375	213
57	1448	0.75	230	16	52.25	1327	4.75	121
57	1448	1	325	22	51.5	1308	5.5	140
57	1448	1.25	470	32	50.875	1292	6.125	156
57	1448	1.5	585	40	50.125	1273	6.875	175
57	1448	1.75	775	53	49.375	1254	7.625	194
57	1448	2	875	60	48.625	1235	8.375	213
59	1499	0.75	225	16	54.25	1378	4.75	121
59	1499	1	320	22	53.5	1359	5.5	140
59	1499	1.25	435	30	52.875	1343	6.125	156
59	1499	1.5	590	41	52.125	1324	6.875	175
59	1499	1.75	720	50	51.375	1305	7.625	194
59	1499	2	850	59	50.625	1286	8.375	213

Continued on the next page.

Table 8B continued.

Shell ins	Shell mm	Bolts ins	Pressure psig	Pressure barg	OTL ins	OTL mm	Clearance ins	Clearance mm
62	1575	0.75	210	14	57.25	1454	4.75	121
62	1575	1	305	21	56.5	1435	5.5	140
62	1575	1.25	420	29	55.875	1419	6.125	156
62	1575	1.5	570	39	55.125	1400	6.875	175
62	1575	1.75	705	49	54.375	1381	7.625	194
62	1575	2	810	52	53.625	1362	8.375	213
64	1626	0.75	205	13	59.125	1502	4.875	124
64	1626	1	300	20	58.375	1483	5.625	143
64	1626	1.25	395	26	57.75	1467	6.25	159
64	1626	1.5	535	36	57	1448	7	178
64	1626	1.75	665	44	56.25	1429	7.75	197
64	1626	2	760	52	55.5	1410	8.5	216
67	1702	0.75	195	13	61.875	1572	5.125	130
67	1702	1	285	20	61.125	1553	5.875	149
67	1702	1.25	375	26	60.5	1537	6.5	165
67	1702	1.5	515	36	59.75	1518	7.25	184
67	1702	1.75	640	44	59	1499	8	203
67	1702	2	745	51	58.25	1480	8.75	222
69	1753	0.75	190	13	63.875	1622	5.125	130
69	1753	1	280	19	63.125	1603	5.875	149
69	1753	1.25	375	26	62.5	1588	6.5	165
69	1753	1.5	485	33	61.75	1568	7.25	184
69	1753	1.75	605	42	61	1549	8	203
69	1753	2	700	48	60.25	1530	8.75	222
72	1829	0.75	185	13	66.875	1699	5.125	130
72	1829	1	265	18	66.125	1680	5.875	149
72	1829	1.25	360	25	65.5	1664	6.5	165
72	1829	1.5	470	32	64.75	1645	7.25	184
72	1829	1.75	595	41	64	1626	8	203
72	1829	2	690	48	63.25	1607	8.75	222
74	1880	0.75	180	12	68.875	1749	5.125	130
74	1880	1	250	17	68.125	1730	5.875	149
74	1880	1.25	355	24	67.5	1715	6.5	165
74	1880	1.5	445	31	66.75	1695	7.25	184
74	1880	1.75	560	39	66	1676	8	203
74	1880	2	660	46	65.25	1657	8.75	222

For welded tubes 2 x 3 = 6 mm (0.25 ins) should be added to the bundle/shell clearances in the above table. This decreases the OTL by the same amount.

TABLE 9 – NOZZLE PROJECTIONS

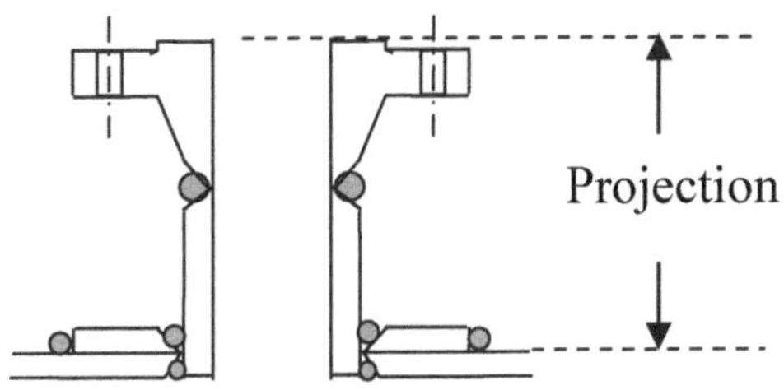

Figure 2.82

The table below gives the nozzle projections in mm for the most commonly used flange ratings.

Nozzle NB ins	150 lb	300 lb	600 lb	900 lb
2	175	175	200	200
4	200	200	200	200
6	200	225	225	250
8	200	225	250	250
10	200	250	250	300
12	225	250	250	300
14	225	250	300	350
16	250	250	300	350
18	250	300	350	400
20	275	300	350	400
24	275	300	350	400

The thickness of any insulation must be added to the values in this table.

TABLE 10 – INLET AND OUTLET SPACES

The development and use of these tables is explained in detail in Chapter 2. A nomenclature and summary follow:

D_s & D_n The shell inside diameter and the nozzle nominal bore.

t The shell thickness for the pressure at the top of the page

TS The tubesheet thickness for the pressure at the top of the page for all types of head except for U tube bundles. For U tubes only increase this value by 25%. The pressure for the tubesheet is the higher of the shell side and tube side values.

F_{stat} The length of the shell girth flange for the pressure at the top of the page.

$F_{S\ head}$ The length required for the split ring and bolts for an S type floating head for the pressure at the top of the table.

$F_{T\ head}$ The length required for the bolts and access for a T type floating head for the pressure at the top of the table

L1 The length between the face of the tubesheet and the centre line of the first baffle for a stationary tubesheet at the pressure at the top of the page. If the tube side pressure is higher than the shell side pressure this length must be increased by the difference in lengths of the shell flanges, F_{stat}, at the two pressures.

L2 S The length between the face of the tubesheet and the centre line of the first baffle for an S type floating tubesheet at the pressure at the top of the page. If the tube side pressure is higher than the shell side pressure this length must be increased by the difference in lengths of the split rings and bolts, $F_{S\ head}$, at the two pressures.

L2 T The length between the face of the tubesheet and the centre line of the first baffle for a T type floating tubesheet at the pressure at the top of the page. . If the tube side pressure is higher than the shell side pressure this length must be increased by the difference in lengths of the bolts and access, $F_{T\ head}$, at the two pressures.

L2 U & L2 T special are estimated as explained in Chapter 2.

SHELL SIDE DESIGN PRESSURE 150 psig, 10.34 barg

D_s ins	D_s mm	D_n ins	t mm	TS mm	F_{stat} mm	$F_{S\ head}$ mm	$F_{T\ head}$ mm	L1 mm	L2 S mm	L2 T mm
12.09	**307**	**2**	**8.4**	**27**	**70**	**67**	**25.4**	**267**	**356**	**279**
		4						318	406	330
		6						368	457	381
13.25	**337**	**2**	**9.5**	**29**	**75**	**76**	**38.1**	**279**	**394**	**305**
		4						305	419	330
		6						356	470	381
15.25	**387**	**2**	**9.5**	**30**	**75**	**76**	**50.8**	**279**	**406**	**318**
		4						305	432	343
		6						356	483	394
		8						406	533	445
17.25	**438**	**2**	**9.5**	**33**	**75**	**76**	**50.8**	**279**	**406**	**318**
		4						305	432	343
		6						356	483	394
		8						406	533	445
19.25	**489**	**2**	**9.5**	**35**	**75**	**76**	**50.8**	**279**	**419**	**318**
		4						305	445	343
		6						356	495	394
		8						406	546	445
22	**559**	**4**	**9.5**	**40**	**81**	**83**	**50.8**	**318**	**457**	**356**
		6						368	508	406
		8						406	546	445
		10						470	610	508
24	**610**	**4**	**9.5**	**41**	**81**	**83**	**50.8**	**318**	**470**	**356**
		6						368	521	406
		8						419	572	457
		10						470	622	508
		12						521	673	559
26	**660**	**4**	**9.5**	**44**	**83**	**89**	**50.8**	**318**	**483**	**356**
		6						368	533	406
		8						406	572	445
		10						483	648	521
		12						533	699	572
28	**711**	**4**	**9.5**	**46**	**84**	**89**	**63.5**	**318**	**495**	**368**
		6						368	546	419
		8						406	584	457
		10						495	673	546
		12						546	724	597
		14						597	775	648
30	**762**	**6**	**11.1**	**49**	**84**	**89**	**63.5**	**368**	**546**	**419**
		8						419	597	470
		10						470	648	521
		12						546	724	597

Continued on the next page.

SHELL SIDE DESIGN PRESSURE 150 psig, 10.34 barg

D_s ins	D_s mm	D_n ins	t mm	TS mm	F_{stat} mm	$F_{S\ head}$ mm	$F_{T\ head}$ mm	L1 mm	L2 S mm	L2 T mm
32	**813**	**6**	**11.1**	**52**	**84**	**94**	**63.5**	**381**	**559**	**424**
		8						432	610	475
		10						483	660	526
		12						533	711	577
		14						584	762	627
		16						622	800	665
34	**864**	**6**	**11.1**	**54**	**87**	**95**	**63.5**	**381**	**572**	**424**
		8						432	622	475
		10						495	686	538
		12						546	737	589
		14						622	813	665
		16						660	851	704
36	**914**	**6**	**11.1**	**57**	**91**	**98**	**63.5**	**394**	**584**	**437**
		8						445	635	488
		10						508	699	551
		12						559	749	602
		14						622	813	665
		16						660	851	704
		18						711	902	754
38	**965**	**6**	**11.1**	**59**	**92**	**98**	**63.5**	**394**	**584**	**437**
		10						533	724	577
		14						622	813	665
		18						737	927	780
40	**1016**	**8**	**12.7**	**62**	**92**	**103**	**63.5**	**432**	**622**	**475**
		12						533	724	577
		16						648	838	691
		20						762	953	805
42	**1067**	**6**	**12.7**	**65**	**94**	**103**	**63.5**	**406**	**597**	**450**
		10						533	724	577
		14						622	813	665
		18						724	914	767
44	**1118**	**8**	**12.7**	**67**	**97**	**108**	**63.5**	**445**	**635**	**488**
		12						533	724	577
		16						673	864	716
		20						775	965	818
46	**1168**	**6**	**12.7**	**70**	**100**	**108**	**63.5**	**432**	**635**	**475**
		10						533	737	577
		14						622	826	665
		18						737	940	780
		20						800	1003	843

Continued on the next page.

SHELL SIDE DESIGN PRESSURE 150 psig, 10.34 barg

D_s ins	D_s mm	D_n ins	t mm	TS mm	F_{stat} mm	$F_{S\ head}$ mm	$F_{T\ head}$ mm	L1 mm	L2 S mm	L2 T mm
48	**1219**	**8**	**12.7**	**73**	**105**	**108**	**63.5**	**508**	**711**	**551**
		12						584	787	627
		16						673	876	716
		20						813	1016	856
		24						914	1118	958
50	**1270**	**6**	**12.7**	**75**	**108**	**111**	**63.5**	**445**	**648**	**488**
		10						546	749	589
		14						622	826	665
		18						762	965	805
		24						940	1143	983
52	**1321**	**8**	**12.7**	**78**	**111**	**111**	**63.5**	**508**	**711**	**551**
		12						584	787	627
		16						699	902	742
		20						826	1029	869
		24						965	1168	1008
54	**1372**	**6**	**12.7**	**79**	**116**	**114**	**63.5**	**457**	**673**	**500**
		10						559	775	602
		14						648	864	691
		18						775	991	818
		24						978	1194	1021
56	**1422**	**8**	**12.7**	**83**	**121**	**119**	**63.5**	**508**	**724**	**551**
		12						610	826	653
		16						724	940	767
		20						864	1080	907
		24						991	1207	1034
58	**1473**	**6**	**14.3**	**86**	**121**	**121**	**63.5**	**457**	**673**	**500**
		10						559	775	602
		14						648	864	691
		18						775	991	818
		24						991	1207	1034
60	**1524**	**8**	**14.3**	**87**	**121**	**125**	**63.5**	**521**	**737**	**564**
		12						622	838	665
		16						724	940	767
		20						864	1080	907
		24						991	1207	1034

SHELL SIDE DESIGN PRESSURE 300 psig, 20.69 barg

D_s ins	D_s mm	D_n ins	t mm	TS mm	F_{stat} mm	$F_{S\ head}$ mm	$F_{T\ head}$ mm	L1 mm	L2 S mm	L2 T mm
12.09	**307**	**2**	**8.4**	**33**	**70**	**67**	**38.1**	**267**	**368**	**292**
		4						**330**	**432**	**356**
		6						**381**	**483**	**406**
13.25	**337**	**2**	**9.5**	**37**	**75**	**76**	**50.8**	**279**	**406**	**318**
		4						**330**	**457**	**368**
		6						**394**	**521**	**432**
15.25	**387**	**2**	**9.5**	**40**	**75**	**76**	**50.8**	**305**	**445**	**343**
		4						**343**	**483**	**381**
		6						**394**	**533**	**432**
		8						**445**	**584**	**483**
17.25	**438**	**2**	**9.5**	**43**	**75**	**76**	**50.8**	**318**	**457**	**356**
		4						**356**	**495**	**394**
		6						**406**	**546**	**445**
		8						**457**	**597**	**495**
19.25	**489**	**2**	**11.1**	**48**	**76**	**76**	**50.8**	**318**	**457**	**356**
		4						**356**	**495**	**394**
		6						**406**	**546**	**445**
		8						**457**	**597**	**495**
22	**559**	**4**	**11.1**	**52**	**83**	**83**	**50.8**	**356**	**508**	**394**
		6						**406**	**559**	**445**
		8						**457**	**610**	**495**
		10						**508**	**660**	**546**
24	**610**	**4**	**12.7**	**56**	**86**	**83**	**50.8**	**356**	**521**	**394**
		6						**406**	**572**	**445**
		8						**457**	**622**	**495**
		10						**521**	**686**	**559**
		12						**572**	**737**	**610**
26	**660**	**4**	**12.7**	**57**	**87**	**84**	**63.5**	**356**	**521**	**406**
		6						**419**	**584**	**470**
		8						**470**	**635**	**521**
		10						**508**	**673**	**559**
		12						**572**	**737**	**622**
28	**711**	**4**	**12.7**	**64**	**92**	**92**	**63.5**	**368**	**546**	**419**
		6						**419**	**597**	**470**
		8						**470**	**648**	**521**
		10						**508**	**686**	**559**
		12						**597**	**775**	**648**
		14						**610**	**787**	**660**
30	**762**	**6**	**14.3**	**67**	**95**	**97**	**63.5**	**432**	**610**	**483**
		8						**483**	**660**	**533**
		10						**521**	**699**	**572**
		12						**597**	**775**	**648**

Continued on the next page.

SHELL SIDE DESIGN PRESSURE 300 psig, 20.69 barg

D_s ins	D_s mm	D_n ins	t mm	TS mm	F_{stat} mm	$F_{S\ head}$ mm	$F_{T\ head}$ mm	L1 mm	L2 S mm	L2 T mm
32	**813**	**6**	**14.3**	**70**	**100**	**100**	**63.5**	**445**	**635**	**488**
		8						495	686	538
		10						533	724	577
		12						610	800	653
		14						635	826	678
		16						699	889	742
34	**864**	**6**	**15.9**	**75**	**100**	**105**	**63.5**	**445**	**635**	**488**
		8						508	699	551
		10						546	737	589
		12						610	800	653
		14						648	838	691
		16						724	914	767
36	**914**	**6**	**15.9**	**78**	**102**	**108**	**63.5**	**445**	**648**	**488**
		8						508	711	551
		10						546	749	589
		12						610	813	653
		14						648	851	691
		16						724	927	767
		18						800	1003	843
38	**965**	**6**	**17.5**	**81**	**105**	**111**	**63.5**	**445**	**660**	**488**
		10						546	762	589
		14						648	864	691
		18						800	1016	843
40	**1016**	**8**	**17.5**	**86**	**106**	**119**	**63.5**	**508**	**737**	**551**
		12						622	851	665
		16						724	953	767
		20						864	1092	907
42	**1067**	**6**	**17.5**	**89**	**113**	**121**	**63.5**	**445**	**686**	**488**
		10						559	800	602
		14						673	914	716
		18						813	1054	856
44	**1118**	**8**	**19.1**	**92**	**111**	**125**	**63.5**	**521**	**762**	**564**
		12						635	876	678
		16						737	978	780
		20						864	1105	907
46	**1168**	**6**	**19.1**	**97**	**116**	**132**	**63.5**	**457**	**711**	**500**
		10						559	813	602
		14						673	927	716
		18						826	1080	869
		20						876	1130	919

Continued on the next page.

SHELL SIDE DESIGN PRESSURE 300 psig, 20.69 barg

D_s ins	D_s mm	D_n ins	t mm	TS mm	F_{stat} mm	$F_{S\ head}$ mm	$F_{T\ head}$ mm	L1 mm	L2 S mm	L2 T mm
48	**1219**	**8**	**20.7**	**100**	**122**	**133**	**63.5**	**521**	**775**	**564**
		12						635	889	678
		16						737	991	780
		20						889	1143	932
		24						1016	1270	1059
50	**1270**	**6**	**20.7**	**105**	**125**	**140**	**63.5**	**470**	**724**	**513**
		10						559	813	602
		14						686	940	729
		18						838	1092	881
		24						1029	1283	1072
52	**1321**	**8**	**20.7**	**108**	**130**	**141**	**63.5**	**533**	**787**	**577**
		12						660	914	704
		16						775	1029	818
		20						927	1181	970
		24						1041	1295	1085
54	**1372**	**6**	**22.2**	**111**	**130**	**144**	**63.5**	**483**	**749**	**526**
		10						572	838	615
		14						686	953	729
		18						838	1105	881
		24						1054	1321	1097
56	**1422**	**8**	**22.2**	**114**	**133**	**152**	**63.5**	**546**	**813**	**589**
		12						673	940	716
		16						775	1041	818
		20						927	1194	970
		24						1054	1321	1097
58	**1473**	**6**	**23.8**	**119**	**138**	**156**	**76.2**	**483**	**762**	**538**
		10						572	851	627
		14						686	965	742
		18						838	1118	894
		24						1067	1346	1123
60	**1524**	**8**	**23.8**	**122**	**144**	**163**	**76.2**	**546**	**838**	**602**
		16						775	1067	831
		20						927	1219	983
		24						1067	1359	1123

SHELL SIDE DESIGN PRESSURE 450 psig, 31.02 barg

D_s ins	D_s mm	D_n ins	t mm	TS mm	F_{stat} mm	$F_{S\ head}$ mm	$F_{T\ head}$ mm	L1 mm	L2 S mm	L2 T mm
12	305	2	9.5	40	70	67	50.8	305	445	343
		4						343	483	381
		6						394	533	432
13.13	334	2	12.0	43	75	76	50.8	305	445	343
		4						356	495	394
		6						406	546	445
15	381	2	12.7	48	75	73	50.8	305	457	343
		4						343	495	381
		6						406	559	445
		8						445	597	483
17	432	2	12.7	51	78	71	63.5	318	470	368
		4						343	495	394
		6						419	572	470
		8						457	610	508
19.25	489	2	14.3	57	83	83	63.5	318	495	368
		4						356	533	406
		6						419	597	470
		8						457	635	508
22	559	4	15.9	64	87	89	63.5	356	533	406
		6						419	597	470
		8						457	635	508
		10						508	686	559
24	610	4	15.9	67	94	102	63.5	356	546	406
		6						432	622	483
		8						483	673	533
		10						521	711	572
		12						584	775	635
26	660	4	15.9	70	95	102	63.5	368	572	419
		6						445	648	495
		8						495	699	546
		10						533	737	584
		12						597	800	648
28	711	4	17.5	76	100	110	63.5	368	584	419
		6						445	660	495
		8						495	711	546
		10						533	749	584
		12						610	826	660
		14						648	864	699
30	762	6	19.1	81	108	119	63.5	445	660	495
		8						495	711	546
		10						533	749	584
		12						622	838	673

Continued on the next page.

SHELL SIDE DESIGN PRESSURE 450 psig, 31.02 barg

D_s ins	D_s mm	D_n ins	t mm	TS mm	F_{stat} mm	$F_{S\ head}$ mm	$F_{T\ head}$ mm	L1 mm	L2 S mm	L2 T mm
32	**813**	**6**	**20.7**	**86**	**110**	**121**	**63.5**	**457**	**673**	**500**
		8						521	737	564
		10						546	762	589
		12						635	851	678
		14						660	876	704
		16						737	953	780
34	**864**	**6**	**20.7**	**91**	**114**	**129**	**63.5**	**470**	**699**	**513**
		8						521	749	564
		10						559	787	602
		12						635	864	678
		14						673	902	716
		16						762	991	805
36	**914**	**6**	**22.2**	**95**	**119**	**133**	**63.5**	**483**	**724**	**526**
		8						521	762	564
		10						572	813	615
		12						660	902	704
		14						673	914	716
		16						762	1003	805
		18						838	1080	881
38	**965**	**6**	**23.8**	**100**	**122**	**137**	**63.5**	**483**	**724**	**526**
		10						572	813	615
		14						699	940	742
		18						838	1080	881
40	**1016**	**8**	**23.8**	**105**	**125**	**144**	**63.5**	**533**	**787**	**577**
		12						660	914	704
		16						787	1041	831
		20						927	1181	970
42	**1067**	**6**	**25.4**	**110**	**132**	**149**	**63.5**	**495**	**749**	**538**
		10						584	838	627
		14						699	953	742
		18						851	1105	894
44	**1118**	**8**	**27.0**	**114**	**132**	**152**	**63.5**	**546**	**813**	**589**
		12						673	940	716
		16						787	1054	831
		20						927	1194	970
46	**1168**	**6**	**27.0**	**119**	**141**	**165**	**63.5**	**508**	**775**	**551**
		10						597	864	640
		14						711	978	754
		18						876	1143	919
		20						940	1207	983

Continued on the next page.

SHELL SIDE DESIGN PRESSURE 450 psig, 31.02 barg

D_s ins	D_s mm	D_n ins	t mm	TS mm	F_{stat} mm	$F_{S\ head}$ mm	$F_{T\ head}$ mm	L1 mm	L2 S mm	L2 T mm
48	**1219**	**8**	**28.6**	**124**	**143**	**168**	**76.2**	**559**	**838**	**615**
		12						673	953	729
		16						800	1080	856
		20						940	1219	996
		24						1067	1346	1123
50	**1270**	**6**	**28.6**	**127**	**146**	**175**	**76.2**	**521**	**813**	**577**
		10						597	889	653
		14						724	1016	780
		18						889	1181	945
		24						1029	1321	1085
52	**1321**	**8**	**30.2**	**132**	**148**	**179**	**76.2**	**572**	**876**	**627**
		12						699	1003	754
		16						800	1105	856
		20						940	1245	996
		24						1067	1372	1123
54	**1372**	**6**	**31.8**	**137**	**154**	**184**	**88.9**	**521**	**838**	**589**
		10						610	927	678
		14						737	1054	805
		18						902	1219	970
		24						1092	1410	1161
56	**1422**	**8**	**31.8**	**143**	**165**	**189**	**88.9**	**584**	**902**	**653**
		12						711	1029	780
		16						838	1156	907
		20						978	1295	1046
		24						1118	1435	1186
58	**1473**	**6**	**28.6**	**146**	**183**	**194**	**88.9**	**572**	**902**	**640**
		10						673	1003	742
		14						800	1130	869
		18						978	1308	1046
		24						1143	1473	1212
60	**1524**	**8**	**30.2**	**151**	**184**	**208**	**88.9**	**622**	**965**	**691**
		16						876	1219	958
		20						1029	1372	1110
		24						1168	1511	1250

SHELL SIDE DESIGN PRESSURE 600 psig, 41.38 barg

D_s ins	D_s mm	D_n ins	t mm	TS mm	F_{stat} mm	$F_{S\ head}$ mm	$F_{T\ head}$ mm	L1 mm	L2 S mm	L2 T mm
11.75	**298**	**2**	**12.7**	**44**	**71**	**67**	**63.5**	**305**	**445**	**356**
		4						356	495	406
		6						406	546	457
13	**330**	**2**	**12.7**	**48**	**75**	**76**	**63.5**	**318**	**457**	**368**
		4						356	495	406
		6						406	546	457
14.69	**373**	**2**	**16.7**	**54**	**78**	**83**	**63.5**	**318**	**483**	**368**
		4						356	521	406
		6						406	572	457
		8						457	622	508
16.88	**429**	**2**	**14.3**	**59**	**89**	**89**	**63.5**	**330**	**508**	**381**
		4						368	546	419
		6						419	597	470
		8						470	648	521
19.25	**489**	**2**	**17.5**	**65**	**89**	**98**	**63.5**	**330**	**508**	**381**
		4						368	546	419
		6						432	610	483
		8						483	660	533
21.9	**556**	**4**	**19.1**	**70**	**97**	**102**	**63.5**	**368**	**559**	**419**
		6						445	635	495
		8						495	686	546
		10						533	724	584
24	**610**	**4**	**20.7**	**78**	**103**	**113**	**63.5**	**381**	**584**	**432**
		6						457	660	508
		8						495	699	546
		10						546	749	597
		12						597	800	648
26	**660**	**4**	**22.2**	**83**	**108**	**125**	**63.5**	**394**	**610**	**445**
		6						457	673	508
		8						508	724	559
		10						559	775	610
		12						622	838	673
28	**711**	**4**	**23.8**	**89**	**116**	**130**	**63.5**	**406**	**635**	**457**
		6						457	686	508
		8						521	749	572
		10						559	787	610
		12						622	851	673
		14						673	902	724
30	**762**	**6**	**25.0**	**94**	**121**	**141**	**63.5**	**470**	**711**	**521**
		8						533	775	584
		10						572	813	622
		12						648	889	699

Continued on the next page.

SHELL SIDE DESIGN PRESSURE 600 psig, 41.38 barg

D_s ins	D_s mm	D_n ins	t mm	TS mm	F_{stat} mm	$F_{S\ head}$ mm	$F_{T\ head}$ mm	L1 mm	L2 S mm	L2 T mm
32	**813**	**6**	**25.4**	**100**	**127**	**146**	**76.2**	**483**	**724**	**538**
		8						533	775	589
		10						572	813	627
		12						673	914	729
		14						699	940	754
		16						775	1016	831
34	**864**	**6**	**27.0**	**105**	**130**	**151**	**76.2**	**508**	**762**	**564**
		8						559	813	615
		10						597	851	653
		12						673	927	729
		14						711	965	767
		16						800	1054	856
36	**914**	**6**	**28.6**	**111**	**133**	**157**	**76.2**	**508**	**775**	**564**
		8						559	826	615
		10						610	876	665
		12						686	953	742
		14						711	978	767
		16						800	1067	856
		18						851	1118	907
38	**965**	**6**	**30.2**	**116**	**141**	**165**	**88.9**	**508**	**787**	**577**
		10						610	889	678
		14						711	991	780
		18						851	1130	919
40	**1016**	**8**	**31.8**	**122**	**148**	**170**	**88.9**	**572**	**864**	**640**
		12						699	991	767
		16						800	1092	869
		20						953	1245	1021
42	**1067**	**6**	**31.8**	**127**	**154**	**176**	**88.9**	**521**	**813**	**589**
		10						635	927	704
		14						749	1041	818
		18						889	1181	958
44	**1118**	**8**	**28.6**	**132**	**168**	**181**	**88.9**	**622**	**927**	**691**
		12						762	1067	831
		16						889	1194	958
		20						1029	1334	1097
46	**1168**	**6**	**30.2**	**138**	**175**	**187**	**88.9**	**559**	**864**	**627**
		10						673	978	742
		14						813	1118	881
		18						940	1245	1008
		20						1041	1346	1110

Continued on the next page.

SHELL SIDE DESIGN PRESSURE 600 psig, 41.38 barg

D_s ins	D_s mm	D_n ins	t mm	TS mm	F_{stat} mm	$F_{S\ head}$ mm	$F_{T\ head}$ mm	L1 mm	L2 S mm	L2 T mm
48	**1219**	**8**	**31.8**	**143**	**176**	**195**	**101.6**	**622**	**940**	**704**
		12						762	1080	843
		16						889	1207	970
		20						1041	1359	1123
		24						1168	1486	1250
50	**1270**	**6**	**33.4**	**149**	**183**	**202**	**101.6**	**559**	**889**	**640**
		10						686	1016	767
		14						826	1156	907
		18						953	1283	1034
		24						1194	1524	1275
52	**1321**	**8**	**33.4**	**154**	**192**	**205**	**101.6**	**648**	**978**	**729**
		12						775	1105	856
		16						914	1245	996
		20						1067	1397	1148
		24						1207	1537	1288
54	**1372**	**6**	**34.9**	**159**	**195**	**211**	**101.6**	**584**	**927**	**665**
		10						711	1054	792
		14						838	1181	919
		18						991	1334	1072
		24						1207	1549	1288
56	**1422**	**8**	**36.5**	**165**	**197**	**221**	**101.6**	**648**	**991**	**729**
		12						787	1130	869
		16						927	1270	1008
		20						1067	1410	1148
		24						1219	1562	1300
58	**1473**	**6**	**36.5**	**170**	**208**	**225**	**101.6**	**610**	**965**	**691**
		10						724	1080	805
		14						851	1207	932
		18						1003	1359	1085
		24						1232	1588	1313
60	**1524**	**8**	**38.1**	**176**	**211**	**246**	**101.6**	**673**	**1054**	**754**
		16						940	1321	1021
		20						1105	1486	1186
		24						1232	1613	1313

SHELL SIDE DESIGN PRESSURE 750 psig, 51.72 barg

D_s ins	D_s mm	D_n ins	t mm	TS mm	F_{stat} mm	$F_{S\ head}$ mm	$F_{T\ head}$ mm	L1 mm	L2 S mm	L2 T mm
11.75	**298**	**2**	**12.7**	**49**	**78**	**79**	**63.5**	**305**	**483**	**356**
		4						343	521	394
		6						419	597	470
12.81	**325**	**2**	**15.1**	**52**	**79**	**87**	**63.5**	**318**	**495**	**368**
		4						356	533	406
		6						406	584	457
14.69	**373**	**2**	**16.7**	**59**	**84**	**92**	**63.5**	**330**	**508**	**381**
		4						368	546	419
		6						419	597	470
		8						457	635	508
16.5	**419**	**2**	**19.1**	**65**	**89**	**95**	**76.2**	**330**	**521**	**394**
		4						368	559	432
		6						432	622	495
		8						470	660	533
19.25	**489**	**2**	**20.7**	**73**	**95**	**105**	**76.2**	**330**	**533**	**394**
		4						368	572	432
		6						432	635	495
		8						483	686	546
22	**559**	**4**	**22.2**	**81**	**108**	**124**	**76.2**	**394**	**610**	**457**
		6						457	673	521
		8						495	711	559
		10						546	762	610
24	**610**	**4**	**23.8**	**87**	**116**	**133**	**76.2**	**406**	**635**	**470**
		6						470	699	533
		8						508	737	572
		10						559	787	622
		12						660	889	724
26	**660**	**4**	**27.0**	**94**	**119**	**140**	**76.2**	**406**	**648**	**470**
		6						470	711	533
		8						533	775	597
		10						572	813	635
		12						660	902	724
28	**711**	**4**	**28.6**	**102**	**130**	**149**	**76.2**	**419**	**660**	**483**
		6						495	737	559
		8						533	775	597
		10						584	826	648
		12						660	902	724
		14						699	940	762
30	**762**	**6**	**30.2**	**108**	**135**	**157**	**88.9**	**508**	**762**	**584**
		8						546	800	622
		10						584	838	660
		12						660	914	737

Continued on the next page.

SHELL SIDE DESIGN PRESSURE 750 psig, 51.72 barg

D_s ins	D_s mm	D_n ins	t mm	TS mm	F_{stat} mm	$F_{S\ head}$ mm	$F_{T\ head}$ mm	L1 mm	L2 S mm	L2 T mm
32	**813**	**6**	**31.8**	**114**	**141**	**163**	**88.9**	**495**	**762**	**564**
		8						559	826	627
		10						610	876	678
		12						686	953	754
		14						737	1003	805
		16						787	1054	856
34	**864**	**6**	**28.6**	**119**	**159**	**170**	**88.9**	**546**	**826**	**615**
		8						622	902	691
		10						660	940	729
		12						737	1016	805
		14						800	1080	869
		16						851	1130	919
36	**914**	**6**	**30.2**	**125**	**162**	**183**	**88.9**	**559**	**851**	**627**
		8						622	914	691
		10						660	953	729
		12						749	1041	818
		14						800	1092	869
		16						864	1156	932
		18						902	1194	970
38	**965**	**6**	**31.8**	**132**	**170**	**189**	**88.9**	**559**	**864**	**627**
		10						660	965	729
		14						800	1105	869
		18						914	1219	983
40	**1016**	**8**	**33.4**	**137**	**175**	**200**	**101.6**	**635**	**953**	**716**
		12						775	1092	856
		16						864	1181	945
		20						1016	1334	1097
42	**1067**	**6**	**34.9**	**143**	**176**	**208**	**101.6**	**572**	**889**	**653**
		10						686	1003	767
		14						826	1143	907
		18						927	1245	1008
44	**1118**	**8**	**36.5**	**149**	**186**	**219**	**101.6**	**660**	**991**	**742**
		12						787	1118	869
		16						889	1219	970
		20						1041	1372	1123
46	**1168**	**6**	**36.5**	**156**	**195**	**214**	**101.6**	**610**	**953**	**691**
		10						724	1067	805
		14						851	1194	932
		18						953	1295	1034
		20						1067	1410	1148

Continued on the next page.

SHELL SIDE DESIGN PRESSURE 750 psig, 51.72 barg

D_s ins	D_s mm	D_n ins	t mm	TS mm	F_{stat} mm	$F_{S\ head}$ mm	$F_{T\ head}$ mm	L1 mm	L2 S mm	L2 T mm
48	**1219**	**8**	**38.1**	**162**	**203**	**225**	**101.6**	**673**	**1029**	**754**
		12						826	1181	907
		16						914	1270	996
		20						1067	1422	1148
		24						1232	1588	1313
50	**1270**	**6**	**39.7**	**168**	**206**	**233**	**101.6**	**610**	**978**	**691**
		10						724	1092	805
		14						851	1219	932
		18						978	1346	1059
		24						1232	1600	1313
52	**1321**	**8**	**41.3**	**175**	**214**	**243**	**101.6**	**686**	**1054**	**767**
		12						826	1194	907
		16						940	1308	1021
		20						1092	1461	1173
		24						1232	1600	1313
54	**1372**	**6**	**42.9**	**181**	**225**	**252**	**101.6**	**648**	**1029**	**729**
		10						762	1143	843
		14						889	1270	970
		18						991	1372	1072
		24						1257	1638	1339
56	**1422**	**8**	**44.5**	**187**	**229**	**264**	**114.3**	**711**	**1105**	**805**
		12						851	1245	945
		16						965	1359	1059
		20						1118	1511	1212
		24						1257	1651	1351
58	**1473**	**6**	**46.1**	**194**	**243**	**286**	**114.3**	**660**	**1067**	**754**
		10						775	1181	869
		14						914	1321	1008
		18						1016	1422	1110
		24						1270	1676	1364
60	**1524**	**8**	**47.6**	**200**	**246**	**289**	**127**	**724**	**1156**	**831**
		12						864	1295	970
		16						978	1410	1085
		20						1130	1562	1237
		24						1270	1702	1377

TABLE 11 – TYPICAL HEAT TRANSFER COEFFICIENTS

11A LIQUIDS

Fluid	Viscosity	Shell side Cooling	Shell side Heating	Tube side Cooling	Tube side Heating
Oil mixtures	50	120 0.0083	180 0.0055	60 0.0167	120 0.0083
	35	240 0.0042	300 0.0033	90 0.0111	150 0.0067
	8	420 0.0023	500 0.002	120 0.0083	200 0.005
	3	480 0.0021	600 0.0017	390 0.0026	450 0.0022
	1	550 0.0018	700 0.0014	500 0.002	550 0.0018
	0.7	720 0.00139	840 0.00119	690 0.00145	750 0.00133
	0.45	900 0.0011	1020 0.00098	840 0.0012	1080 0.00092
	0.3	1080 0.00092	1200 0.00083	1050 0.00095	1170 0.00085
Diesel		690 0.00145	780 0.00128	570 0.00175	690 0.00145
Kerosene		870 0.00115	930 0.00107	840 0.0012	900 0.0011
Heavy Naphtha		870 0.00115	930 0.00107	780 0.00128	840 0.0012
Light Naphtha		1080 0.00092	1080 0.00092	1080 0.00092	1080 0.00092
Alcohols Organic solvents		1080 0.00092	1080 0.00092	1080 0.00092	1080 0.00092
Gasoline		1200 0.00083	1200 0.00083	1200 0.00083	1200 0.00083
Light hydrocarbons		1500 0.00067	1500 0.00067	1500 0.00067	1500 0.00067
Brine 75% water		3500 0.00028	3500 0.00028	3500 0.00028	3500 0.00028
Water, ammonia		5000 (0.0002)	5000 (0.0002)	5000 (0.0002)	5000 (0.0002)

Heat transfer coefficients are in W/m^2K; below them are the thermal resistances in m^2K/W.

The viscosity values are absolute viscosities in cps.

11B GASES & VAPOURS

	0.7 bar	3.5 bar	7 bar	20 bar	35 bar
Air	78 0.01282	150 0.00667	300 0.00333	510 0.00196	600 0.00167
Light inorganic	81 0.01235	170 0.00588	320 0.00312	550 0.00182	630 0.00159
Ammonia	84 0.0122	180 0.00555	330 0.00303	570 0.00175	660 0.00152
Steam	90 0.01111	190 0.00526	350 0.00286	650 0.00154	750 0.00133
Light HC's	150 0.00667	360 0.00278	600 0.00167	1020 0.00098	1200 0.00083
Medium HC's Organic solvents	180 0.00555	420 0.00238	630 0.00159	1080 0.00093	1320 0.00076
Hydrogen – 100%	240 0.00417	630 0.00159	1140 0.00088	2100 0.00048	2540 0.00039
Hydrogen – 75%	210 0.00476	480 0.00208	900 .00111	1670 0.00059	2040 0.00049
Hydrogen – 50%	180 0.00555	420 0.00238	780 0.00128	1480 0.00068	1860 0.00054
Hydrogen – 25%	150 0.00667	330 0.00303	600 0.00167	1080 0.00093	1620 0.00062

Heat transfer coefficients are in W/m^2K; below them are the thermal resistances in m^2K/W.

For hydrogen the % figures are by volume/mol.

11C DURING PHASE CHANGES

CONDENSING		BOILING	
Steam	10000 0.0001	Water	10000 0.0001
Steam 5% non condensables	6000 0.00017		
Steam 10% non condensables	4000 0.00025		
Ammonia	8000 0.00013	Ammonia	4000 0.00025
Propane, Butane	2500 0.0004	Propane, Butane	2500 0.0004
Pure light HC's	1800 0.00056	Pure light HC's	1800 0.00056
Mixed light HC's	1300 0.00077	Mixed light HC's	1500 0.00067
Mixed medium HC's	700	Mixed medium HC's	1200 0.00083
Gasoline	1200 0.00083		
Pure organic solvents	1500 0.00067		

Heat transfer coefficients are in W/m^2K; below them are the thermal resistances in m^2K/W.

For steam the percentage values for the non condensables are by volume/mol.

For the boiling coefficients it is assumed that there is a high enough temperature difference for full nucleate boiling without there being such a high a temperature difference that there is a danger of film boiling.

These tables should only be used for the preliminary sizing of heat exchangers.

TABLE 12 – EQUIVALENT LENGTHS FOR FITTINGS

TABLE 12A

The equivalent lengths in this table are in meters.

NB	BENDS			VALVES		T JUNCTIONS	
ins	45°	Long	Short	Globe	Gate	Hard	Soft
4	1.2	1.5	2.1	35	1.2	7	2.1
6	1.8	2.1	3.0	52	2.1	10	3.0
8	2.4	2.7	4.3	70	2.7	13	4.3
10	2.7	3.4	5.2	89	3.4	16	5.2
12	3.4	4.0	6.1	102	4.0	19	6.1
14	3.7	4.6	6.7	115	4.6	21	7
16	4.3	5.2	7.9	133	5.2	24	8
18	4.9	5.8	8.8	150	5.8	27	9
20	5.5	6.4	10	166	6.4	30	10
24	6.4	7.6	12		7.6	36	12
30	8.2	9.8	15		9.8	45	15
36	10	12	18		12	54	18
42	12	14	21		14	63	21
48	13	16	24		16	72	24

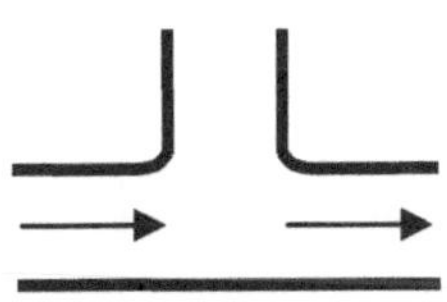
Soft T

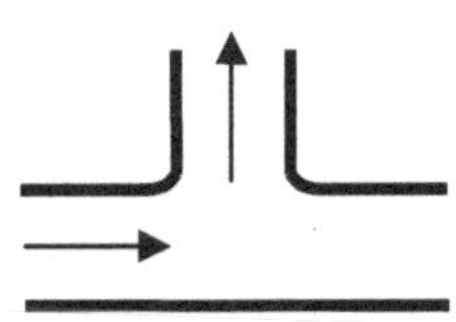
Hard T

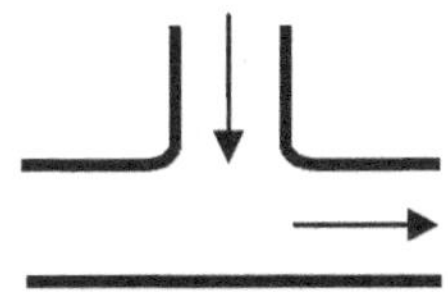
Hard T

The long radius bend has a radius of 1.5 x nominal bore diameter.

The short radius bend has a radius equal to the nominal bore diameter

The values for the valves are when they are fully open. A globe valve is used for fine control of the flow rate of fluid. Gate valves are used to isolate the exchanger and are therefore either fully open or shut. A butterfly valve had a similar resistance to a gate valve when fully open. It may be used to give a crude limited manual control of the fluid flow rate.

TABLE 12B

The values in this table are Le/NB ratios. The table value must be multiplied by the nominal bore of the smallest pipe, d in meters to give the equivalent length of pipe in meters.

Nominal pipe sizes ins		DIAMETER INCREASE		DIAMETER DECREASE	
d	D	Abrupt	Expander	Abrupt	Reducer
4	6	4	4	3	2
	8	8	10	5	3
	10	12	14	7	5
6	8	4	4	4	1
	10	11	12	6	2
	12	12	14	7	4
	14	17	19	8	4
8	10	4	4	3	3
	12	11	11	6	3
	14	15	18	8	4
	16	16	19	9	5
10	12	5	5	4	2
	14	12	12	7	3
	16	14	17	9	4
	18	18	24	10	5
12	14	5	5	5	2
	16	8	8	7	2
	18	12	12	10	3
	20	17	20	12	5
14	16	6	6	5	2
	18	8	8	6	2
	20	11	12	9	4
16	18	6	6	6	2
	20	6	6	6	2
18	20	6	6	6	2
	24	11	12	12	5
20	24	7	7	7	2

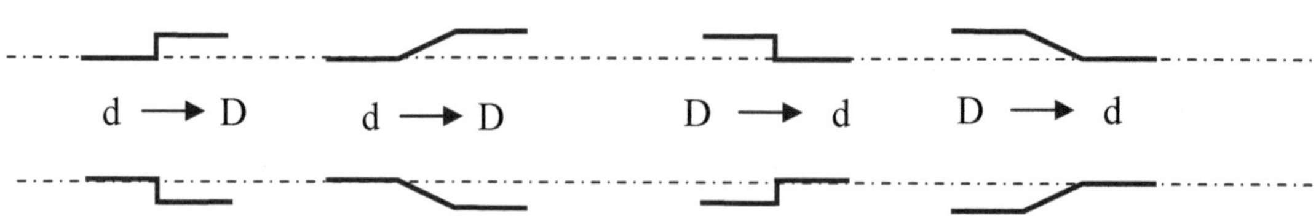

Pipe size increase

Pipe size decrease

CHECK LISTS & DATA SHEET

This section contains the following information:

Company	SHELL-AND-TUBE HEAT EXCHANGER DATA SHEET (SI UNITS)		Engineering contractor
P.O. No.:	Doc. No.:		Page 1 of

No.						
1	Client:		Location:			
2	Process unit:		Item No.:			
3	Job No.:		Fabricator:			
4	Service of unit:		No. of units:			
5	Size:	TEMA Type:	Connected in:	Parallel	Series	
6	Effective surface per unit (plain/finned):	m²	Shells/unit:	Effective surface per shell:	m²	
7	**Performance of one unit**		Shell side		Tube side	
8			Inlet	Outlet	Inlet	Outlet
9	Fluid name					
10	Fluid quantity, total:	kg/h				
11	Vapour:	kg/h				
12	Liquid:	kg/h				
13	Steam:	kg/h				
14	Water:	kg/h				
15	Non-condensable/relative molecular mass:	kg/h	/		/	
16	Temperature:	°C				
17	Density (vapour/liquid):	kg/m³	/	/	/	/
18	Viscosity (vapour/liquid):	mPa·s	/	/	/	/
19	Specific heat (vapour/liquid):	kJ/(kg·K)	/	/	/	/
20	Thermal conductivity (vapour/liquid):	W/(m·K)	/	/	/	/
21	Specific latent heat:	kJ/kg @ °C	@		@	
22	Inlet pressure:	kPa (ga)				
23	Velocity:	m/s				
24	Pressure drop (allowable/calculated):	kPa	/		/	
25	Fouling resistance:	m²·K/W				
26	Average film coefficient:	W/(m²·K)				
27	Heat exchanged:	kW	Mean temperature difference (MTD):	°C	corrected/weighted	
28	Heat transfer rate:	W/(m²·K)	Required:	Fouled:	Clean:	
29	ρv²:	kg/(m·s²)	Inlet nozzle:	Bundle entrance:	Bundle exit:	
30	**Construction per shell**					
31	Tube number:	OD: mm	Nozzles – Number, size and rating			
32	Tube wall thickness: mm	minimum wall/average wall		Shell side	Tube side	
33	Tube pitch: mm	Tube layout:	Inlet:			
34	Tube length: m	Tube type: plain/finned/other	Outlet:			
35	Tube to tubesheet joint:		Intermediate:			
36	Shell diameter: mm	(ID/OD) /	Vent:			
37	Cross-baffle type:		Drain:			
38	Baffle spacing: c/c mm	No. of cross passes:	Pressure relief:			
39	Baffle Cut: %	vertical/horizontal	Design pressure: kPa (ga)			
40	Tube support type:		Vacuum: kPa (abs)			
41	Longitudinal baffle seal type:		Design temp. (Max/MDMT): °C	/	/	
42	Bypass seal type:		Number of passes per shell:			
43	Impingement protection: (Y/N)	Type:	Corrosion allowance: mm			
44	**Materials of construction**					
45	Shell:	Tubes:	Gaskets			
46	Shell cover:		Shell side:			
47	Channel or bonnet:		Tube side:			
48	Channel cover:		Floating head:			
49	Floating head cover/bolts:		Spare gaskets required:			
50	Tubesheet: Stationary:	Floating:	Test ring required:	(Y/N)		
51	Baffles: Cross:	Longitudinal:	Insulation - shell:	mm		
52	Tube support material:		Insulation - channel:	mm	inlet/outlet	
53	Expansion joint type:		Expansion joint material:			
54	Pressure design code:	Stamp:	Calculate MAWP: (Y/N)	TEMA Class:		
55	Remarks:					
56						
57						

Rev. No.	Description	Date	Prepared by	Reviewed by

Company	**SHELL-AND-TUBE HEAT EXCHANGER CONNECTIONS AND MATERIALS DATA SHEET (SI UNITS)**	Engineering contractor
P.O. No.:	**Doc. No.:**	**Page 2 of**

1	Client:	Location:
2	Process unit:	Item No.:
3	Job No.:	Fabricator:
4	**Connection schedule (optional)**	

	Mark	Number required	Size	Rating	Facing	Description
5						
6						
7						
8						
9						
10						
11						
12						
13						
14						
15						
16						

			Corrosion allowance
17	**Materials of construction (optional)**		Corrosion allowance
18	Shell:		mm
19		Head:	mm
20		Pipe/stub ends:	mm
21		Nozzle necks:	mm
22		Nozzle flanges:	mm
23		Body flanges:	mm
24		Expansion joint:	mm
25		Support:	mm
26		Bolting (internal):	mm
27		Bolting (external):	mm
28		Nozzle reinforcement:	mm
29	Tubes:		
30	Tubesheets:		mm
31	Bonnet/channel:		mm
32		Bonnet head(s):	mm
33		Channel cover(s):	mm
34		Body flanges:	mm
35		Pipe/stub ends:	mm
36		Bolting (internal):	mm
37		Bolting (external):	mm
38		Nozzle reinforcement:	mm
39		Nozzle necks:	mm
40		Nozzle flanges:	mm
41	Baffles, spacers, tie rods:		mm
42			
43			
44			
45			
46			
47			
48			

49	**Gaskets (optional)**				**Mechanical data (optional)**					
50	Shell side:	Thickness:	mm		MAWP (hot and corroded):		kPa (ga)			
51	Y =	Pa		m =	MAP (new and cold):		kPa (ga)			
52	Tube side:	Thickness:	mm		Hydrotest pressure:					
53	Y =	Pa		m =		Field:	kPa (ga)	Shop:	kPa (ga)	
54	Floating head:	Thickness:	mm		Mass:	Empty:	kg	Bundle:	kg	
55	Y =	Pa		m =		Full of water:	kg			

Rev. No.	Revision	Date	Prepared by	Reviewed by

Company	SHELL-AND-TUBE HEAT EXCHANGER ADDITIONAL INFORMATION DATA SHEET (SI UNITS)	Engineering contractor
P.O. No.:	Doc. No.:	Page 3 of

1	Client:	Location:
2	Process unit:	Item No.:
3	Job No.:	Fabricator:

Additional remarks, sketches, etc. (optional)

Rev. No.	Revision	Date	Prepared by	Reviewed by

EN ISO 16812:2007

Company	SHELL-AND-TUBE HEAT EXCHANGER ADDITIONAL PROCESS INFORMATION DATA SHEET (SI UNITS)	Engineering contractor
P.O. No.:	Doc. No.:	Page 4 of

Client:	Location:
Process unit:	Item No.:
Job No.:	Fabricator:

Fluid name:		Ref. Pressure 1:	kPa (abs)
Pressure kPa (abs)	Temp. °C	Enthalpy kJ/kg	Vapour mass fraction

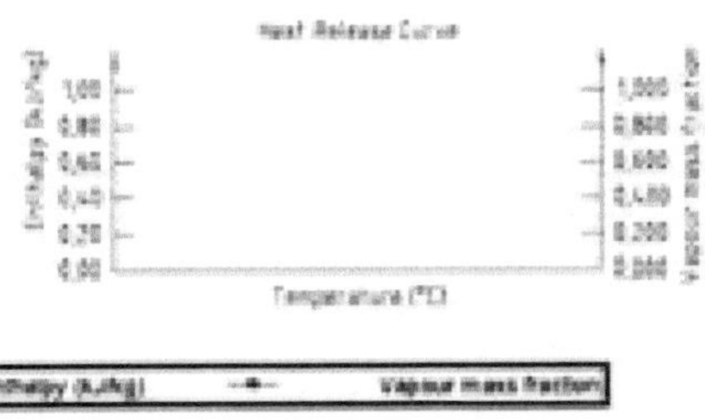

Density vapour	Density liquid	Viscosity vapour	Viscosity liquid	Thermal conductivity vapour	Thermal conductivity liquid	Specific heat vapour	Specific heat liquid	Surface tension	Liquid critical pressure	Liquid critical temp.
kg/m3	kg/m3	mPa·s	mPa·s	W/m·K	W/m·K	kJ/(kg·K)	kJ/(kg·K)	N/m	kPa (abs)	°C

Fluid name:		Ref. Pressure 2:	kPa (abs)
Pressure kPa (abs)	Temp. °C	Enthalpy kJ/kg	Vapour mass fraction

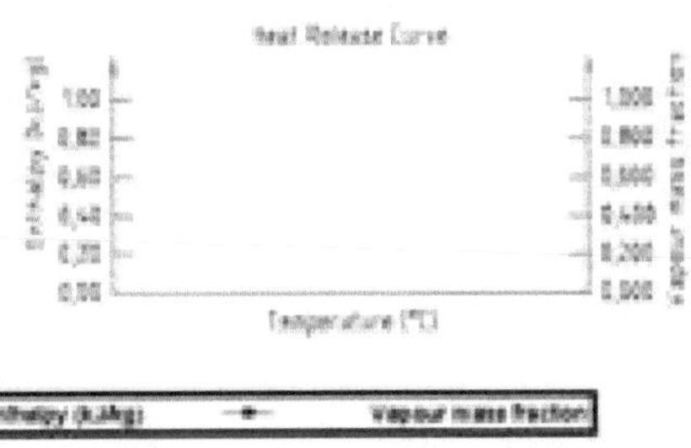

Density vapour	Density liquid	Viscosity vapour	Viscosity liquid	Thermal conductivity vapour	Thermal conductivity liquid	Specific heat vapour	Specific heat liquid	Surface tension	Liquid critical pressure	Liquid critical temp.
kg/m3	kg/m3	mPa·s	mPa·s	W/m·K	W/m·K	kJ/(kg·K)	kJ/(kg·K)	N/m	kPa (abs)	°C

Rev. No.	Revision	Date	Prepared by	Reviewed by

Company	SHELL-AND-TUBE HEAT EXCHANGER ADDITIONAL PROCESS INFORMATION DATA SHEET (SI UNITS)	Engineering contractor
P.O. No.:	**Doc. No.:**	**Page 5 of**

Client:	Location:
Process unit:	Item No.:
Job No.:	Fabricator:

Design conditions for expansion joint (optional)

Case[a]	Shell side: Flow condition[b]	Shell side: Fluid temperature Inlet	Shell side: Fluid temperature Outlet[c]	Shell side: Pressure[d]	Shell side: Mean metal temp.[e]	Shell side: Number of cycles[f]	Tube side: Flow condition[b]	Tube side: Fluid temperature Inlet	Tube side: Fluid temperature Outlet[c]	Tube side: Pressure[d]	Tube side: Mean metal temp.[e]	Tube side: Number of cycles[f]
		°C	°C	kPa (ga)	°C			°C	°C	kPa (ga)	°C	

Determine the mean shell and tube metal temperatures at the listed cases. Evaluate the need for an expansion joint based on the metal temperatures at these conditions with either or both sides clean or with specified fouling.

Unless otherwise stated, operation in accordance with the recommendations of the TEMA Standards, paragraph E3.2, "Operating Procedures", is assumed.

- a A case which may affect design (e.g. steam-out, upset, etc.).
- b F = flowing (specify flow rate), S= stagnant, E = empty.
- c Outlet temperature if known, thermal designer determines other conditions.
- d Specify design pressure for operating conditions. Use maximum actual pressure at other conditions.
- e Mean metal temperature to be specified by thermal designer.
- f Number of cycles that occur during life of the exchanger for each case.

Rev. No.	Revision	Date	Prepared by	Reviewed by

PROCESS DATA REQUIRED for THE THERMAL DESIGN OF HEAT EXCHANGERS

TEMPERATURES - At inlet and outlet

FLOW RATES

These should be on a mass basis. Volume flow rates are ambiguous unless a reference temperature and pressure are also given.

PRESSURES

The pressures at inlet with a value for the allowable pressure drop for both fluids. For feed/effluent type units a total pressure drop for both fluids is more appropriate and may lead to a more economic design.

DUTY (Heat flow rate)

PHYSICAL PROPERTIES

Specific Heat Capacity
Absolute viscosity
Thermal conductivity
Density
Surface Tension (only required for two-phase fluids)
Critical pressure (only required for boiling fluids)

The critical pressure is for the ***boiling liquid***. It should ideally be the actual value but a pseudo value will be safe. All other properties should be given ***at two temperatures***. An exception is a high viscosity fluid where a viscosity value at a ***third*** temperature may be necessary. For viscous liquids a viscosity value at a third temperature – close to that of the cooling medium coldest temperature – should be provided.

Physical properties are not necessary for fresh water/steam (but should be provided for sea water). They may also not be necessary for common fluids which are in most standard property data banks e.g. propane. Using agreed data banks saves time for both the process engineer and the thermal design engineer.

HEATING/COOLING COOLING CURVE DATA

These should be provided for all fluids undergoing a phase change unless the process is isothermal or the fluid involved is a standard fluid such as water/steam or a pure fluid such as a refrigerant where the heat exchanger design program being used contains the same data bank as the one process are using.

The data required should be for a series of equilibrium states covering the full temperature range. The equilibrium states must include ***all turning points particularly bubble or dew*** points. There is no need for anything other than end points for any sections which are a straight line e.g. desuperheating.

The data required for each state is:

Temperature
Mass fraction (not mol fraction)
Specific enthalpy, (or a duty from which a specific enthalpy can be calculated.)

For reboilers two sets of data should be provided. One set should be at the tower pressure (or above the liquid level for kettles) and the second set at a higher pressure. The pressure for the second set of data should be sufficient to allow for all possible static head effects to be taken into account without the need for extrapolation.

For condensers a single curve which starts at the inlet pressure and ends at the anticipated outlet pressure is normally sufficient and always safe. However for low pressure processes accompanied by a low MTD providing data at both the inlet pressure and at a lower pressure (inlet pressure – allowable pressure drop) may result in a significant saving in surface area.

FOULING FACTORS

These should be specified for both fluids. Alternatively an overall percentage oversurface can be specified.

CLEANING REQUIREMENTS

Mechanical or chemical cleaning may be necessary for many fluids. This will affect the type of heat exchanger selected.

MATERIALS

At least generic materials for both sides of the heat exchanger..

PLOT LIMITATIONS

In general long thin exchangers are more economical than short fat ones. It is important to know if there are any length (or height for vertical units) limitations.

FLUID ALLOCATION

Is it acceptable to switch the fluids?

HEAT EXCHANGER JOB NOTES

Whenever a designer starts work on a project for the first time he is required to become familiar with the relevant specifications before he even looks at a process data sheet. This can be very onerous and boring. It is not unusual to be faced with a plethora of pages that could cover the Great Wall of China and still have enough left over to light the bonfire on Guy Forks night. It seems to be a modern trend that everyone involved will have contributed a specification. A document may have hundreds of pages with only a handful being relevant to the thermal designs and of course those few pages will not necessarily be in sequence.

Normally specifications evolve over many years and are products of a committee. Inputs from people with excellent credentials in other fields but limited thermal design knowledge stand out like sore thumbs. Then there are ambiguities – what is meant by 'clean' for example.

It is not unusual to have two experienced thermal design engineers reaching a different conclusion on particular aspects of the way the heat exchangers should be designed. For good consistent designs throughout a job any possible ambiguities should be sorted out before any designs are started. To do this the following sequence of events is recommended:

(1) The job leader for the thermal design work should go through all the specifications and extract relevant details. From these details he/she should compile a shell and tube thermal design specification for the job.

(2) This should be checked and agreed by another experienced designer.

(3) The document should then be circulated to everyone involved in the project who might comment on future designs – client, projects, process, licensure etc.

An example of typical job notes follows:

These notes are intended to clarify and expand the contents of: - list all the standards and specifications relevant to the job. Where a particular unit is required to deviate from these notes it should be written into the Process Data Sheet. For example a unit may have a length restriction lower than the maximum below.

(1) Relevant codes e.g. TEMA R.

(2) Design Programs. The XXXX program Rev Y will be used for the thermal designs. If there are concerns about a particular unit it will be checked on ZZZZ. (It is important to state the Rev of the program. It is not unusual for updates to occur part way through a job. Frequently updates will change the fourth significant figure of lots of the output whilst having no overall effect on the result. This can waste a lot of time. It is better to stick with one Rev of a program unless there is a good reason not to.)

(3) For exchangers with removable bundles the maximum tube length is 6 m. With straight tubes shorter lengths should be in one meter increments. U tubes may be any length up to the maximum. (If lengths are required in the old US customary units it is important to clear up the details. An example is the maximum tube length; the nearest conversion from 20 ft is 6.096 m, (or is it to be rounded off to 6.1 m)

(4) For fixed tubesheet exchangers the maximum length is 8 m. shorter lengths should be in one meter increments.

(5) Tube diameters (OD), gauges and wall thickness to be input to the computer program are as follows:

CS and low alloy	19.05 mm	14 BWG min	2.3 mm
	25.4 mm	12 BWG min	3.1 mm
SS & none ferrous	19.05 mm	16 BWG avg	1.65 mm
	25.4 mm	14 BWG avg	2.11 mm
Titanium	19.05 mm	20 BWG avg	0.89 mm

(6) Shells are to be specified in pipe sizes up to 16 inch. Larger diameters should be in 10 mm increments. For exchangers with removable bundles the maximum diameter is limited only by a maximum mass of 20 tonne.

(7) A fluid shall be considered to be clean if the fouling factor is no more than 0.0003 m^2K/W. Clean fluids may be used inside U tubes. For clean fluids flowing outside tubes a triangular or rotated triangular pitch will be considered. Fluids flowing outside tubes with a fouling factor greater than 0.0003 m^2K/W require a square or rotated square pitch.

(8) Where required by TEMA impingement protection will be specified. This may be a plate or rods. Where rods are used a single row is sufficient for a 90° tube layout but two rows will be used for all other tube orientations.

S & T THERMAL DESIGN CHECK LIST

	DESIGN	CHECK
PROCESS DATA SHEET REVIEW		
If TEMA type is specified is it correct?		
Overcapacity/oversurface required?		
Any alternative conditions?		
Are fluids 'normal' – do the physical properties look OK?		
Extra temperature for properties for viscous fluids		
Heat release curves – all turning points covered? Extra curves?		
Pure 2 phase fluids – can data bank be used?		
Temperature cross – F type shell allowed?		
Any surface temperature warnings?		
Unusual design? Discuss with another engineer.		
Any similar items on another plant?		
Are line sizes given?		
CHECK P & I DIAGRAM		
Control of unit		
Either stream through adjacent units?		
AFTER THE FIRST COMPUTER RUN		
Recheck the process data input		
Is the duty and mass balance OK?		
Are multi shells required or below $20m^2$ – double pipe?		
Impingement protection – plate or rods?		
AFTER THE FINAL COMPUTER RUN		
Check flow directions, nozzle arrangement		
Have the allowable pressure drops been fully utilized?		
Are the heat transfer coefficients in the anticipated range?		
Length, shell diameter & bundle mass limits		
Velocity & ρu^2 limits		
Stream analysis – cross flow/long flow balance, by pass streams		
Should switching fluids be considered?		
Could there be any problems with the supports?		
S type head OK for tubeside pressure or T – check with mech.		
Alternative design with enhanced surface tubes?		
VIBRATION CHECK		
Limits OK?		
Extra supports for revamp?		
Acoustic baffles used?		
TUBE LAYOUT DIAGRAM		
Check tube count, tubes/pass (modified by tie rods etc)		
Bundle/shell clearance		
U tubes – correct bend/diameter ratio for material used?		
Entrance & Exit flow areas compared with nozzle flow area		
Number & position of tie rods		
Number & position of sealing strips - compare with C stream		
Partition lane width, expanded or welded tubes, alignment		
Partition lane sealing – compare with F stream		

Baffle cut on tube centre line or in between tubes		
Bundle runners – position, type, alignment		
Acoustic baffles position, height & thickness		
Flow paths around acoustic baffles		
OTHER CALCULATIONS		
In house programs		
Hand calculations		
F SHELLS ONLY		
Type of seal		
Uninsulated / Insulated longitudinal baffle (drain holes)		
Pressure & Temperature differences/shell		
FTS ITEMS ONLY		
Construction of joints for bundle/shell clearance		
Table of metal temperatures – all cases covered?		
KETTLES ONLY		
Flux limits, recirculation, vapor escape from ports, lanes		
Shell fluid inlet arrangement		
Liquid level, weir, hold up calculations		
Vapour generation – even or variable?		
Extra calculations for single pass kettles		
Cone or dished ends?		
Entrainment calculations – positions of vapour outlet nozzles		
THERMOSYPONS ONLY		
Tower drawings – type of system		
Calculations for static head and outlet vertical height		
Pipework assumptions, valve in inlet line? - Note on status		
Type of heating medium (Single phase co-current)		
Boiling fluid – type and operating pressure		
Balance and proportions of system pressure drops.		
Design vaporization rate compared with PDS target value		
Instability check		
System control & part load operation		
System sensitivity		
Warning messages on TDS depending on last five items		

NOTES

DETAILED DRAWINGS CHECK THERMAL CHECK

	DESIGN	CHECK
TUBES		
Length = baffled + end spaces + tubesheets + protrusion		
Number, gauge, pitch & orientation		
Position of top & bottom rows		
U tubes – min bends, any rows of heavier gauge?		
Alignment – cleaning lanes unobstructed.		
BAFFLES & SUPPORTS		
Thicknesses, circulation spaces around supports.		
Ears		
Any 'special anti vibration supports'?		
Any deflecting baffles for flow to potential dead spaces?		
Tie rods picking up all edges, number of tie rods		
Any tops 'clipped off' – particularly steam items		
Drain notches – any required above standard size?		
Cut positions – on tube centre lines or with min 2 mm metal		
Acoustic – positions, height, thickness, & flow areas past		
IMPINGEMENT PLATES		
Size, shape & thickness		
Proximity to first tube row		
Method of support		
SEALING STRIPS		
Full length for horizontal baffles, baffle/baffle for vertical		
Positions & number		
BUNDLE RUNNERS		
No interference with flow to or from a nozzle		
Any drain notches required?		
NOZZLES		
ID compared with ID on computer run.		
Proximity to first baffle		
Distance to tubesheet for vertical items		
Any chemical cleaning connections required?		
Orientation for vertical units		
PASS PARTITION LANES		
Width of horizontal & vertical lanes		
Lane sealing – diameter of tubes/rods/trips, clearances		
Any sealing tubes open at one end only		
CLEARENCES		
Bundle/shell & tube/baffle (particularly if 'tight' required)		
VENTS & DRAINS		
All compartments		
Any extra vents required?		
'Specials' for vertical thermosyphon units		
F TYPE ONLY		
Position of baffle cuts adjacent to turnaround space		
Length of turnaround space		

Type of seal		
Construction and drainage of insulated horizontal baffle		
KETTLES ONLY		
Do any supports interfere with the circulation paths?		
Weir height/liquid level		
Length of the space beyond any weir		
Port space – particularly for stab-in type		
Position of all outlet nozzles – particularly one pass units		
Fluid inlet distribution system – particularly one pass units		
Any internal two phase distributors/deflector plates		
Any calming section for instrument connections		
NTIW DEIGNS ONLY		
Special attention to supports, runners & tie rods		
Number & length of any rows of impingement rods		
Position of impingement plate – well clear of nearest tubes		
FLOW PATH AREAS – ALL UNITS		
Check throughout – particularly around inlets & outlets		

REFERENCES

(1) Practical Thermal Design of Shell-and-Tube Heat Exchangers
R Mukherjee – Begell House inc. 2004
(2) Heat Exchangers – Selection, Rating and Thermal Design
Sadik Kakac & Hongtan Liu – CRC Press 2002
(3) TEMA – Ninth Edition 2007
Tubular Exchanger Manufacturers Association
(4) Mechanical Services for Buildings
T D Eastop & W E Watson – Longman 1992
(5) Heat Transmission
William H McAdams – McGraw-Hill 3rd Edition
(6) Heat Exchangers – Selection Design and Construction
E A D Saunders – Longman 1988
(7) Vaporisers – Selection, Design & Operation
R A Smith – Longman 1986
(8) Heat Transfer – A Basic Approach
M Necati Ozisik - McGraw-Hill 1985
(9) Shell and Tube Heat Exchangers
Stanley Yokell - McGraw-Hill 1990
(10) Boilers Evaporators & Condensers
Sadik Kakac – John Wiley & Sons 1991
(11) Heat Transfer
F J Bayley J M Owen AB Turner –Nelson and Sons 1972
(12) Engineering Heat Transfer
William S Janna – Van Nostrand Reinhold 1988
(13) Chemical Engineering
Coulson and Richardson – Pergamon Press 1954
(14) Process Heat Transfer
Kern - McGraw-Hill 1950
(15) Thermodynamic and Transport Properties of Fluids
G F C Rogers and Y R Mayhew – Third Edition
(16) Fouling Notebook
T R Bott – Institution of Chemical Engineers 1990
(17) Mechanics of fluids
B S Massey - Van Nostrand Reinhold 6th Ed 1989
(18) Fluid Mechanics
Francis – Edward Arnold 3rd Ed 1969
(19) Fluid Mechanics
Douglas/Gasiorek/Swaffield – Pitman 1979
(20) Heat Exchange Institute Standards
(21) Physical Properties of Hydrocarbons
Robert W Gallant.
(22) Fluid Mechanics for Engineers
P S Barna –Butterworth 1969
(23) Handbook of Hydraulics
W H King
(24) Two Phase Flow and Heat Transfer
D Butterworth & G F Hewitt – Oxford University Press 1977

(25) Heat Exchanger Design
A P Fraas – 2nd Edition John Wiley & Sons
(26) Shell & Tube Reboilers
J W Palen – Hemisphere Publishing Corporation 1987
(27) Liquid Entrainment
D M Newitt, D Drombrowski & F Knelman – I Chem. E 1954
(28) Flooding in a vertical updraft partial condenser.
K G English et Al – Chem. Eng. Progress1963
(29) Engineering Dynamics
S V A Swanson – English Universities Press 1963
(30) Natural frequency of heat exchanger U tubes
C D Nieh and Zang – Hydrocarbon Processing May 1991
(31) Experience with acoustic vibrations in tubular exchangers
E A Barrington – Chem. Eng. Progress July 1973
(32) Cure exchanger acoustic vibrations
E A Barrington – Chem. Eng. Progress July 1978
(33) Acoustic resonance in heat exchanger tube bundles
R D Blevins – Third Keswick International Conf. 1982
(34) Flow induced vibrations in heat exchangers
J S Fitz-Hugh – Symposium of Vib. Probs Keswick . 1973
(35) Acoustic resonance in banks of tubes
R Parker – Journal of Sound & Vibrations Nov 1977
(36) Fouling in Heat Exchangers
N Epstein – Heat Transfer Volume 6 1978
(37) Fouling on surfaces
Gudmundsson PhD University of Birmingham UK 1977
(38) Reboiler System Design
AJ McCarthy & BR Smith – Chem. Eng Progress May 1995
(39) ECMD Trays & High Flux Tubing
UOP LLC & Chevron AIChE March 1999
(40) BS EN ISO 16812
British Standards – London 2007 (Available worldwide)

EPILOGUE

All the technical writing including the index has been completed a few weeks ago. All that remains is to overcome the technical hurdles involved in publishing. If you are reading this I must have managed that as well.

It will have taken just over two years when my original target was one. During that period I have taken just one weeks holiday and I took a laptop with me so that I could carry on working. (Not very successfully; I was plagued with IT problems and with no IT crutch to hand I finally gave in and became a more normal tourist. Lanzarote is a very small but interesting volcanic island with lots of fascinating thermal features and an incredible history. I was welcome in all their bars, just as I once was in my own country.) It will be nice to take another holiday without having to worry about work.

Finishing off with a few pages which anyone can read has made this project into a very short book enveloping the main technical content. That can be no bad thing since there will at least be something that my children and grandchildren can read. I doubt if any of them will ever delve into the depths of the main text. My next book may be something which they can read from cover to cover.

I am finding that publishing a book involves a lot of 'chicken and egg' situations. There are a lot of options to consider when making hard copies available. Hard copies are essential if libraries and technical institutions are to be able to stock the book. It is very attractive to be able to sub contract this work but a PDF of the whole book, including this section is required before I can do this. I have been quoted prices of under £15 per copy for printing only if I then distribute the book myself. This might work in the UK where I am familiar with many of the organisations which would be interested but certainly not worldwide. Using more conventional and efficient channels of distribution will increase this cost; I have no idea how the cost will escalate if I choose this option. I do know that since I purchased my copy of Mr Mukherjee's book for £376.27 the cost has increased to £494.96 and as far as I know it remains the only other book devoted to the thermal design of shell and tube heat exchangers.

One thing that is certain is that I will not be making any money out of this venture. If I select an option that involves me receiving the usual 12.5% royalties I will send these off to one of the many wildlife charities that I support.

Making no money at all from publishing this book may not please many technical authors particularly Mr Mukherjee but I hope that many of them will appreciate my motive. It would be nice if I could meet Mr Mukherjee some time and discuss our books over a drink – first round on me.

I may also have upset some Americans with my regular snide remarks about their insistence on maintaining an outdated system of units. On the other hand I have worked with many Americans who agree with me. In president Obama the Americans now have one of the most enlightened and internationally educated presidents in their history. I have great hopes that he may see the need for America to join the rest of the world and ditch their old system of units. Unfortunately he has far more pressing issues to deal with first.

The main design service providers of heat transfer programs, particularly HTRI may not welcome this book. (Begell House are the publisher for Mr Mukherjee's book but only the HTRI program is mentioned.) My book contains criticisms of many aspects of their programs both historically and currently. These are intended to be constructive and if taken in the right spirit could lead to improved programs and improved thermal designs. I have already made many contributions to their programs by being one of the first to point out mistakes when new versions have been introduced. I was not always credited with these since many were channelled through the head of the section I was working in at the time. The most notable of these was pointing out the errors in the default value for γ, the ratio of the specific heats at constant pressure and constant volume. The default value has now been improved but should be unnecessary and I have explained how to correct this in chapter 15 on vibrations.

I will be sending free inspection copies of this book to both HTRI and HTFS. It would be nice if they could recommend it to their client companies.

Some individuals may also be upset. In order to demonstrate the many pitfalls in thermal designs I have used real situations and reported them as I saw and experienced them. This is especially true during sections on the historical development of computer programs. Where references have not been complimentary I have never mentioned individuals. An example of this is the use of fixed shell/port diameter ratios for kettles long after more sophisticated calculation methods became available. Despite not mentioning individuals there will be readers who will be able to identify them. I am sure that most of them will be OK with this. The next time we meet it will probably be "You son of...... you owe me a drink"; then the conversation will drift to the attributes of the barmaid or sport. Of course there may be the odd one who will not be so amicable.

Anything which might upset an individual could also upset the company involved but I know that everything that I have reported actually took place. I also know that there have been significant improvements to the design standards of these companies since then.

We live in a world where even the remotest possibility of upsetting an individual or a company could result in litigation. The craziest of cases are reported in the newspapers every day. Litigation and the measures taken to avoid it have significantly added to the cost of the Nation Health Service in the UK. Lawyers like bankers continue to amass ridiculous incomes at the expense of those employed in useful productive professions.

If my company, Shell and Tube Consultancy and Publications Ltd became involved in litigation nonsense there would be nothing forthcoming to make their efforts worthwhile. I would never consider employing a lawyer to defend us.

I have some limited personal experience of the law and their archaic ways. I once undertook the selling of one property and the buying of another without any legal assistance. It was back in the early seventies when I was very short of cash and the obstacles they raised to try and stop me could fill a few pages in another book. (For example the refusal to accept a final transfer document without giving any reason – just advice to get legal assistance. I finally nailed that one – the wrong colour of tape binding the document; it was red when it should have been green!) They failed to stop me and I learned a lot.

I also carried out my own legal work during the divorce from my first wife. There was never any problem with my main concern – access to the children. We came to an amicable arrangement about money, which her lawyer disapproved of since he wanted to boost his fees by taking out a court order against me. This aspect of the divorce rather backfired on me when a year later I realised that money paid out to a court order was tax deductible whereas money paid through an amicable arrangement was not. I soon found myself involved in legal matters again. The solution was very simple – *I took out a court order against myself!* I can still picture the faces on the clerks as I filled in all their forms – it was a first for them as well as for me.

Many years after that I had problems being granted probate when my father died. He had made out a very simple will leaving all he had to be divided equally between me and my brother. It was a very modest estate and all the legal requirements had been followed when the will was drawn up. Unfortunately the will had been drawn up in England where my father had lived all his life but he died in Scotland while staying with my brother who has lived there for most of his life. The legal system in Scotland is not the same as in England and I was told I needed a solicitor to reconcile the two systems. I knew that the last thing my father would have wanted was to leave money to a solicitor. I eventually found a minor official who was prepared to sign the necessary papers in return for a couple of bottles of the best Scotch whisky. This may not have been strictly legal but it was justice and a victory for common sense – a rare commodity these days.

I know my father would have approved of my actions concerning his will but he has left me with one further problem – how to dispose of his remains. We only ever discussed such matters during a good session on the beer. I knew he had no problems with cremation but there was never a definitive answer as to what should come next. His simplest response would be "the compost heap" a more elaborate one would be to scatter a few ashes in each of the pubs and clubs he had frequented – an impossible task. I am still undecided about that one and his tasteful oak urn is still on the top of the bookcase behind me as I am writing this. I suppose I am passing on the problem to my children for when they decide what to do with me.

Throughout the technical sections of this book I have always tried to investigate the extreme limits of both variables and correlations. I have done the same with the possibility of litigation. It is a limit which I hope I do not have to experience because I have so much more to pack into a limited time span. There are other books to write and a lot of wildlife projects to keep me occupied. However if my company was ever taken to court I would defend it myself and go to jail rather than pay any fine. I have been prepared to go to jail for my principle on other occasions during my life. The last time was in July 2007 when the smoking ban was introduced.

For months the talk in all the bars I used had been about opposing the ban. I can not recall anyone being in favour of it – even those who had given up smoking years ago. When the fateful day came I went to my local bar, ordered my usual drink, downed it very quickly and then paid for another one. I then lit up a cigar as I normally did. I refused to put it out and suggested that they call the police. They rang the police but they clearly did not see me as a priority call out. By now I had more bar space than normal as everyone moved as far away from me as possible. The scene was reminiscent of the bar in a western when everyone moves away when the bad guy enters. There was silence – not a single word of support from people who had been far more vocal than I had against the ban.

The landlord found a novel way of cooling down the situation; I was soon drenched with soda water from a pressurised bottle and the last part of a good cigar was wasted. I downed the last few millimetres of my pint, sarcastically thanked everyone for their support and left. I have never been back since.

Now people huddle in shelters (which must be exposed to the elements in three directions to maximise discomfort) and doorways - which is technically illegal. When I walk past I see pensioners and young bar staff, who have not had time to put on a coat, shivering. This must give the politicians who introduced the law a lot of satisfaction. The lies they used during their campaign would have come easy to them after all their practice when filling in their expense claims.

They may also be very pleased about the predictable side effect – closing down pubs and clubs at an unprecedented rate which destroys the social life of none-smokers and smokers alike, particularly for residents of small villages who have lost their only pub. Perhaps the main instigators are members of a temperance society – not too far fetched since we did have a closet Catholic as prime minister for about ten years.

I now find myself extolling the virtues of Germany where local referenda were held to find out what their people wanted. This has resulted in pubs where smoking is banned and pubs which anyone may use. They may even have separate areas in the same building – common sense?

We now have a new coalition government which is 'talking about' more power to the people. Halting the closure of pubs would also help the economy so it will be interesting to see how they deal with this issue. I contacted my MP early in the campaign and was pleased to find that he agreed with my views and thanked me for making him aware of the situation in Germany.

Until there is a change I will remain an antisocial, stay at home drinker except for the rare occasions when the UK weather permits outdoor drinking.

Whilst writing this book I thought that just giving it away might be a world first. Now as I am writing the final words I have been brought down to earth – religious organisations have been doing it for years. I remember a Coptic Christian who I once met in Egypt insisting that I have a small bible containing their variation of Christianity; interesting company for an atheist approaching the end of his days?

INDEX

9781844268443
SD - #0074 - 060826 - C0 - 297/210/37 - PB - 9781844268443 - Gloss Lamination